Ralf Lindau

Climate Atlas of the Atlantic Ocean

Springer
Berlin
Heidelberg
New York
Barcelona
Hong Kong
London
Milan
Paris
Singapore
Tokyo

Ralf Lindau

CLIMATE ATLAS of the *ATLANTIC OCEAN*

Derived from the Comprehensive Ocean
Atmosphere Data Set (COADS)

With 428 Plates, 19 Figures and 2 Tables

 Springer

Dr. Ralf Lindau
University of Kiel
Department of Marine Sciences
Düsternbrooker Weg 20
24105 Kiel
Germany
E-mail: rlindau@ifm.uni-kiel.de

Additional material to this book can be downloaded from
http://extra.springer.com

ISBN-13:978-3-642-64009-4   e-ISBN-13:978-3-642-59526-4
DOI:10.1007/978-3-642-59526-4

Library of Congress Catatloging-in-Publication Data
Climate atlas of the Atlantic Ocean: derived from the Comprehensive Ocean Atmosphere Data Set
(COADS) / Ralf Lindau. p. cm. Rev. ed. of: The Bunker climate atlas of the North Atlantic Ocean /
Hans-Jörg Isemer, Lutz Hasse. c1985-1987.
Includes bibliographical references.
ISBN-13:978-3-642-64009-4
1. Atlantic Ocean-Climate-Charts, diagrams, etc. 2. Climatic changes-Atlantic Ocean-Charts, dia-
grams, etc. 3. Ocean-atmosphere interaction-Atlantic Ocean Maps. I. Isemer, Hans-Jörg. Bunker
climate atlas of the North Atlantic Ocean. II. Title.

Springer -Verlag Berlin Heidelberg New York
a member of BertelsmannSpringer Science+Business Media GmbH

© Springer -Verlag Berlin Heidelberg 2001

The use of general descriptive names, registered names, trademarks, etc. in this publication does not
imply, even in the absence of a specific statement, that such names are exempt from the relevant
protective laws and regulations and therefore free for general use.

Product liability: The publishers cannot guarantee the accuracy of any information about dosage
and application contained in this book. In every individual case the user must check such information by consulting the relevant literature.

Typesetting: Camera-ready by author
Cover Design: design & production, Heidelberg
Production: PRO EDIT GmbH, Heidelberg
Printed on acid-free paper   SPIN: 10643444   32/3136 Re – 5 4 3 2 1 0

Acknowledgement

I greatly appreciate the support and faith of Prof. Dr. Lutz Hasse. It were the productive discussions with him and Dr. Hans-Joerg Isemer which laid the foundations for this atlas.

Preface

The presented climate atlas follows with regard to structure and volume the Bunker Climate Atlas of the North Atlantic Ocean published 1985 by H.-J. Isemer and L. Hasse, which has become a standard work for the climate study community.

A new edition was appropriate, since during the last decade much progress was achieved in the correct interpretation of the wind speed reported by merchant ships. As the wind speed is an essential parameter for the fundamental air-sea interactions, namely the exchange of momentum and energy between ocean and atmosphere, considerable impact on resulting climate key parameters as Sverdrup transport or meridional heat transport could be expected.

Furthermore, individual ship observations are nowadays available from COADS, the Comprehensive Ocean Atmosphere Data Set, and, due to the grown computer power, these data are processable in reasonable time. Thus, improving the calculations of air-sea fluxes, as performed by Isemer and Hasse, is no longer a complex and tricky venture, since flux parameterizations can be applied directly to the ship observations.

Compared to the Bunker Altas we expanded the considered sea area and included also the South Atlantic. Considering the entire Atlantic Ocean reveals not only interesting features of the southern hemisphere but allows also a finer assessment of the resulting energy fluxes.

Kiel, August 2000   Ralf Lindau

Contents

1   Introduction

The growing interest in climate and climate change has renewed the demand for climate descriptions of the world oceans. Fields of climate variables are used as boundary conditions or sea truth for atmospheric and oceanic models and coupled models of the atmosphere-ocean system. The fast improvements that modeling has made in recent years also call for more detailed descriptions of processes at the air-sea interface, especially of the air-sea fluxes of momentum, heat and matter. The ocean heat transport is understood as a major branch of the global climate system. The large scale ocean circulations depend on wind stress and thermohaline forcing, that both would likely be influenced by climate changes.

Today marine weather observations are still the major source for the study of longterm climate at the ocean. Since area covering data of direct flux measurements are not available, air-sea fluxes have to be derived by parameterisations using the basic meteorological observations.

The climate atlas presented here is restricted to the Atlantic Ocean. The Atlantic is of special interest for two reasons. The cross-equatorial ocean heat transport of about 1 petawatt is an outstanding climate feature, that contributes to the relatively high sea surface temperatures in the Northeastern North Atlantic and the mild climate of Europe. From a more technical view, the data coverage of the Atlantic is much denser than for other parts of the world oceans. Hence the Atlantic can well serve as a reliable test area of sea truth for both modelling and satellite remote sensing.

2   Data and Data Treatment

Individual ship reports from the Comprehensive Ocean-Atmosphere Data Set (COADS) of the Atlantic Ocean from 1940 to 1979 are used for this climatological study. Maps of monthly mean distribution are calculated not only for the basic ship observations as wind speed or air pressure but also for the indirectly derived climate variables such as evaporation or wind stress, which are obtained by parameterisations. For the derived variables the individual method is used that conserves the effects of cross-correlations. The classical climatological method, that uses averaged values of the meteorological parameters, cannot account for these contributions. For the North East Atlantic JOSEY & al. (1995) showed that the latent heat flux is overestimated by the crude climatological method. The negative correlation between wind speed and sea-air humidity difference causes an overestimation from 1-2 Wm^{-2} in winter to 7 Wm^{-2} in summer.

The distribution of ship observations in the Atlantic Ocean is very inhomogeneous. Especially in the South Atlantic reports are concentrated along narrow shipping routes, whereas broad areas remain nearly unsampled (fig.1). However, coastal regions, where we suppose the strongest gradients, are in general well frequented by ships. In the open ocean the relatively small number of available observations is tolerable, because strong gradients are not expected. Thus, we developped an adaptive interpolation scheme to account for the inhomogeneous data density.

The individual reports or values of derived climate variables are averaged on a grid of 1°latitude by 1°longitude. A minimum of 300 observations around the respective grid point is required. If this number is reached in a surrounding of 111 km by 111 km, no interpolation is performed and the arithmetic mean is taken as the grid point value. Otherwise the radius of influence is enlarged step by step from 1°to 5°(measured as great circle distance) until 300 observations are enclosed. Then a least square fit of a two-dimensional quadratic polynomial yields the value for the considered grid point. There is the possible problem that the reports are non-uniformly distributed around a grid point and the polynominal fit would perform an extrapolation rather than an interpolation. An algorithm was used to ensure that in all quadrants of a given grid point at least 10 observations were available. For nearly all grid

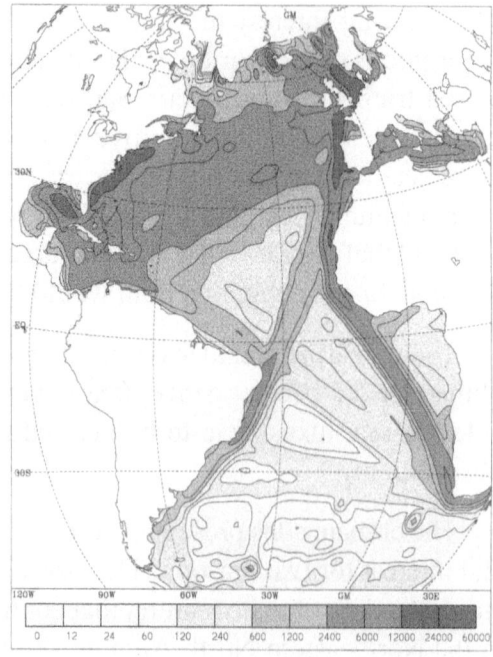

Figure 1: *Number of observations per* $10^{10}m^2$ *in the period 1940 to 1979.*

points these conditions are fulfilled, exept small regions in the central South Atlantic. To avoid gaps on the climate maps we allowed moderate linear extrapolations in such areas.

3   Derivation of Air Sea Fluxes by Parameterisations

Air-sea interaction is brought about by fluxes of momentum, heat and matter. Since these are not measured routineously, they are determined by parameterisations. In the following, we give a short account of the formulations that we selected.

3.1   Net Shortwave Radiation

Net shortwave radiation SR depends on the albedo of the sea surface α and on Q_s, the portion of solar radiation, which reaches the surface.

$$SR = Q_s \left(1 - \alpha\right) \tag{1}$$

For the albedo α, we used the latitude dependent monthly mean values derived by PAYNE (1972). Q_s is a function of solar altitude, transmissivity of the atmosphere and cloud cover. For Q_s we used a parameterisation derived by MALEVSKII et al. (1992). His scheme is based on about 50,000 ship-borne radiation measurements well distributed over the world

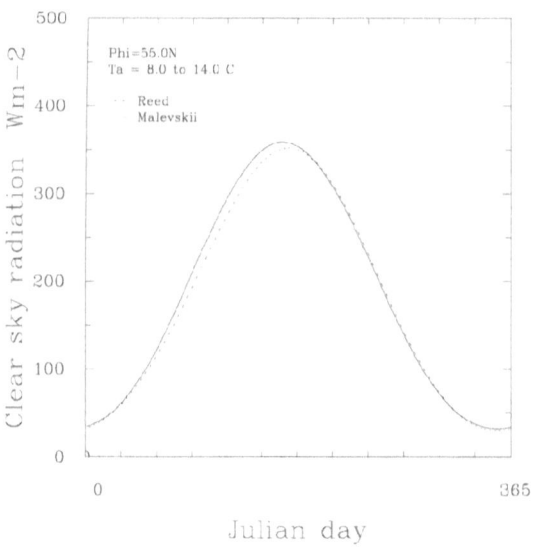

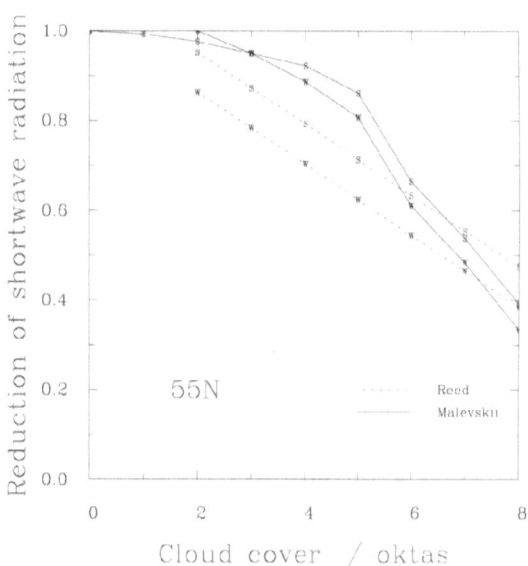

Figure 2: *Clear sky radiation in 55°N of Malevskii and of Reed. For Malevskii's parameterisation, which requires the air temperature, a sinusoidal function (8 °C in February and 14 °C in August) is supposed*

Figure 3: *Reduction of shortwave radiation by clouds. The functions for summer (21. July) and winter (21. December) of Reed and Malevskii are compared for a latitude of 55°N.*

ocean (NIEKAMP, 1992). His scheme is preferred to the often used formula of REED (1977), because MALEVSKII parameterized the variation of transmissivity, which is a constant in the formulation of REED. Following MALEVSKII, in a first step the clear sky radiation Q_0 (fig. 2) is obtained by

$$Q_0 = 1000 \, C \, (sin\gamma)^D \, Wm^{-2} \qquad , with : \gamma = sun \; altitude \qquad (2)$$

where the coefficients C and D depend on the transmission P

$$
\begin{aligned}
C &= 1.20 - 1.7P + 2P^2 & &(3) \\
D &= 0.38 + 2.5P - 2P^2 & , for \; P > 0.75 &\quad (4)
\end{aligned}
$$

or

$$
\begin{aligned}
C &= 1.50 - 2.1P + 2P^2 & &(5) \\
D &= 1.808 - 0.9P & , for \; P \leq 0.75 &\quad (6)
\end{aligned}
$$

The transmission factor P is given as a function of air temperature T_a, because air temperature is strongly correlated to the water vapor content, which tends to decrease the transmissivity of the atmosphere.

$$P = 0.79 - 0.003T_a \qquad (7)$$

Table 1 Cloud reduction of the shortwave radiation

γ	Total cloud cover n_t										
.	0.00	0.10	0.20	0.30	0.40	0.50	0.60	0.70	0.80	0.90	1.00
5	1.00	1.02	1.02	1.00	0.95	0.91	0.85	0.78	0.70	0.57	0.37
10	1.00	1.02	1.02	1.00	0.96	0.92	0.85	0.78	0.69	0.56	0.36
20	1.00	1.01	1.01	0.99	0.96	0.92	0.86	0.78	0.68	0.54	0.35
30	1.00	1.00	1.00	0.98	0.96	0.93	0.87	0.80	0.70	0.57	0.38
40	1.00	1.00	0.99	0.98	0.96	0.94	0.88	0.81	0.72	0.60	0.42
50	1.00	0.99	0.98	0.97	0.96	0.94	0.89	0.83	0.75	0.64	0.45
60	1.00	0.98	0.97	0.97	0.96	0.94	0.90	0.84	0.77	0.68	0.49
70	1.00	0.97	0.96	0.96	0.96	0.95	0.92	0.86	0.80	0.71	0.52
80	1.00	0.96	0.96	0.96	0.96	0.96	0.93	0.89	0.83	0.75	0.55
90	1.00	0.96	0.96	0.96	0.96	0.96	0.95	0.91	0.86	0.78	0.59

In a second step the effects of clouds are taken into account by

$$Q_s = Q_0 F(n_t, \gamma) \tag{8}$$

where the factor of cloud reduction F depends on total cloud cover n_t and solar altitude γ (fig.3). This factor was tabulated by MALEVSKII (1992) and is reproduced as table 1.

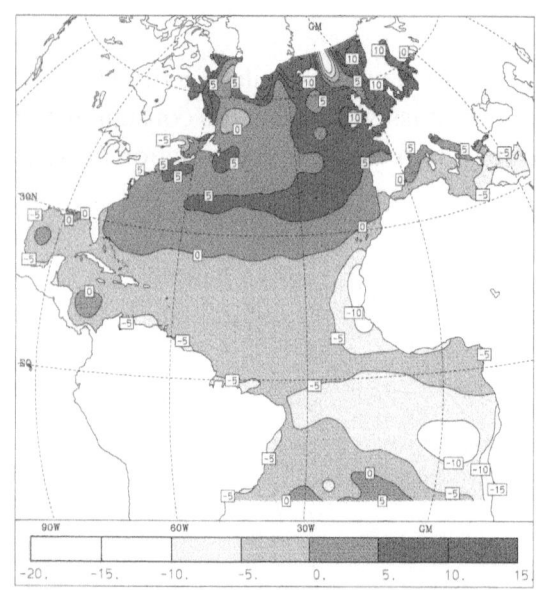

Figure 4: *Annual mean difference of shortwave radiation (Malevskii minus Isemer & Hasse)*

The MALEVSKII scheme may be compared to other parameterisations of shortwave radiation, e.g. REED (1977). He introduced an often applied formula, which has been modified by ISEMER & HASSE (1987). They proposed to increase the cloud coefficient c_n from 0.620 to 0.637 and to decrease the transmissivity of the atmosphere from 0.70 to 0.69. Compared to the results of ISEMER & HASSE, MALEVSKII's scheme yields more radiation in high latitudes and smaller values in low latitudes (fig. 4). But the differences are in general less than $5\ Wm^{-2}$. Averaged over the Atlantic between 70°N and 25°S, which is the southern limit of definition for REED's formula, both parameterisations give about $183\ Wm^{-2}$ and differ less than $1\ Wm^{-2}$.

The parameterisation of MALEVSKII yields hourly means of the radiation. Since the merchant ship observations are carried out at fixed hours referred to GMT, the local time of ship reports is a function of geographic longitude. To avoid errors due to this effect, for each individual report a fictitious daily mean radiation is computed by averaging over all solar altitudes of the individual day and position, assuming the same cloud cover throughout the day.

3.2 Net Longwave Radiation

ISEMER & HASSE (1987) calculated the net heat loss of the sea surface due to longwave radiation by a formula of EFIMOVA (1961), which has been modified by BUNKER (1976). It was in good agreement with a radiation transfer model of FUNG (1984), but only clear sky conditions were tested.

Recent calculations in the quasi-enclosed Mediterranean Sea show, that net longwave radiation LR needs to be considerably larger to balance the well measured heat flux through the strait of Gibraltar. BUNKER et al. (1982) proposed a new scheme including the effects of different cloud types. Compared to his previous formula the mean LR for the Mediterranean Sea is increased from 52 Wm^{-2} to 67 Wm^{-2}.

For our calculations we used an algorithm derived by BIGNAMI & al. (1991), which recently has been slightly revised by BIGNAMI (1995). The parameterization bases on measurements of radiative fluxes in the Tyrrhenian Sea (SCHIANO & al., 1993).

$$LR = \epsilon \, \sigma \, T_s^4 \left(0.344 - 6.66 \, 10^{-3} \, e_a \right) \left(1 - 0.42 \, C \right) \tag{9}$$

with $\epsilon = 0.98$ is the sea surface emissivity, σ the Stefan-Boltzmann constant, T_s the water temperature, e_a the vapor pressure in hPa and C the fractional cloud cover.

Applying this formula to COADS, and considering additionally that in the Mediterranean the transmissivity of air is strongly reduced by aerosols, GILMAN & GARRETT (1994) obtained much better results for the longterm heat budget. The BIGNAMI formula diminishes the conventionally resulting heat gain of about 30 Wm^{-2} to a realistic value near zero.

3.3 Latent and Sensible Heat Fluxes

Turbulent heat fluxes are individually calculated by the well-known bulk formulae

$$LE \;=\; \rho \, L \, C_E \, (q_a - q_s) \, W \tag{10}$$
$$H \;=\; \rho \, c_p \, C_H \, (T_a - T_s) \, W \tag{11}$$

with ρ: air density using individual observations of pressure and virtual temperature

L: temperature dependent heat of vaporisation

W: scalar wind speed

c_p: specific heat of air at constant pressure

$q_a - q_s$:   the difference between specific humidity of air and saturation humidity at sea surface temperature

$T_a - T_s$:   the difference between air and sea temperature

C_E, C_H: exchange coefficients for water vapor and heat

Three major uncertainties are associated with these formulae:

1.   The determination of exchange coefficients C_E and C_H.
2.   Systematic errors in the ship-based measurements of $T_a - Ts$ and $q_a - q_s$
3.   Conversion of the original Beaufort estimates into wind speeds.

3.3.1 Exchange Coefficients

In this study we used the wind-speed and stability-dependent exchange coefficients of BUNKER (1976), modified by ISEMER & HASSE (1987). Their comparison with several open ocean measurements showed, that BUNKER's exchange coefficients are too high. Consequently, they reduced C_E by 13% and C_H by 17%, equivalent to a neutral Dalton-number C_{EN} near 1.25 10^{-3} in the range of low to moderate winds.

The stability dependence of C_E and C_H is usually expressed in terms of the air-sea temperature difference. Following ISEMER & HASSE we used the virtual temperature difference instead of the actual values, to take into account the influence of water vapor on the density gradient.

3.3.2 Systematic Errors of Temperature Observations

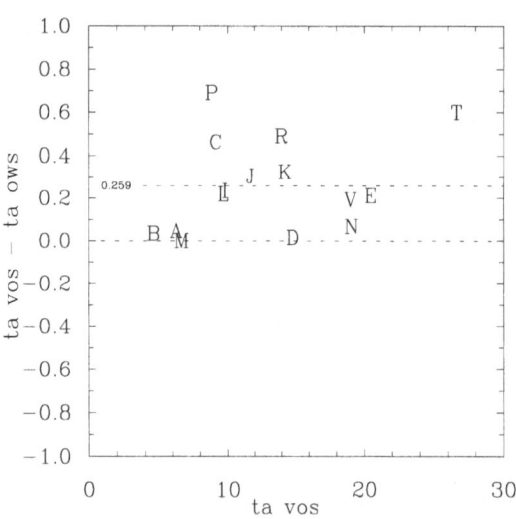

It has been known for a long time, that temperature measurements on bord of merchant ships have considerable systematic errors. Their consequences on heat flux calculations were already pointed out by DIETRICH (1950). Important for sensible and latent heat fluxes are the small differences between air temperature T_a and wet bulb temperature T_w respectively against the sea surface temperature T_s. To quantify the systematic errors, we compared temperature measurememts of Ocean Weather Ships (OWS), which are considered to be correct, with reports of merchant ships (VOS) in the vicinity of the OWS. Data were available from 15 Stations, 11 in the North Atlantic and 4 in the North Pacific, operating during different periods between 1945 and 1989. Simultaneous merchant ship observations within a radius of 300 km were extracted from COADS. In this way 700,000 pairs of observations were found. To avoid errors due to an unsymmetric distribution of merchant ships around

Figure 5: *The mean difference of air temperature measurements (Merchant ships minus Ocean Weather Ships OWS) at Atlantic OWSs A, B, C, D, E, I, J, K, L, M, R, and Pacific OWSs P, N, T, V. The dashed line shows the mean of all stations.*

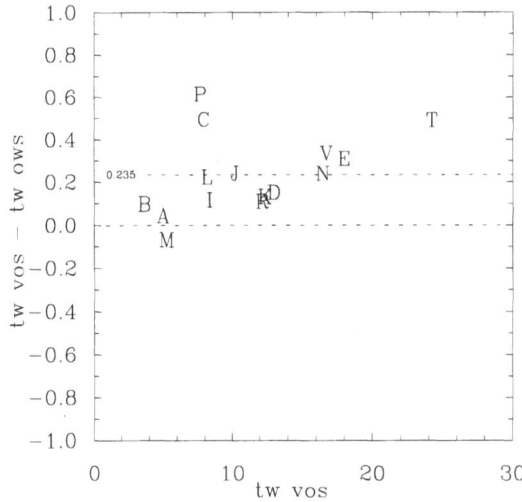

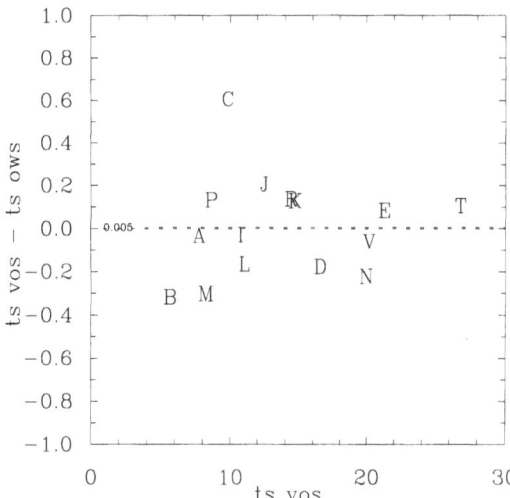

Figure 6: *As fig.5, but for wet bulb temperature.*

Figure 7: *As fig.5, but for sea surface temperature.*

the OWS, data were interpolated on a two-dimensional linear field for each month and each station.

The air temperature measured on merchant ships is higher than OWS-reports for all stations, except OWS M (fig. 5). The mean difference $T_{aVOS} - T_{aOWS}$ varies considerably from - 0.00 °C at OWS M to 0.69 °C at OWS P. Because no relation to climate zones is obvious, we averaged over all stations and obtained an overestimation of 0.259°C. The comparison of wet bulb temperatures yields similar results. T_w is generally overestimated by merchant ships, though the mean bias is only 0.235 °C (fig 6). However, important for heat fluxes is mainly the error of the air-sea temperature differences. Averaging the results from all stations we found only a very small bias of 0.005 °C between merchant ships and OWS-measured T_s (fig 7).

Thus, merchant ships are reporting a too warm atmosphere, containing therefore too much water vapor. As values at sea surface are fairly well determined, vertical temperature and humidity gradients are too small, so that sensible and latent heat fluxes are underestimated. Consequently, we corrected the merchant ship observation by subtracting the following biases:

$$\Delta T_a = 0.259°C$$
$$\Delta T_f = 0.235°C$$
$$\Delta T_s = 0.005°C$$

3.3.3 Beaufort Equivalents

Even nowadays the majority of wind observations at sea is obtained as Beaufort estimates, whereas bulk formulae require the wind speed. Therefore, a correct equivalent scale is of

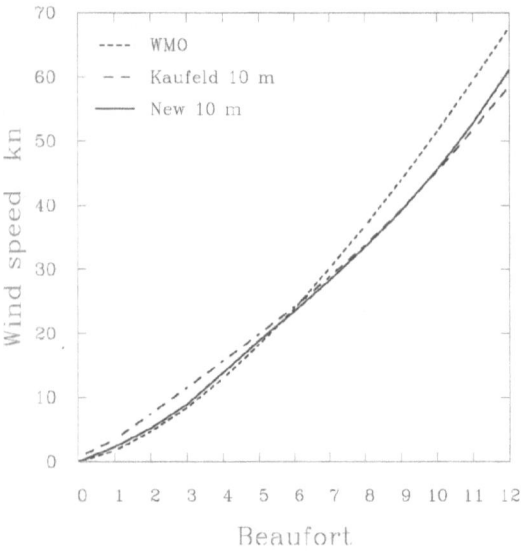

Figure 8: *Different Beaufort equivalent scales. For a better comparability the Kaufeld scale is reduced from 25m to 10m.*

Figure 9: *New 10m-equivalent scale. Isolines are indicating data density. Outside of the -2-isoline, for instance, a portion of 10^{-2} of the total data is found.*

Bft	0	1	2	3	4	5	6	7	8	9	10	11	12
WMO	0.0	1.7	4.7	8.4	13.0	18.3	23.9	30.2	36.8	44.0	51.4	59.4	67.7
New	0.0	2.3	5.2	8.9	13.9	18.9	23.5	28.3	33.5	39.2	45.5	52.7	61.1
N	6	378	2287	8441	17197	11598	8870	4655	2068	597	122	15	1

Table 2: *New 10m-equivalent values compared to the WMO Code 1100. N gives the number of data pairs, which consists of daily means for OWS measurements and spatial means for Voluntary Observing Ships (VOS)*

great importance. KAUFELD (1981) derived a new scale by comparing wind measurements at OWSs with Beaufort estimates of nearby passing merchant ships (fig. 8). KAUFELD's scale has been used by ISEMER & HASSE (1987) and in the following years by several authors.

But also KAUFELD's scale has proved to be wrong. Applying his scale for example on COADS, it does not reproduce monthly mean wind speeds at OWSs (fig. 10), although OWS data were used for its derivation. On average, wind speed is overestimated by KAUFELD's scale. A possible explanation is that german ships, which are found to report too weak winds, were overrepresented in his data set. Furthermore, the slope of KAUFELD's scale is too low. The reason is that the different error variances of OWS and of merchant ship observations have not been taken into account when deriving the scale.

Therefore, we developed a new Beaufort equivalent scale (table 2), by comparing daily means of OWS wind speeds with spatial means of surrounding COADS observations. The averaging radius and the number of included VOS observations were selected such that both the error variance and the natural variability were equal to the OWS data. (fig. 9). This new

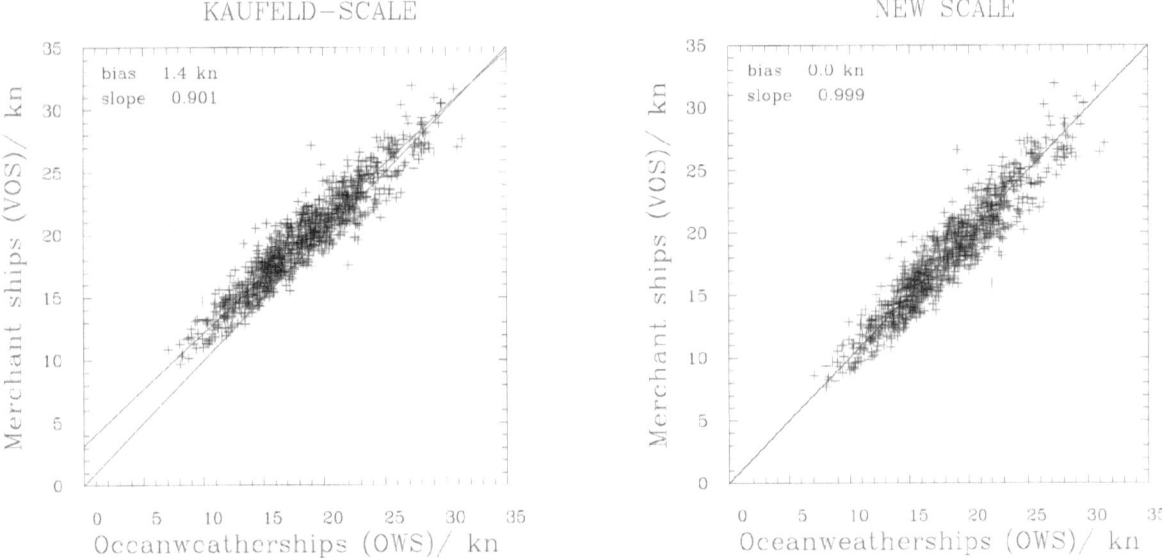

Figure 10: *Application of the Kaufeld scale (left) and the new scale (right). Monthly means of converted Beaufort estimates from merchant ships compared to OWS-measurements of the North Atlantic. Merchant ship observation are taken from a 5° x 7°-surrounding of OWSs.*

scale converts Beaufort estimates in wind speeds that correspond to OWS measurements (fig. 10).

It is not surprising that our new scale is rather similar to the old WMO scale Code 1100, originally derived at the turn of the century (fig. 8). In developing this scale, the measurements at the reference stations were taken as true, and the random variations totally ascribed to imperfections of the Beaufort estimates. This is in good agreement with the error variances which we obtained from COADS as 2.86 $m^2 s^{-2}$ for OWS and 6.22 $m^2 s^{-2}$ for voluntary observing ships.

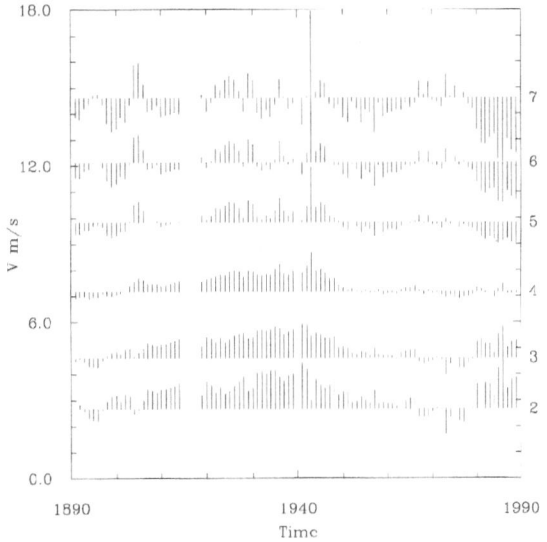

Figure 11: *Time dependent equivalent scale of Beaufort 2-7.*

3.3.4 Time Dependent Beaufort Scale

As the definition of the Beaufort scale changed during the times, it is likely that an equivalent scale is not constant in time, too. We checked North Atlantic wind observations of the last hundred years against simultaneous individual pressure differences between reporting ships. Assumming a constant relationship between wind speed and pressure gradient throughout the years, a time-dependent equivalent scale is obtained (fig.11). Applying this scale on Beaufort wind series, much of the inferred interdecadal trend is removed (fig. 12).

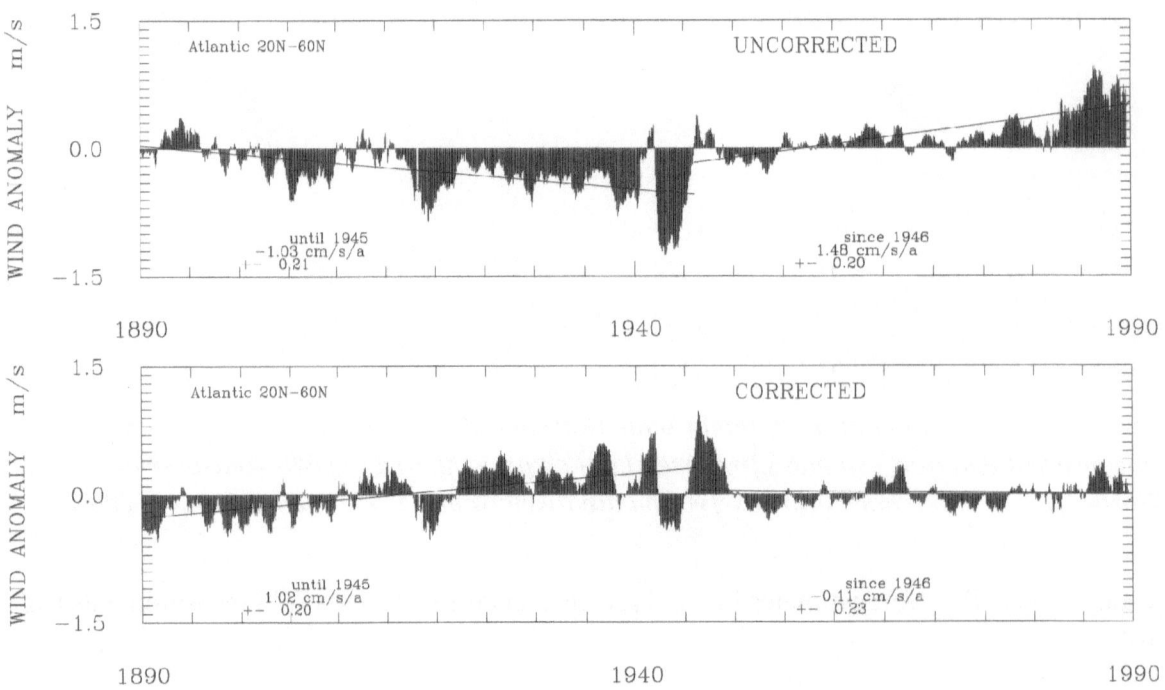

Figure 12: *Time series of North Atlantic wind anomaly according to uncorrected Beaufort estimates of COADS and after applying the time-dependent scale of fig.11. Trends and their errors for the period until 1945 and since 1946 are given in cm per second per year.*

3.4   Wind Stress

The new Beaufort equivalent scale is applied to compute the mean wind stress from individual COADS reports. Further we used the stability- and wind speed dependent drag coefficients proposed by ISEMER & HASSE (1987), which were selected to represent open ocean conditions and are reduced by 21% compared to BUNKER's (1976) coefficients. Even with unbiased Beaufort scale and drag coefficients observational errors and subgrid-scale variability in wind speed and direction need to be considered. The wind stress components are defined as

$$\tau_x = \rho \, C_D \, u \, W \tag{12}$$
$$\tau_y = \rho \, C_D \, v \, W \tag{13}$$

with ρ air density and C_D drag coefficient. u,v denote the east and north component of the wind and W the scalar wind speed, with $W^2 = u^2 + v^2$.

Introducing the wind steadiness s,

$$s = \frac{\sqrt{\overline{uW}^2 + \overline{vW}^2}}{\overline{WW}} \tag{14}$$

the vector mean pseudo stress can be expressed as a function of squared wind speed and wind steadiness.

$$\frac{\overline{\tau}}{\rho C_D} = s\,\overline{WW} \tag{15}$$

Both factors on the right hand side are liable to observational errors.
Obviously, the observation error ΔW of the wind speed has to be taken into account, because stress depends on W^2, so that error effects are not compensated by averaging. The mean stress is overestimated by the factor

$$1 + \frac{\overline{\Delta W \Delta W}}{\overline{WW}} \tag{16}$$

where $\overline{\Delta W \Delta W}$ denotes the error variance.

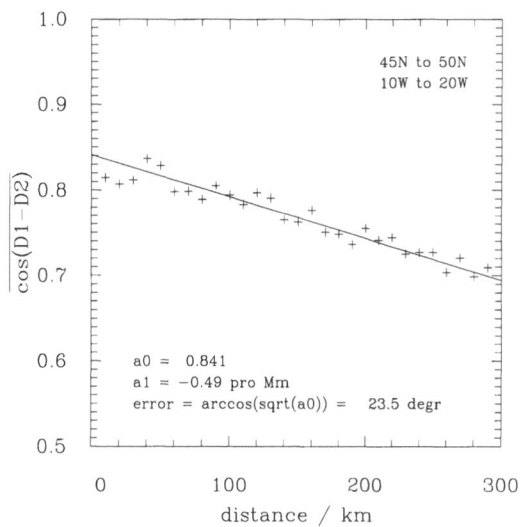

Figure 13: *Example for concluding the error of the wind direction. The mean value of $\cos D_1 - D_2$ where D_1 and D_2 denote tewo individual simultaneous ship observations of the wind direction, is plotted against the ship distance s. A linear fit with $a_0 + a_1 s$ yields a value for $s = 0$*

Not only the observational errors of the wind speed affect the mean stress values, but also errors of the wind direction. These errors, denoted here as Δd, cause a spurious decrease of wind steadiness, consequently the mean wind stress is underestimated. It is easy to show that this effect can be quantified by the factor $\overline{\cos \Delta d}$.
Thus, the ratio of calculated and true wind stress is

$$\frac{\overline{\tau_{obs}}}{\overline{\tau}} = \left(1 + \frac{\overline{\Delta W \Delta W}}{\overline{WW}}\right)\,\overline{\cos \Delta d} \tag{17}$$

To determine the error effects, the mean value of $\cos \Delta d$ and the relative error variance of W are evaluated for different monthly 5°x 10°- fields in the North Atlantic. For this purpose pairs of simultaneous wind reports were extracted from COADS. Their differences in the observed wind speed were computed and sorted into groups of ship distance (fig.13). In this way it is possible to extrapolate lineary to a hypothetical value at zero distance, where only observational errors take effect. If (1) more than 100,000 pairs of observations were found and (2) the linear fit explained more than 95% of the variance, the relative error $\overline{\Delta W \Delta W}/\overline{WW}$ is plotted as a function of $\overline{WW}$. A relative error of about 7% results, decreasing with higher wind speeds (fig.14). An analogous procedure is applied to determinate the mean directional

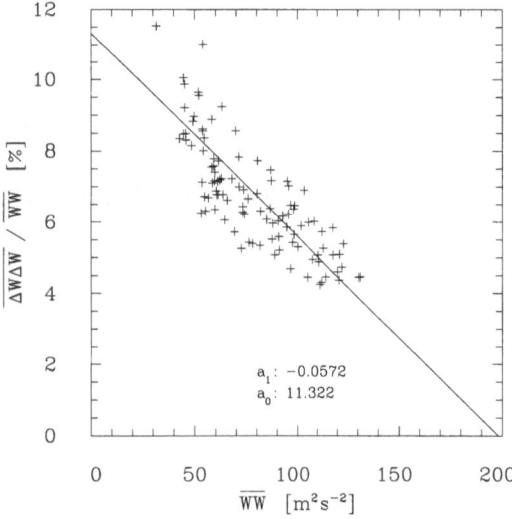

Figure 14: *Relative error of the wind speed as a function of squared wind speed WW. Individual crosses depict monthly 5°x 10°averages. A linear fit of $a_0 + a_1WW$ is applied.*

Figure 15: *As fig.14, but for the wind direction. Two linear fits for WW less than and greater than $50m^2s^{-2}$ are applied*

error $\overline{\cos \Delta d}$ (fig.15). For weak winds, corresponding to mean squared wind speeds of less than 50 m^2s^{-2}, the directional error is

$$\overline{\cos \Delta d} = 0.700 + c\,\overline{WW} \qquad\qquad , c = 0.00331\ s^2m^{-2} \qquad\qquad (18)$$

and for stronger winds

$$\overline{\cos \Delta d} = 0.837 + c\,\overline{WW} \qquad\qquad , c = 0.00058\ s^2m^{-2} \qquad\qquad (19)$$

Summarizing both, strength and steadiness error, the latter dominates, so that the wind stress is underestimated by 5% (for $\overline{WW} = 150\ m^2s^{-2}$) to 10% (for $\overline{WW} = 40\ m^2s^{-2}$).

4   Net Air-Sea Heat Flux and Meridional Heat Transport

The net heat gain of the ocean is determined by summing the fluxes of shortwave (SR) and longwave radiation (LR) and the turbulent fluxes of sensible (H) and latent heat (LE) at the sea surface.

$$NHF = SR + LR + LE + H \qquad\qquad (20)$$

The annual average of net heat flux (NHF) is small in most parts of the Atlantic (fig.16). Of course, the Gulf Stream region is an exeption where the ocean looses large amounts of energy mainly due to strong evaporation. Also the warm water of Agulhas Current provides for a considerable heat loss off South Africa. On the other hand the ocean gains energy in the equatorial zone due to strong downward radiation and weak evaporation, caused by low wind speeds in this region. Net heat uptake occurs also over the cold waters of the Labrador current near Newfoundland, at latitudes south of 45°S and in upwelling regions.

Considering longterm averages and neglecting decadal changes of the mean ocean temperature, any net heat loss at the sea surface has to be compensated by an oceanic heat transport into this region. These horizontal transports are sufficiently well determined from oceanographic sections to serve as an independent constraint for the computed net air-sea heat flux. As the Atlantic is laterally enclosed only meridional heat transports (MHTs) contribute to a heat exchange with other oceans. The heat transport across 65°N, where the northern boundary of the Atlantic may be defined, is small. According to GULEV & TICHONOV (1989) the net heat loss into the Arctic Ocean is equal to 0.275 PW, whereas AARGAARD & GREISMAN (1975) estimated a smaller value of 0.098 PW. Integrating the net air-sea heat flux from 65°N southward, yields the MHT as function of latitude:

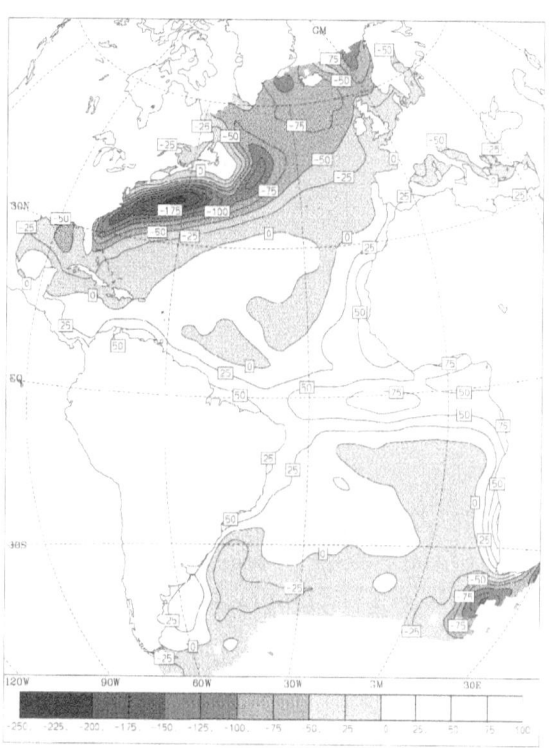

Figure 16: *Annual net air sea heat exchange*

$$MHT_{(\varphi)} = \int\limits_{65°}^{\varphi} \int\limits_{\lambda_{(\varphi_{east})}}^{\lambda_{(\varphi_{west})}} NHF \, d\lambda d\varphi \; + \; MHT_{(65°)} \qquad (21)$$

Beginning with 0.098 PW at 65°N a maximum northward heat transport of 1.09 PW results at 23°N (fig.18) and an energy rate of 0.78 PW is crossing the equator. Southward of 10°S the MHT remains at a nearly constant northward flow of about 0.5 PW, because the heat budget in this region is approximately balanced. The vertical bars in fig.18 indicate the oceanographic estimates of BRYDEN & HALL (1980) at 25°N, of WUNSCH (1984) for the equator and of HOLFORT (1994) at 30°S. Since our results are in good agreement with these independent estimates no additional adjustment of parameterisations was necessary.

Our results may be compared with the two studies of HASTENRATH & LAMB (1978) and of ISEMER & HASSE (1987), denoted in the following as H&L and I&H, respectively. H&L used the climatological method, i.e. averaged values of meteorological parameters are taken for the heat flux evaluation, whereas I&H's results are derived from the individually computed fluxes of BUNKER. In both studies only parts of the Atlantic Ocean are considered.

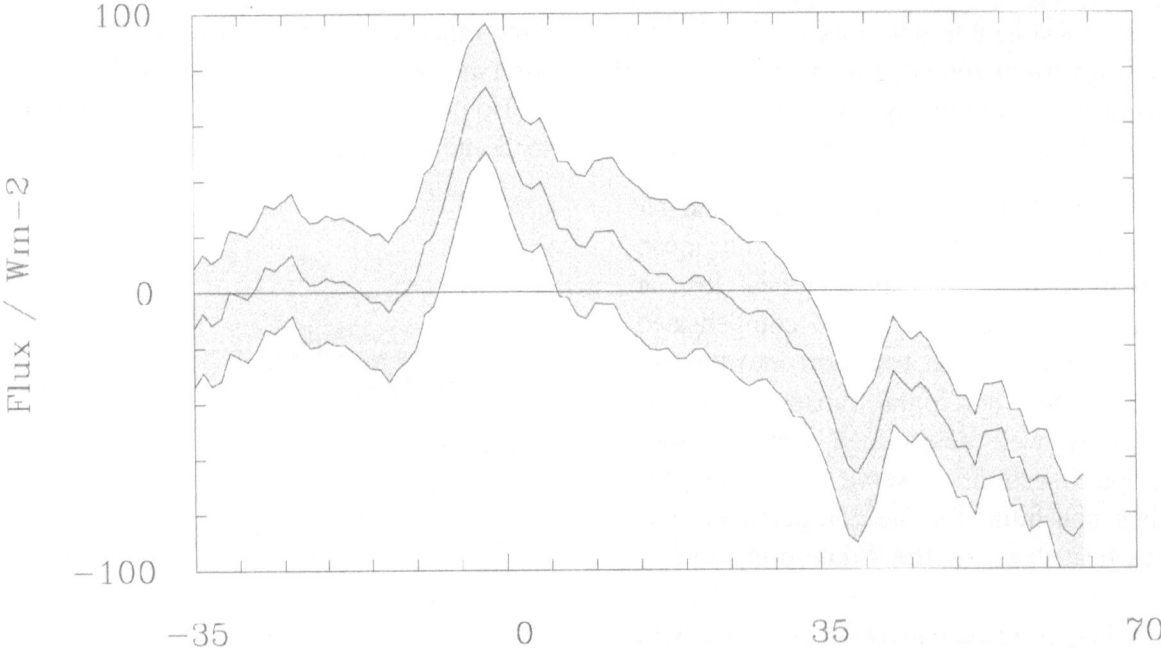

Figure 17: *Zonal mean of net air-sea heat exchange. Shaded area gives errors using the estimates from Isemer & Hasse (1987), which yields relative errors of 7.8 % for shortwave radiation, 14.7 % for longwave radiation, 12.6 % for latent heat flux and 31.3 % for sensible heat flux.*

For better comparison we repeated their calculations with COADS reports by applying their parameterisations of the four heat flux components and integrated the results to obtain the meridional heat transport.

Because the integration is started at the northern edge of the Atlantic, differences between the three studies increase southward. At least at 30°S it becomes obvious that the parameterisations of ISEMER & HASSE and HASTENRATH & LAMB provide too weak (0.17 PW) and too strong (0.92 PW) heat transports, respectively. Our calculation yields a transport of 0.47 PW across 30°S, in good agreement with the oceanographic estimate by HOLFORT (1994). His result is confirmed by other investigations. An overview is given by MACDONALD (1993), who compiled the results of several oceanic heat transport studies. For the Atlantic between 28°S and 32°S he listed eight values from different authors with an average of 0.51 PW.

I&H constrained their parameterisations such, that the MHT at 25°N became equal to 1 PW. Our simulation using exactly the same formulae but another data set (COADS) yields a transport of only 0.87 PW across this latitude. There are three possible reasons for this difference:

The first is of course the use of a different data set, secondly, only mean results of the individually computed fluxes were available for I&H. Thus, crude assumptions were necessary to change the wind speed scale from WMO to KAUFELD scale. Thirdly, the Mediterranean Sea is included in our evaluations, while in the original I&H study the heat transport through

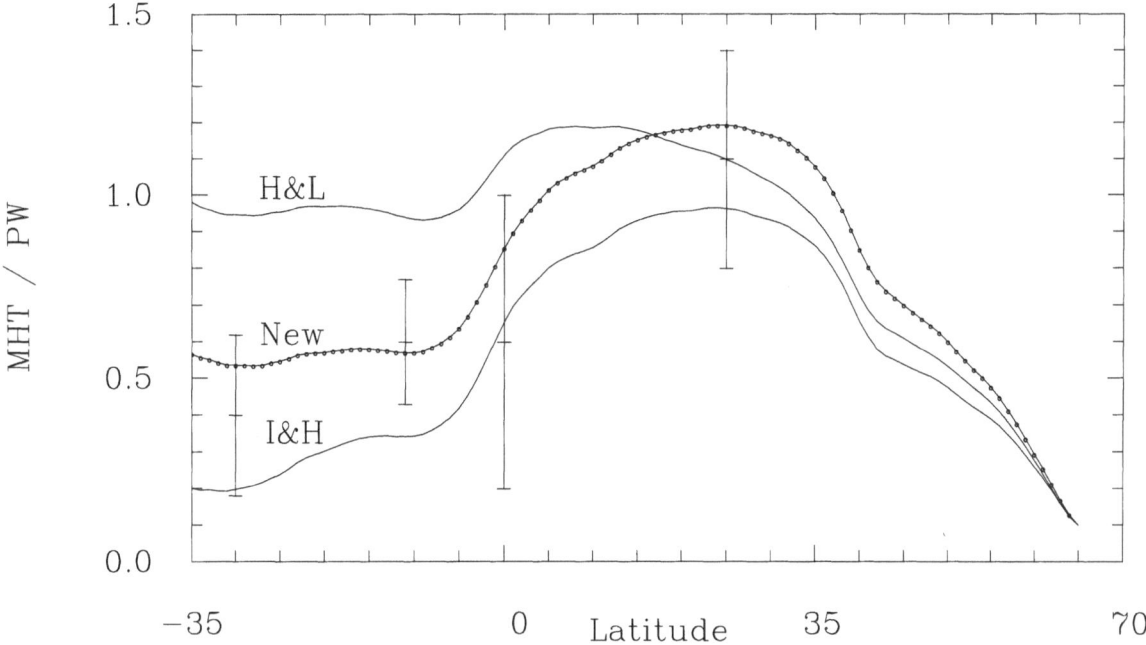

Figure 18: *Atantic meridional heat transport compared to the results of I & H: Isemer & Hasse (1987) and H & L: Hastenrath & Lamb (1978). These authors used different data bases. In order to ensure comparability, their parameterisations have been applied to COADS. Results of oceanographic sections are indicated at the relevant latitude together with their error bars: 30°S: Holfort (1994), 11°S: Speer & al. (1996), 0°: Wunsch (1984), 25°N: Bryden & Hall (1980).*

the strait of Gibraltar was neglected. Thus, all in all the difference in MHT between I&H's original result and our simultion is plausible to explain.

The computed mean fluxes of the Mediterranean Sea are an indication for the consistency of the the applied parameterisations. MAC DONALD & al.(1994) show that the heat transport through the Strait of Gibraltar is directed into the Mediterranean Sea, implying a net annual heat loss from sea surface of about 5 Wm^{-2}. Using I&H parameterisations with the COADS a heat gain of 20 Wm^{-2} is obtained. The parameterisations of HASTENRATH & LAMB give a gain of 15 Wm^{-2} and our own result for the mean net gain is 5 Wm^{-2}. Thus all three studies give a heat gain of the Mediterranean Sea instead of the actual heat loss. It is likely, that parameterisations for the short wave radiation derived at the open ocean are not transferable to the Mediterranean Sea, where stronger concentrations of aerosols cause a lower atmospheric transmissivity (GILMAN & GARRETT, 1994). Nevertheless our result is near to a balanced state, and appears more realistic compared to the two other studies.

The comparison of MHT derived from net heat flux determinations with MHT from oceanographic estimates is a sharp instrument to detect errors of the flux parameterisations. The ocean heat budget is dominated by the balance between incoming shortwave radiation and outgoing latent heat flux. Consequently, small errors in these fluxes cause large errors in the resulting net heat flux, which would become obvious in comparison to the oceanographic results. On the other hand only the over-all effect of the combined four parameterisations can be confirmed but the separate parameterisations may still have errors, which are compensating each other.

To give an impression how reliable the resulting MHT becomes, we assumed an conservative error of only 2% in each parameterisation, which yield an error of 0.3 PW at 30°S, of the same order of magnitude as MHT itself. A variation of the vapor transfer coefficient CE by 5%, to give another example, changes the MWT across 30°S by about 100%, because the mean Atlantic evaporation is 115.8 Wm^{-2} whereas the mean net heat flux, which determines the heat transport into the basin, is only -5.6 Wm^{-2}. By a slight change of the parameterisations nearly any desired MHT can be obtained, without transgressing meteorological constraints. Thus, a realistic MHT does not prove the correctness of the applied parameterisations, but indicates merely, that the used combination is a possible solution.

5   Concluding Remarks

The climate data and air-sea fluxes presented in this work are based on the most comprehensive collection of marine meteorological observations and on the best available parameterisations for the open ocean.

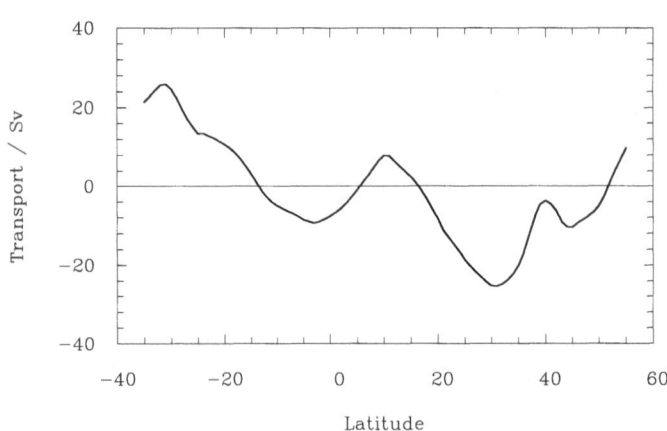

Figure 19: *Meridional Sverdrup transport*

In this study wind and wind stress data were improved in three aspects: First, an improved equivalent scale is used to convert the Beaufort estimates, which is the commonly used wind observation technique at sea. Second, a time-dependent scale is used to obtain temporally consistent wind speeds. Third, it is shown that the inevitable random observational errors cause systematic errors in the mean wind stress. The error variances of wind speed and direction are determined and their systematic effects are removed.

Before computing heat fluxes, systematic errors of temperature measurements were determined and removed. Uncorrected data would give erroneously small fluxes of sensible and latent heat.

The heat flux parameterisations proposed in this study provide results that are consistent with independent oceanographic data. While we cannot assure that the parameterisations of the radiative and turbulent heat flux components are correct in every detail, they are consistent with each other and can be recommended for further use with properly treated marine meteorological data.

6   References

AARGAARD, K. & P.GREISMANN, 1975: *Towards new mass and heat budgets for the Arctic Ocean.* J. Geophys. Res., **80,** 3821-3827.

BIGNAMI, F., R.SANTOLERI, M.SCHIANO & S.MARULLO, 1991: *Net longwave radiation in the western Mediterranean Sea.* Poster session at the 20th General Assembly of the International Union of Geodesy and Geophysics, IAPSO, Wien, August 1991.

BIGNAMI, F., S.MARULLO, R.SANTOLERI & M.E. SCHIANO, 1995: *Longwave radiation budget in the Mediterranean Sea.* J. Geophys. Res., **100,** C2 2501-2514.

BUNKER, A.F., 1976: *Computions of the surface energy flux and annual air-sea interaction cycles of the North Atlantic Ocean.* Mon. Wea. Rev., **108,** 720-732.

BUNKER, A.F., H.CHARNOCK & R.GOLDSMITH, 1982: *A note on the heat balance of the Mediterranean and Red Seas.* J. Mar. Research, **40,** supplement, 73-84.

BRYDEN H.L. & M.M.HALL, 1980: *Heat tranport by currents across 25°N in the Atlantic Ocean.* Science, **207,** 884-886.

DIETRICH G., 1950: *über systematische Fehler in den beobachteten Wasser- und Lufttemperaturen auf dem Meere und ihre Auswirkungen auf die Bestimmung des Wärmeumsatzes zwischen Ozean und Atmosphäre.* Deutsche Hydrogr. Zeitschr. **3,** 314-324.

EFIMOVA, N.A., 1961: *On methods of calculating monthly values of net longwave radiation.* Meteorol. Gidrol. **10,** 28-33.

FUNG I.Y., D.E.HARRISON & A.A.LACIS, 1984: *On the variability of the net longwave radiation at the sea surface.* Rev. Geophys. Space Phys., **22,** 177-193.

GILMAN C. & C.GARRETT, 1994: *Heat flux parameterizations for the Mediterranean Sea: The role of atmospheric aerosols and constraints from the water budget.* J. Geophys. Res., **99,** C3, 5119-5134.

GULEV S.K. & V.A.TICHONOV, 1989: *Interannual variations of the ocean heat balance and meridional heat transport.* Atmosphere-Ocean-Space, **7,** 332-341.

HASTENRATH & LAMB, 1978: *Heat budget atlas of the tropical Atlantic and eastern Pacific Ocean.* University of Wisconsin Press, Madison, 90pp.

HOLFORT J., 1994: *Großräumige Zirkulation und meridionale Transporte im Südatlantik.* Berichte aus dem Institut für Meereskunde an der Christian-Albrechts-Universität, Kiel, **260,** 96 pp.

ISEMER, H.J. & L.HASSE, 1987: *The Bunker climate atlas of the North Atlantic Ocean.* Volume 2: Air-sea interactions. Springer Verlag Berlin, 252 pp.

JOSEY, S.A., E.C. KENT & P.K.TAYLOR, 1995: *Seasonal variations between sampling and classical mean turbulent heat flux estimates in the eastern North Atlantic* Ann. Geophysicae **13**, 1054-1064.

KAUFELD, L., 1981: *The development of a new Beaufort equivalent scale.* Meteorol. Rundschau, **34,** 17-23.

MACDONALD, A., 1993: *Property fluxes at 30oS and their Implications for the Pacific-Indian throughflow and the global heat budget.* J. Geophys. Res., **98,** C4 6851-6868.

MACDONALD, A., J.CANDELA & H.L.BRYDEN, 1994: *An estimate of the net heat transport through the Strait of Gibraltar.* Coastal Estuarine Stud., **46,** edit. P.E. Violette, AGU, Wash., D.C.

MALEVSKII S.P., G.V.GIRDUK & B.EGOROV, 1982: *Calculation of radiative fluxes on the ocean surface.* Leningrad, MGO-Press, 92 pp.

MALEVSKII S.P., G.V.GIRDUK & B.EGOROV, 1992: *Radiation balance of the ocean surface,* Hydrometeoizdat, Leningrad, 148 pp.

NIEKAMP, K.P., 1992: *Untersuchung zur Güte der Parameterisierung von Malevskii-Malevich zur Bestimmung der solaren Einstrahlung an der Ozeanoberfläche.* Diploma Degr., IfM, Kiel, 108 pp.

PAYNE, R.E. 1972: *Albedo of the sea surface.* Journal of Atmosph. Science, **29,** 959-970.

REED, R.K., 1977: *On estimating insolation over the ocean.* J. Phys. Ocean., **7,** 482-485.

SCHIANO M., R.SANTOLERI, F.BIGNAMI, R.M.LEONARDI, S.MARULLO & E. BÖHM, 1993: *Air-sea interaction measurements in the West Mediterranean Sea during the Tyrrhenian Eddy Multi-Platform Observations Experiment.* J. Geoph. Res., **98,** C2 2461-2474.

SPEER, K.G., J.HOLFORT, T.REYNAUD, G.SIEDLER, 1996: *South Atlantic Heat Transport at 11°S.* In: The South Atlantic present and past circulation, G. Welen &G. Siedler eds.,Springer Verlag, Berlin, in press.

WUNSCH, C., 1984: *An eclectic Atlantic Ocean circulation model.* Part 1: The meridional flux of heat. J. Phys. Ocean., **14,**1712-1733.

Part I

Observations

Tim-Latitude Diagrams

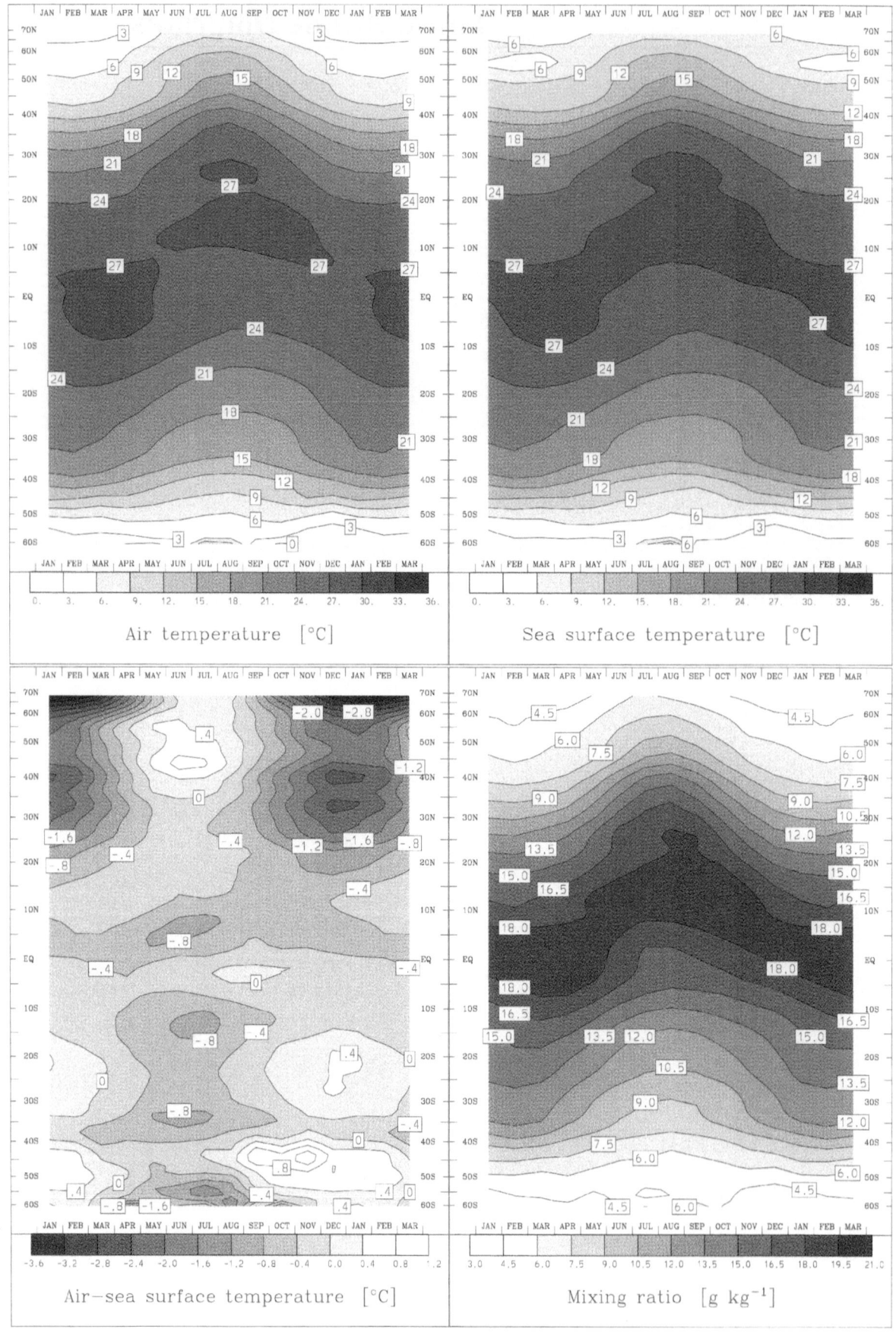

Air temperature [°C]

Sea surface temperature [°C]

Air−sea surface temperature [°C]

Mixing ratio [g kg^{-1}]

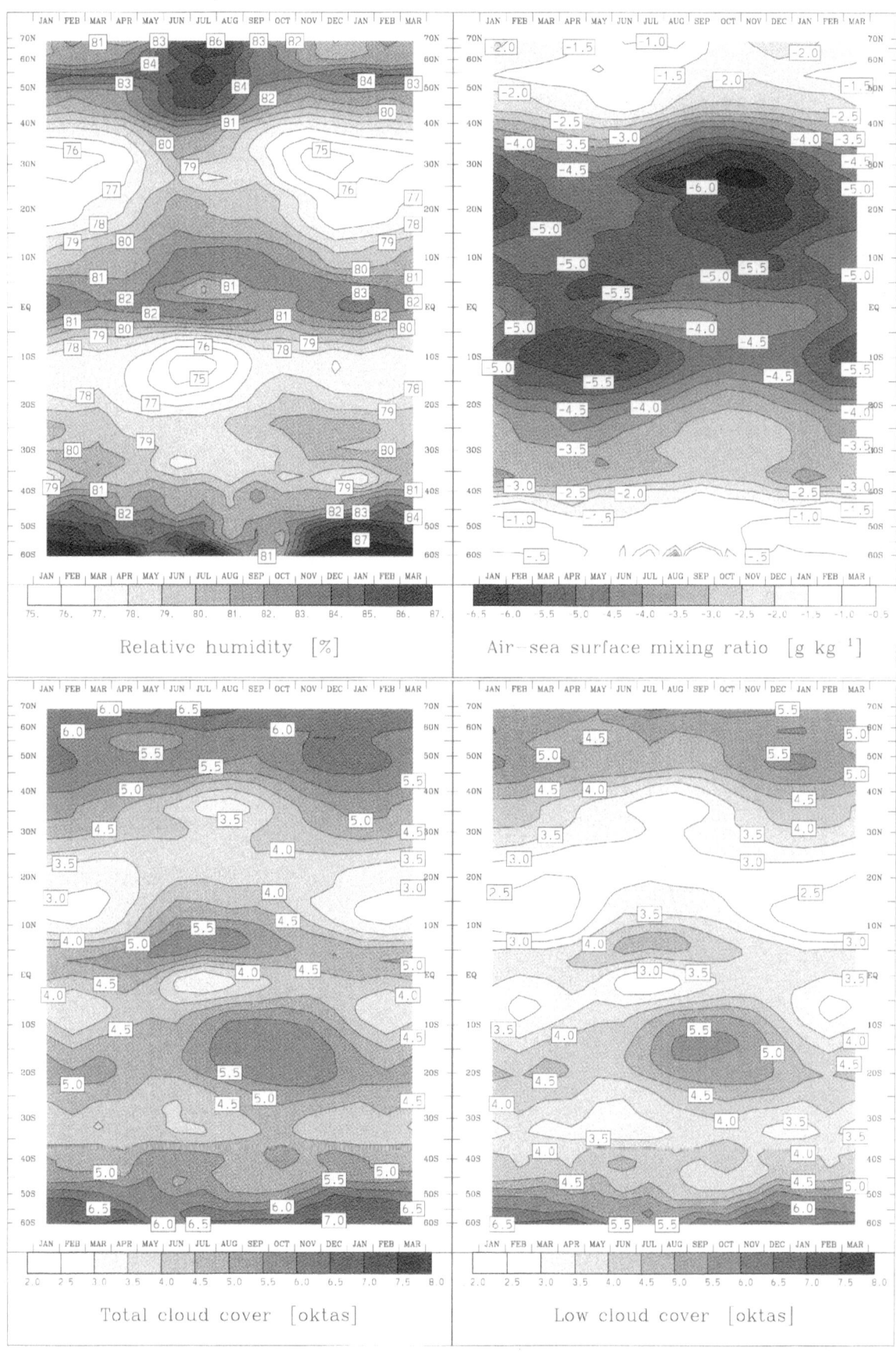

Relative humidity [%]

Air—sea surface mixing ratio [g kg⁻¹]

Total cloud cover [oktas]

Low cloud cover [oktas]

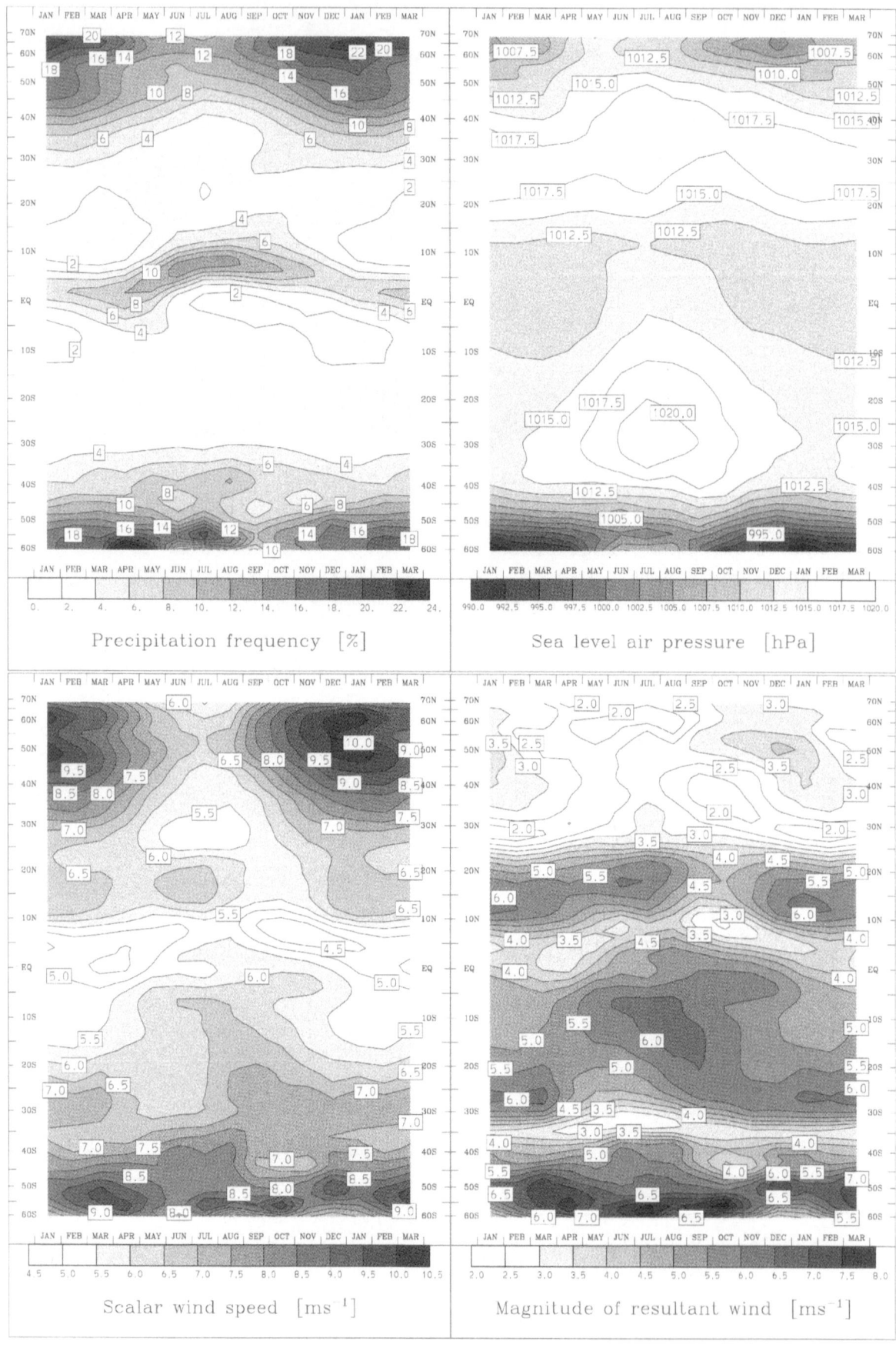

Precipitation frequency [%]

Sea level air pressure [hPa]

Scalar wind speed [ms⁻¹]

Magnitude of resultant wind [ms⁻¹]

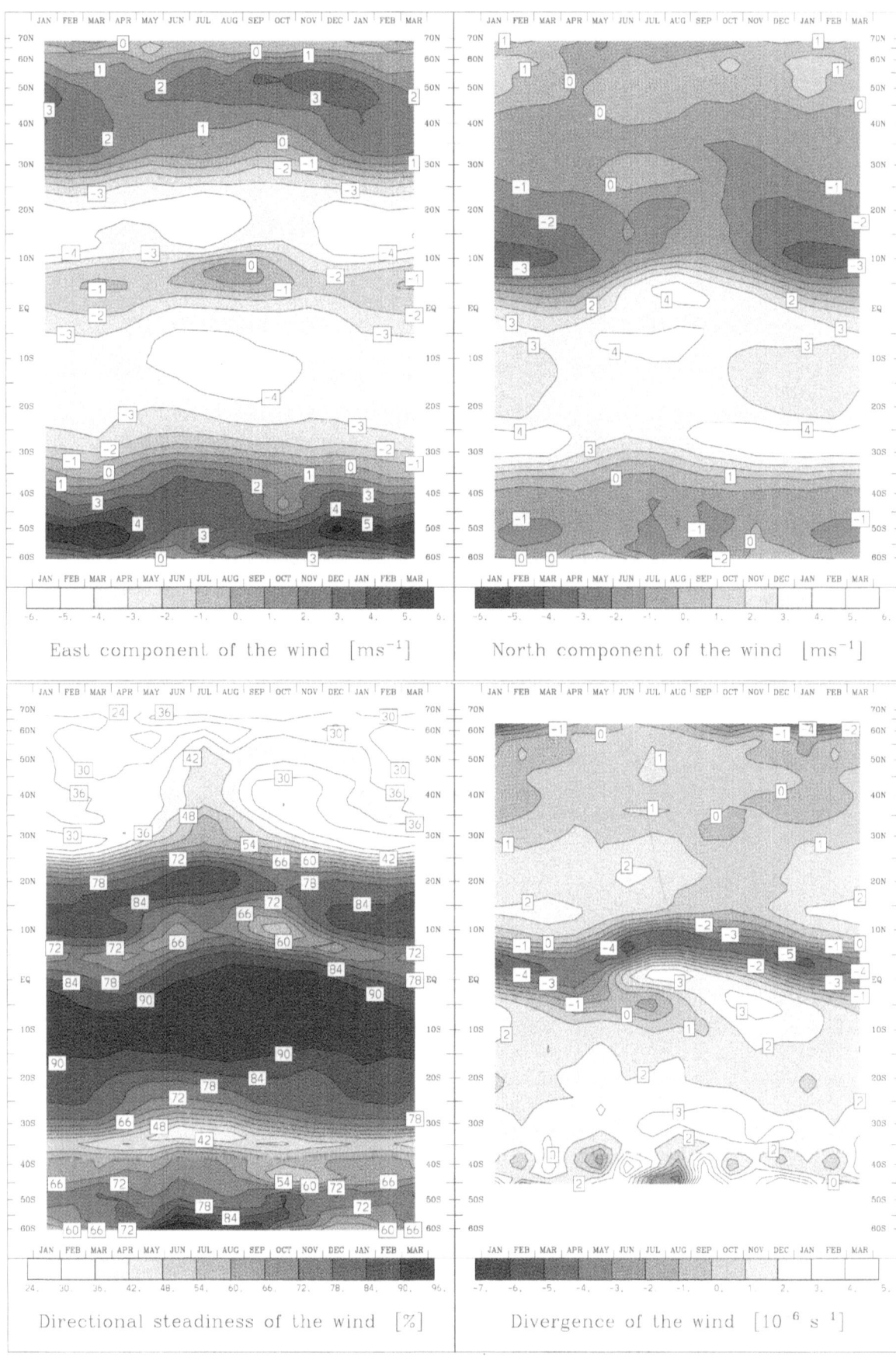

East component of the wind $[\mathrm{ms}^{-1}]$

North component of the wind $[\mathrm{ms}^{-1}]$

Directional steadiness of the wind $[\%]$

Divergence of the wind $[10^{-6}\,\mathrm{s}^{-1}]$

Annual Cycles at Selected Locations

Baltic Sea

54N to 60N   16E to 22E

Air temperature	TA	7.5	°C
Sea surface temp.	TW	8.2	°C
Dew point temp.	DP	5.2	°C
Air − SST	DT	−0.8	°C
Air − dewpoint temp.	DD	2.5	°C
Mixing ratio	MX	6.0	g kg^{-1}
Mixing ratio difference	DQ	−1.2	g kg^{-1}
Total cloud cover	CT	5.0	oktas
Low cloud cover	CL	4.3	oktas
Precipition frequency	PF	7.3	%
Sea level air pressure	PR	1013.7	hPa
Scalar wind speed	FF	7.0	ms^{-1}
Vector wind speed	VV	1.3	ms^{-1}
Mean wind direction	DI	231.1	°

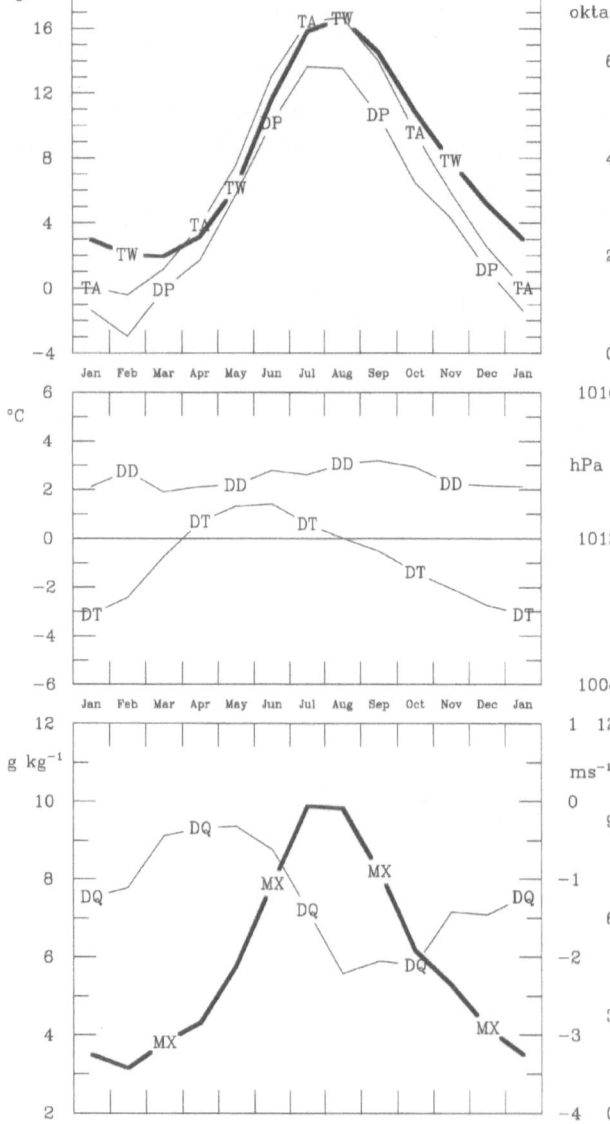

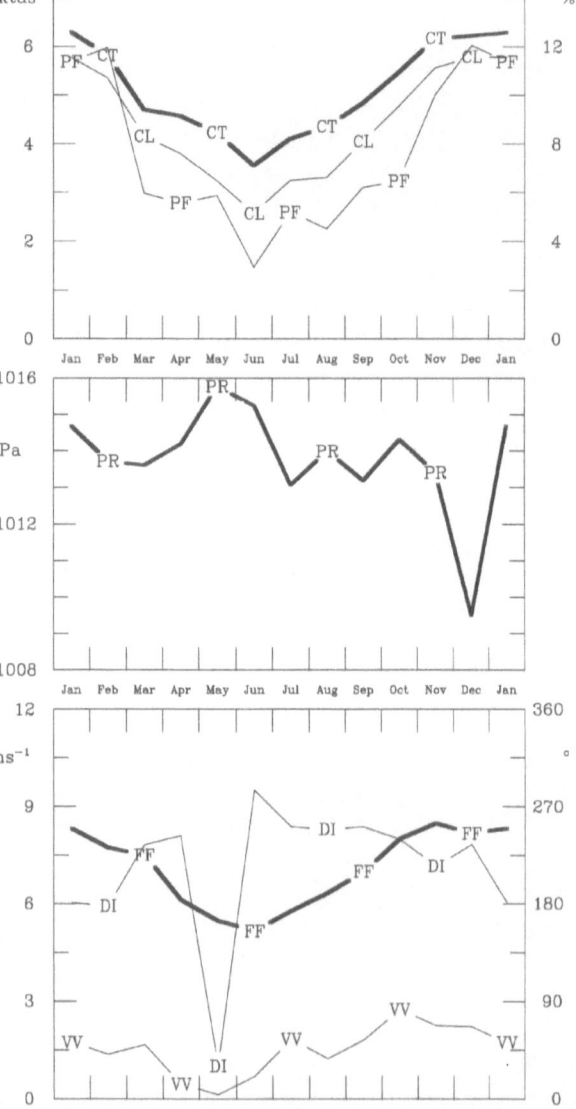

Gulf Stream

38N to 40N   63W to 61W

Air temperature	TA	18.0	°C
Sea surface temp.	TW	21.4	°C
Dew point temp.	DP	13.8	°C
Air – SST	DT	−3.4	°C
Air – dewpoint temp.	DD	4.2	°C
Mixing ratio	MX	10.7	g kg⁻¹
Mixing ratio difference	DQ	−5.9	g kg⁻¹
Total cloud cover	CT	5.7	oktas
Low cloud cover	CL	4.9	oktas
Precipition frequency	PF	11.3	%
Sea level air pressure	PR	1015.7	hPa
Scalar wind speed	FF	8.9	ms⁻¹
Vector wind speed	VV	2.9	ms⁻¹
Mean wind direction	DI	276.4	°

Middle Atlantic Bight

40N to 41N   72W to 71W

Air temperature	TA	11.9	°C
Sea surface temp.	TW	12.8	°C
Dew point temp.	DP	8.9	°C
Air − SST	DT	−0.9	°C
Air − dewpoint temp.	DD	3.3	°C
Mixing ratio	MX	8.0	g kg⁻¹
Mixing ratio difference	DQ	−2.0	g kg⁻¹
Total cloud cover	CT	4.5	oktas
Low cloud cover	CL	3.7	oktas
Precipition frequency	PF	6.7	%
Sea level air pressure	PR	1016.6	hPa
Scalar wind speed	FF	7.1	ms⁻¹
Vector wind speed	VV	1.6	ms⁻¹
Mean wind direction	DI	284.3	°

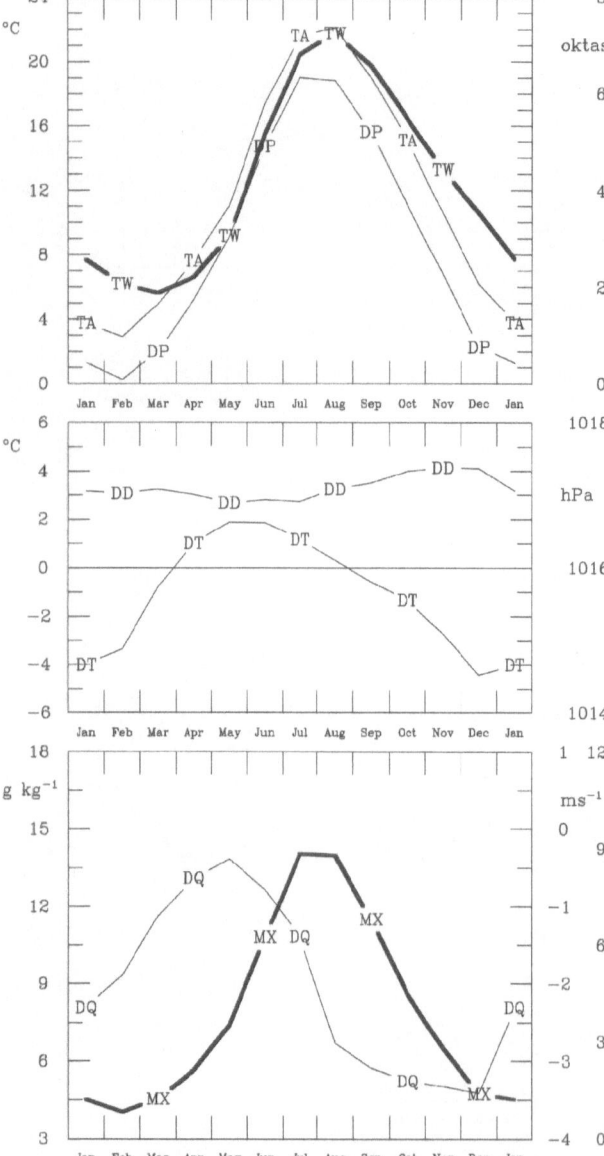

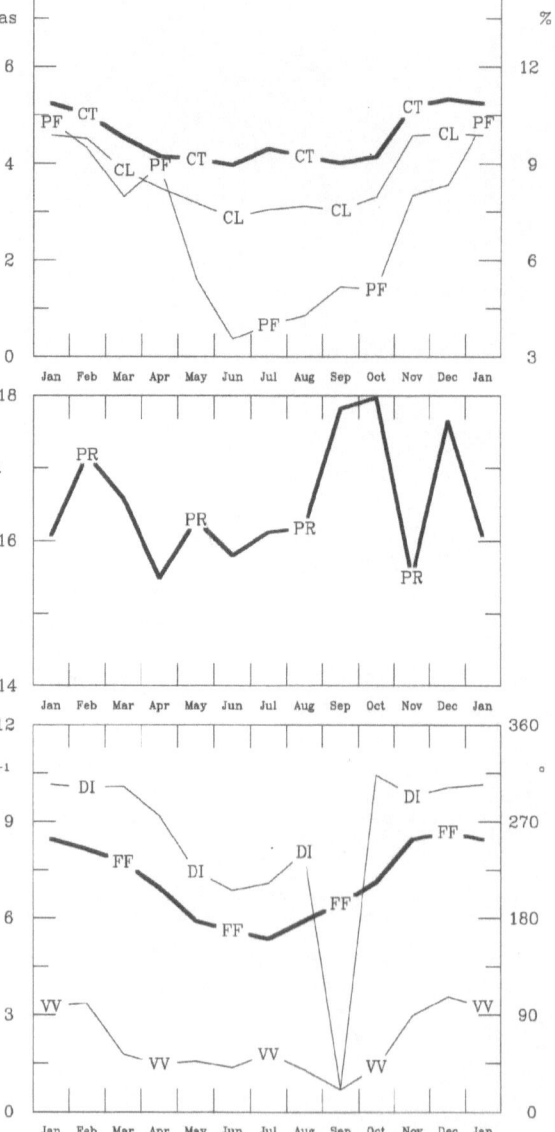

Banks of Newfoundland

Air temperature	TA	4.2	°C
Sea surface temp.	TW	4.3	°C
Dew point temp.	DP	2.3	°C
Air − SST	DT	0.0	°C
Air − dewpoint temp.	DD	2.0	°C
Mixing ratio	MX	4.9	g kg^{-1}
Mixing ratio difference	DQ	−0.5	g kg^{-1}
Total cloud cover	CT	5.8	oktas
Low cloud cover	CL	5.1	oktas
Precipition frequency	PF	15.4	%
Sea level air pressure	PR	1011.7	hPa
Scalar wind speed	FF	8.5	ms^{-1}
Vector wind speed	VV	2.6	ms^{-1}
Mean wind direction	DI	256.9	°

48N to 50N   53W to 51W

Midocean westwind drift

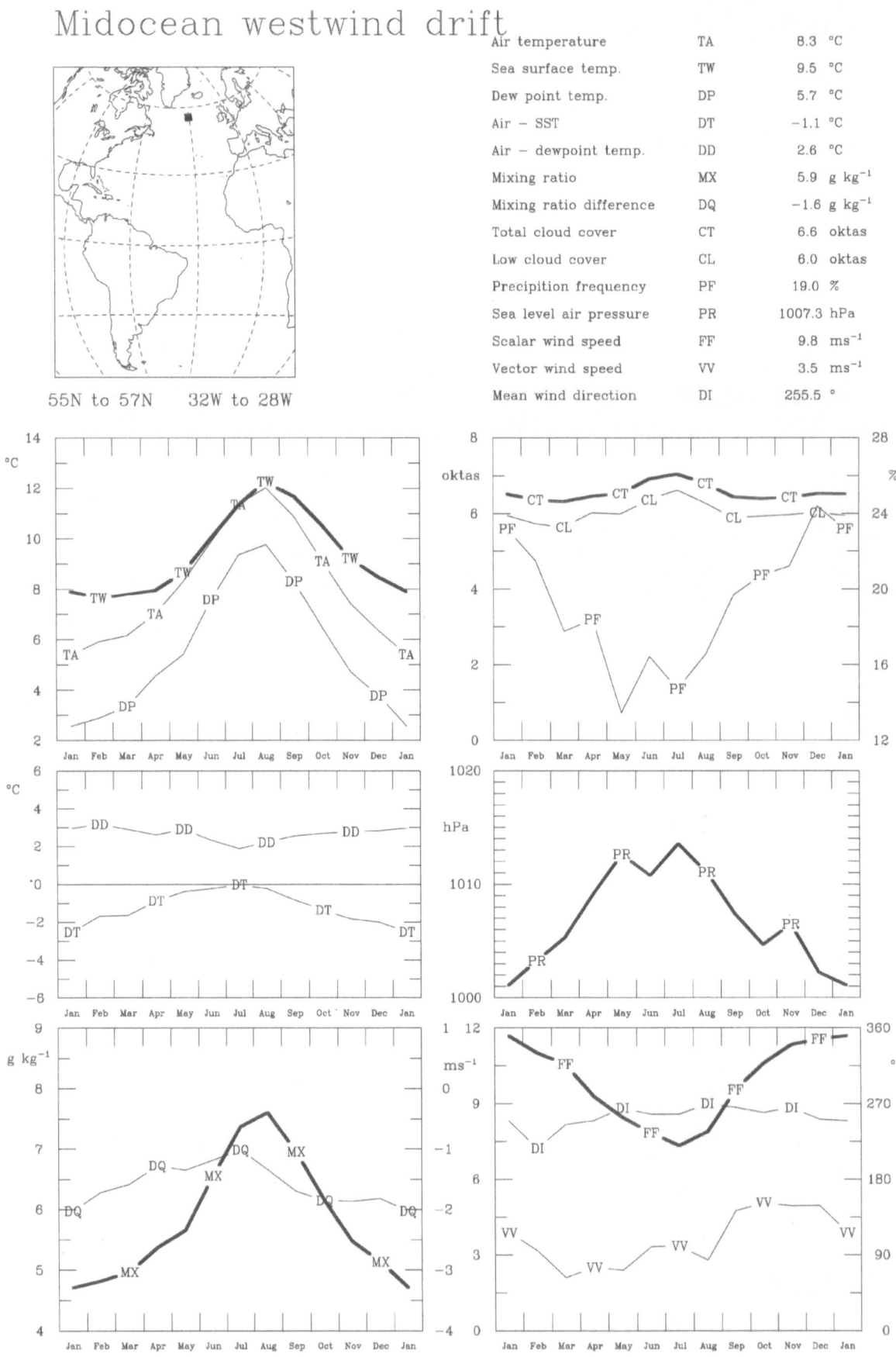

55N to 57N   32W to 28W

Air temperature	TA	8.3	°C
Sea surface temp.	TW	9.5	°C
Dew point temp.	DP	5.7	°C
Air − SST	DT	−1.1	°C
Air − dewpoint temp.	DD	2.6	°C
Mixing ratio	MX	5.9	g kg^{-1}
Mixing ratio difference	DQ	−1.6	g kg^{-1}
Total cloud cover	CT	6.6	oktas
Low cloud cover	CL	6.0	oktas
Precipition frequency	PF	19.0	%
Sea level air pressure	PR	1007.3	hPa
Scalar wind speed	FF	9.8	ms^{-1}
Vector wind speed	VV	3.5	ms^{-1}
Mean wind direction	DI	255.5	°

North Sea

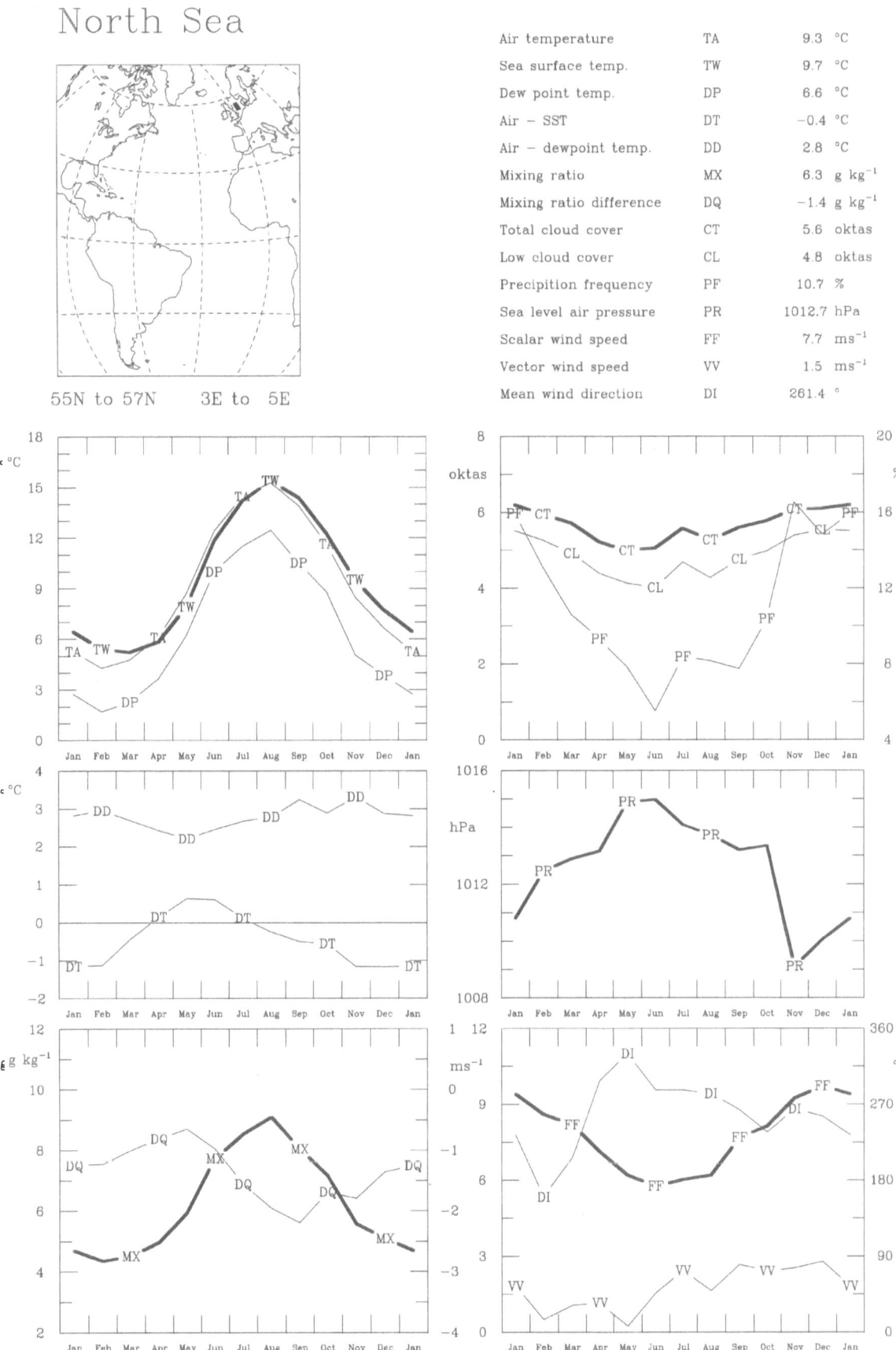

55N to 57N   3E to 5E

Air temperature	TA	9.3	°C
Sea surface temp.	TW	9.7	°C
Dew point temp.	DP	6.6	°C
Air − SST	DT	−0.4	°C
Air − dewpoint temp.	DD	2.8	°C
Mixing ratio	MX	6.3	g kg^{-1}
Mixing ratio difference	DQ	−1.4	g kg^{-1}
Total cloud cover	CT	5.6	oktas
Low cloud cover	CL	4.8	oktas
Precipition frequency	PF	10.7	%
Sea level air pressure	PR	1012.7	hPa
Scalar wind speed	FF	7.7	ms^{-1}
Vector wind speed	VV	1.5	ms^{-1}
Mean wind direction	DI	261.4	°

Subtropical convergence

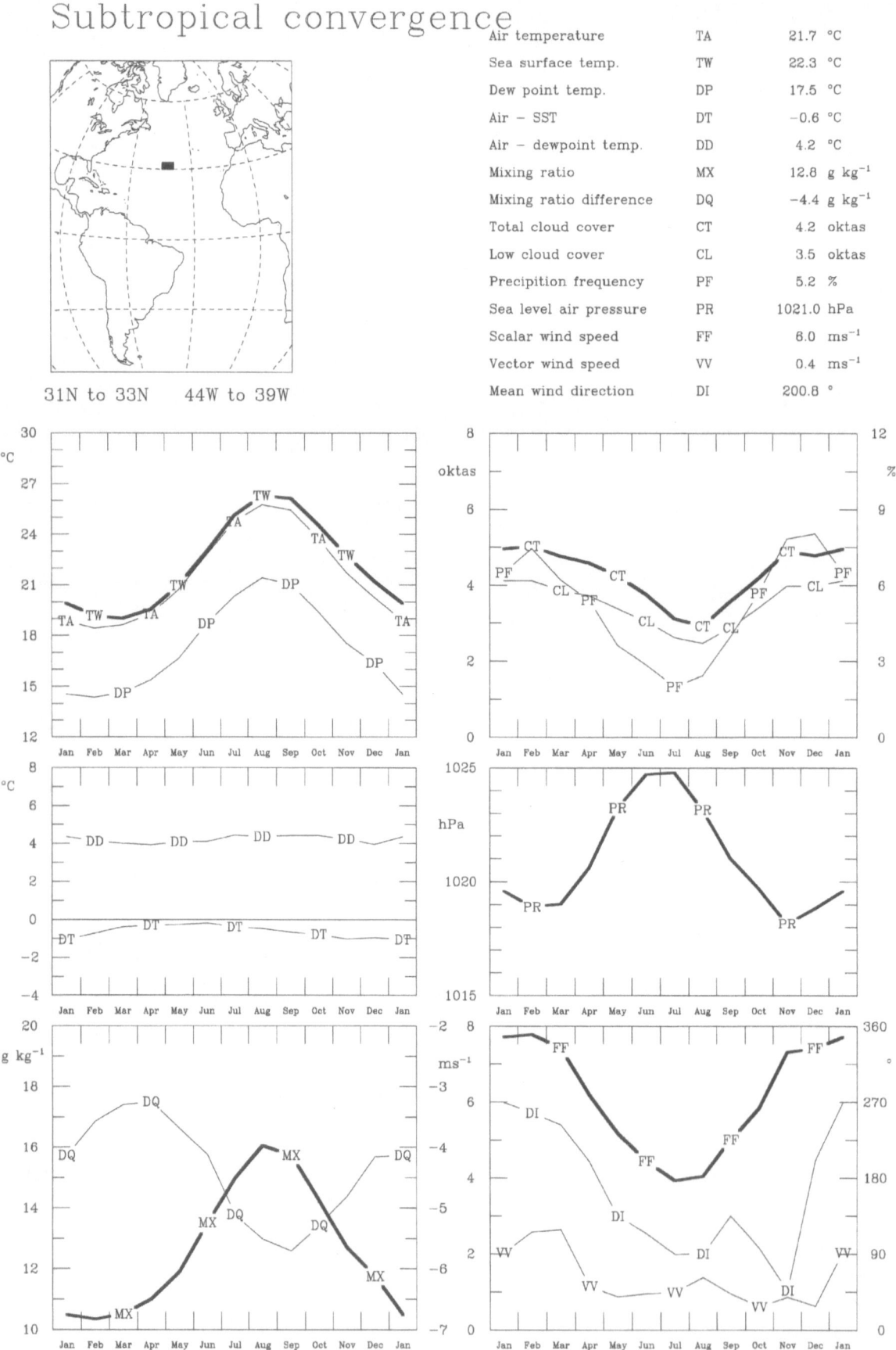

31N to 33N   44W to 39W

Air temperature	TA	21.7	°C
Sea surface temp.	TW	22.3	°C
Dew point temp.	DP	17.5	°C
Air − SST	DT	−0.6	°C
Air − dewpoint temp.	DD	4.2	°C
Mixing ratio	MX	12.8	g kg^{-1}
Mixing ratio difference	DQ	−4.4	g kg^{-1}
Total cloud cover	CT	4.2	oktas
Low cloud cover	CL	3.5	oktas
Precipition frequency	PF	5.2	%
Sea level air pressure	PR	1021.0	hPa
Scalar wind speed	FF	6.0	ms^{-1}
Vector wind speed	VV	0.4	ms^{-1}
Mean wind direction	DI	200.8	°

Coast of Senegal

22N to 23N   19W to 18W

Air temperature	TA	20.8	°C
Sea surface temp.	TW	20.7	°C
Dew point temp.	DP	17.3	°C
Air − SST	DT	0.1	°C
Air − dewpoint temp.	DD	3.6	°C
Mixing ratio	MX	12.5	g kg^{-1}
Mixing ratio difference	DQ	−3.0	g kg^{-1}
Total cloud cover	CT	3.0	oktas
Low cloud cover	CL	2.2	oktas
Precipition frequency	PF	0.4	%
Sea level air pressure	PR	1016.3	hPa
Scalar wind speed	FF	7.3	ms^{-1}
Vector wind speed	VV	6.5	ms^{-1}
Mean wind direction	DI	29.8	°

Trade wind region

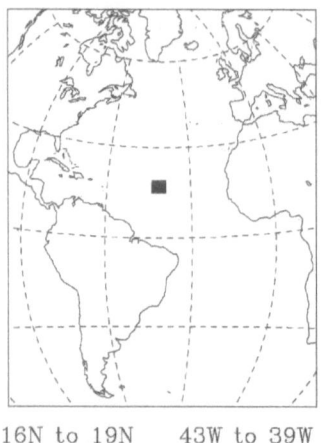

16N to 19N   43W to 39W

Air temperature	TA	25.0 °C
Sea surface temp.	TW	25.4 °C
Dew point temp.	DP	20.7 °C
Air − SST	DT	−0.4 °C
Air − dewpoint temp.	DD	4.4 °C
Mixing ratio	MX	15.5 g kg⁻¹
Mixing ratio difference	DQ	−5.2 g kg⁻¹
Total cloud cover	CT	4.1 oktas
Low cloud cover	CL	3.3 oktas
Precipition frequency	PF	2.8 %
Sea level air pressure	PR	1016.6 hPa
Scalar wind speed	FF	7.0 ms⁻¹
Vector wind speed	VV	6.1 ms⁻¹
Mean wind direction	DI	72.0 °

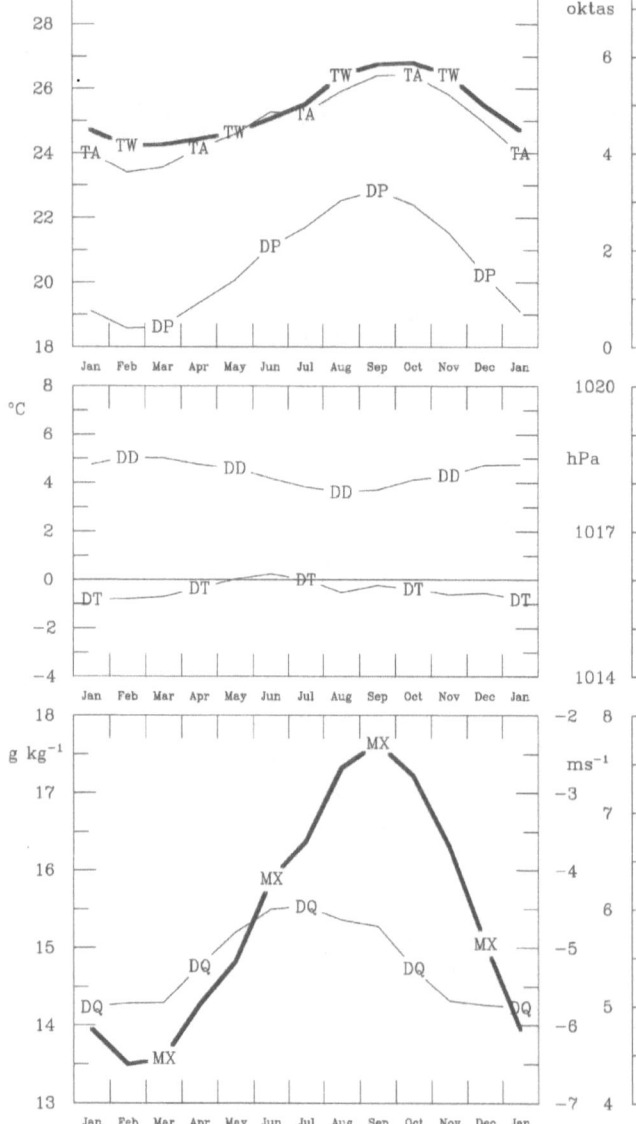

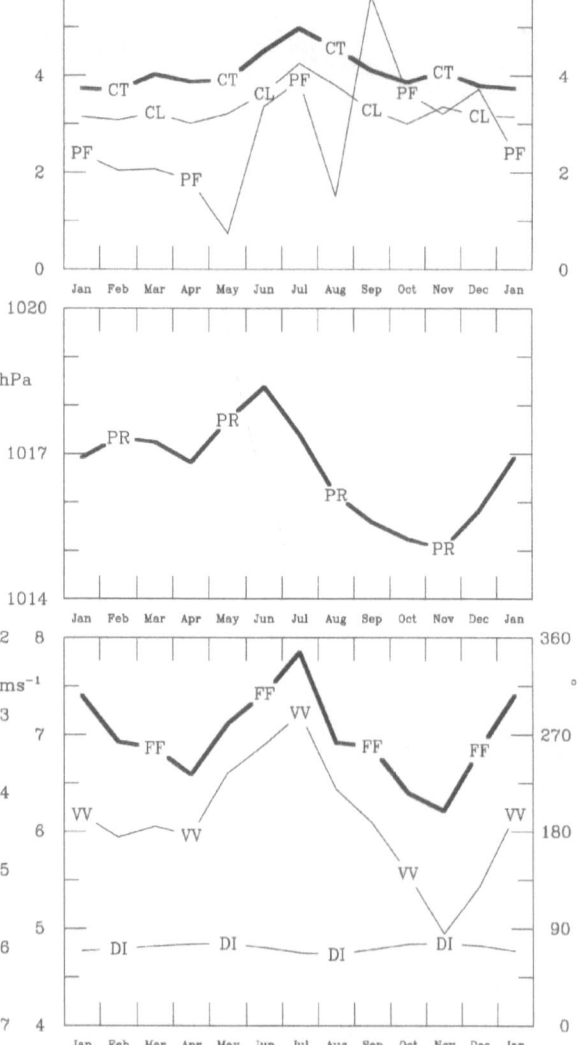

Caribbean Sea

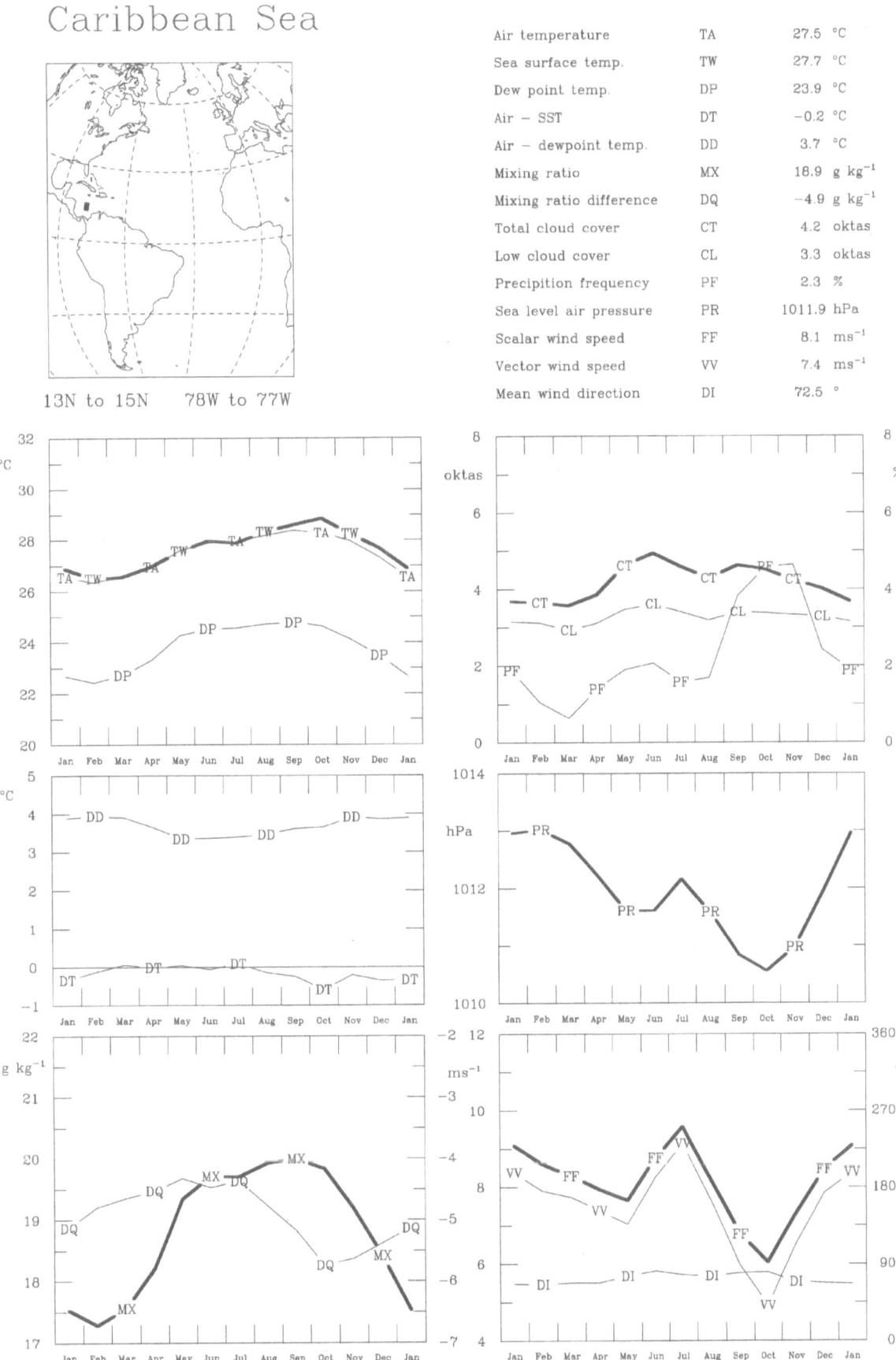

13N to 15N   78W to 77W

Air temperature	TA	27.5	°C
Sea surface temp.	TW	27.7	°C
Dew point temp.	DP	23.9	°C
Air − SST	DT	−0.2	°C
Air − dewpoint temp.	DD	3.7	°C
Mixing ratio	MX	18.9	g kg^{-1}
Mixing ratio difference	DQ	−4.9	g kg^{-1}
Total cloud cover	CT	4.2	oktas
Low cloud cover	CL	3.3	oktas
Precipition frequency	PF	2.3	%
Sea level air pressure	PR	1011.9	hPa
Scalar wind speed	FF	8.1	ms^{-1}
Vector wind speed	VV	7.4	ms^{-1}
Mean wind direction	DI	72.5	°

Cape Hoorn

57S to 52S   70W to 65W

Air temperature	TA	7.0	°C
Sea surface temp.	TW	7.0	°C
Dew point temp.	DP	3.7	°C
Air − SST	DT	−0.1	°C
Air − dewpoint temp.	DD	3.5	°C
Mixing ratio	MX	5.1	g kg^{-1}
Mixing ratio difference	DQ	−1.4	g kg^{-1}
Total cloud cover	CT	5.6	oktas
Low cloud cover	CL	4.6	oktas
Precipition frequency	PF	12.8	%
Sea level air pressure	PR	999.5	hPa
Scalar wind speed	FF	8.2	ms^{-1}
Vector wind speed	VV	4.3	ms^{-1}
Mean wind direction	DI	278.4	°

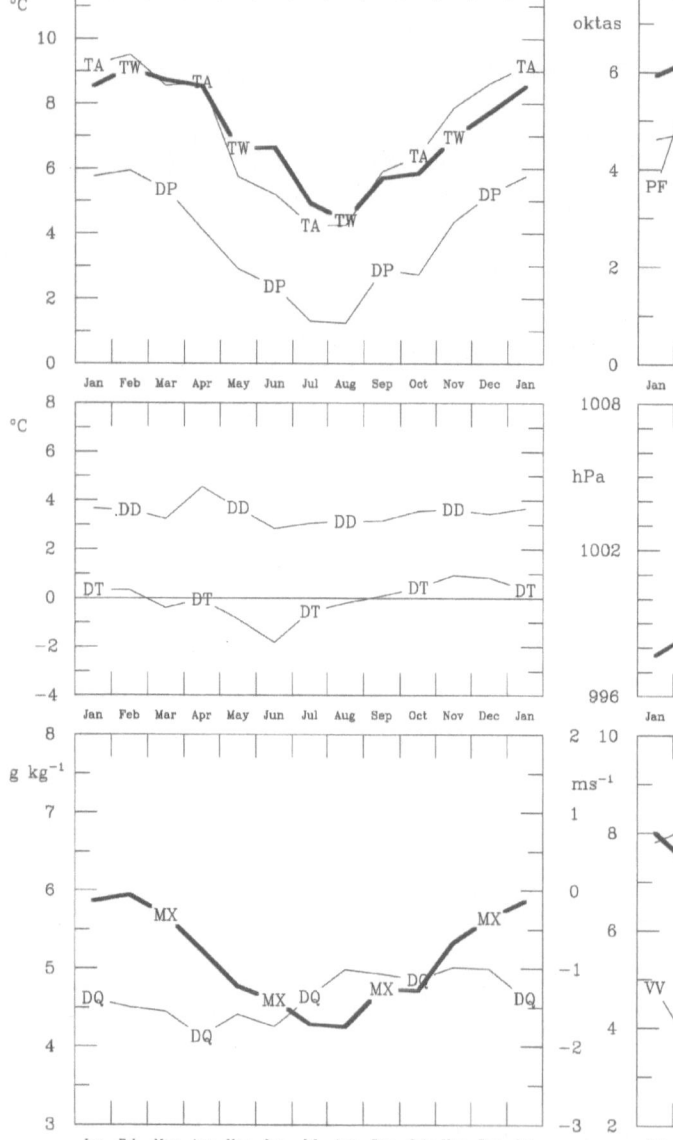

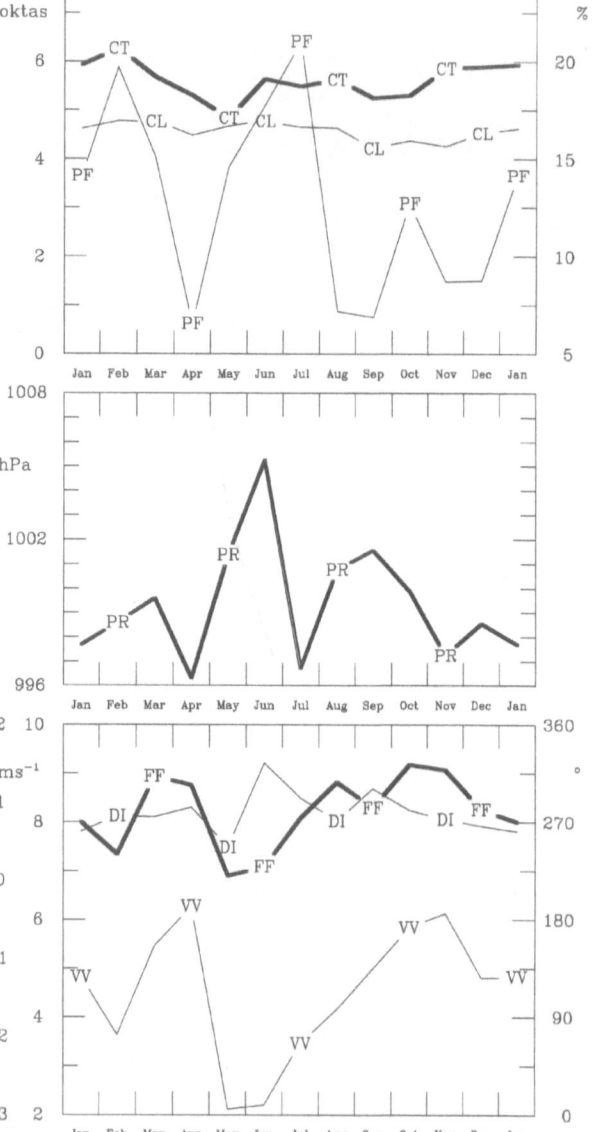

South Georgia

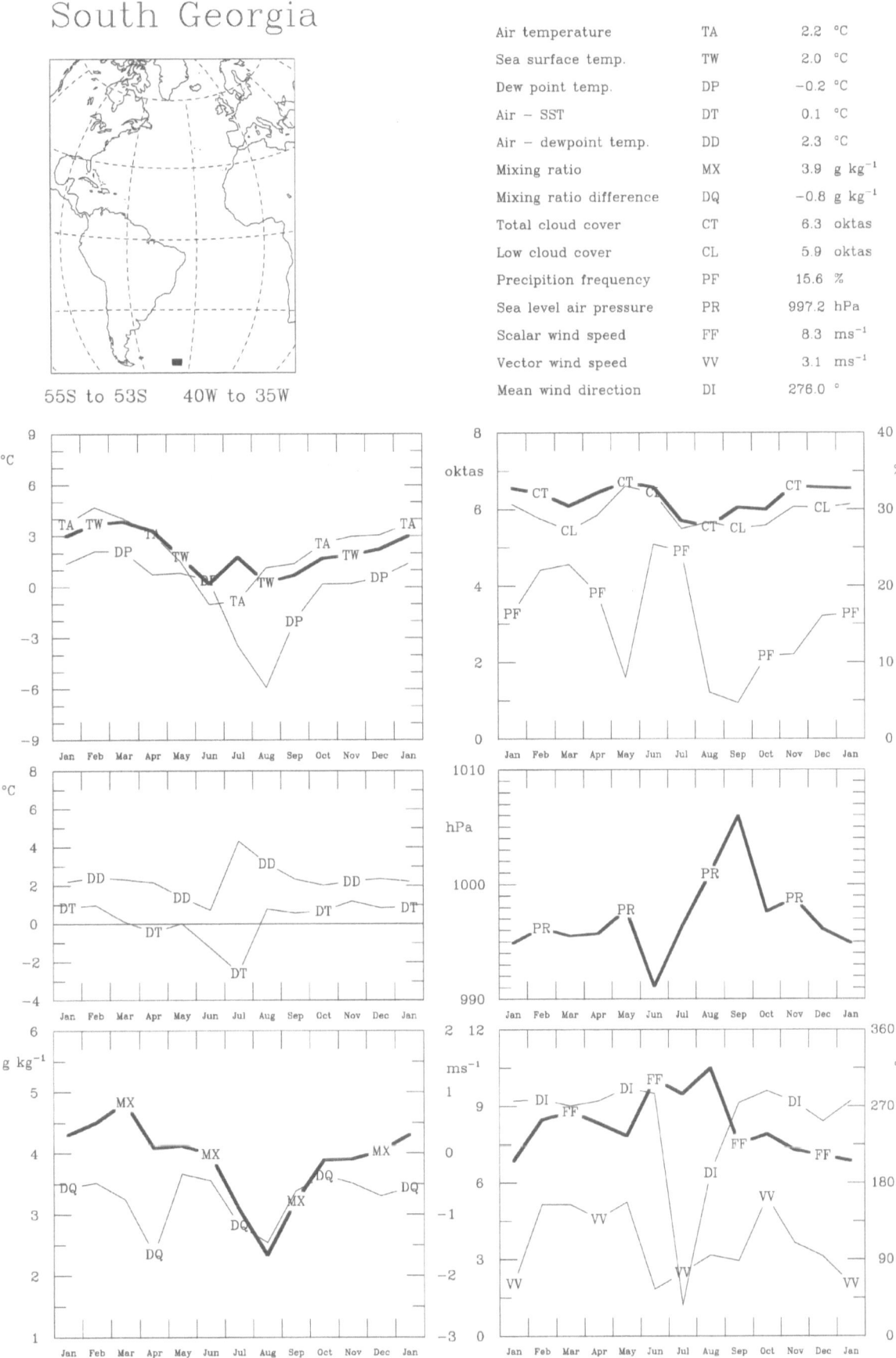

55S to 53S   40W to 35W

Air temperature	TA	2.2	°C
Sea surface temp.	TW	2.0	°C
Dew point temp.	DP	-0.2	°C
Air − SST	DT	0.1	°C
Air − dewpoint temp.	DD	2.3	°C
Mixing ratio	MX	3.9	g kg⁻¹
Mixing ratio difference	DQ	-0.8	g kg⁻¹
Total cloud cover	CT	6.3	oktas
Low cloud cover	CL	5.9	oktas
Precipition frequency	PF	15.6	%
Sea level air pressure	PR	997.2	hPa
Scalar wind speed	FF	8.3	ms⁻¹
Vector wind speed	VV	3.1	ms⁻¹
Mean wind direction	DI	276.0	°

Labrador Sea

58N to 60N   57W to 52W

Air temperature	TA	2.3	°C
Sea surface temp.	TW	4.7	°C
Dew point temp.	DP	0.0	°C
Air − SST	DT	−2.3	°C
Air − dewpoint temp.	DD	2.2	°C
Mixing ratio	MX	4.0	g kg^{-1}
Mixing ratio difference	DQ	−1.3	g kg^{-1}
Total cloud cover	CT	6.8	oktas
Low cloud cover	CL	6.2	oktas
Precipition frequency	PF	22.7	%
Sea level air pressure	PR	1007.5	hPa
Scalar wind speed	FF	9.3	ms^{-1}
Vector wind speed	VV	2.4	ms^{-1}
Mean wind direction	DI	327.9	°

Irminger Sea

59N to 61N   38W to 33W

Air temperature	TA	5.9 °C
Sea surface temp.	TW	7.1 °C
Dew point temp.	DP	3.2 °C
Air − SST	DT	−1.1 °C
Air − dewpoint temp.	DD	2.7 °C
Mixing ratio	MX	4.9 g kg^{-1}
Mixing ratio difference	DQ	−1.3 g kg^{-1}
Total cloud cover	CT	6.6 oktas
Low cloud cover	CL	5.9 oktas
Precipition frequency	PF	18.6 %
Sea level air pressure	PR	1005.2 hPa
Scalar wind speed	FF	9.1 ms^{-1}
Vector wind speed	VV	1.6 ms^{-1}
Mean wind direction	DI	265.8 °

Agulhas Current

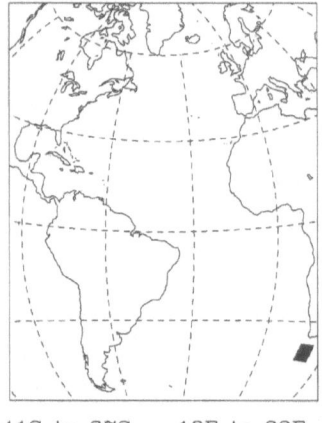

41S to 37S   18E to 23E

Air temperature	TA	16.7	°C
Sea surface temp.	TW	18.8	°C
Dew point temp.	DP	12.2	°C
Air − SST	DT	−2.0	°C
Air − dewpoint temp.	DD	4.7	°C
Mixing ratio	MX	9.2	g kg^{-1}
Mixing ratio difference	DQ	−5.0	g kg^{-1}
Total cloud cover	CT	5.2	oktas
Low cloud cover	CL	4.4	oktas
Precipition frequency	PF	8.7	%
Sea level air pressure	PR	1016.2	hPa
Scalar wind speed	FF	9.1	ms^{-1}
Vector wind speed	VV	4.0	ms^{-1}
Mean wind direction	DI	257.5	°

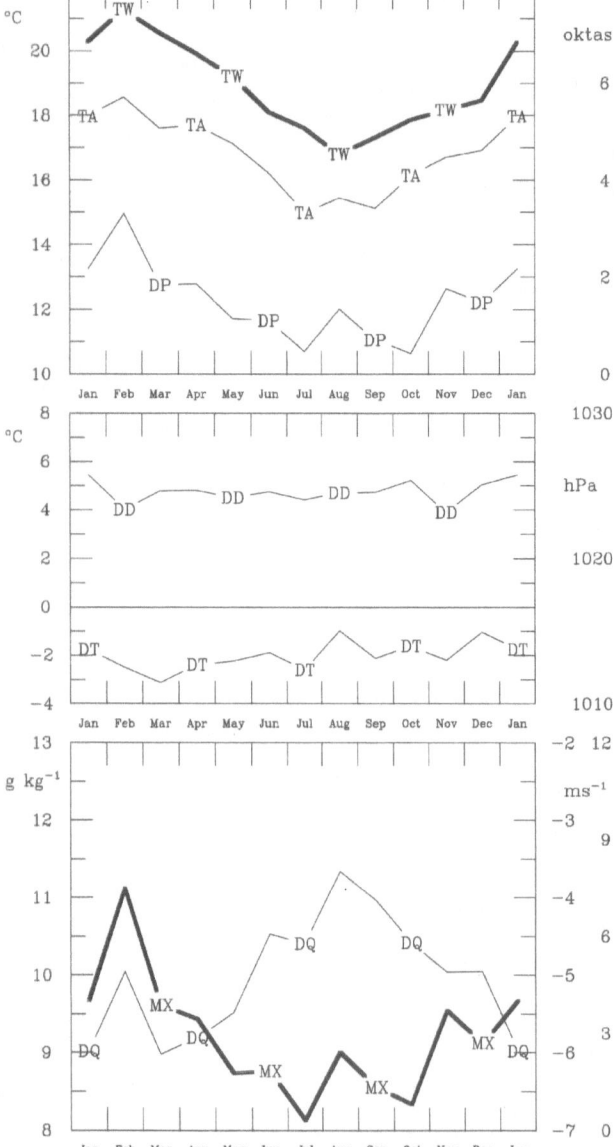

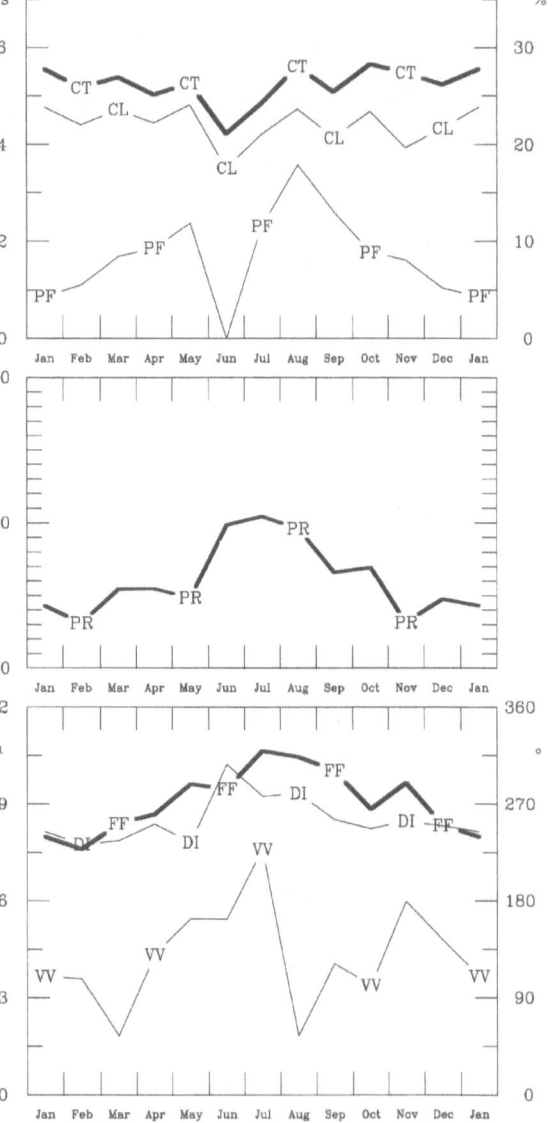

Subtropic convergence

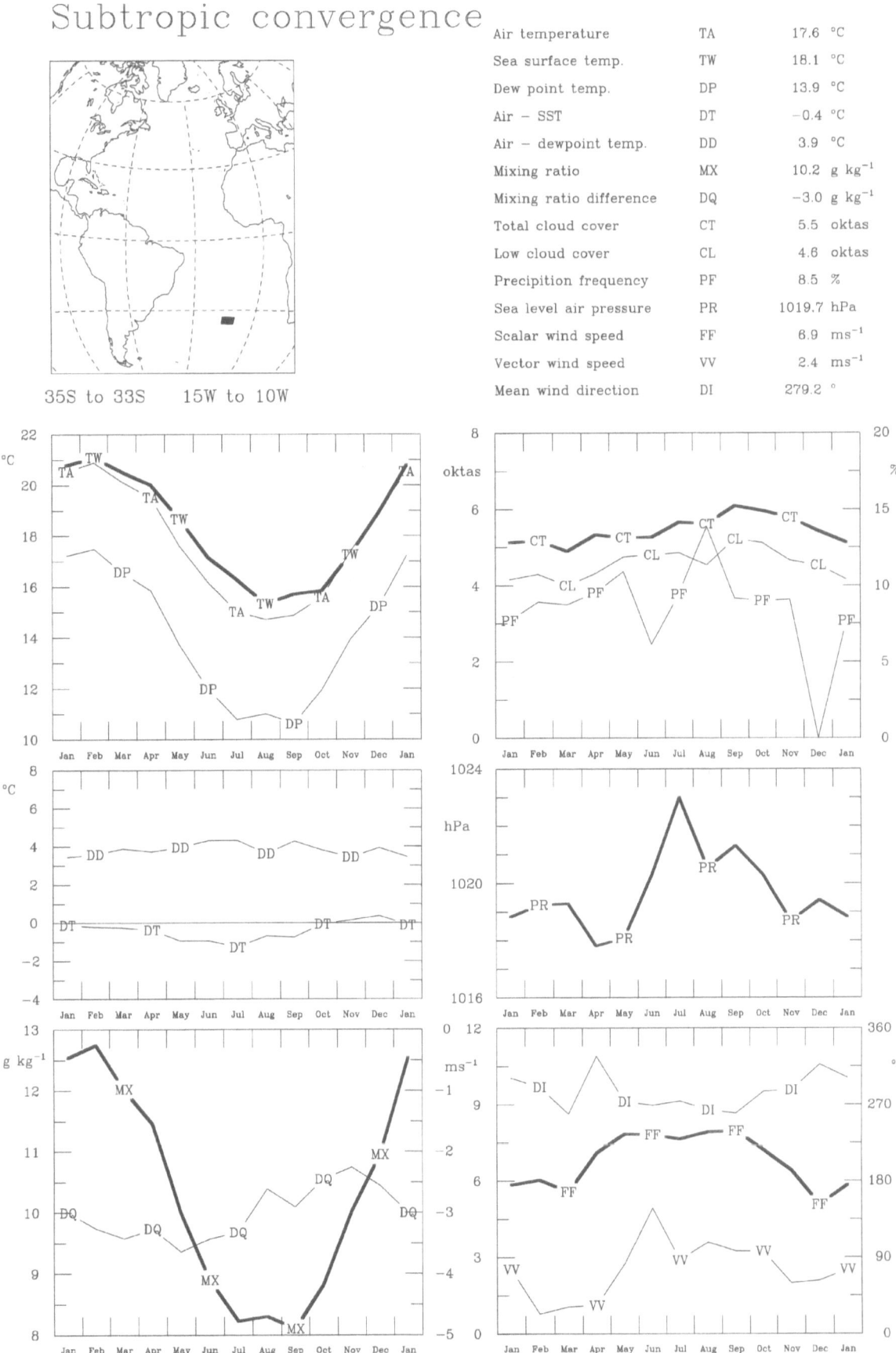

Air temperature	TA	17.6	°C
Sea surface temp.	TW	18.1	°C
Dew point temp.	DP	13.9	°C
Air − SST	DT	−0.4	°C
Air − dewpoint temp.	DD	3.9	°C
Mixing ratio	MX	10.2	g kg^{-1}
Mixing ratio difference	DQ	−3.0	g kg^{-1}
Total cloud cover	CT	5.5	oktas
Low cloud cover	CL	4.6	oktas
Precipition frequency	PF	8.5	%
Sea level air pressure	PR	1019.7	hPa
Scalar wind speed	FF	6.9	ms^{-1}
Vector wind speed	VV	2.4	ms^{-1}
Mean wind direction	DI	279.2	°

35S to 33S   15W to 10W

Benguela Current

26S to 24S   10E to 15E

Air temperature	TA	17.8	°C
Sea surface temp.	TW	17.6	°C
Dew point temp.	DP	14.5	°C
Air − SST	DT	0.2	°C
Air − dewpoint temp.	DD	3.4	°C
Mixing ratio	MX	10.5	g kg^{-1}
Mixing ratio difference	DQ	−2.5	g kg^{-1}
Total cloud cover	CT	4.4	oktas
Low cloud cover	CL	4.0	oktas
Precipition frequency	PF	0.9	%
Sea level air pressure	PR	1016.7	hPa
Scalar wind speed	FF	7.4	ms^{-1}
Vector wind speed	VV	6.7	ms^{-1}
Mean wind direction	DI	157.7	°

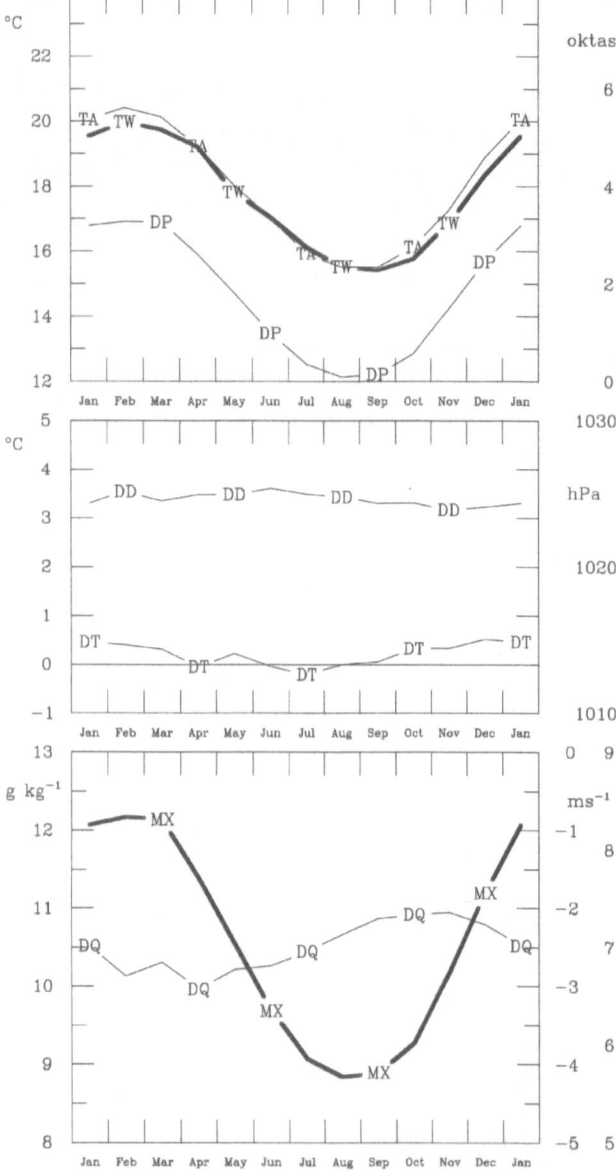

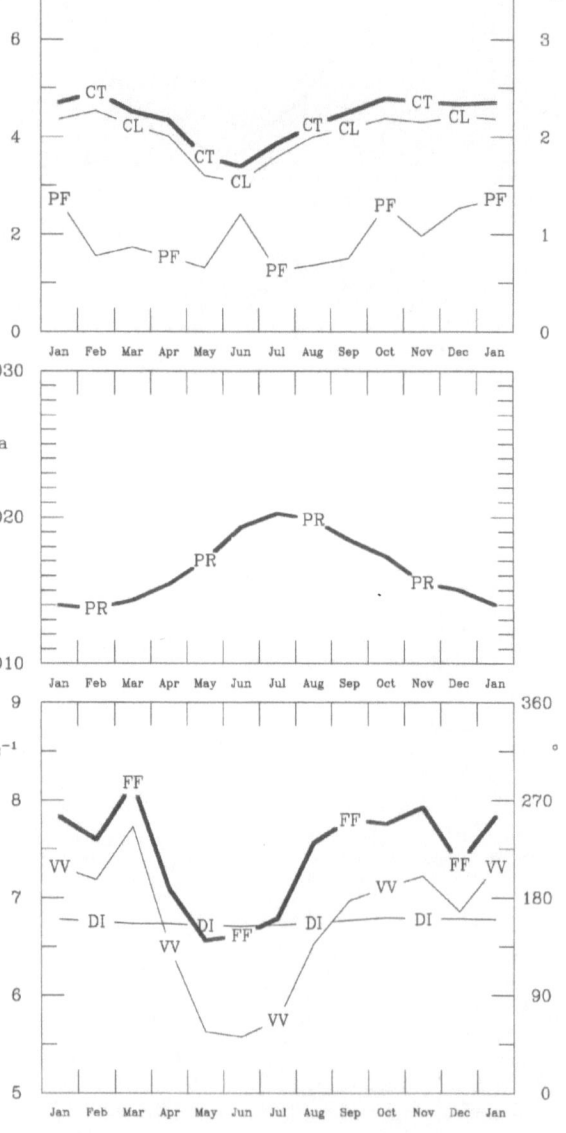

South East Trade Wind

Air temperature	TA	21.2 °C
Sea surface temp.	TW	21.5 °C
Dew point temp.	DP	16.8 °C
Air − SST	DT	−0.3 °C
Air − dewpoint temp.	DD	4.5 °C
Mixing ratio	MX	12.1 g kg^{-1}
Mixing ratio difference	DQ	−4.2 g kg^{-1}
Total cloud cover	CT	6.2 oktas
Low cloud cover	CL	5.9 oktas
Precipition frequency	PF	1.6 %
Sea level air pressure	PR	1015.7 hPa
Scalar wind speed	FF	6.2 ms^{-1}
Vector wind speed	VV	5.9 ms^{-1}
Mean wind direction	DI	140.7 °

16S to 14S   1E to 3E

Mediterranean Sea

34N to 36N   17E to 19E

Air temperature	TA	19.6	°C
Sea surface temp.	TW	20.2	°C
Dew point temp.	DP	15.4	°C
Air − SST	DT	−0.5	°C
Air − dewpoint temp.	DD	4.3	°C
Mixing ratio	MX	11.5	g kg^{-1}
Mixing ratio difference	DQ	−3.9	g kg^{-1}
Total cloud cover	CT	3.1	oktas
Low cloud cover	CL	2.4	oktas
Precipition frequency	PF	3.0	%
Sea level air pressure	PR	1015.5	hPa
Scalar wind speed	FF	5.8	ms^{-1}
Vector wind speed	VV	1.6	ms^{-1}
Mean wind direction	DI	298.1	°

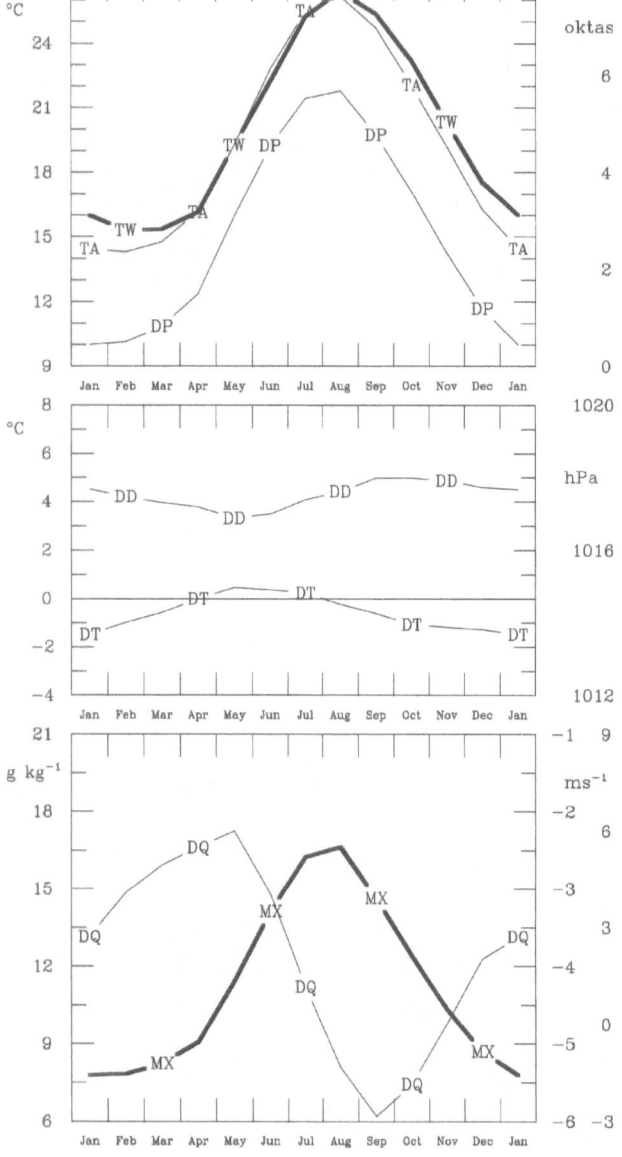

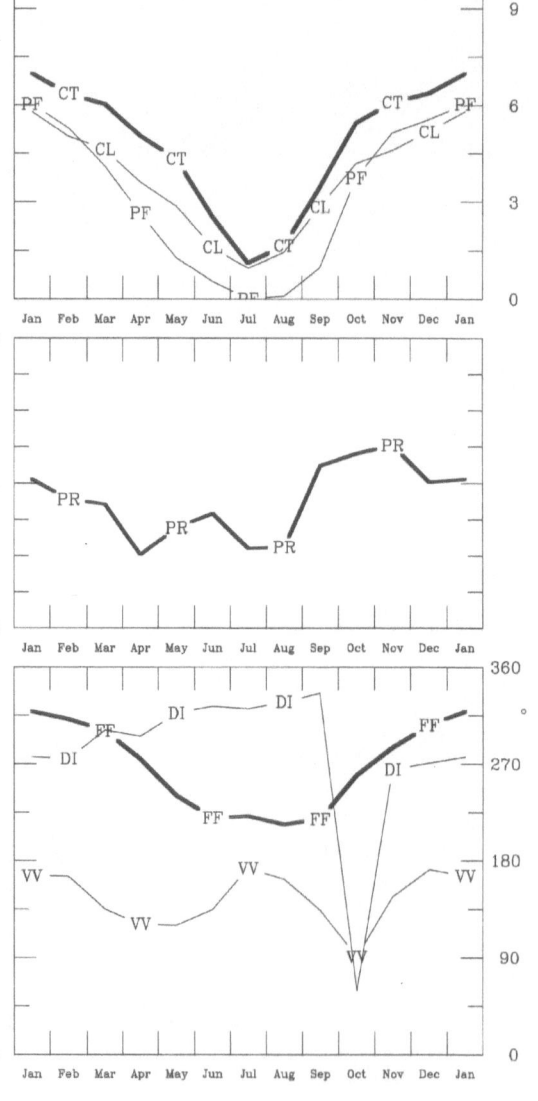

Tropical convergence

Air temperature	TA	26.5 °C
Sea surface temp.	TW	27.1 °C
Dew point temp.	DP	22.8 °C
Air − SST	DT	−0.7 °C
Air − dewpoint temp.	DD	3.7 °C
Mixing ratio	MX	17.7 g kg^{-1}
Mixing ratio difference	DQ	−5.4 g kg^{-1}
Total cloud cover	CT	5.4 oktas
Low cloud cover	CL	4.3 oktas
Precipition frequency	PF	10.3 %
Sea level air pressure	PR	1012.3 hPa
Scalar wind speed	FF	5.3 ms^{-1}
Vector wind speed	VV	2.3 ms^{-1}
Mean wind direction	DI	90.0 °

4N to 6N   30W to 25W

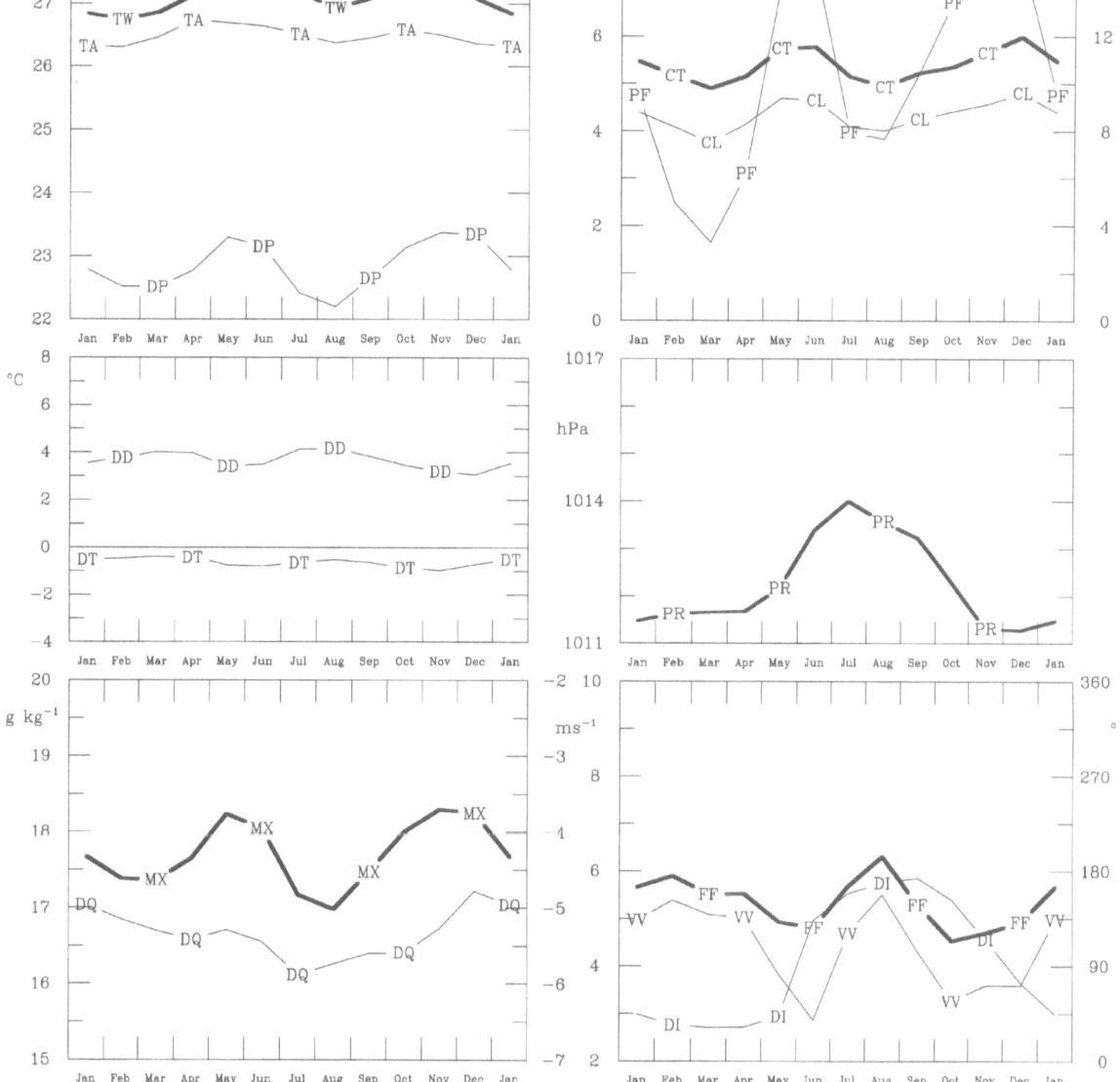

Observation Density

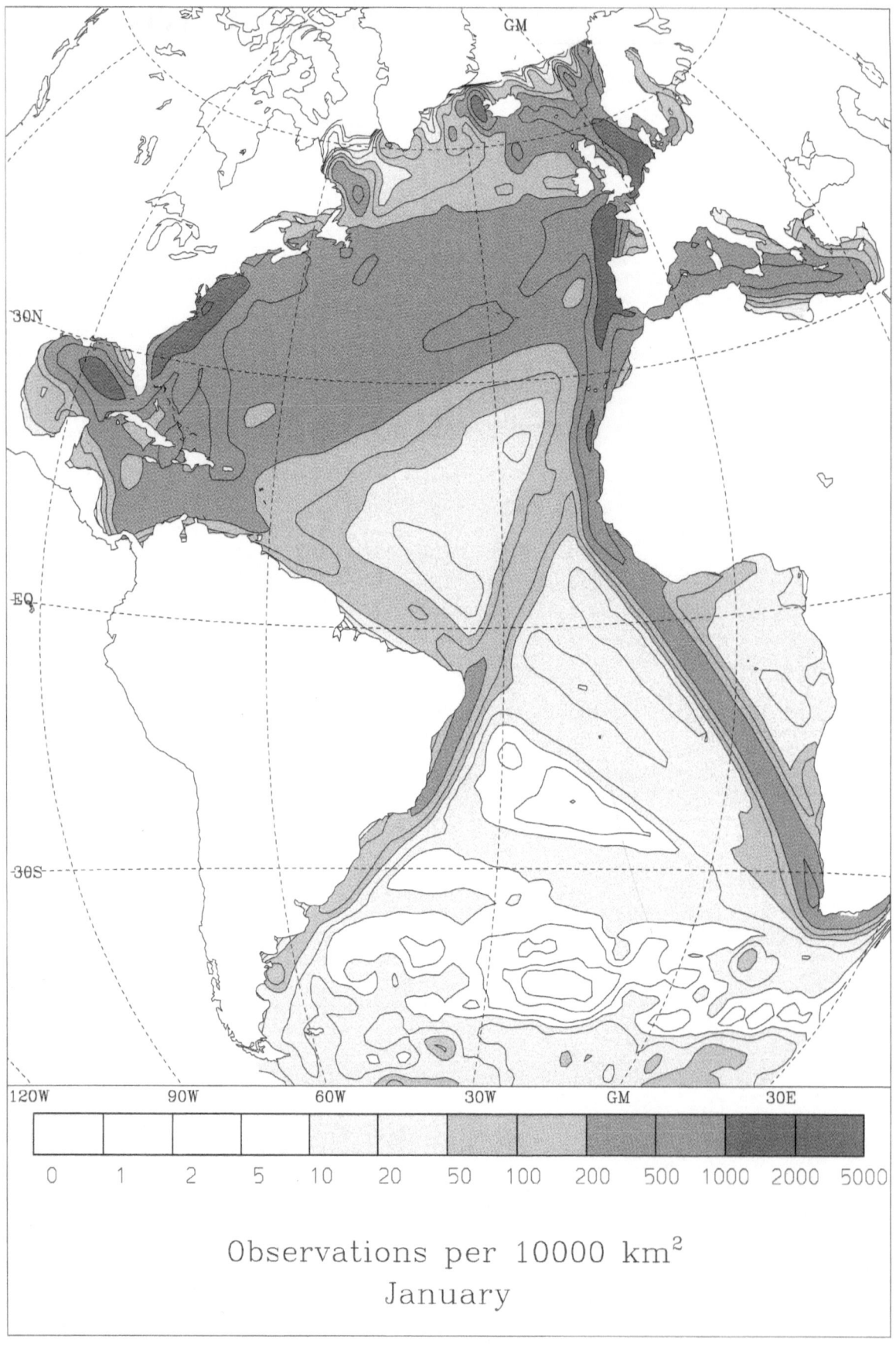

Observations per 10000 km²
January

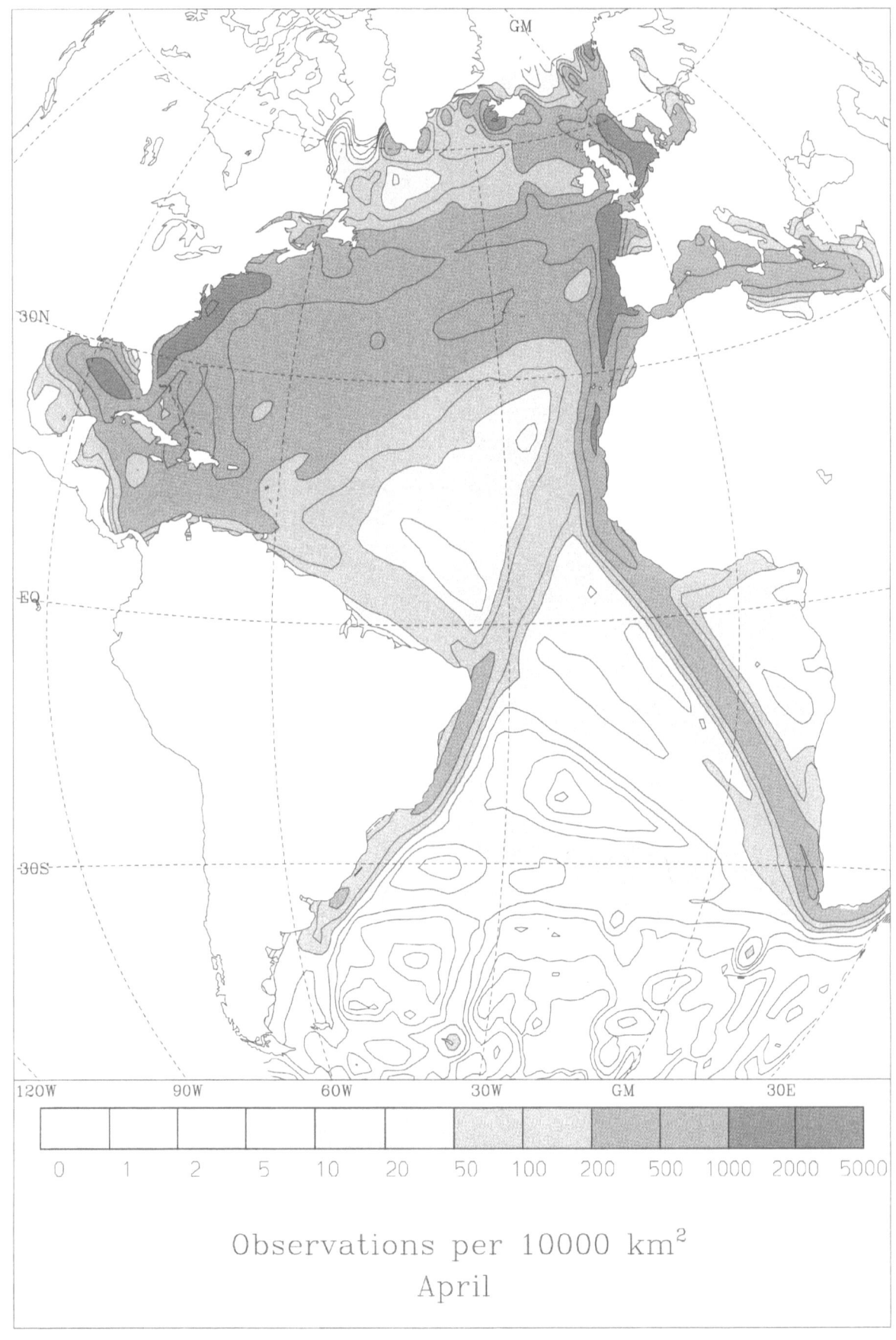

Observations per 10000 km²
April

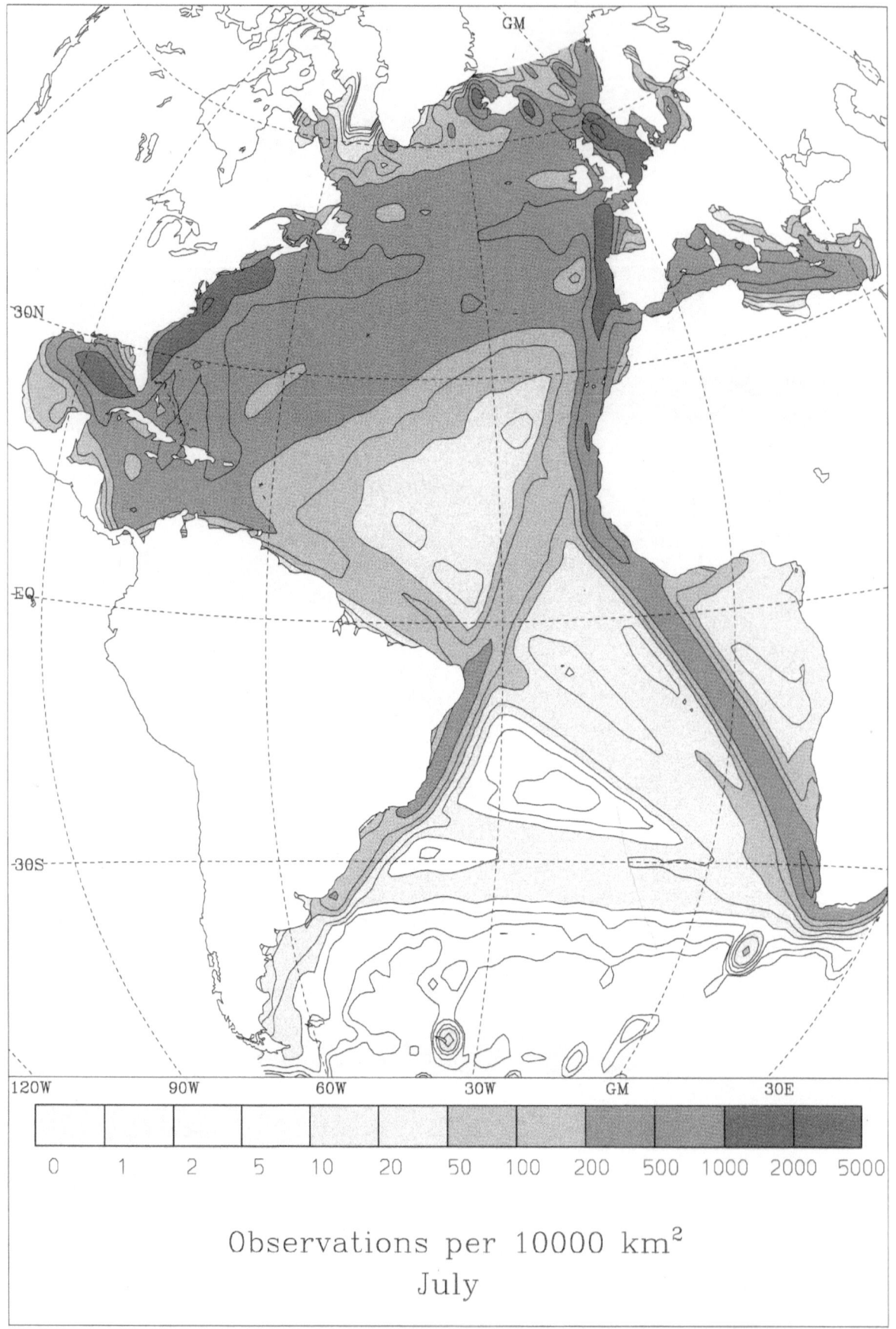

Observations per 10000 km²
July

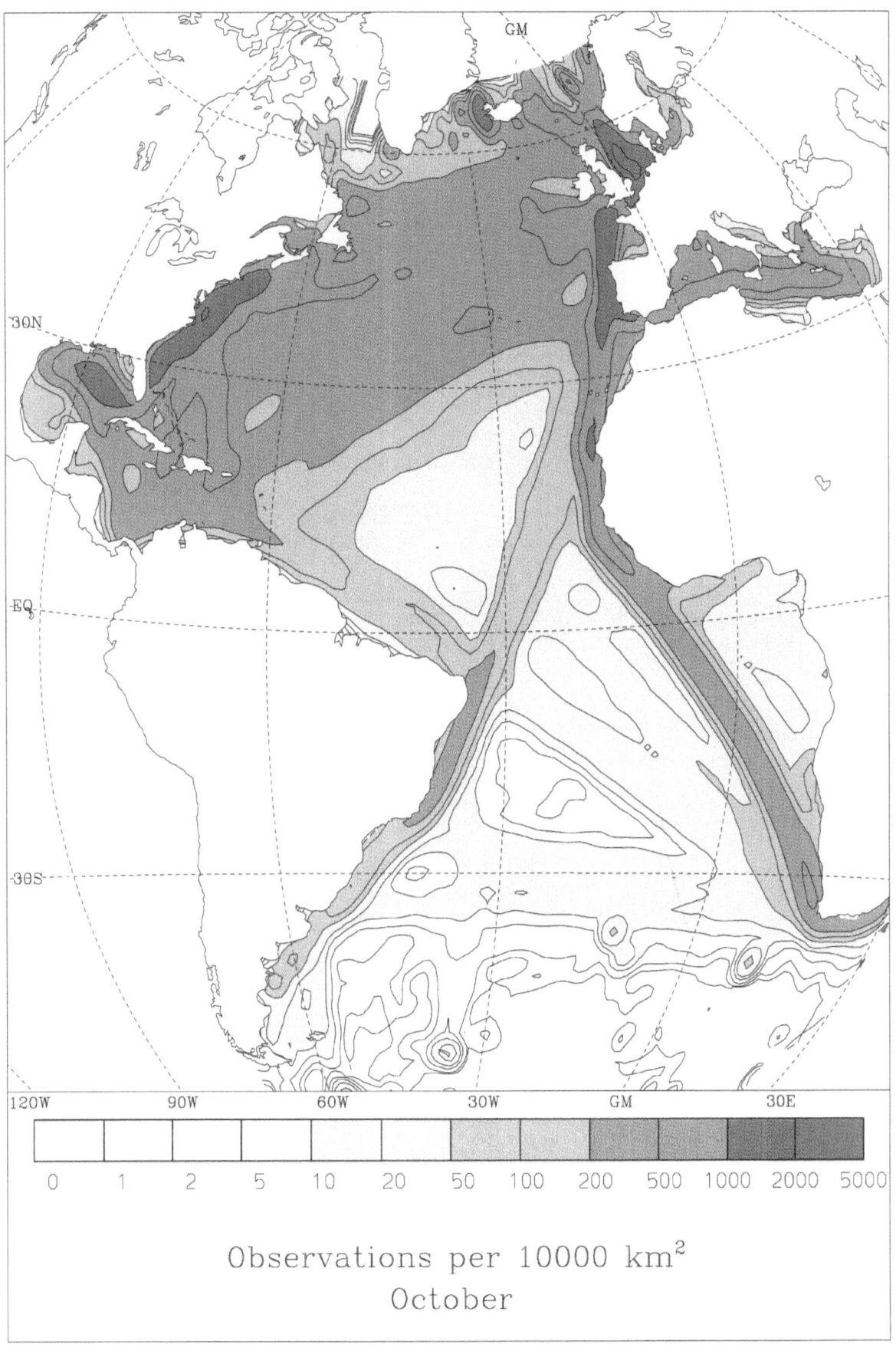

Observations per 10000 km²
October

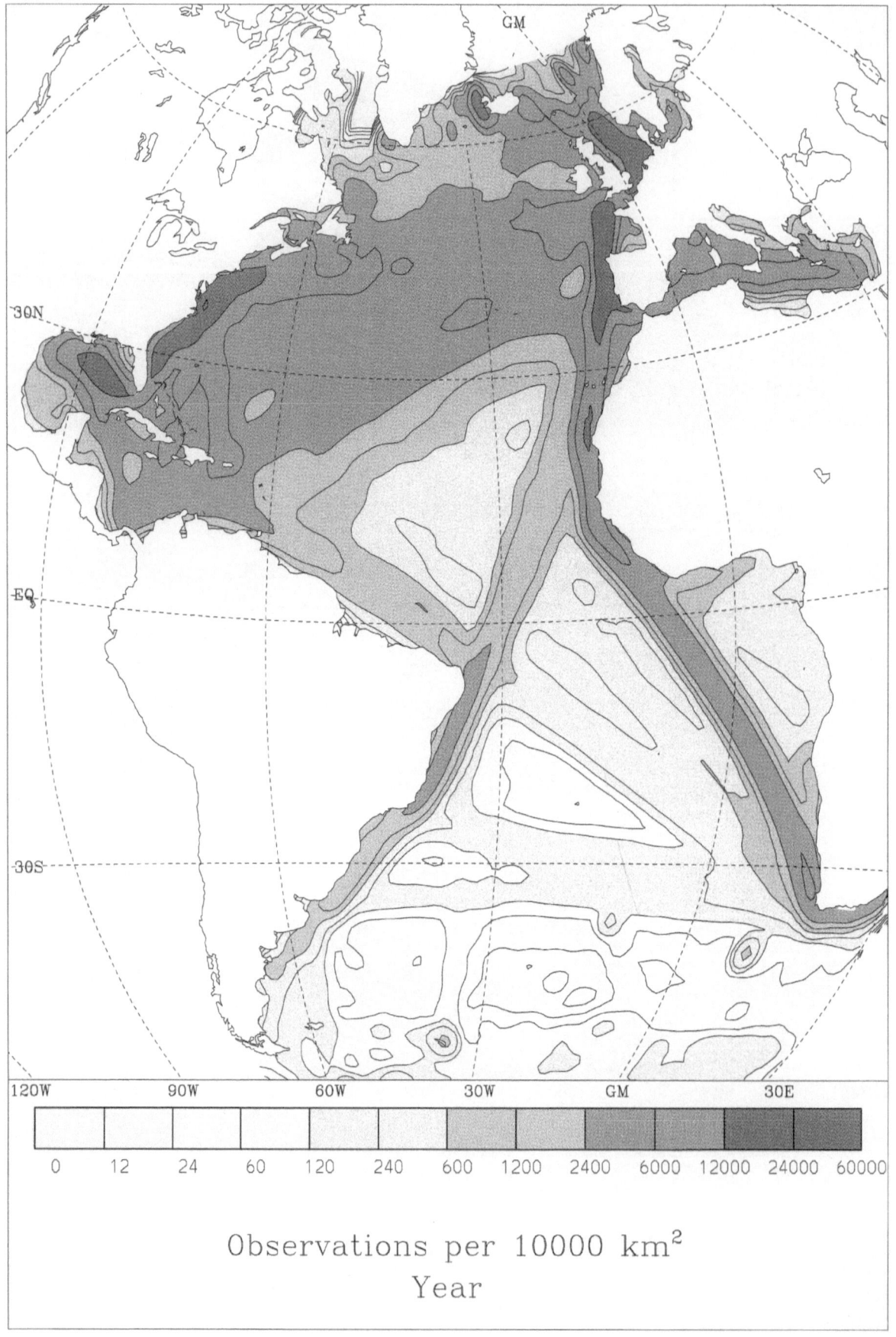

Observations per 10000 km²
Year

Sea Surface Temperature

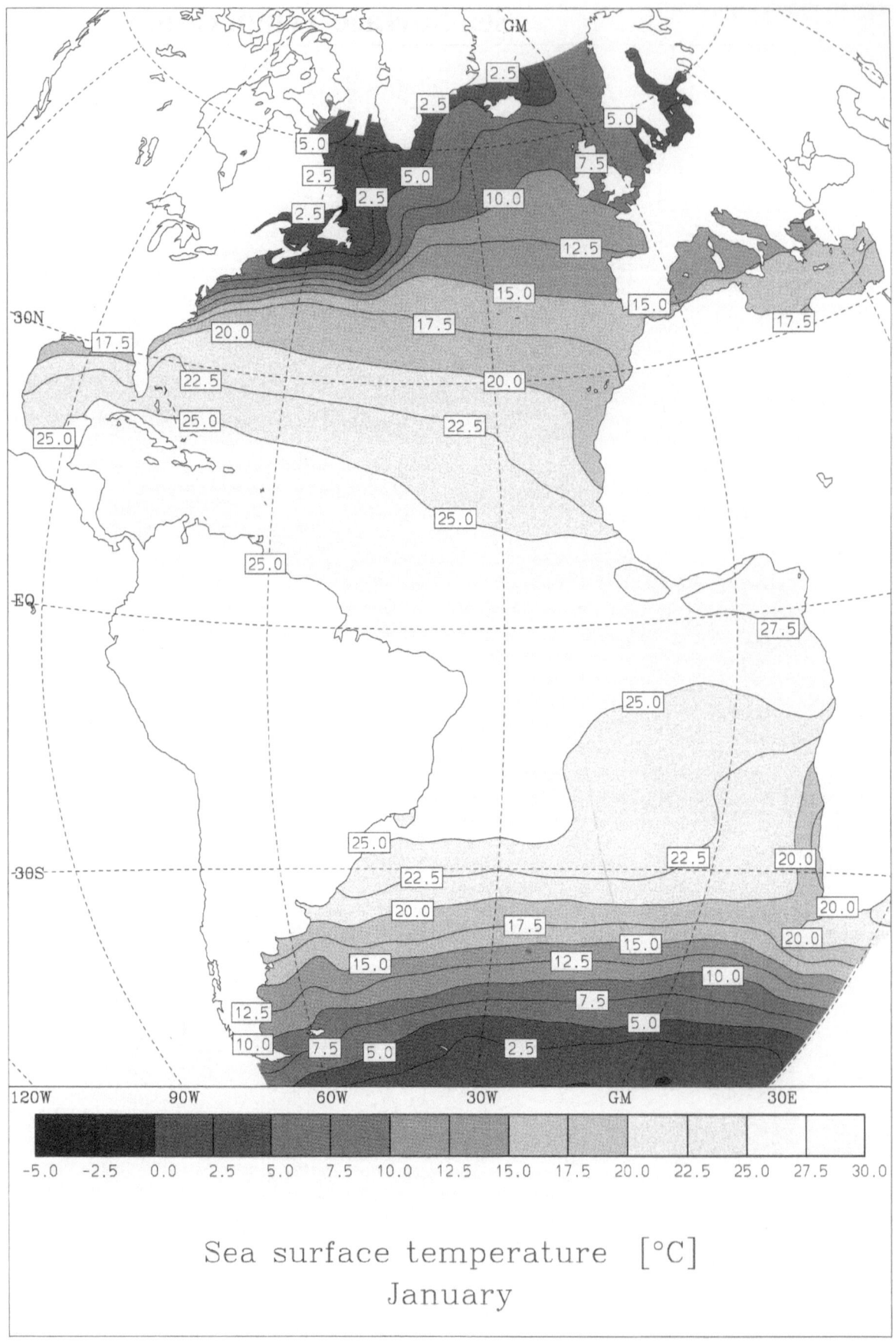

Sea surface temperature [°C]
January

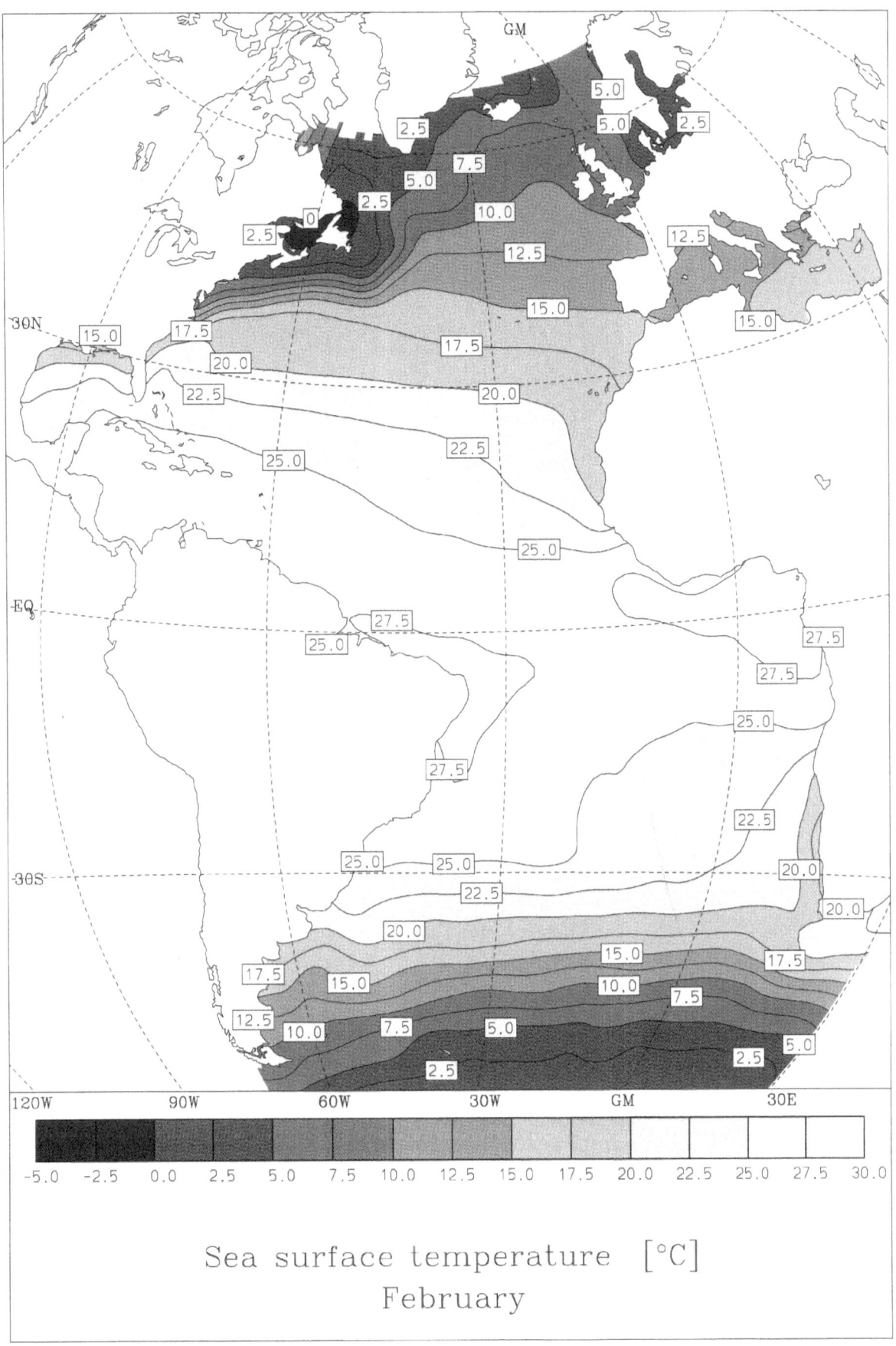

Sea surface temperature [°C]
February

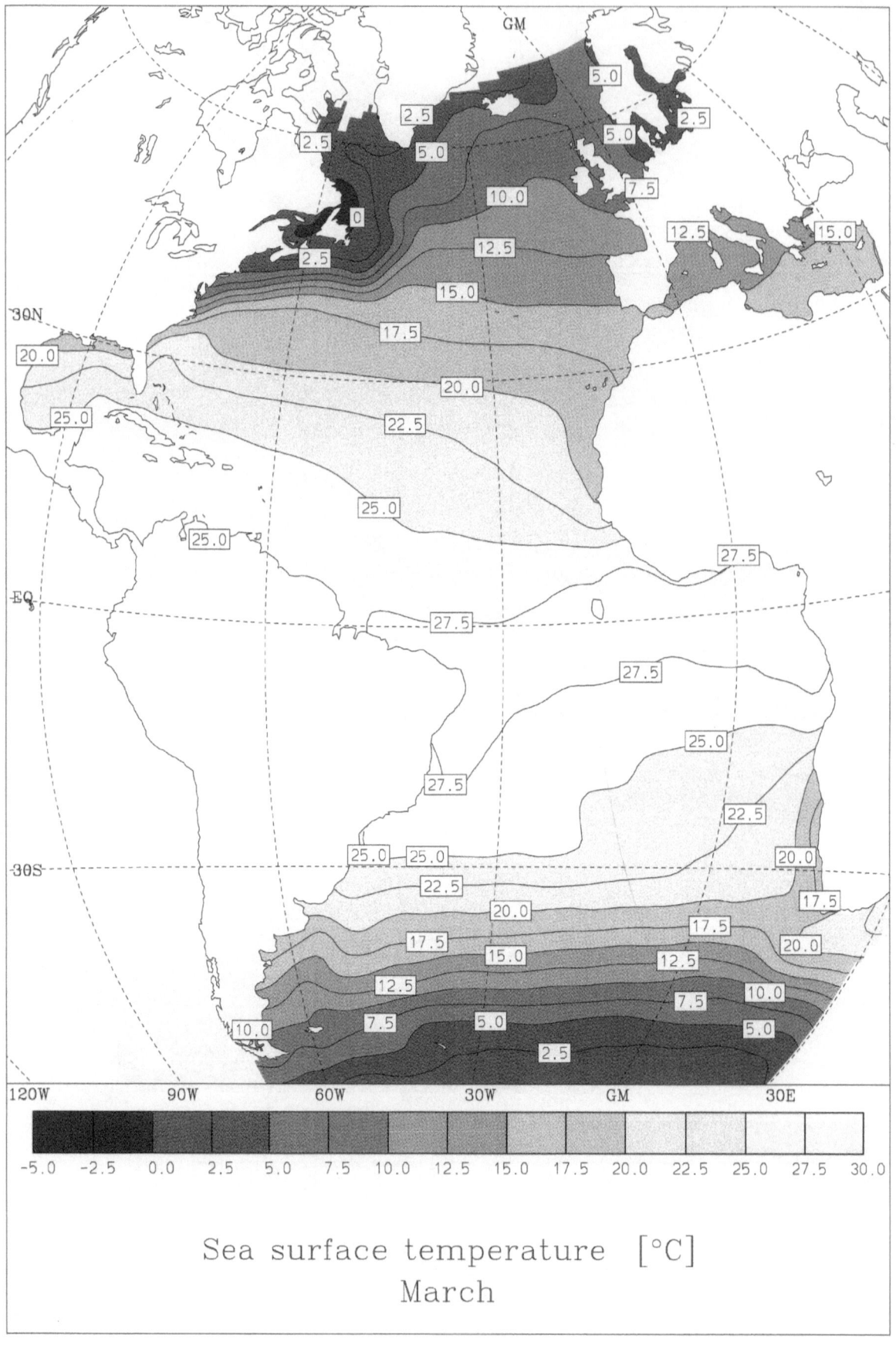

Sea surface temperature [°C]
March

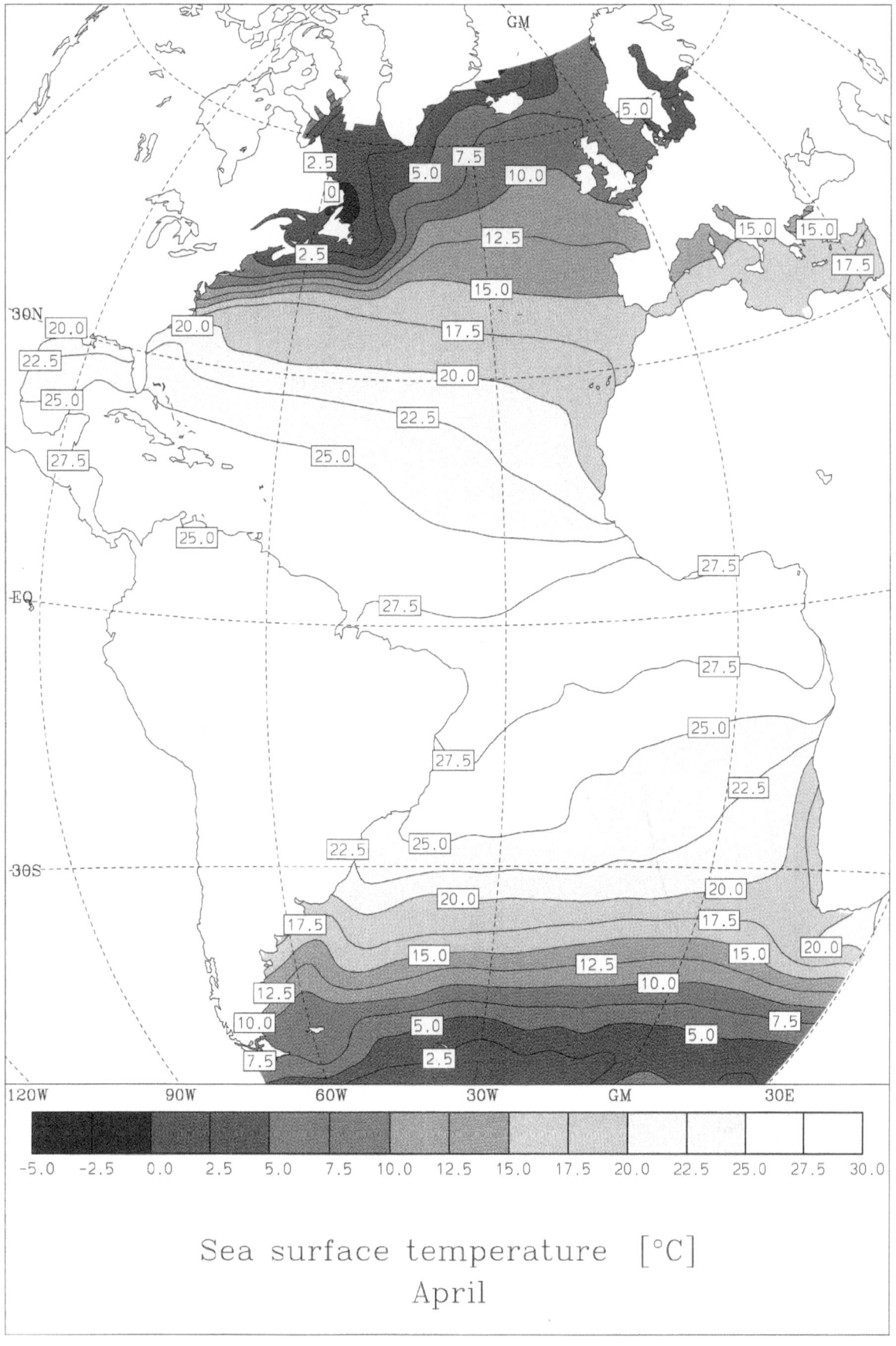

Sea surface temperature [°C]
April

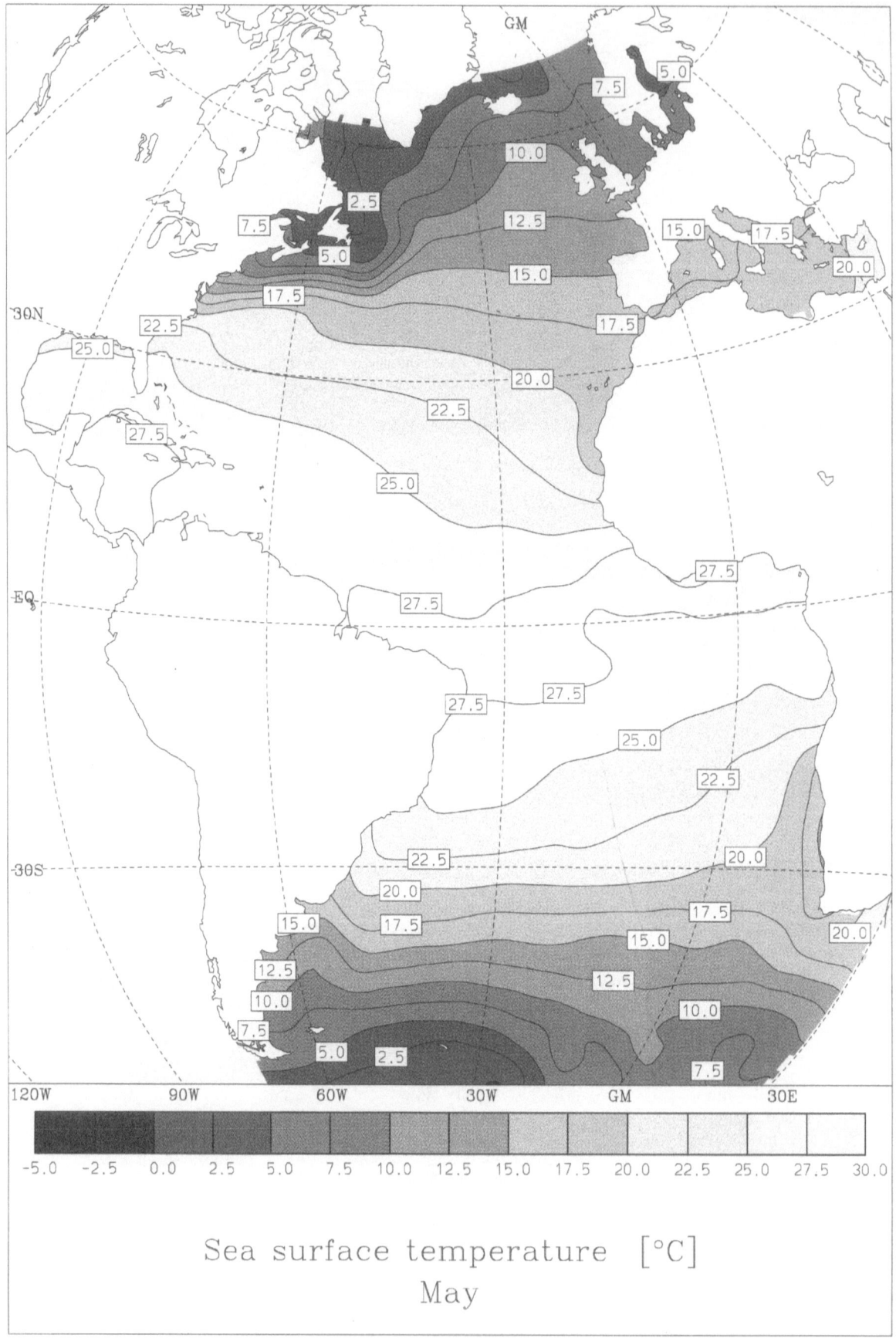

Sea surface temperature [°C]
May

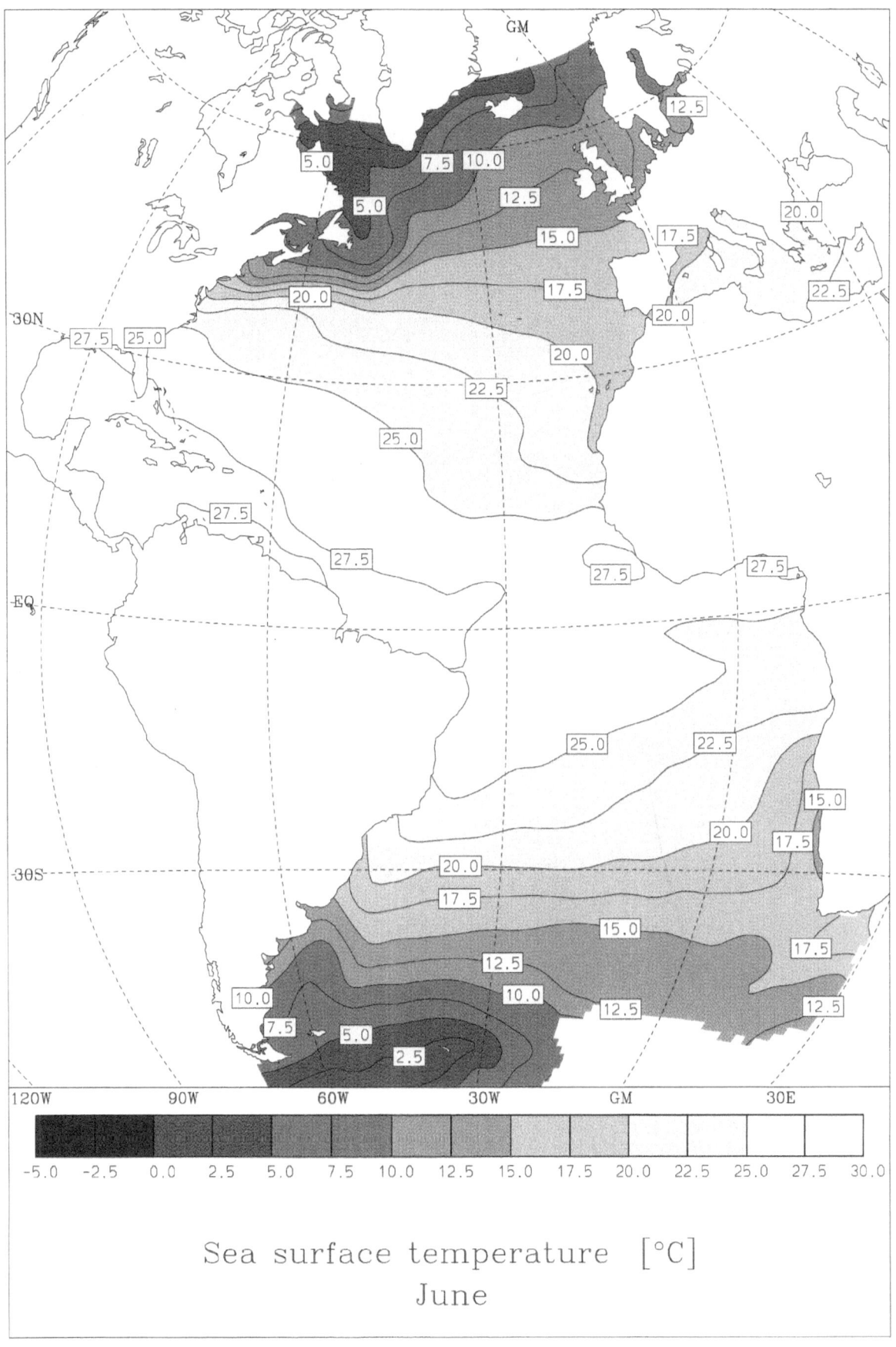

Sea surface temperature [°C]
June

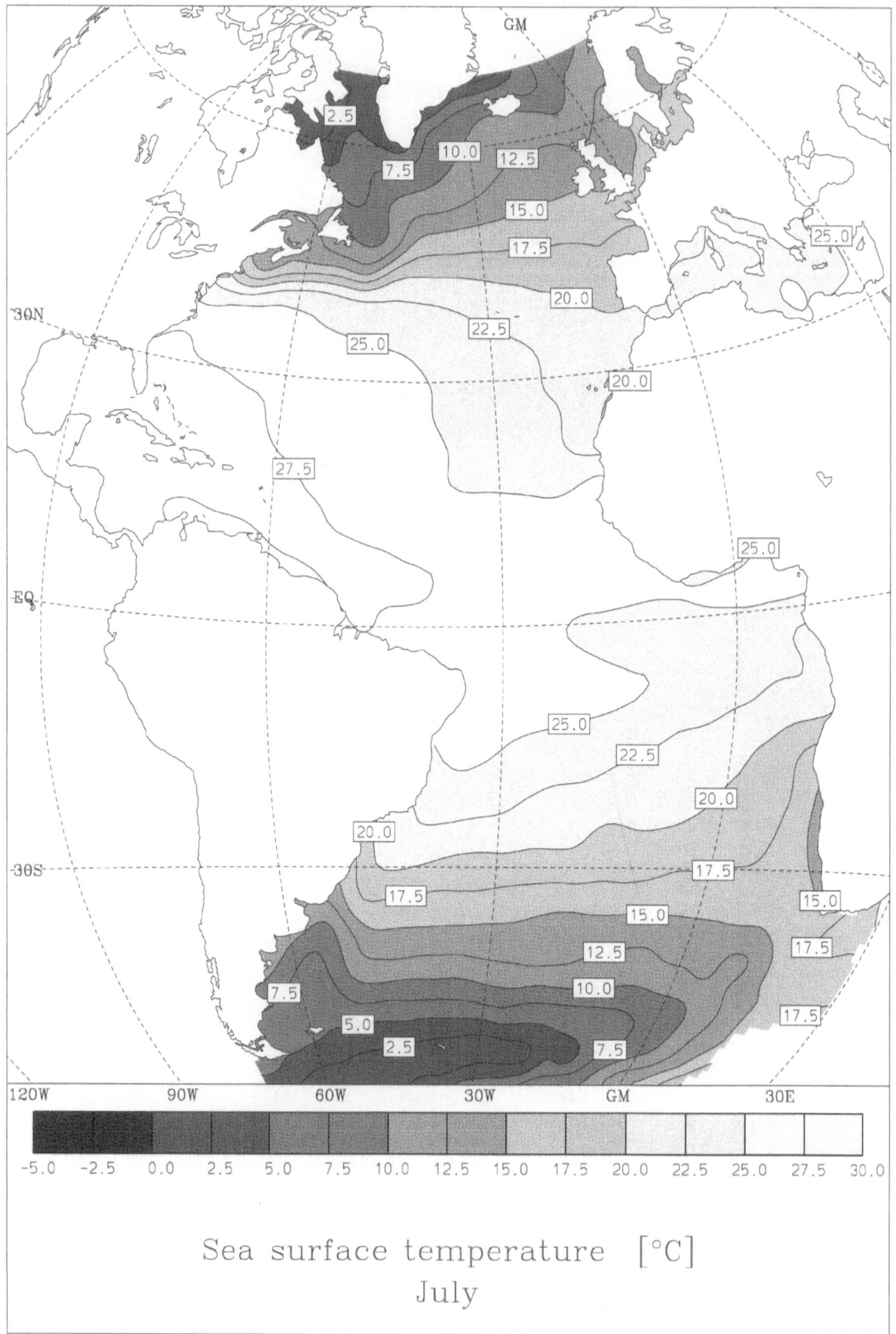

Sea surface temperature [°C]
July

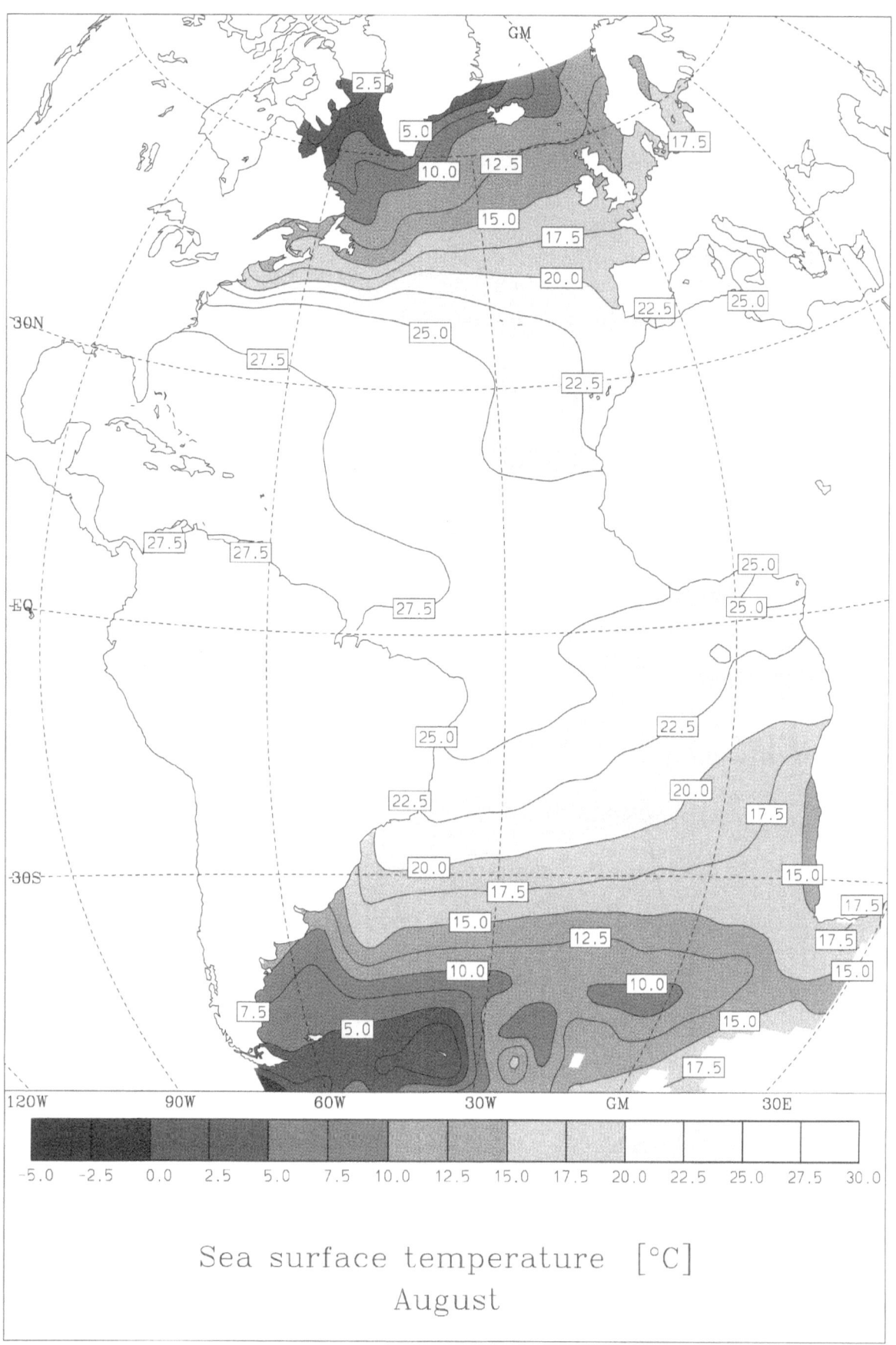

Sea surface temperature [°C]
August

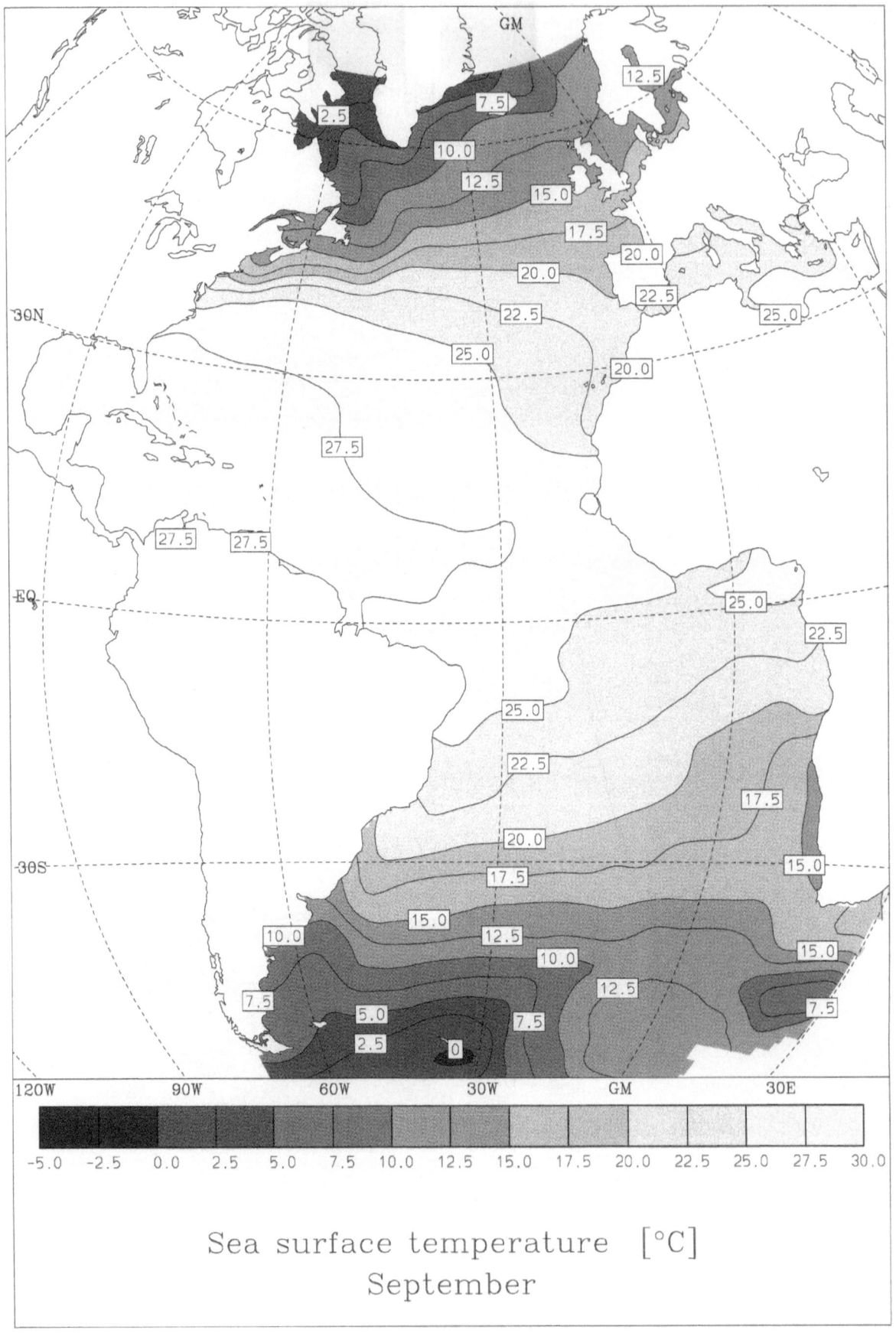

Sea surface temperature [°C]
September

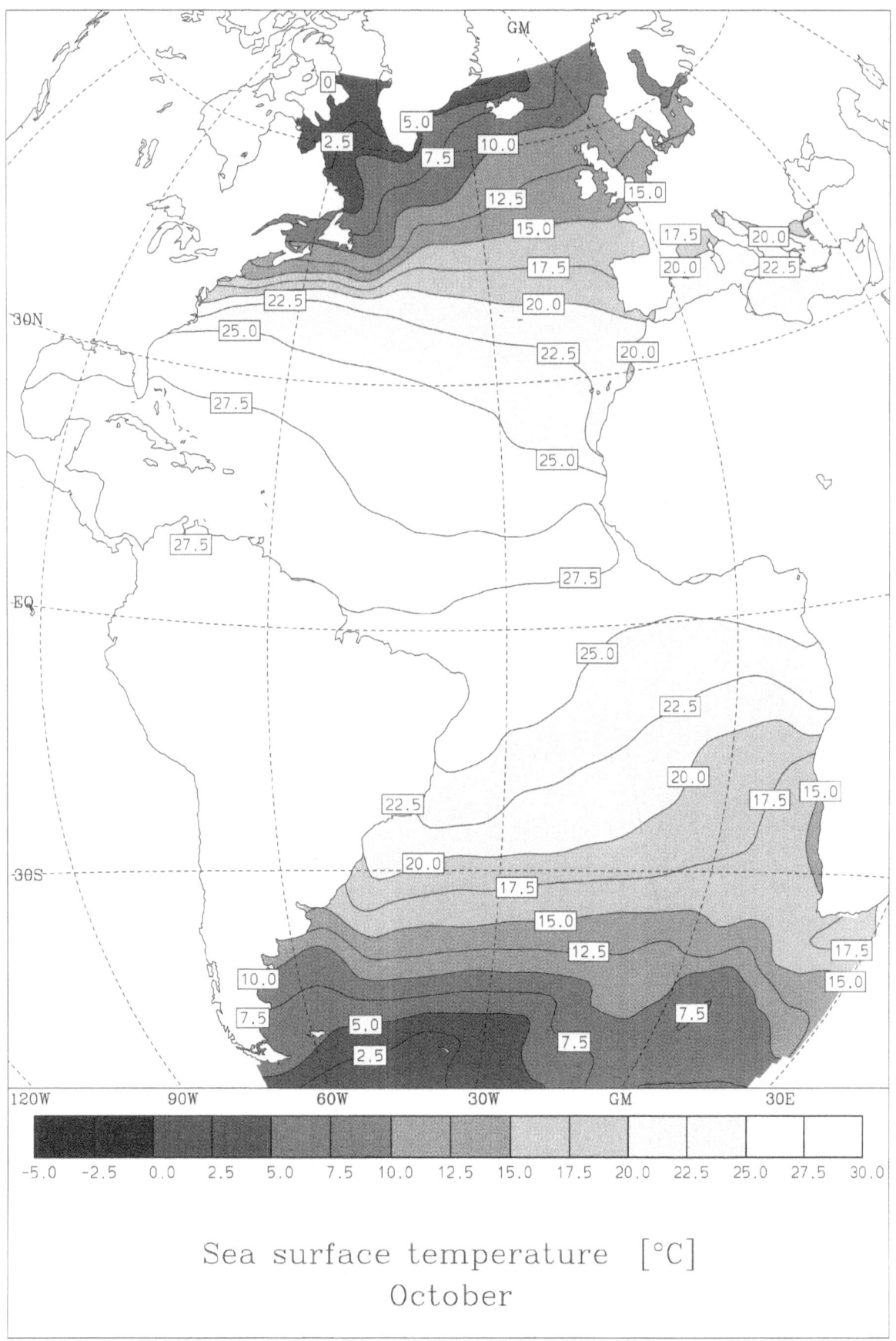

Sea surface temperature [°C]
October

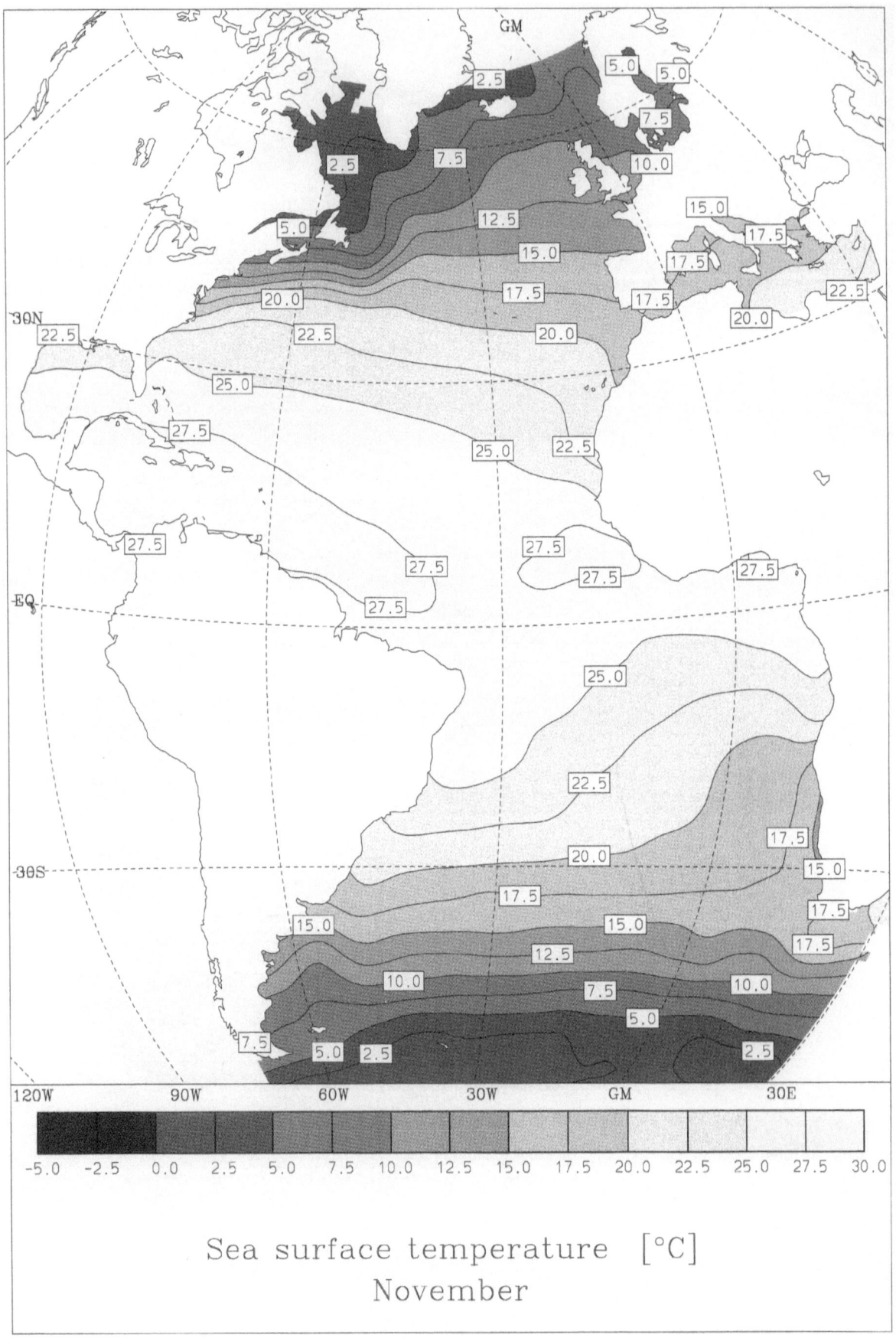

Sea surface temperature [°C]
November

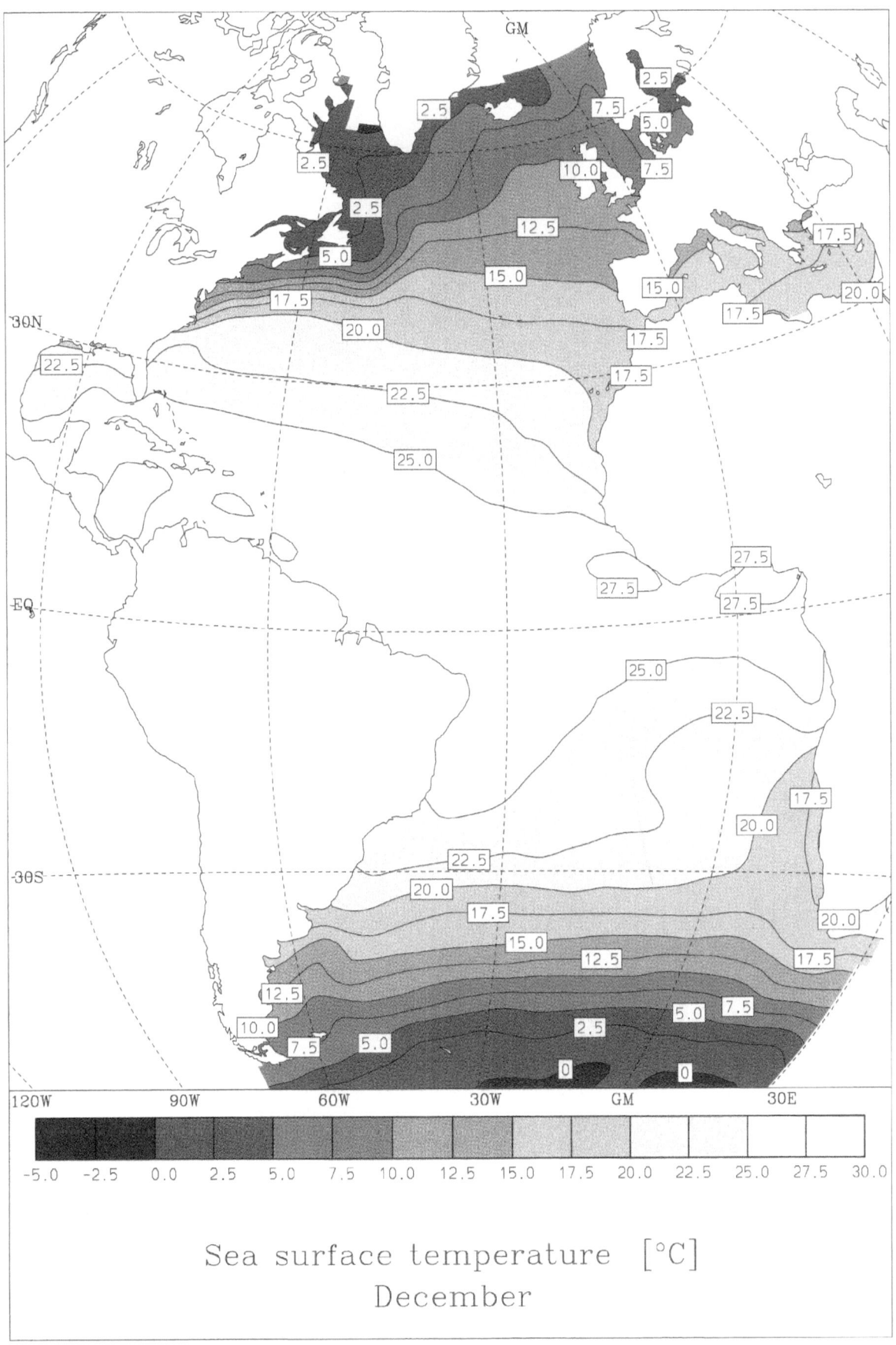

Sea surface temperature [°C]
December

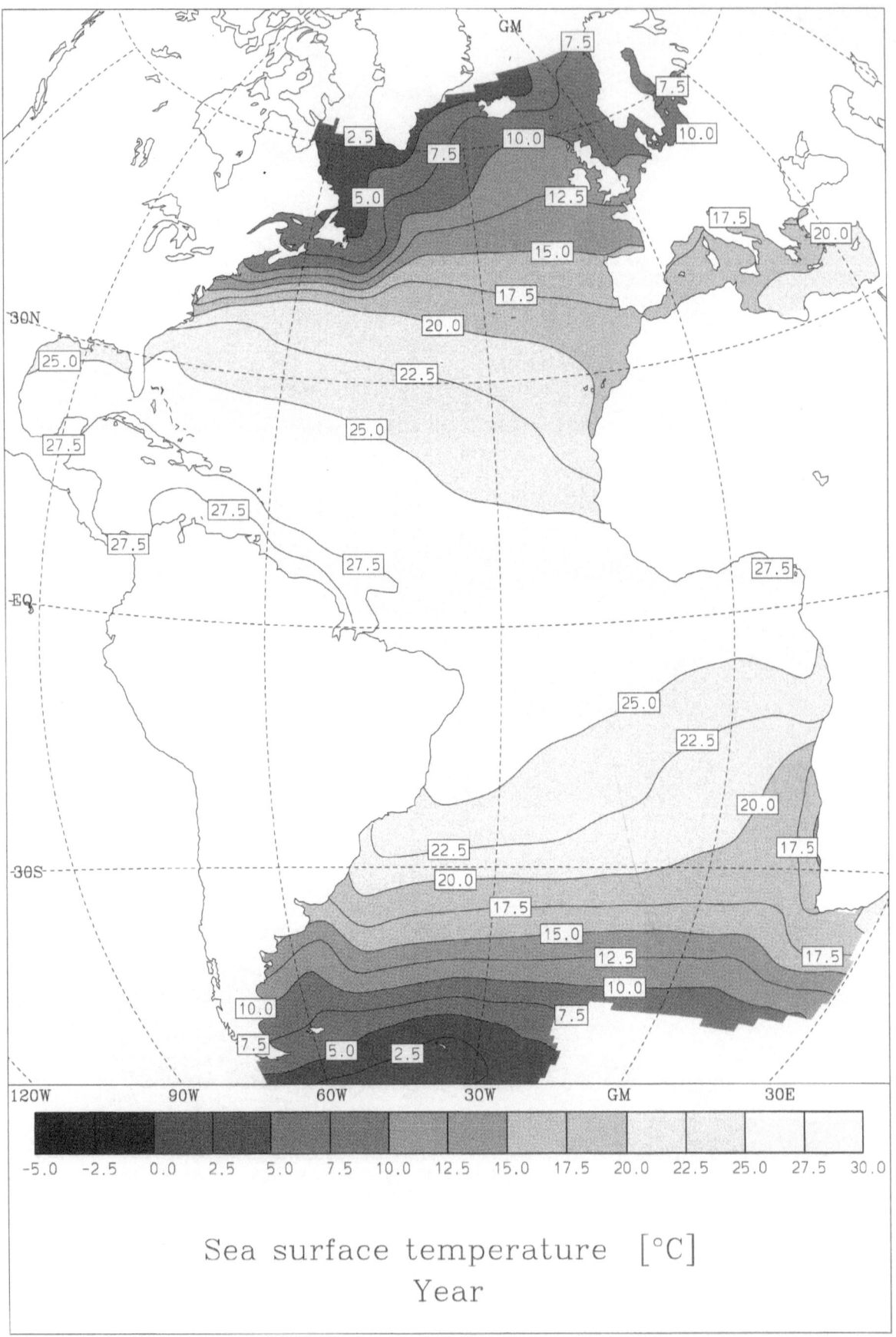

Sea surface temperature [°C]
Year

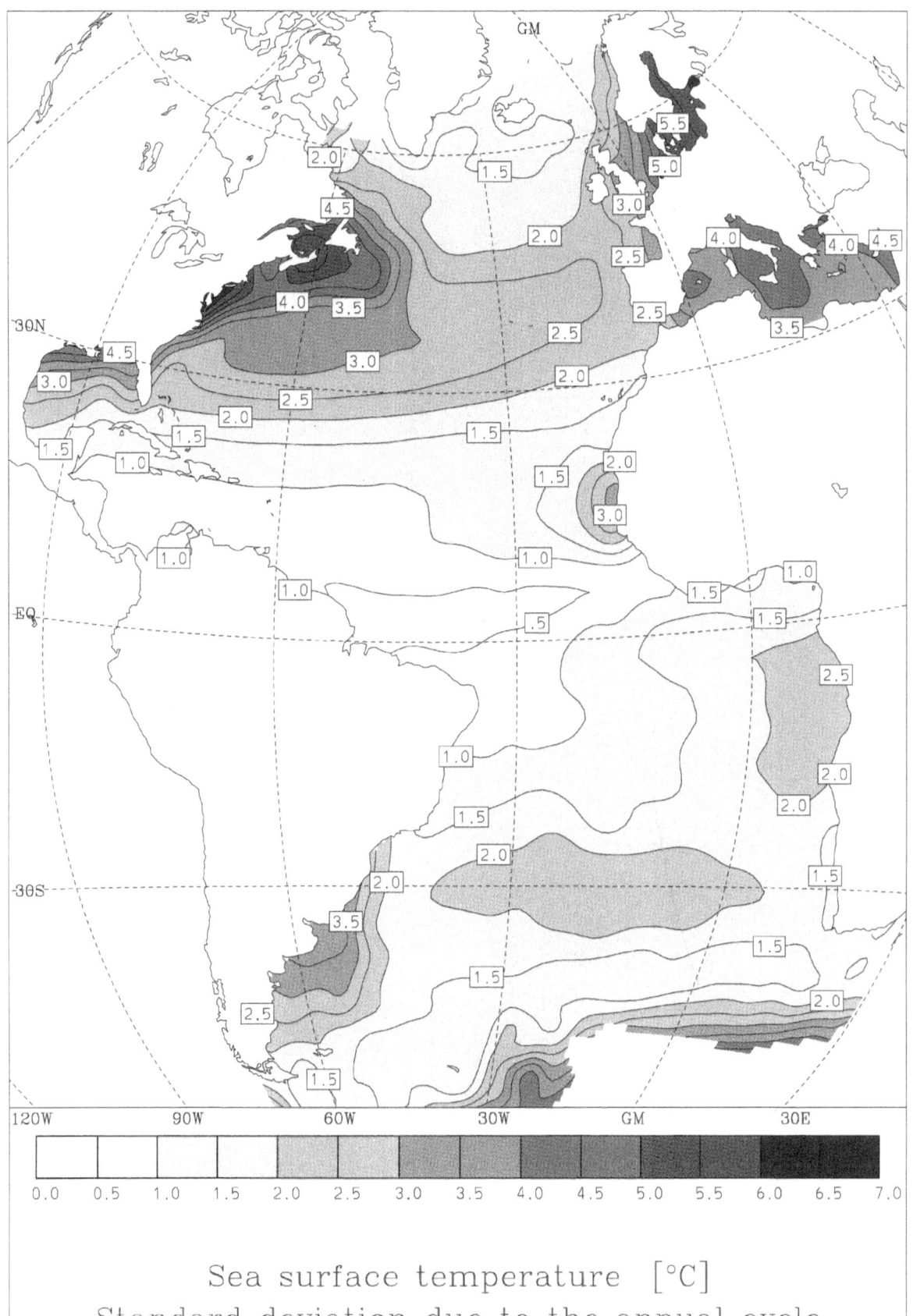

Sea surface temperature [°C]
Standard deviation due to the annual cycle

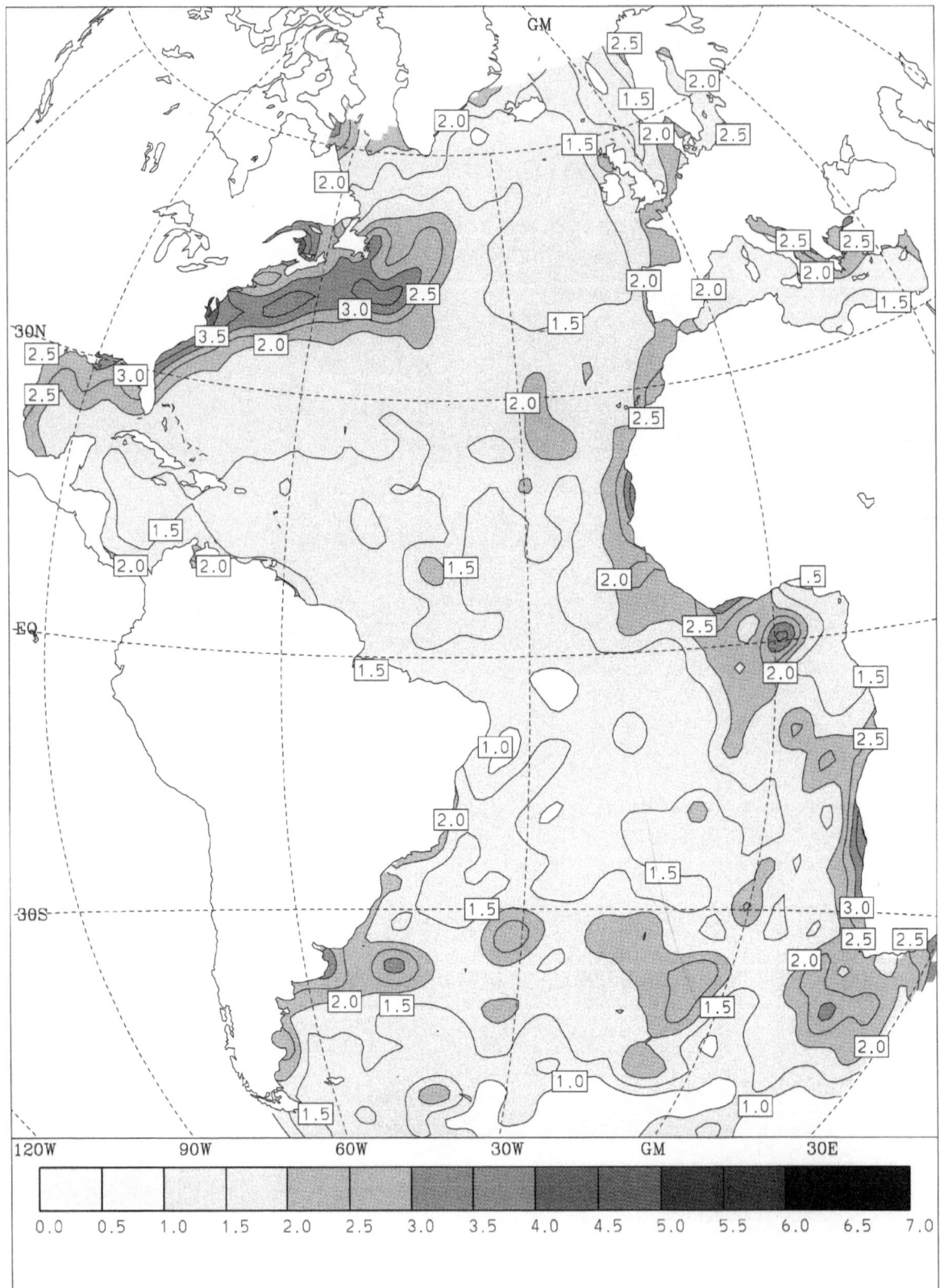

Standard deviation of sea surface temperature [°C]
January

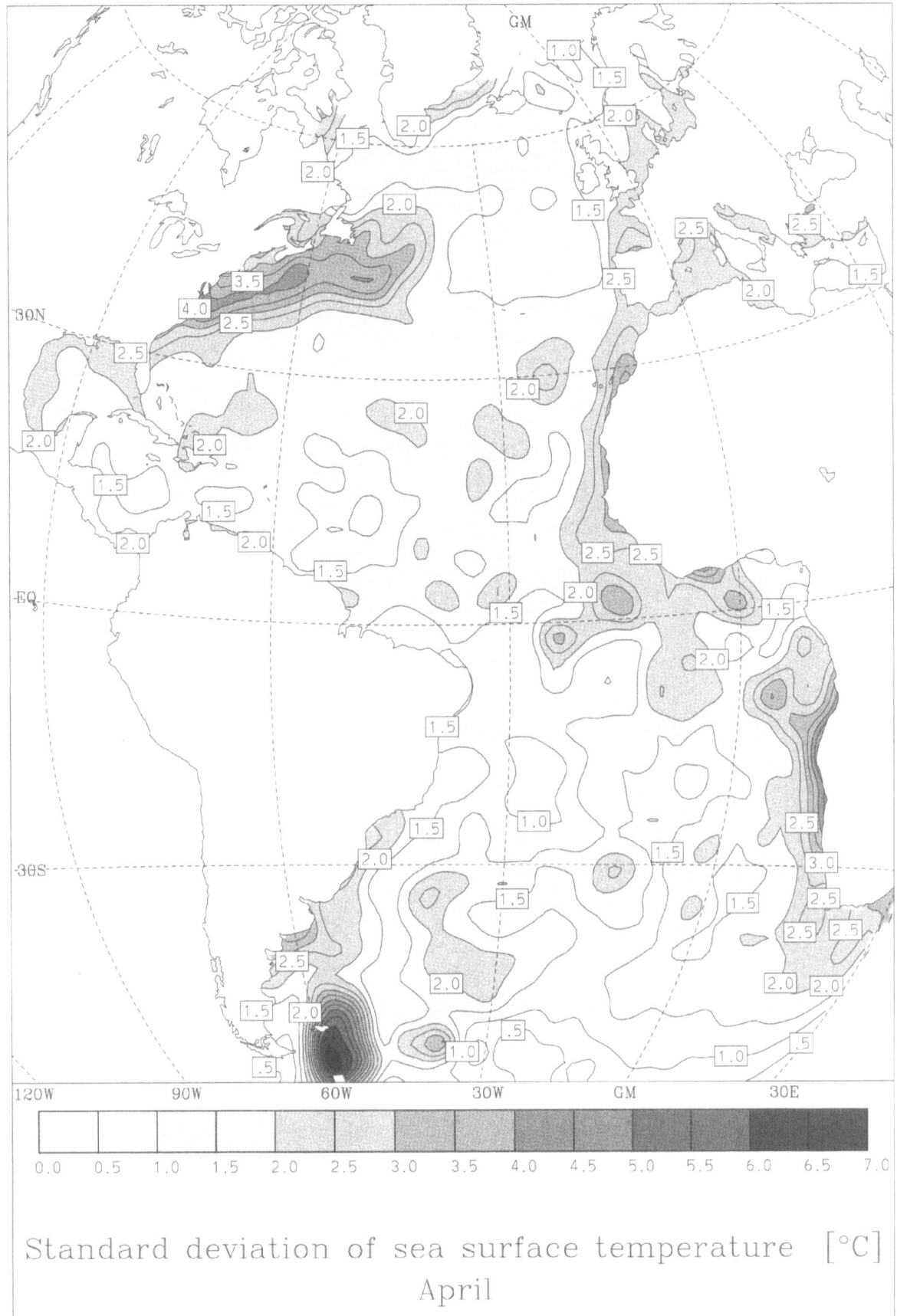

Standard deviation of sea surface temperature [°C]
April

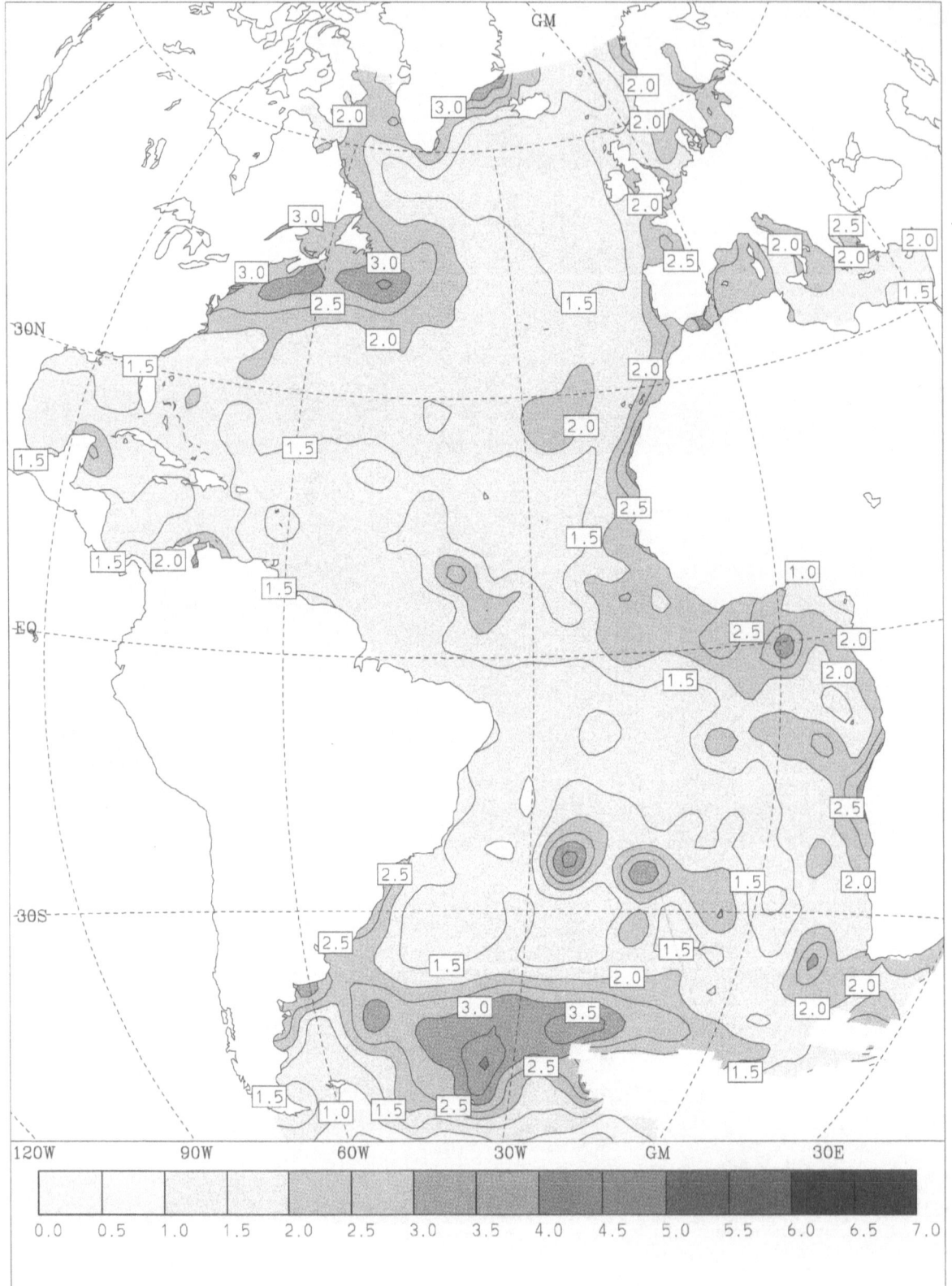

Standard deviation of sea surface temperature [°C]
July

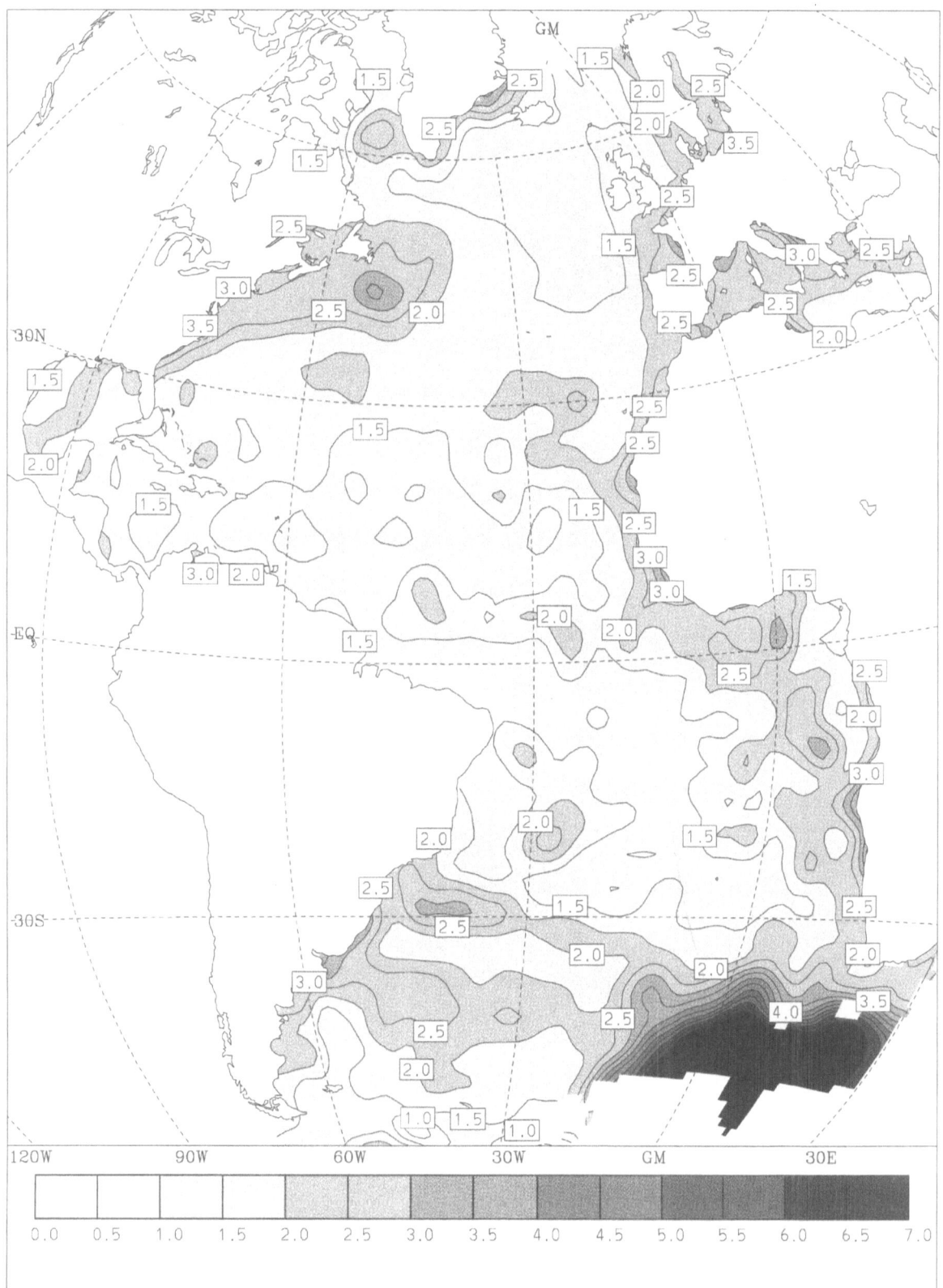

Standard deviation of sea surface temperature [°C]
October

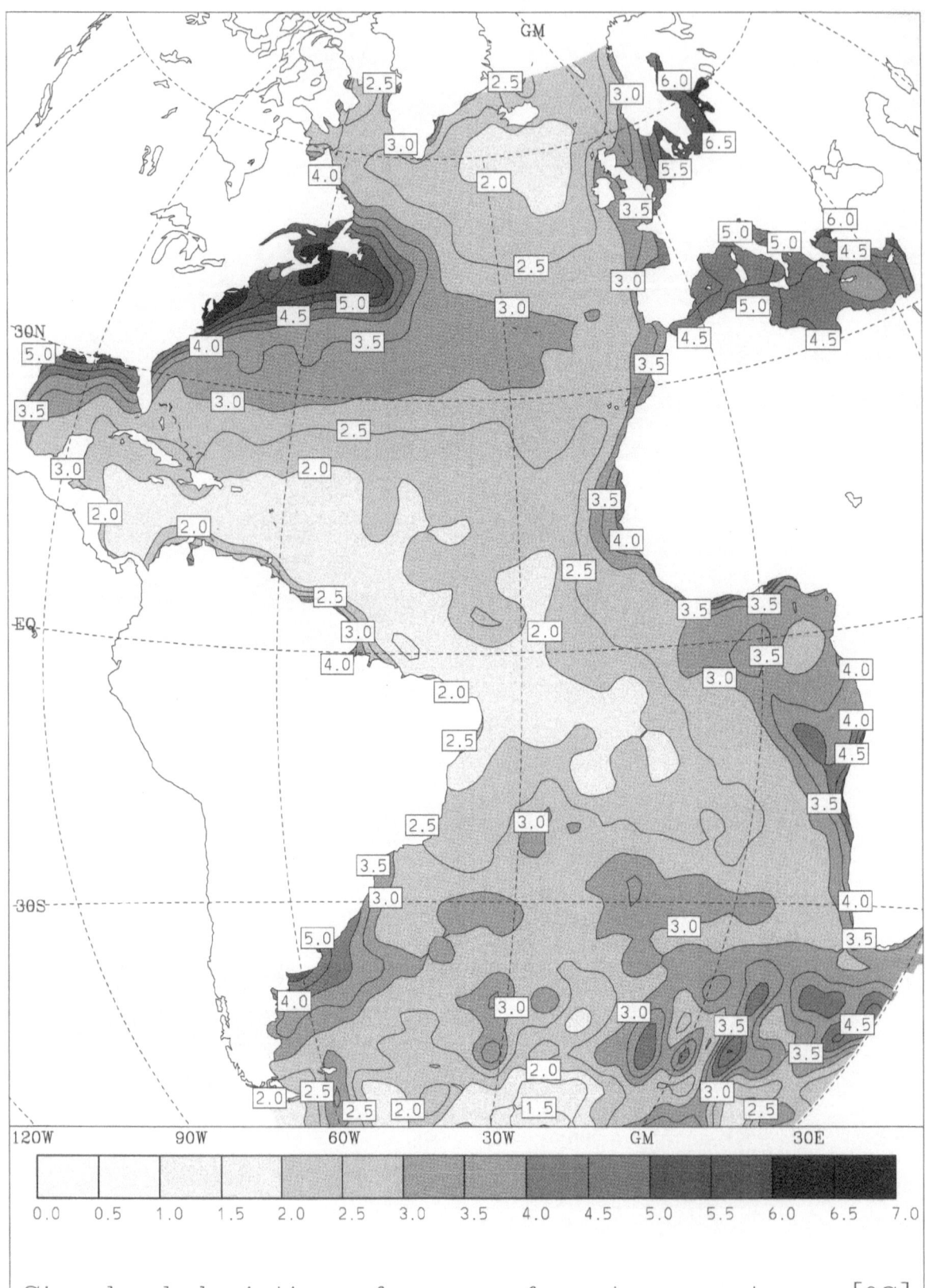

Standard deviation of sea surface temperature [°C]
Year

Air Temperature

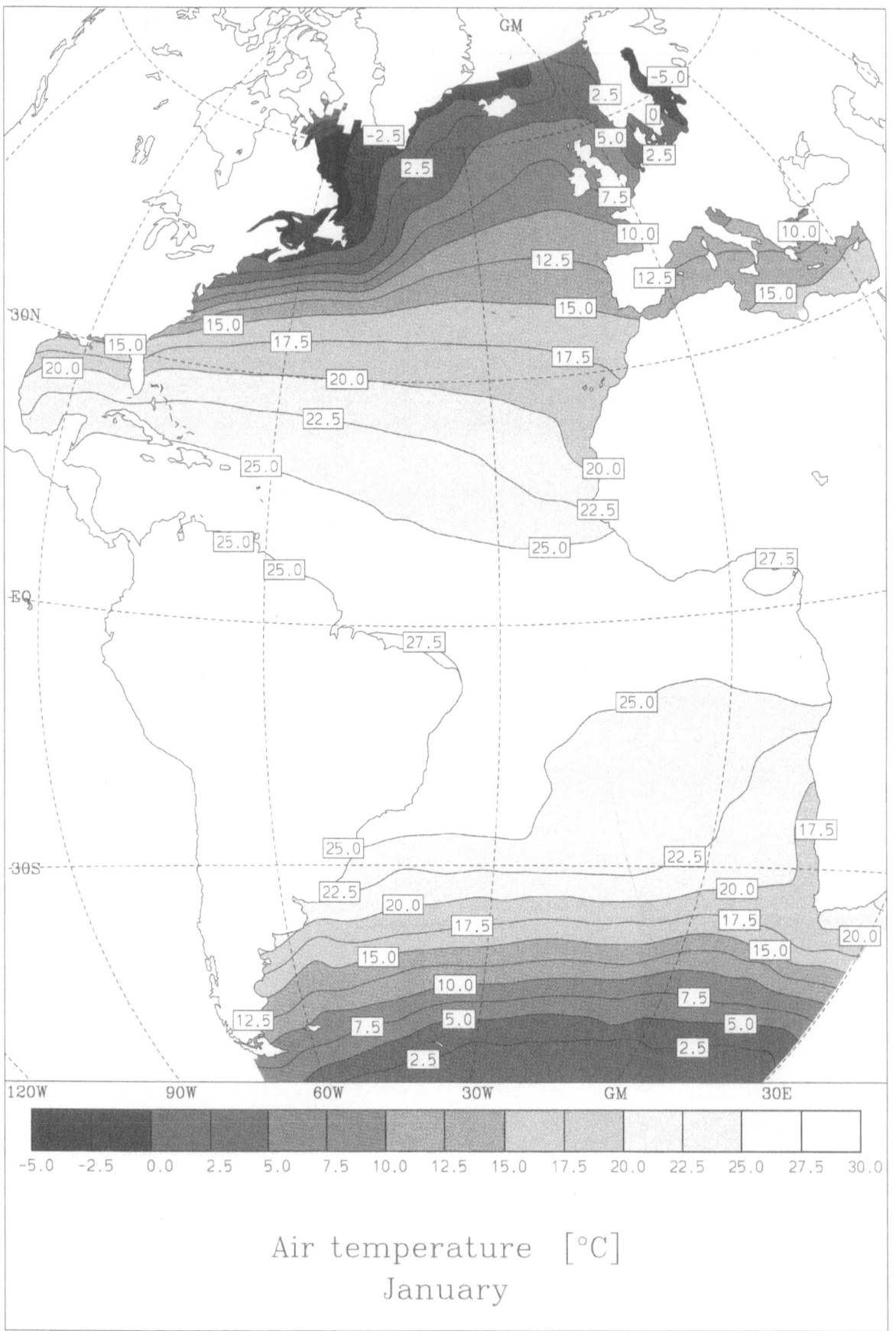

Air temperature [°C]
January

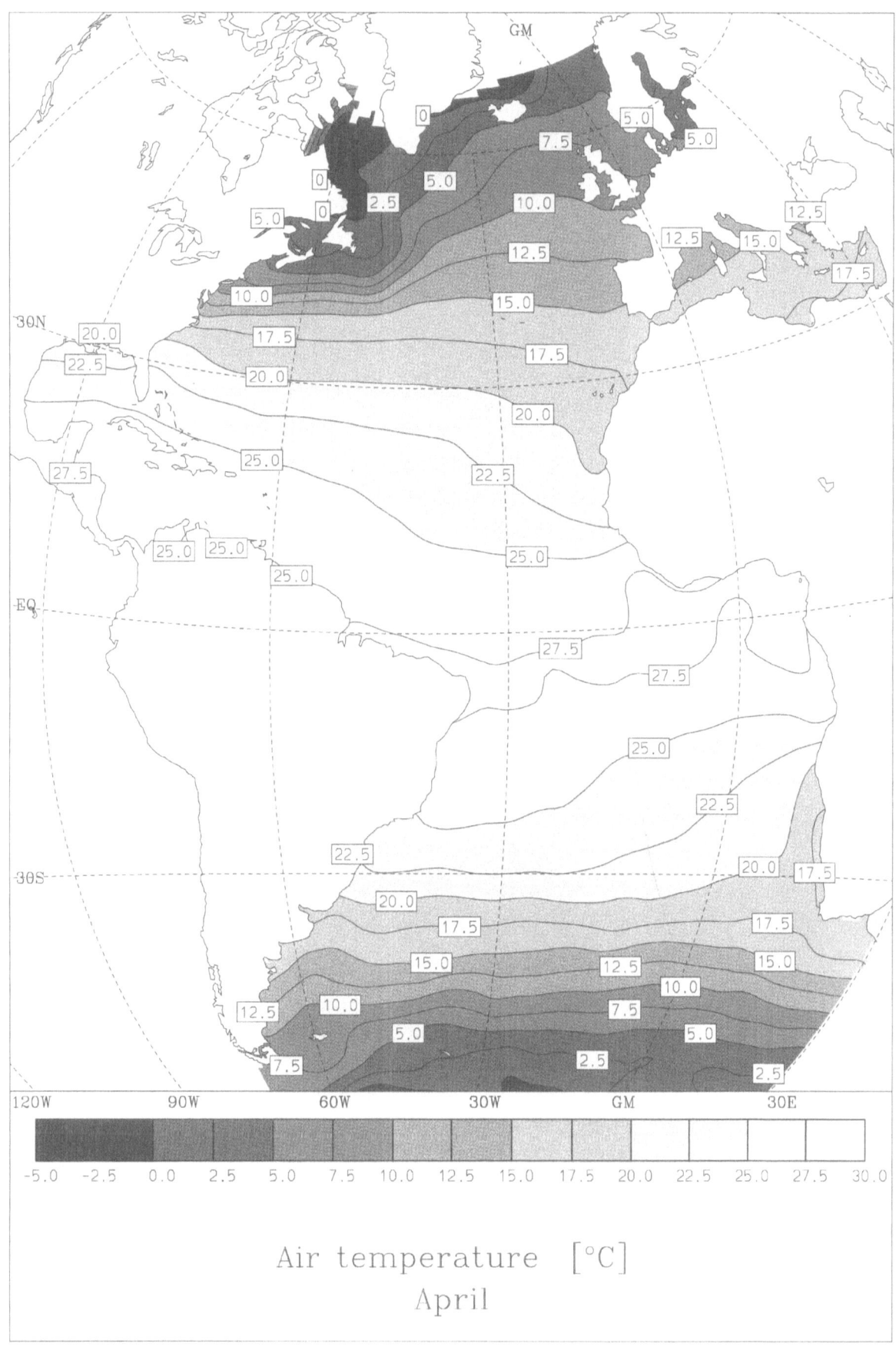

Air temperature [°C]
April

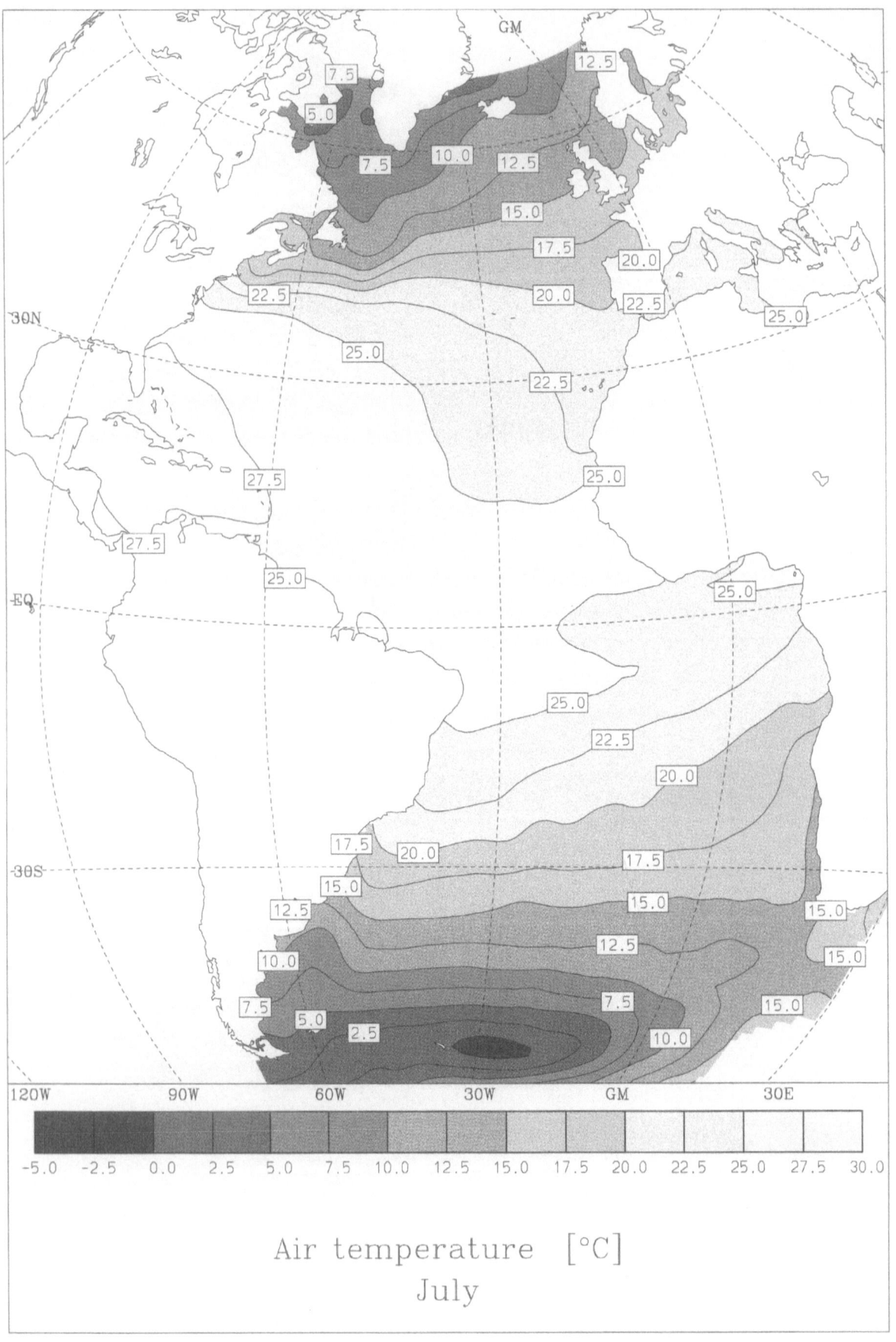

Air temperature [°C]
July

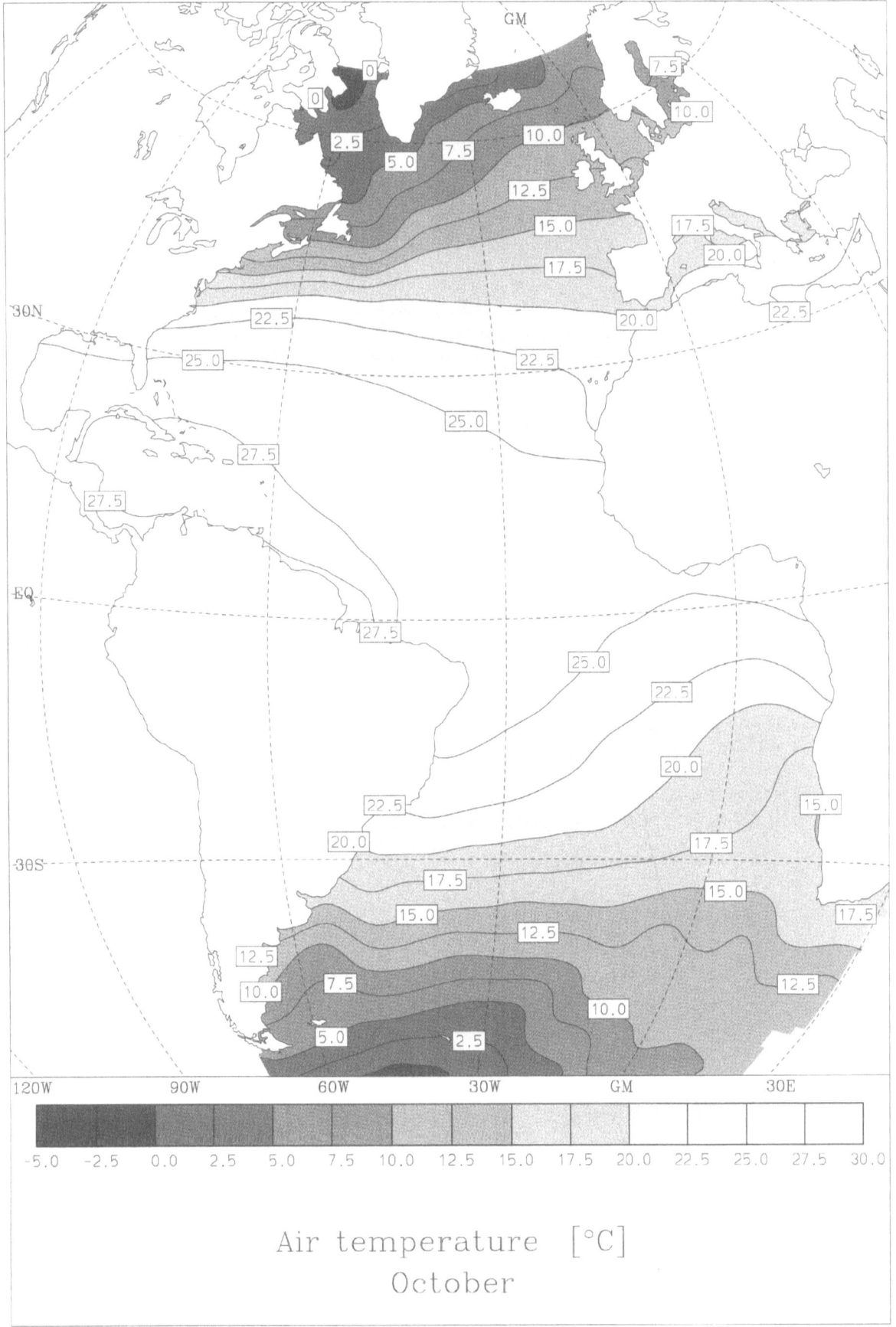

Air temperature [°C]
October

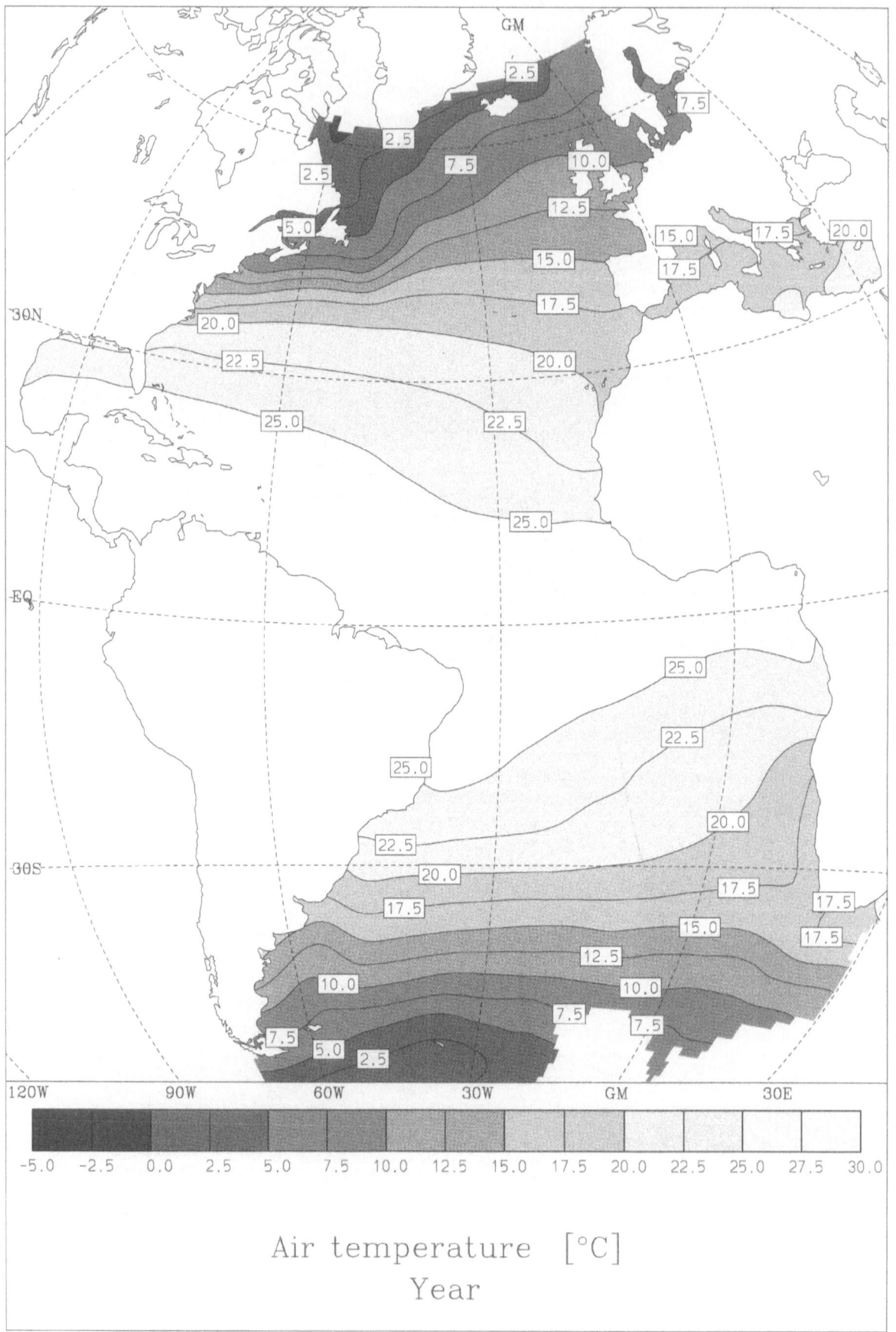

Air temperature [°C]
Year

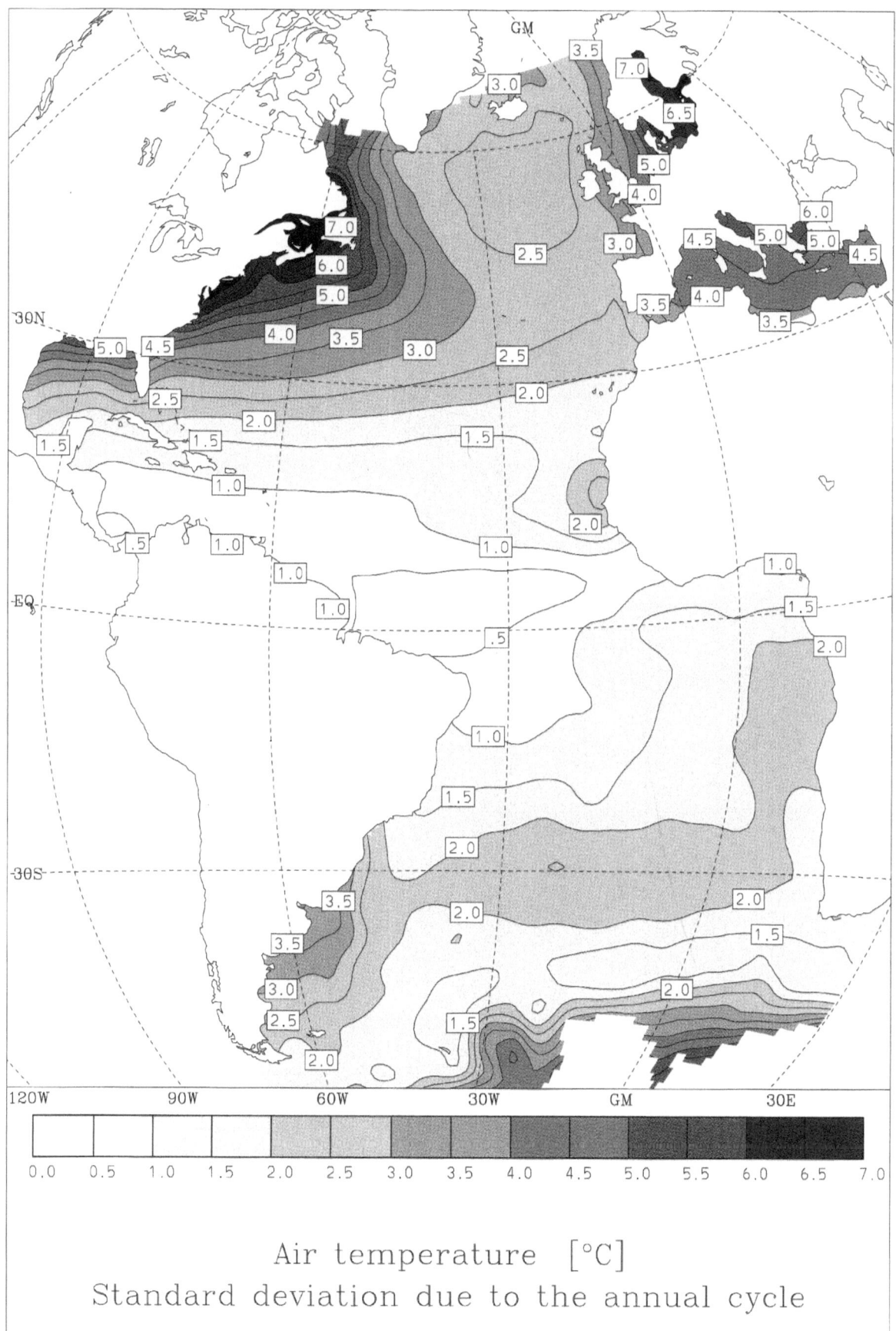

Air temperature [°C]
Standard deviation due to the annual cycle

Air minus Sea Surface Temperature

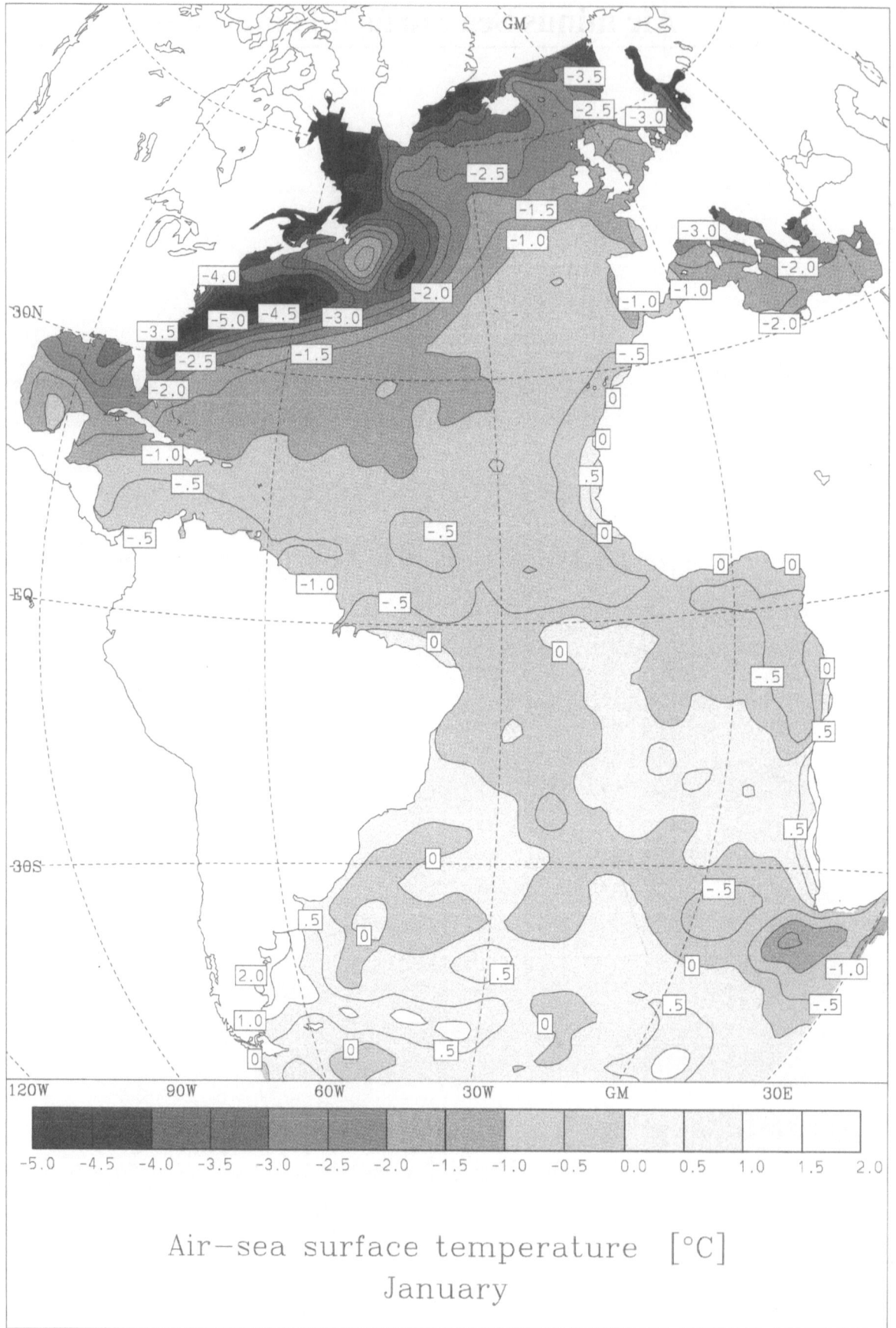

Air−sea surface temperature [°C]
January

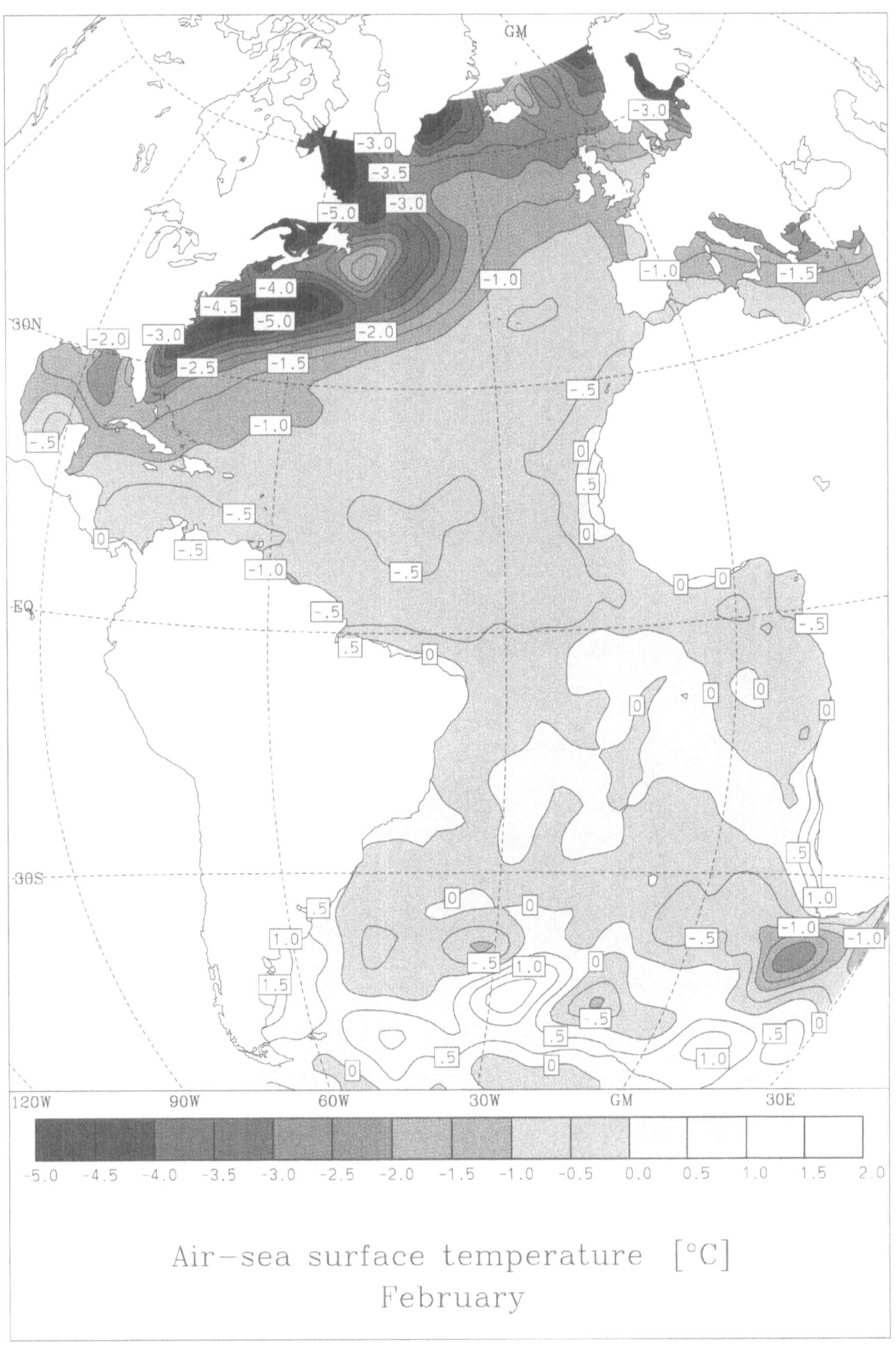

Air–sea surface temperature [°C]
February

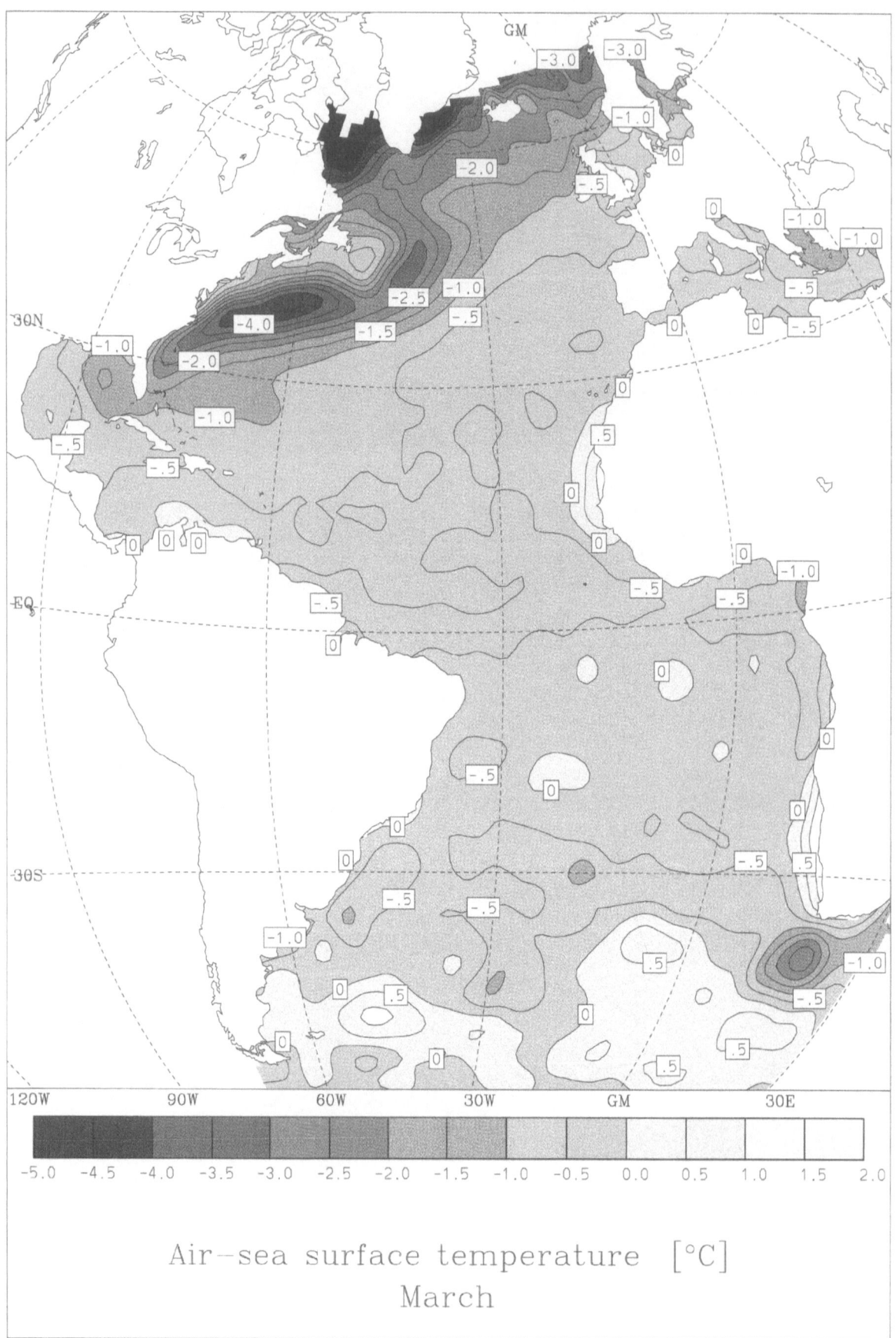

Air−sea surface temperature [°C]
March

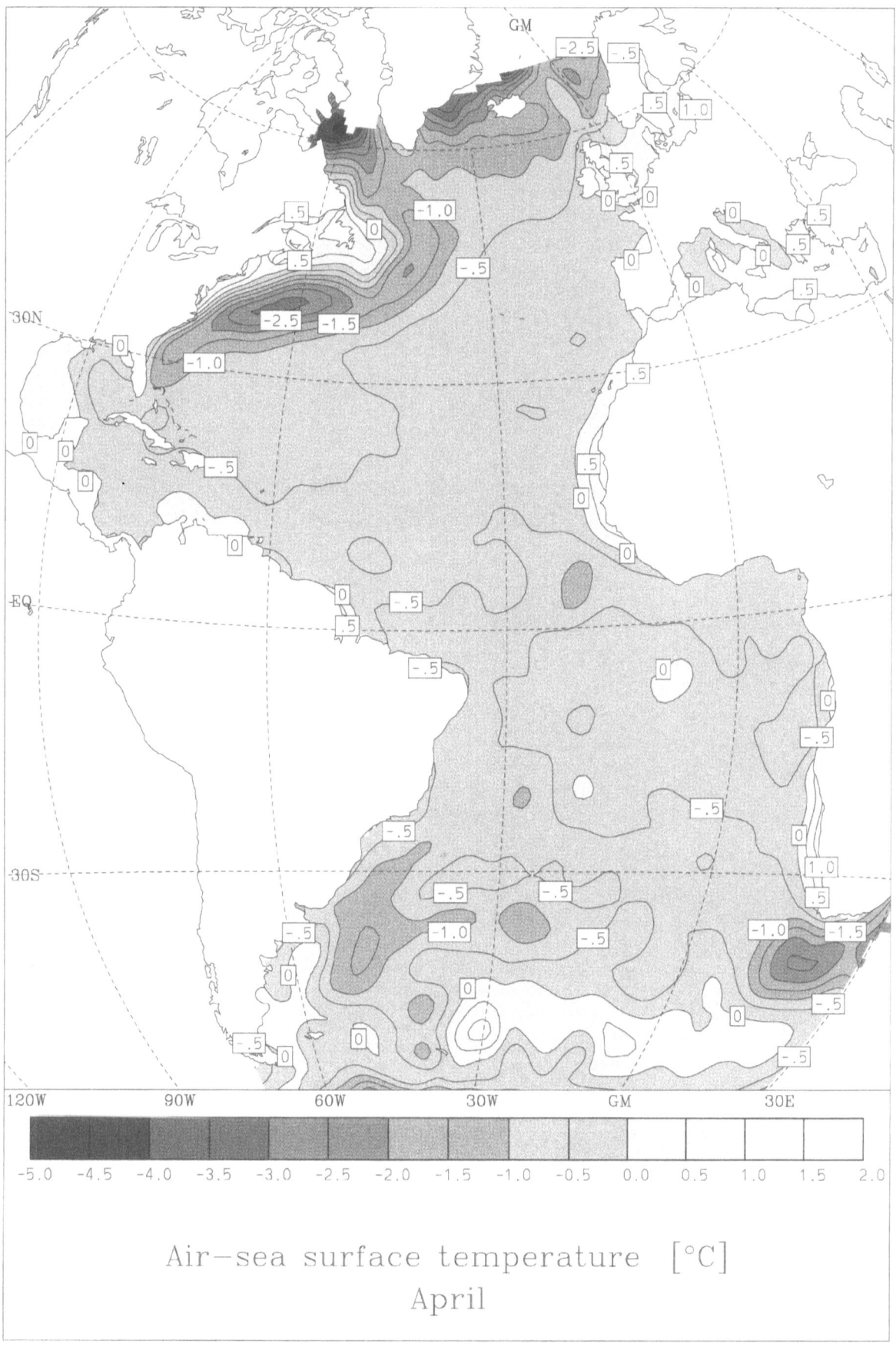

Air−sea surface temperature [°C]
April

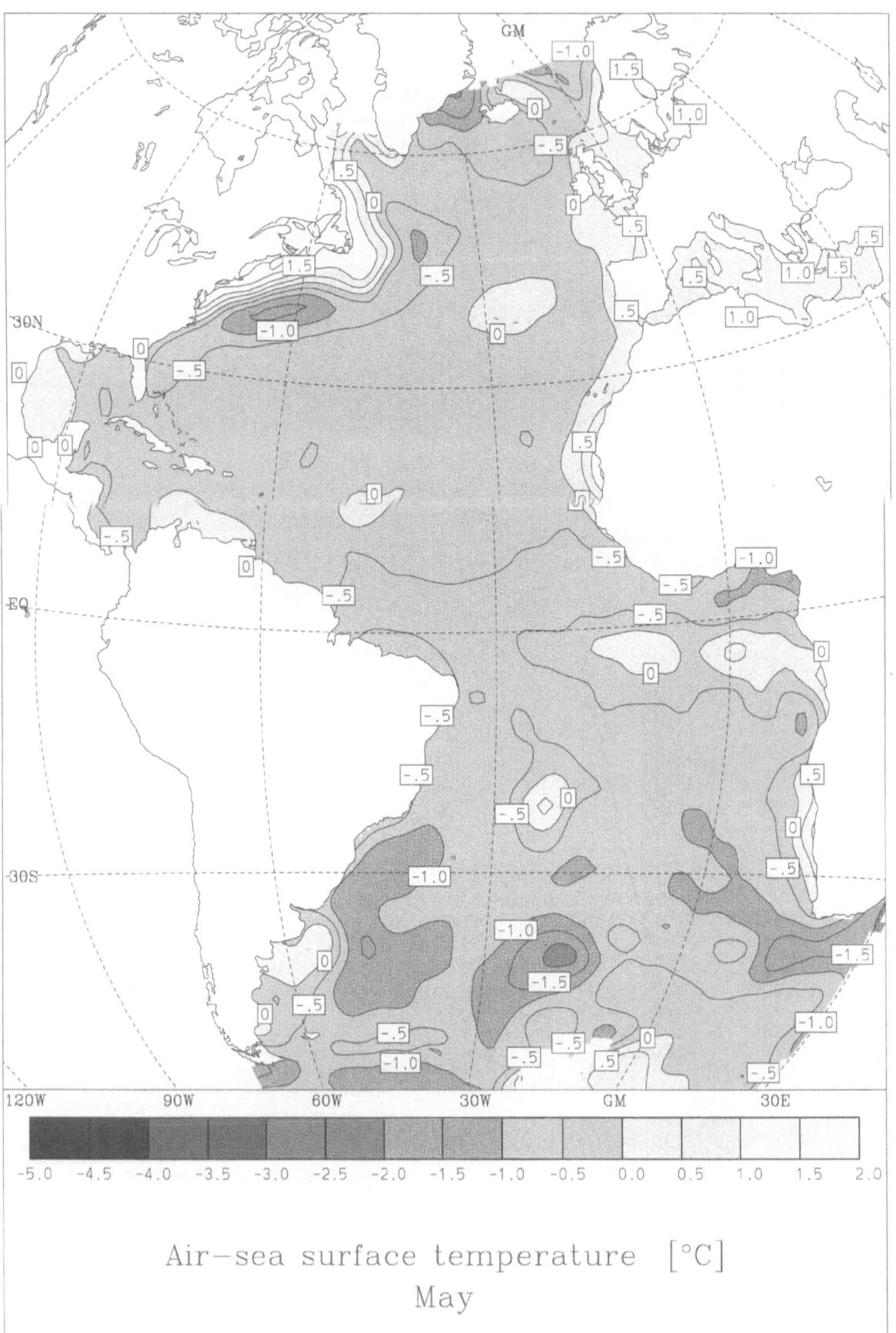

Air–sea surface temperature [°C]
May

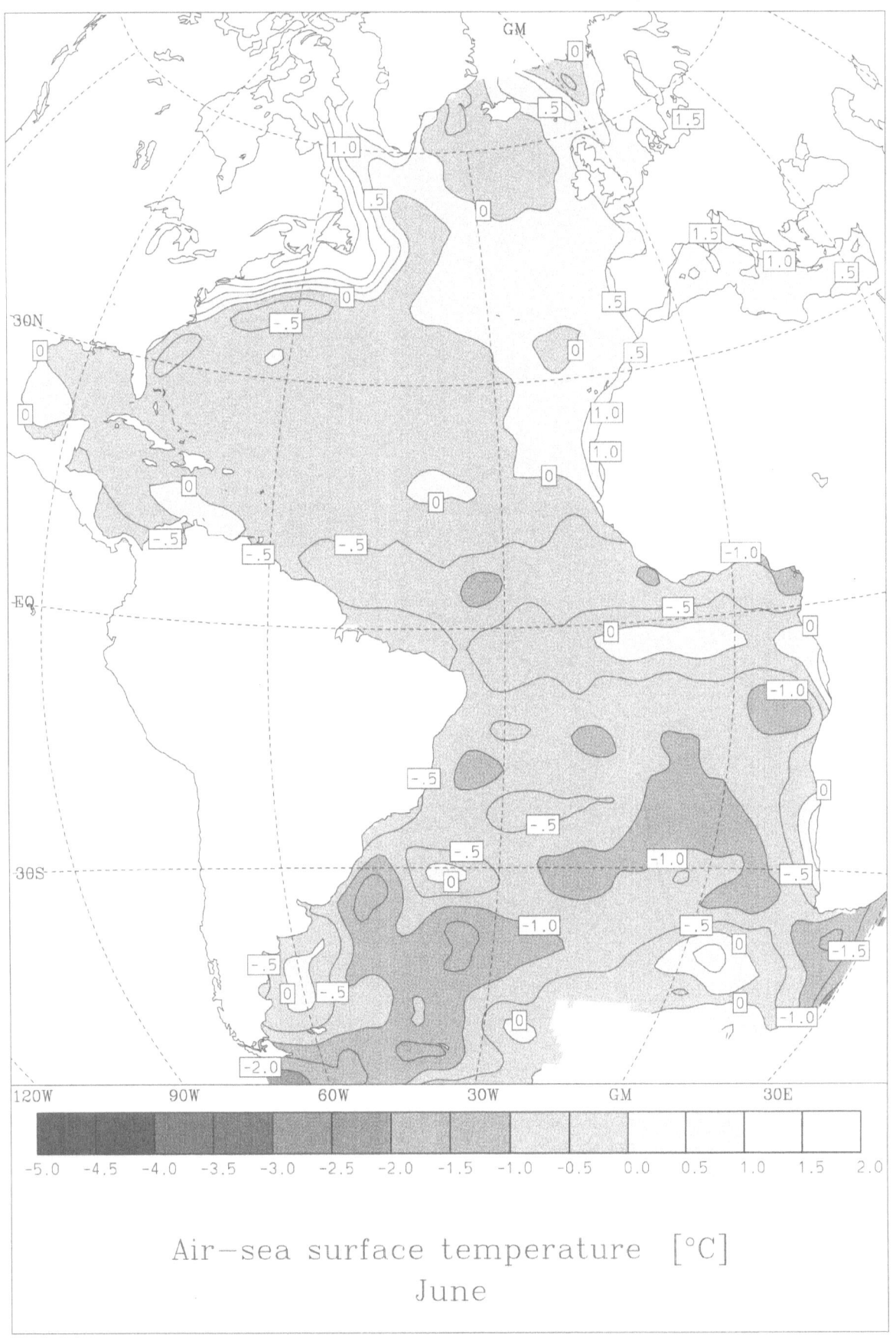

Air−sea surface temperature [°C]
June

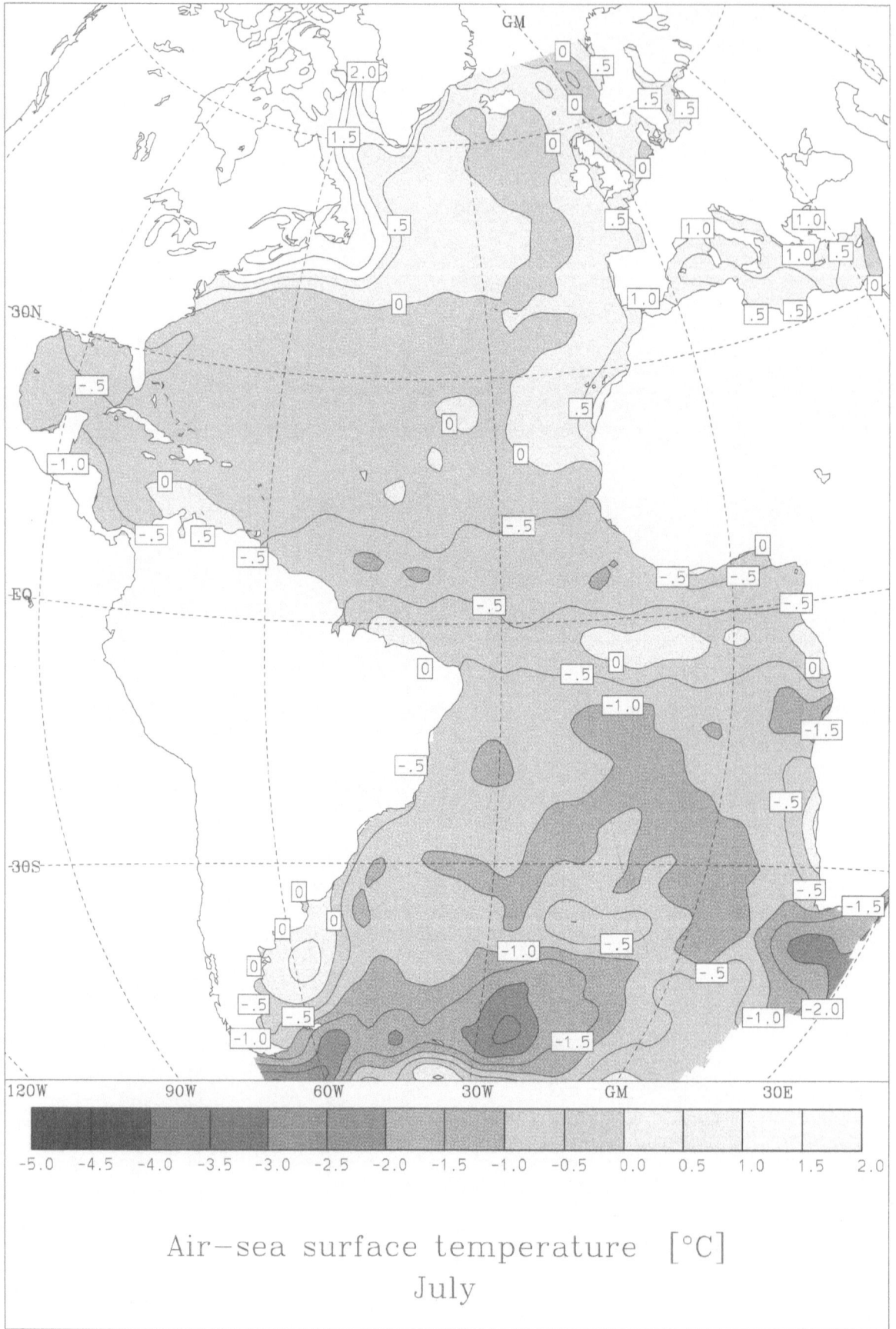

Air−sea surface temperature [°C]
July

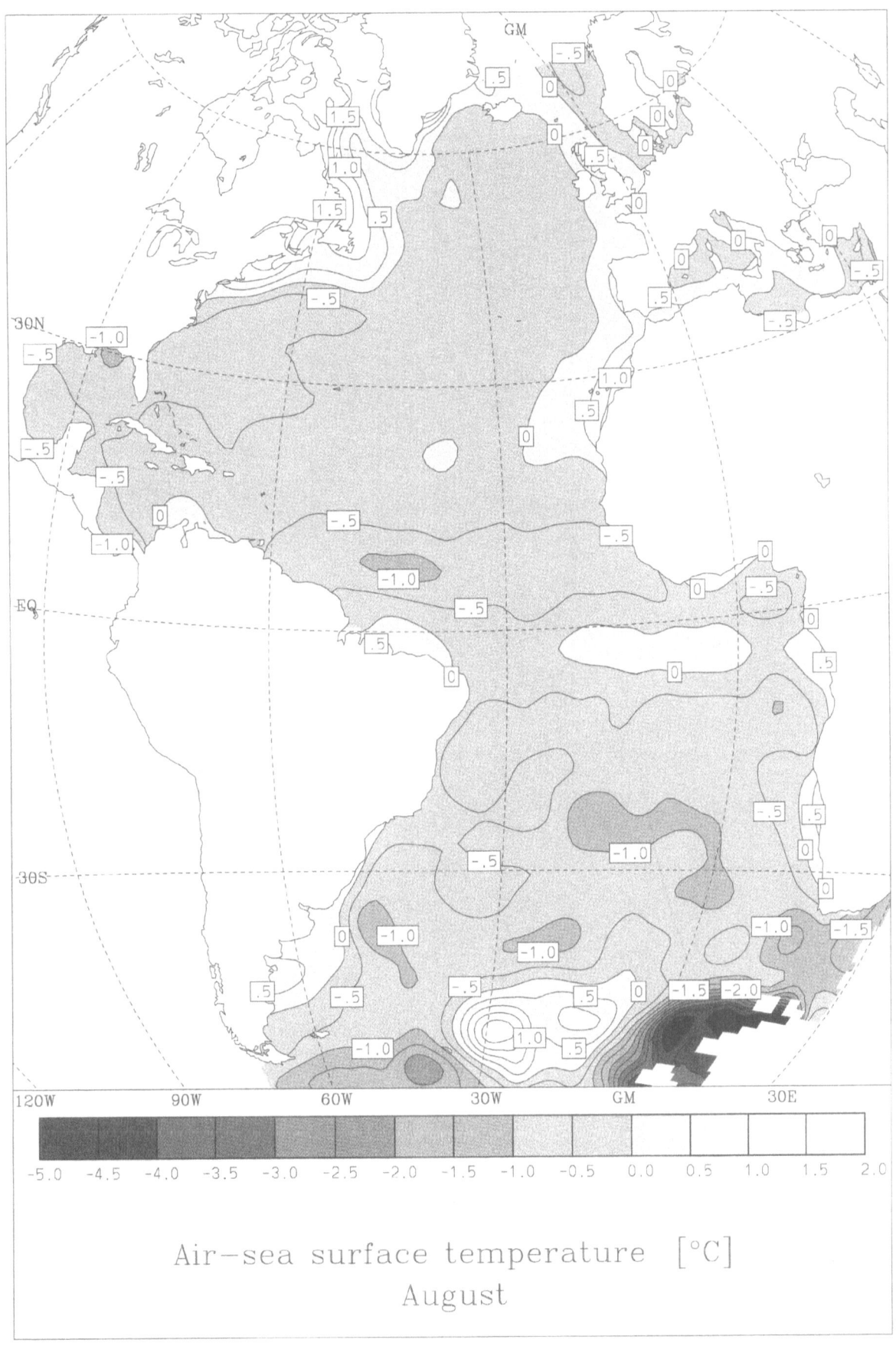

Air–sea surface temperature [°C]
August

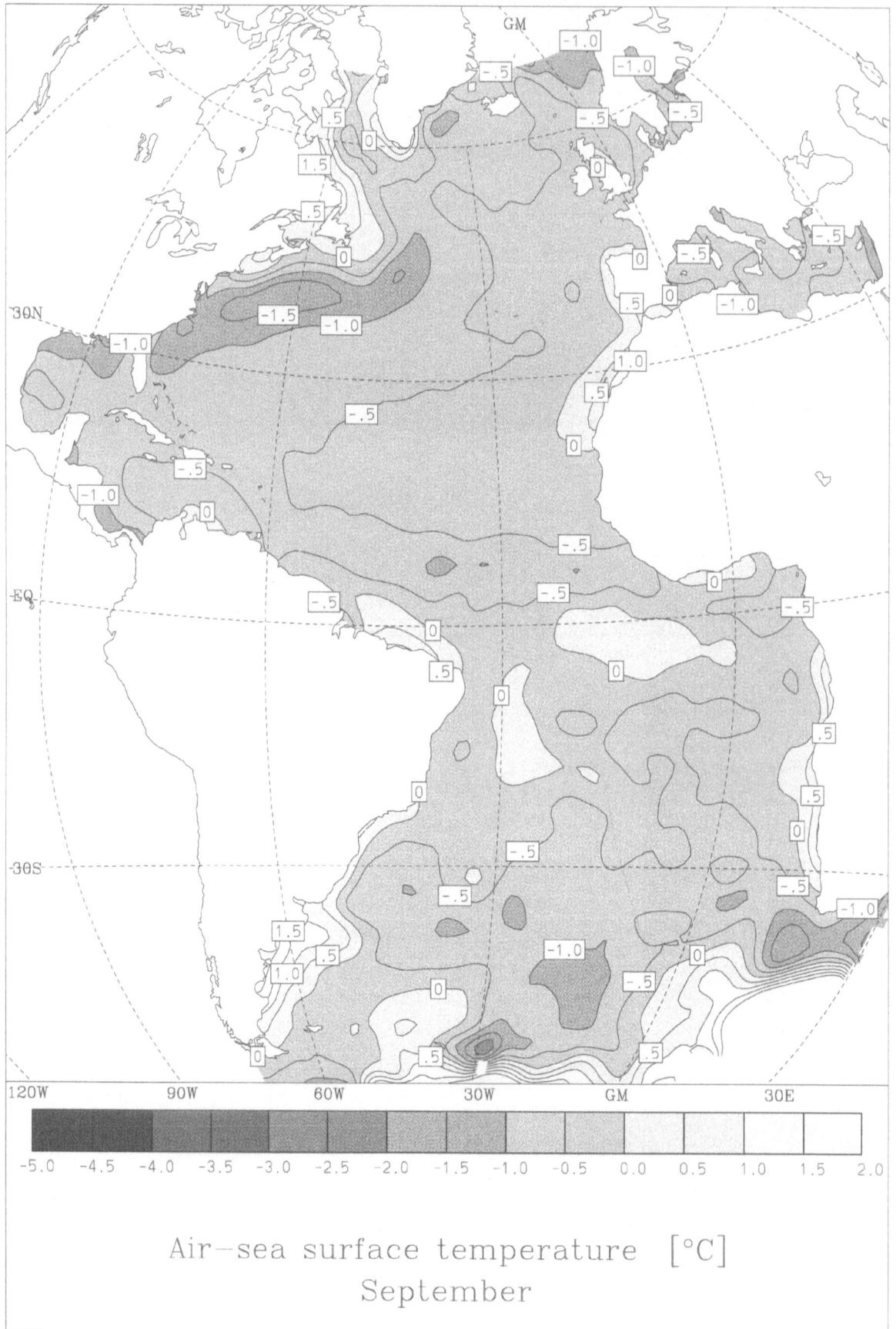

Air—sea surface temperature [°C]
September

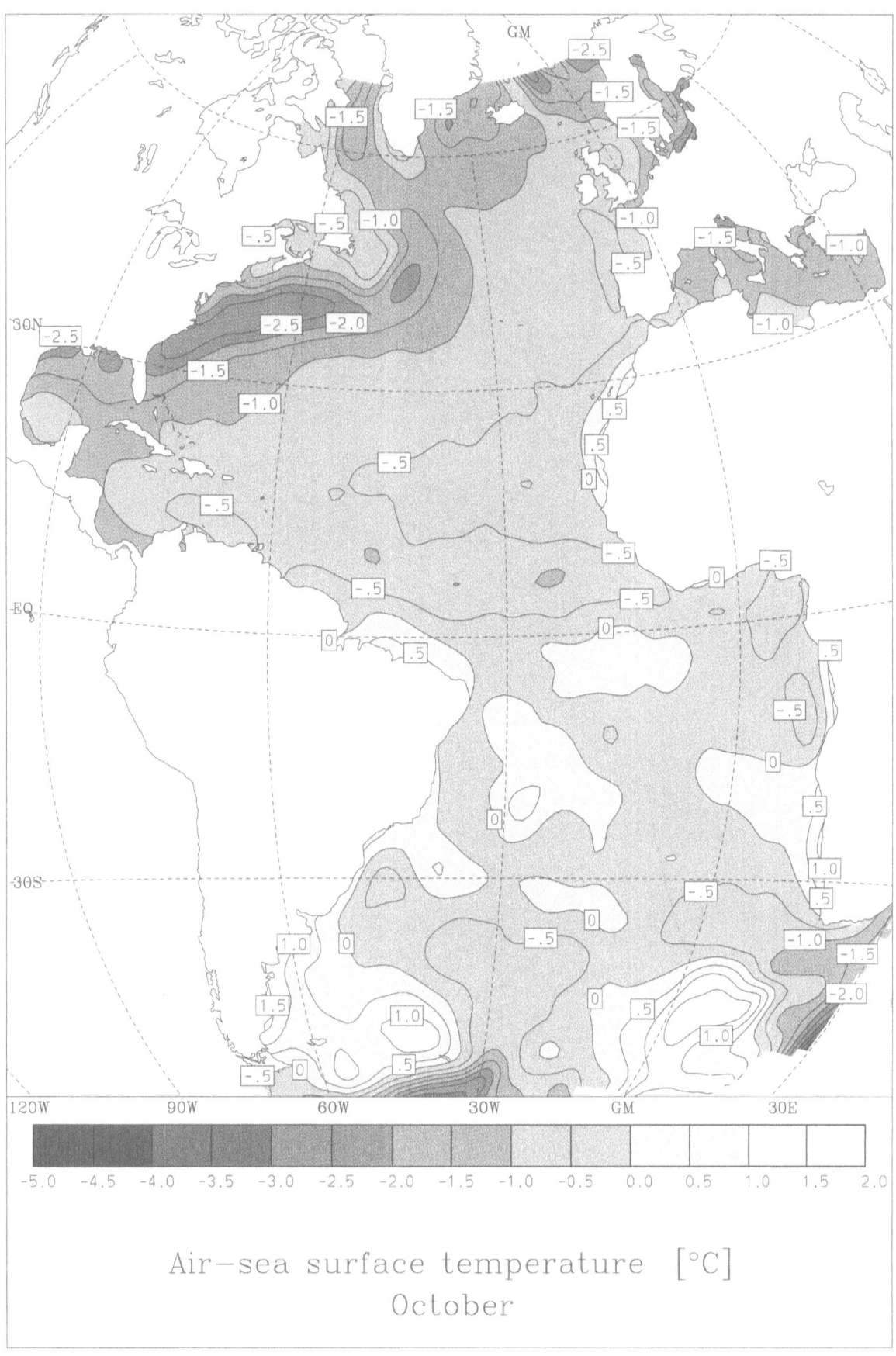

Air−sea surface temperature [°C]
October

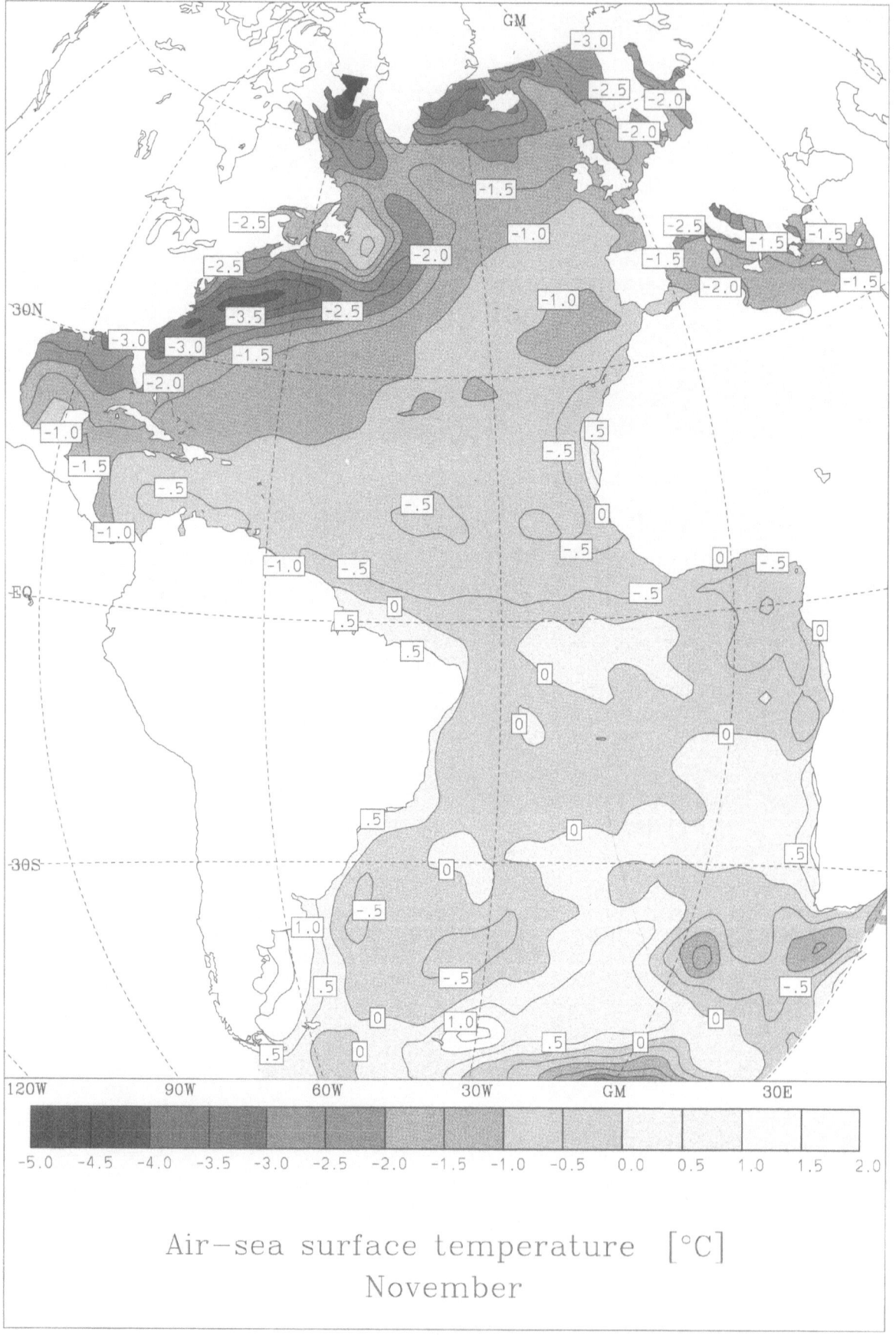

Air–sea surface temperature [°C]
November

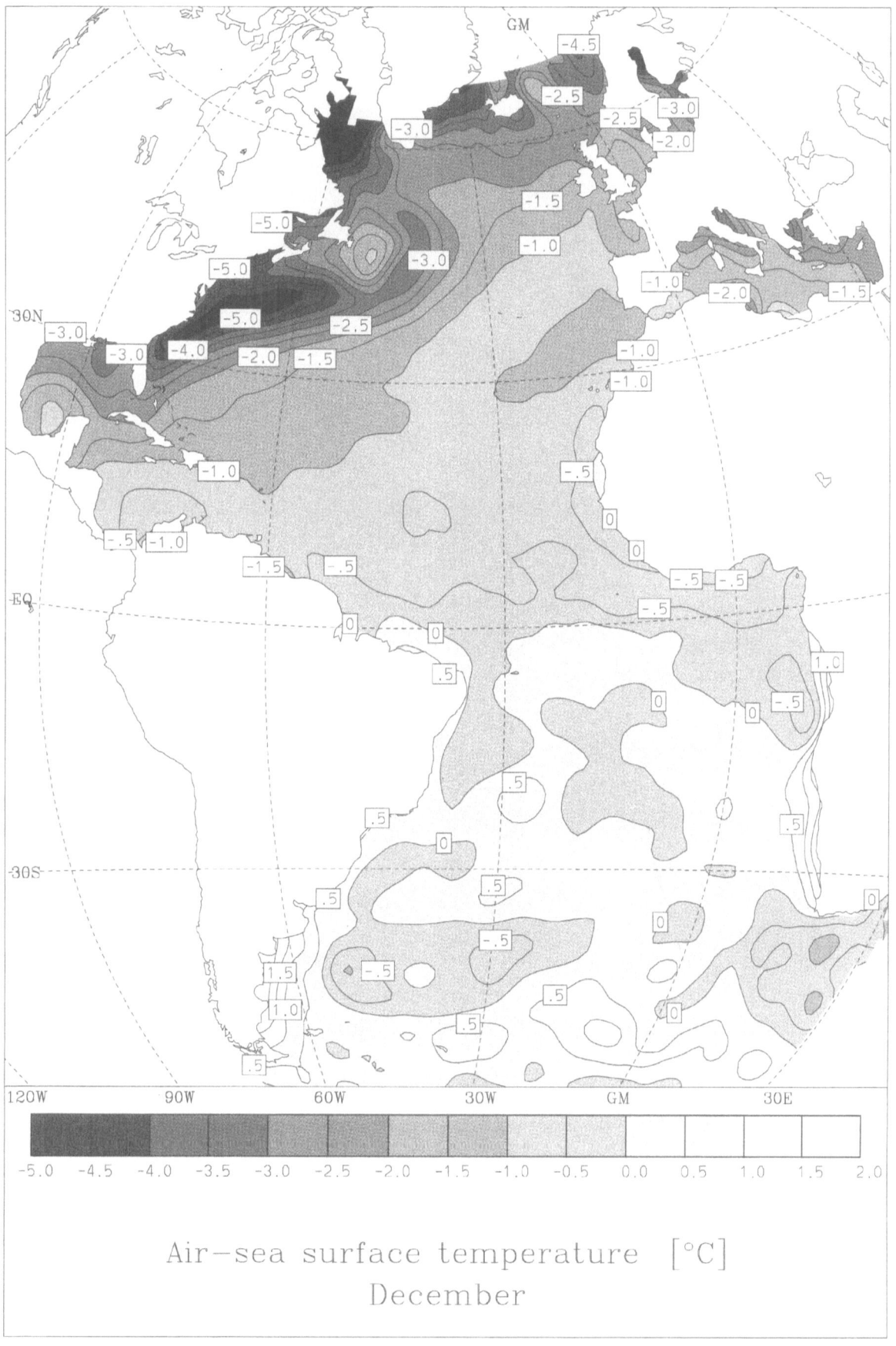

Air−sea surface temperature [°C]
December

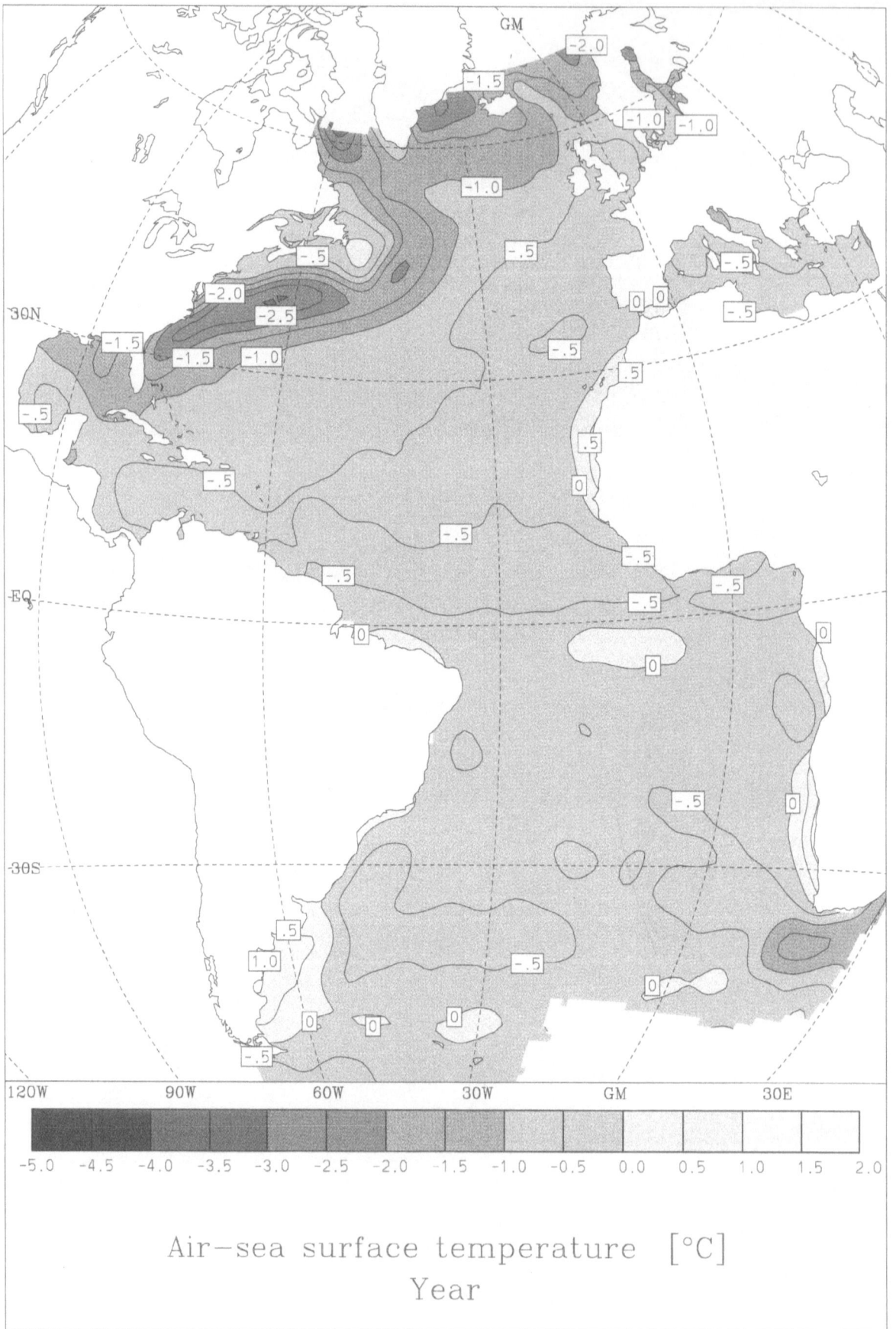

Air−sea surface temperature [°C]
Year

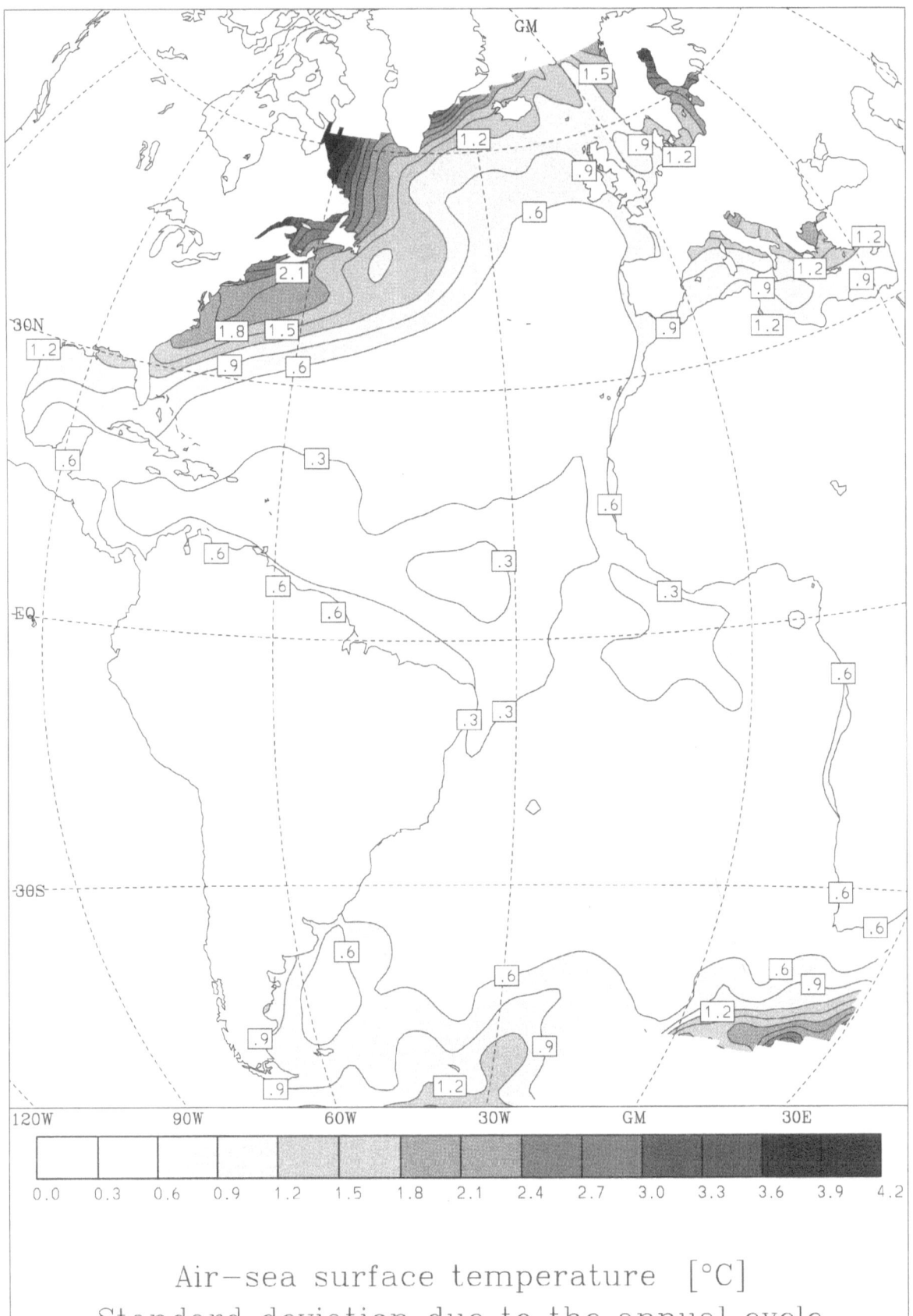

Air-sea surface temperature [°C]
Standard deviation due to the annual cycle

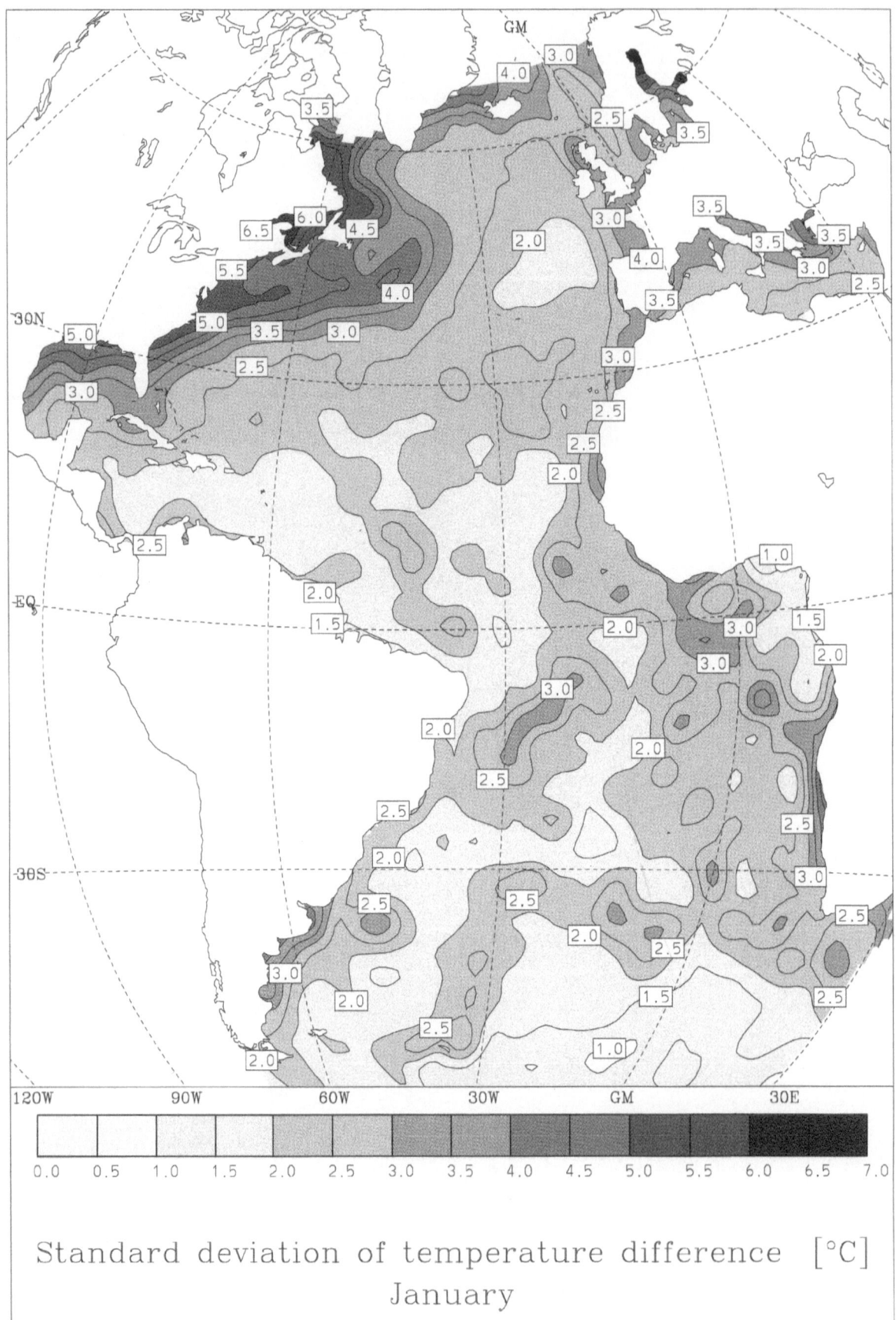

Standard deviation of temperature difference [°C]
January

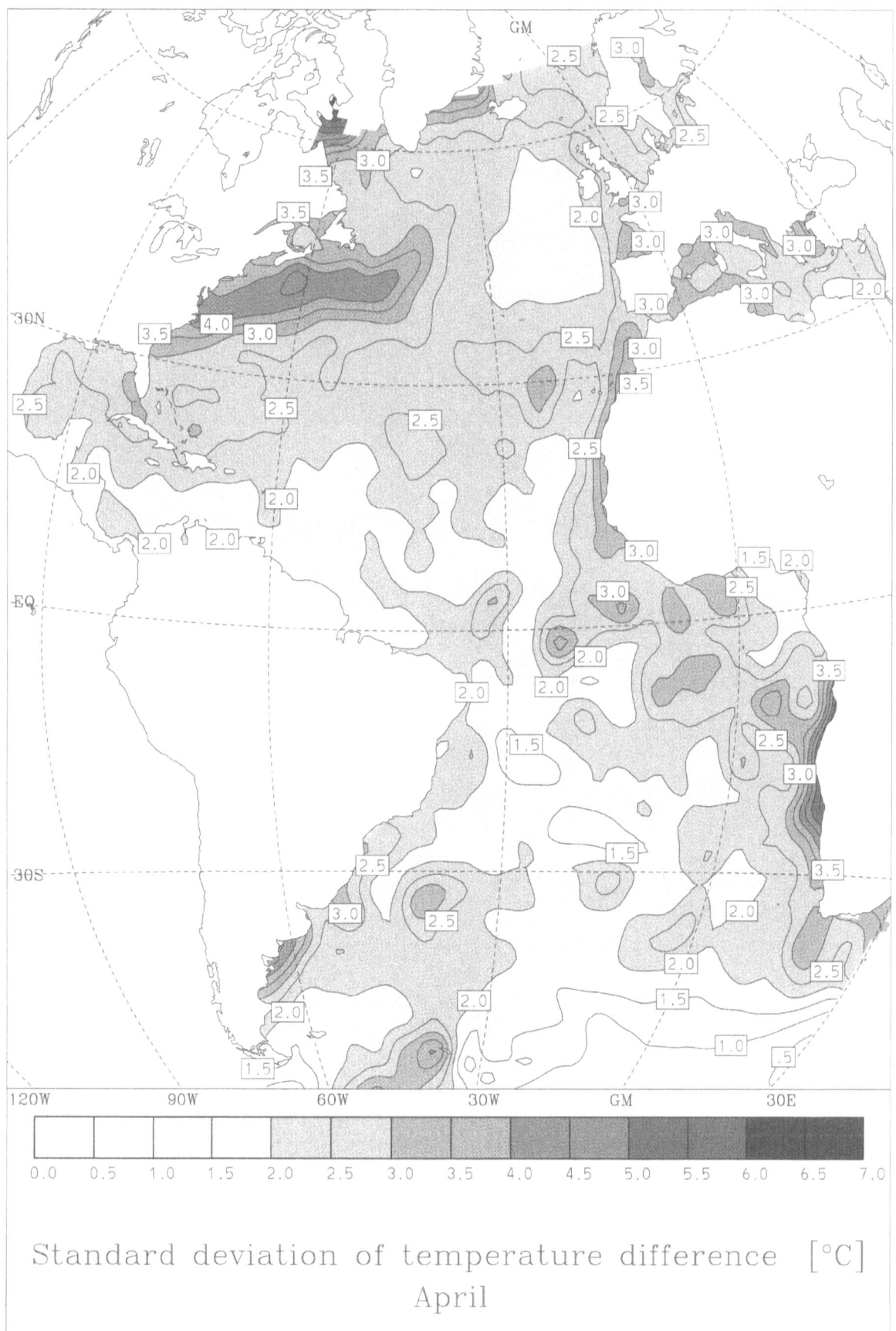

Standard deviation of temperature difference [°C]
April

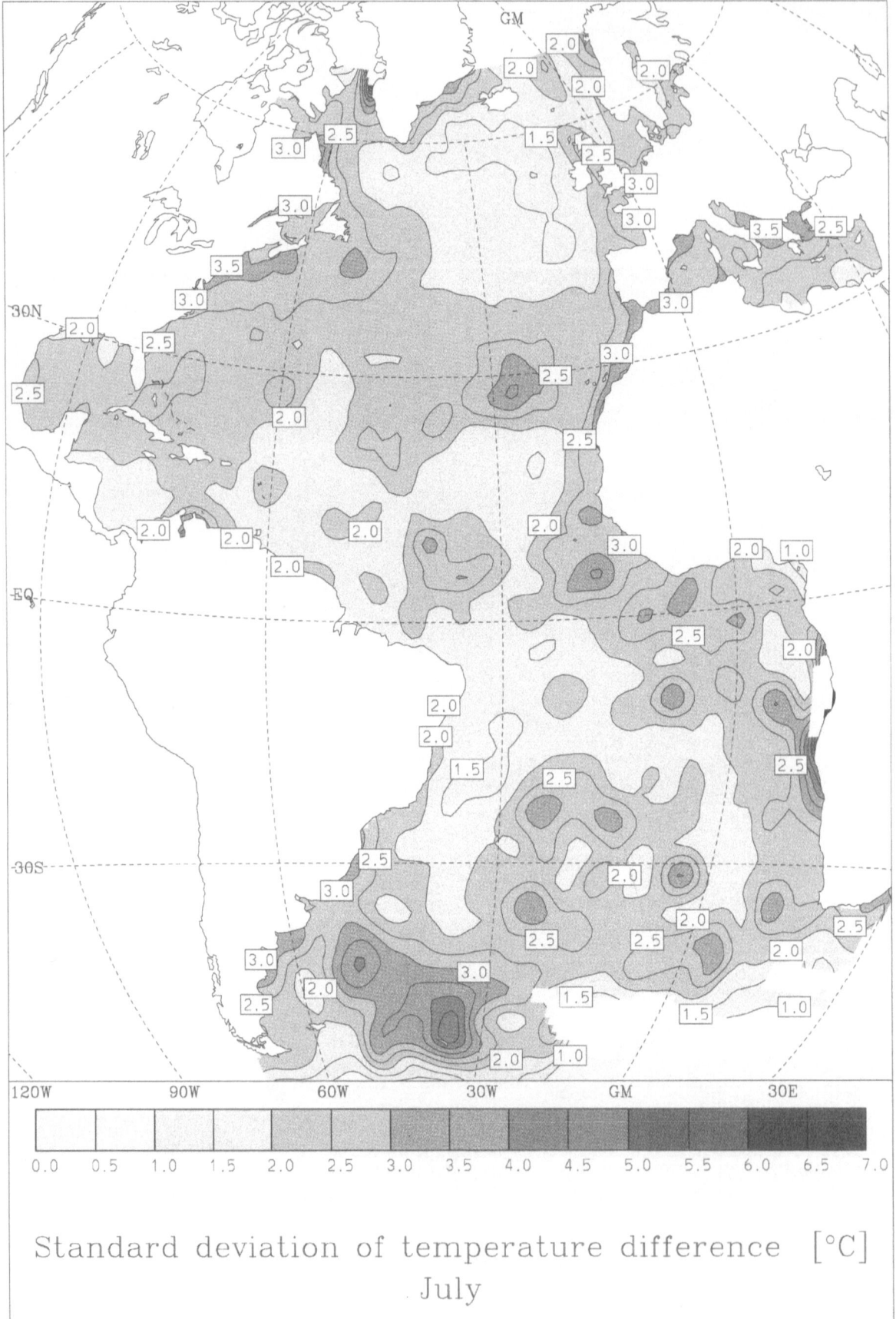

Standard deviation of temperature difference [°C]
July

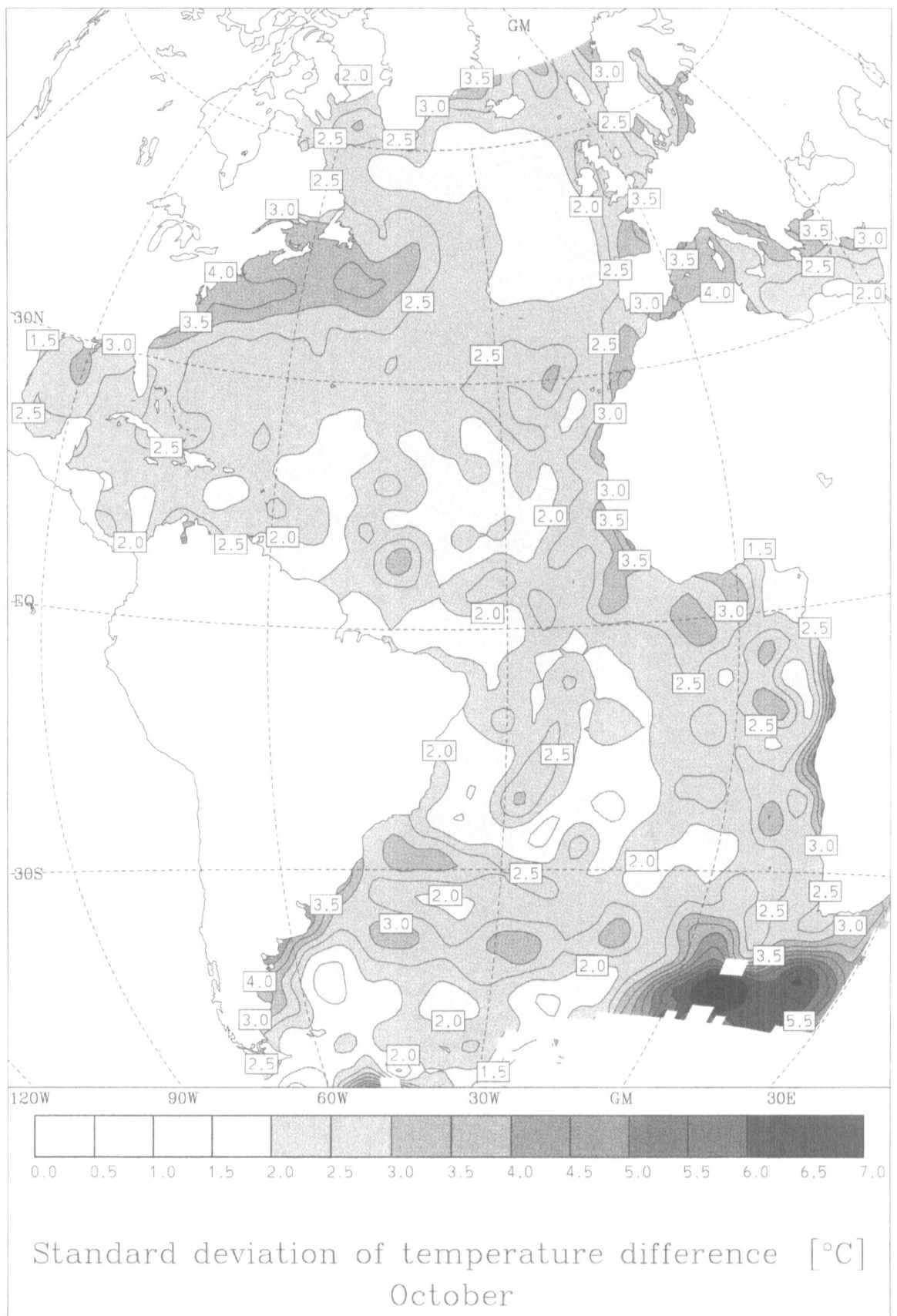

Standard deviation of temperature difference [°C]
October

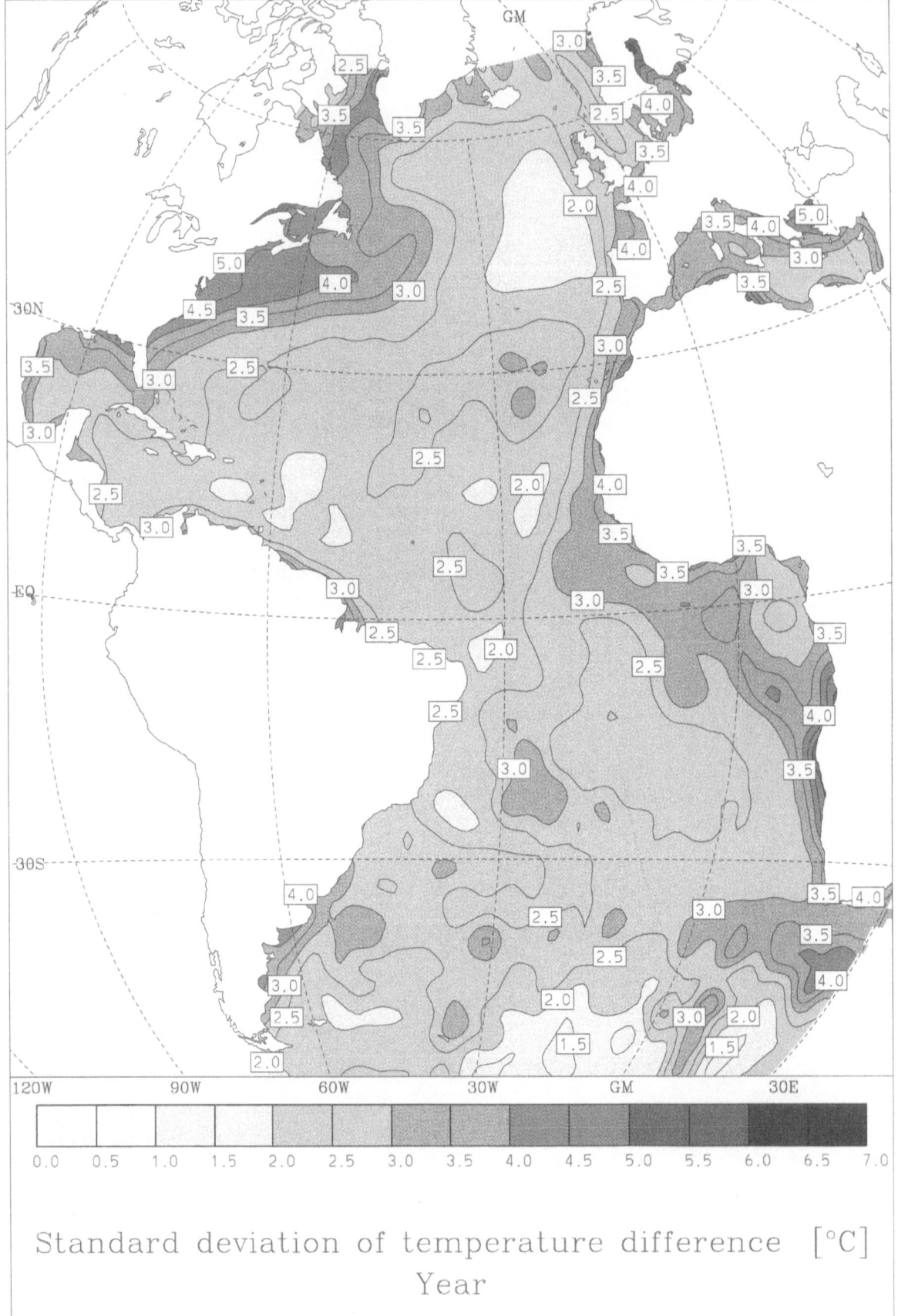

Standard deviation of temperature difference [°C]
Year

Mixing Ratio

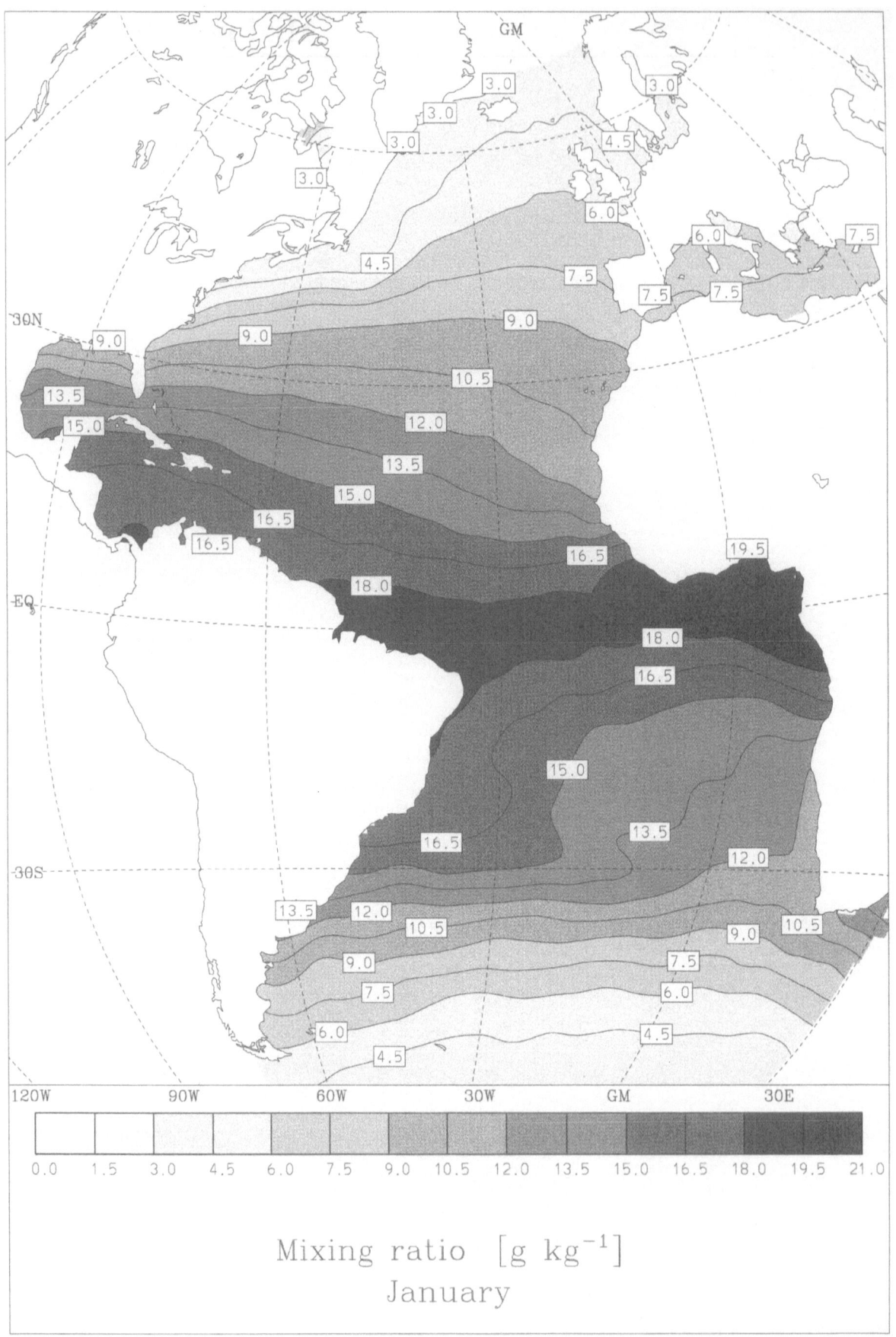

Mixing ratio [g kg^{-1}]
January

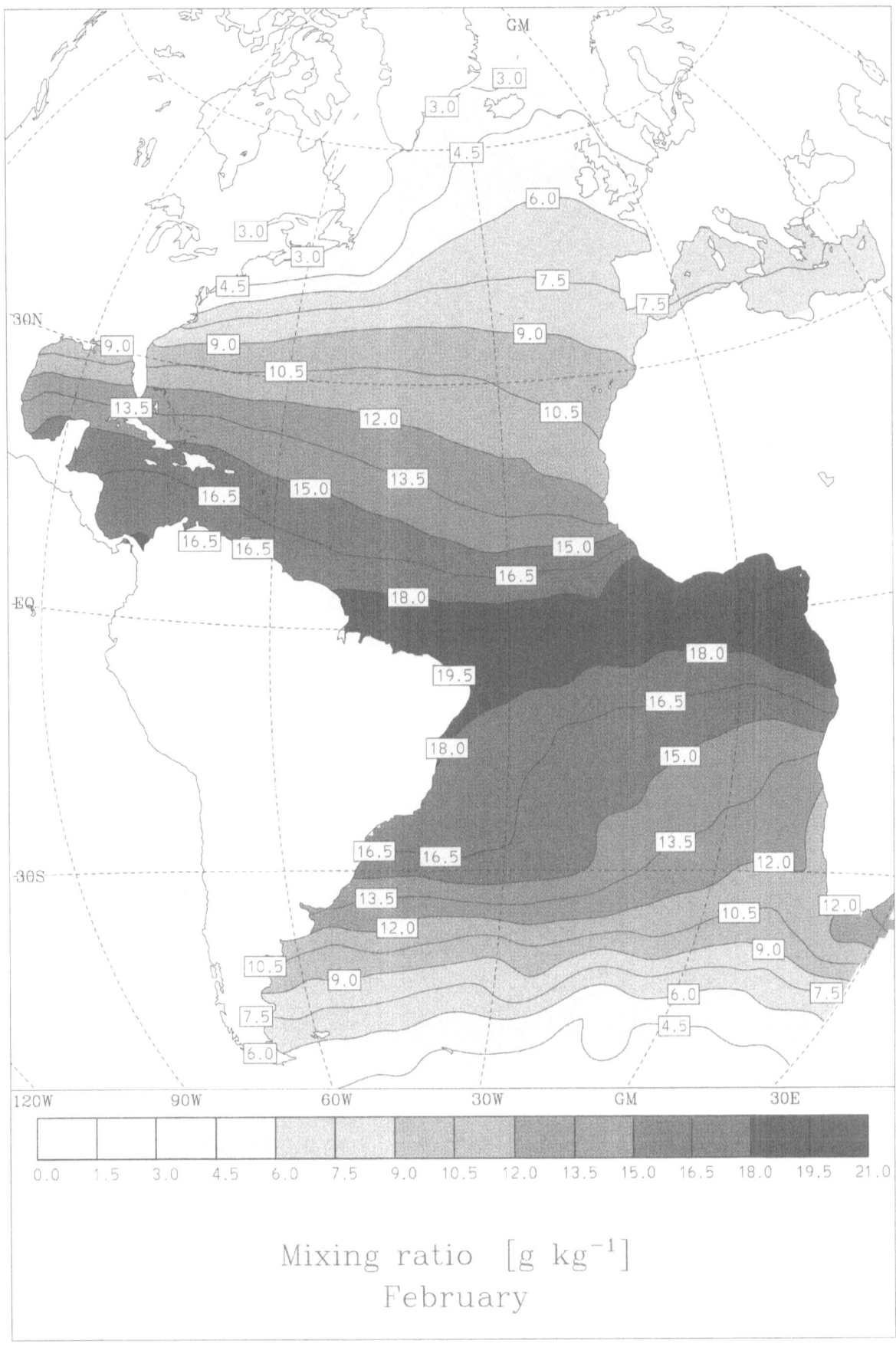

Mixing ratio $[g\ kg^{-1}]$
February

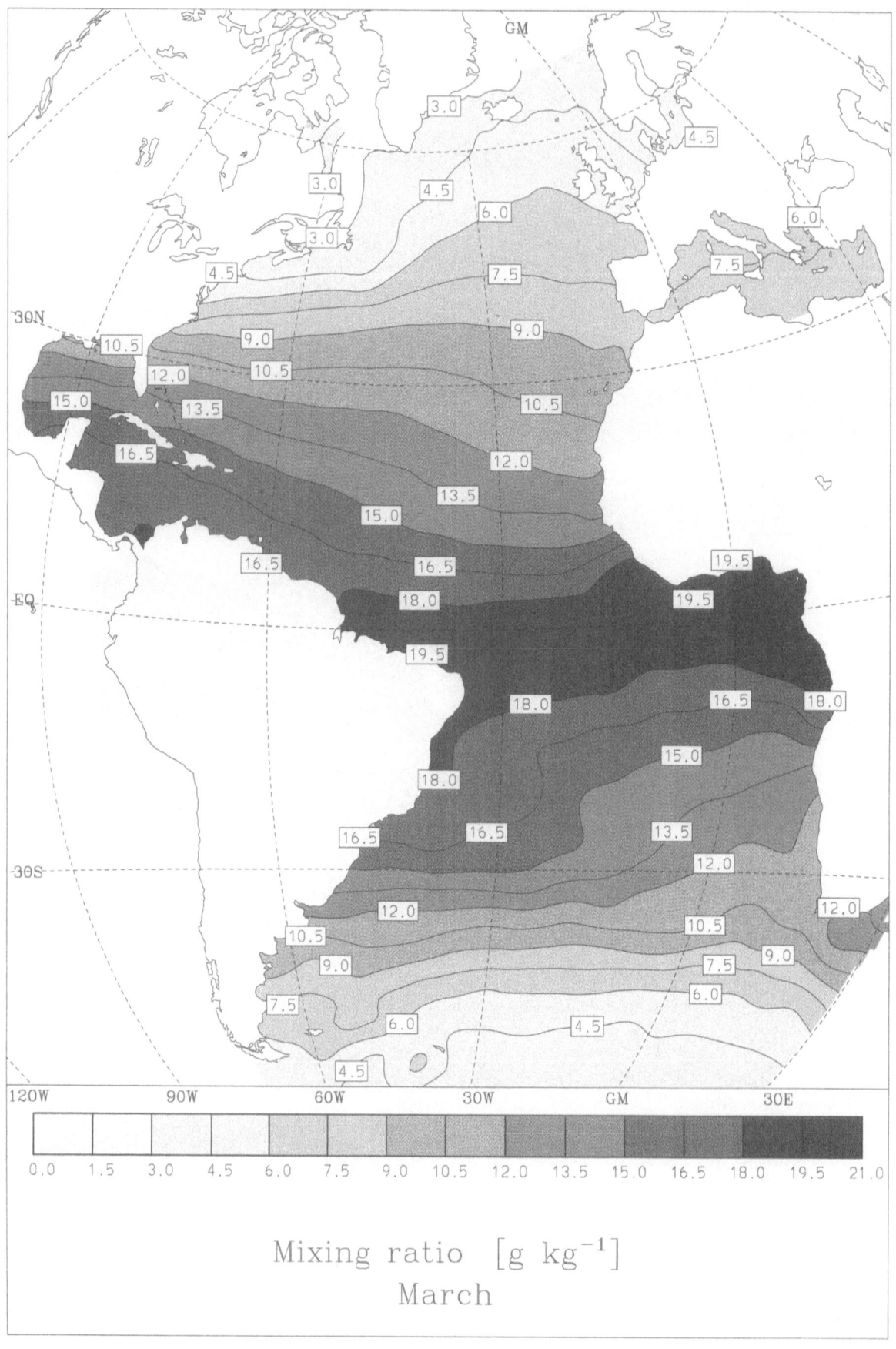

Mixing ratio [g kg^{-1}]
March

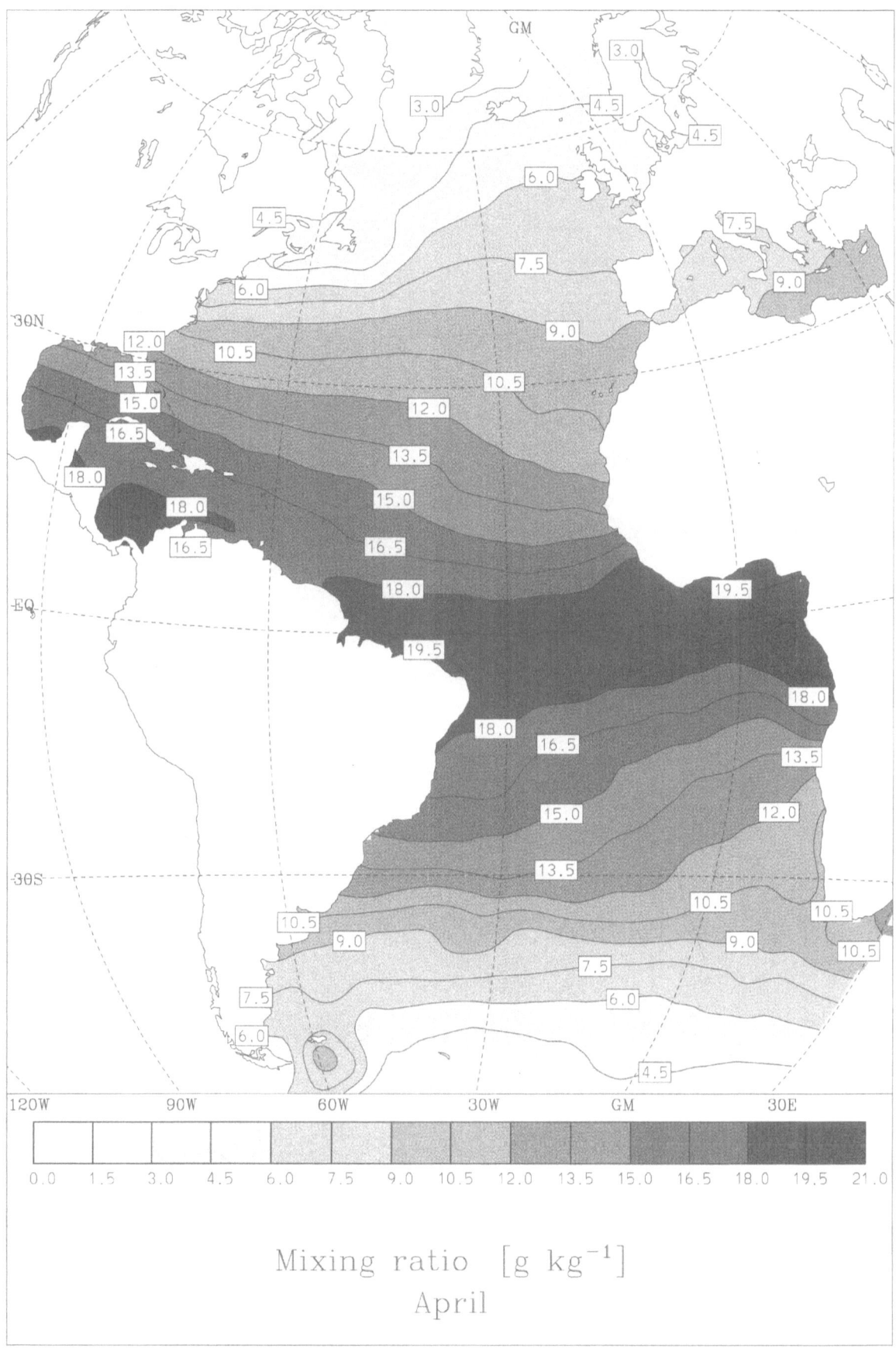

Mixing ratio [g kg^{-1}]

April

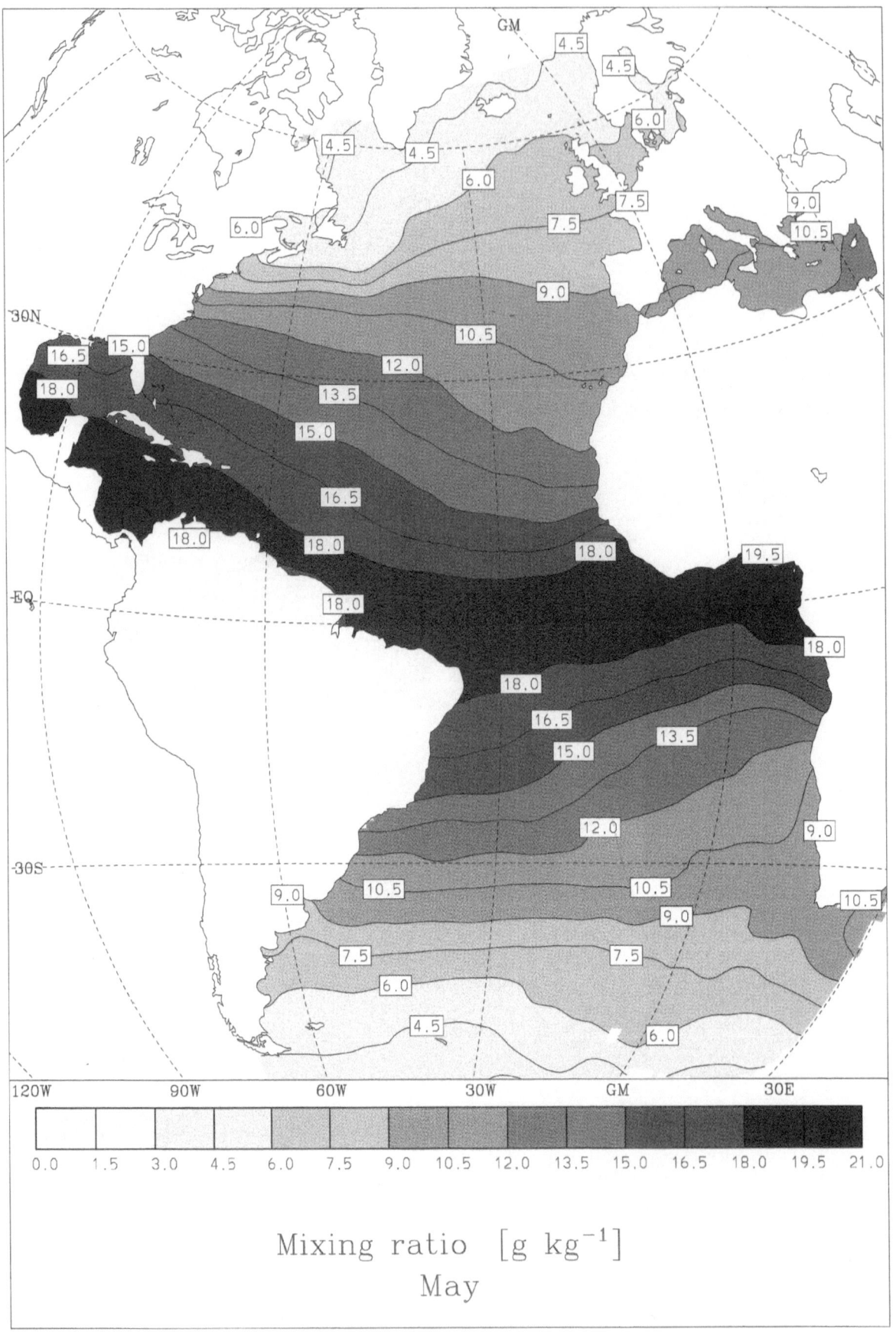

Mixing ratio [g kg^{-1}]
May

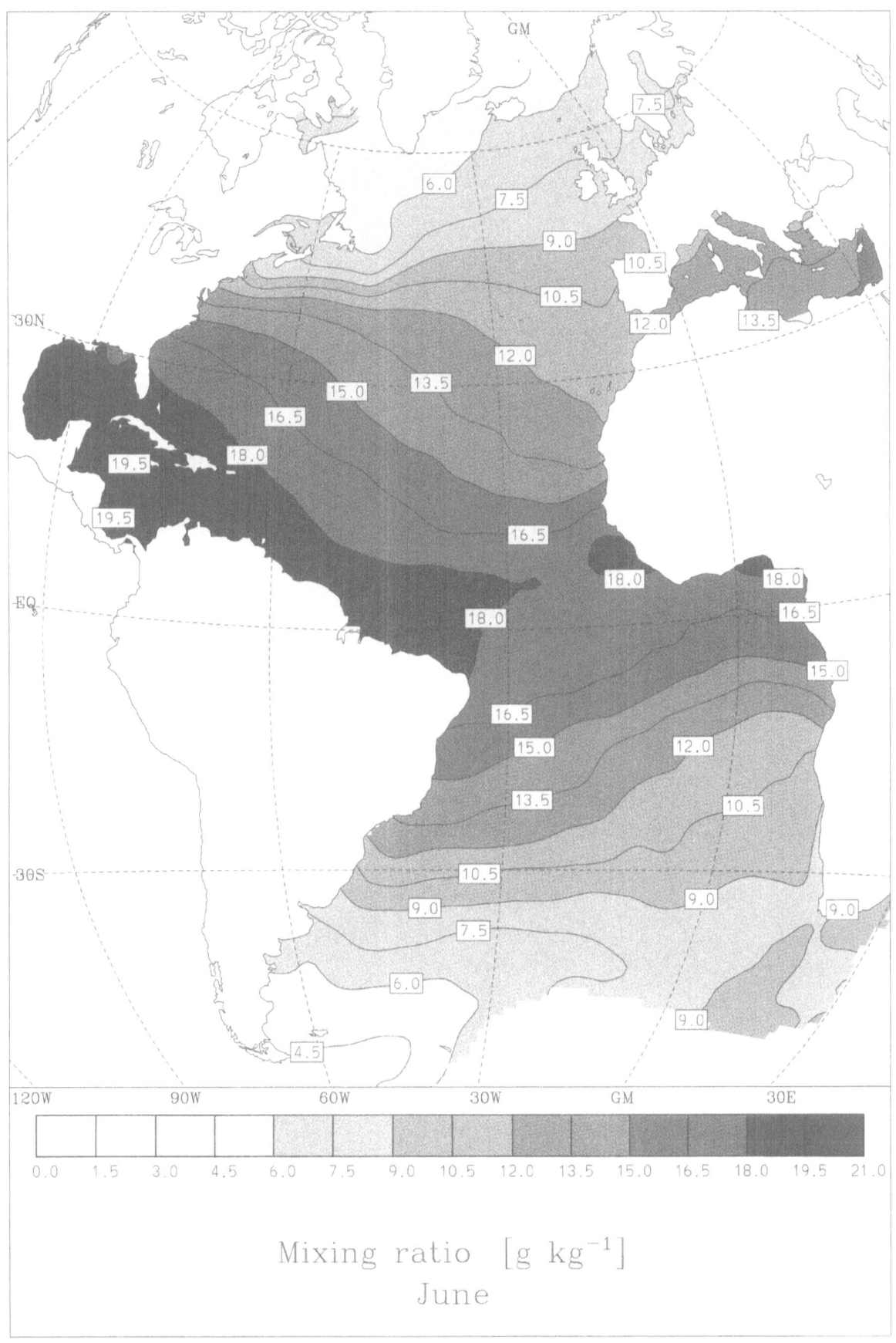

Mixing ratio [g kg^{-1}]
June

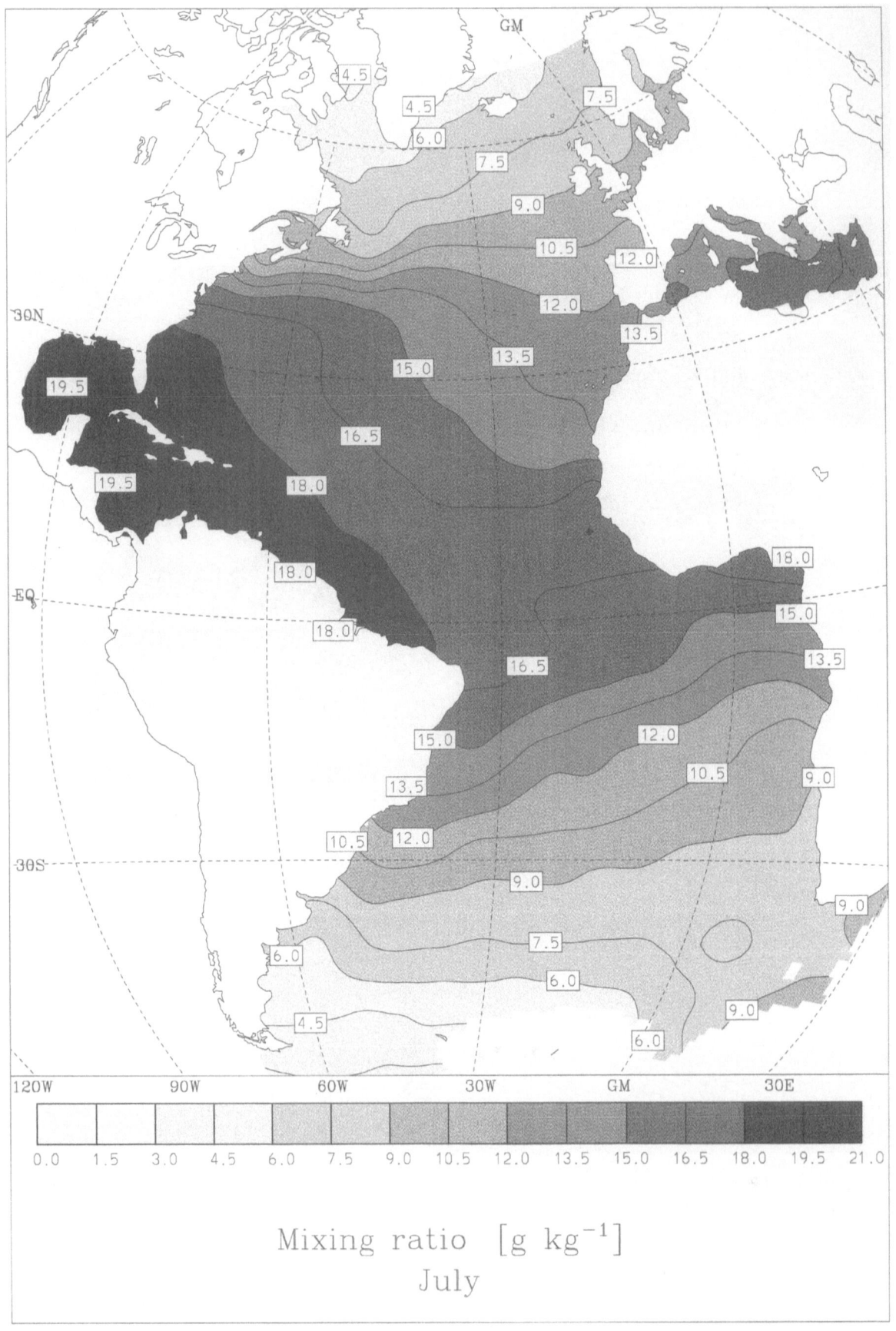

Mixing ratio [g kg^{-1}]
July

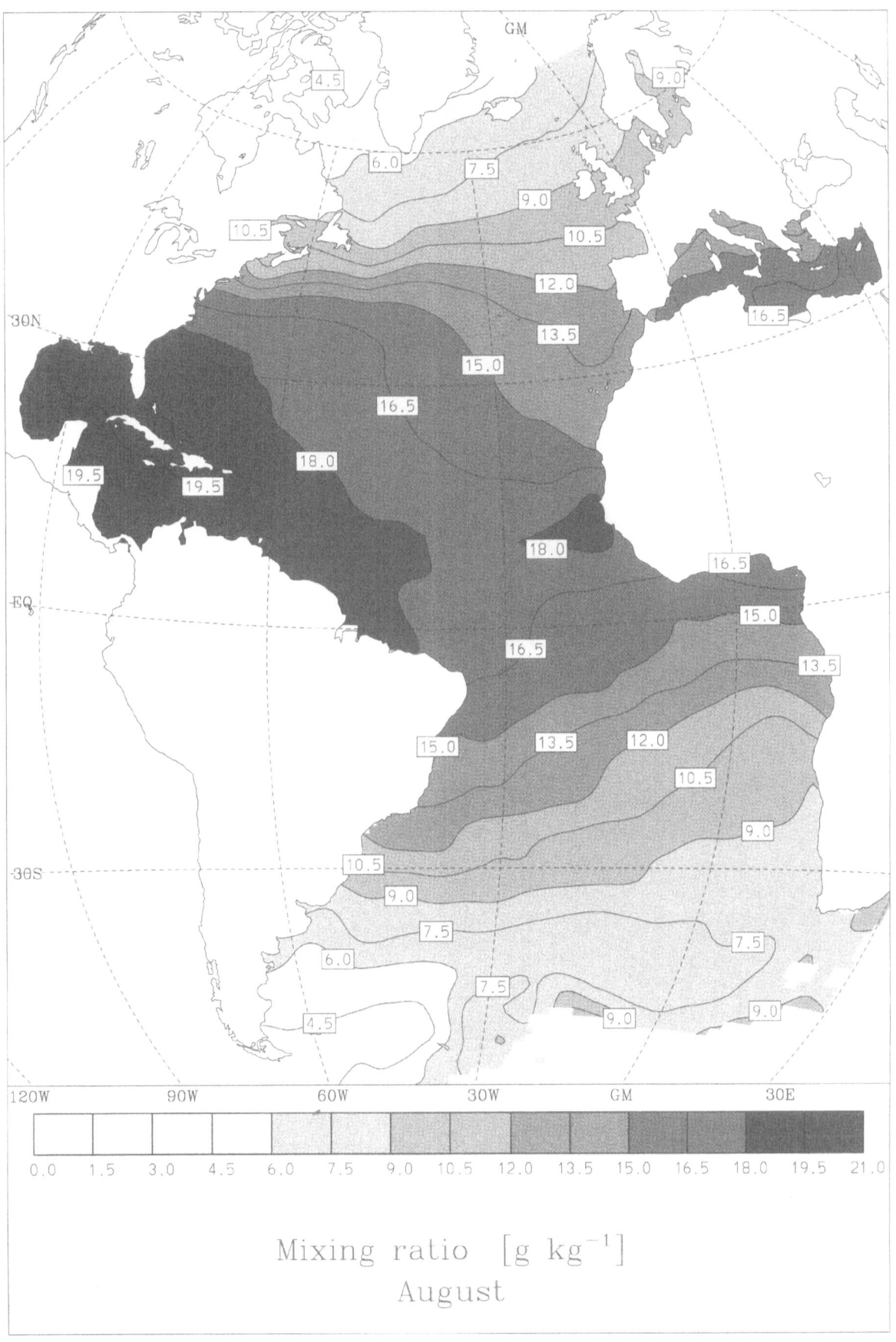

Mixing ratio [g kg^{-1}]

August

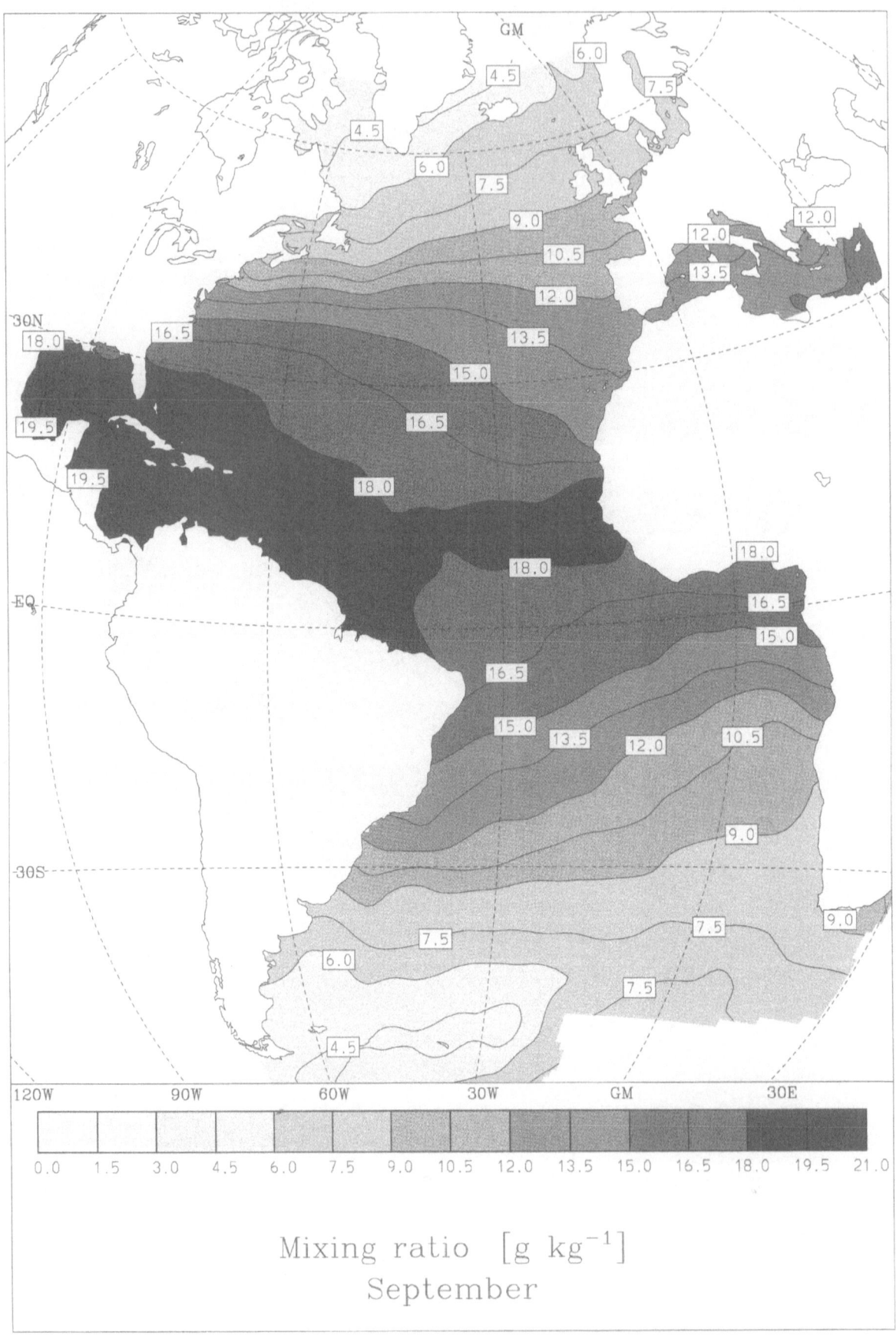

Mixing ratio [g kg^{-1}]
September

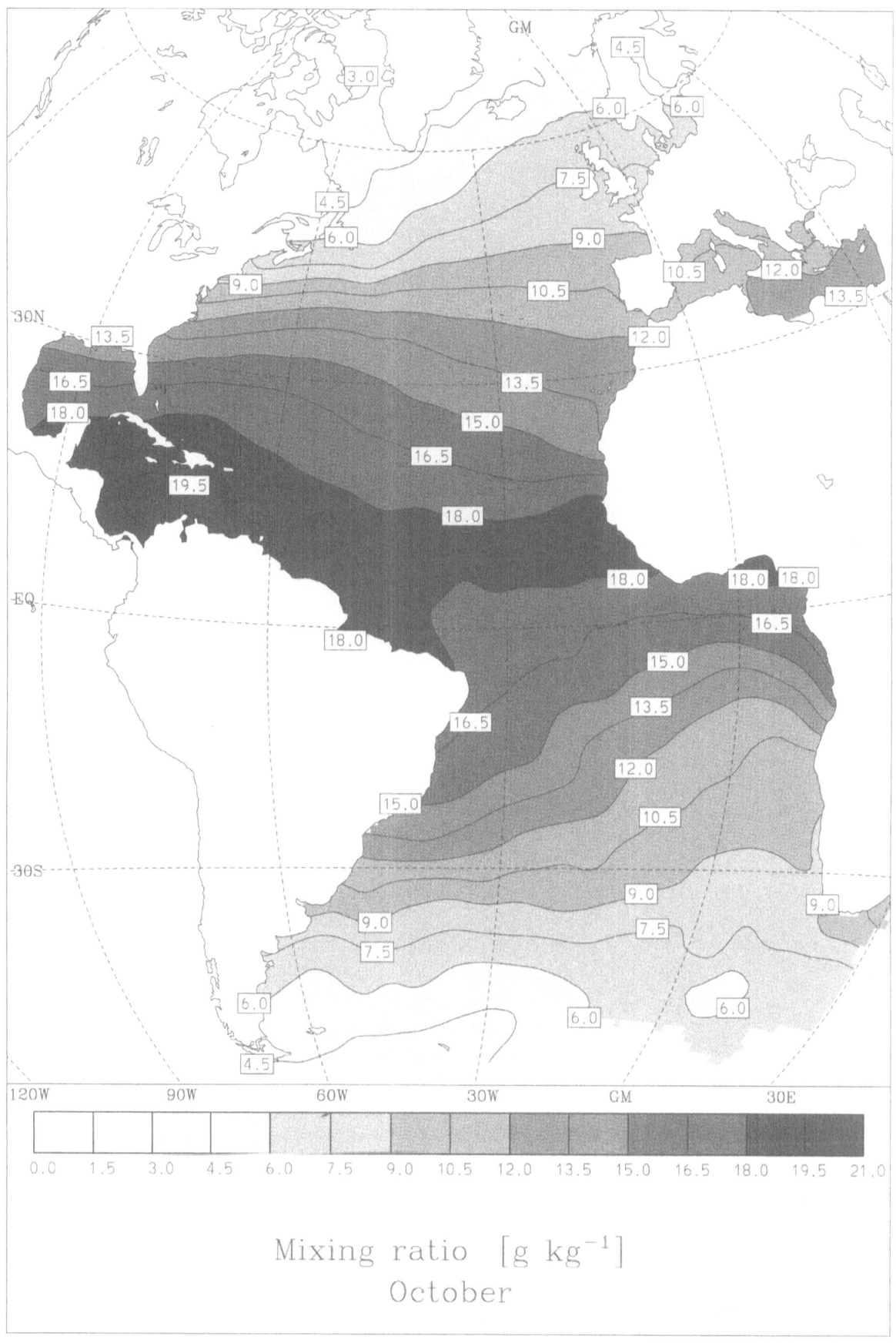

Mixing ratio [g kg^{-1}]
October

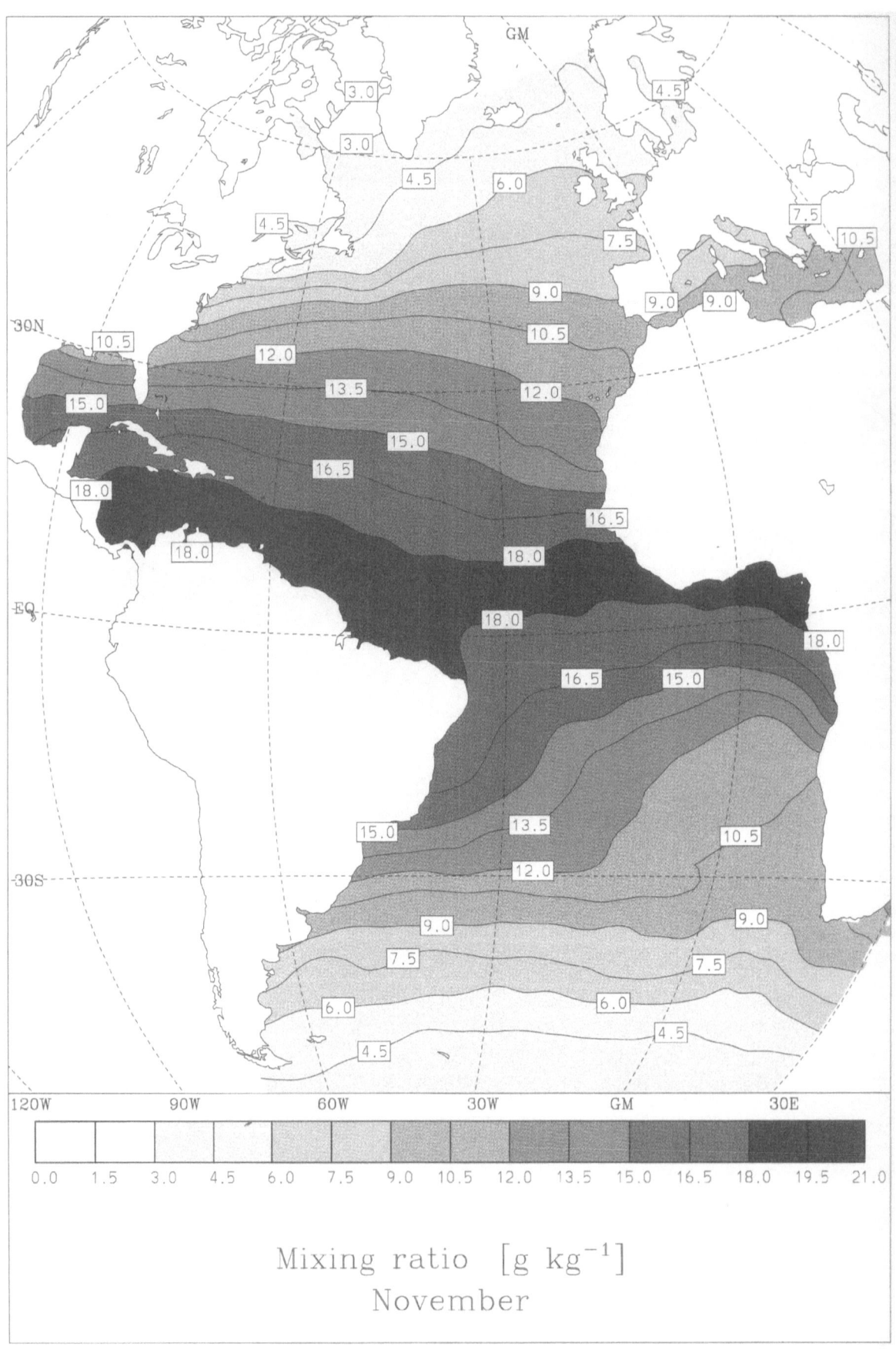

Mixing ratio [g kg^{-1}]
November

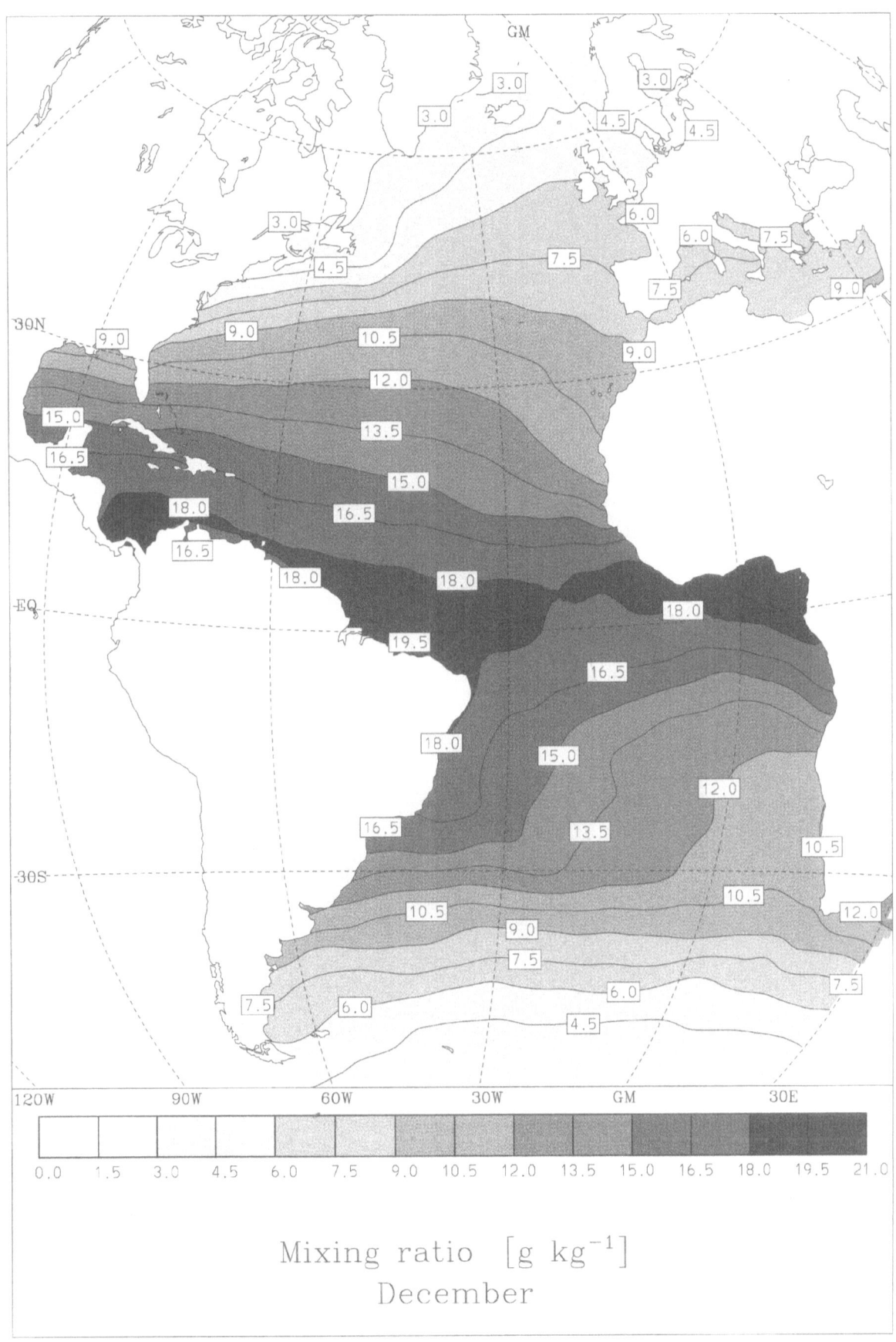

Mixing ratio [g kg^{-1}]
December

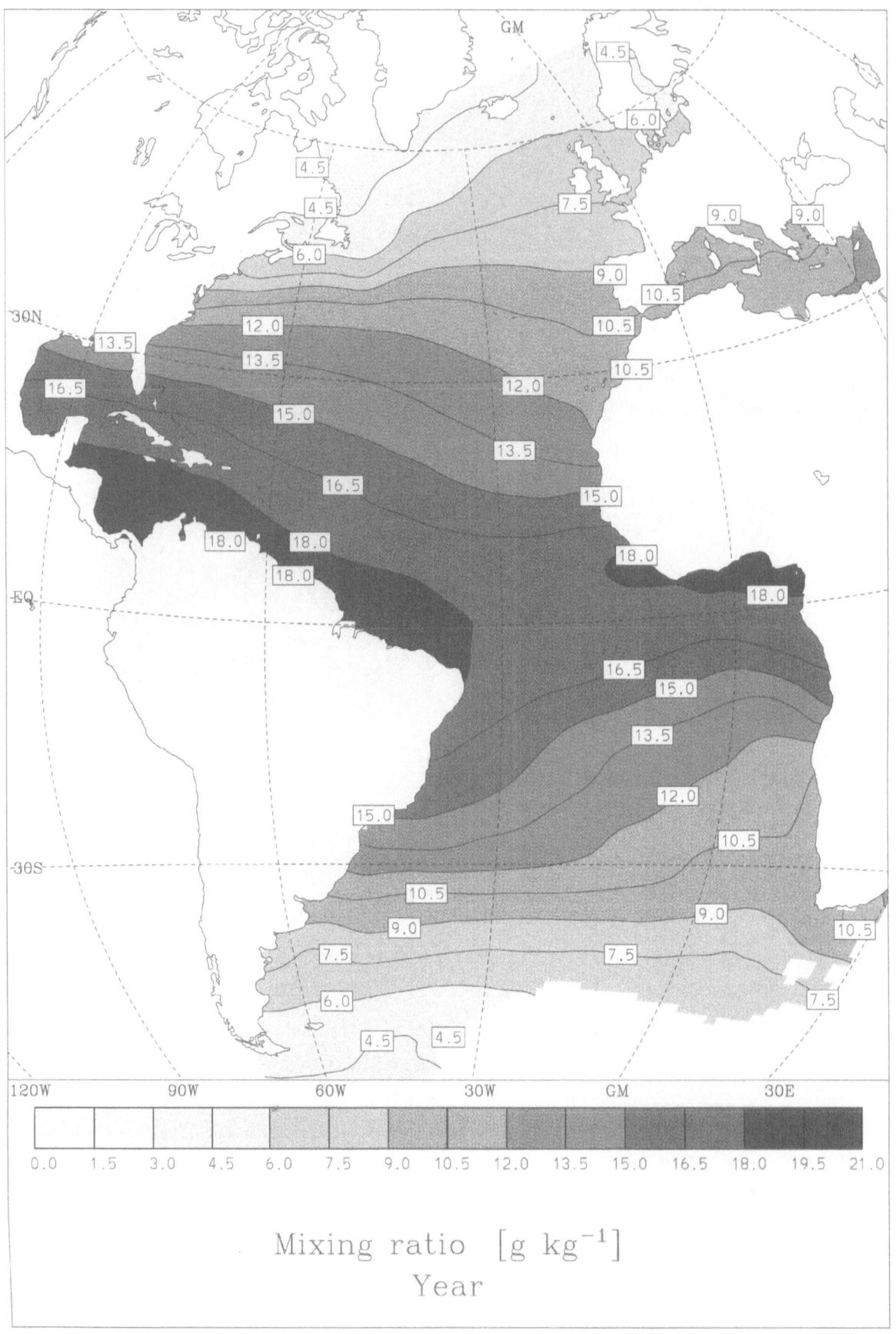

Mixing ratio [g kg^{-1}]
Year

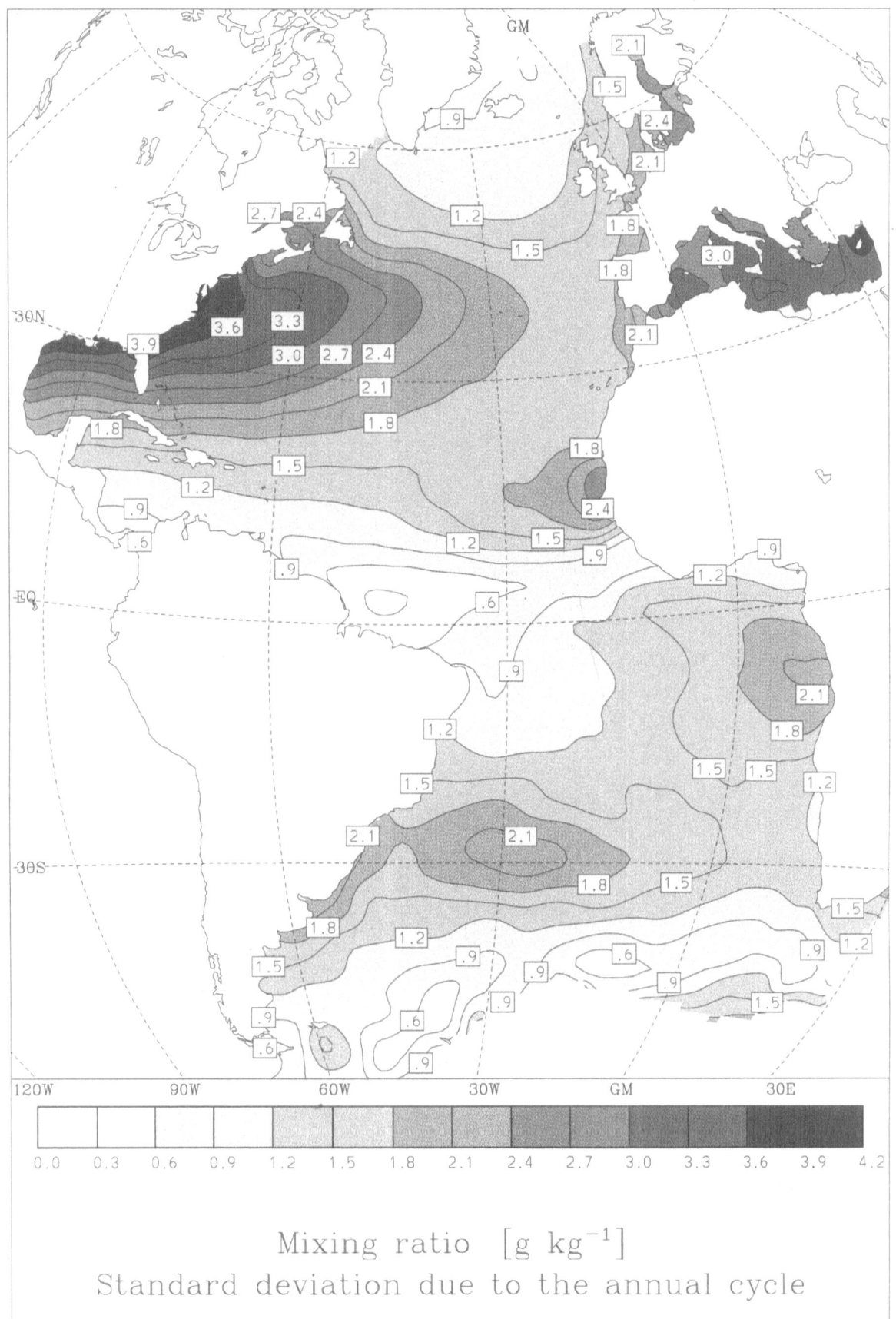

Mixing ratio [g kg^{-1}]
Standard deviation due to the annual cycle

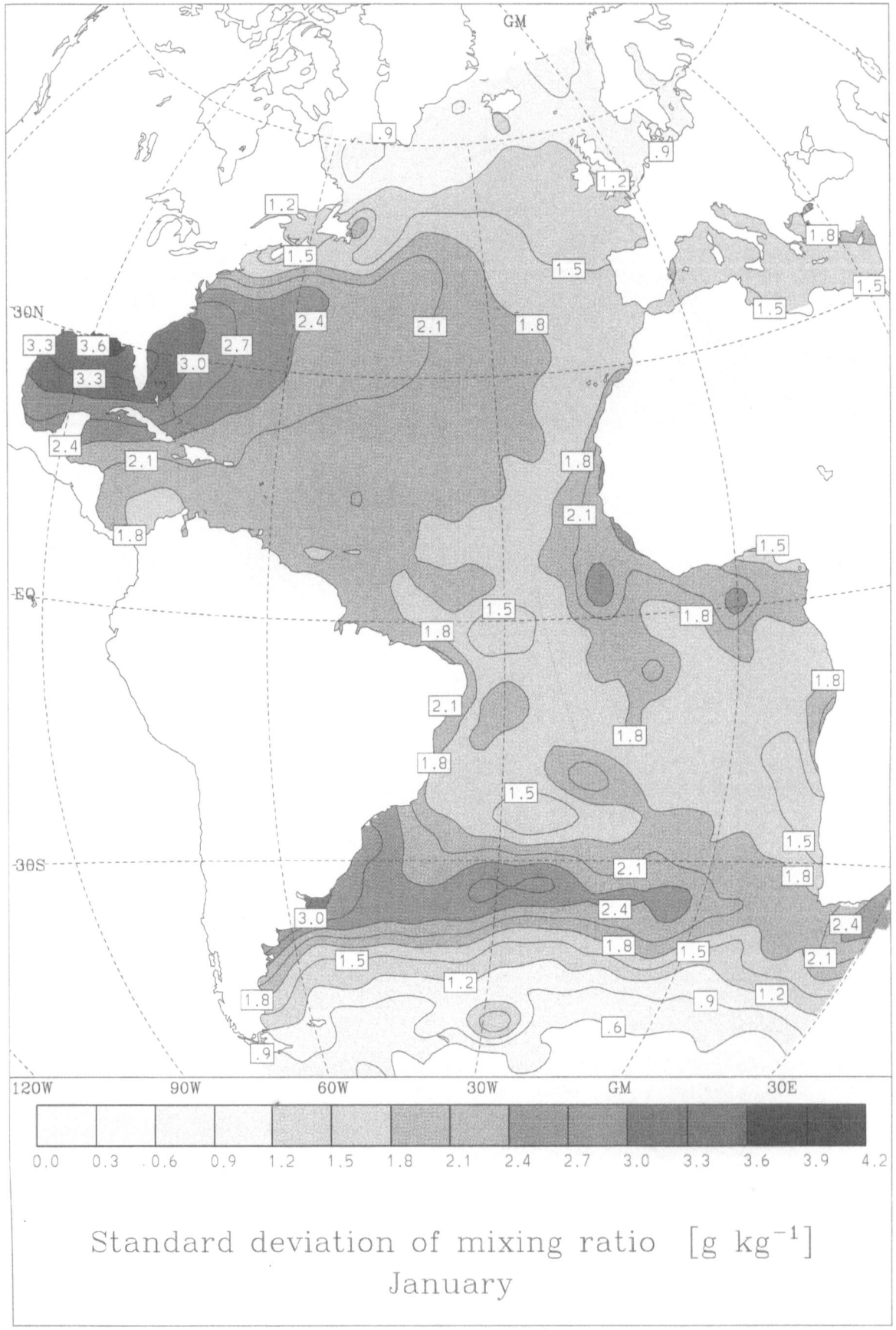

Standard deviation of mixing ratio [g kg^{-1}]
January

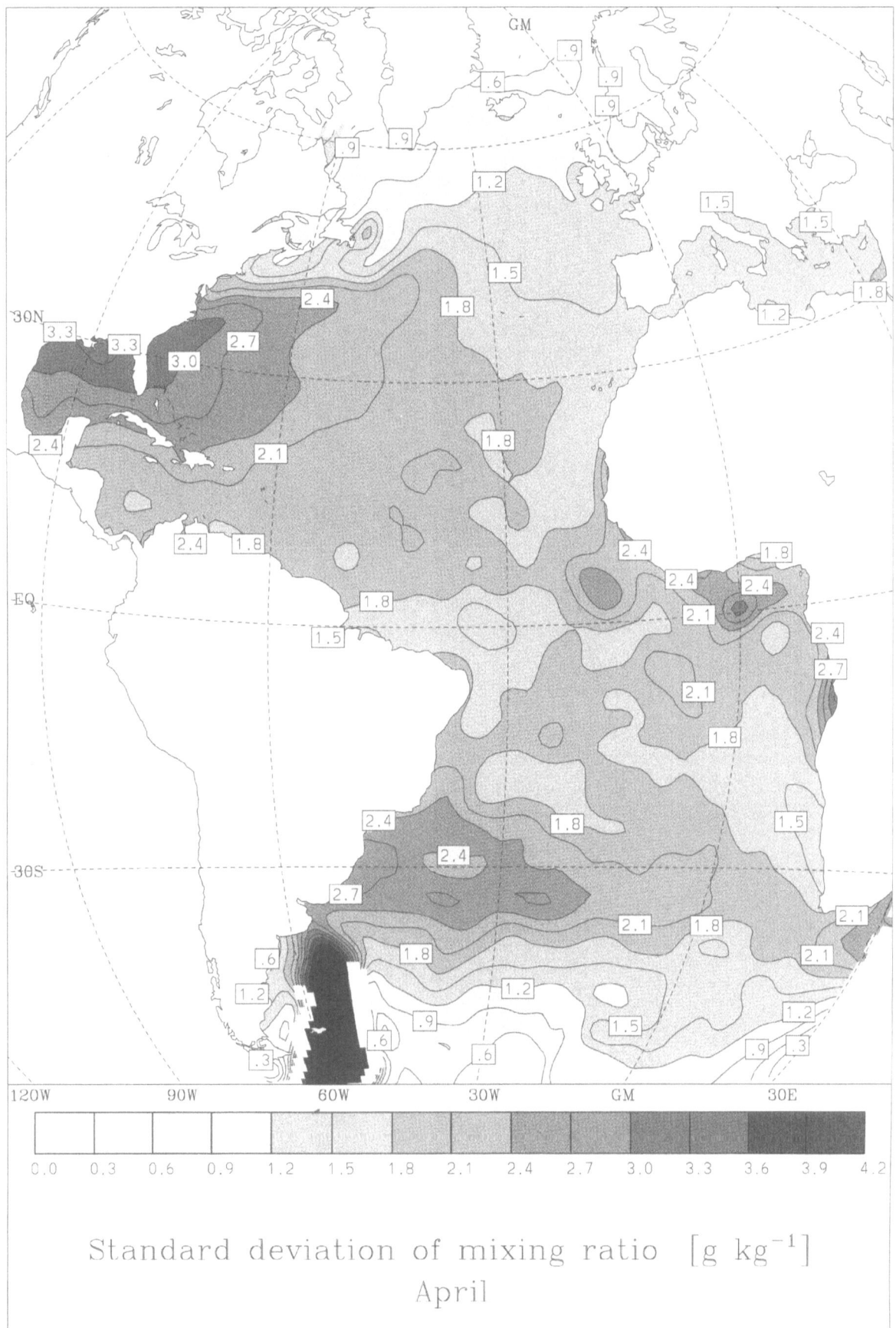

Standard deviation of mixing ratio $[\text{g kg}^{-1}]$
April

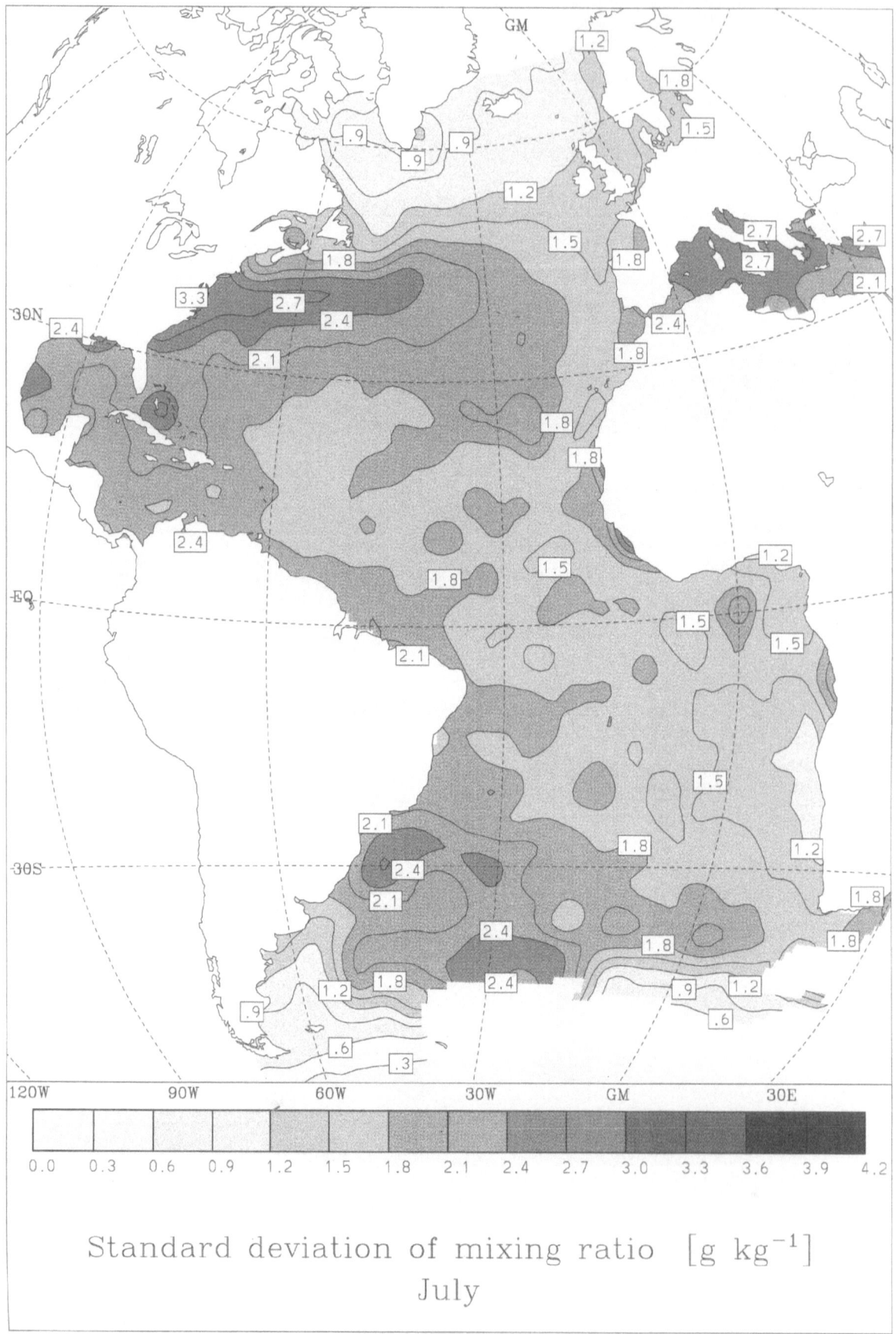

Standard deviation of mixing ratio [g kg⁻¹]
July

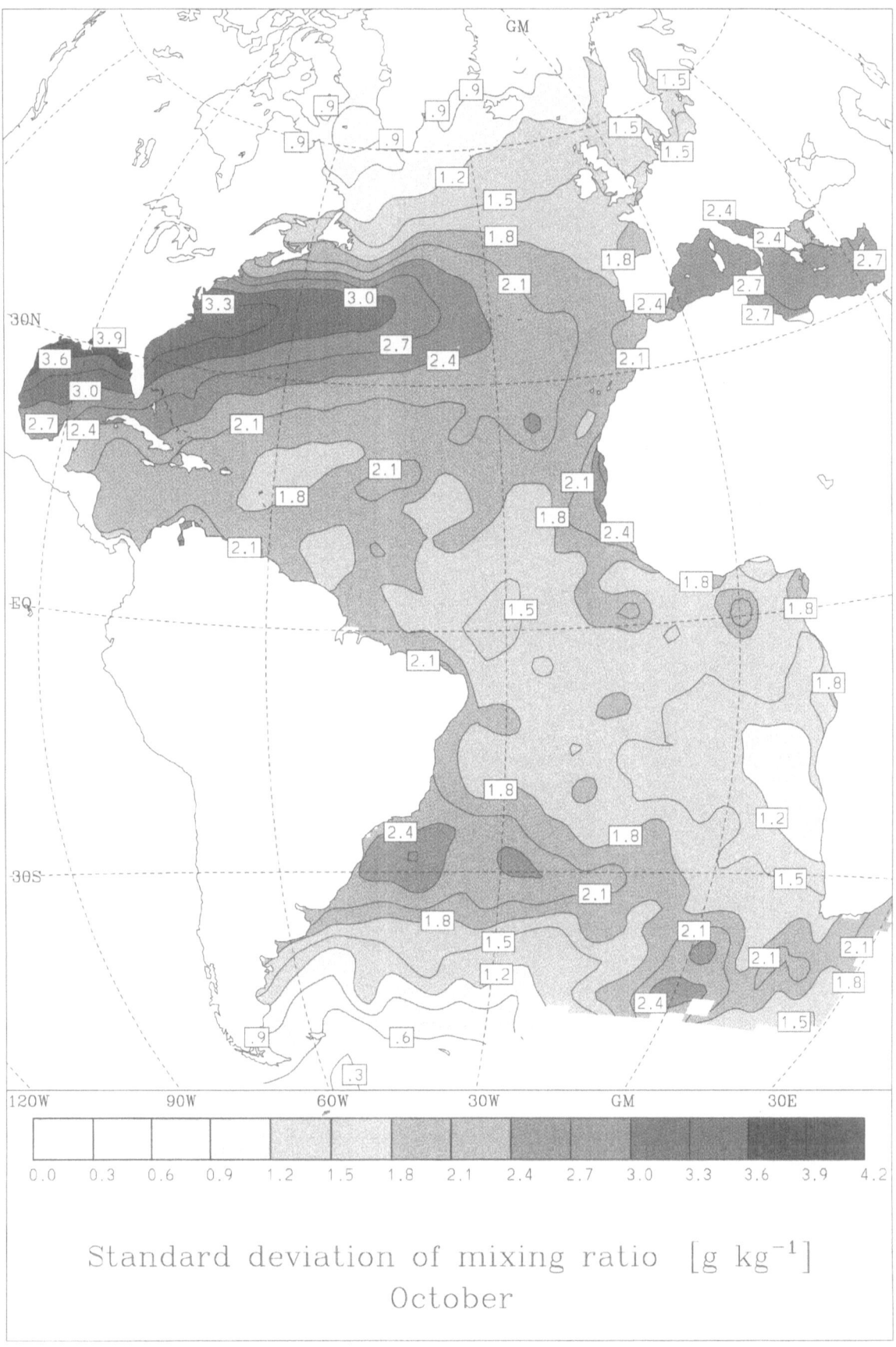

Standard deviation of mixing ratio [g kg^{-1}]
October

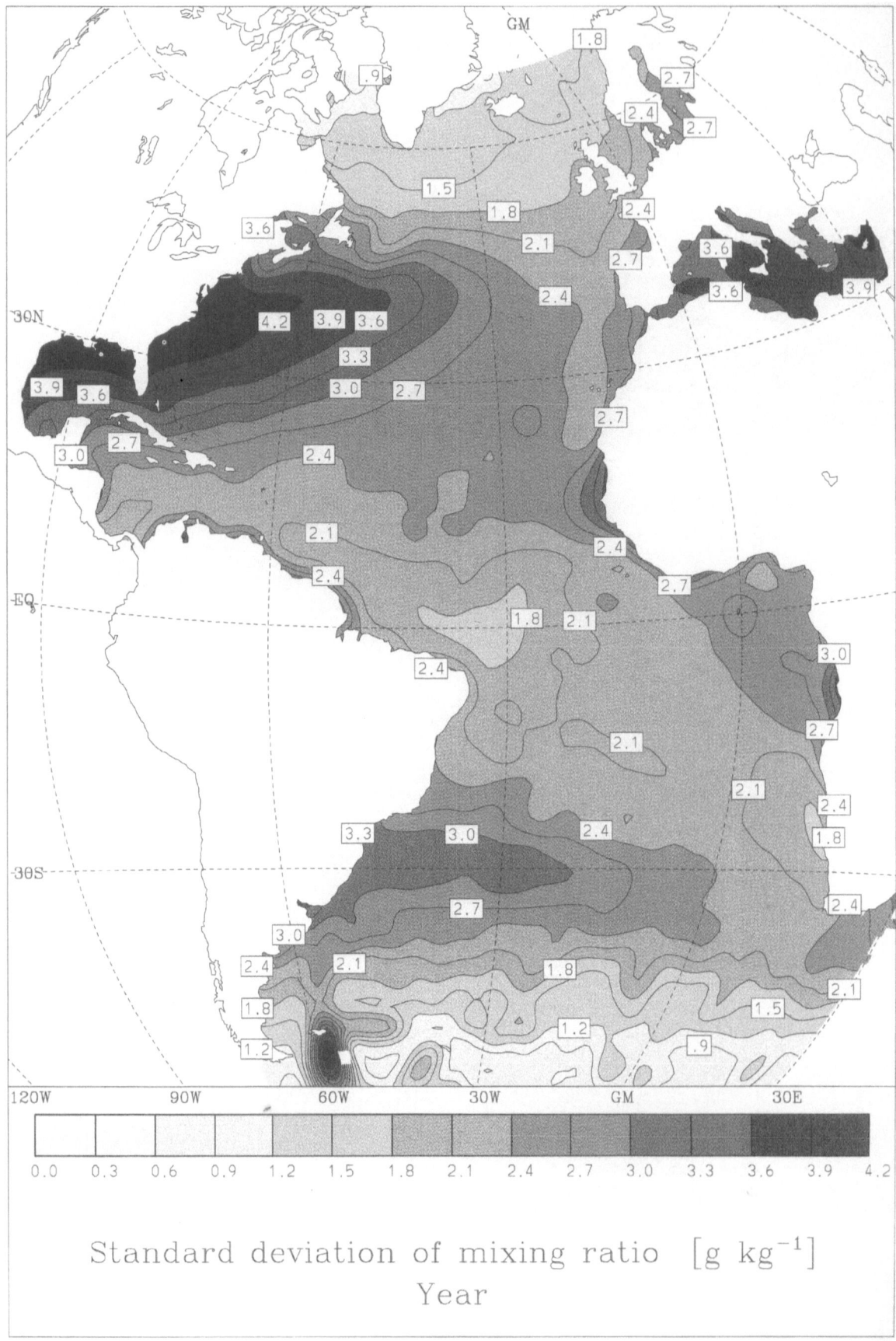

Standard deviation of mixing ratio [g kg^{-1}]
Year

Relative Humidity

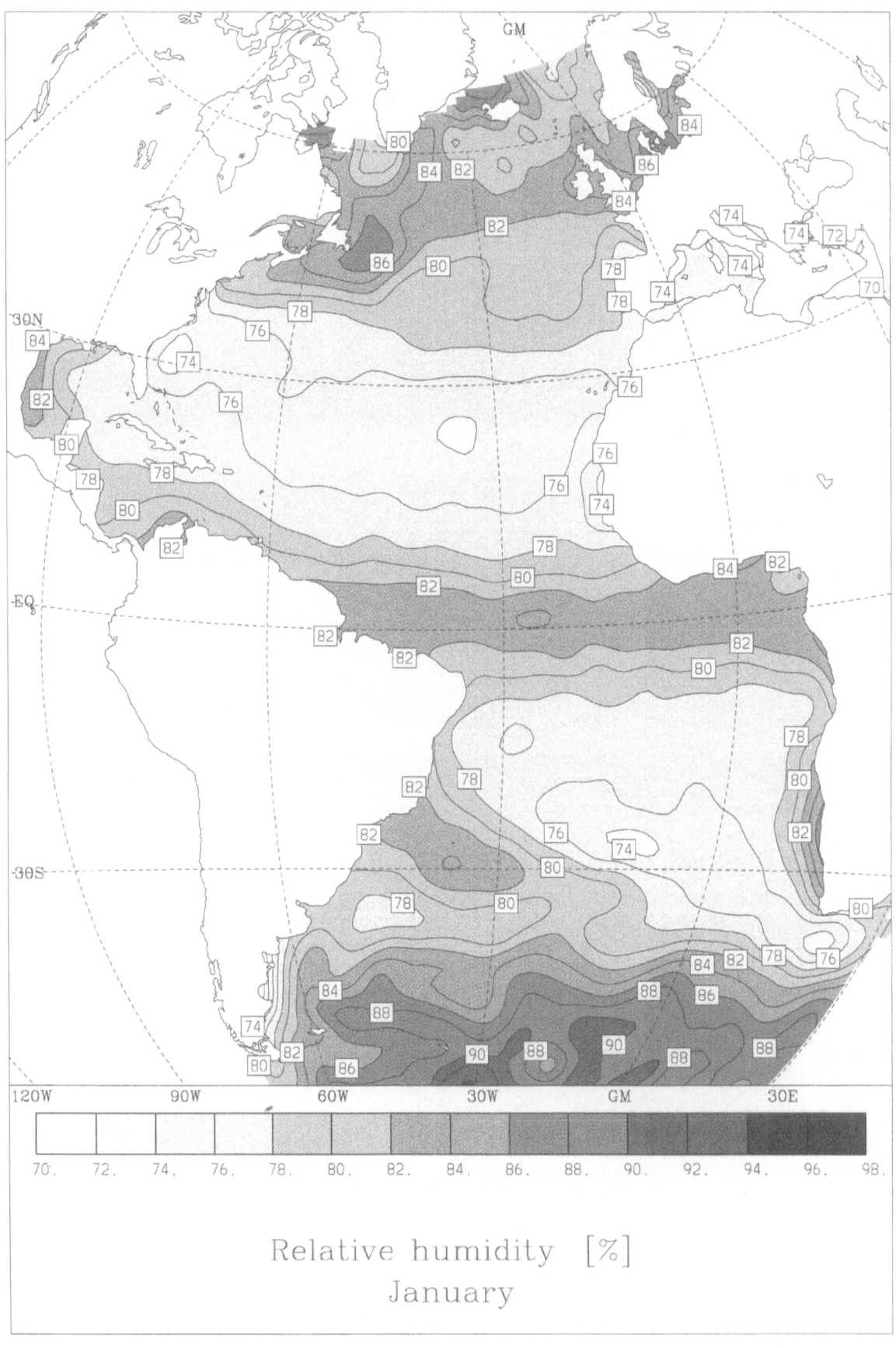

Relative humidity [%]
January

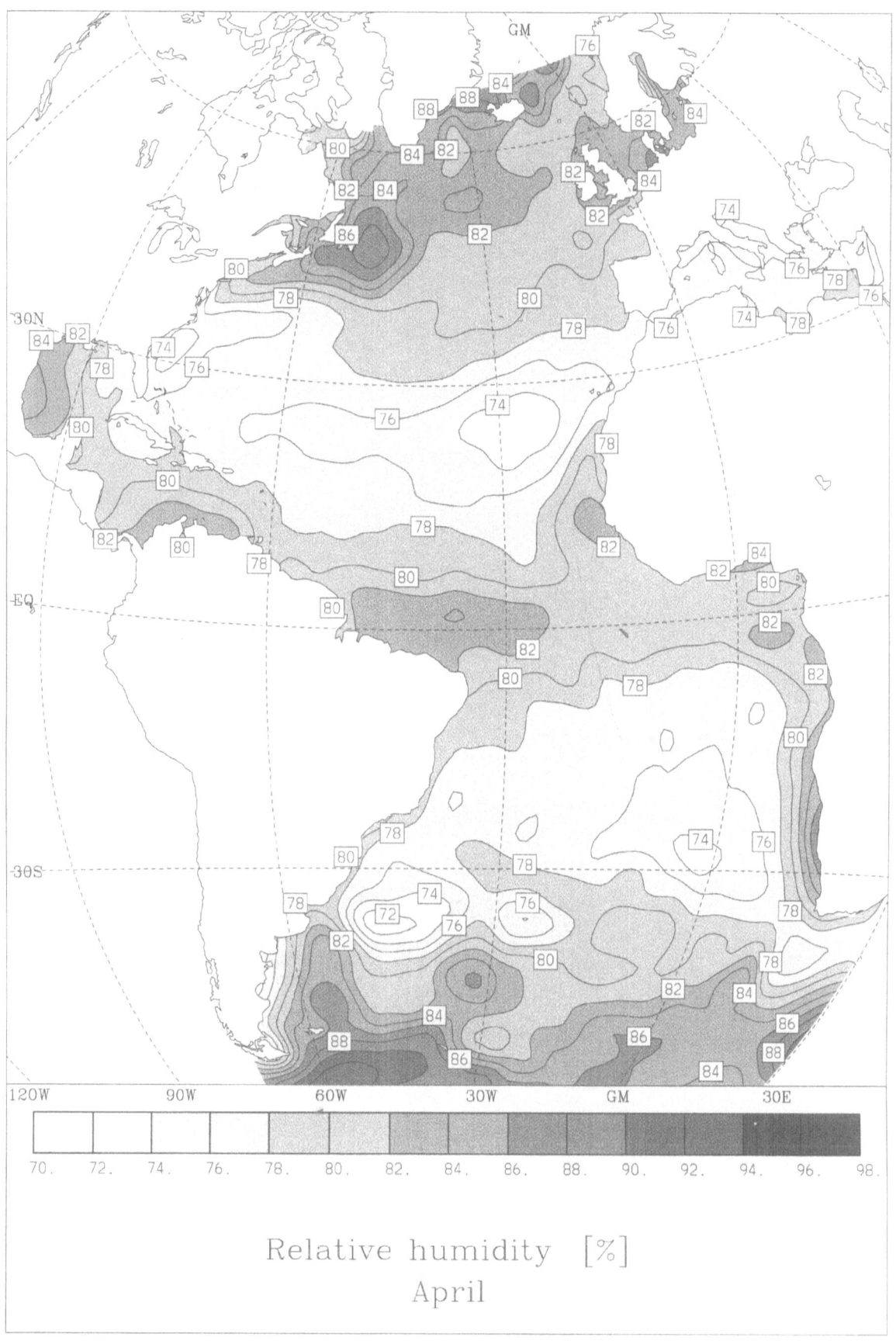

Relative humidity [%]
April

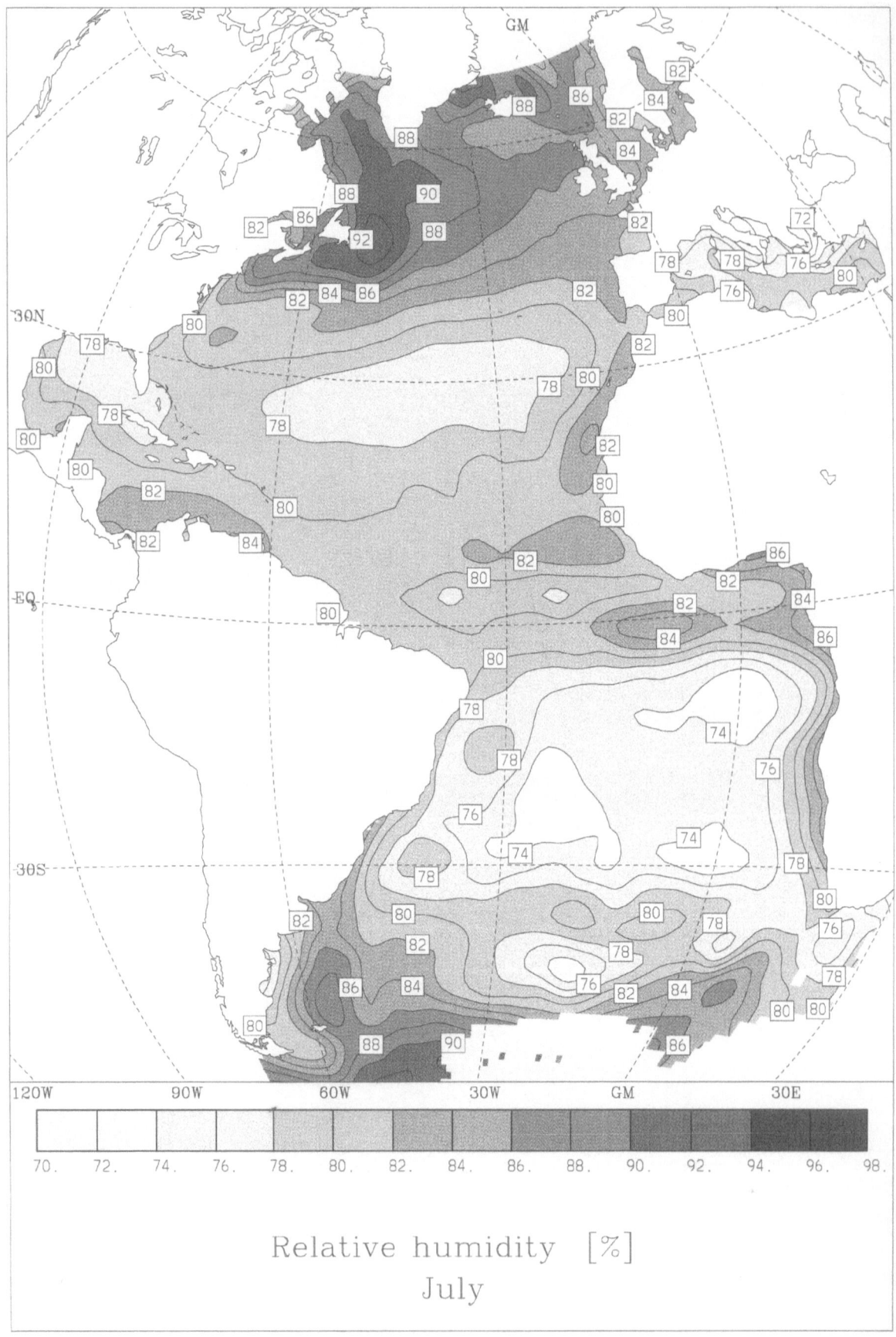

Relative humidity [%]
July

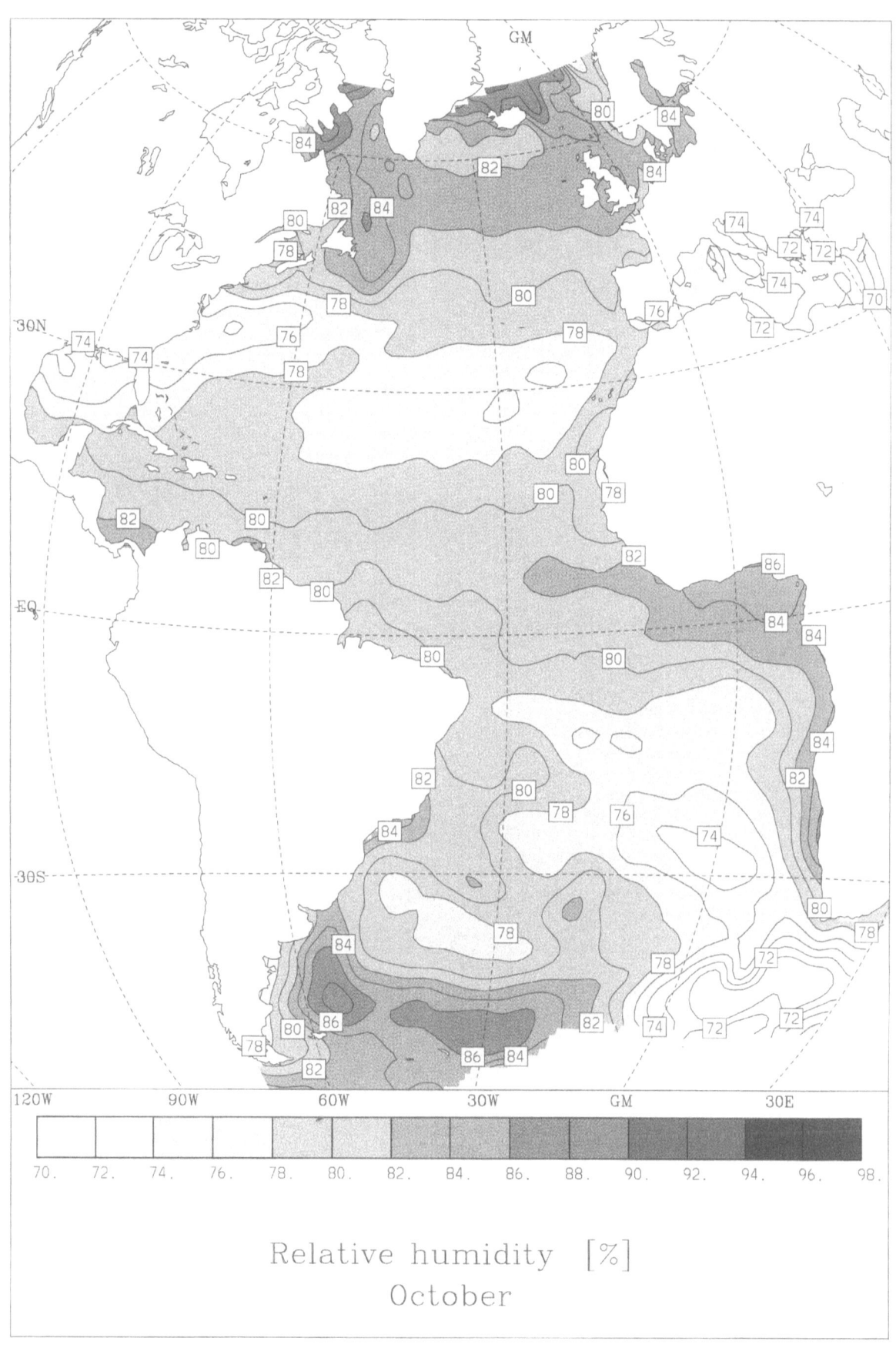

Relative humidity [%]
October

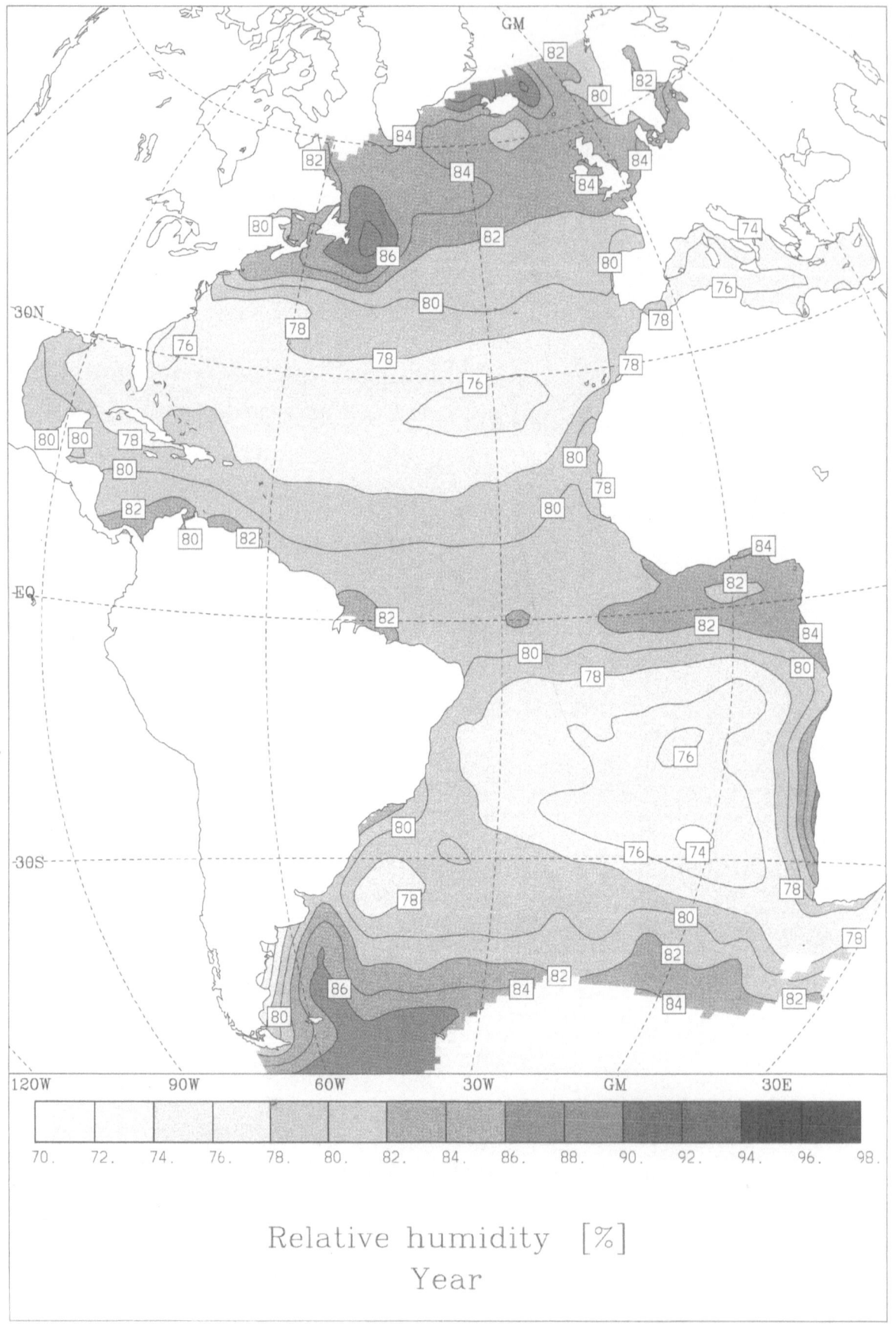

Relative humidity [%]
Year

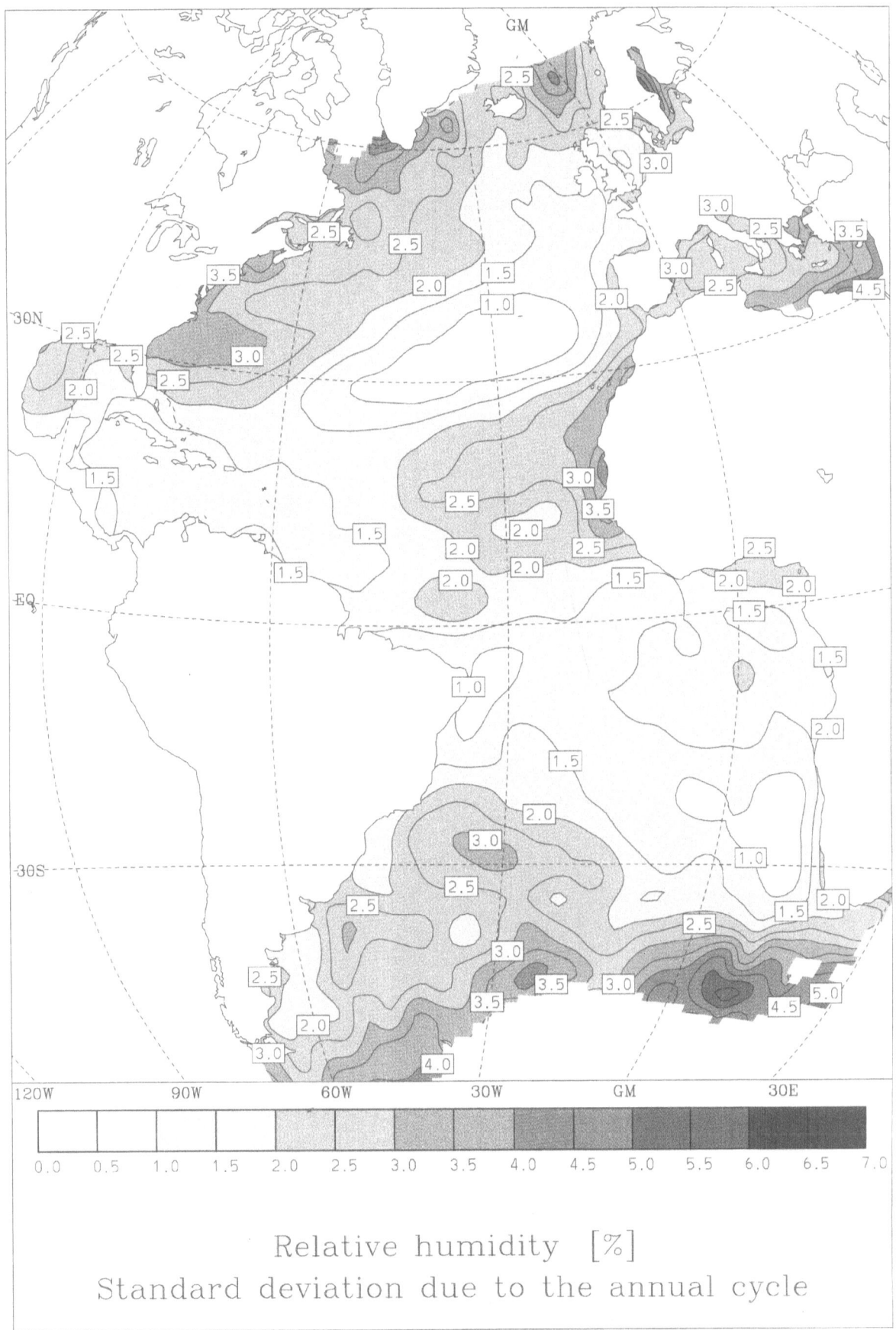

Relative humidity [%]
Standard deviation due to the annual cycle

Air minus Sea Surface Mixing Ratio

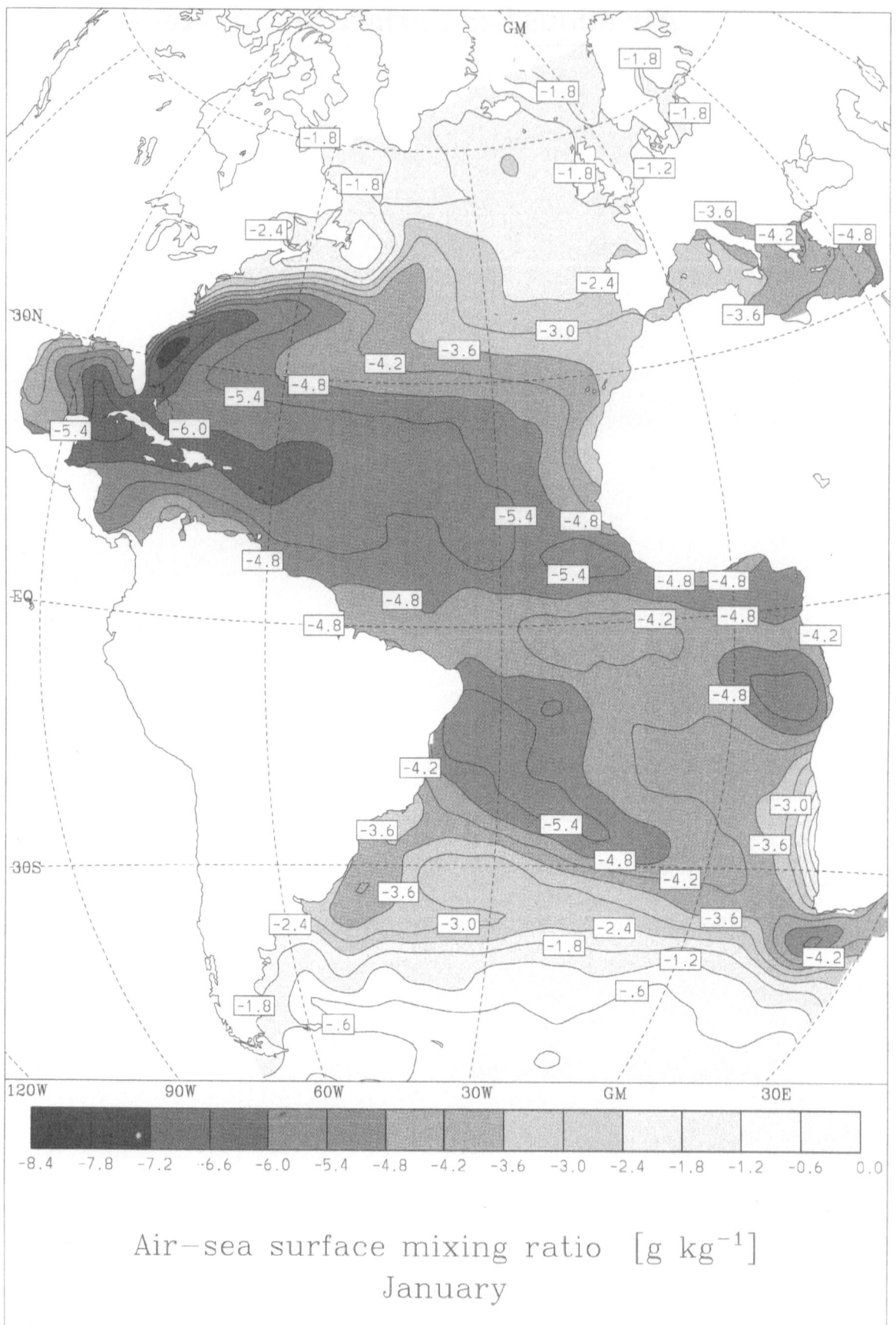

Air−sea surface mixing ratio [g kg^{-1}]
January

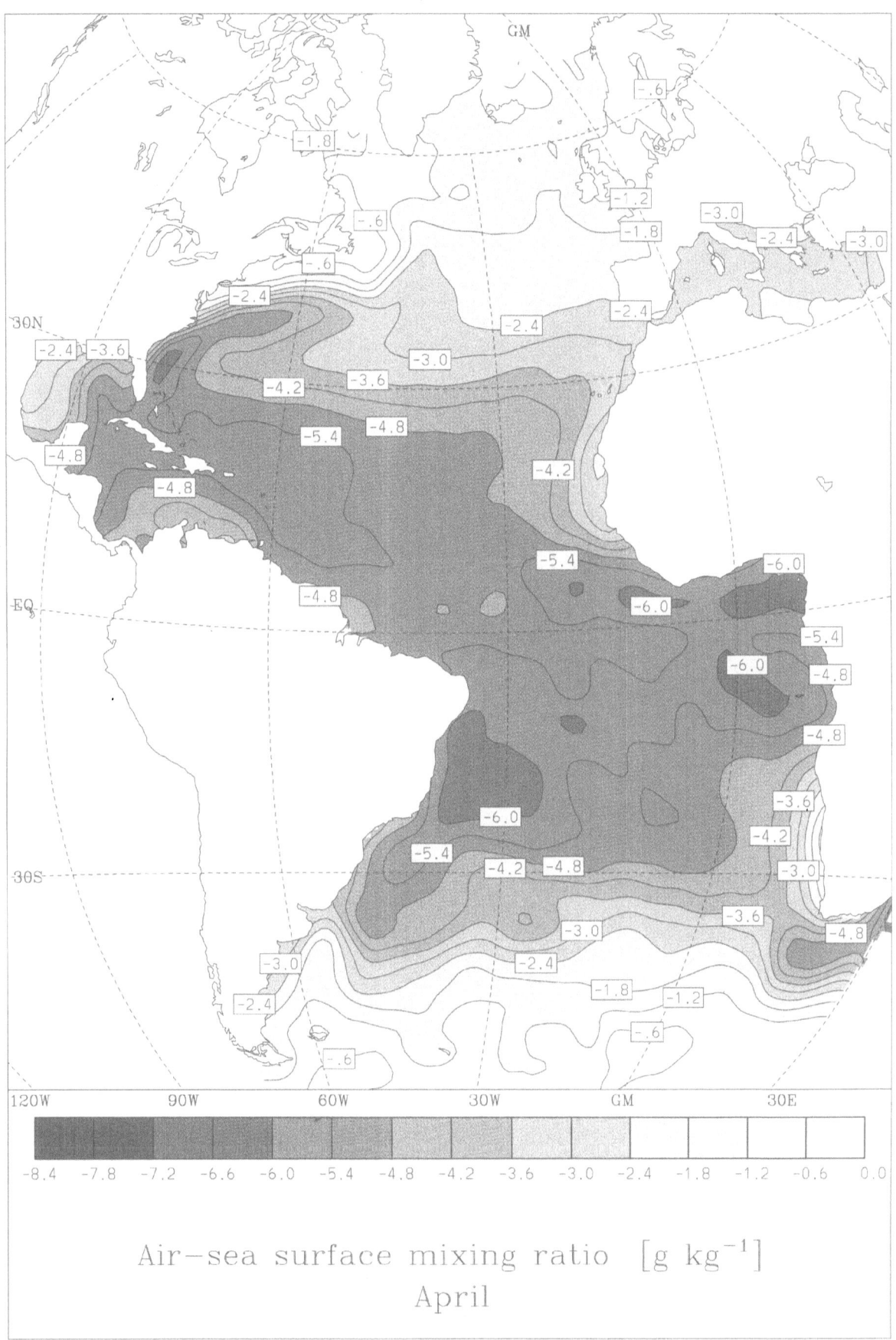

Air—sea surface mixing ratio [g kg^{-1}]
April

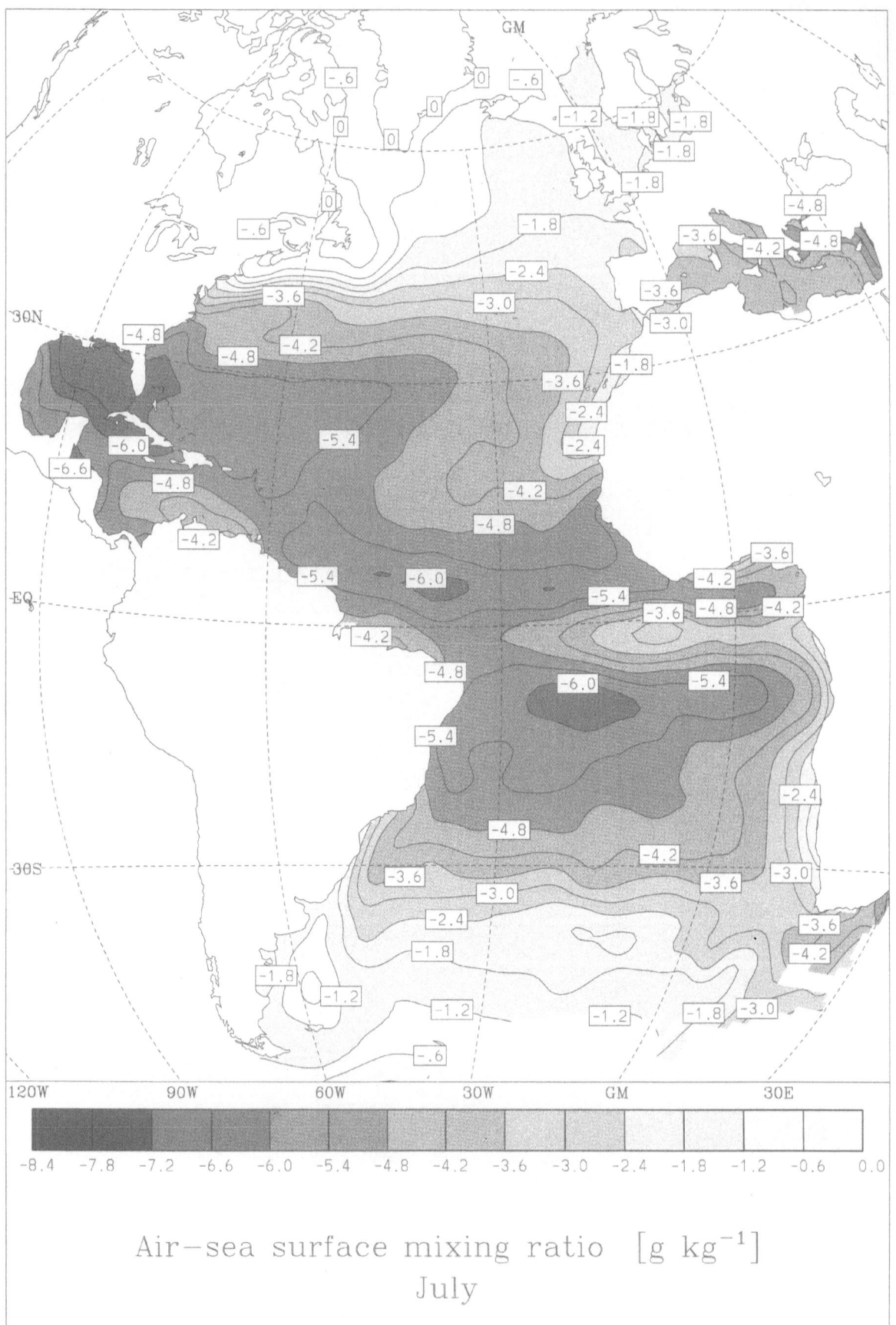

Air−sea surface mixing ratio [g kg^{-1}]
July

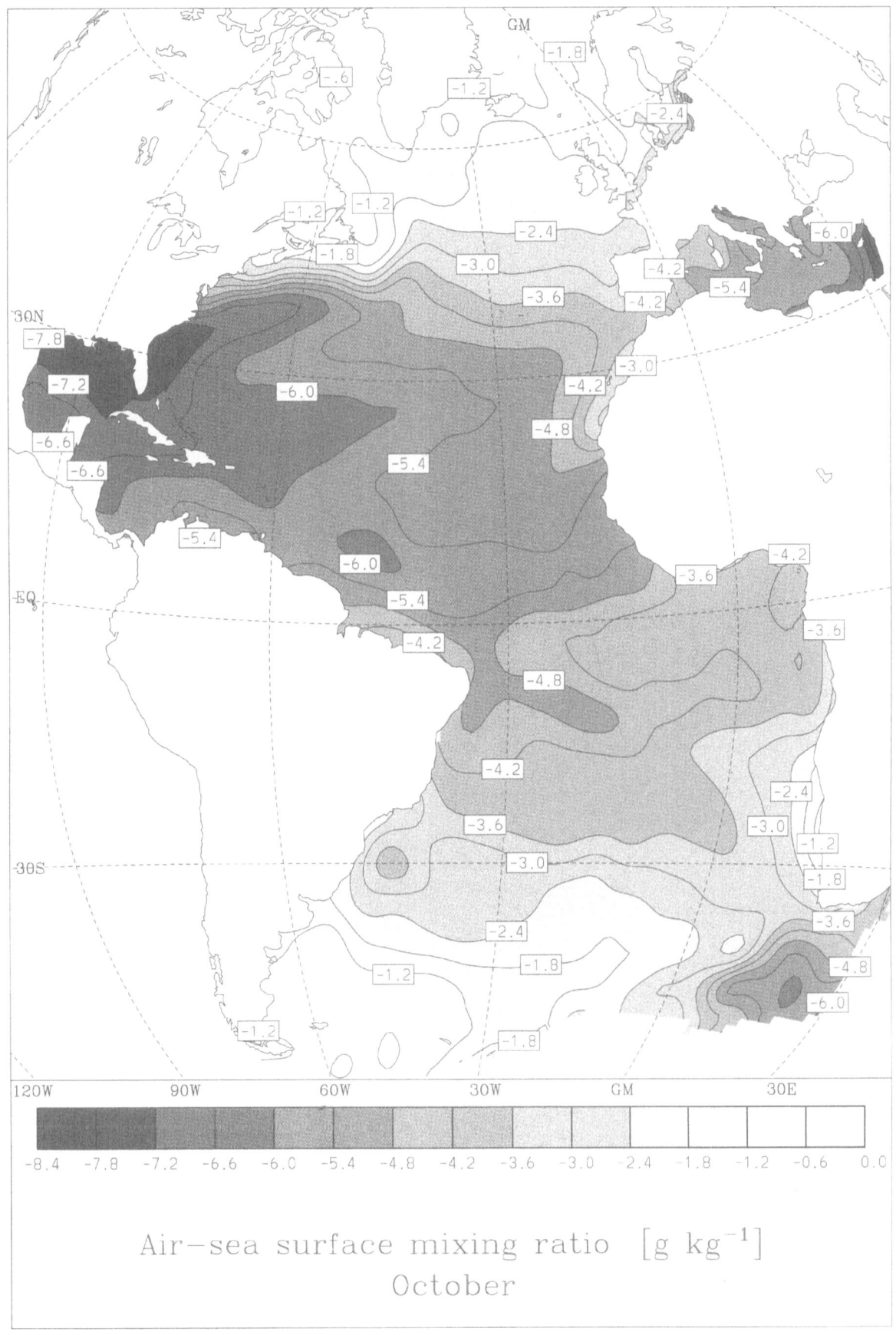

Air–sea surface mixing ratio [g kg^{-1}]
October

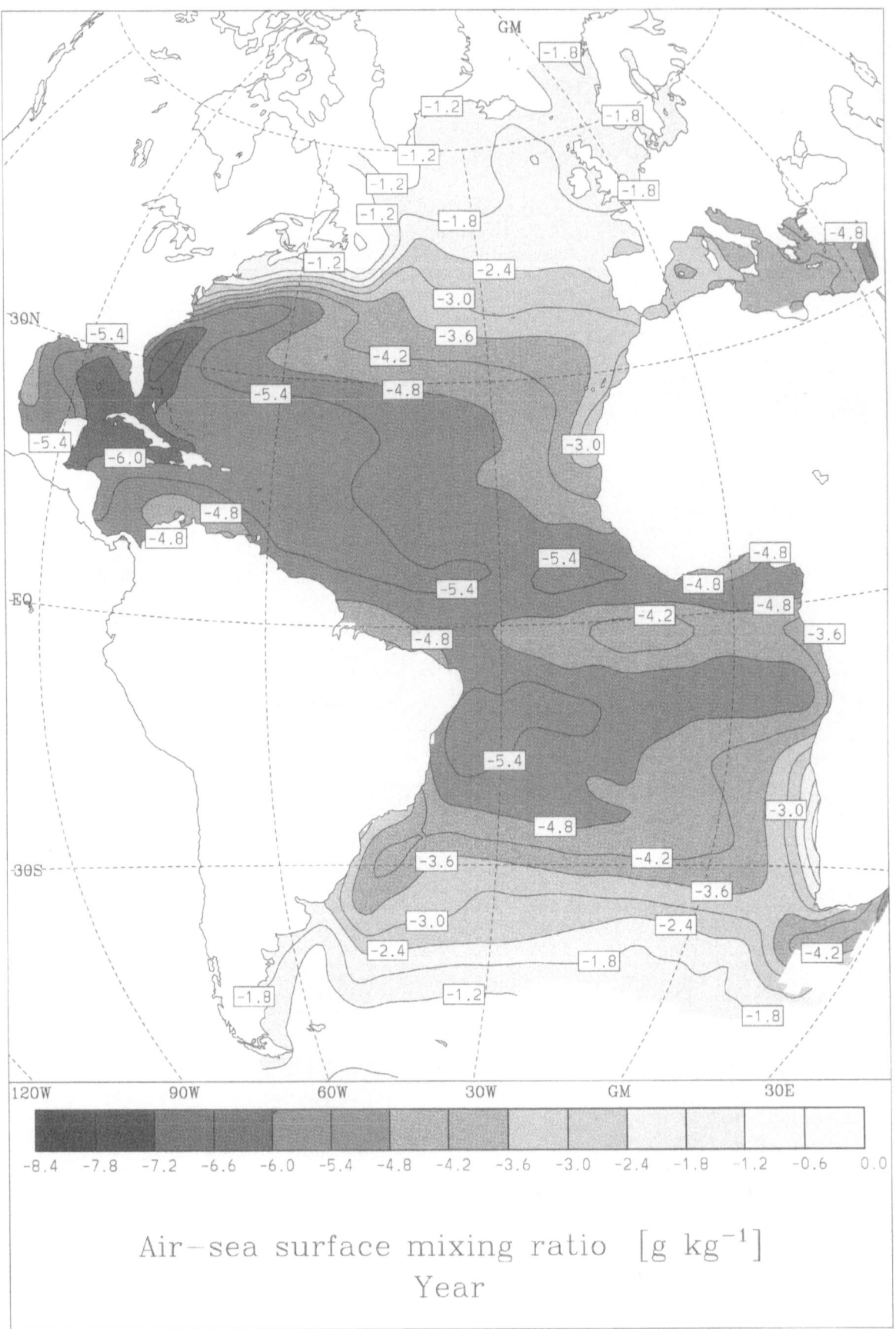

Air-sea surface mixing ratio [g kg^{-1}]
Year

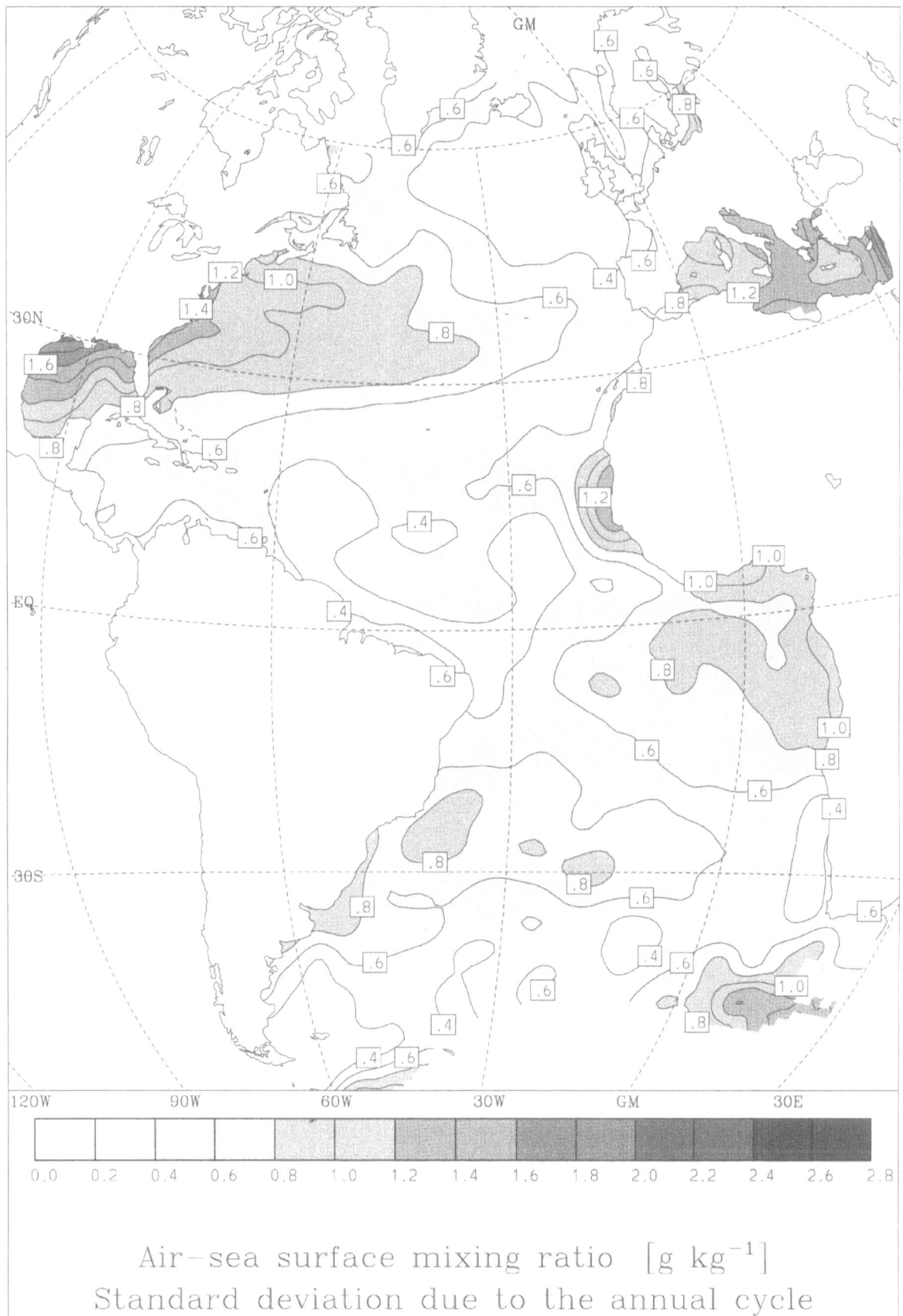

Air−sea surface mixing ratio [g kg^{-1}]
Standard deviation due to the annual cycle

Total Cloud Cover

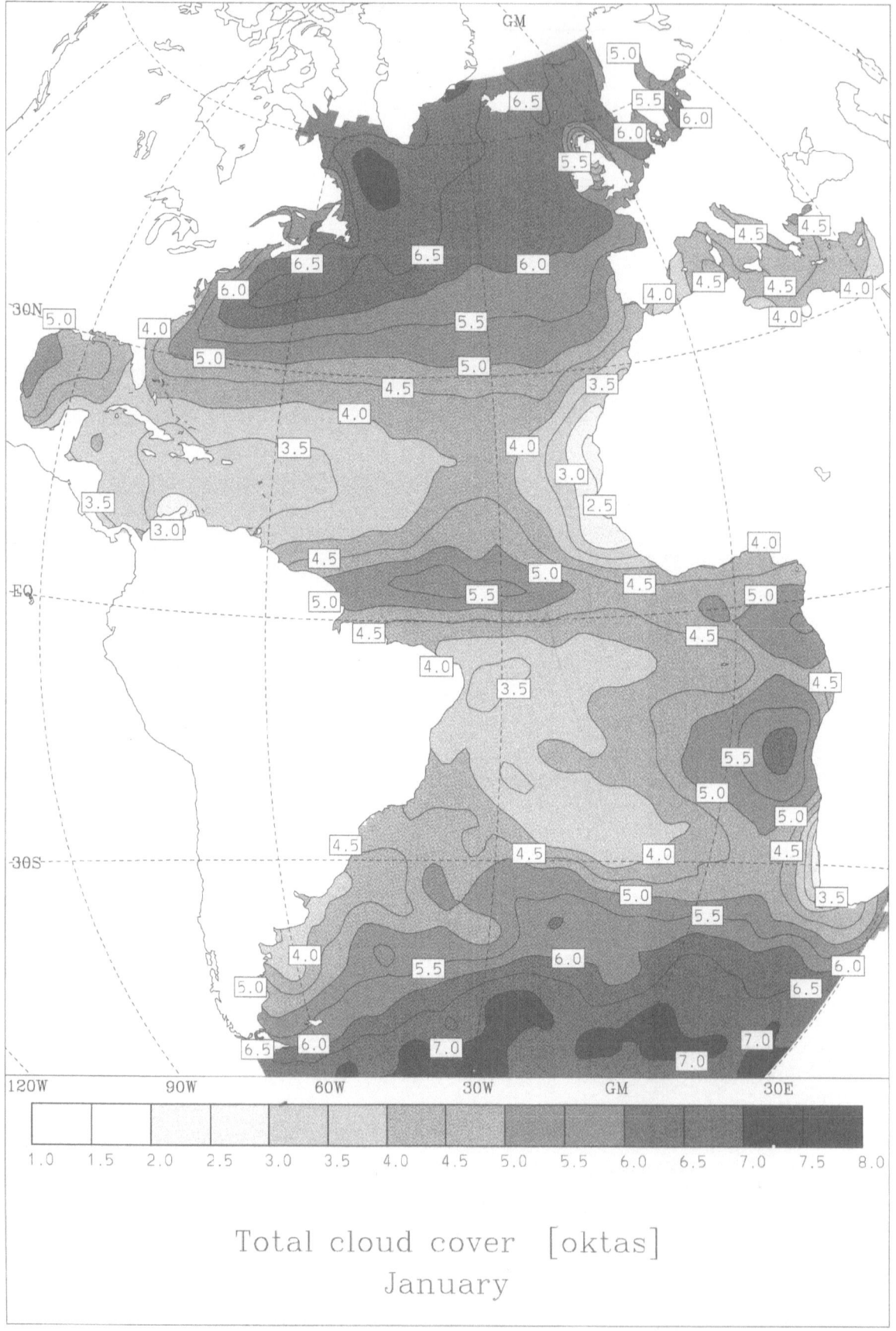

Total cloud cover [oktas]
January

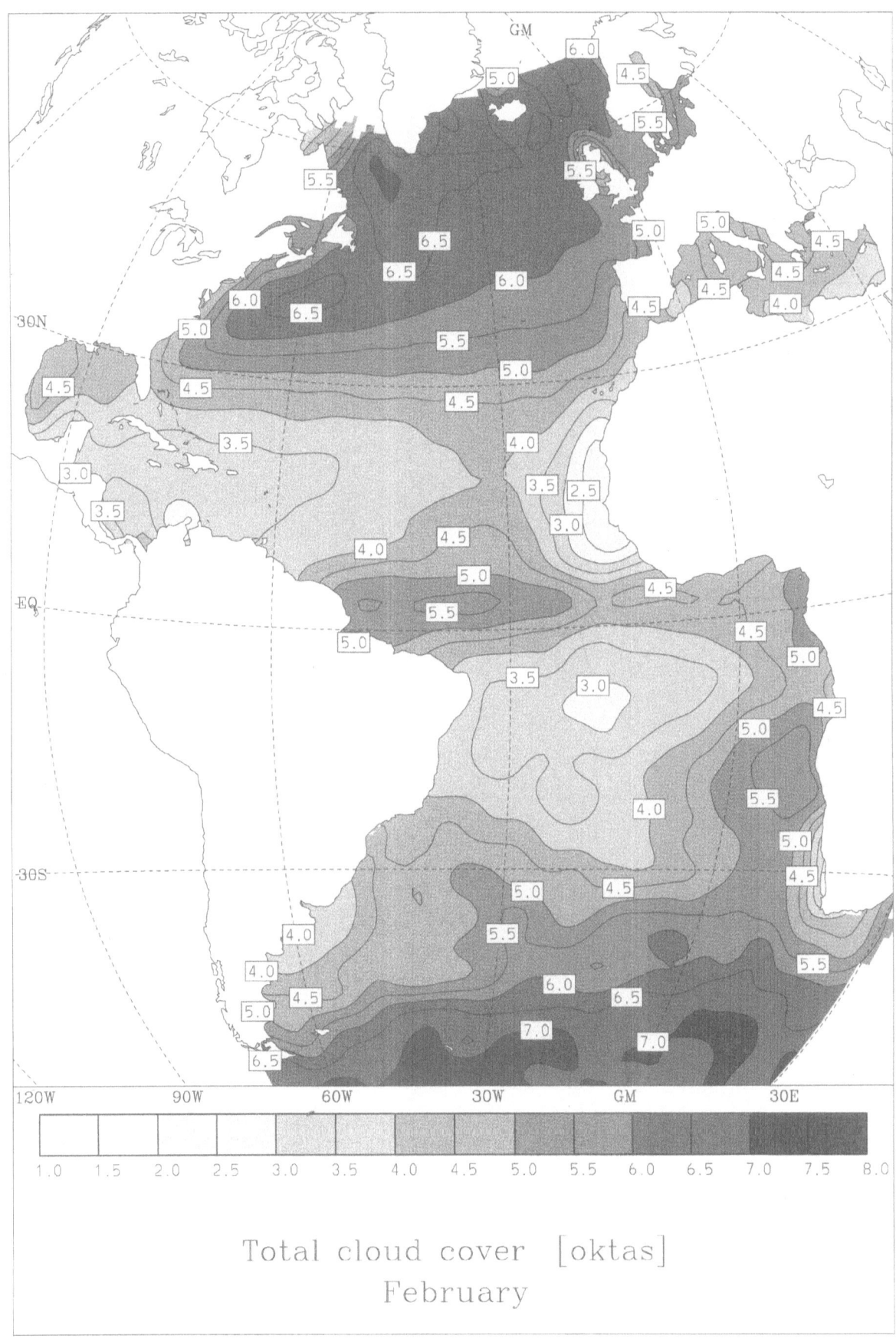

Total cloud cover [oktas]
February

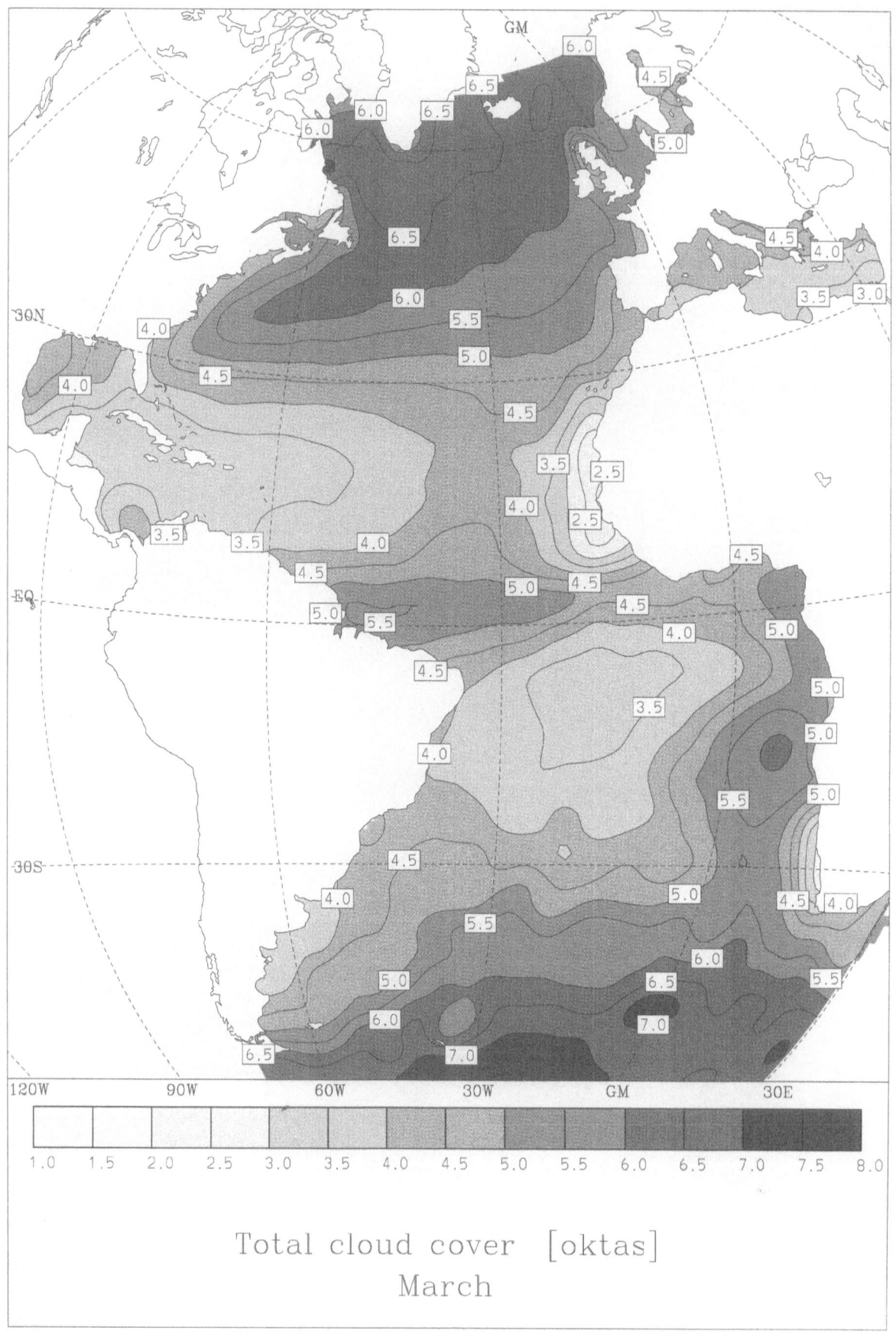

Total cloud cover [oktas]
March

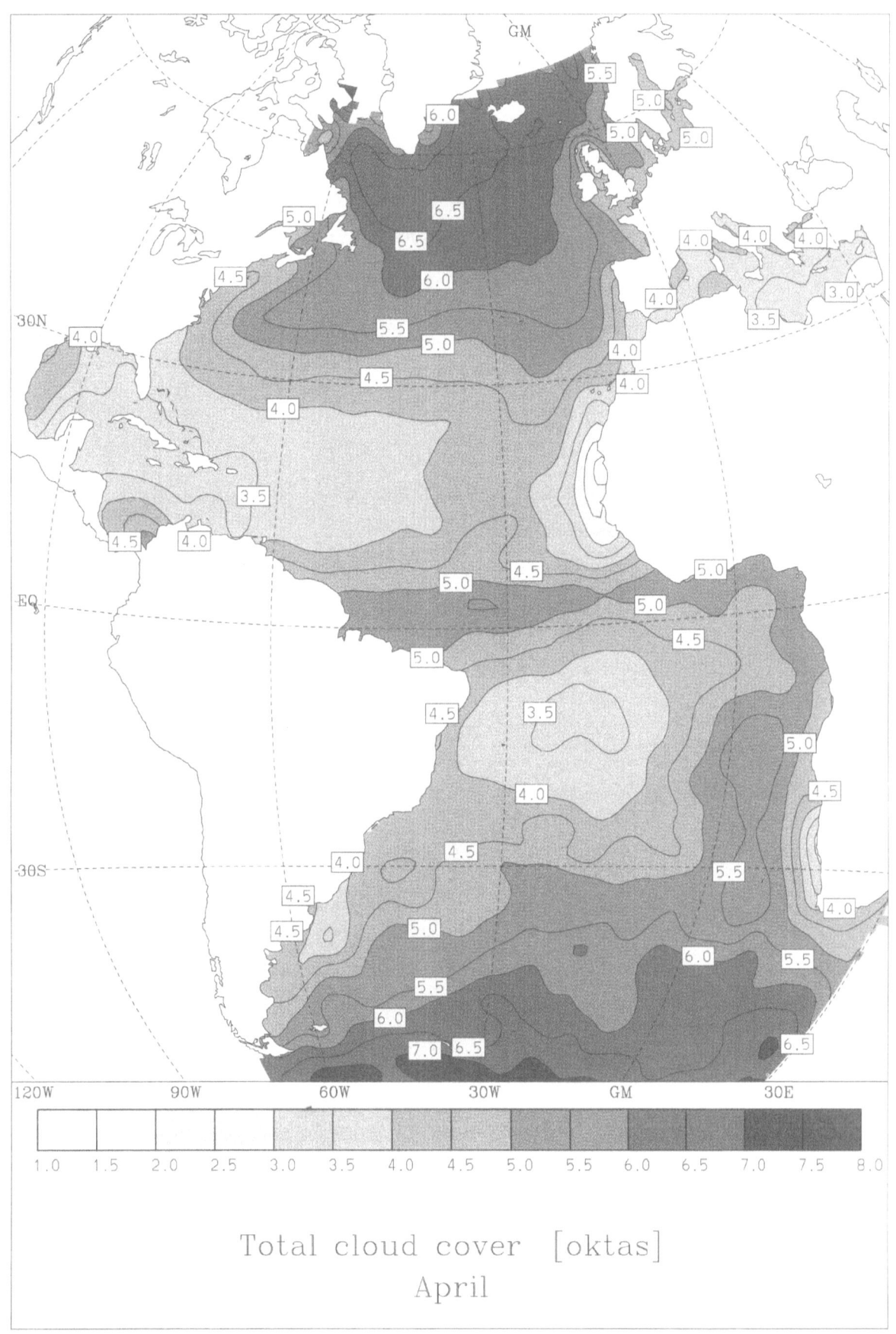

Total cloud cover [oktas]
April

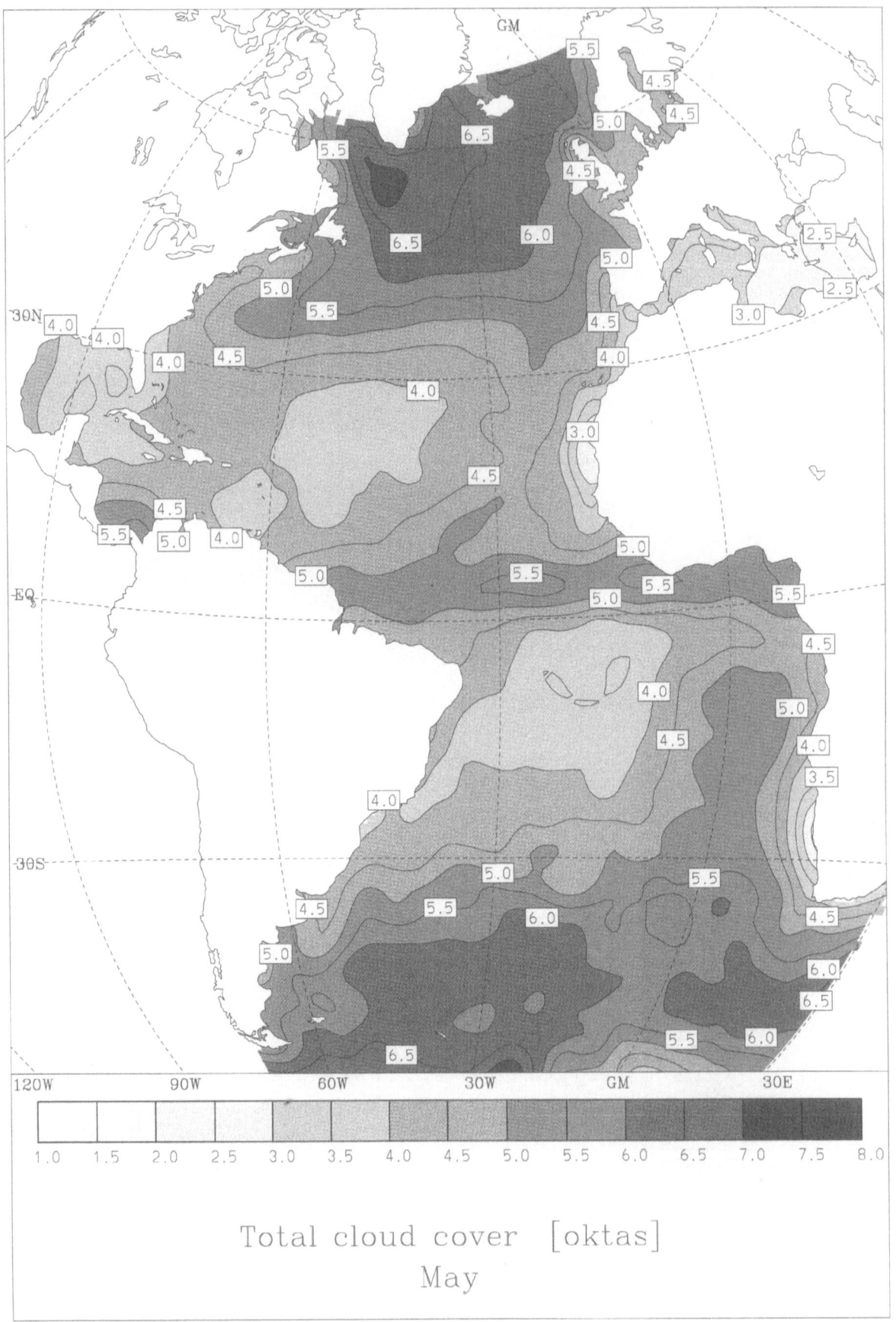

Total cloud cover [oktas]
May

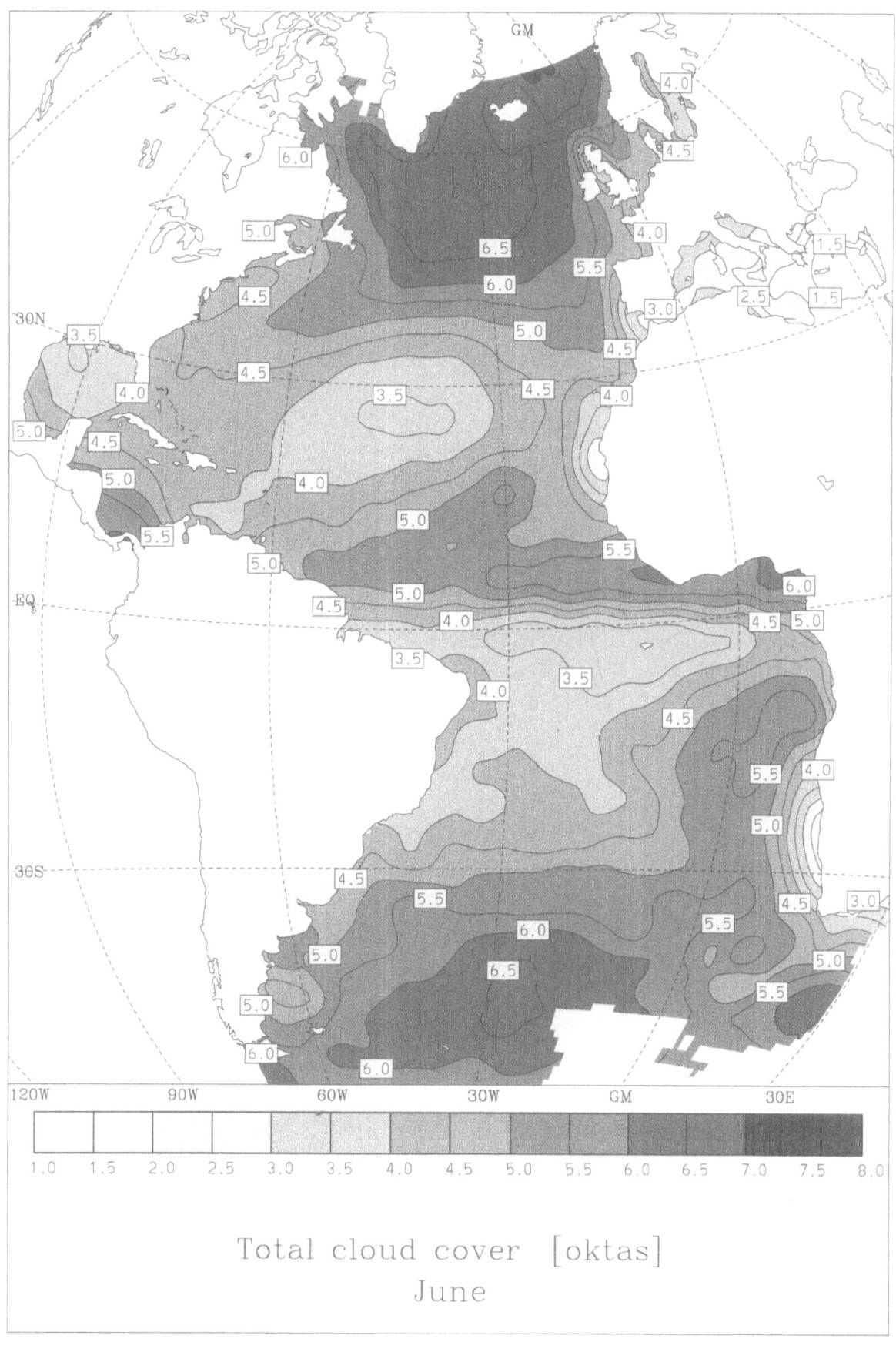

Total cloud cover [oktas]

June

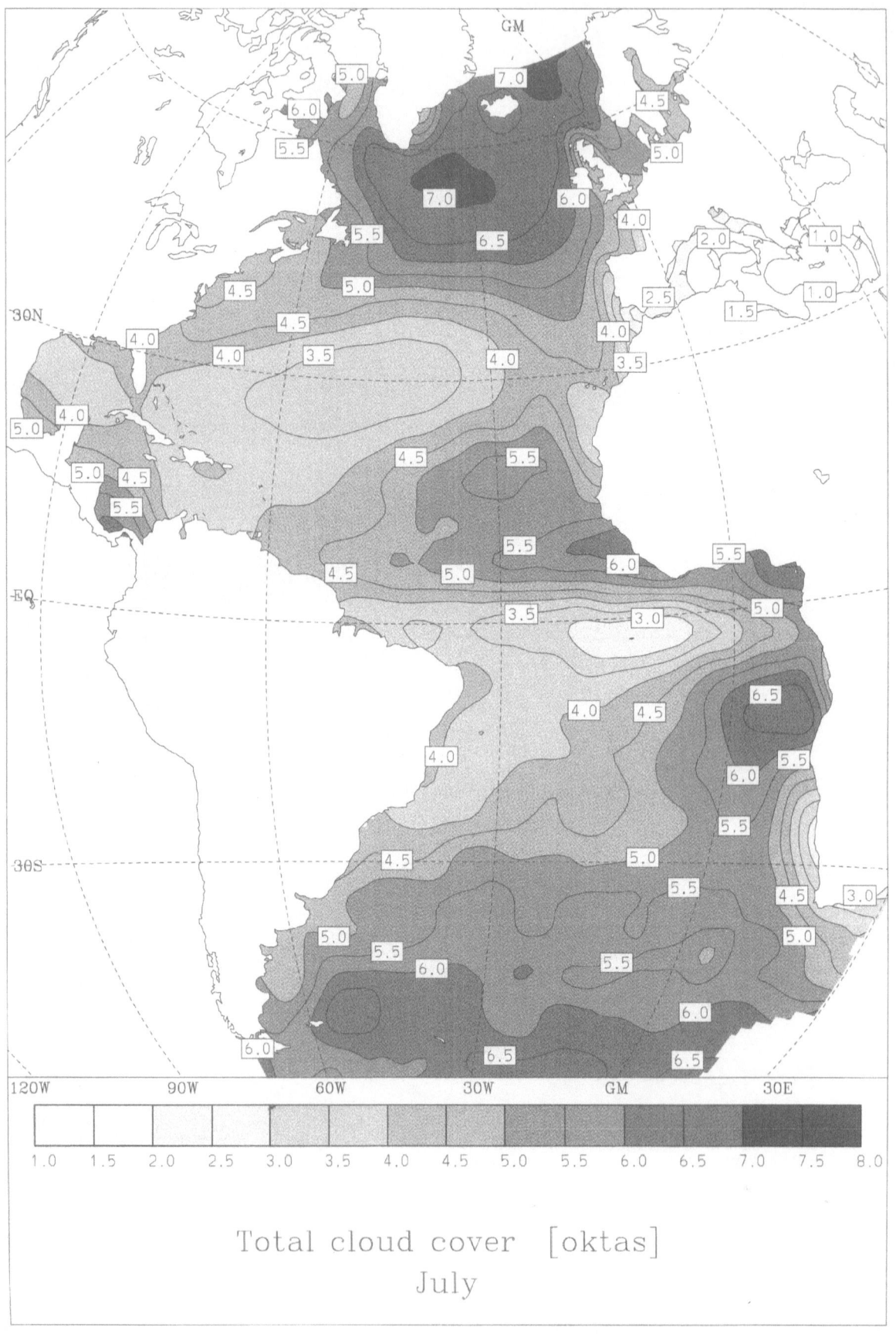

Total cloud cover [oktas]
July

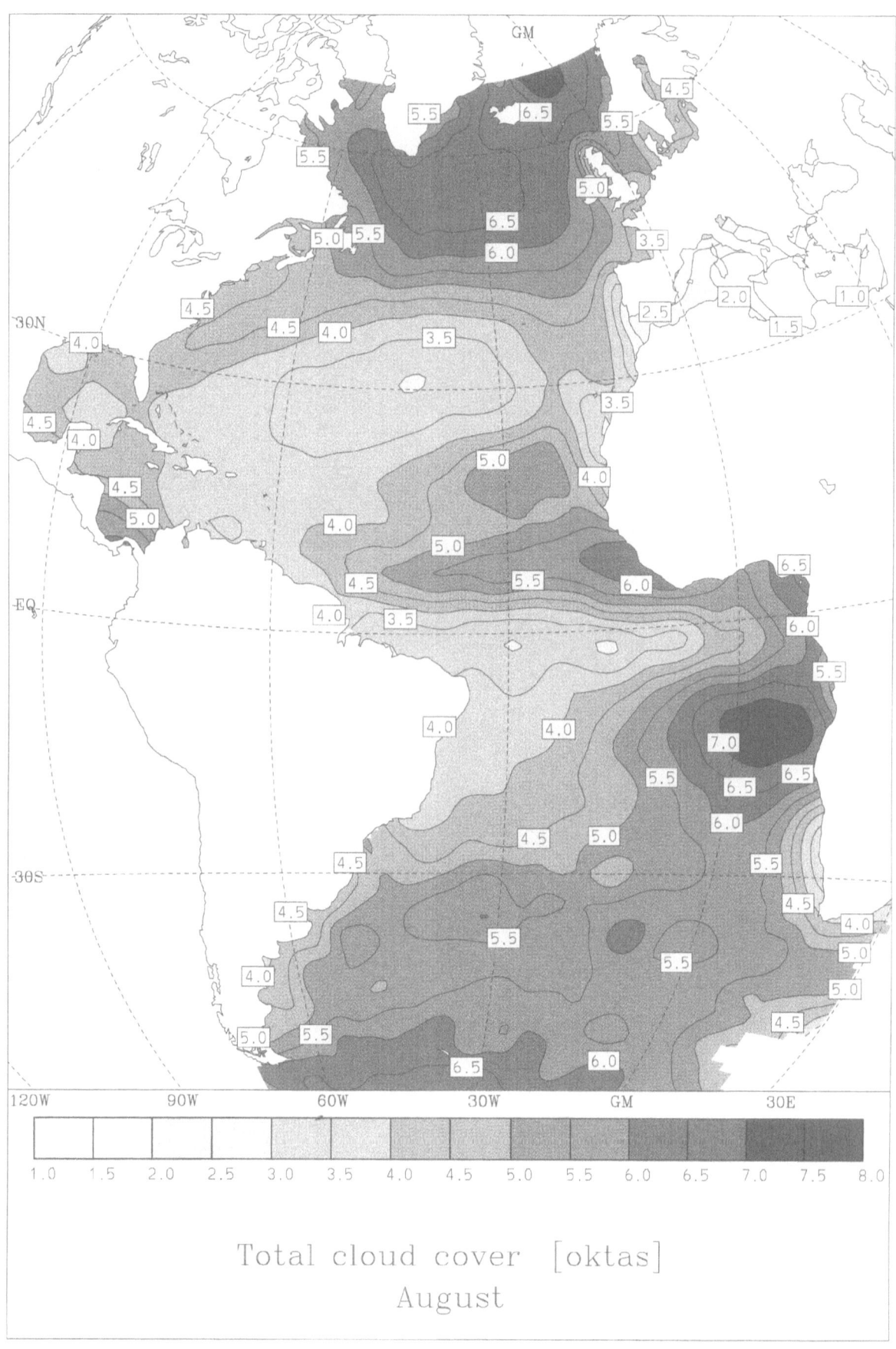

Total cloud cover [oktas]
August

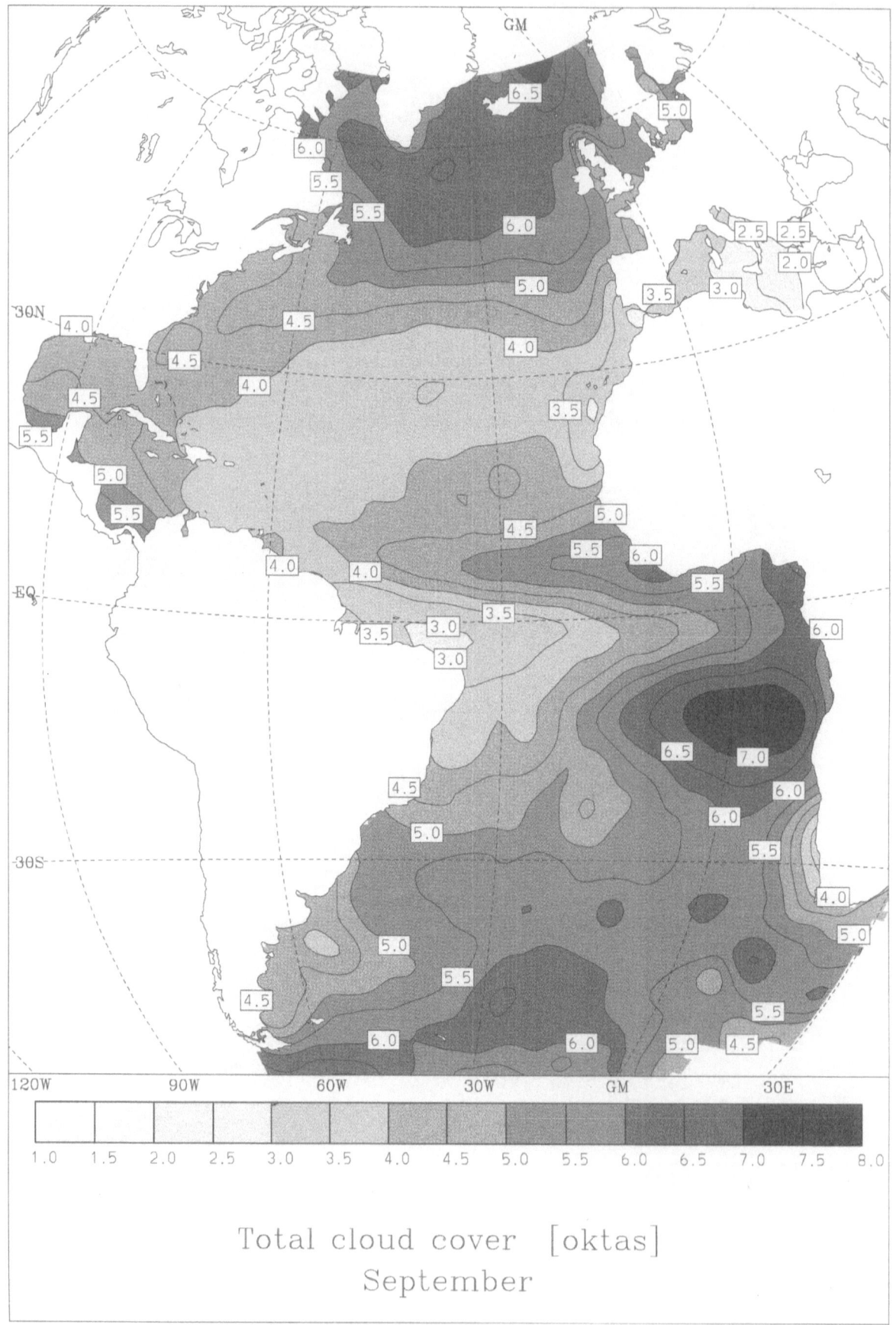

Total cloud cover [oktas]
September

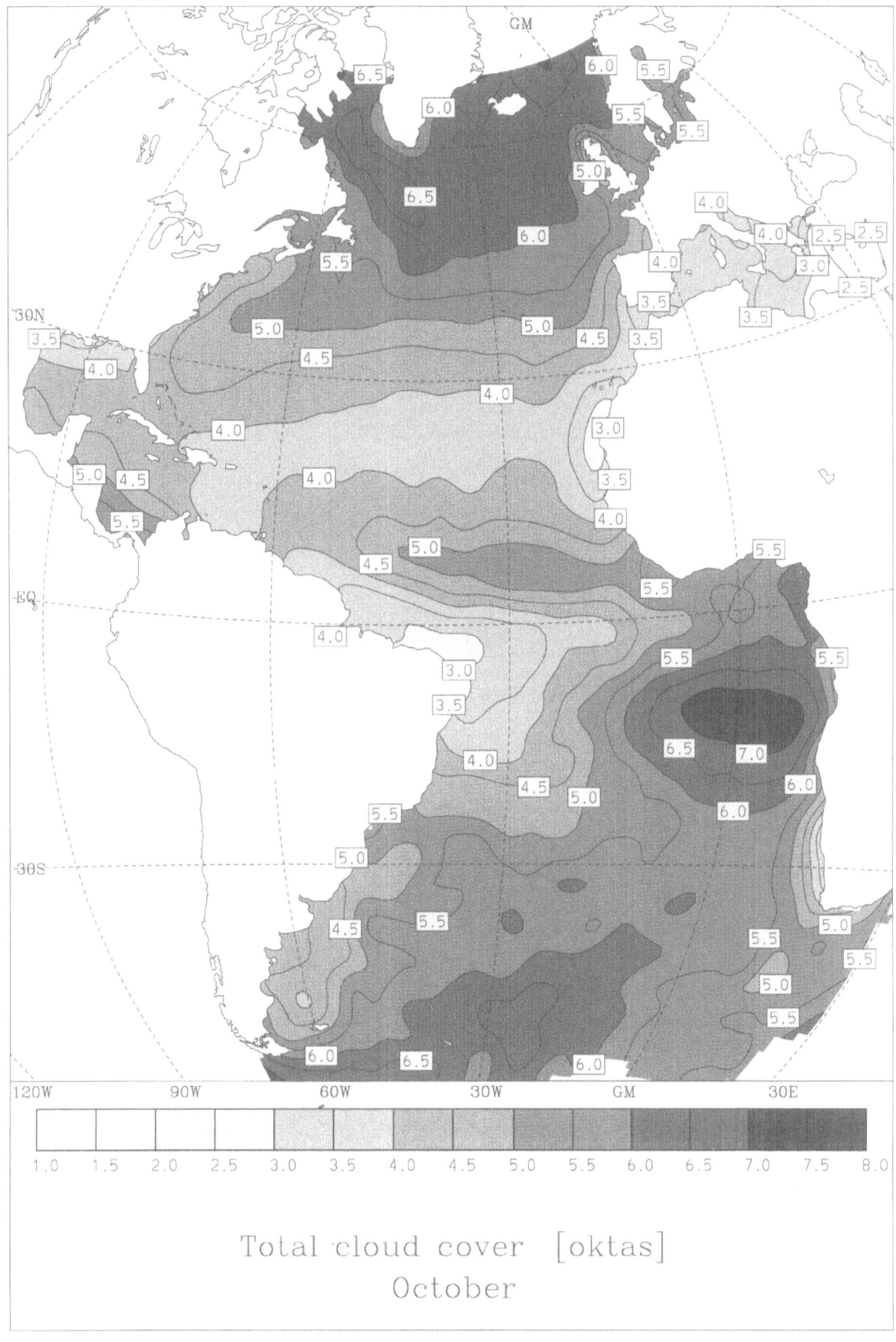

Total cloud cover [oktas]
October

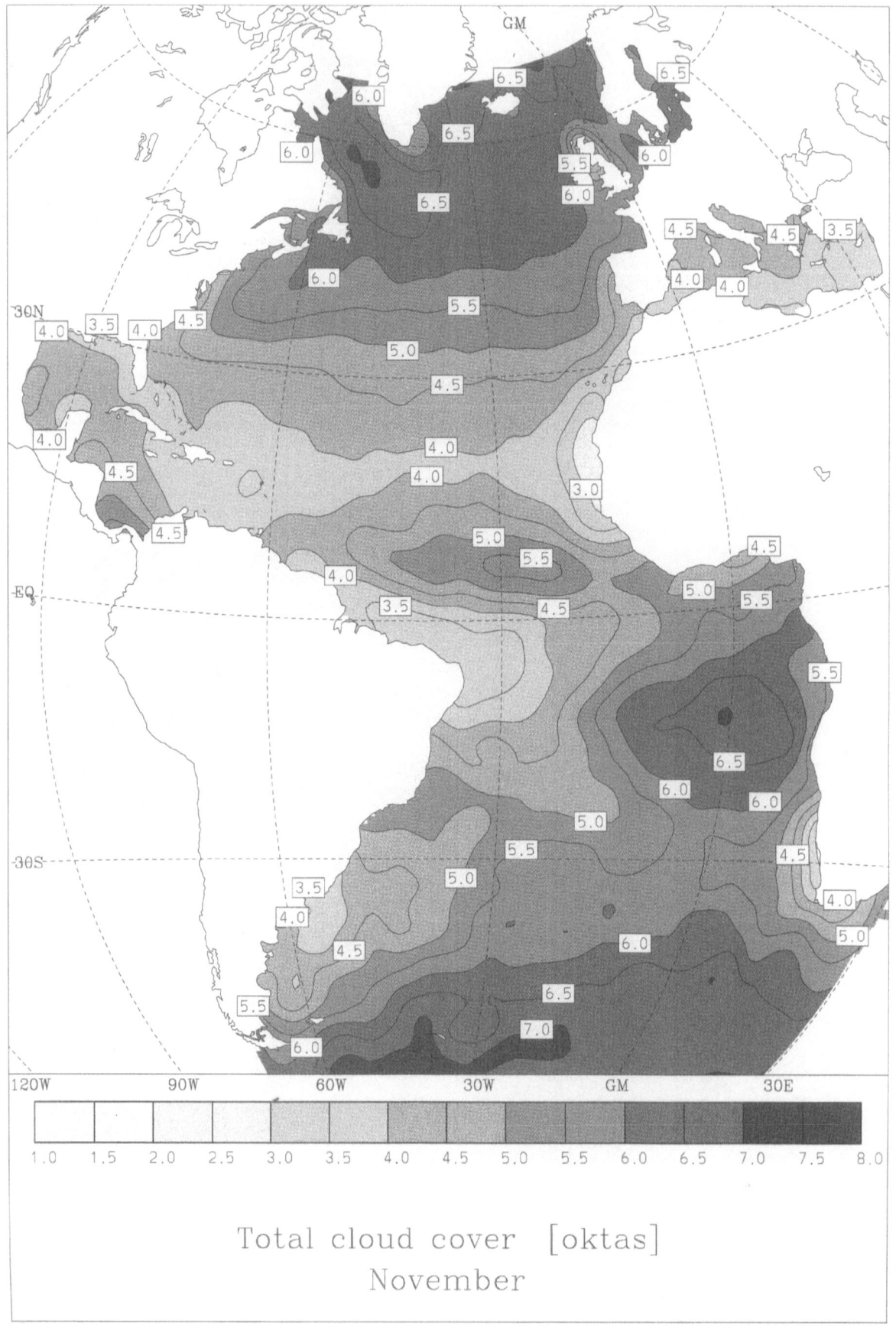

Total cloud cover [oktas]
November

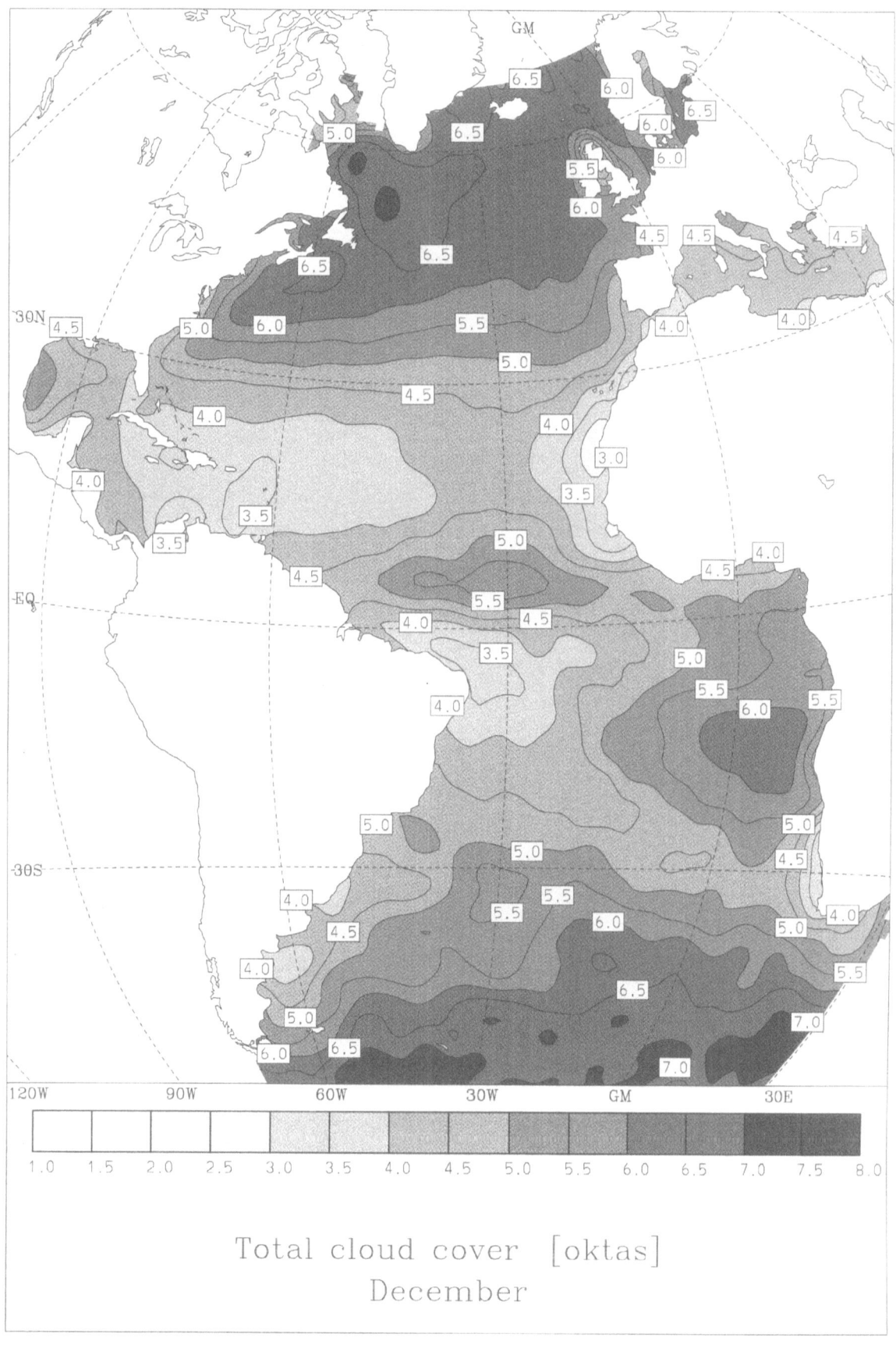

Total cloud cover [oktas]
December

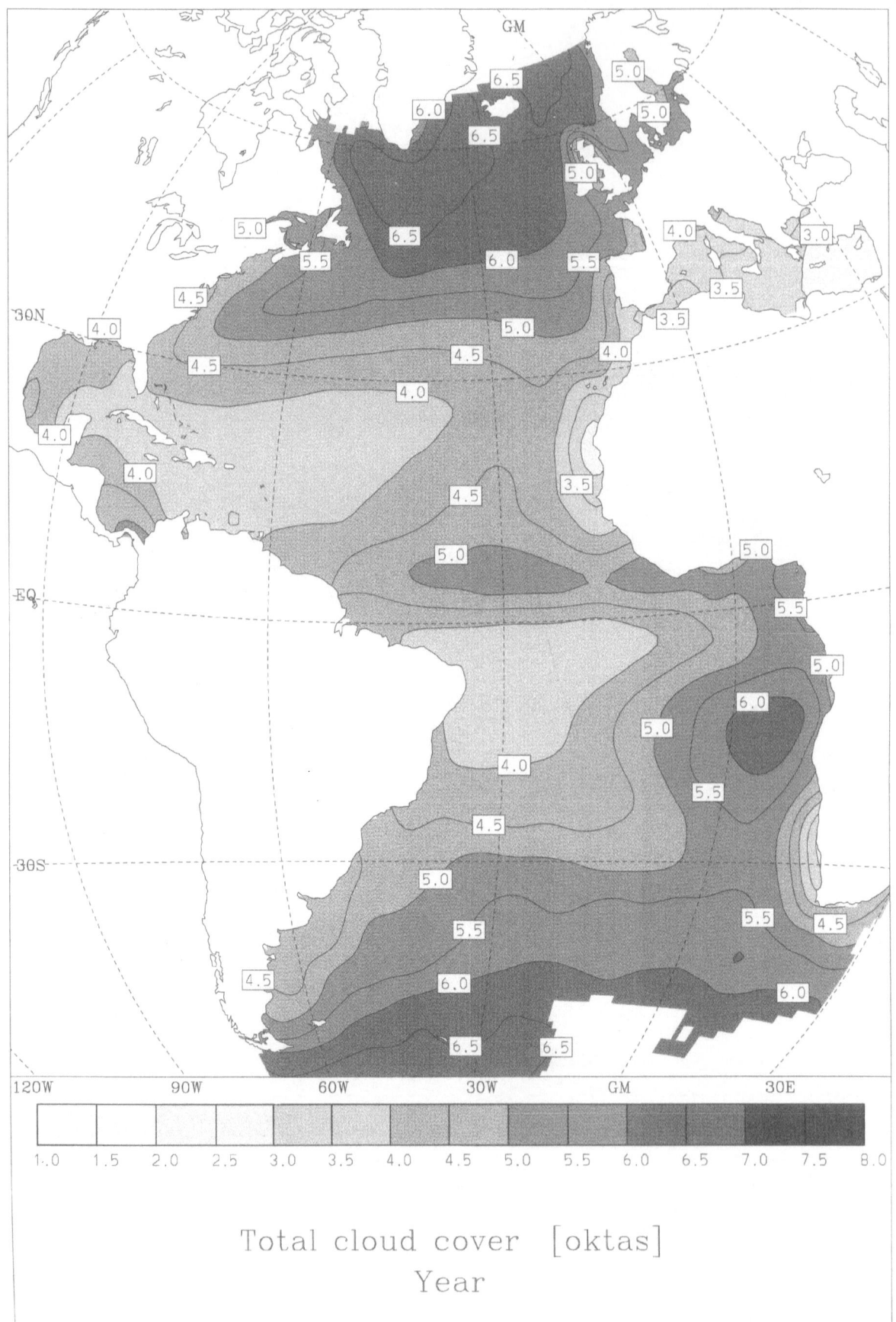

Total cloud cover [oktas]
Year

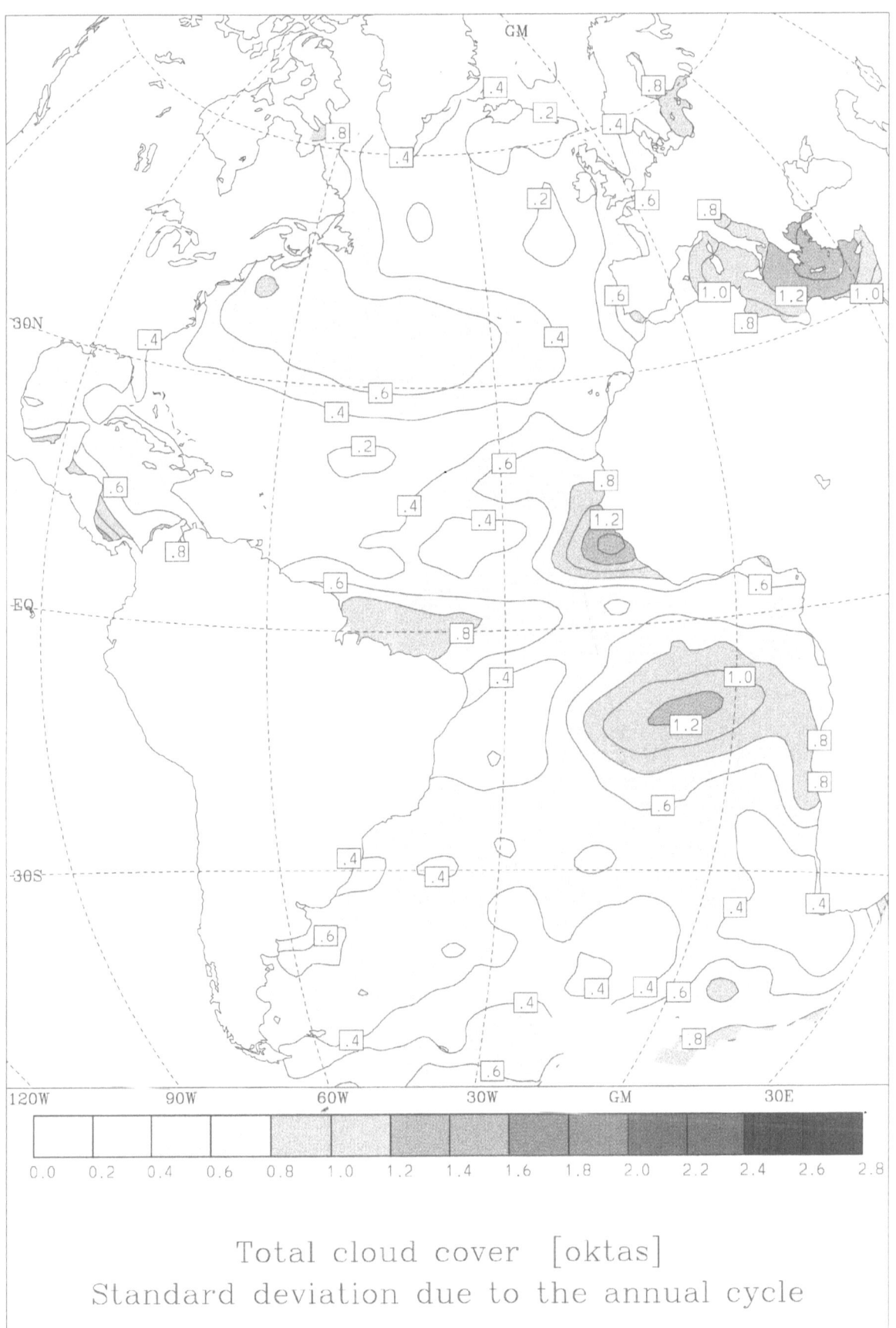

Total cloud cover [oktas]
Standard deviation due to the annual cycle

Low Cloud Cover

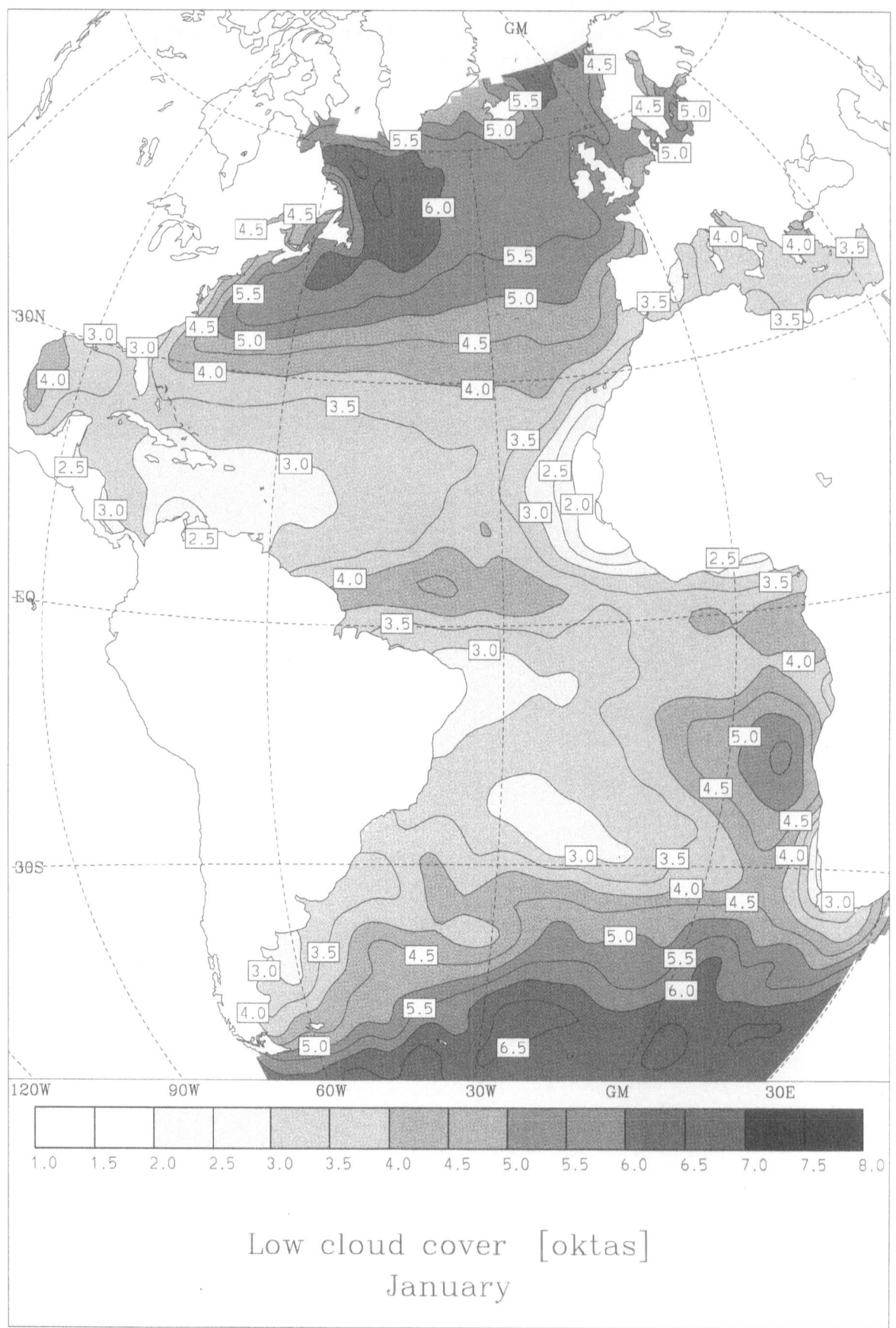

Low cloud cover [oktas]
January

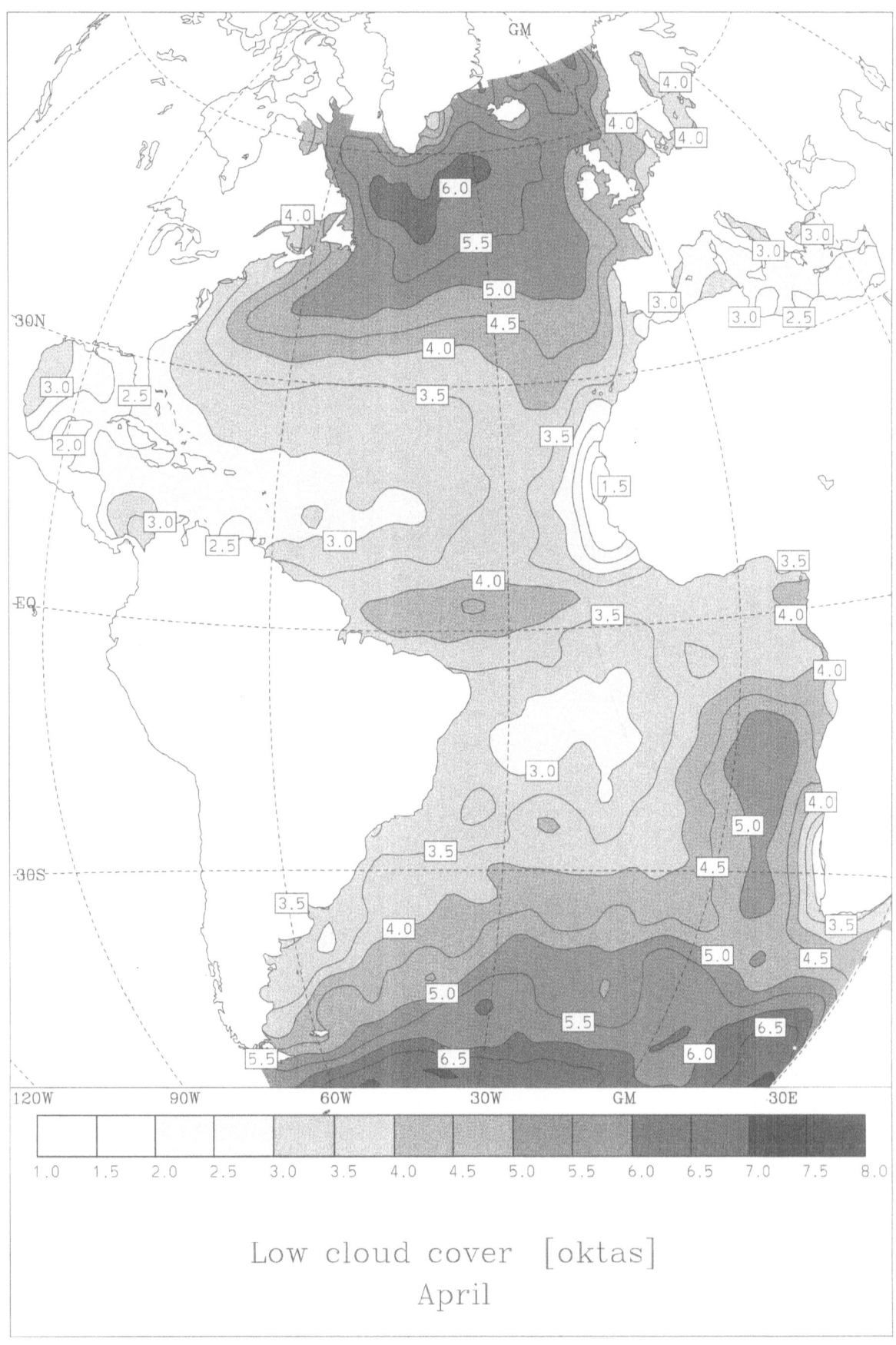

Low cloud cover [oktas]
April

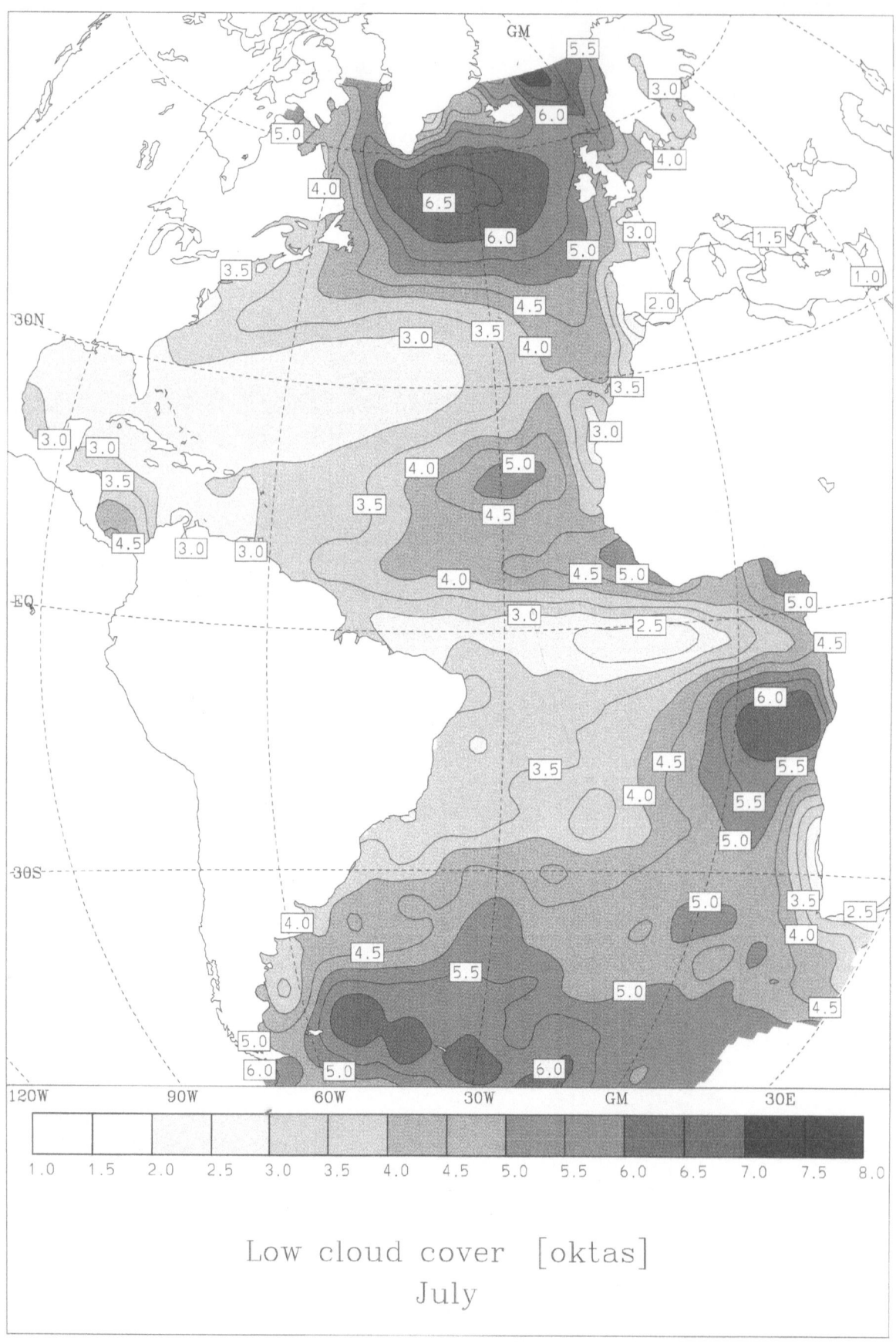

Low cloud cover [oktas]
July

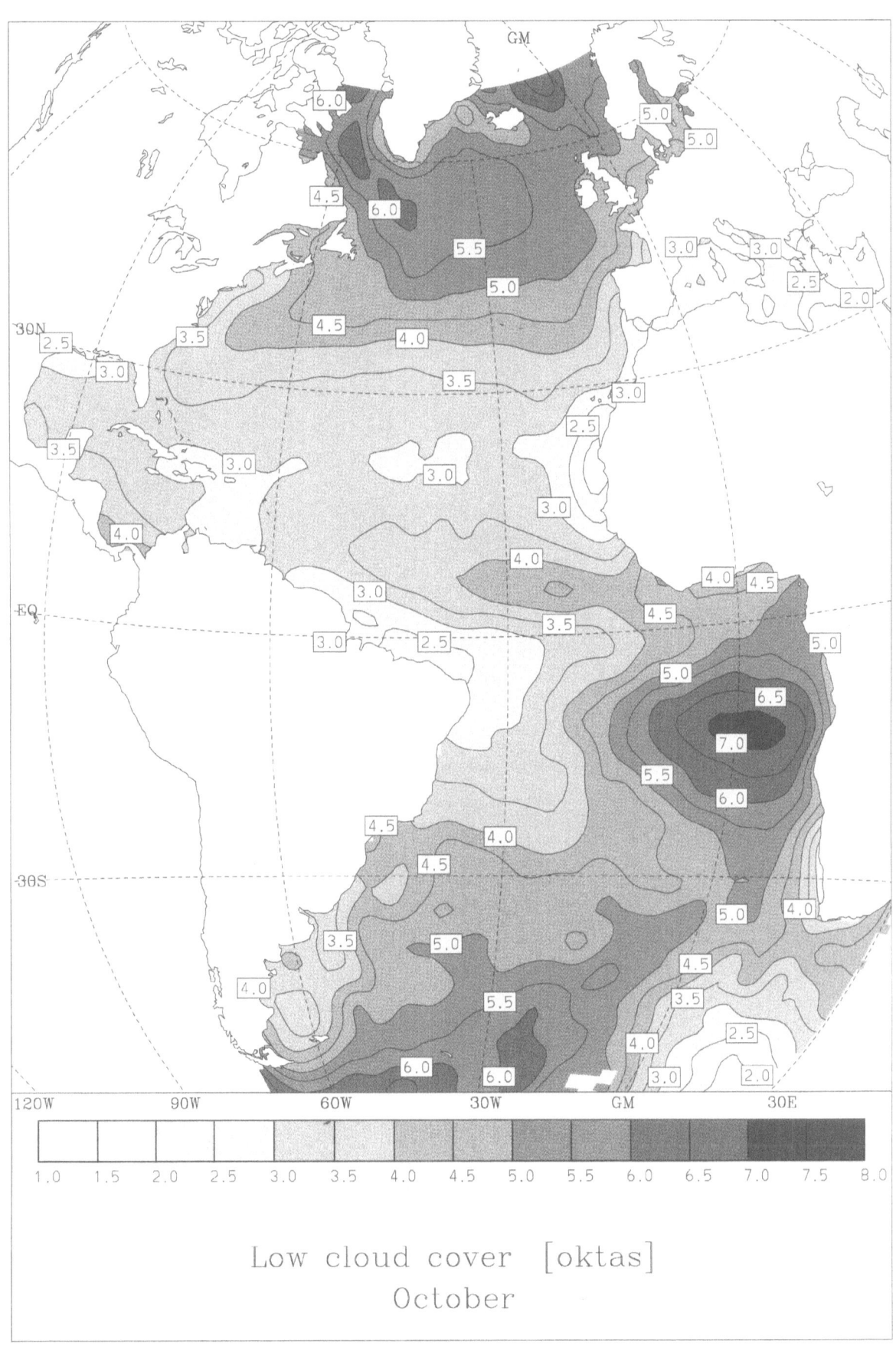

Low cloud cover [oktas]
October

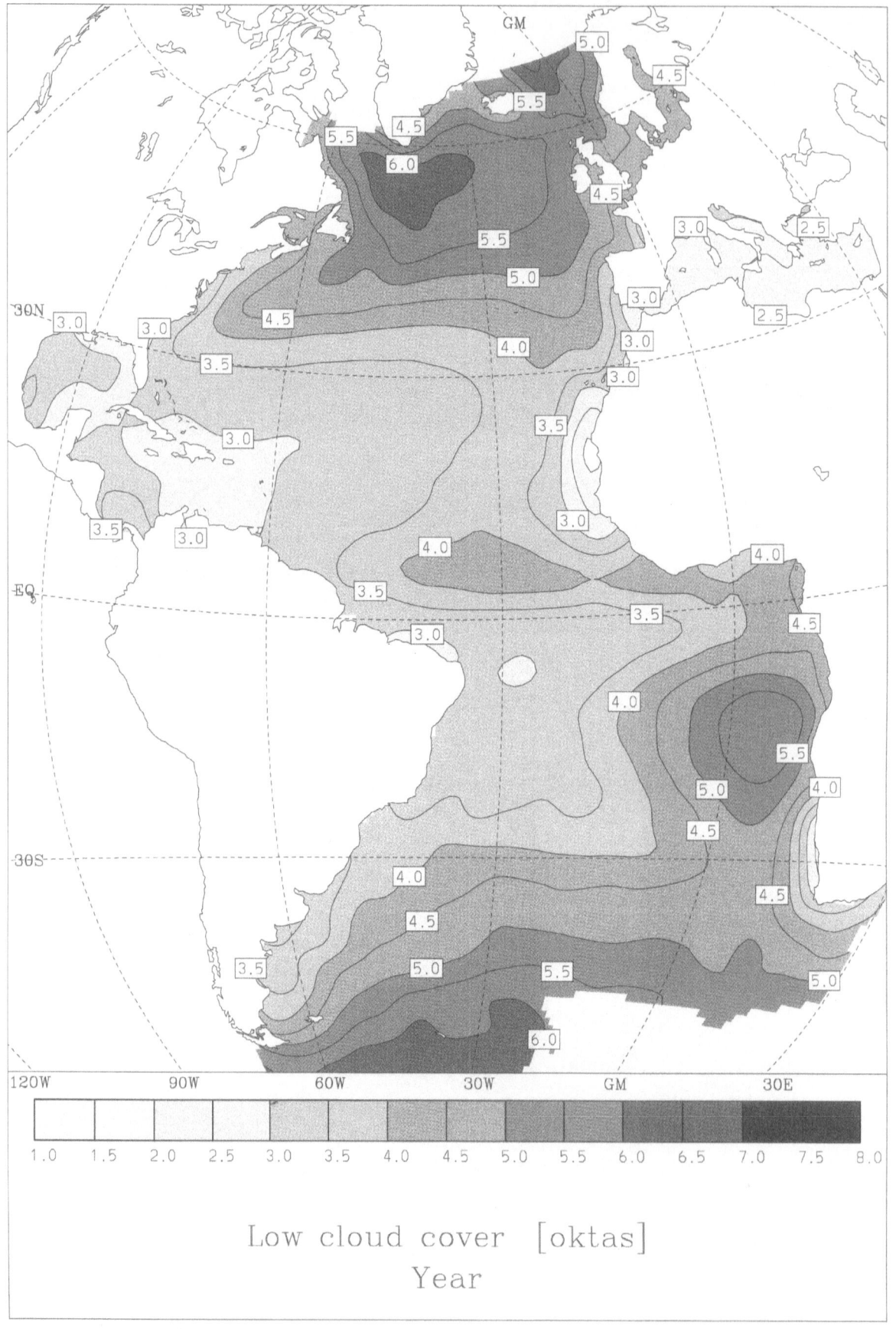

Low cloud cover [oktas]
Year

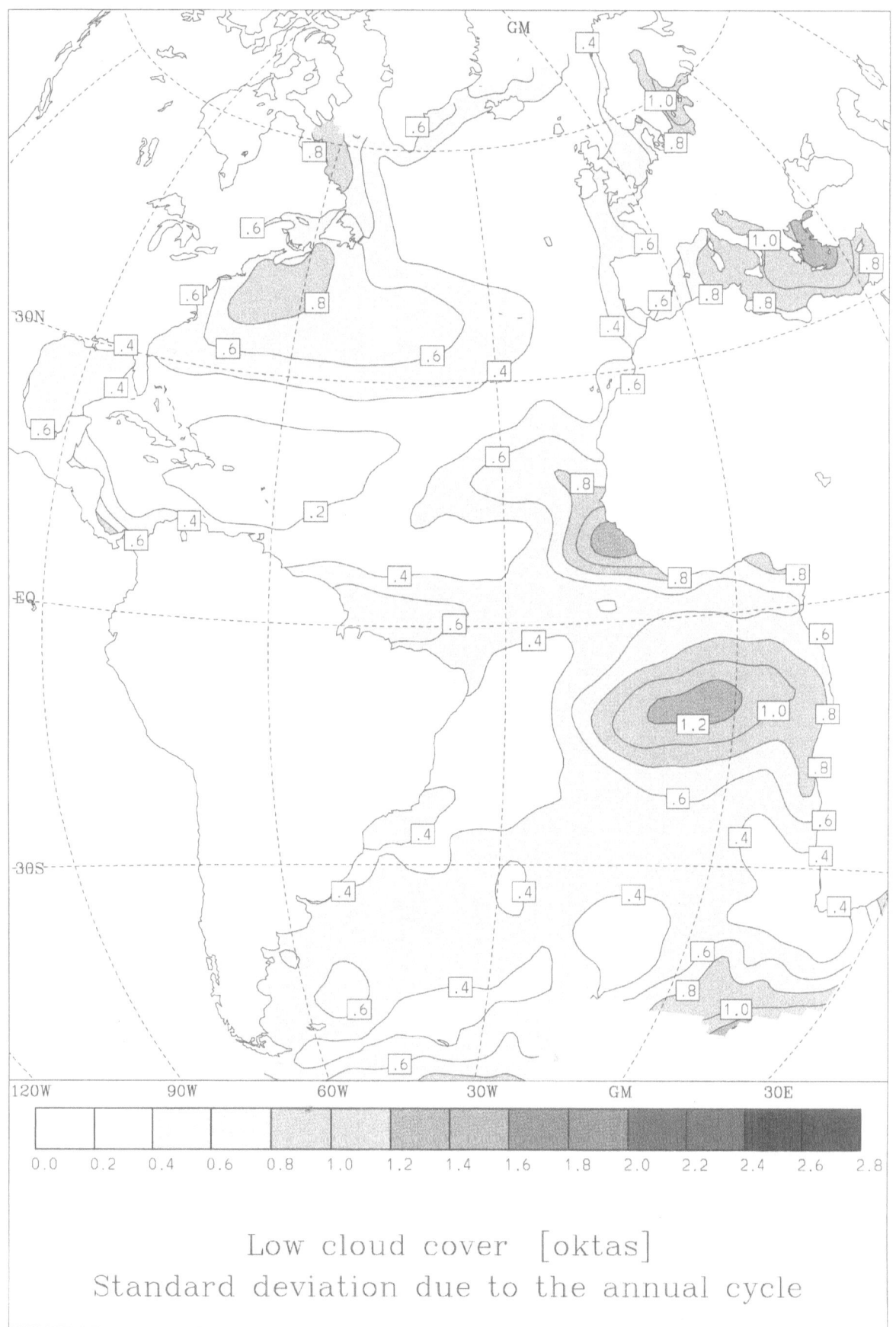

Low cloud cover [oktas]
Standard deviation due to the annual cycle

Precipitation Frequency

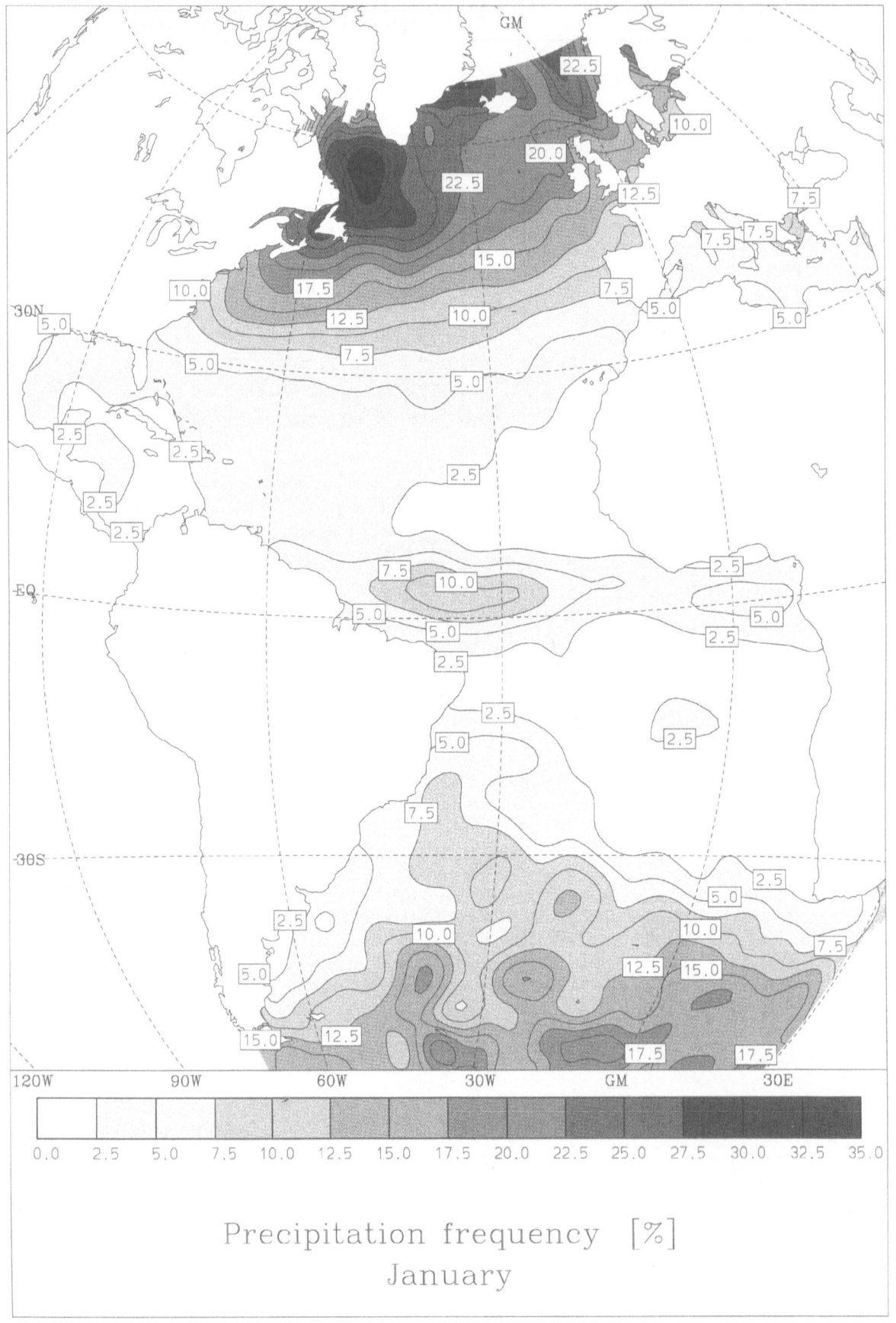

Precipitation frequency [%]
January

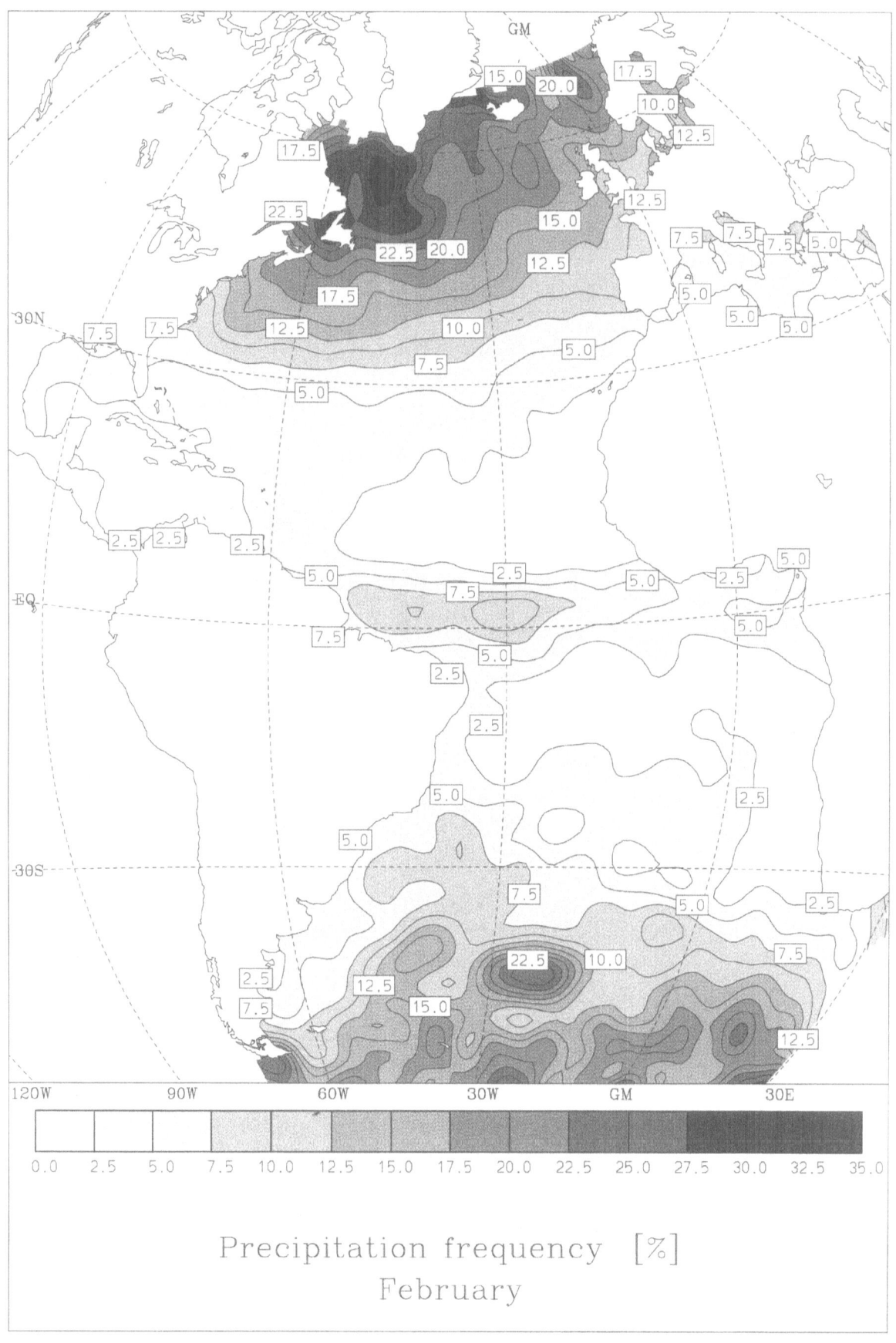

Precipitation frequency [%]
February

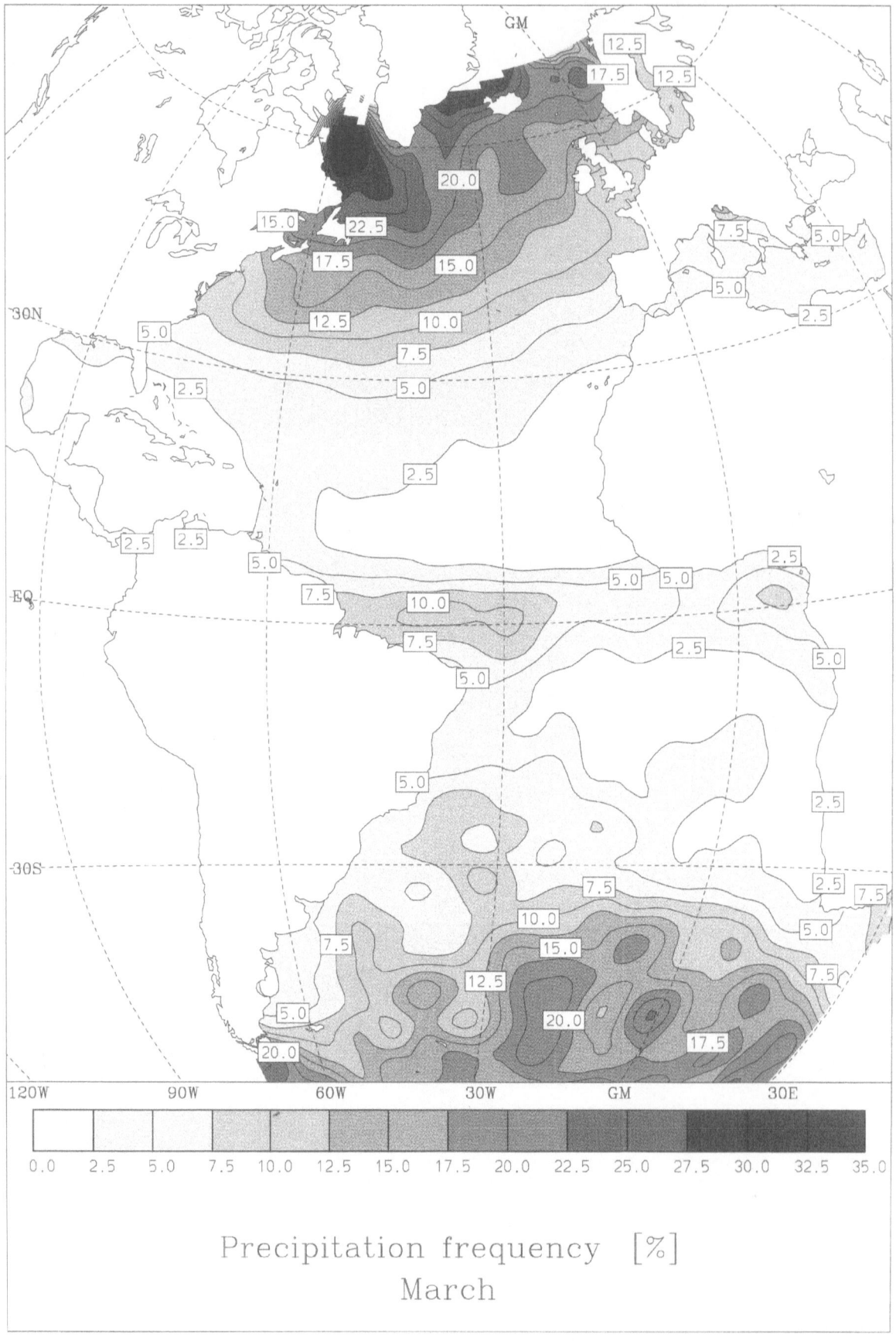

Precipitation frequency [%]
March

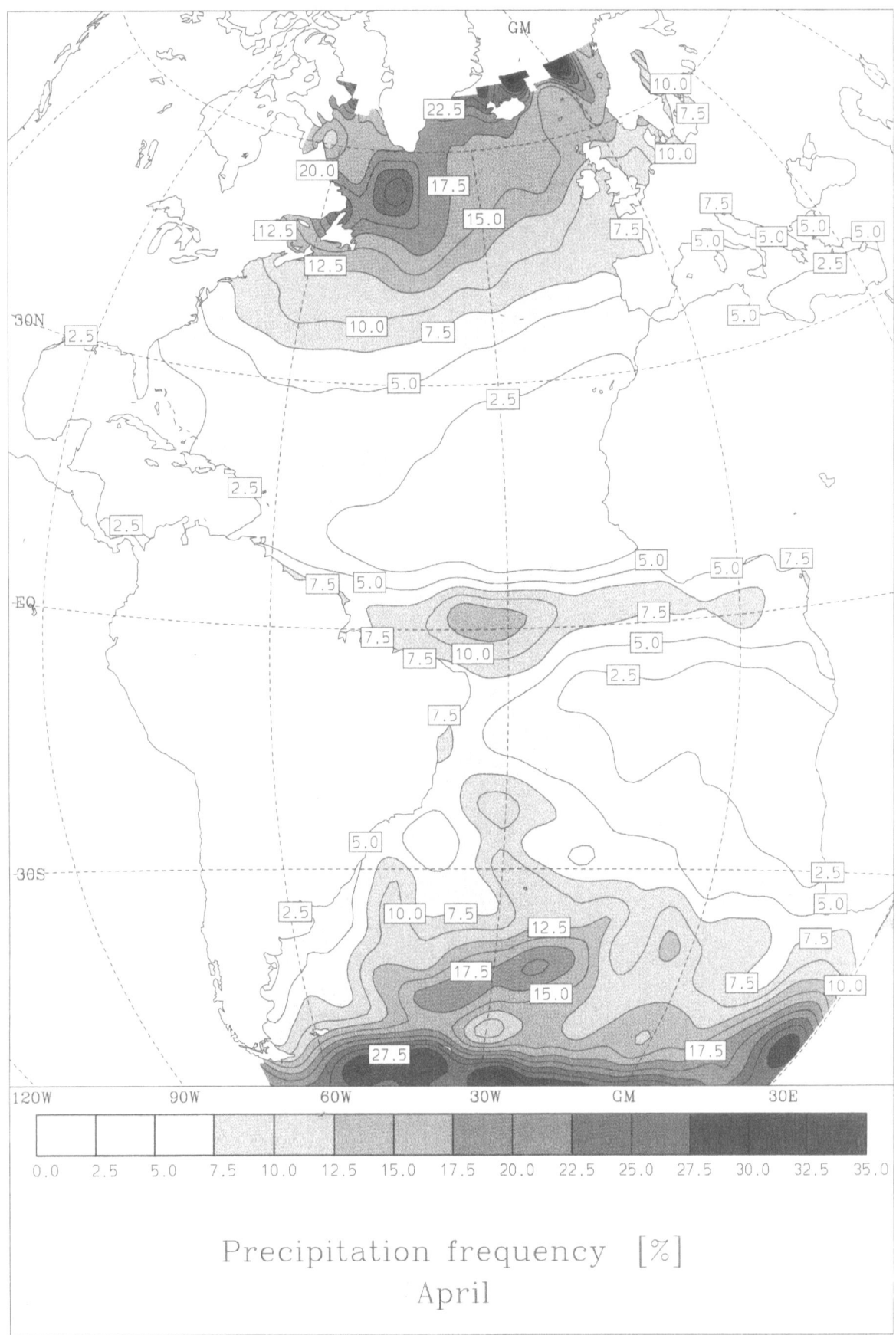

Precipitation frequency [%]
April

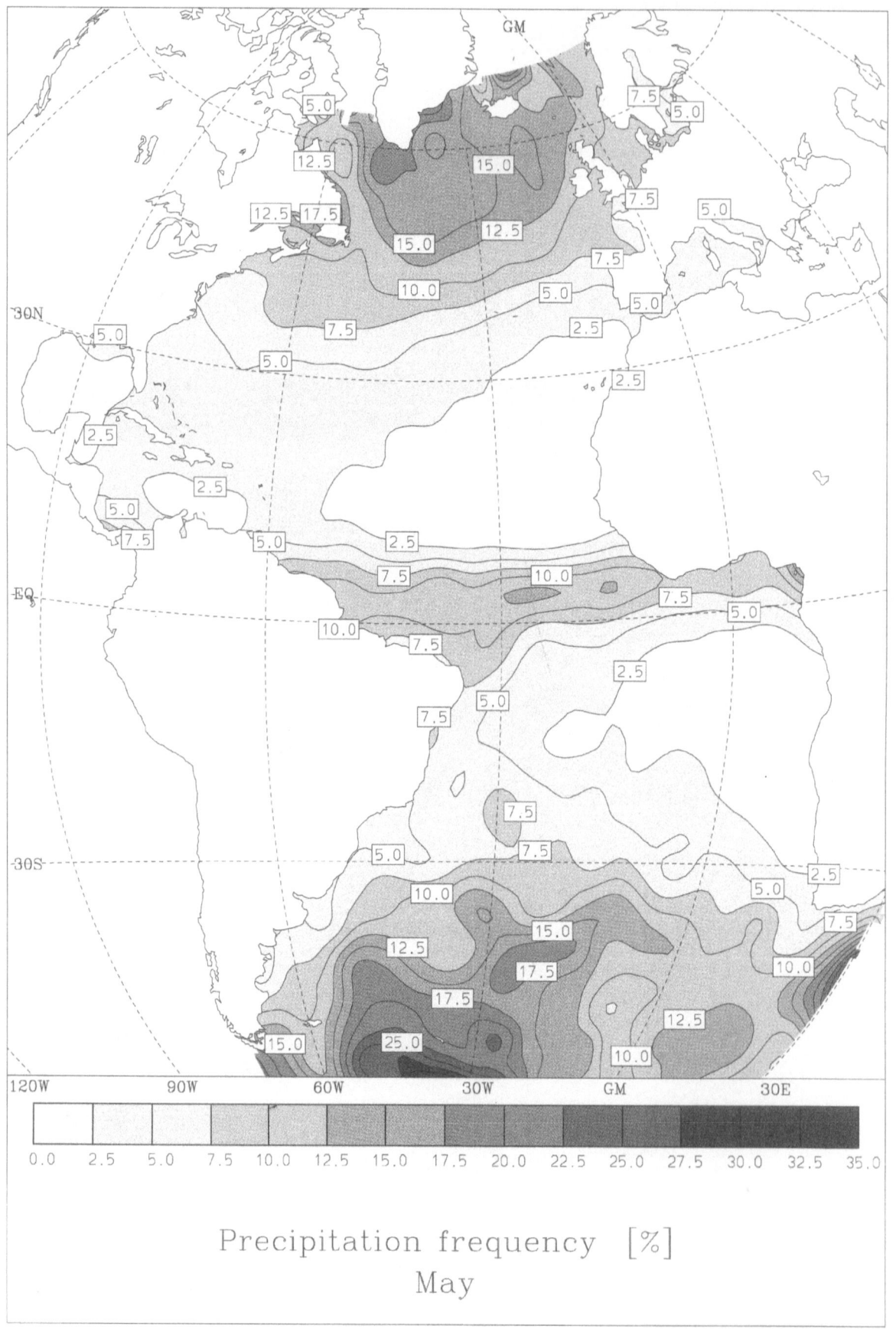

Precipitation frequency [%]
May

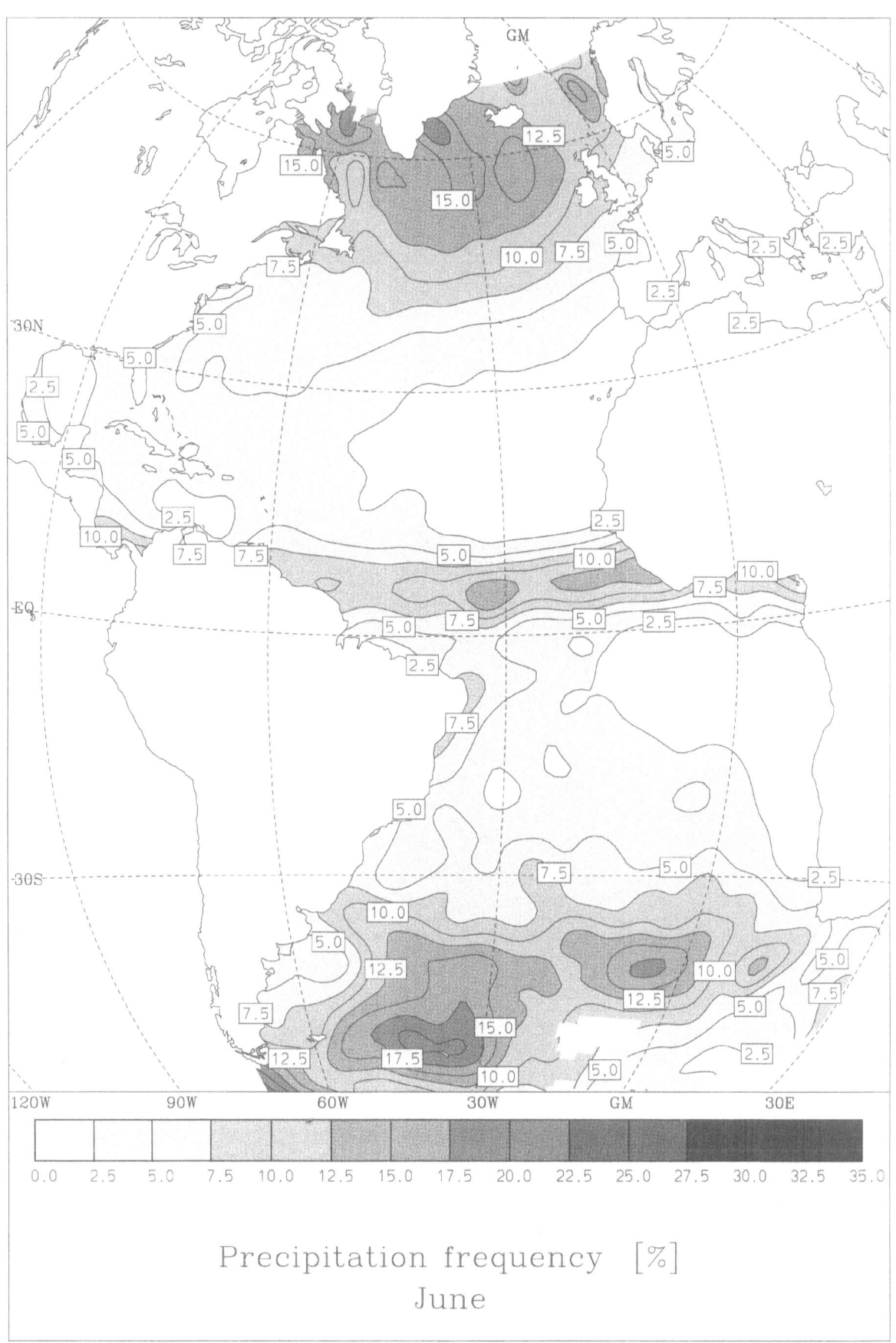

Precipitation frequency [%]
June

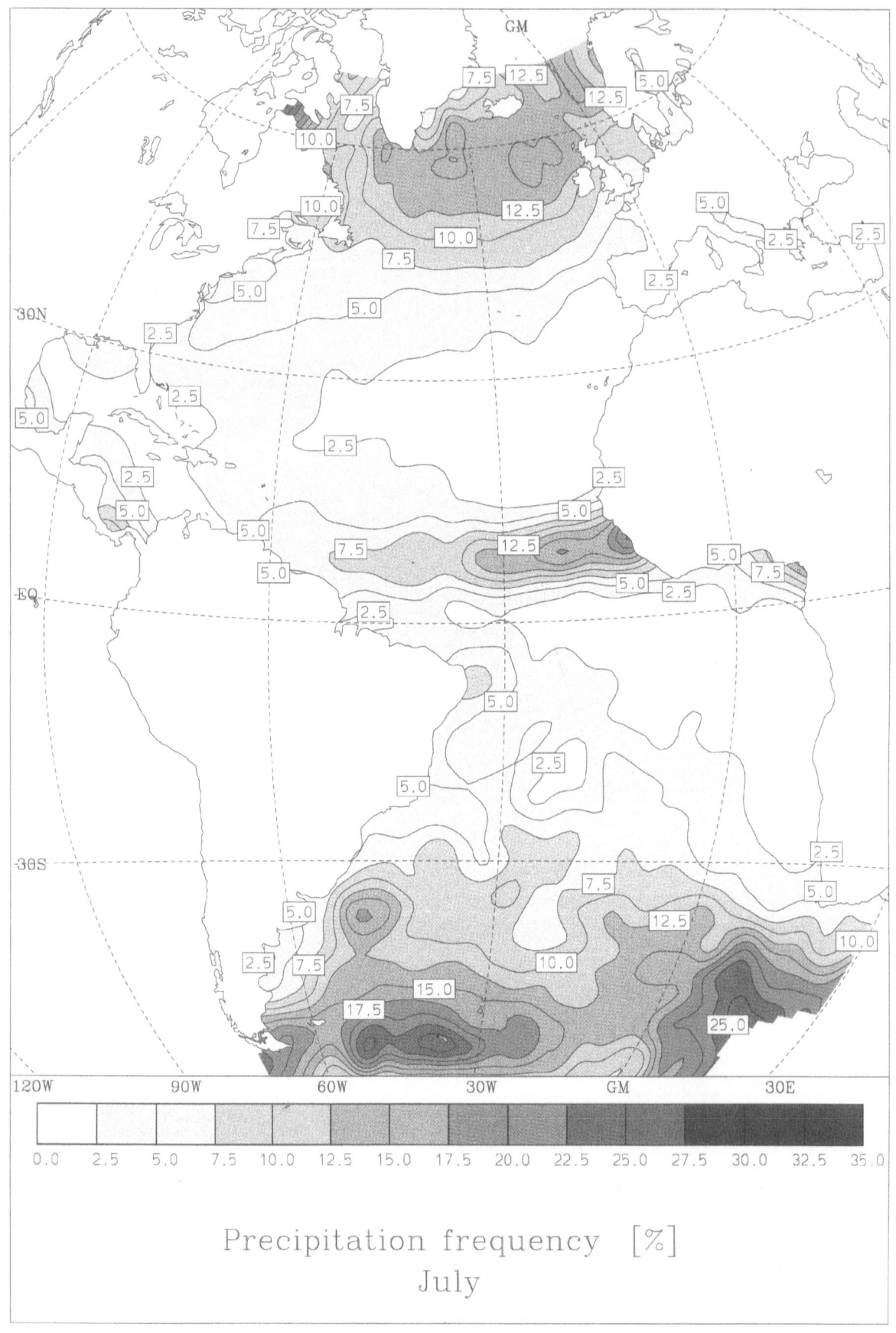

Precipitation frequency [%]
July

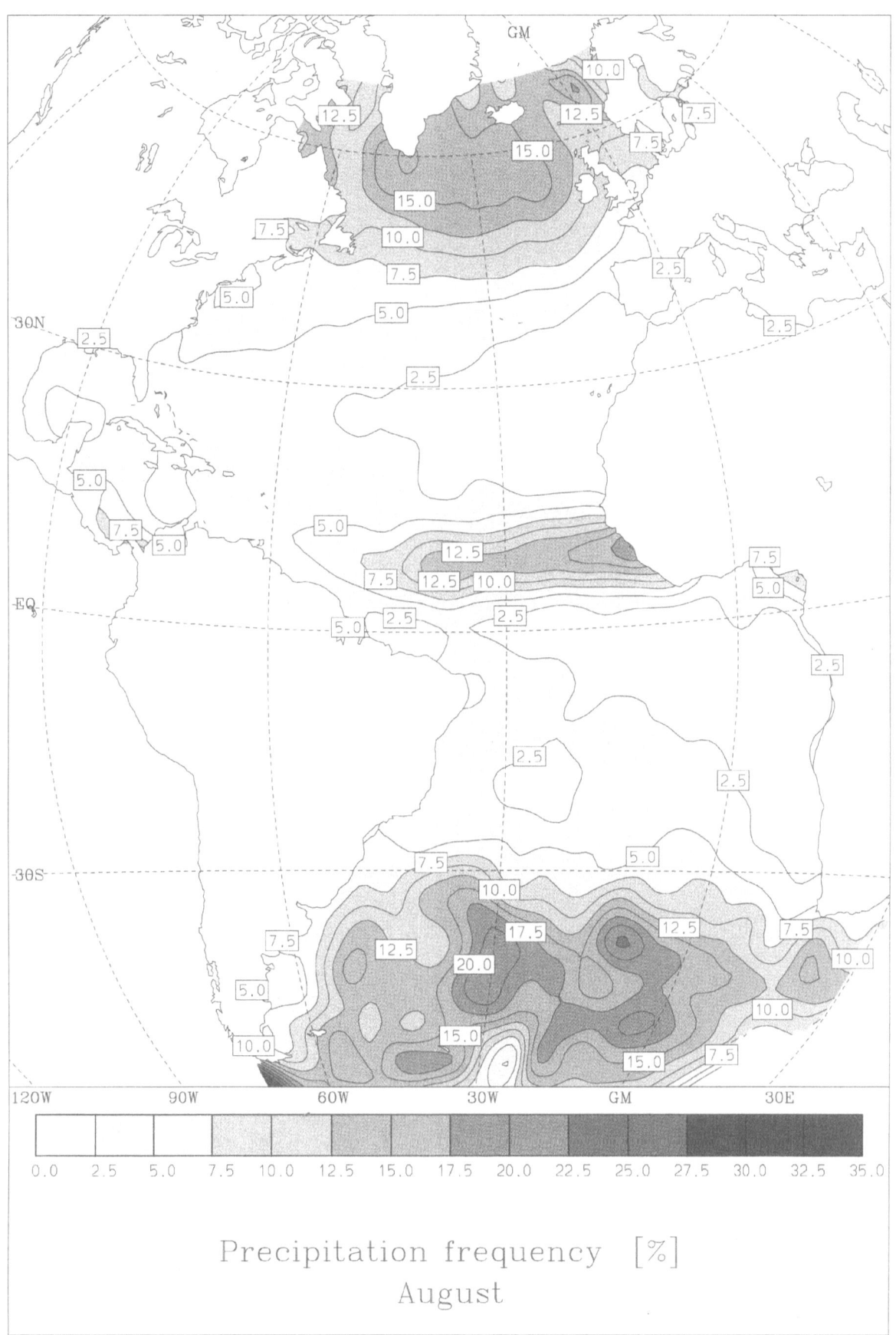

Precipitation frequency [%]
August

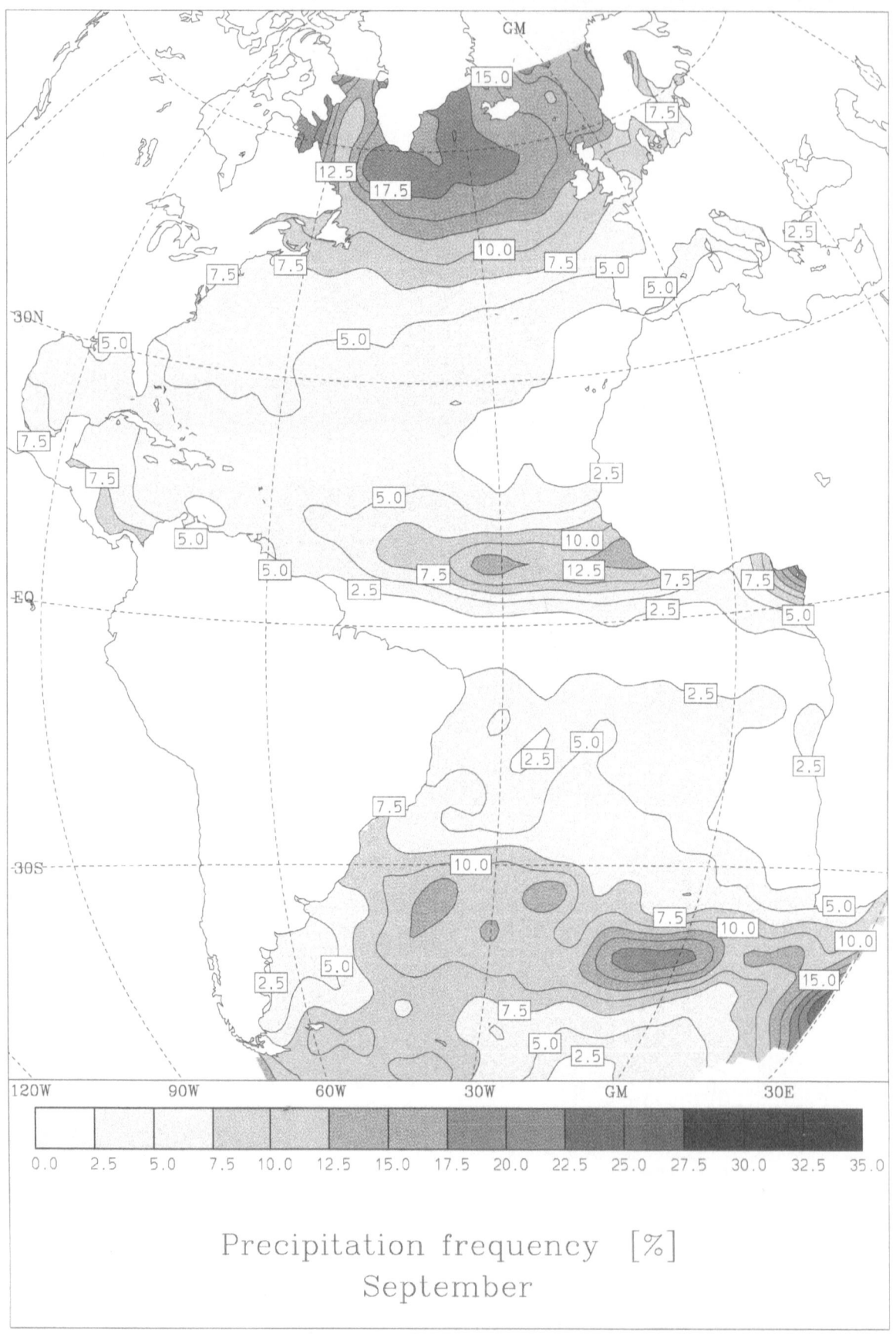

Precipitation frequency [%]
September

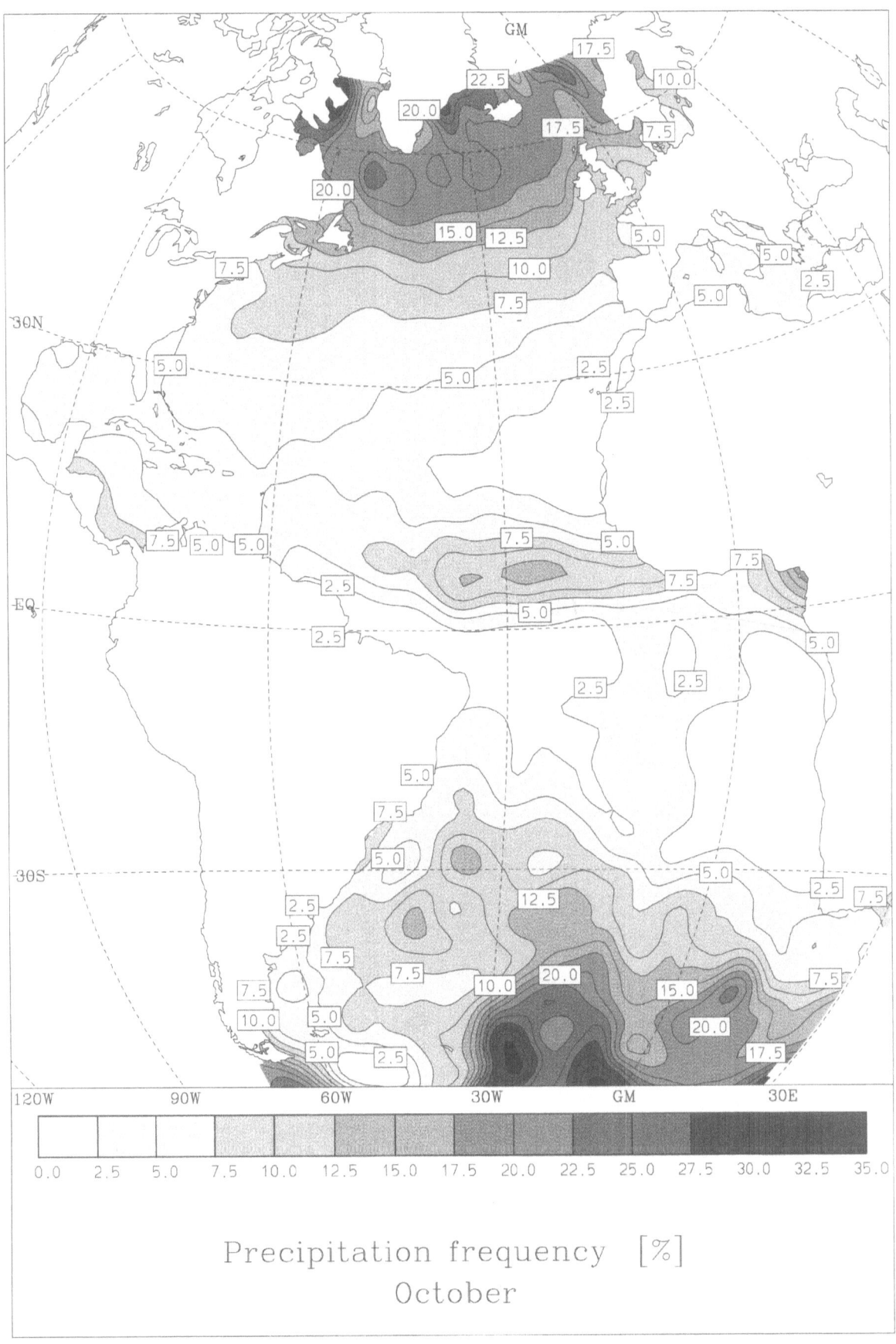

Precipitation frequency [%]
October

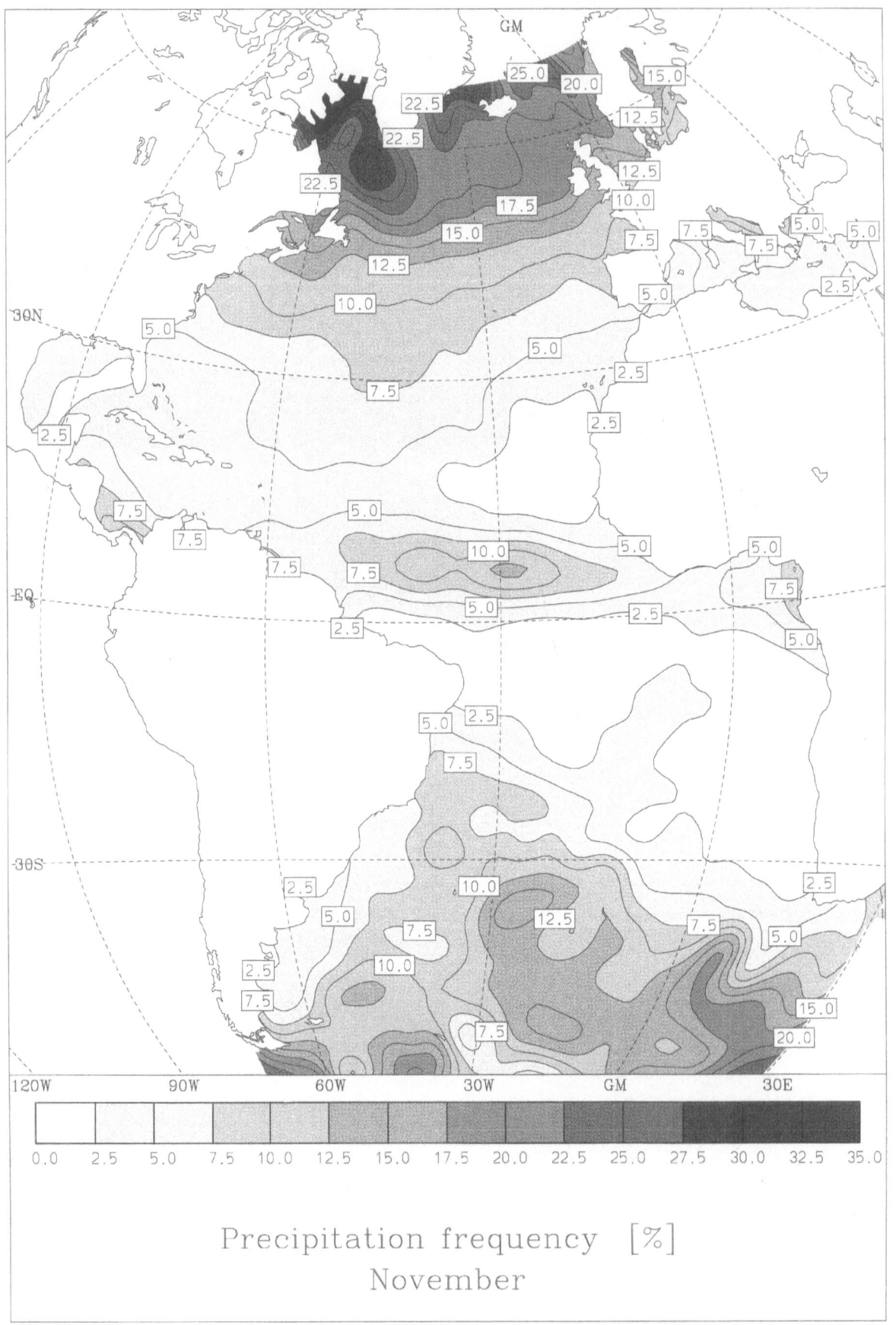

Precipitation frequency [%]
November

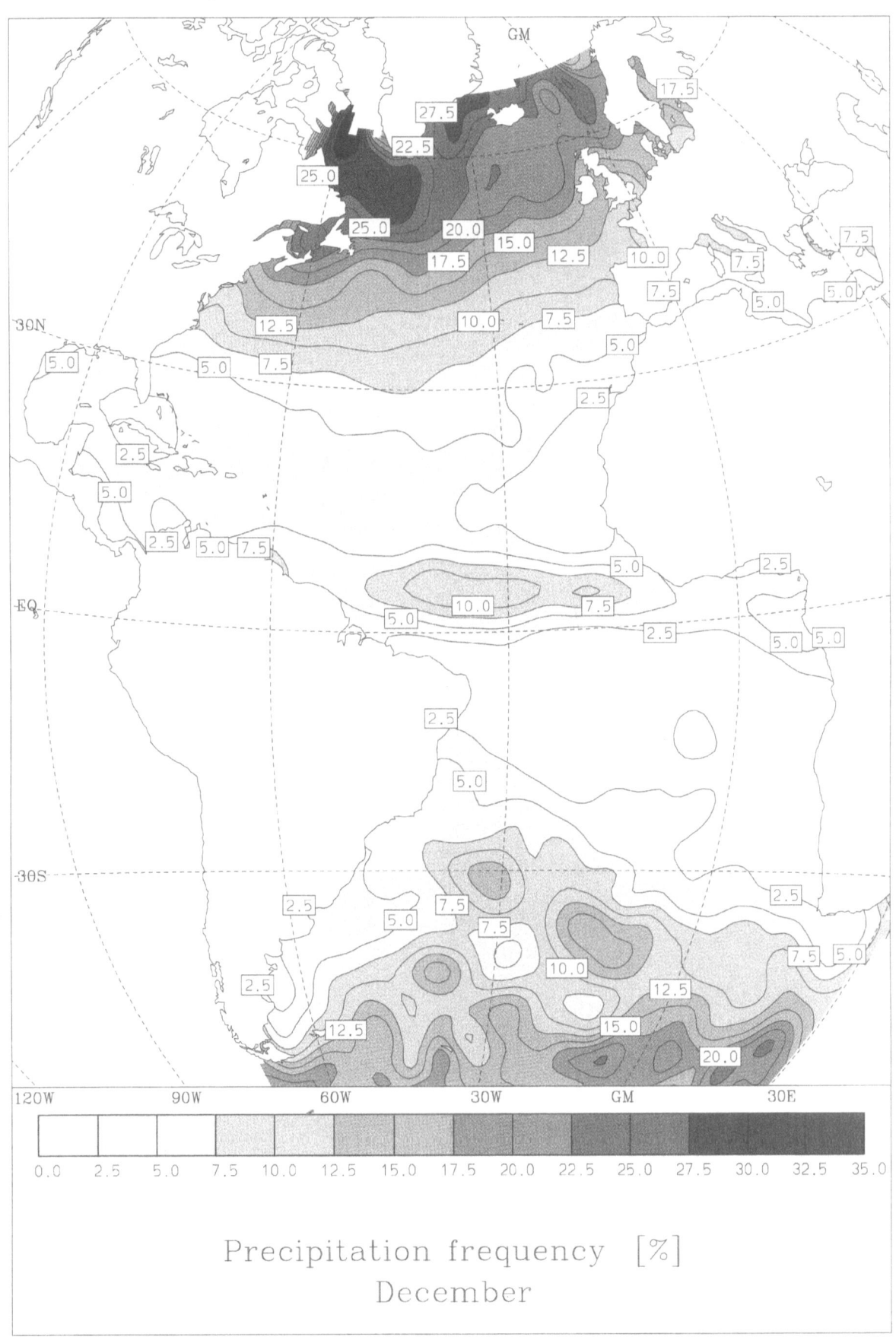

Precipitation frequency [%]
December

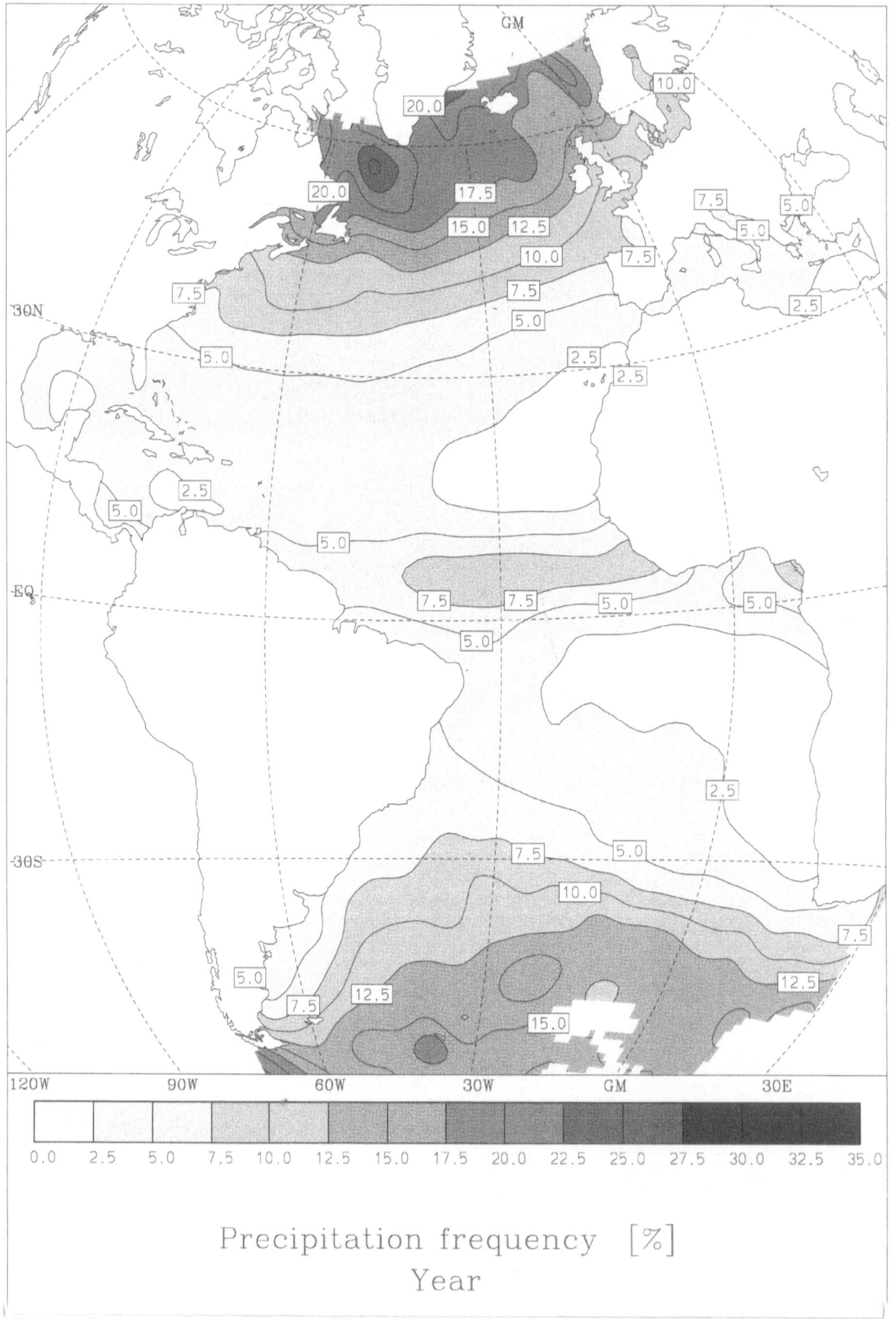

Precipitation frequency [%]
Year

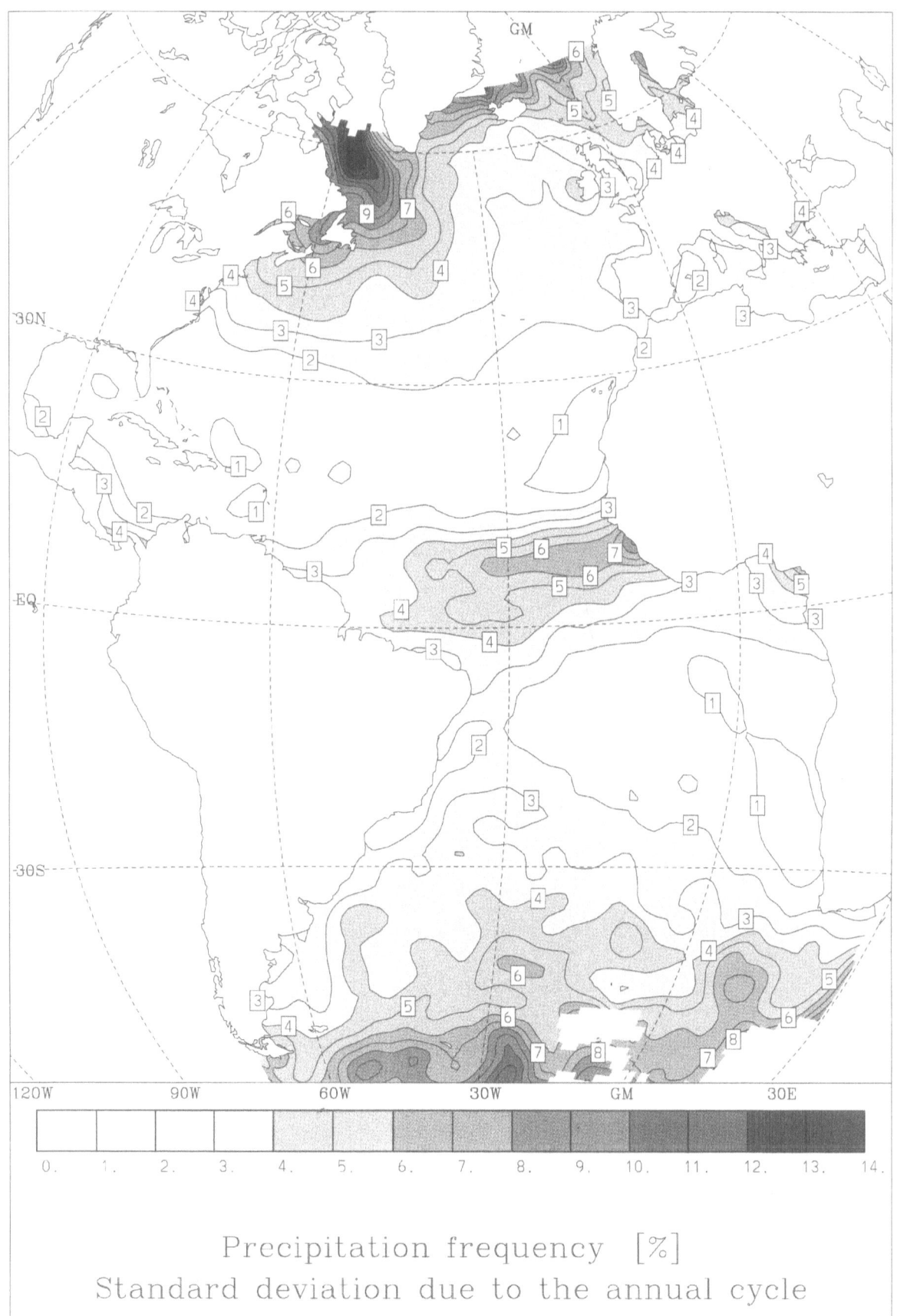

Precipitation frequency [%]
Standard deviation due to the annual cycle

Sea Level Air Pressure

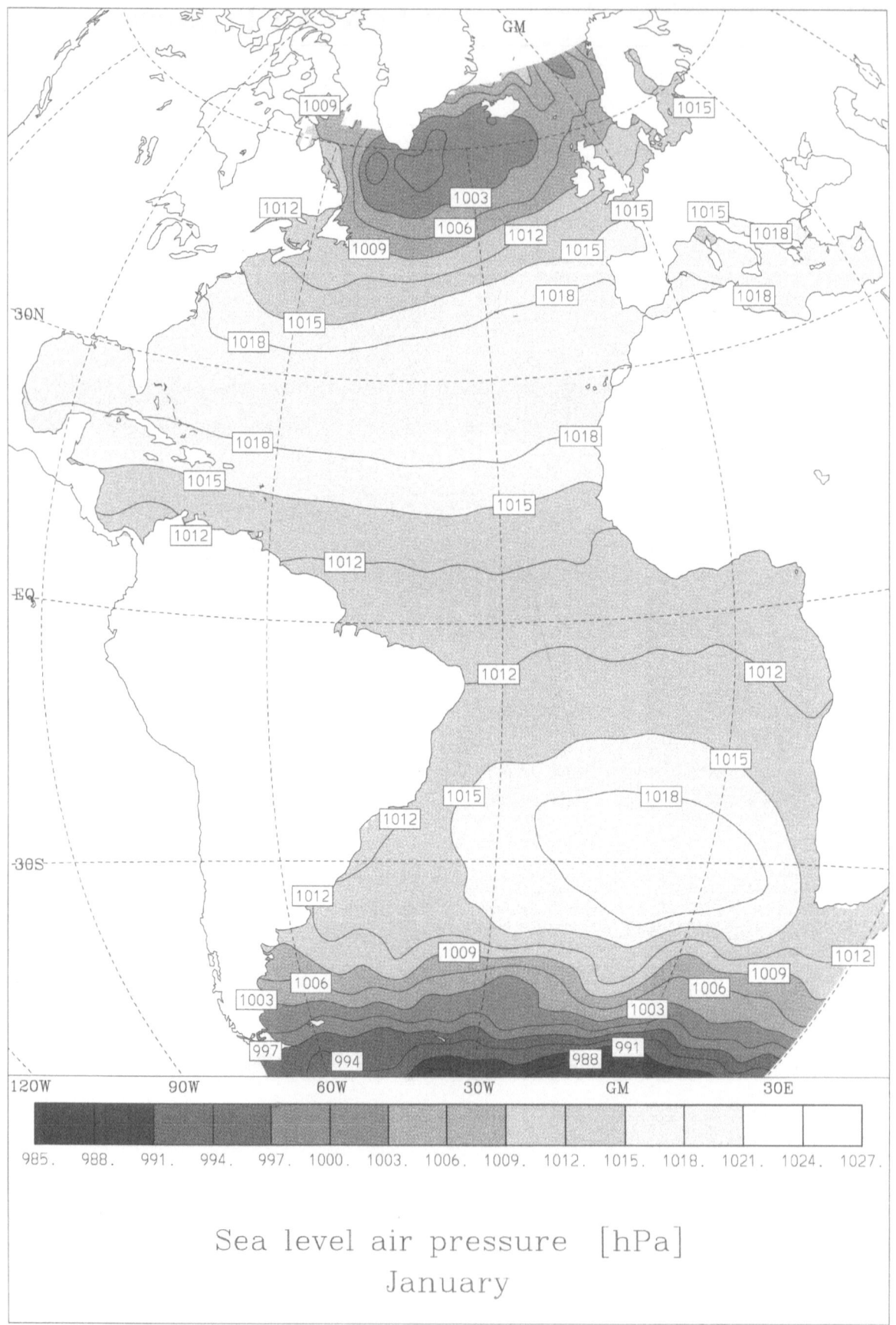

Sea level air pressure [hPa]
January

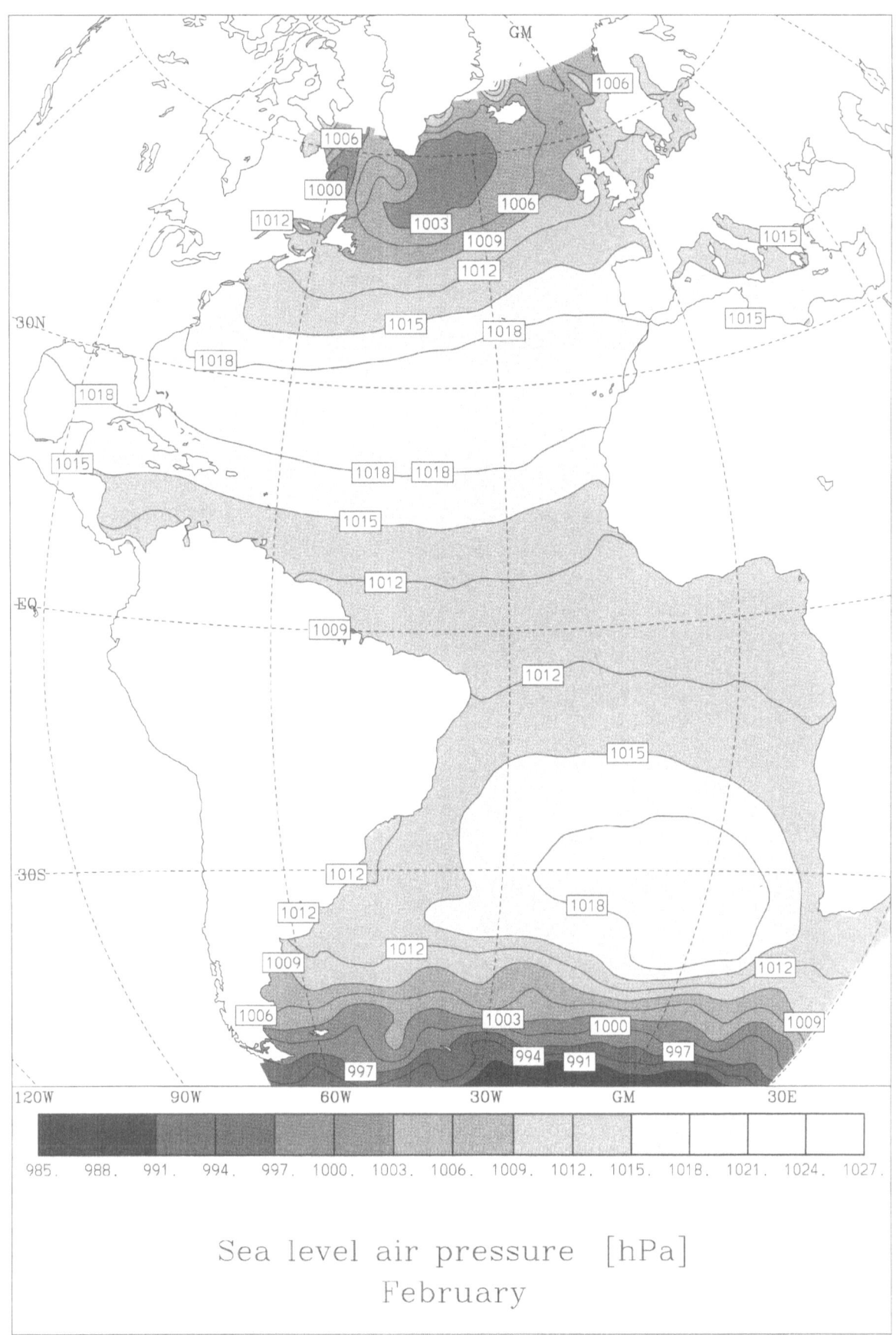

Sea level air pressure [hPa]
February

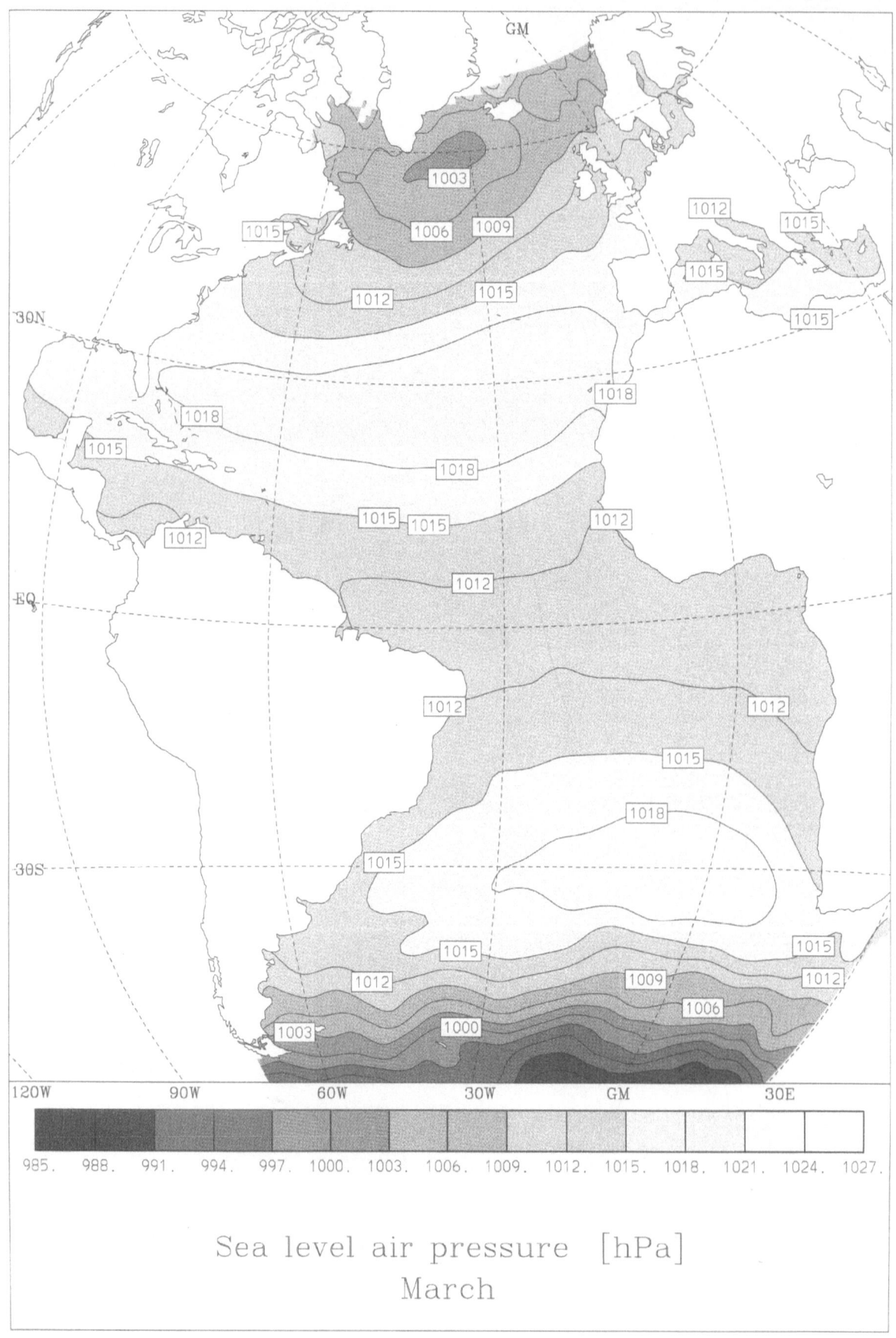

Sea level air pressure [hPa]
March

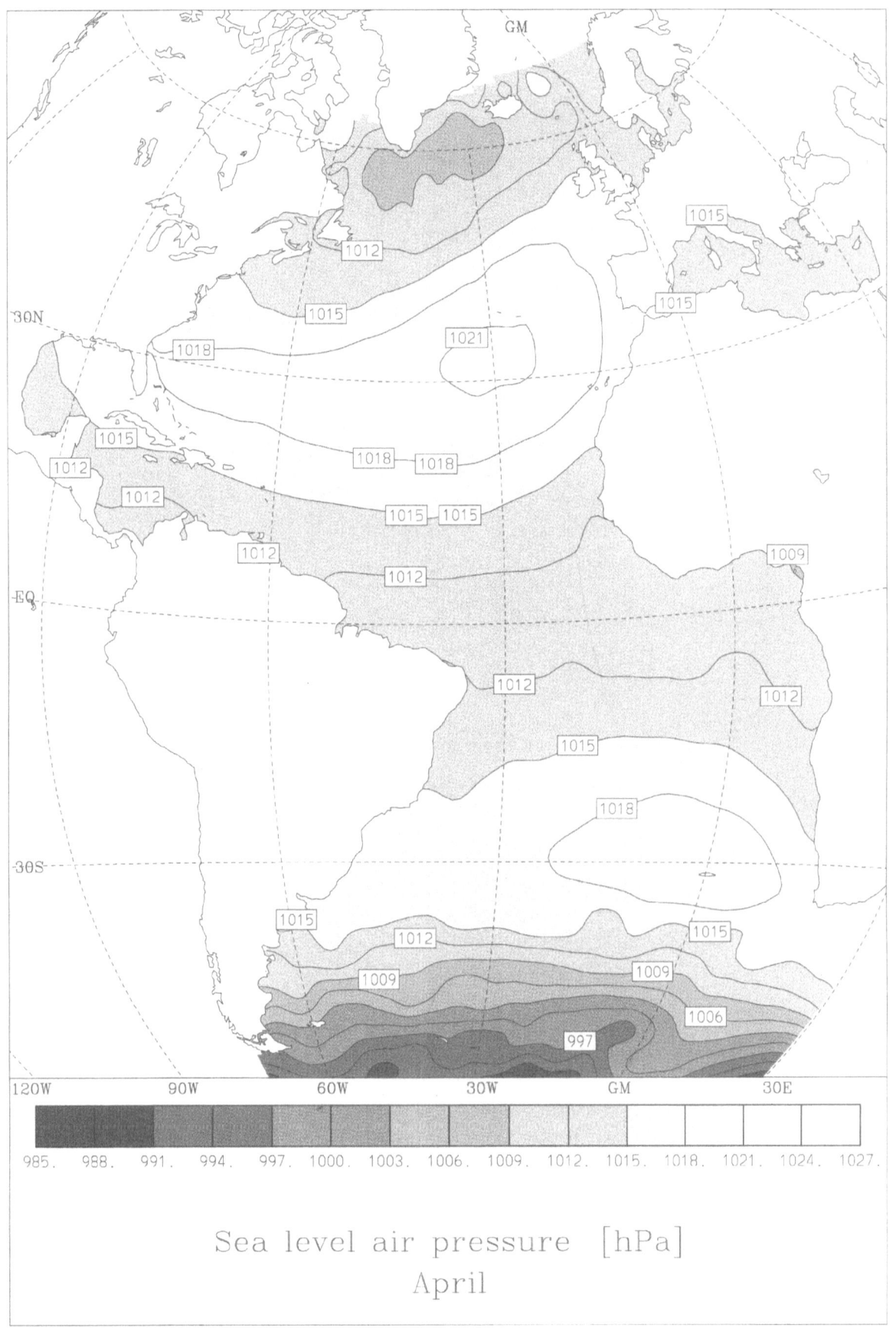

Sea level air pressure [hPa]
April

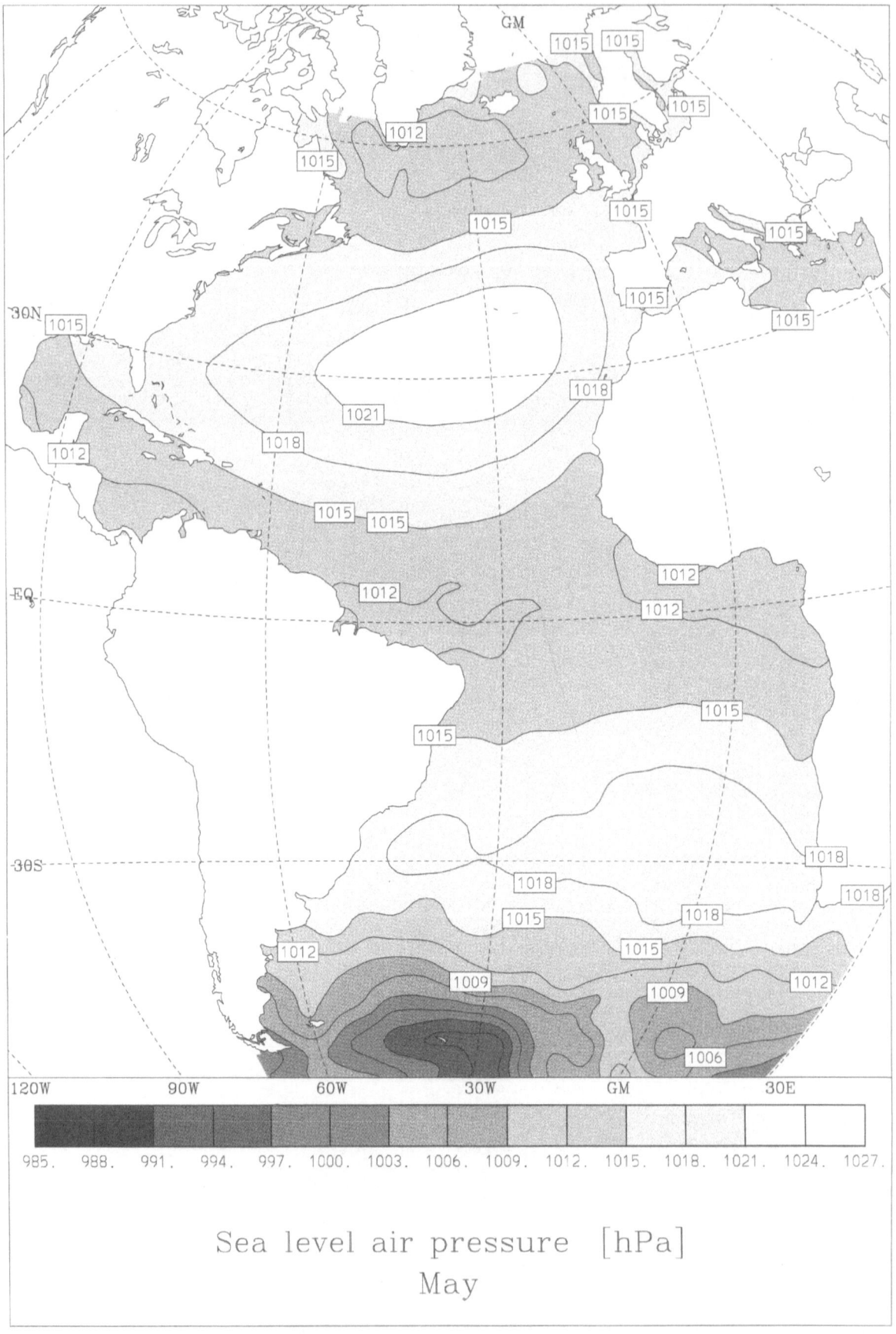

Sea level air pressure [hPa]
May

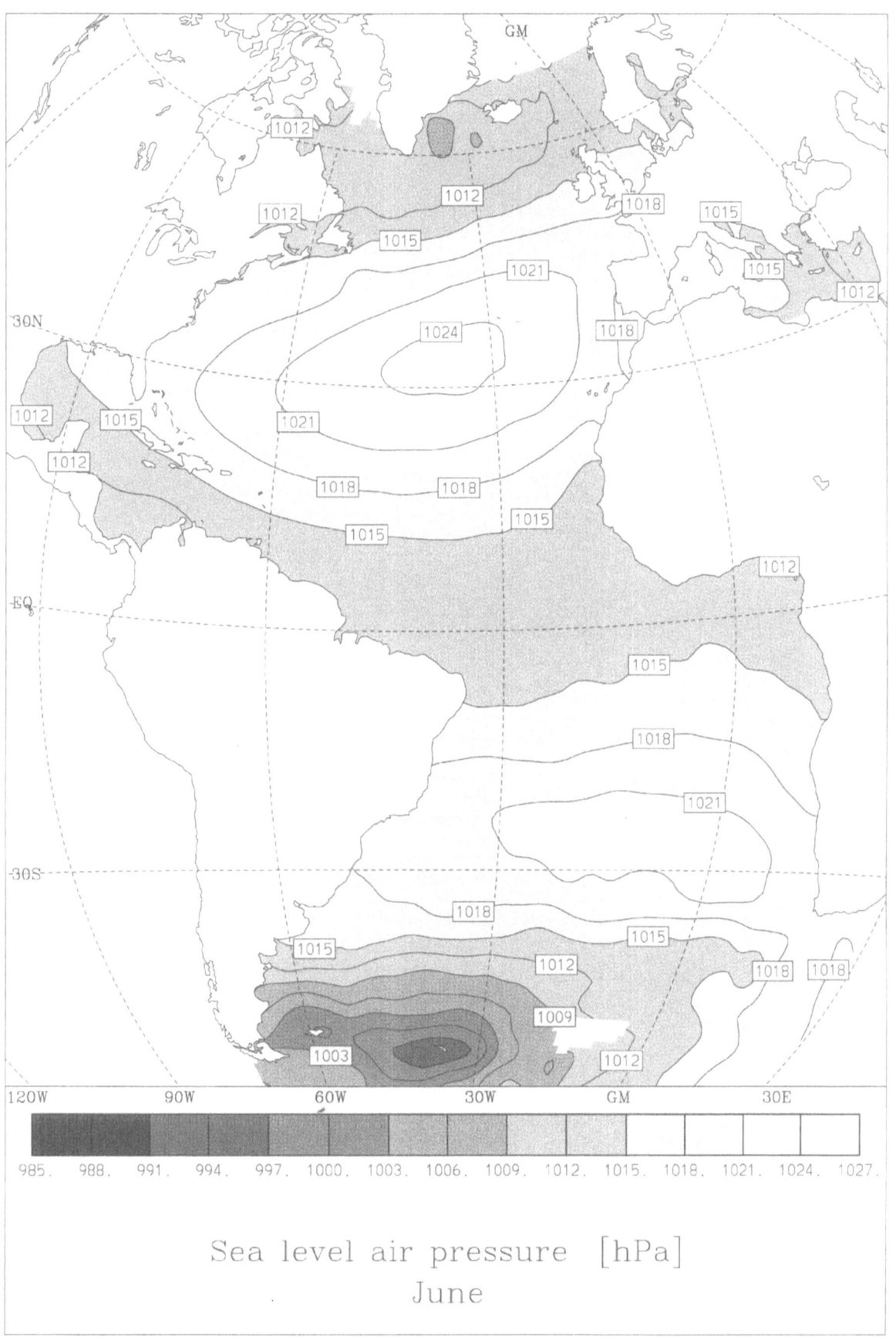

Sea level air pressure [hPa]
June

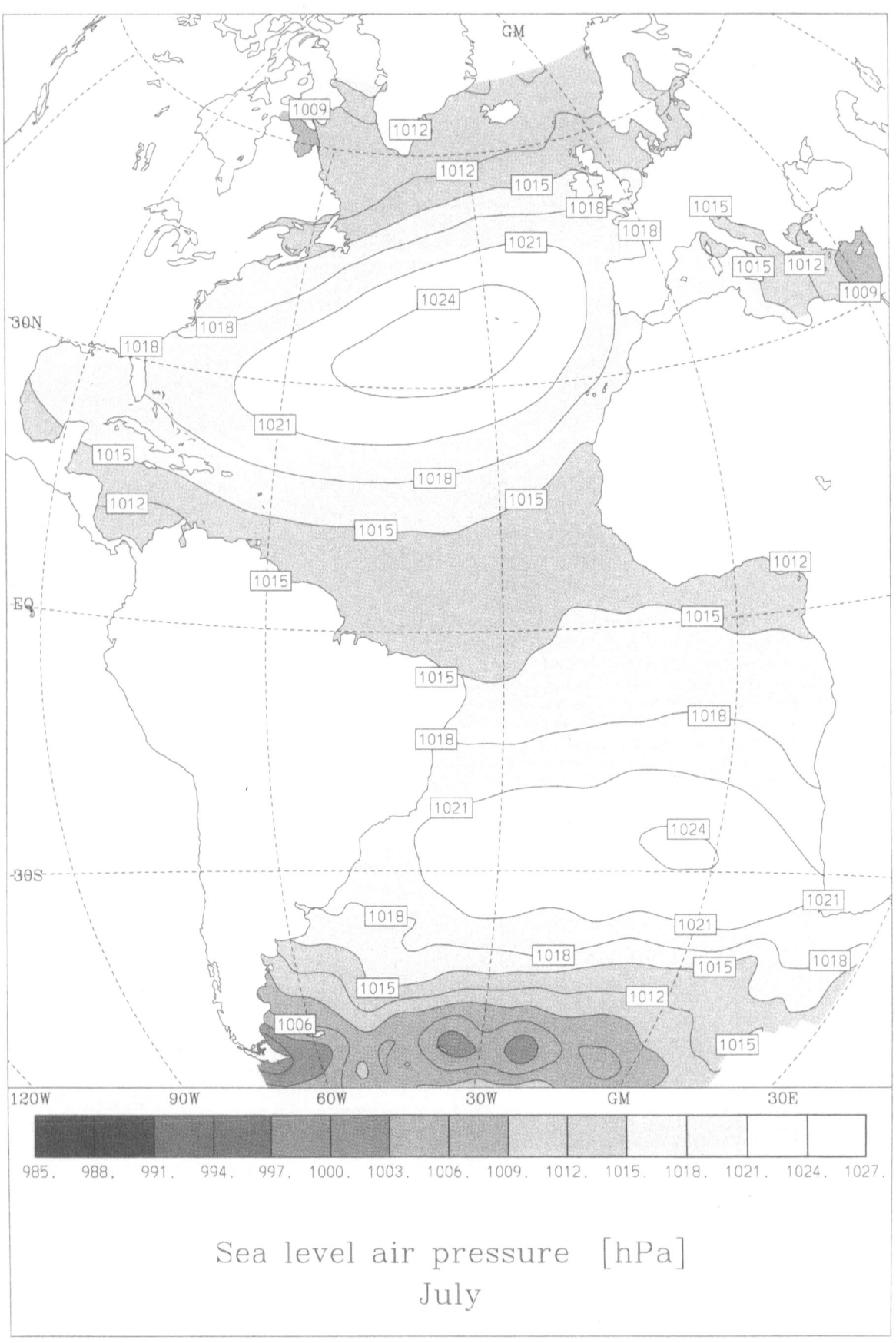

Sea level air pressure [hPa]
July

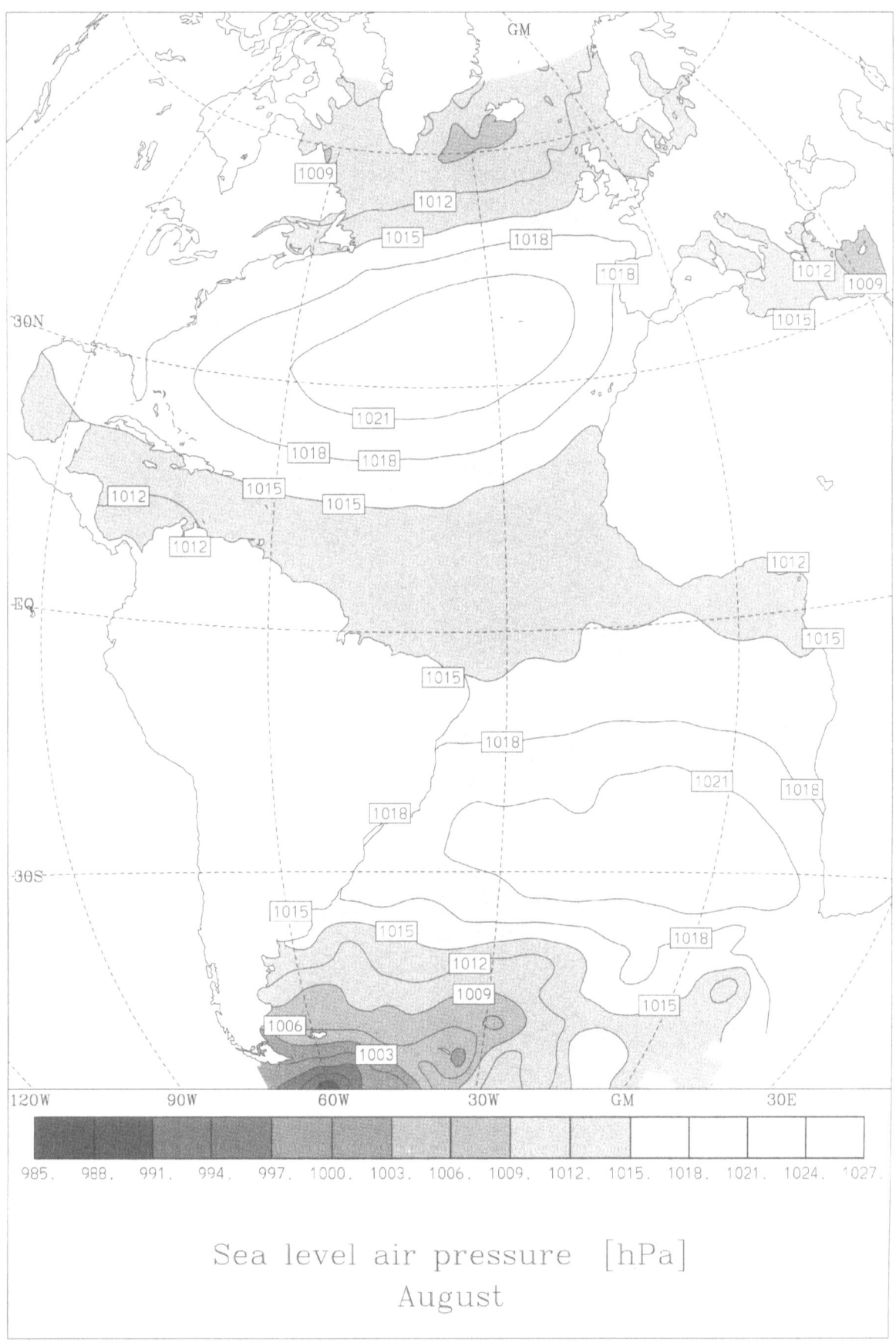

Sea level air pressure [hPa]
August

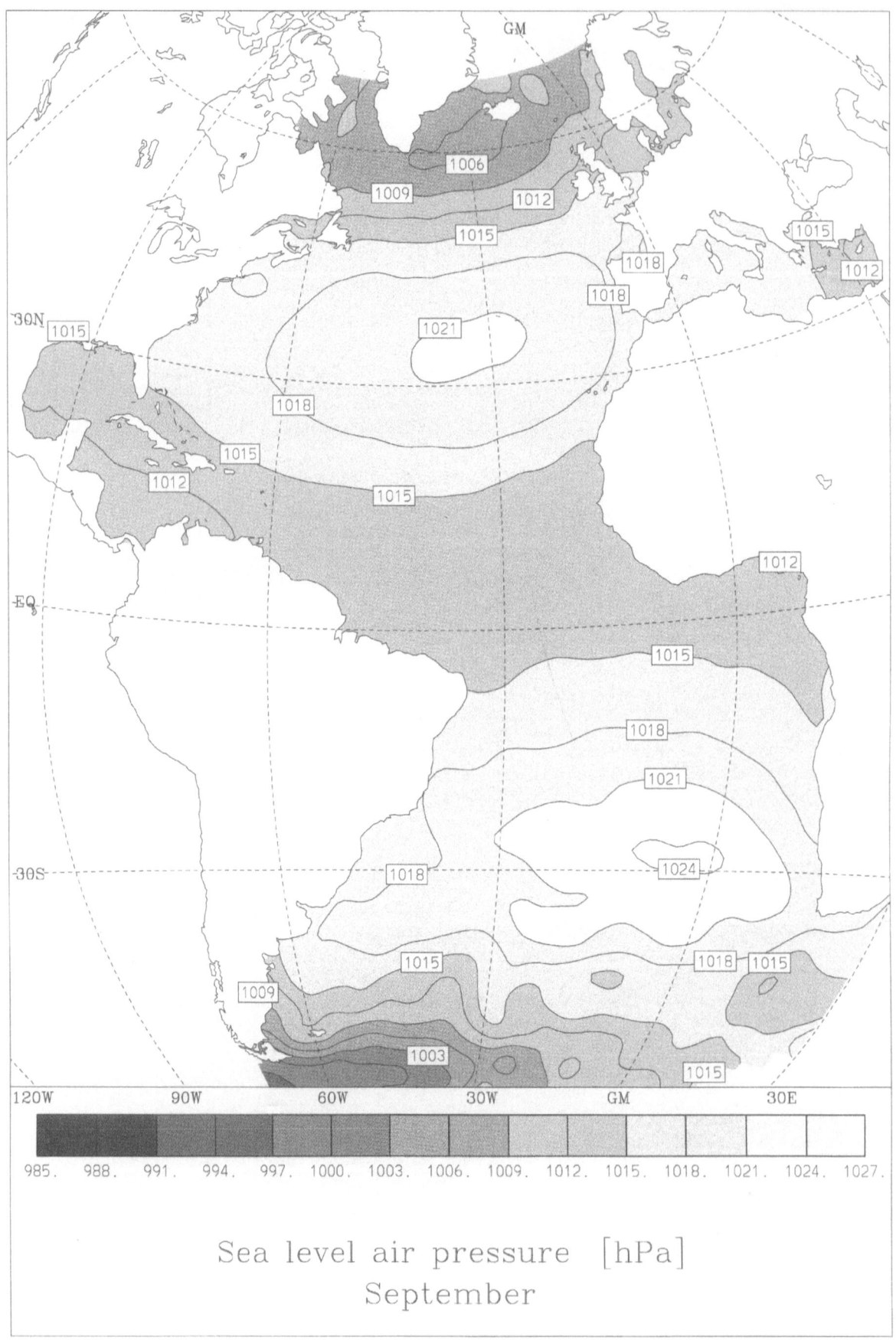

Sea level air pressure [hPa]
September

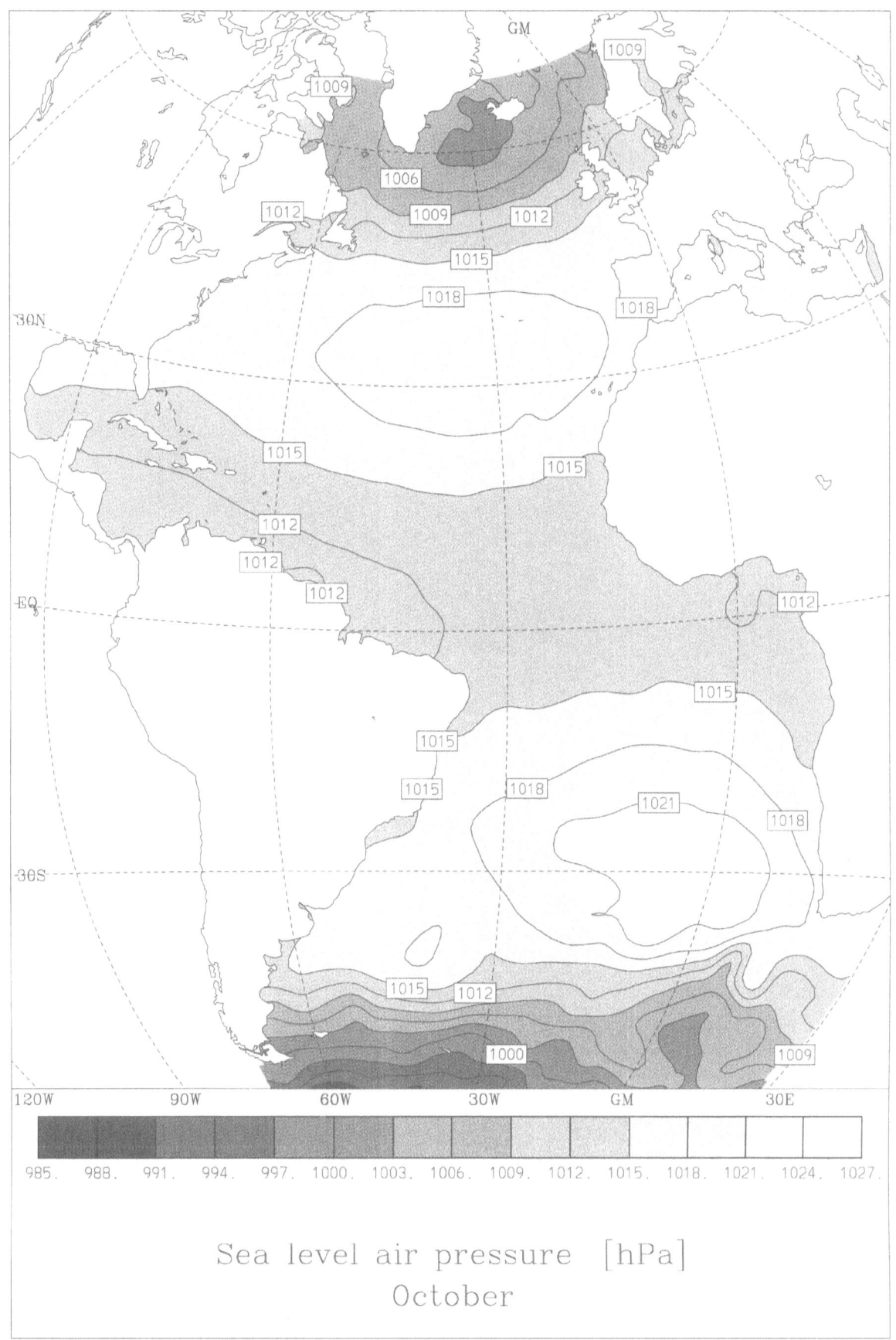

Sea level air pressure [hPa]
October

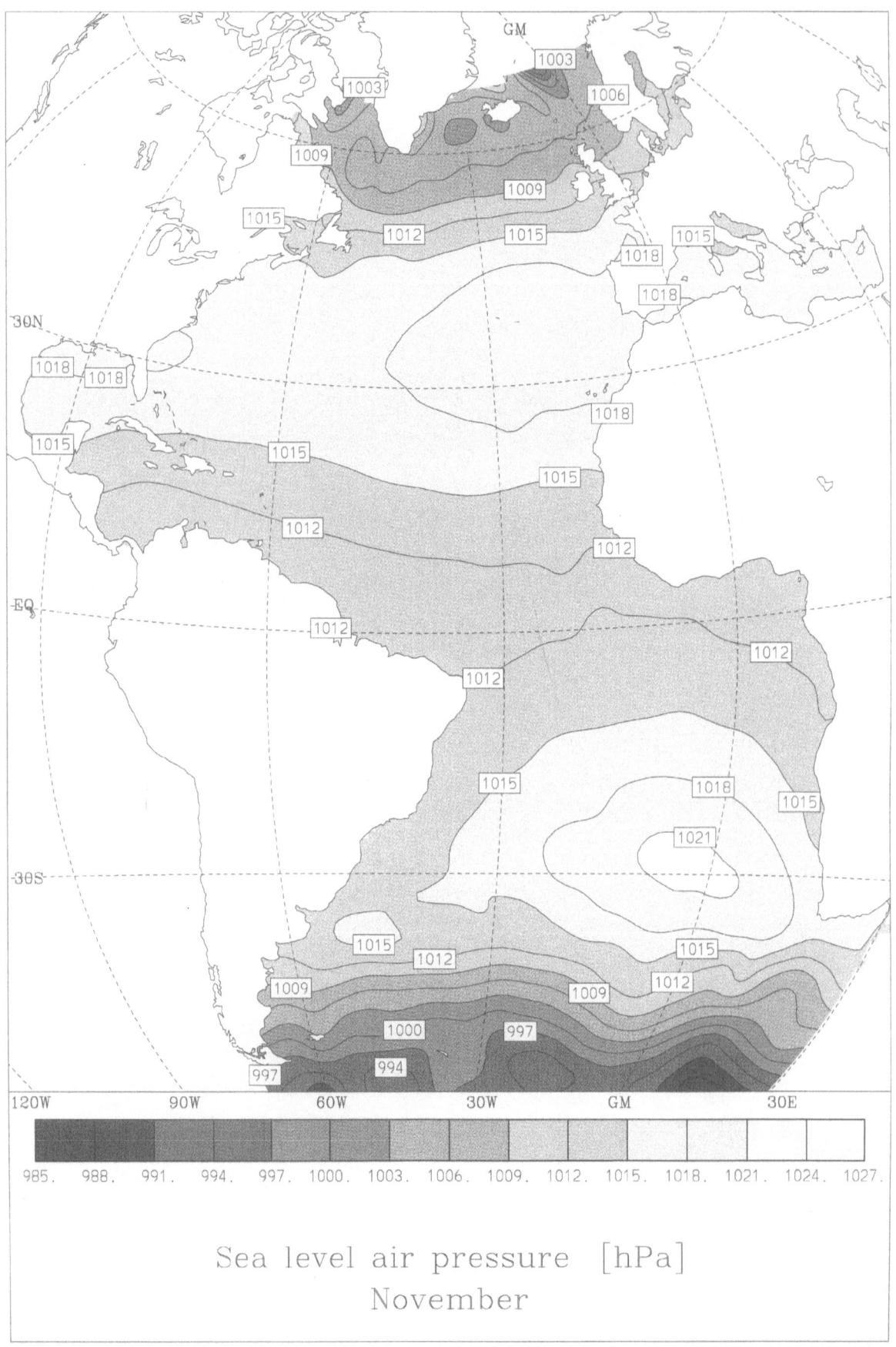

Sea level air pressure [hPa]
November

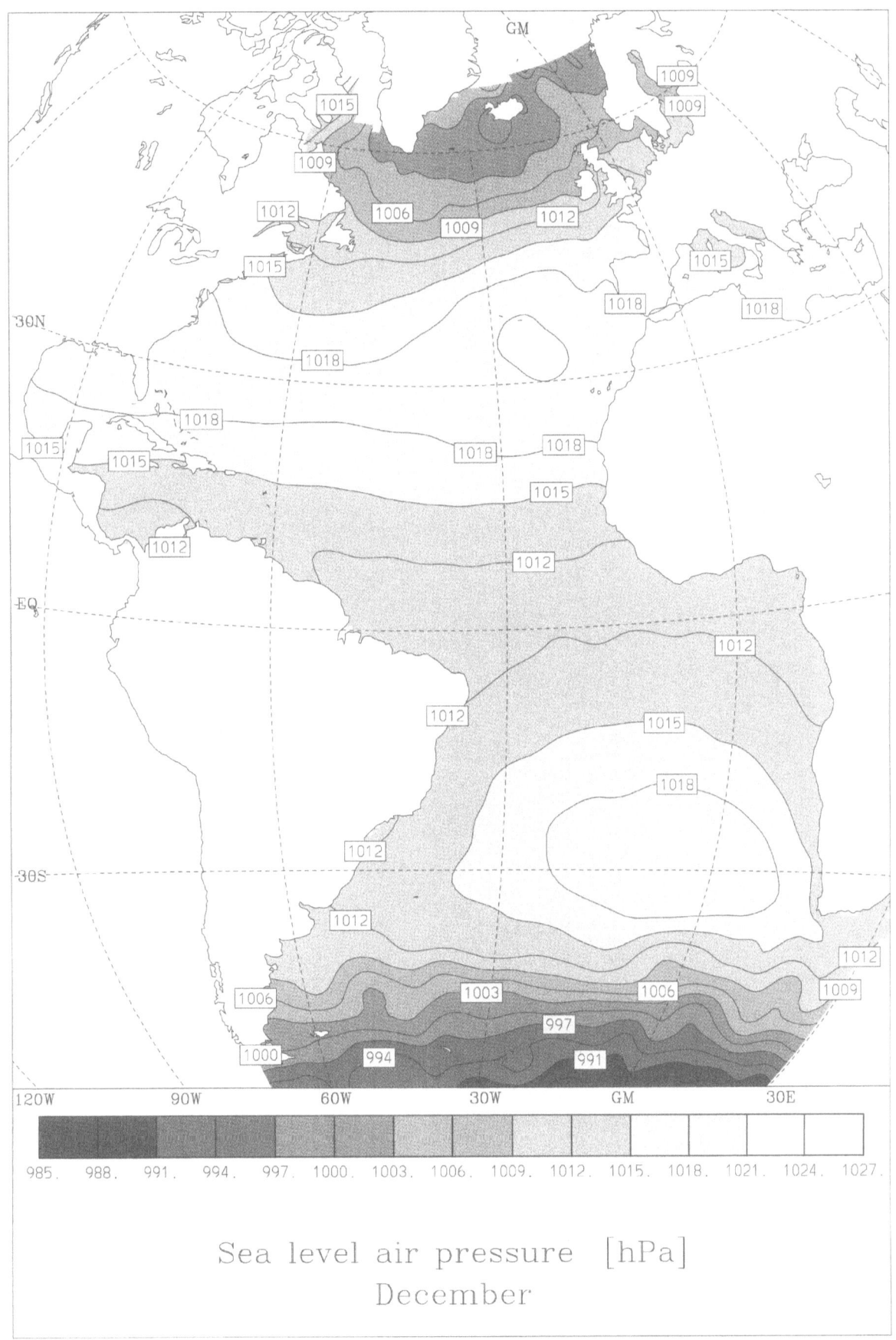

Sea level air pressure [hPa]
December

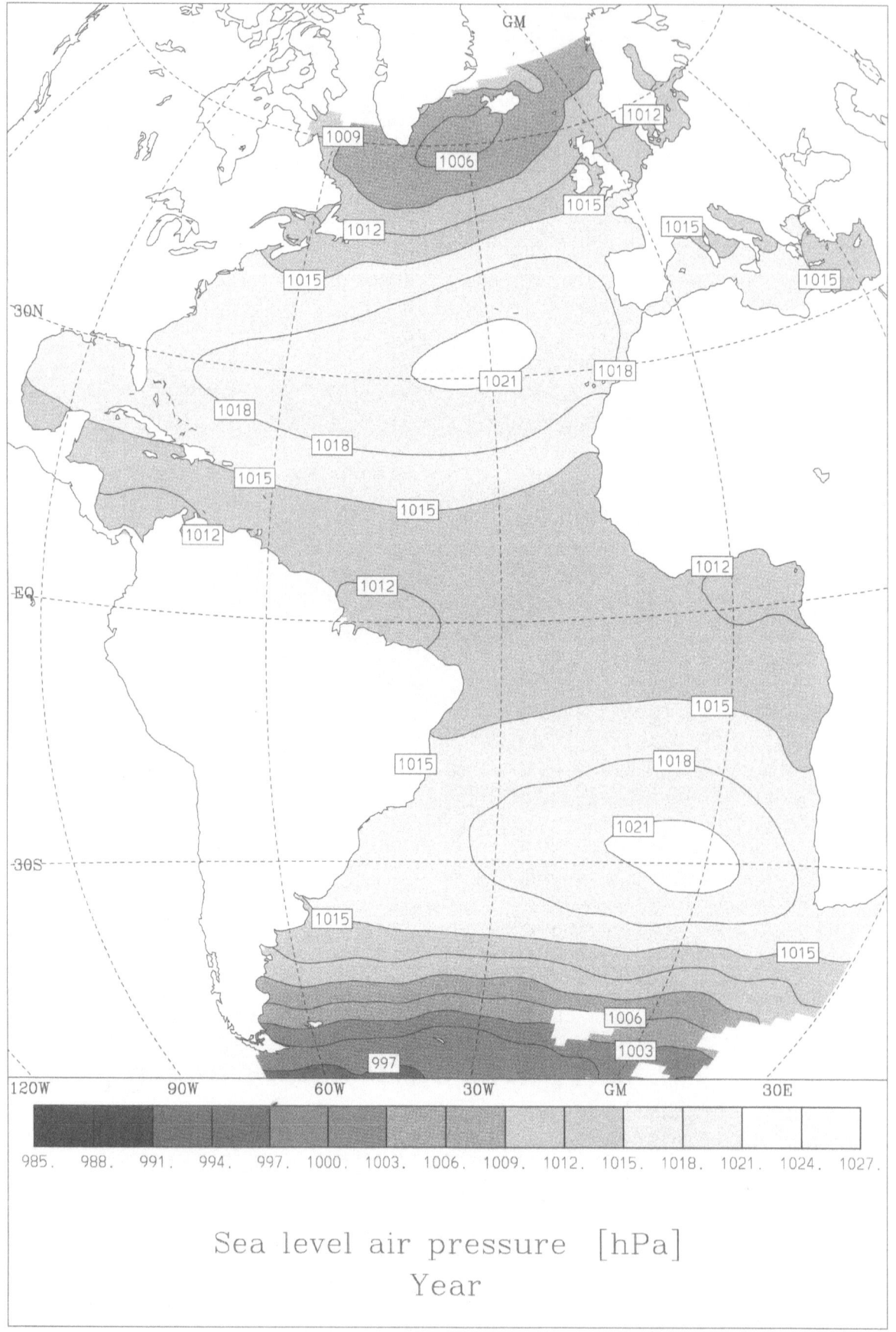

Sea level air pressure [hPa]
Year

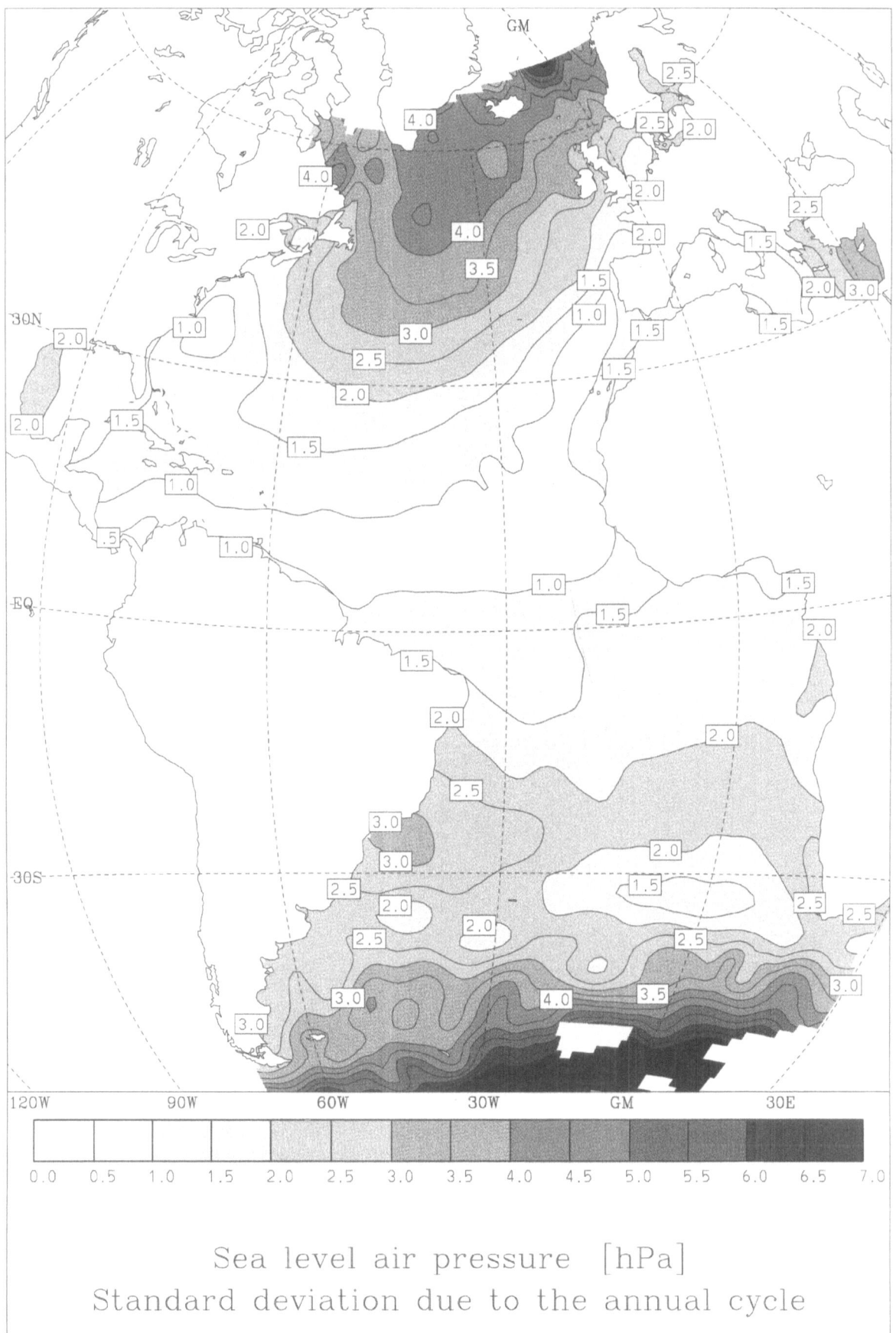

Sea level air pressure [hPa]
Standard deviation due to the annual cycle

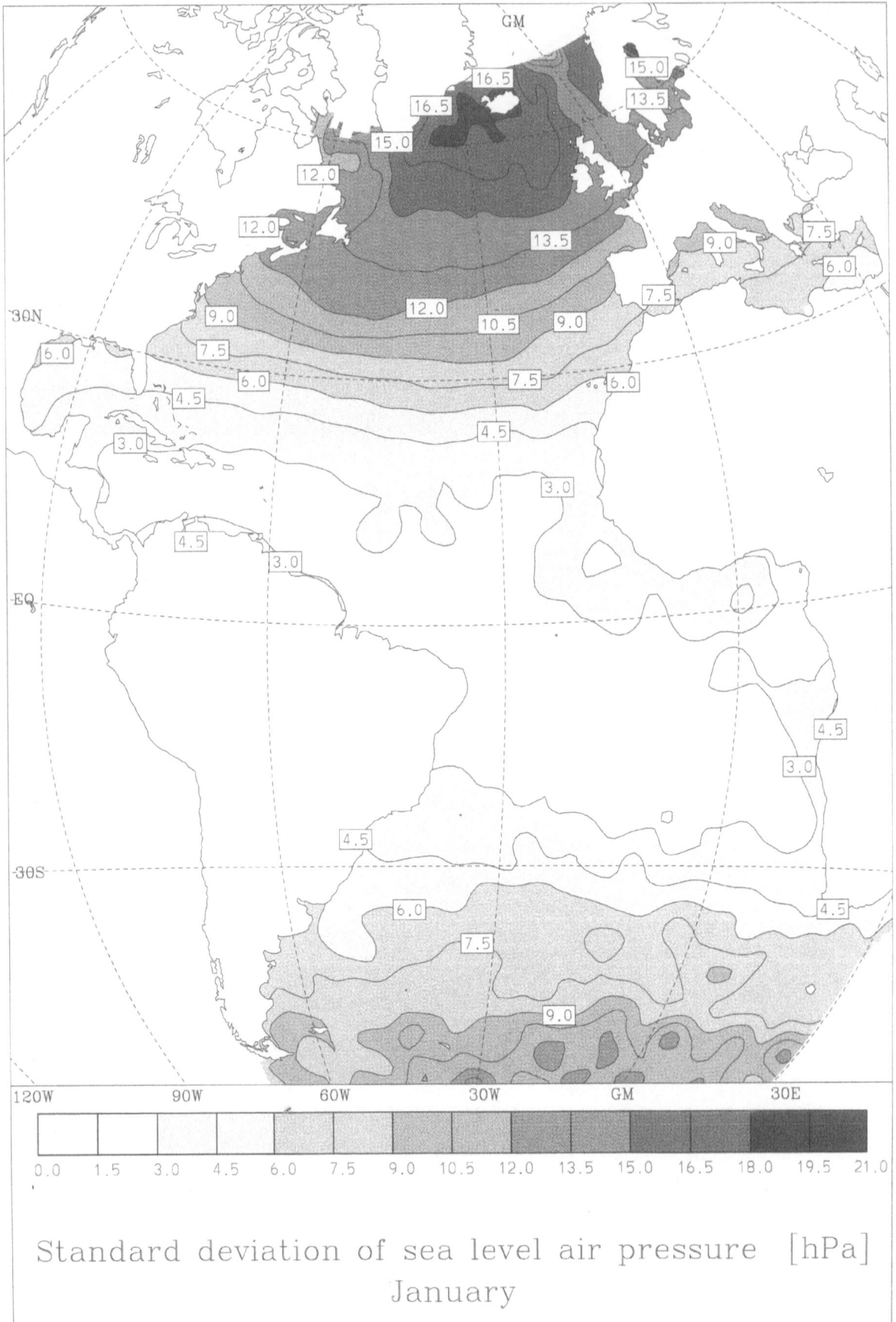

Standard deviation of sea level air pressure [hPa]
January

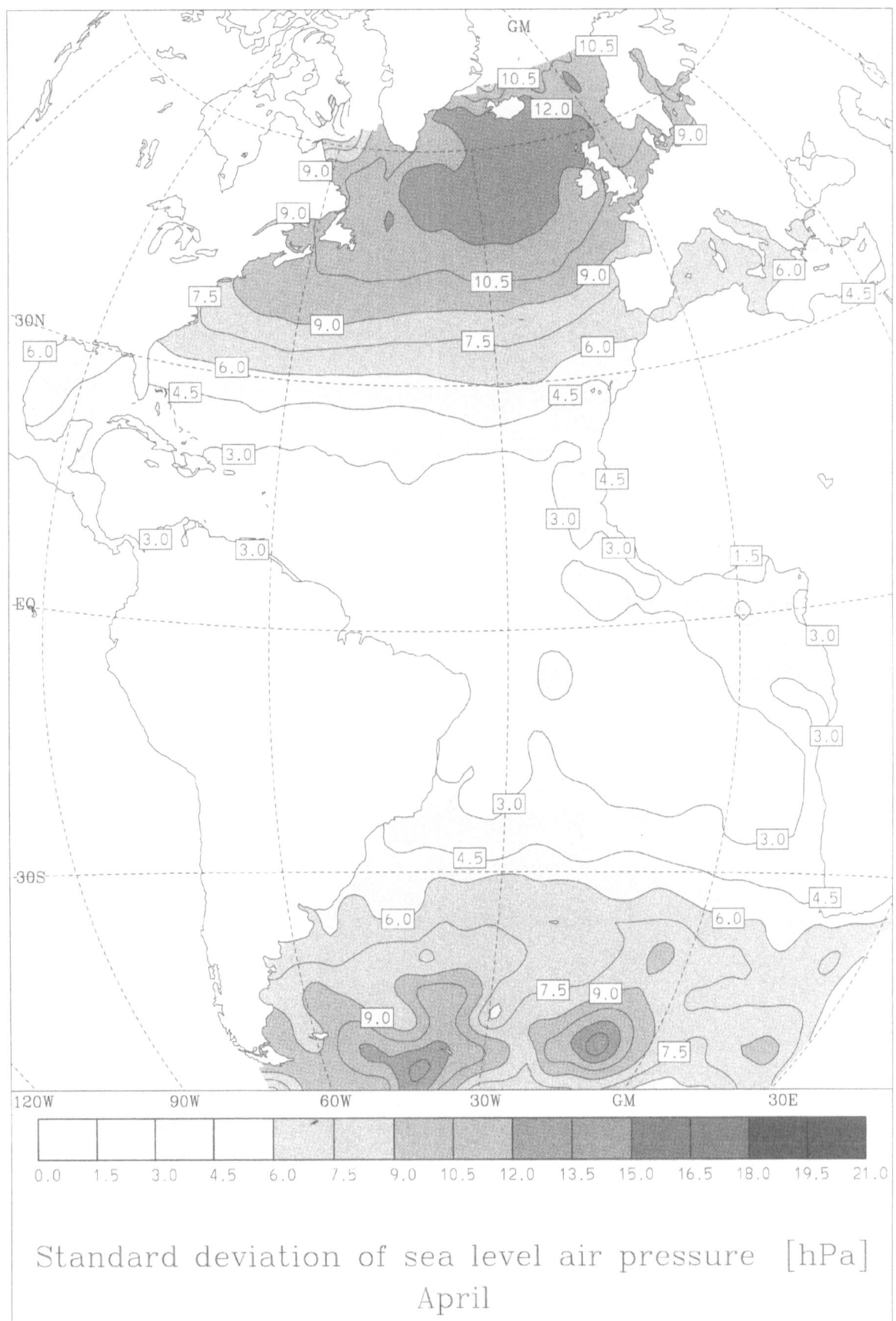

Standard deviation of sea level air pressure [hPa]
April

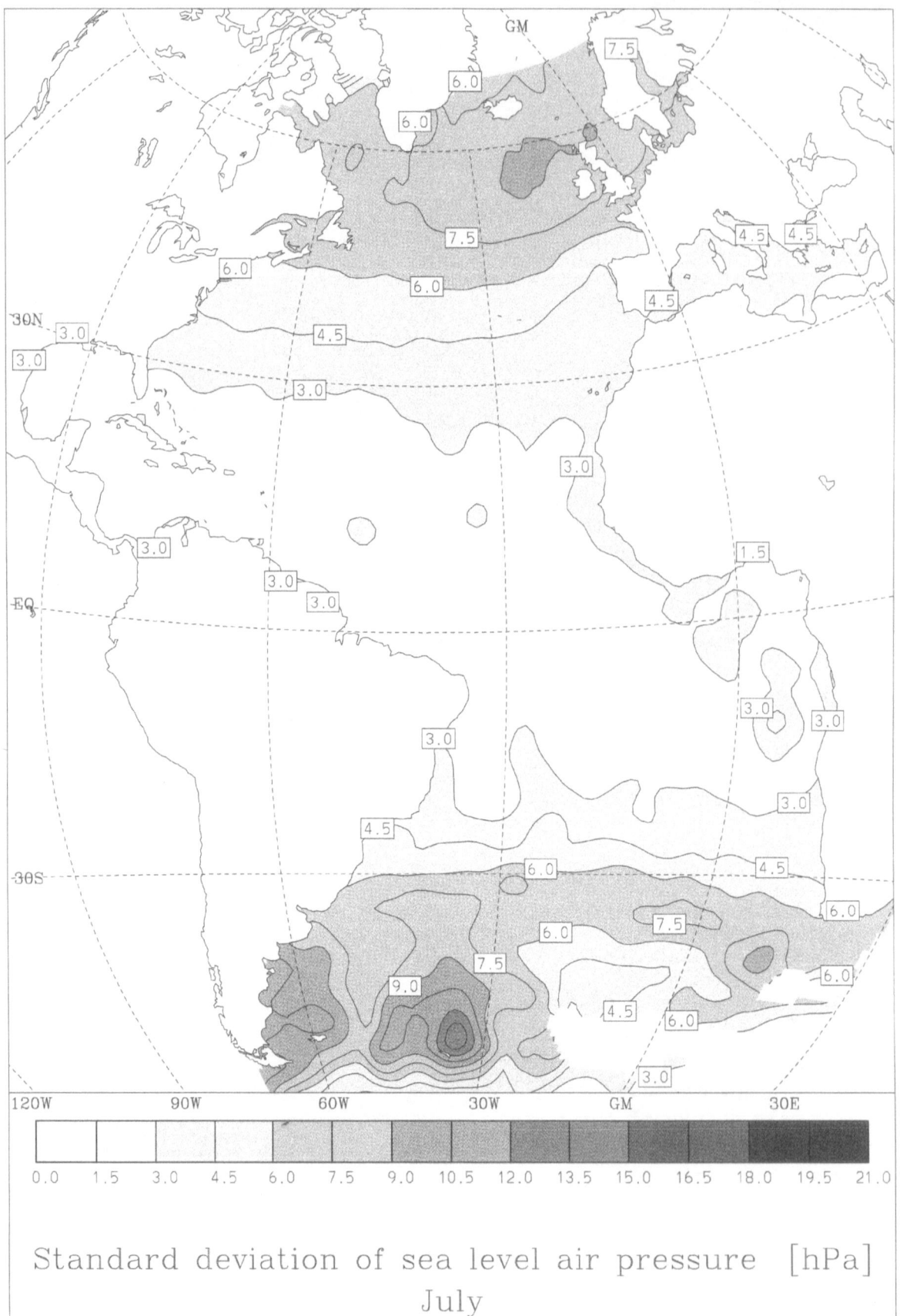

Standard deviation of sea level air pressure [hPa]
July

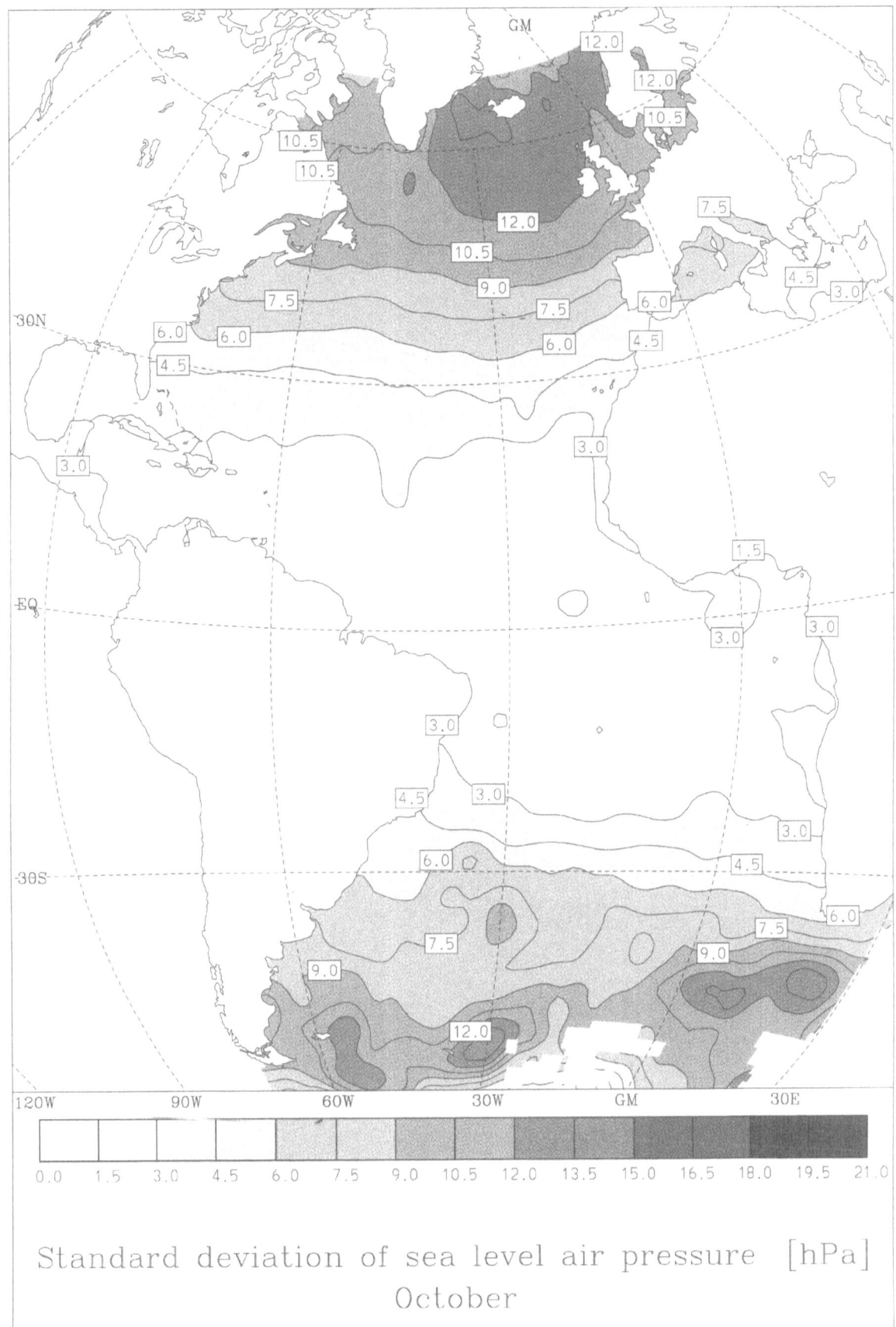

Standard deviation of sea level air pressure [hPa]
October

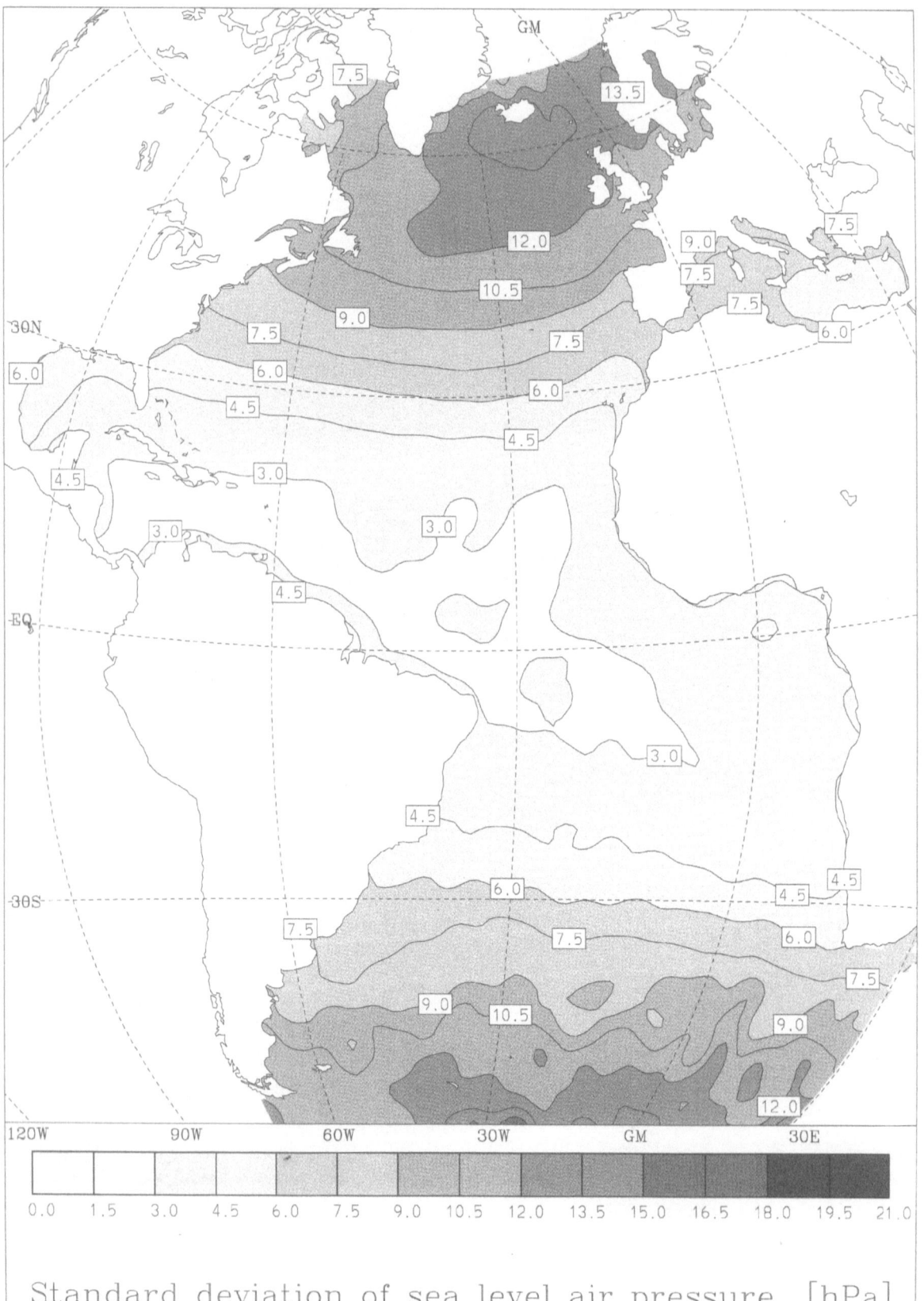

Standard deviation of sea level air pressure [hPa]
Year

Scalar Wind Speed
plus Resultant Wind Vector

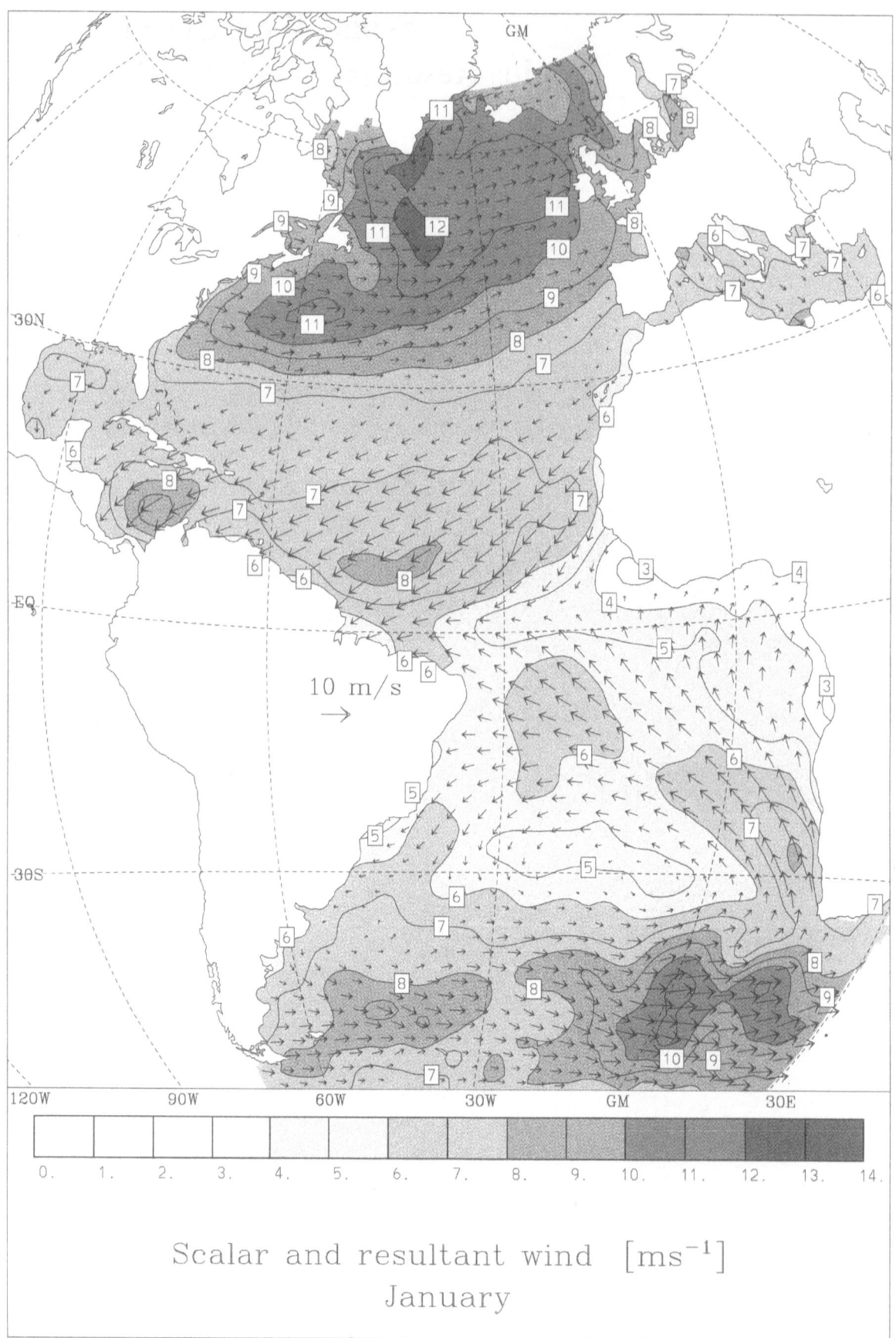

Scalar and resultant wind [ms^{-1}]
January

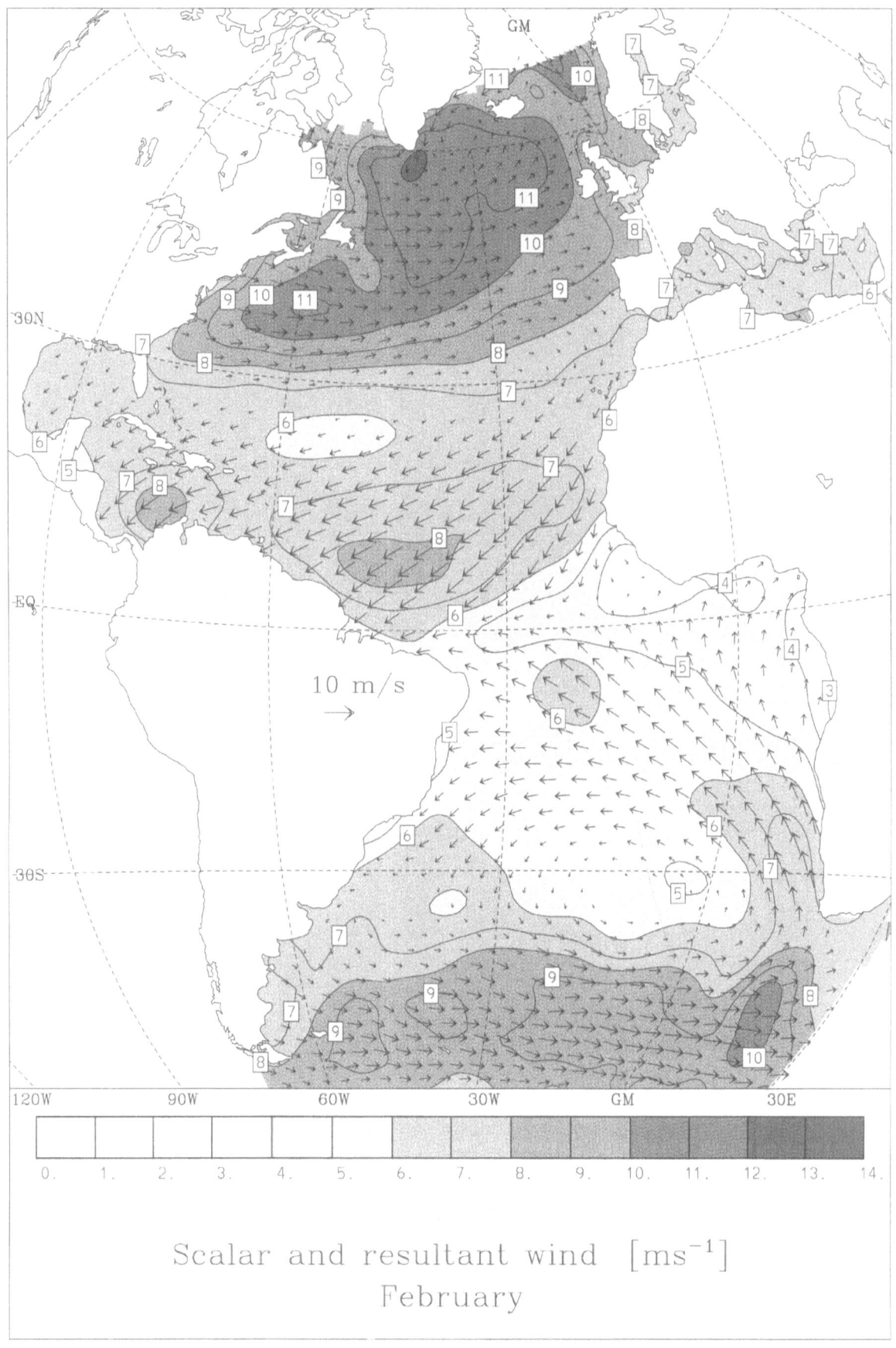

Scalar and resultant wind [ms⁻¹]
February

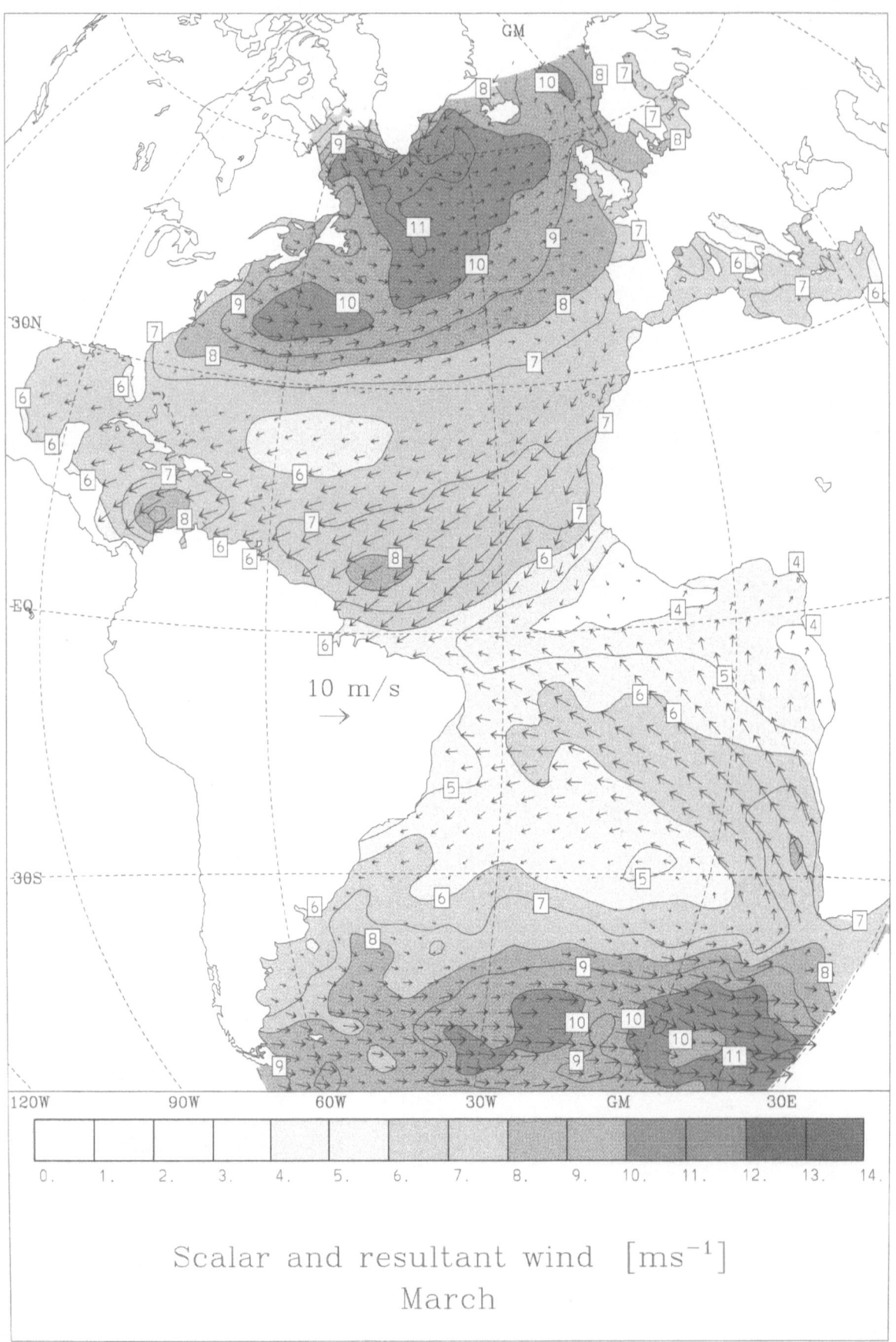

Scalar and resultant wind [ms⁻¹]
March

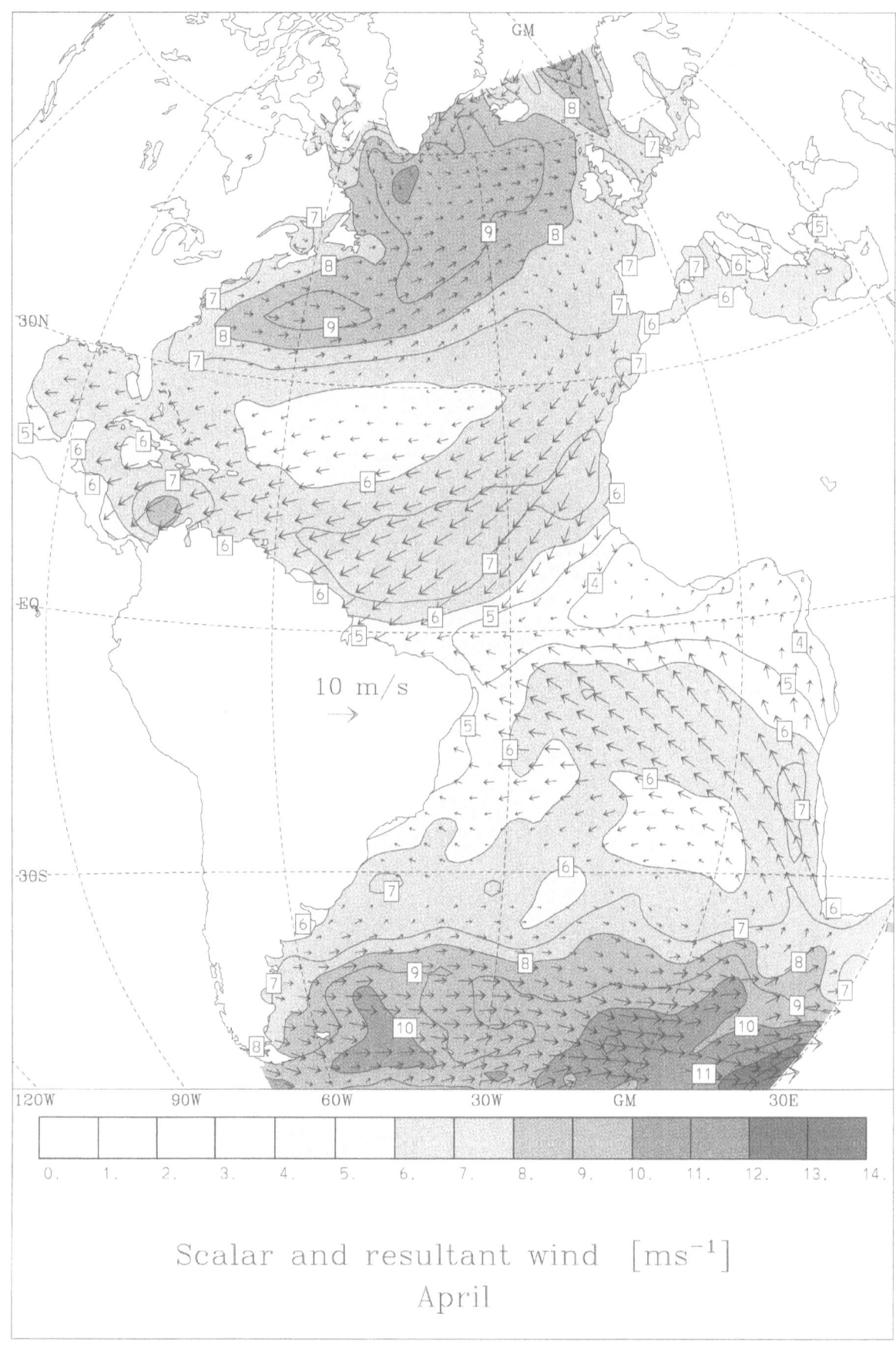

Scalar and resultant wind [ms^{-1}]
April

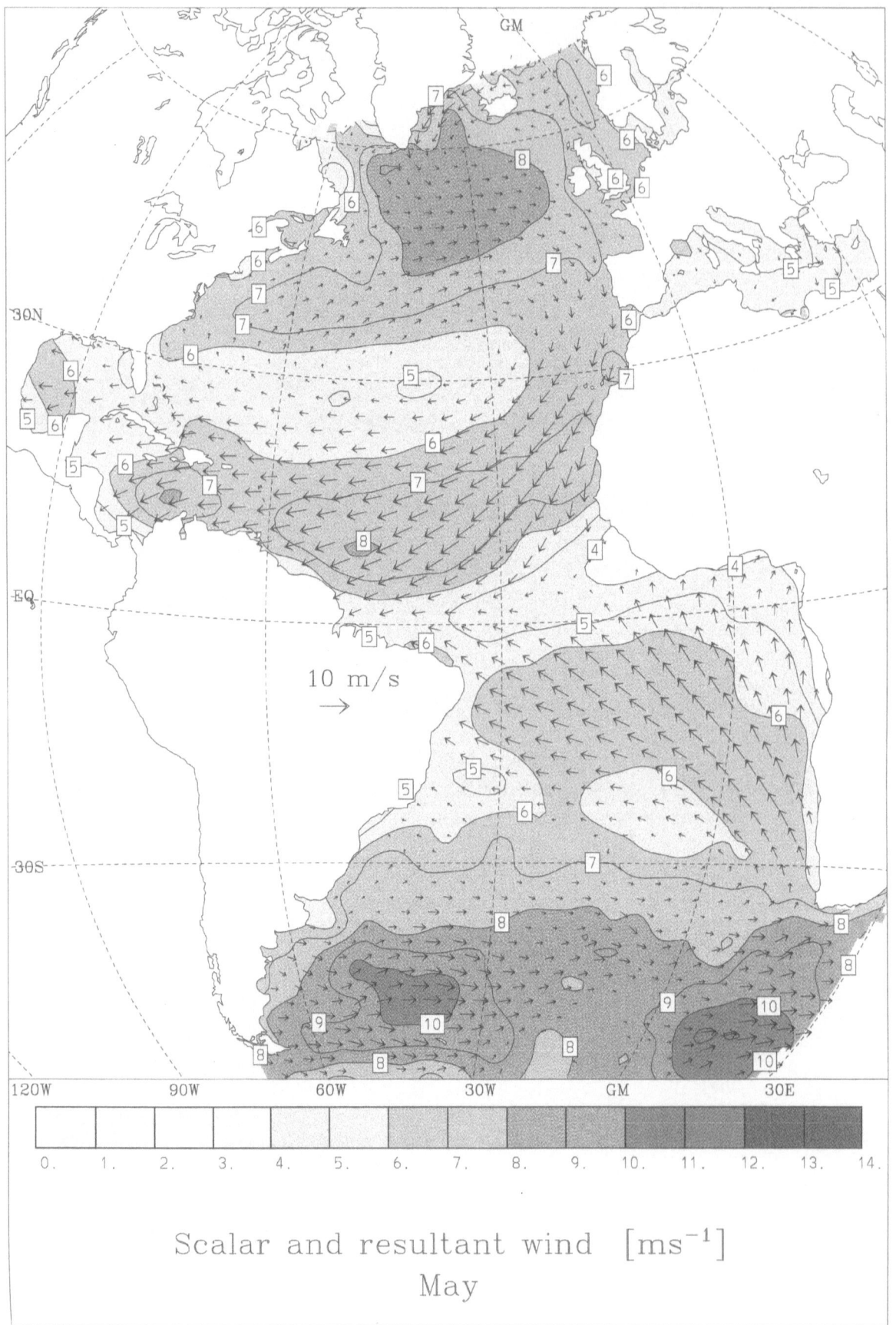

Scalar and resultant wind [ms^{-1}]
May

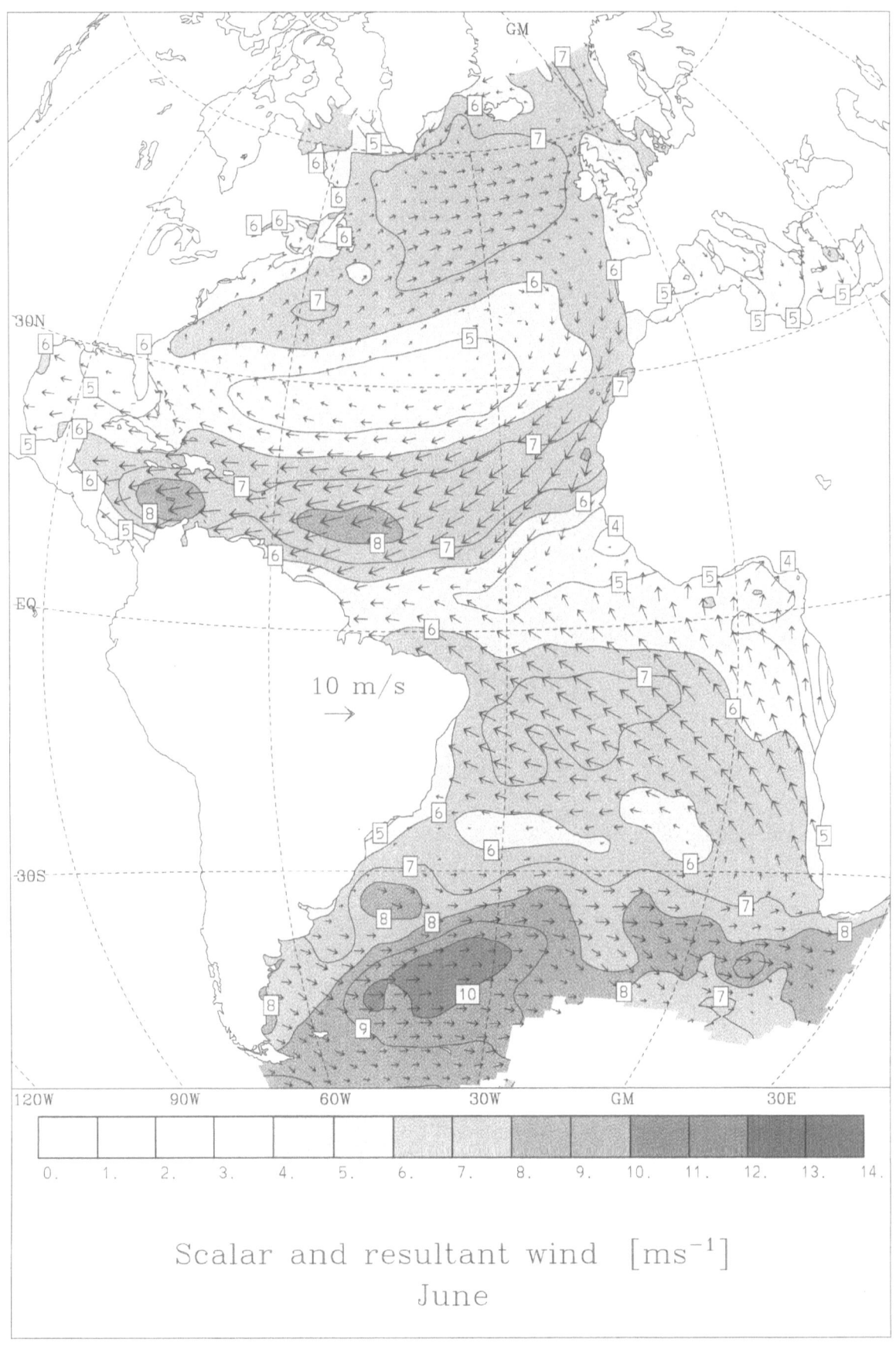

Scalar and resultant wind [ms^{-1}]
June

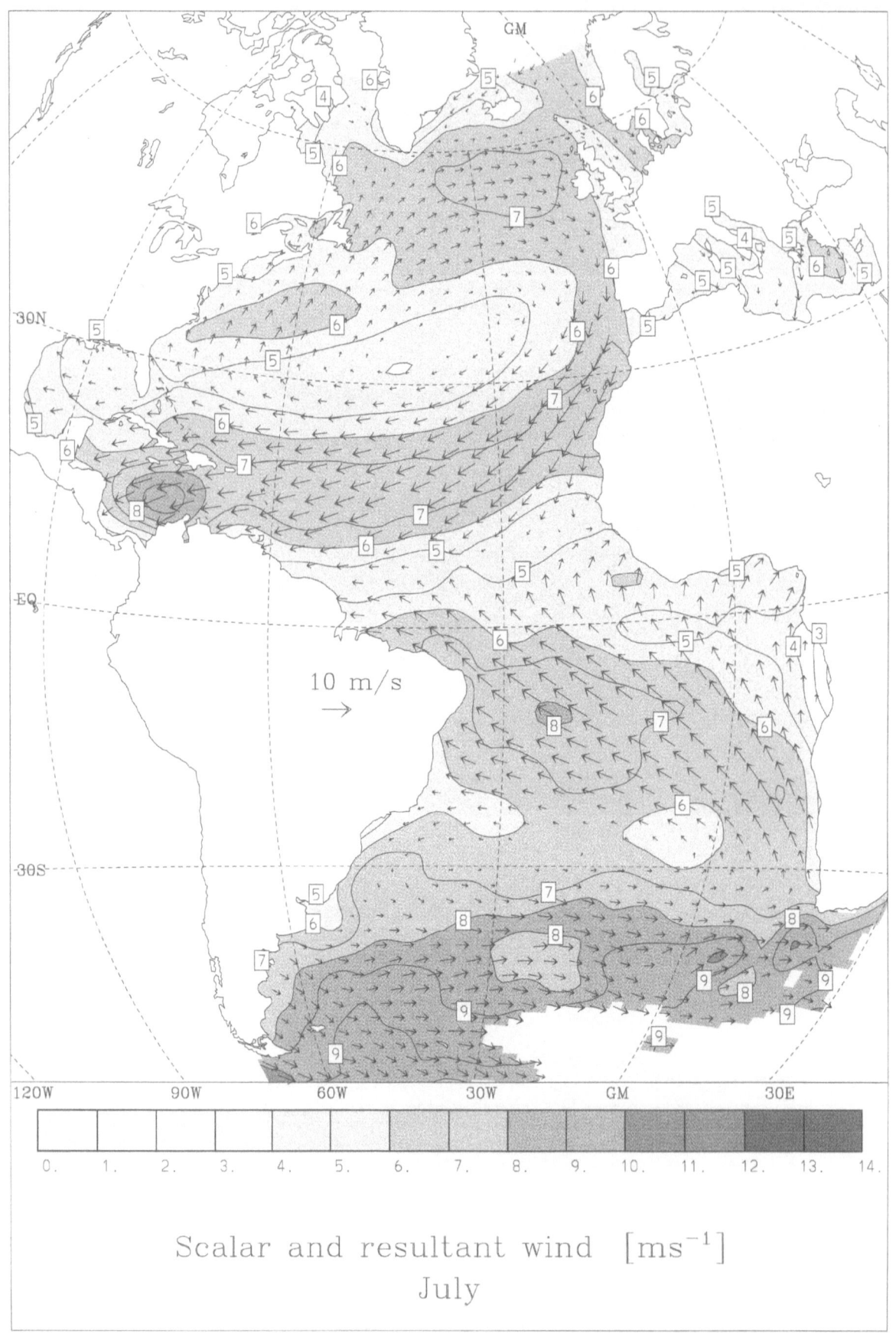

Scalar and resultant wind [ms^{-1}]
July

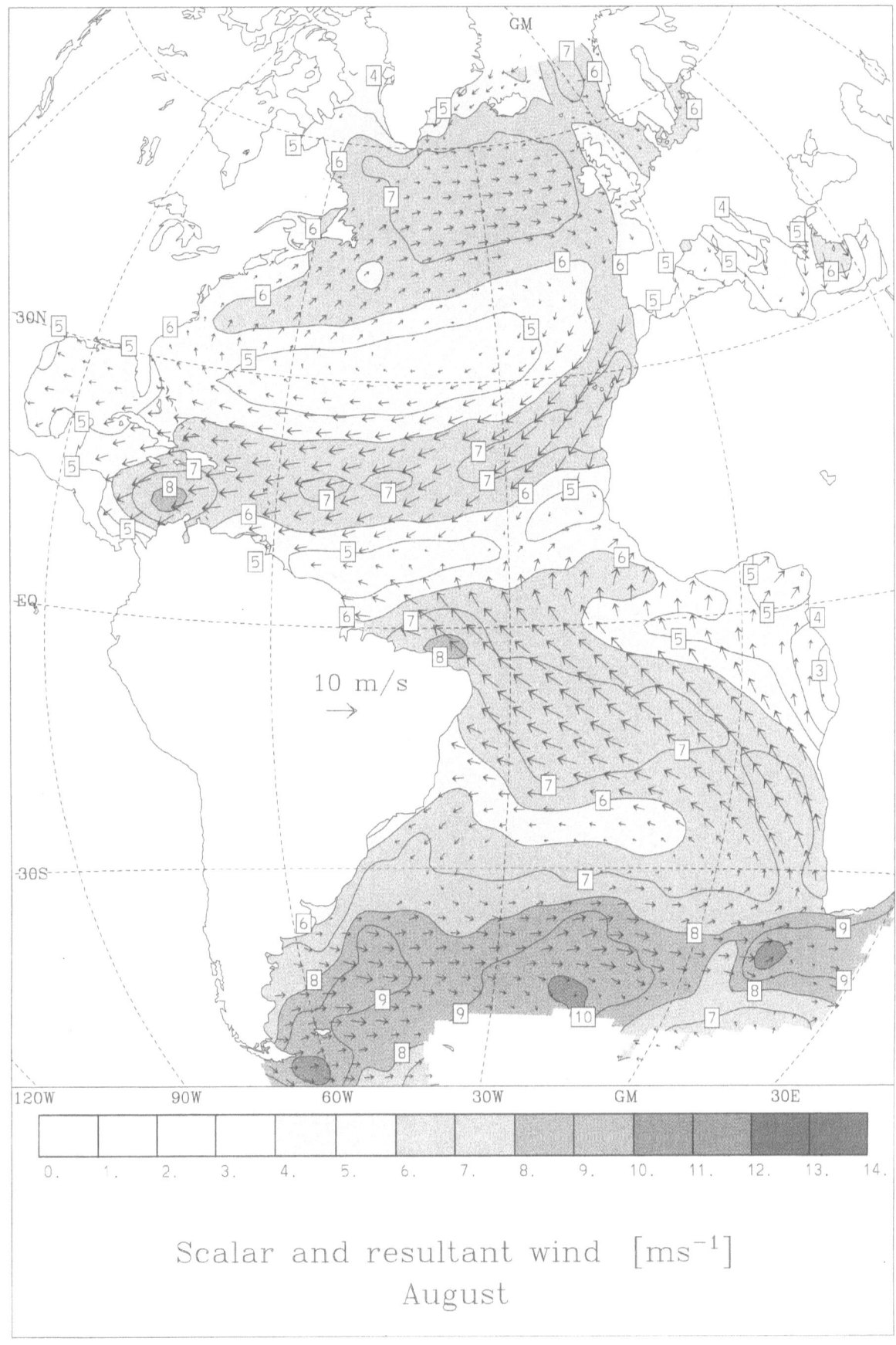

Scalar and resultant wind [ms^{-1}]
August

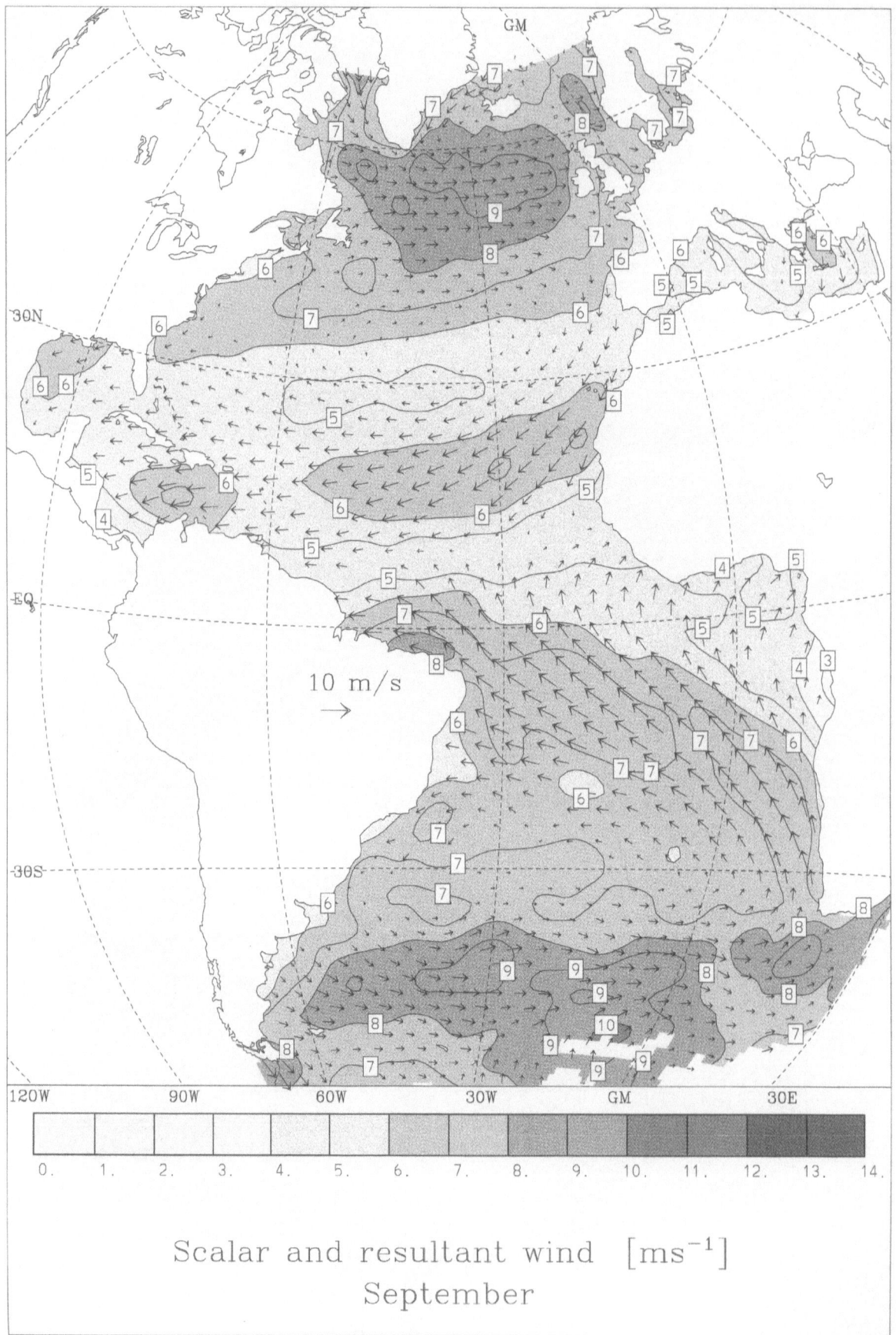

Scalar and resultant wind [ms^{-1}]
September

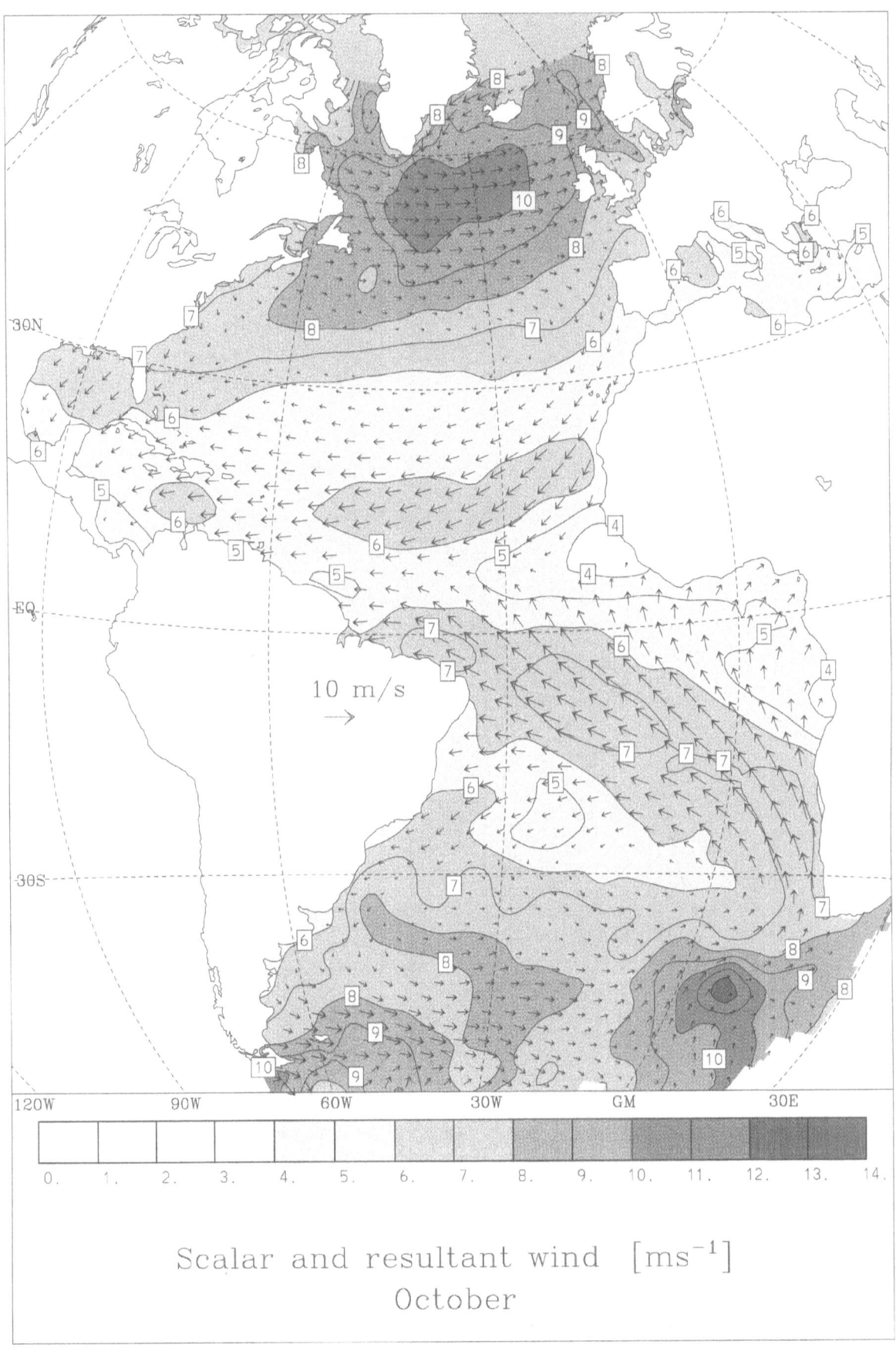

Scalar and resultant wind [ms^{-1}]
October

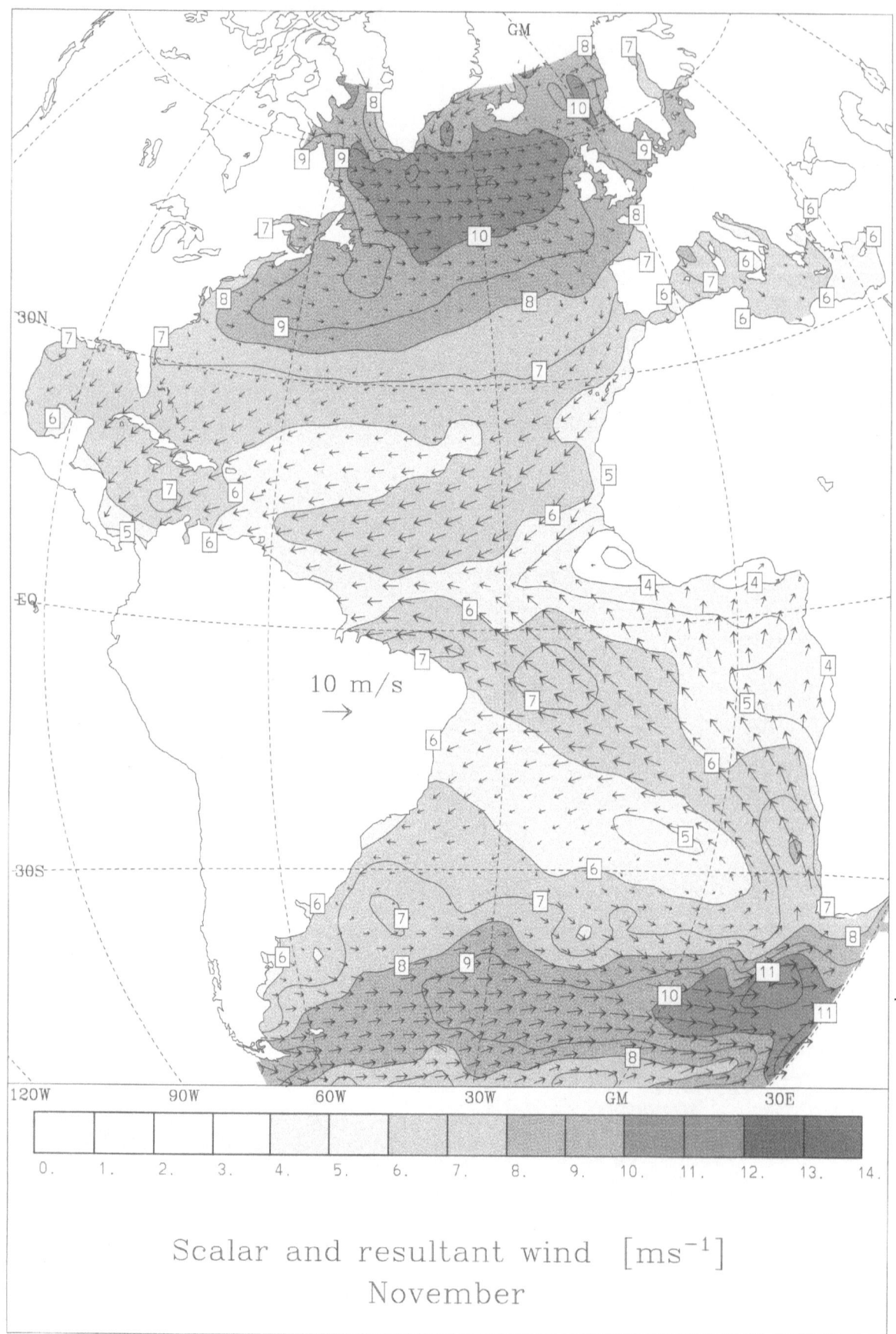

Scalar and resultant wind [ms^{-1}]
November

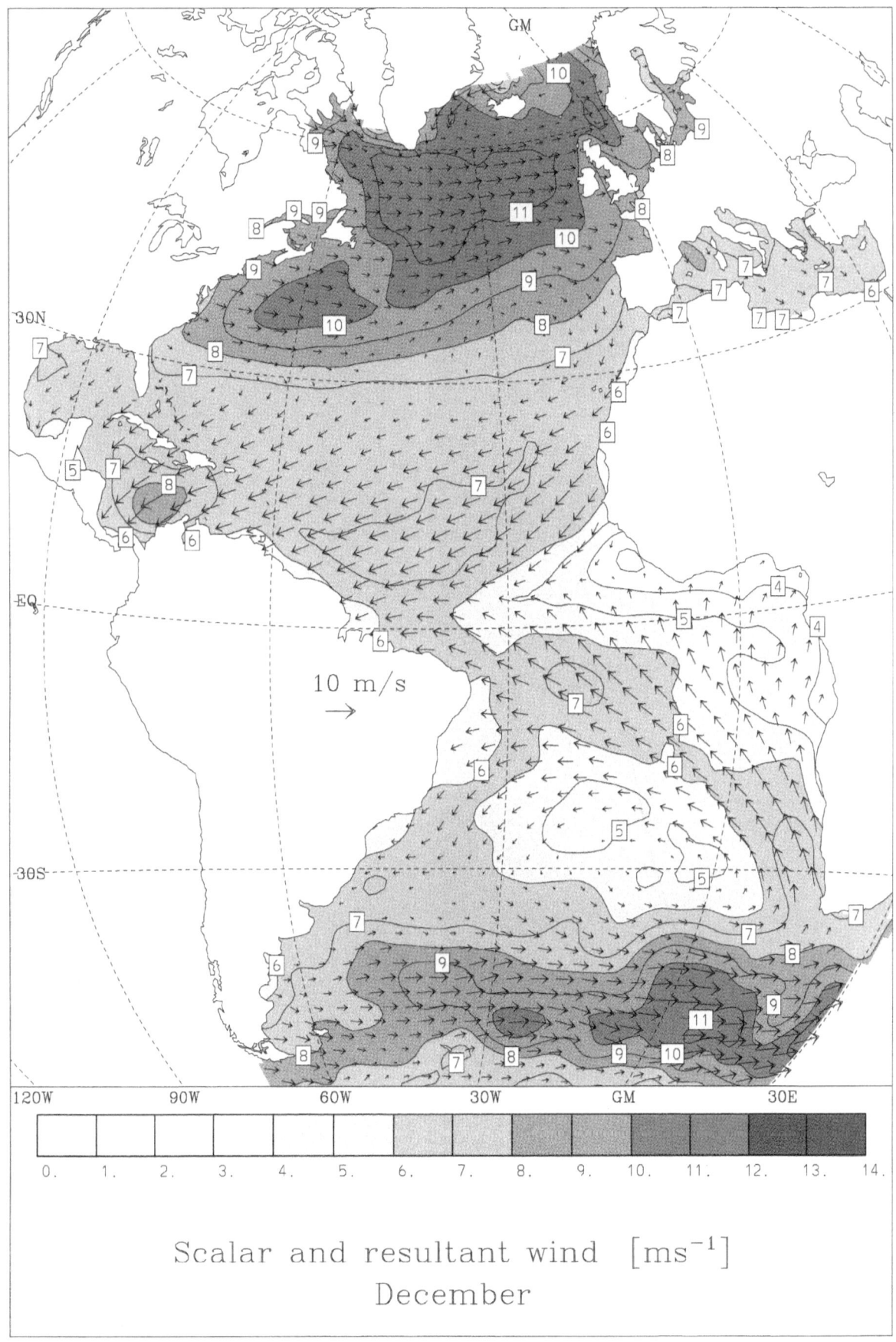

Scalar and resultant wind [ms^{-1}]
December

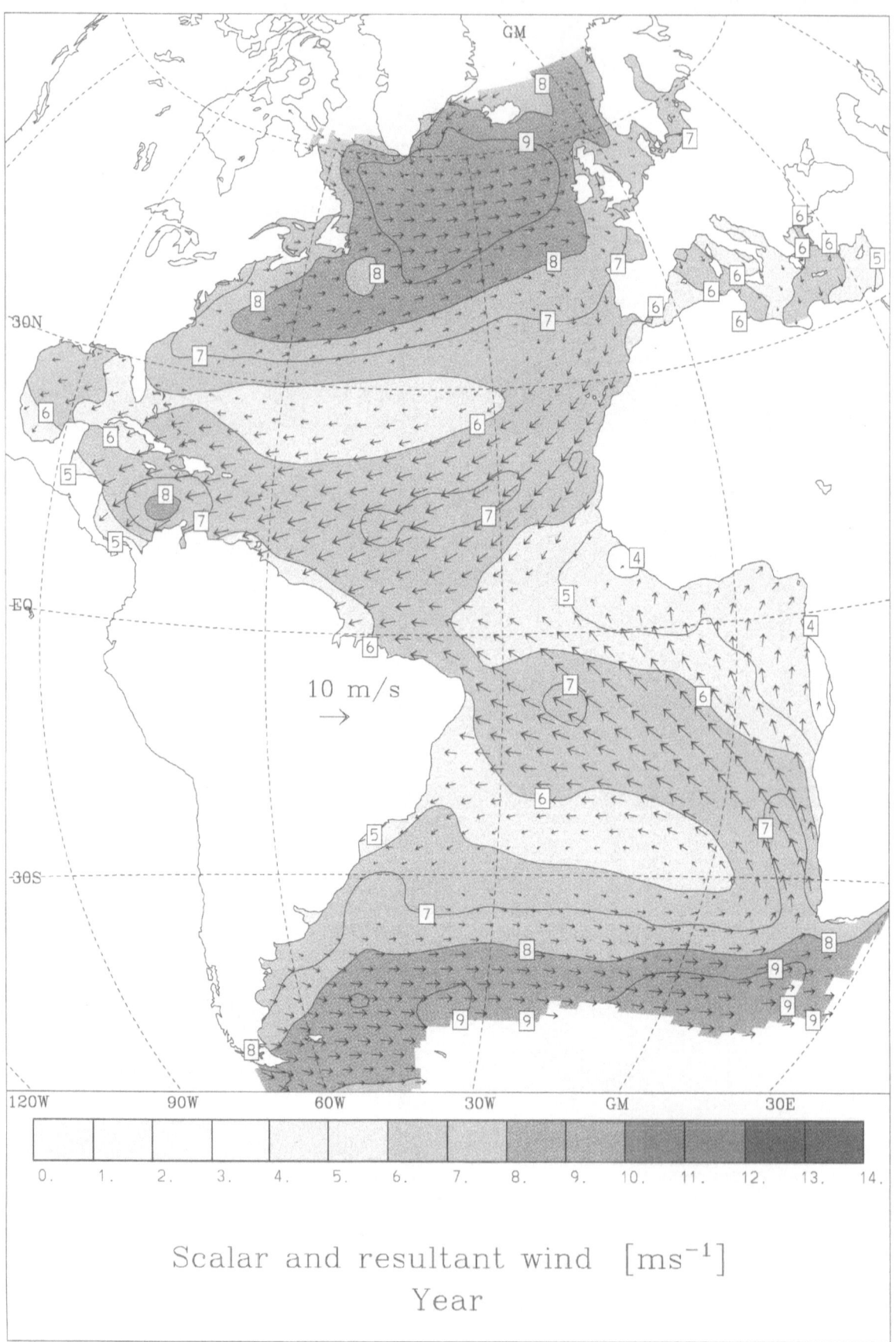

Scalar and resultant wind [ms^{-1}]
Year

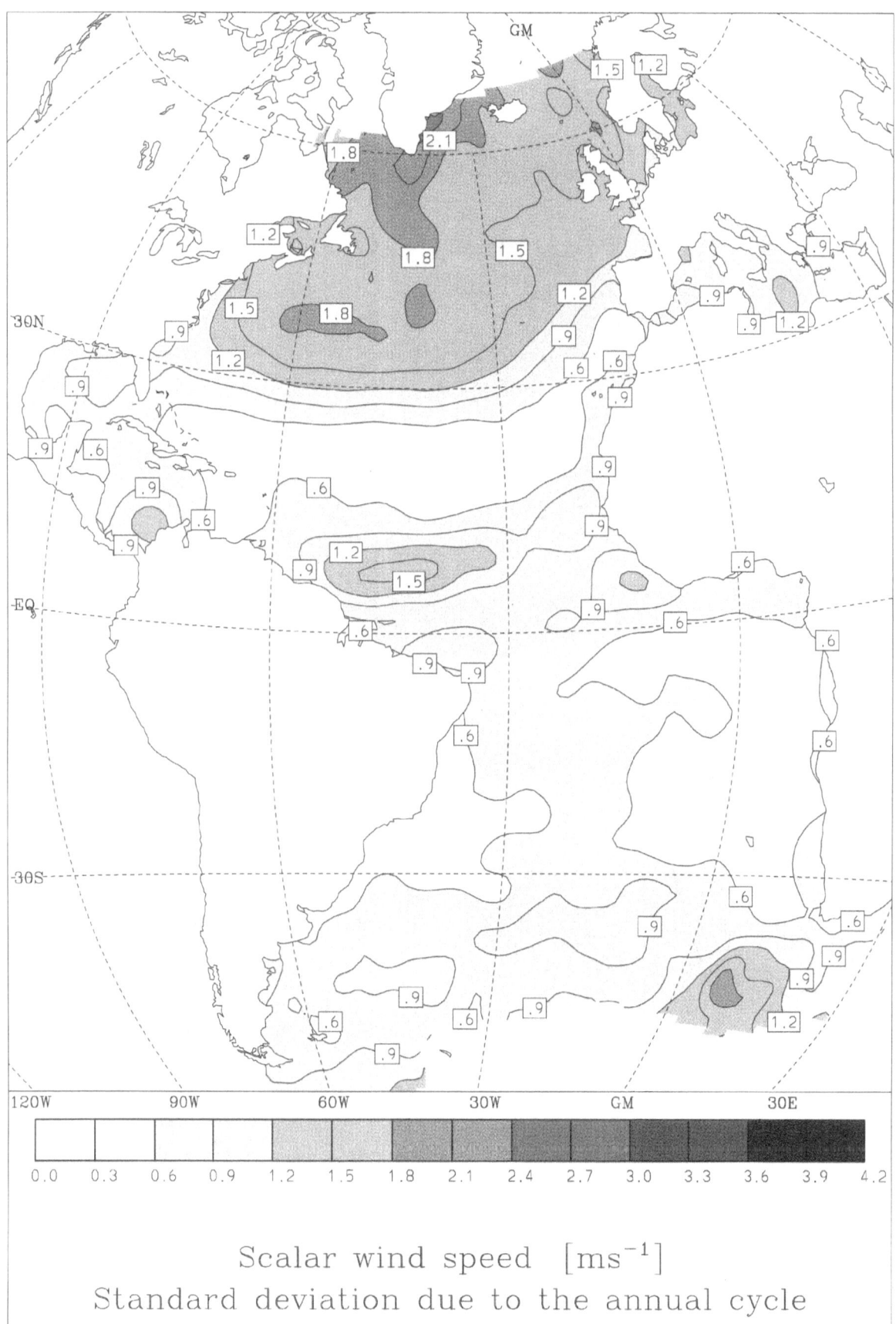

Scalar wind speed $[\mathrm{ms}^{-1}]$
Standard deviation due to the annual cycle

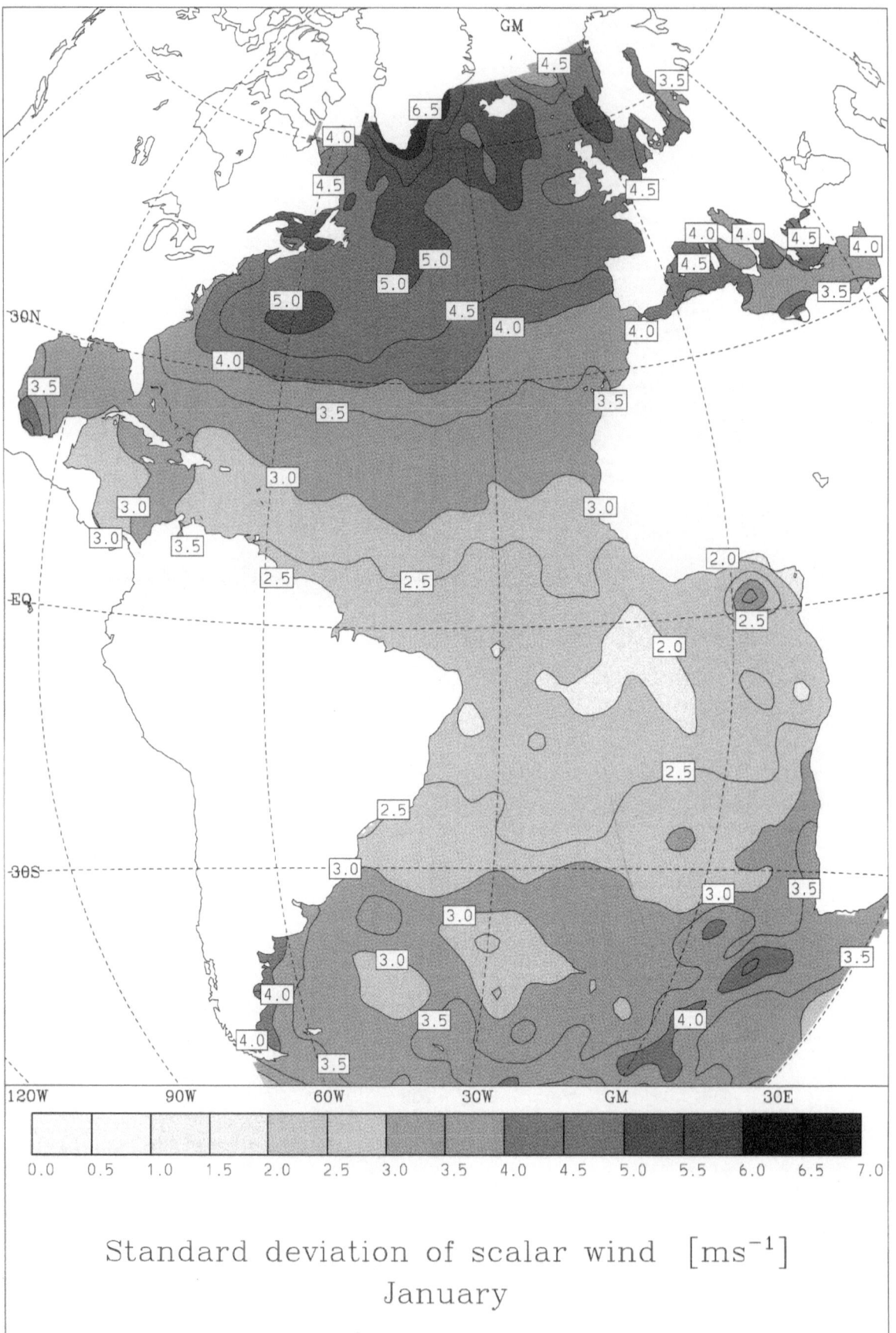

Standard deviation of scalar wind [ms^{-1}]
January

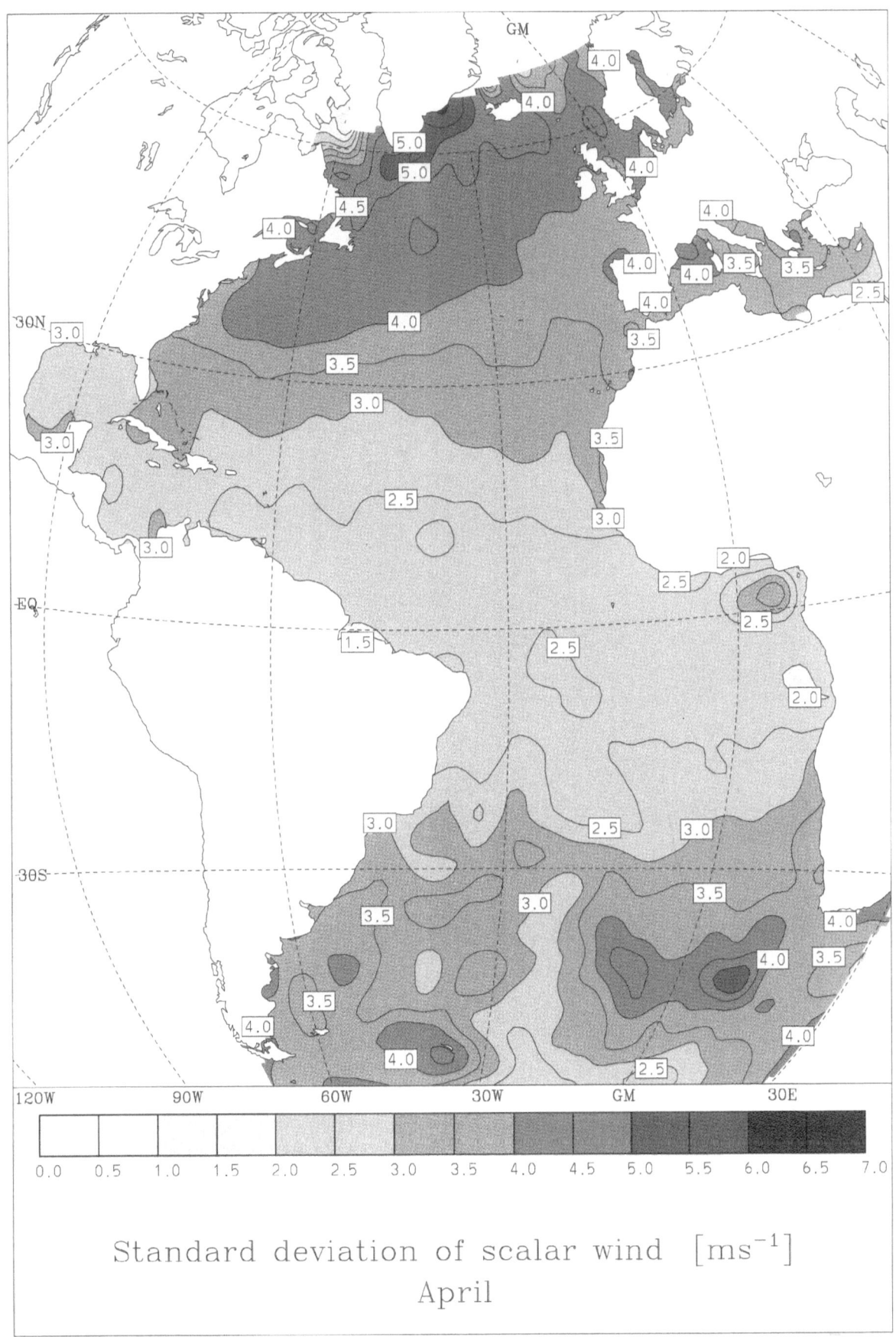

Standard deviation of scalar wind [ms⁻¹]

April

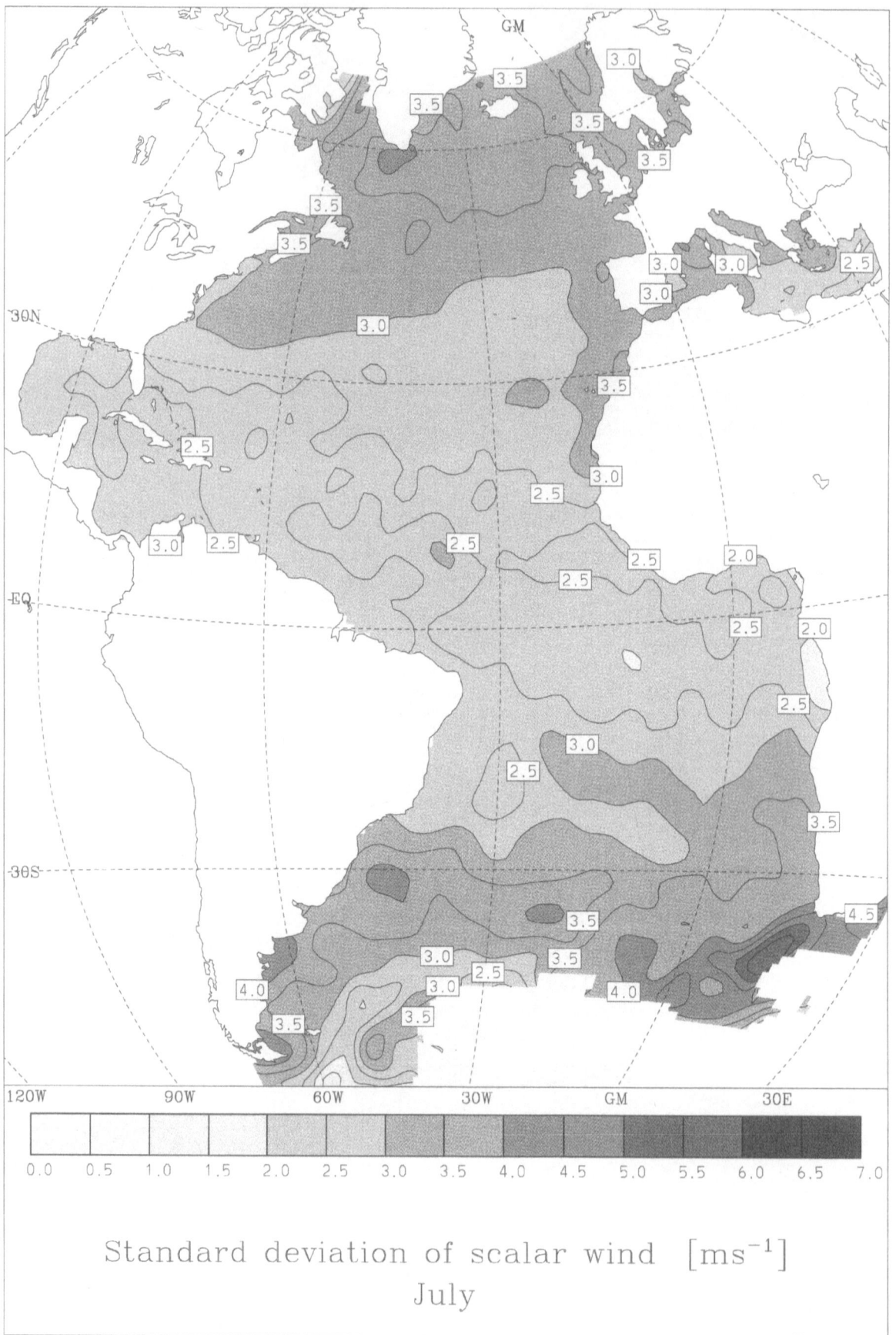

Standard deviation of scalar wind [ms⁻¹]
July

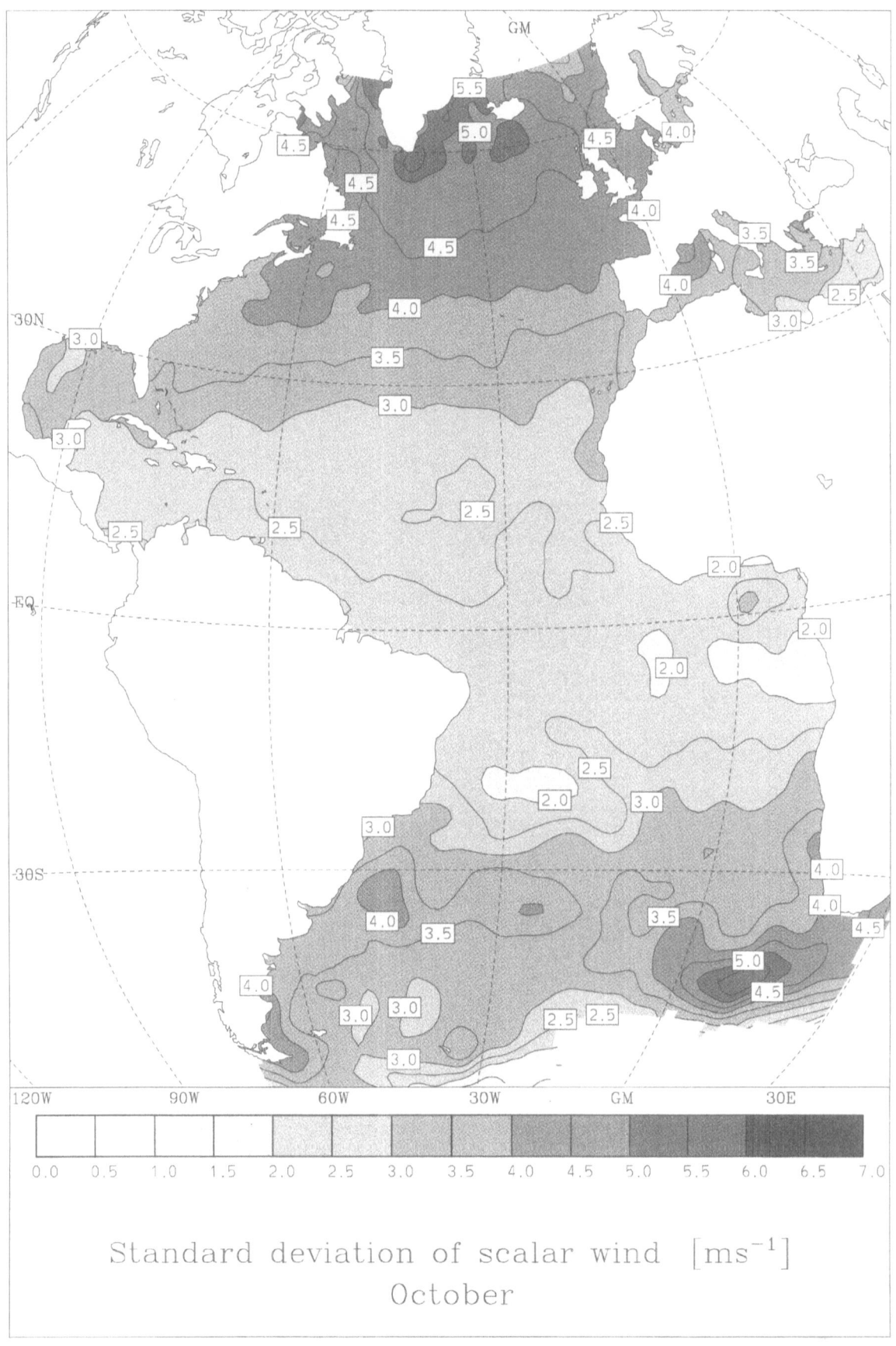

Standard deviation of scalar wind [ms^{-1}]
October

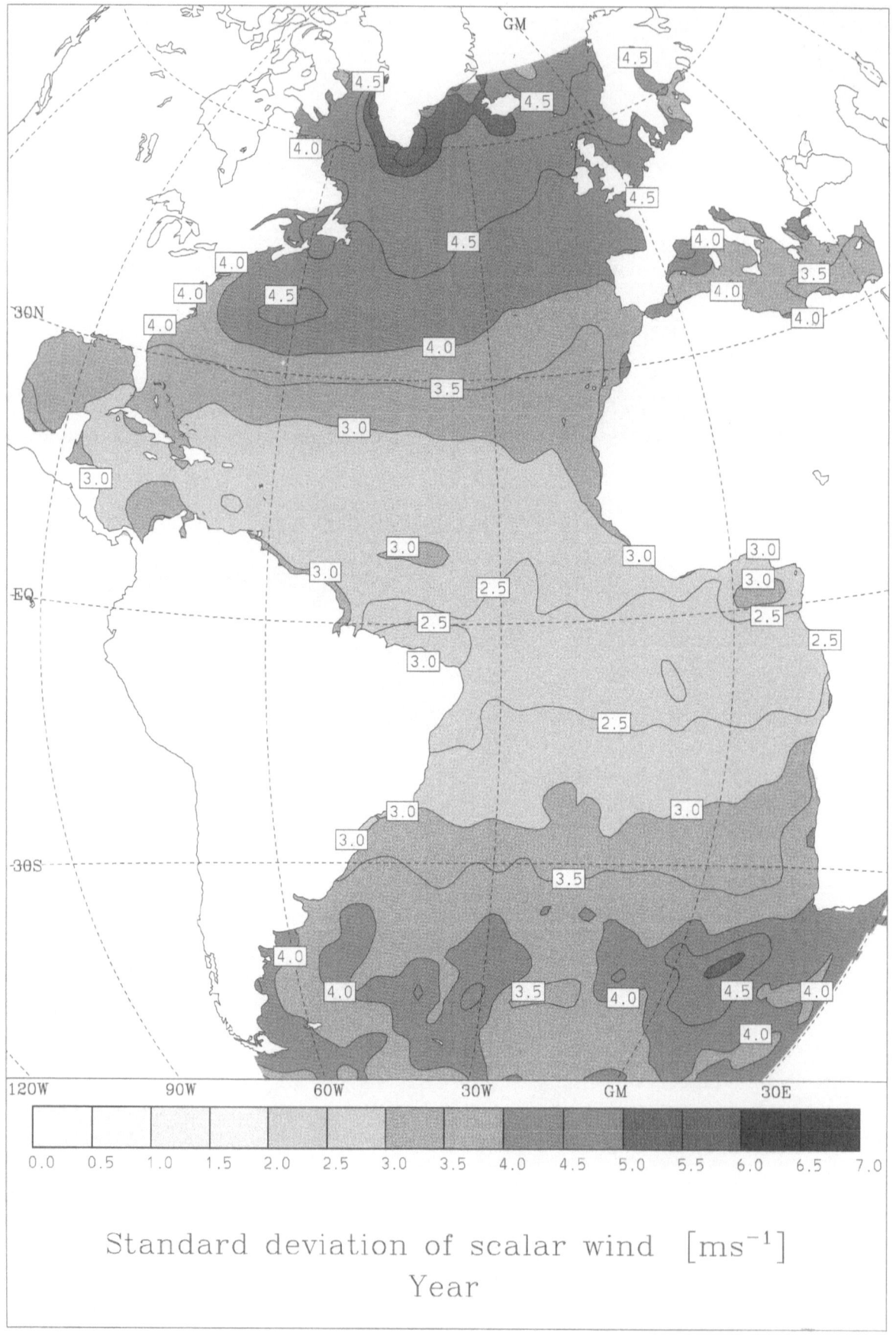

Standard deviation of scalar wind [ms^{-1}]
Year

East Component of the Wind

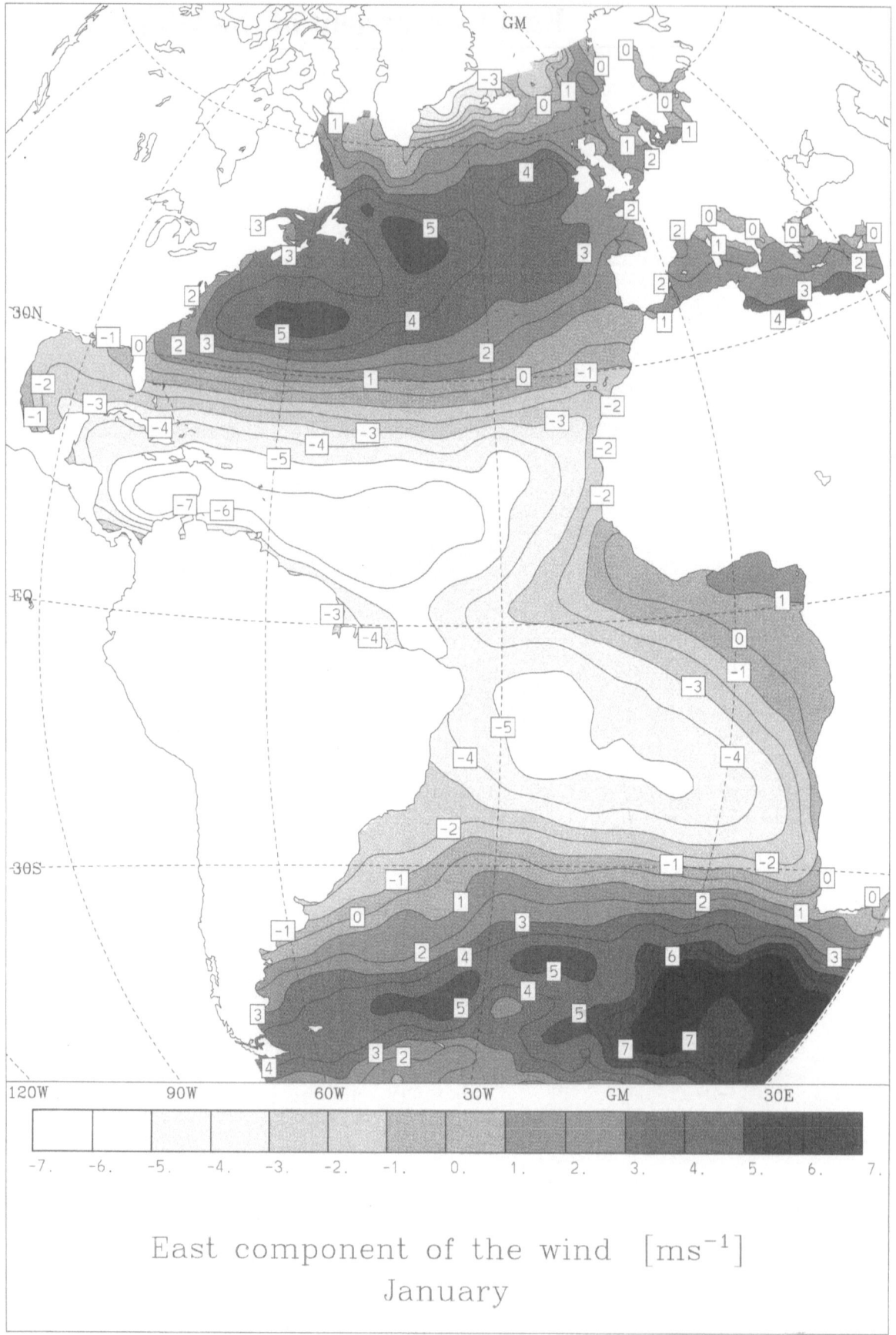

East component of the wind [ms^{-1}]
January

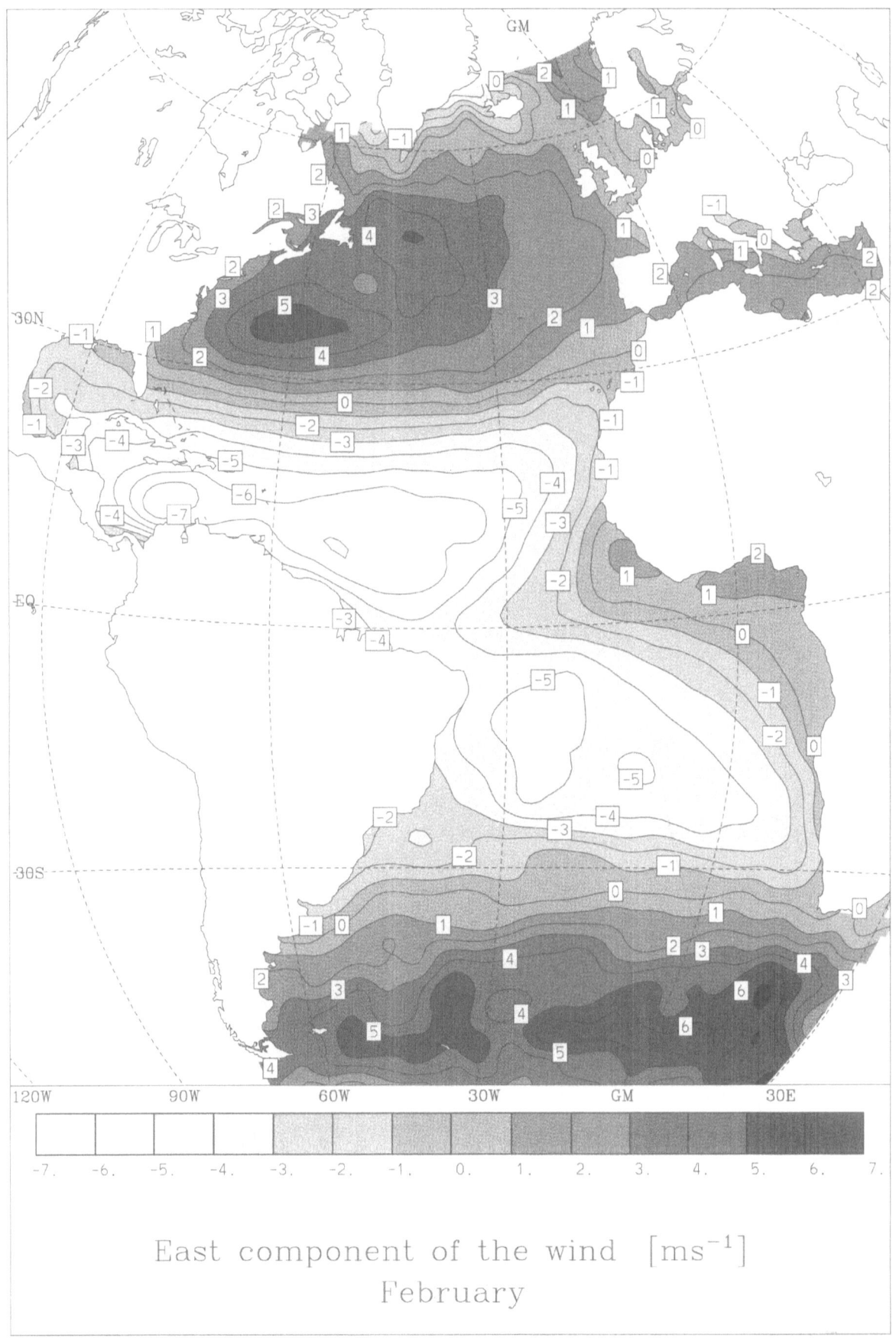

East component of the wind [ms^{-1}]
February

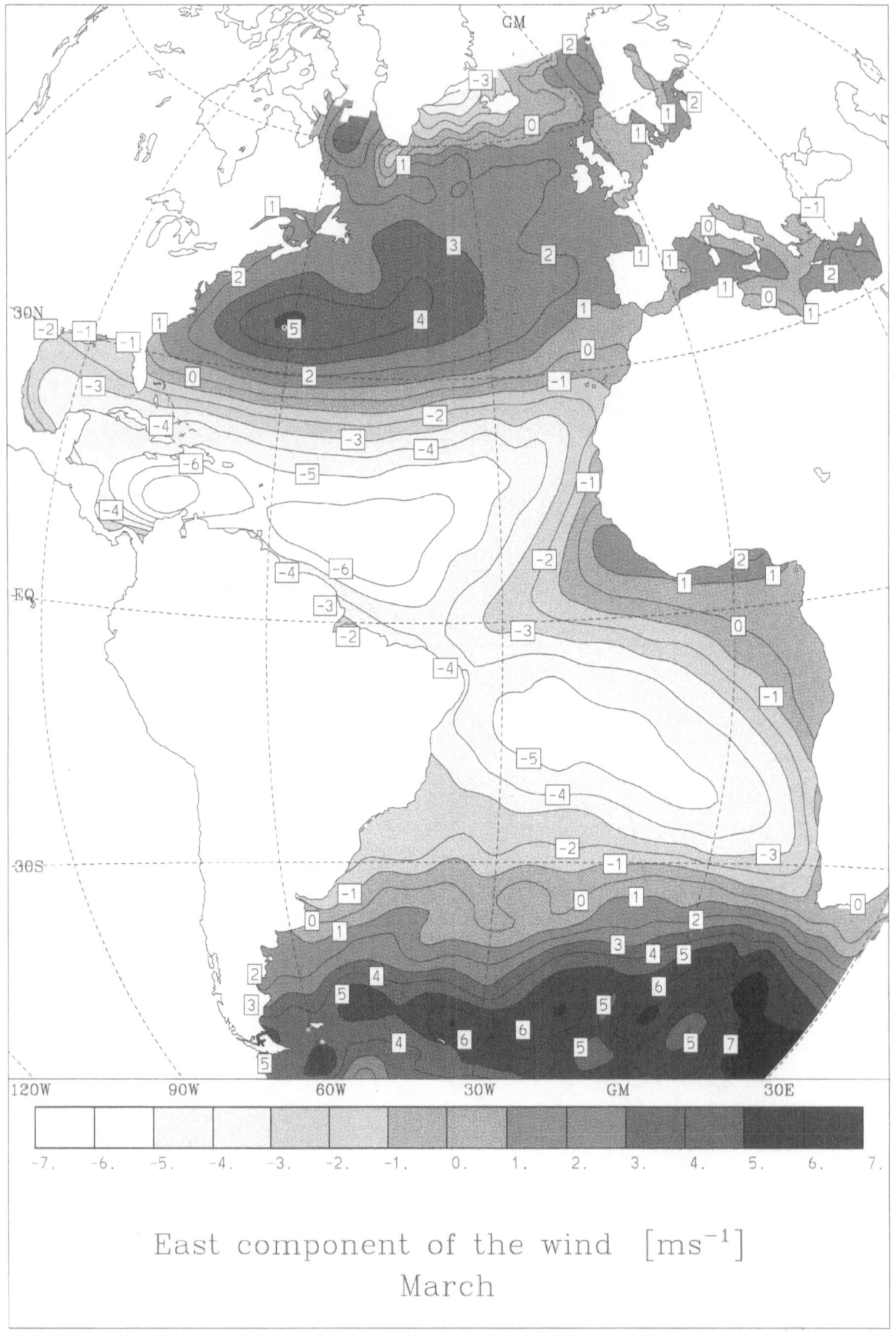

East component of the wind [ms⁻¹]
March

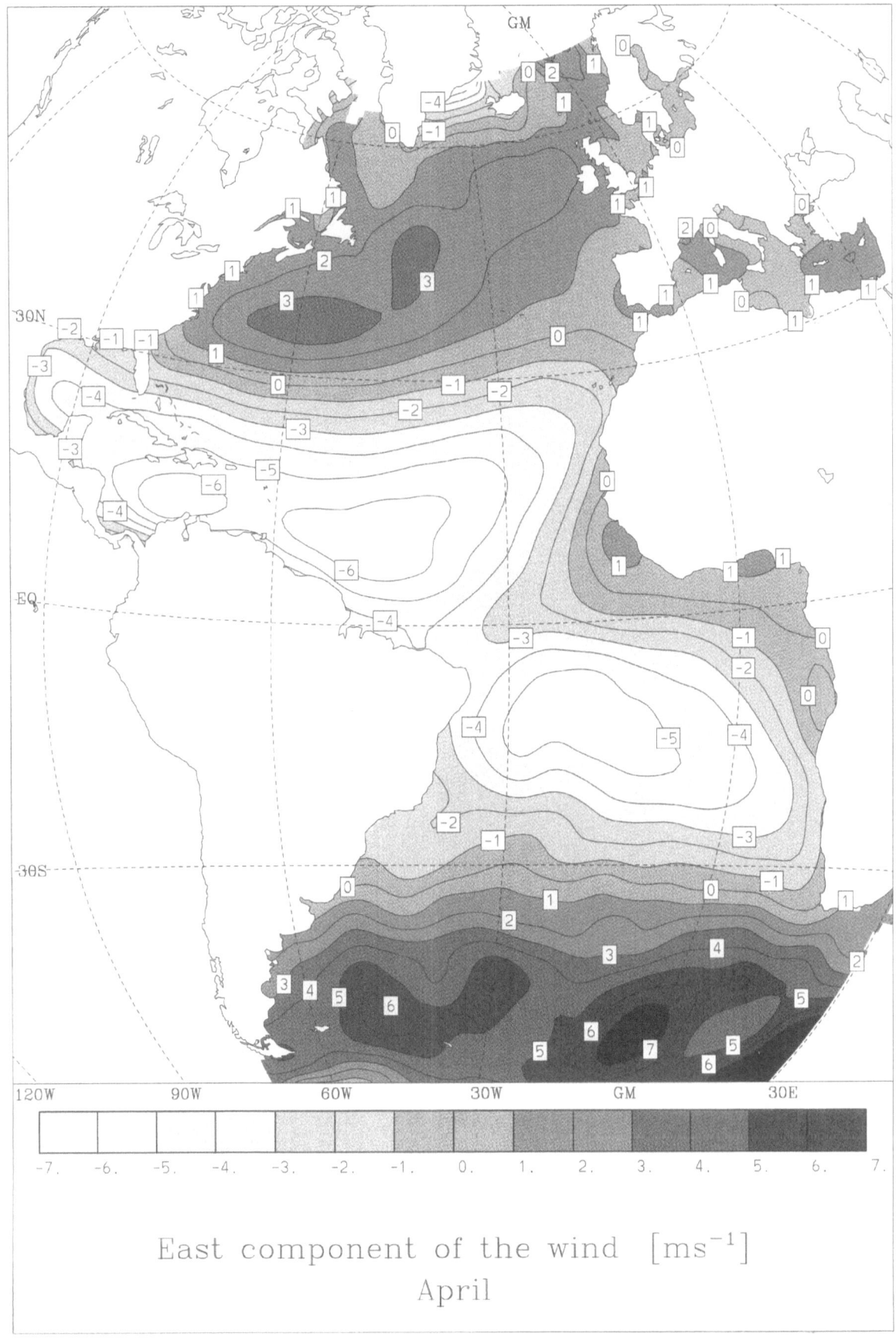

East component of the wind [ms^{-1}]
April

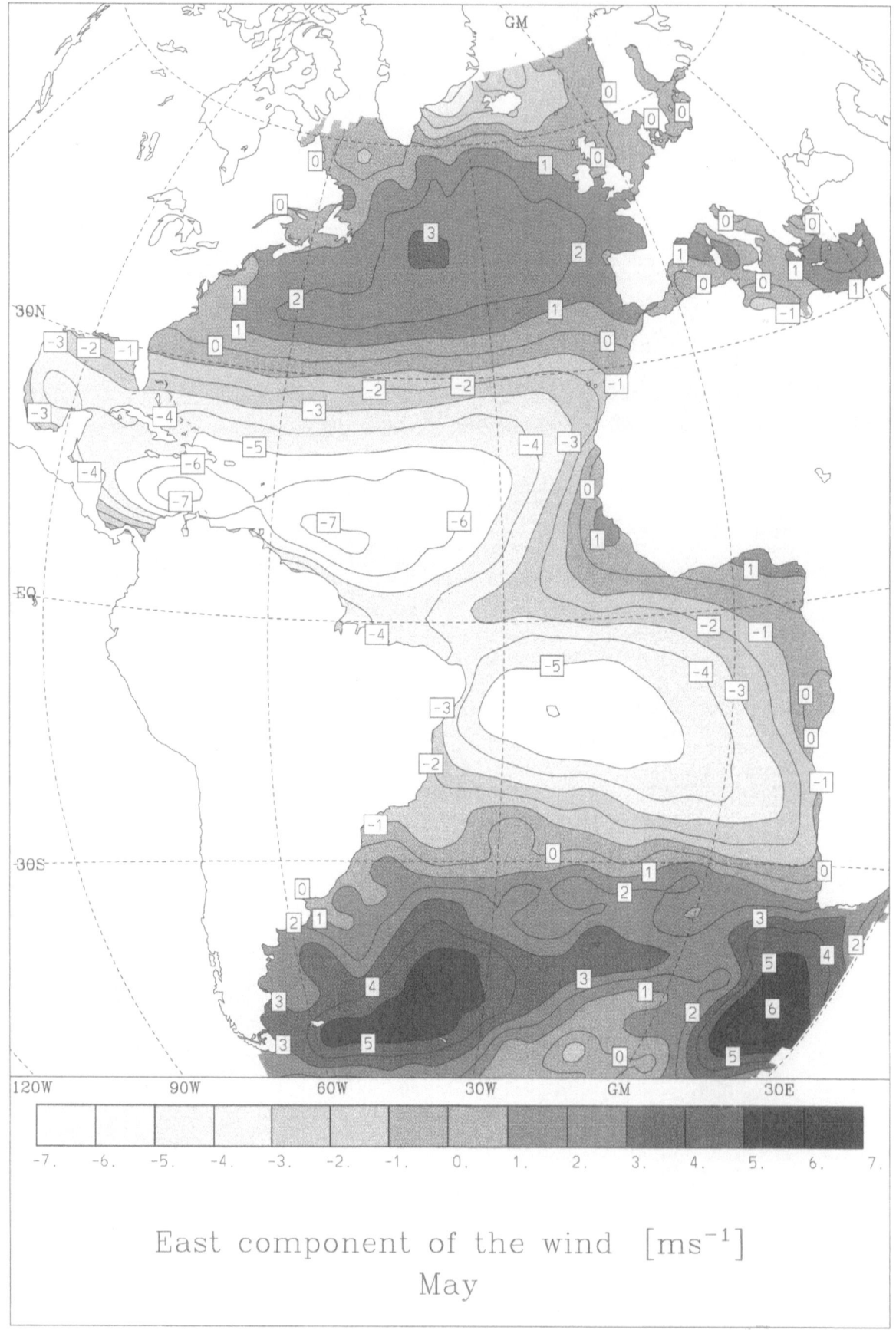

East component of the wind [ms⁻¹]
May

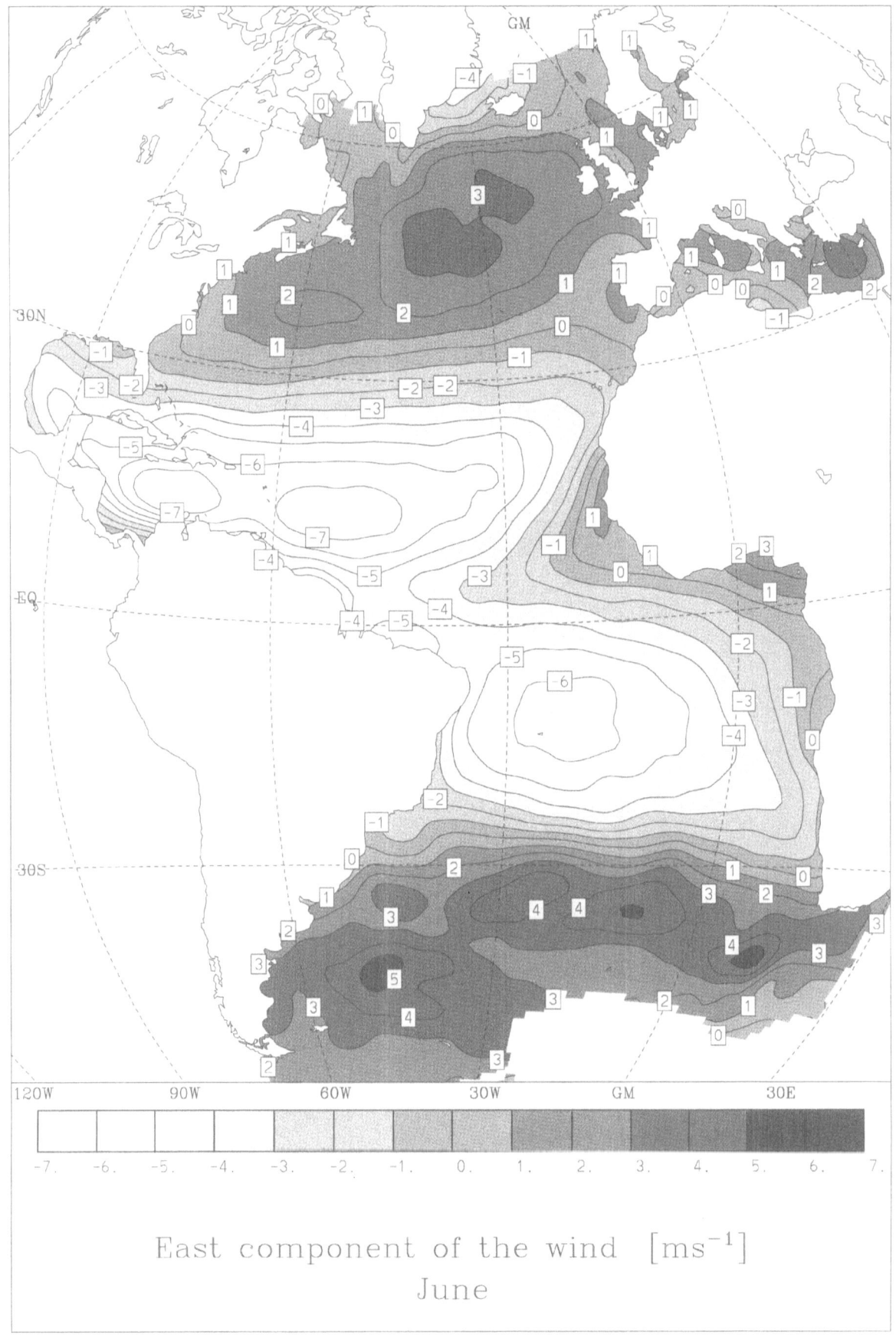

East component of the wind [ms^{-1}]
June

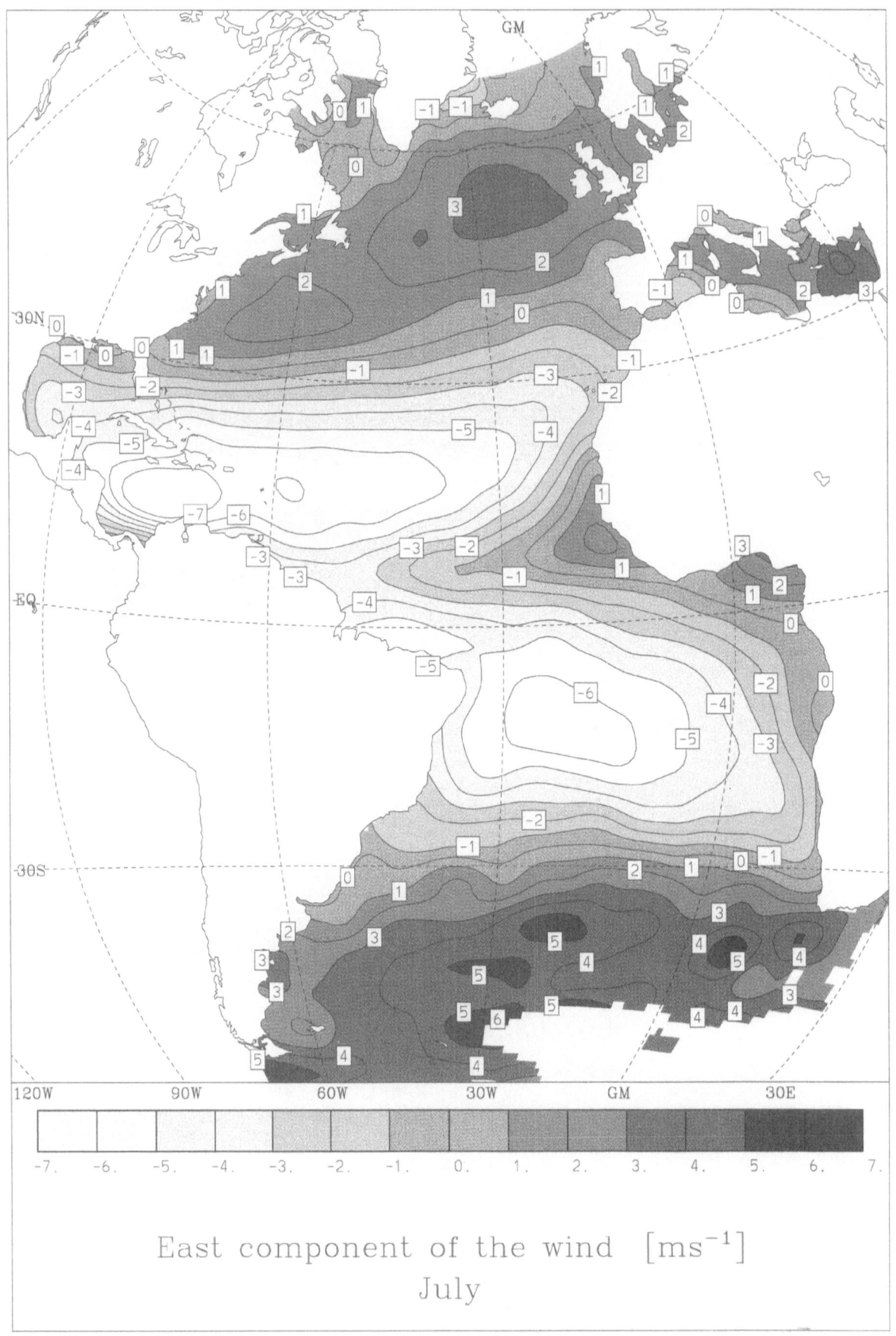

East component of the wind [ms^{-1}]
July

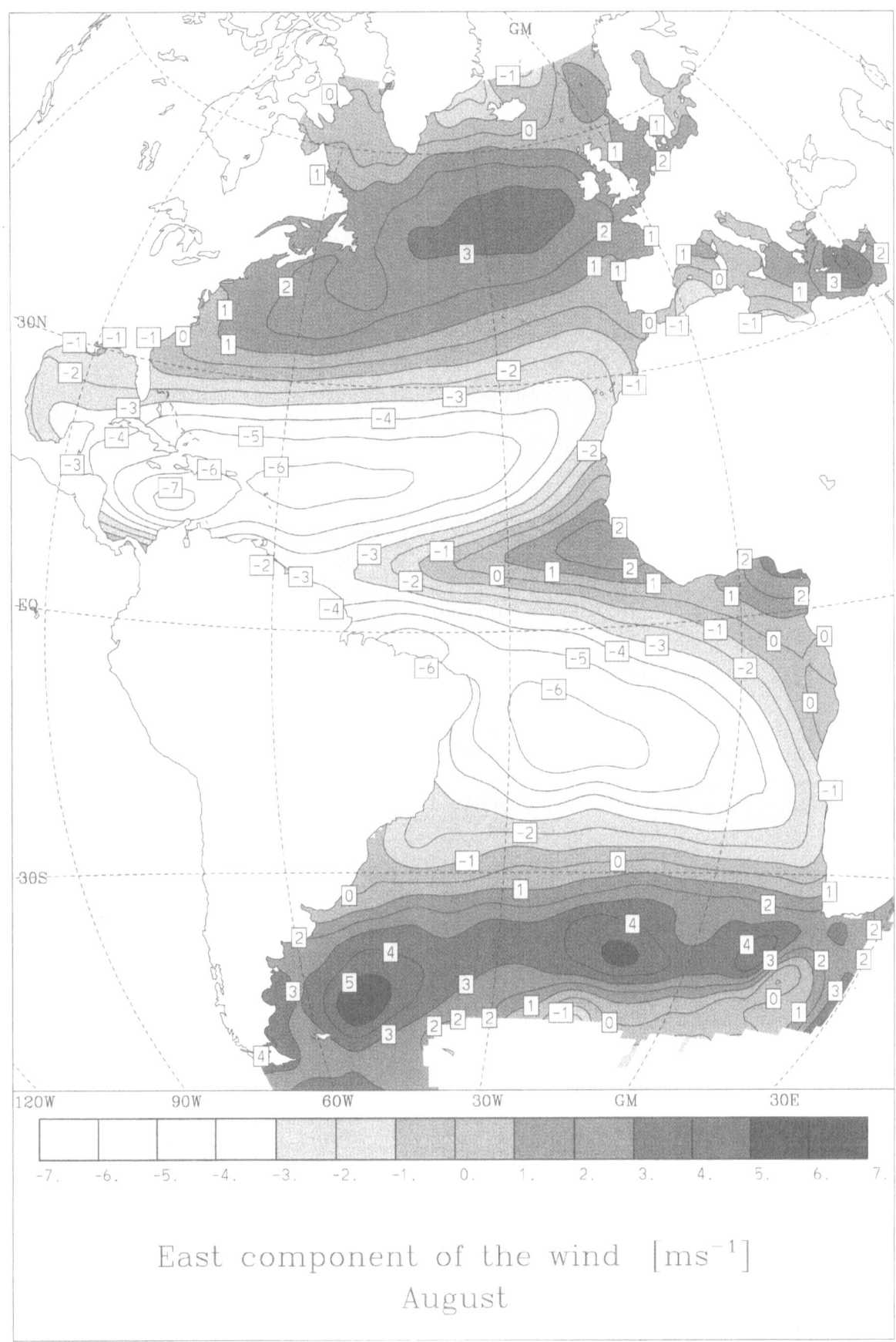

East component of the wind [ms^{-1}]
August

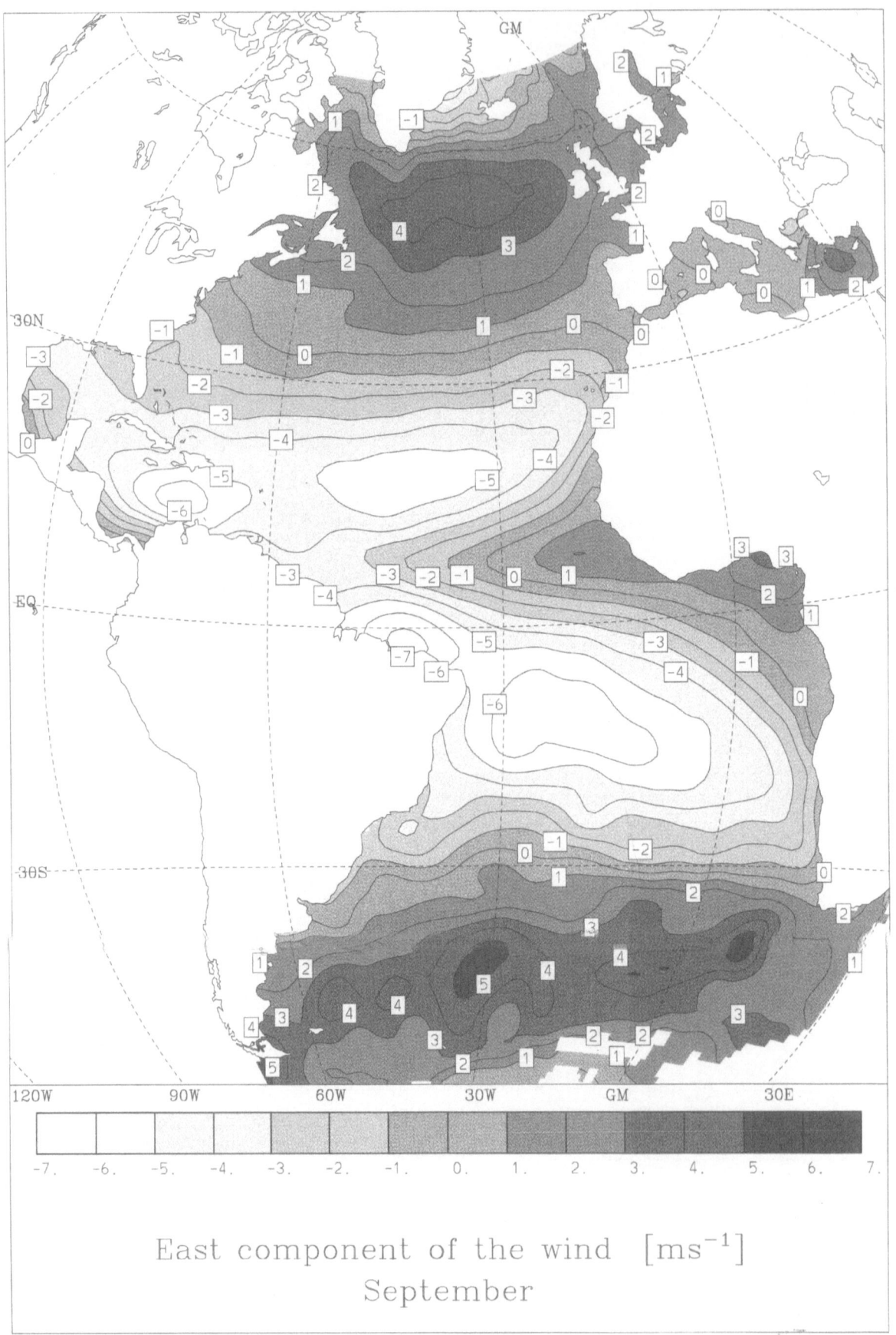

East component of the wind [ms^{-1}]
September

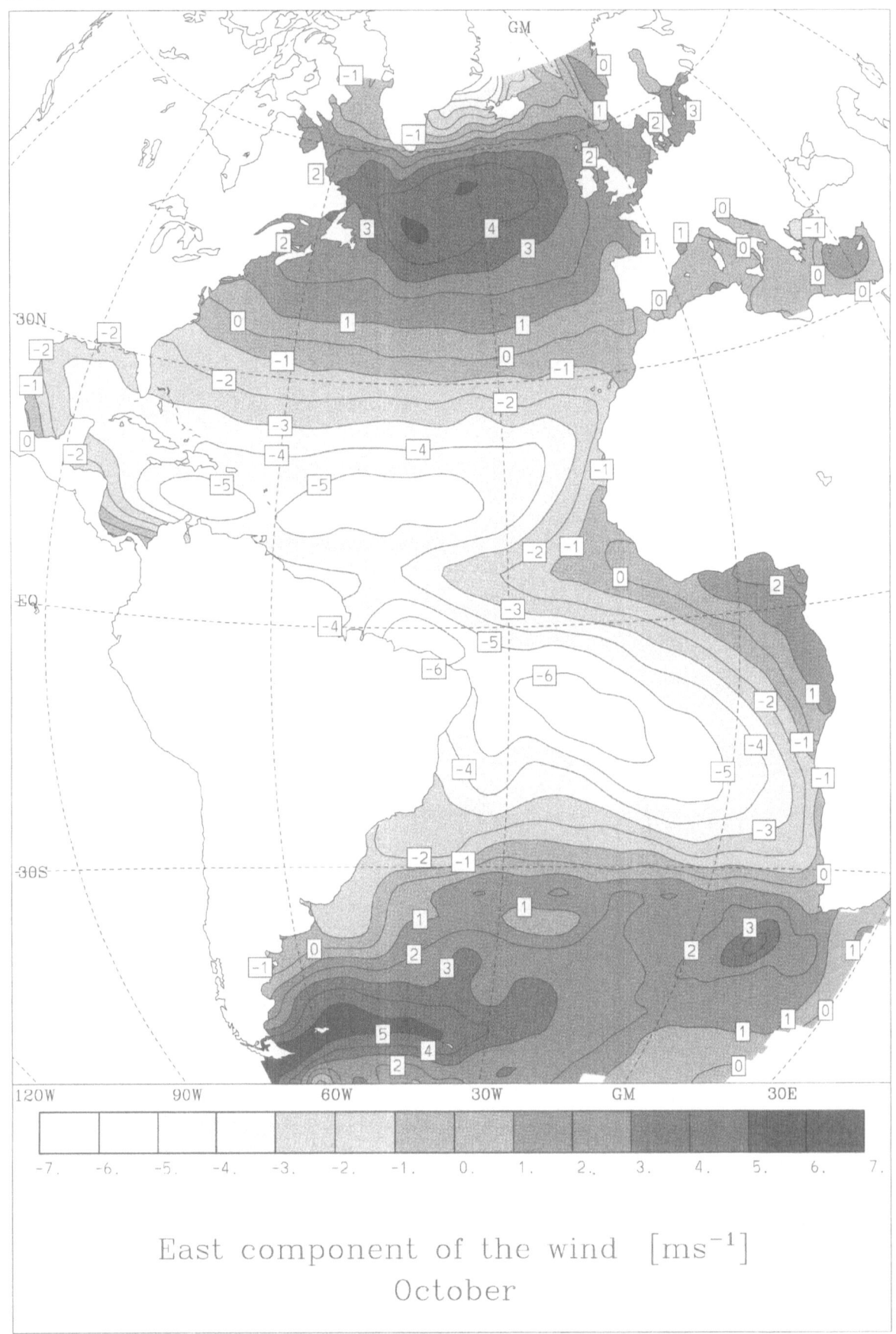

East component of the wind [ms^{-1}]
October

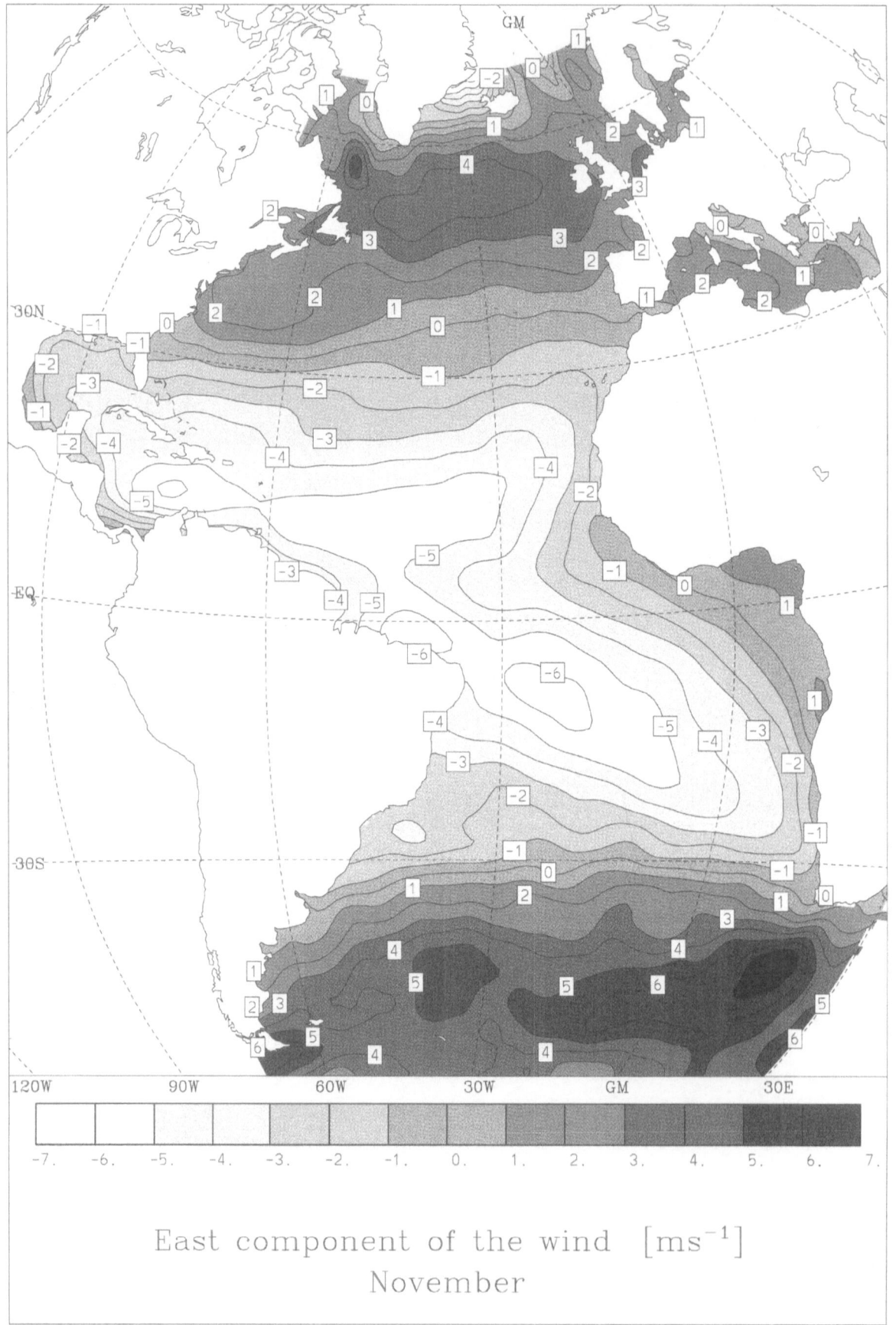

East component of the wind [ms⁻¹]
November

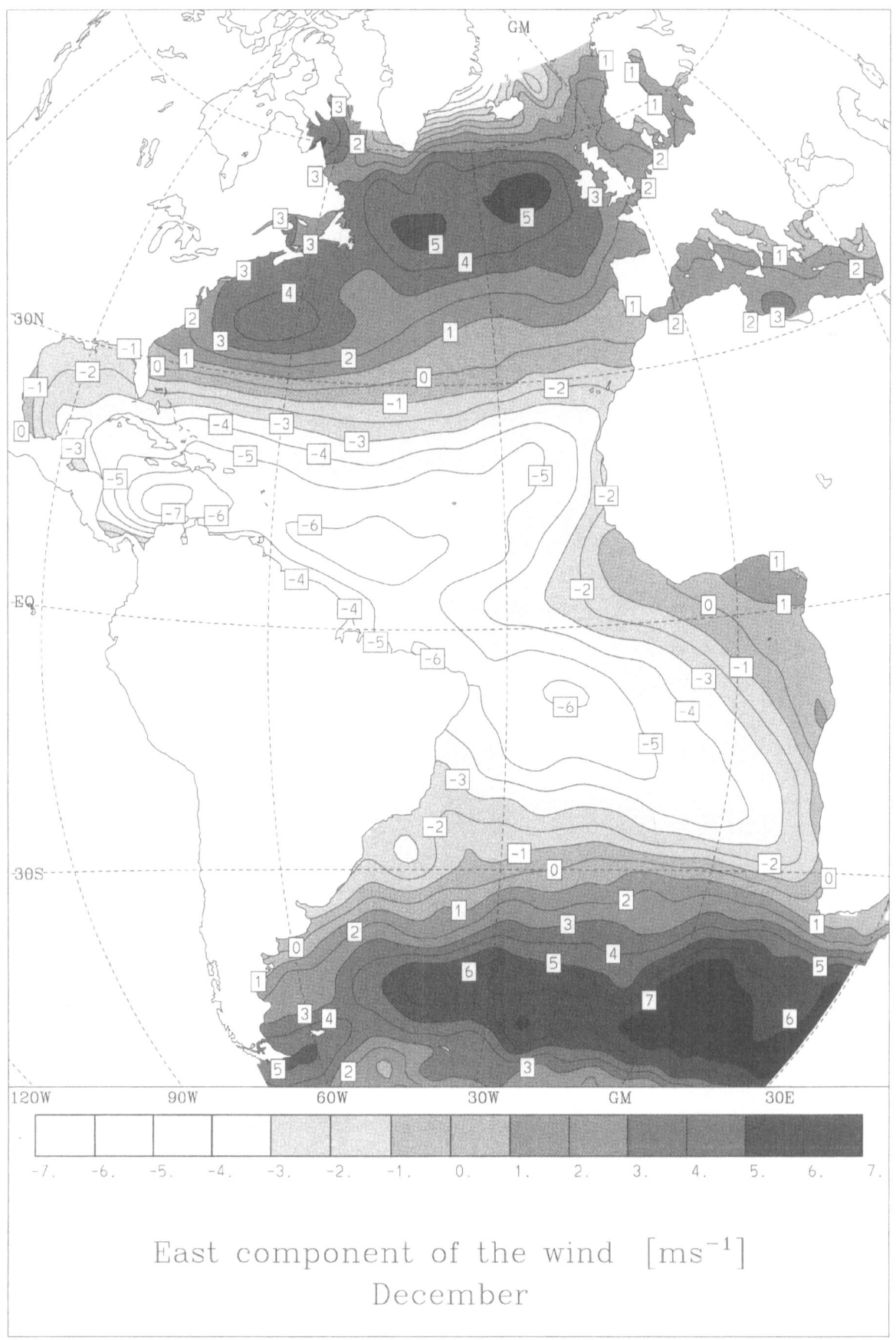

East component of the wind [ms^{-1}]
December

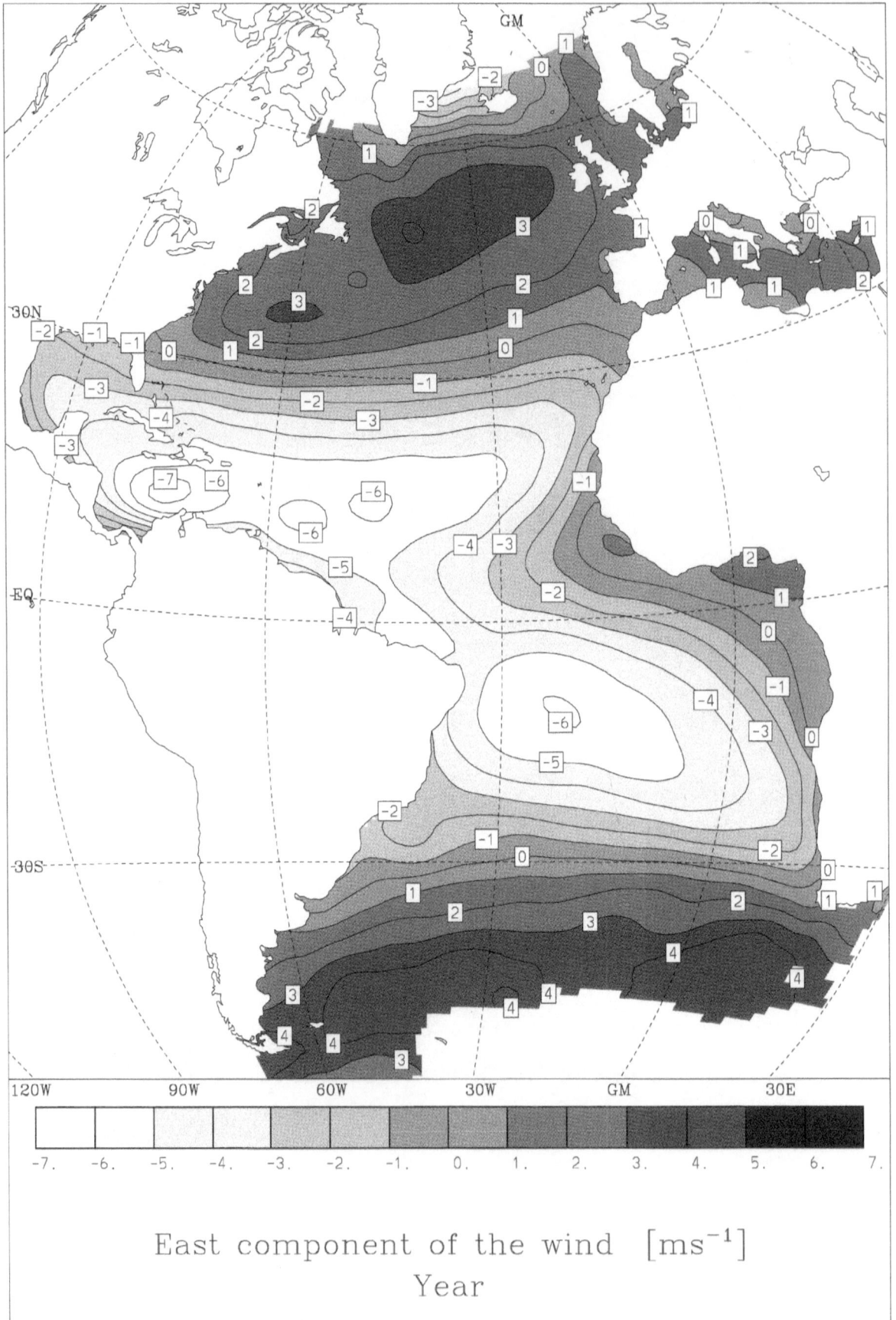

East component of the wind [ms^{-1}]
Year

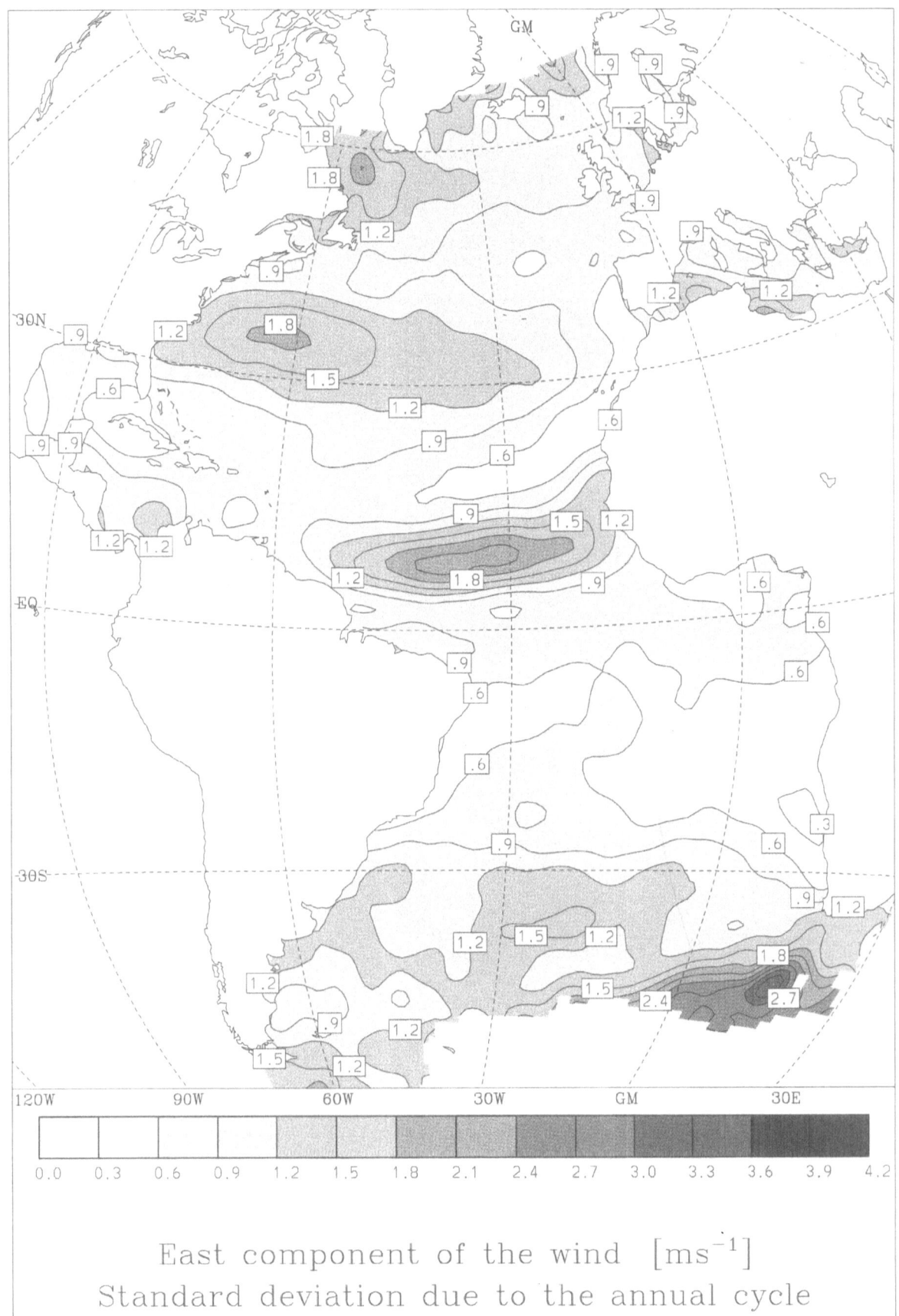

East component of the wind $[ms^{-1}]$
Standard deviation due to the annual cycle

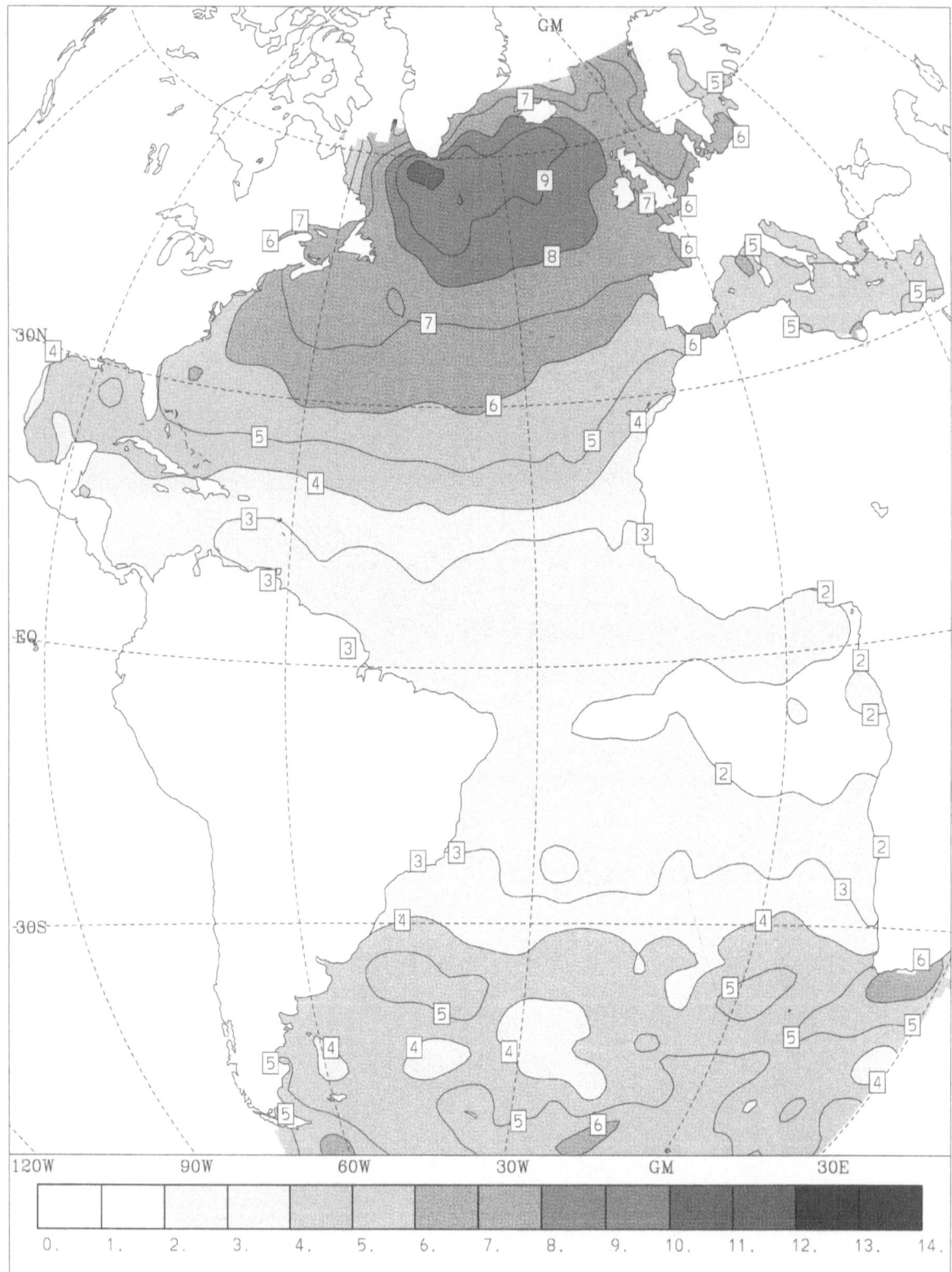

Standard deviation of east wind component $[ms^{-1}]$
January

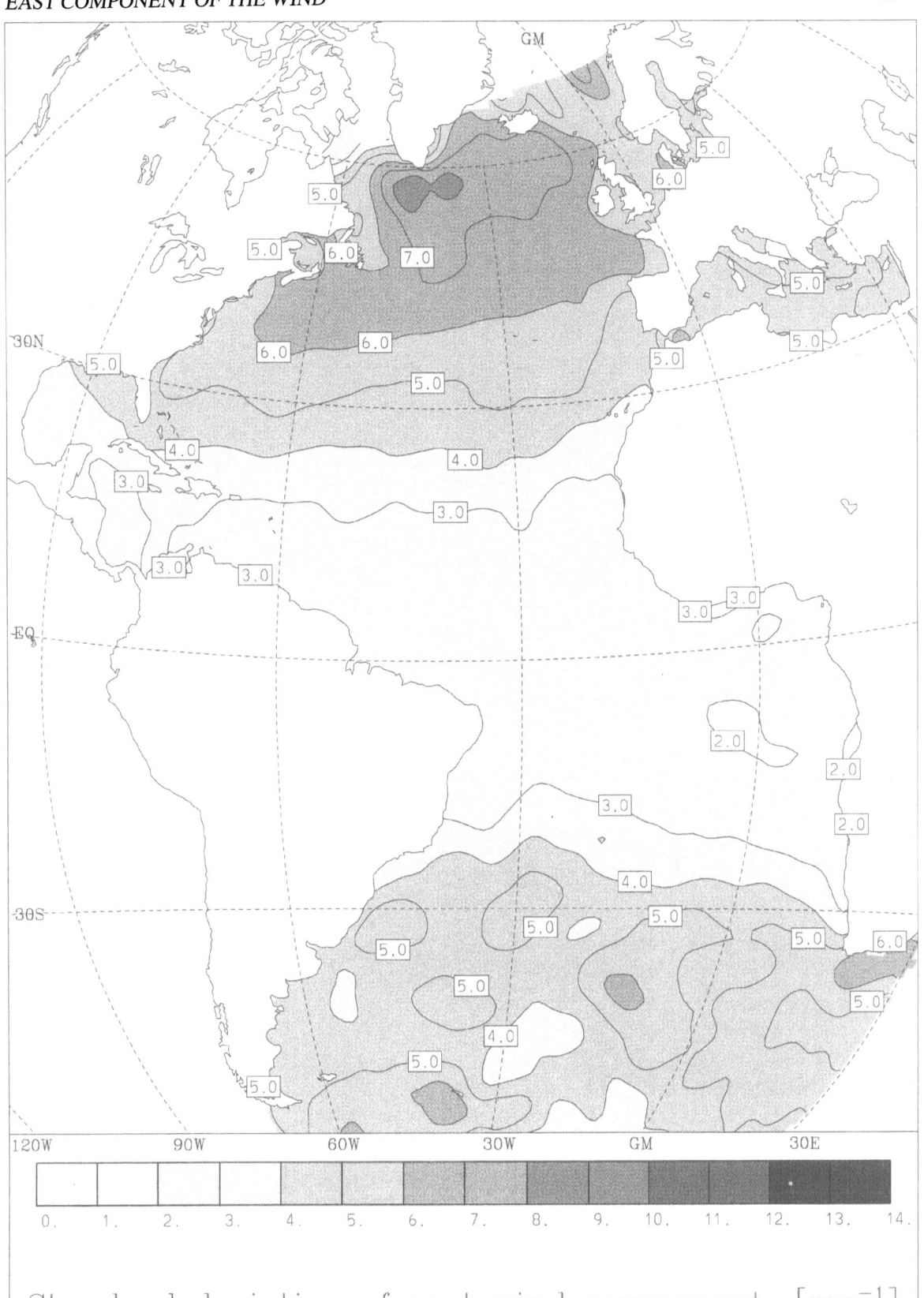

Standard deviation of east wind component [ms^{-1}]
April

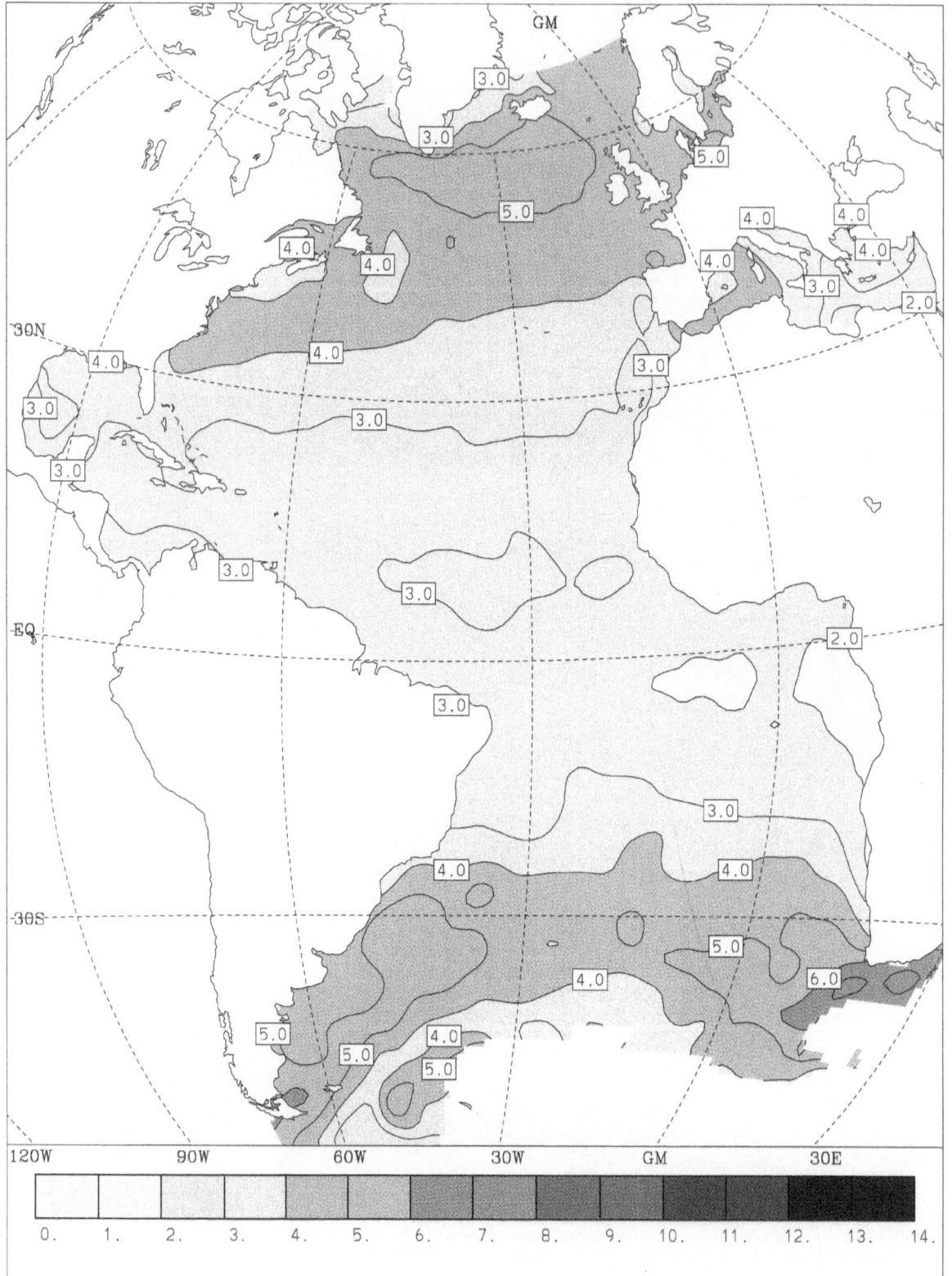

Standard deviation of east wind component [ms^{-1}]
July

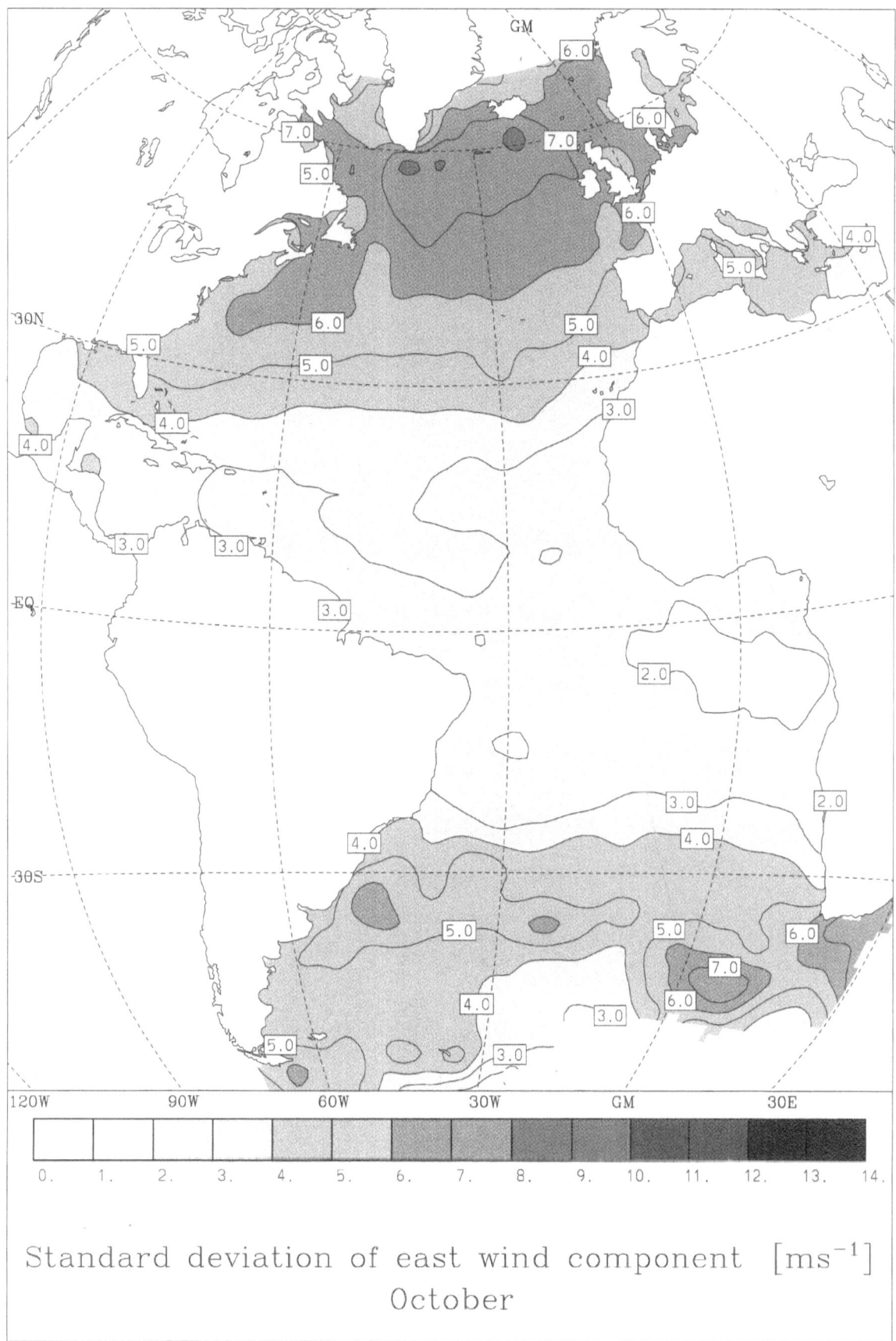

Standard deviation of east wind component [ms^{-1}]
October

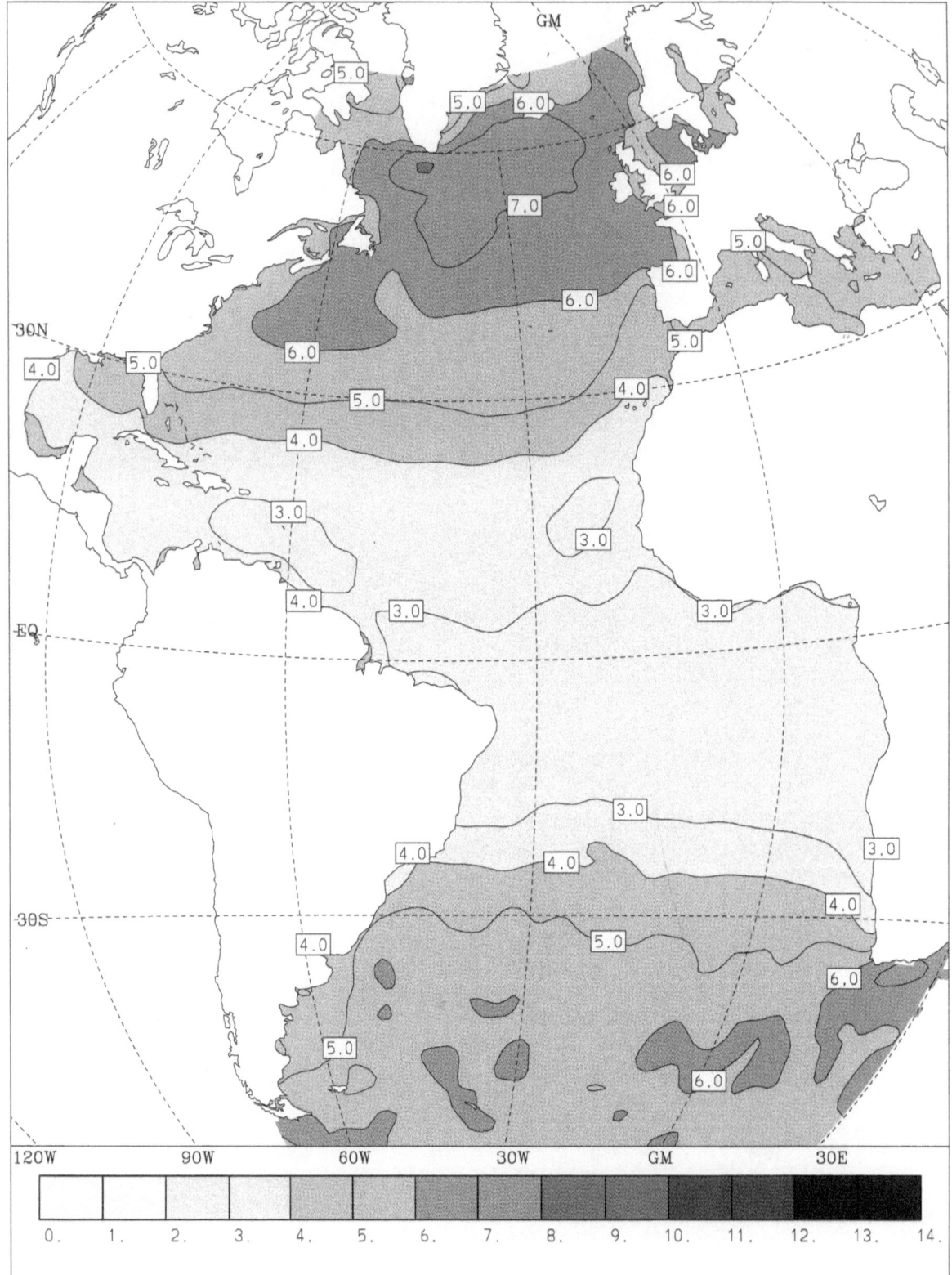

Standard deviation of east wind component [ms^{-1}]
Year

North Component of the Wind

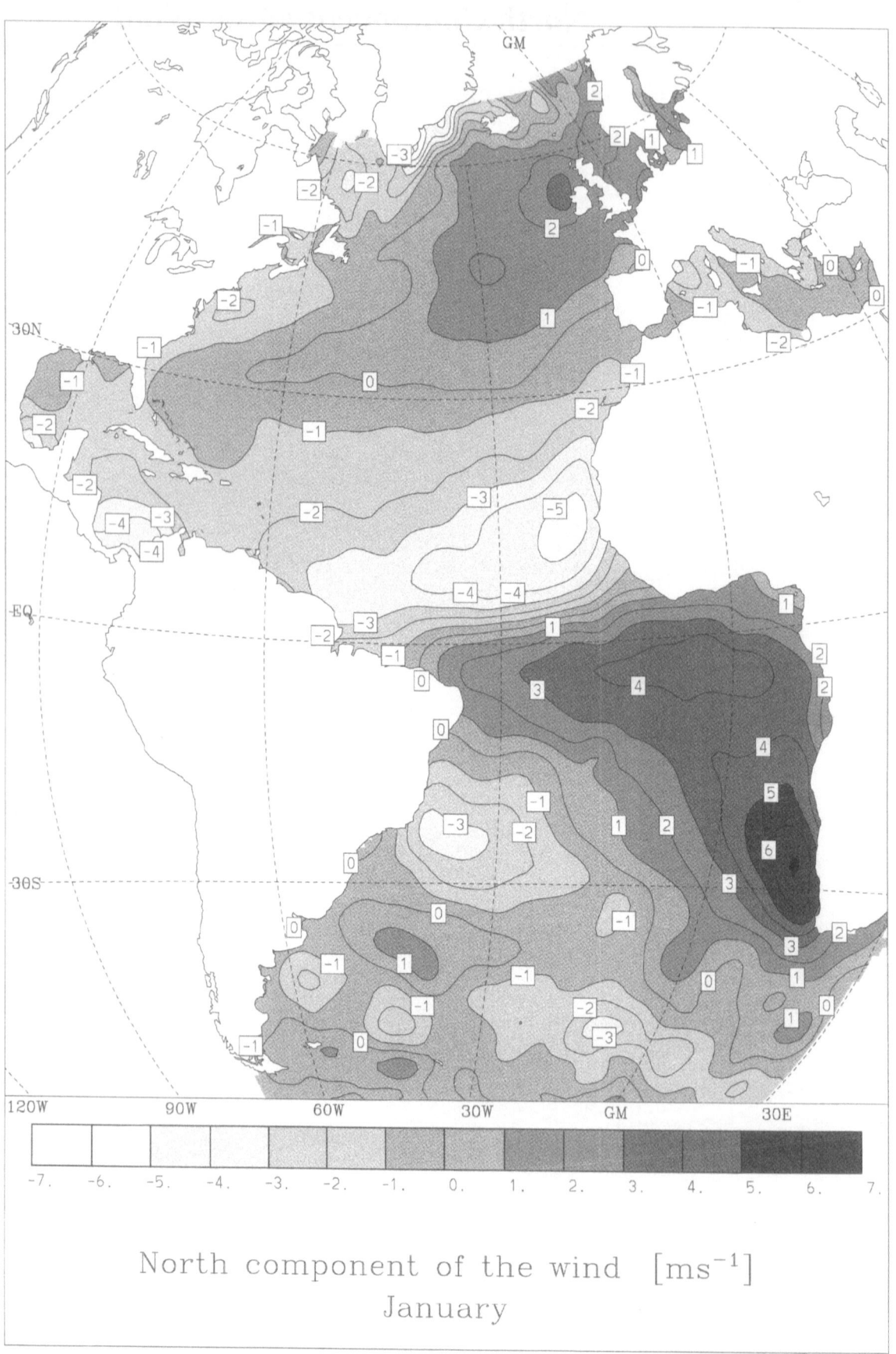

North component of the wind [ms^{-1}]
January

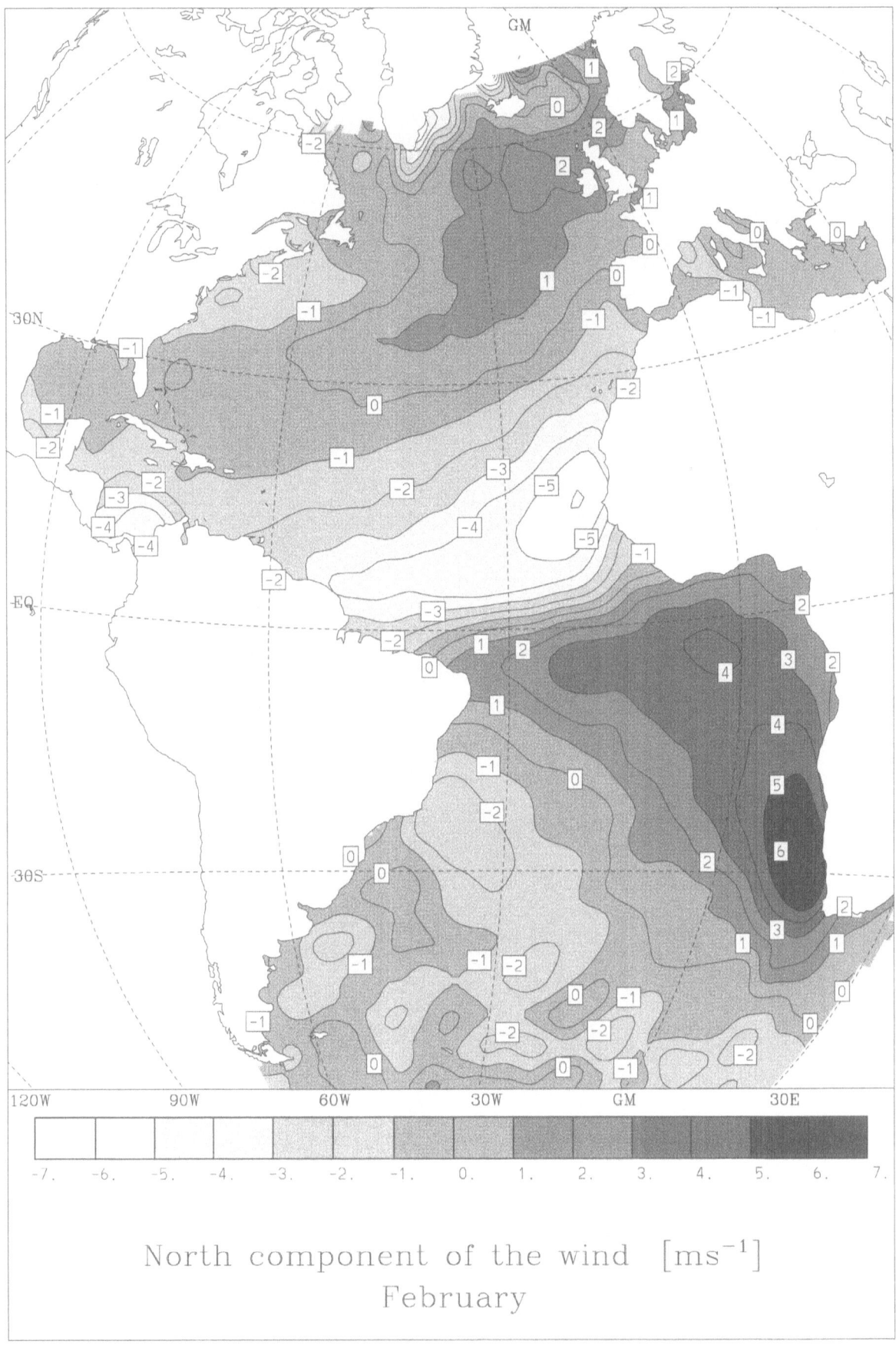

North component of the wind [ms^{-1}]
February

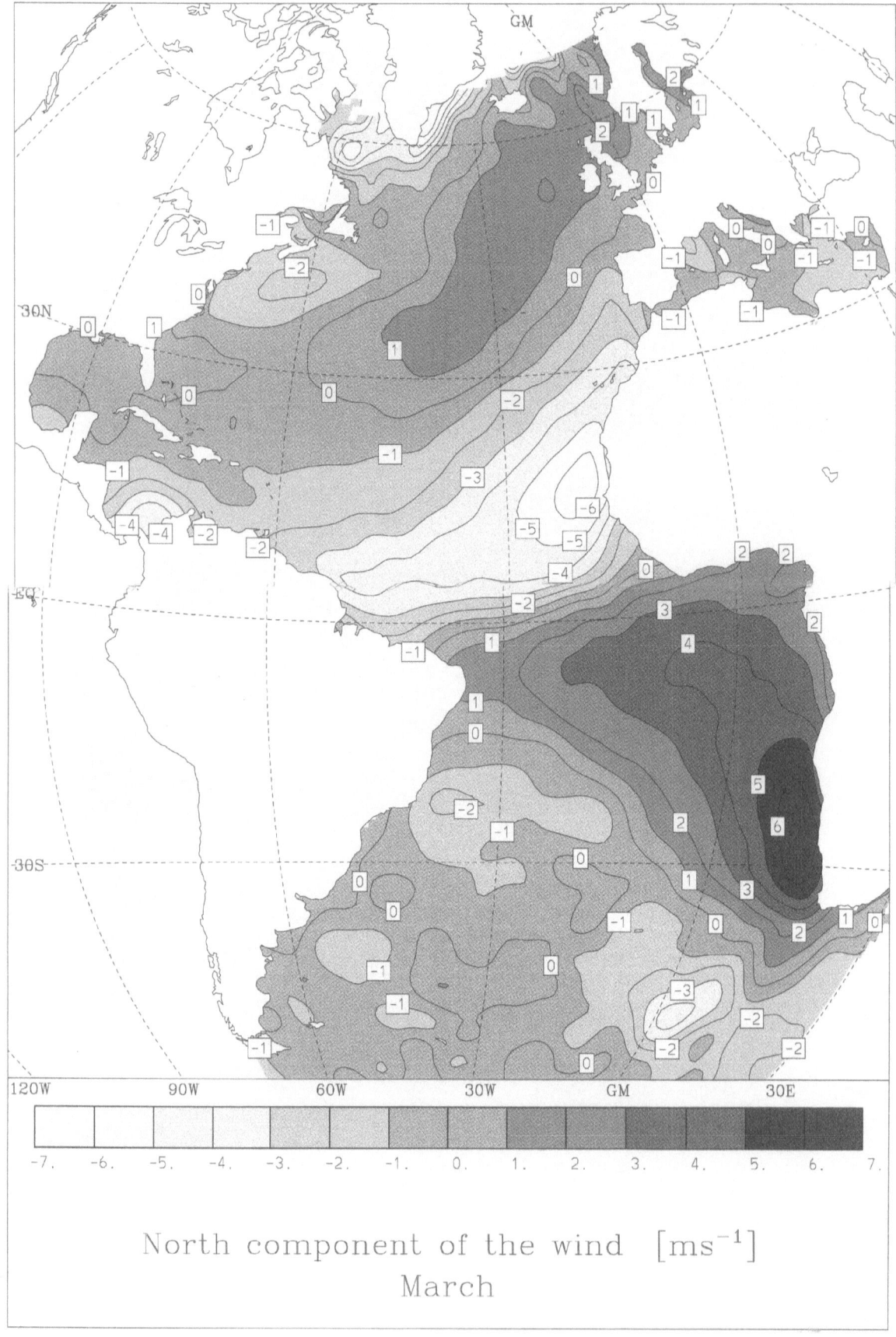

North component of the wind [ms^{-1}]
March

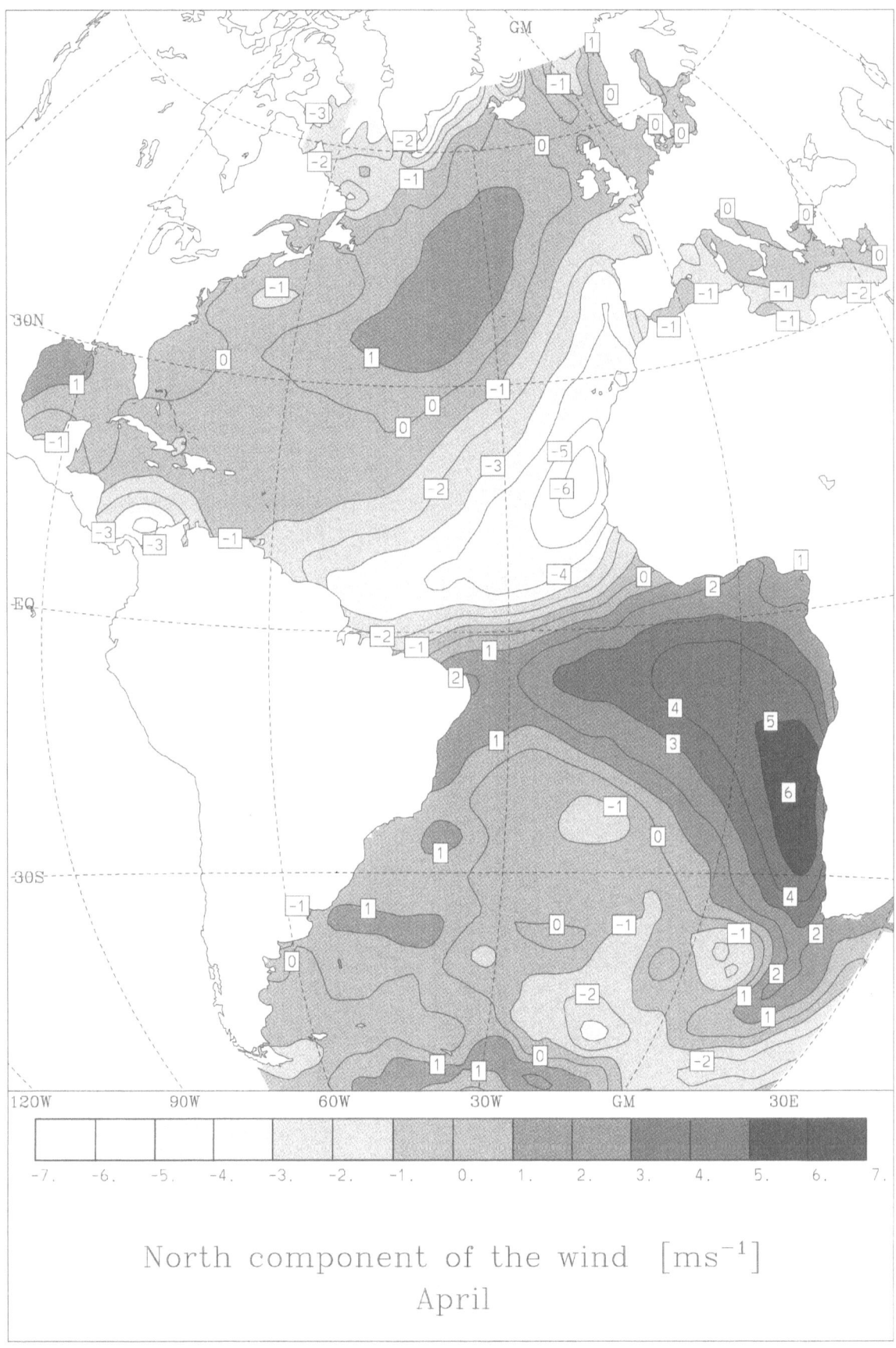

North component of the wind [ms⁻¹]
April

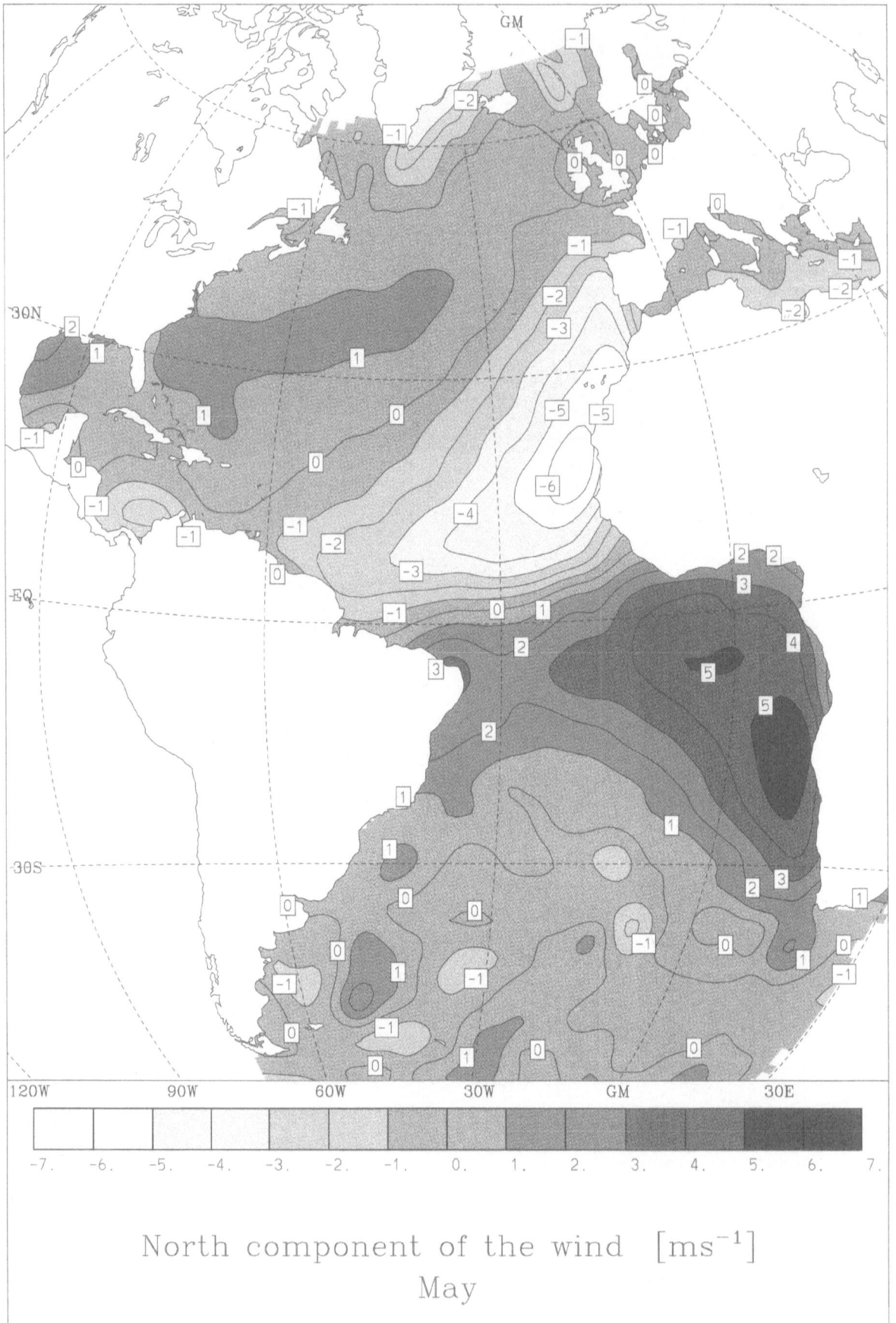

North component of the wind [ms^{-1}]
May

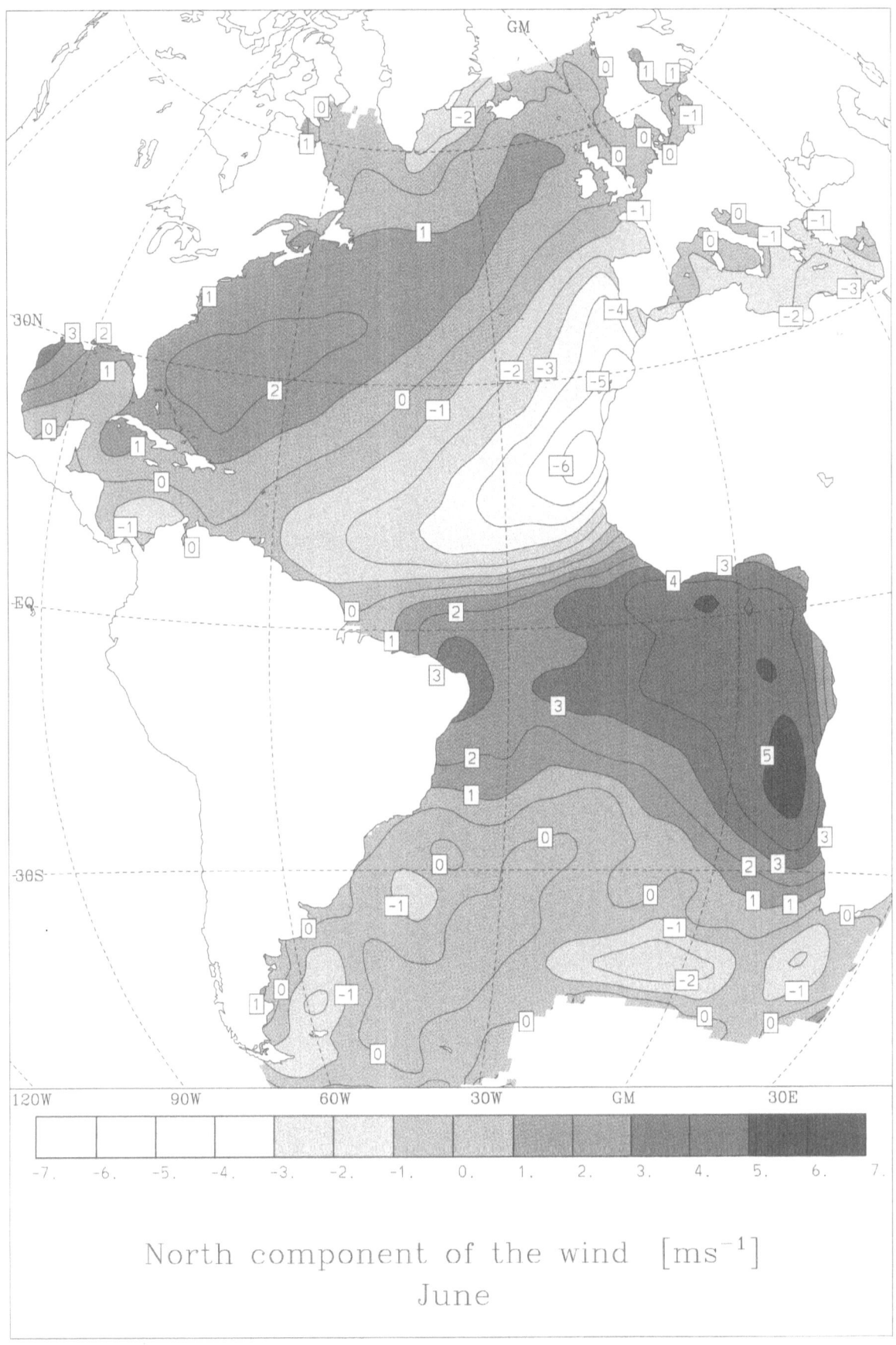

North component of the wind [ms^{-1}]
June

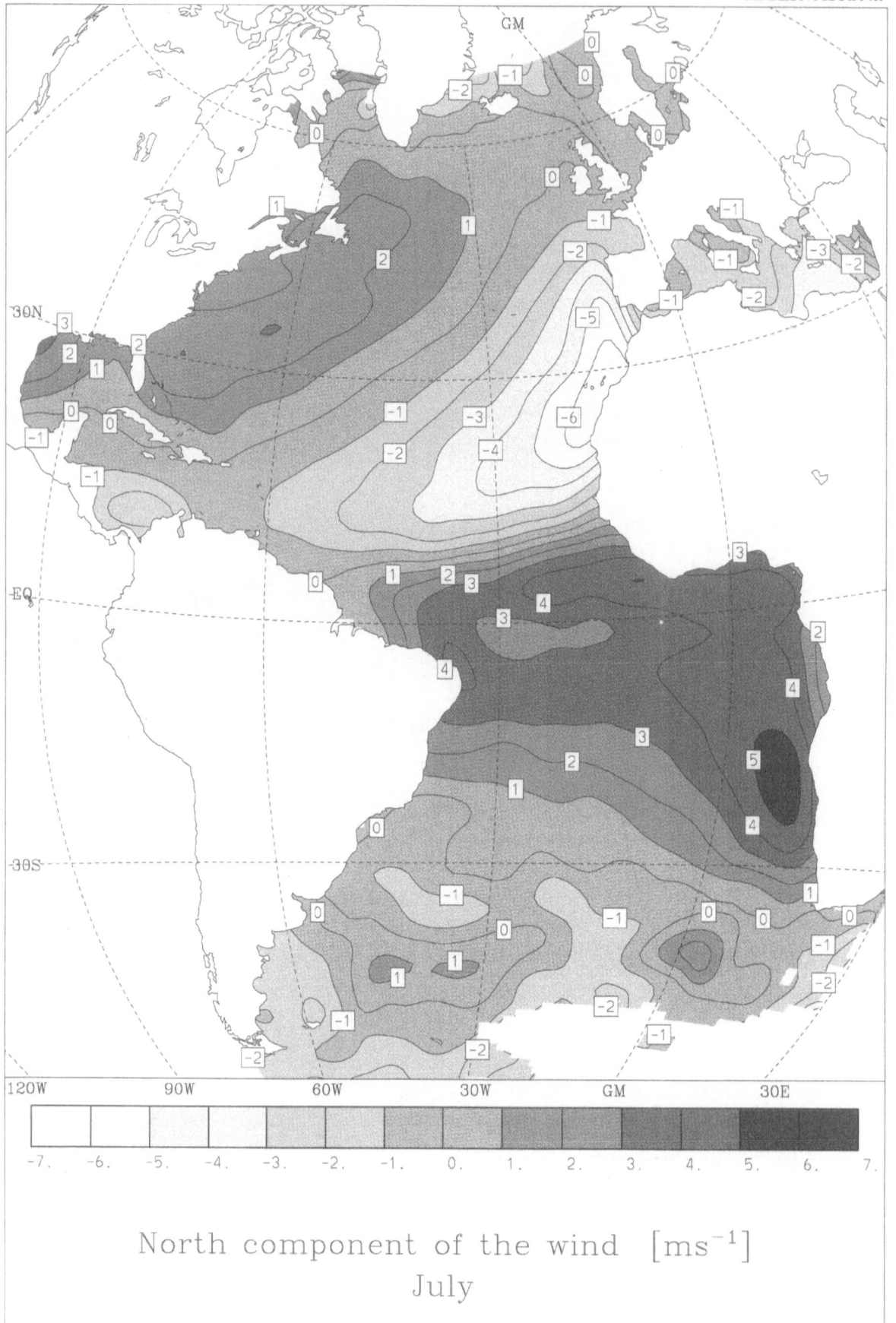

North component of the wind [ms^{-1}]
July

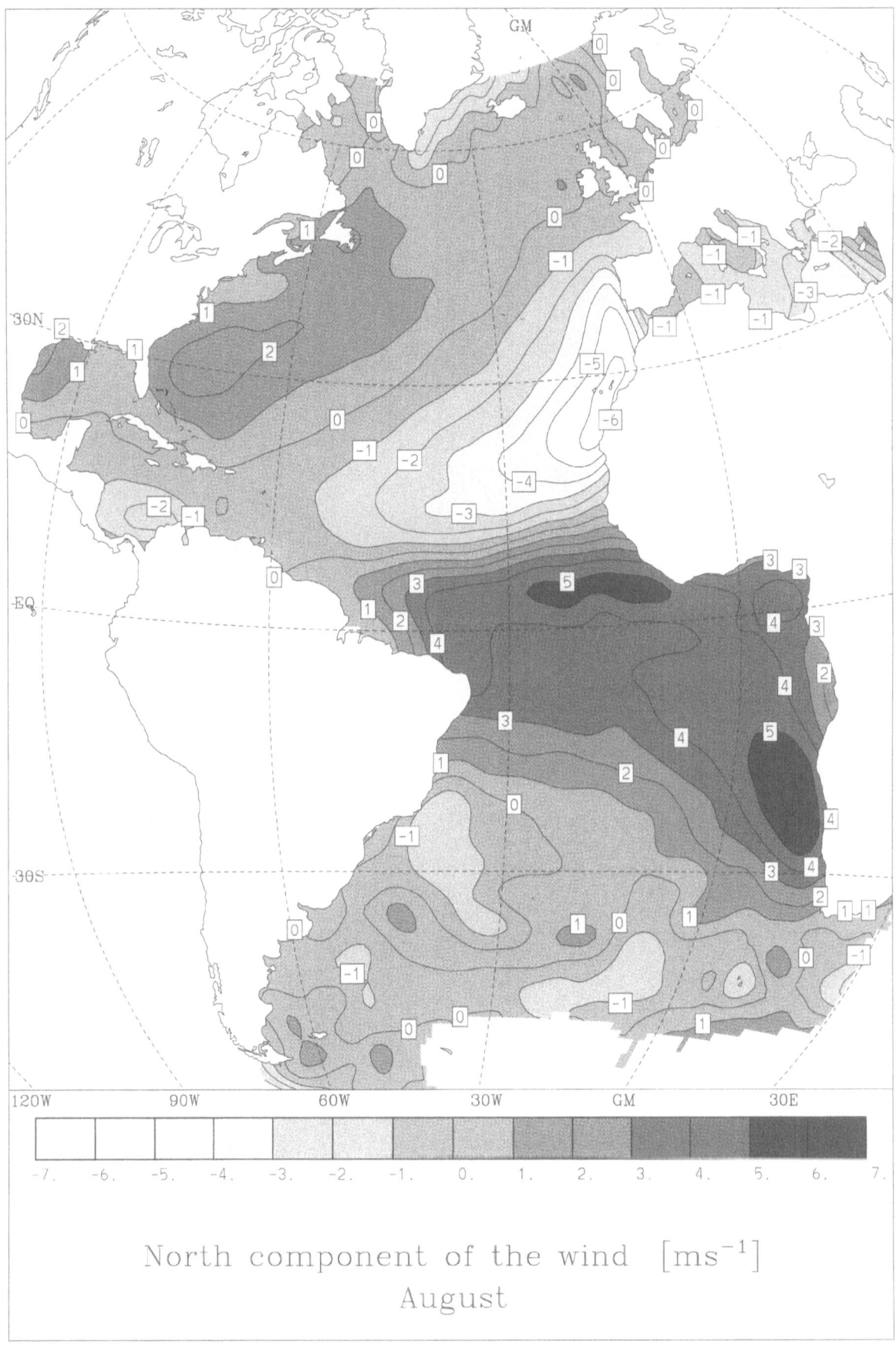

North component of the wind [ms^{-1}]

August

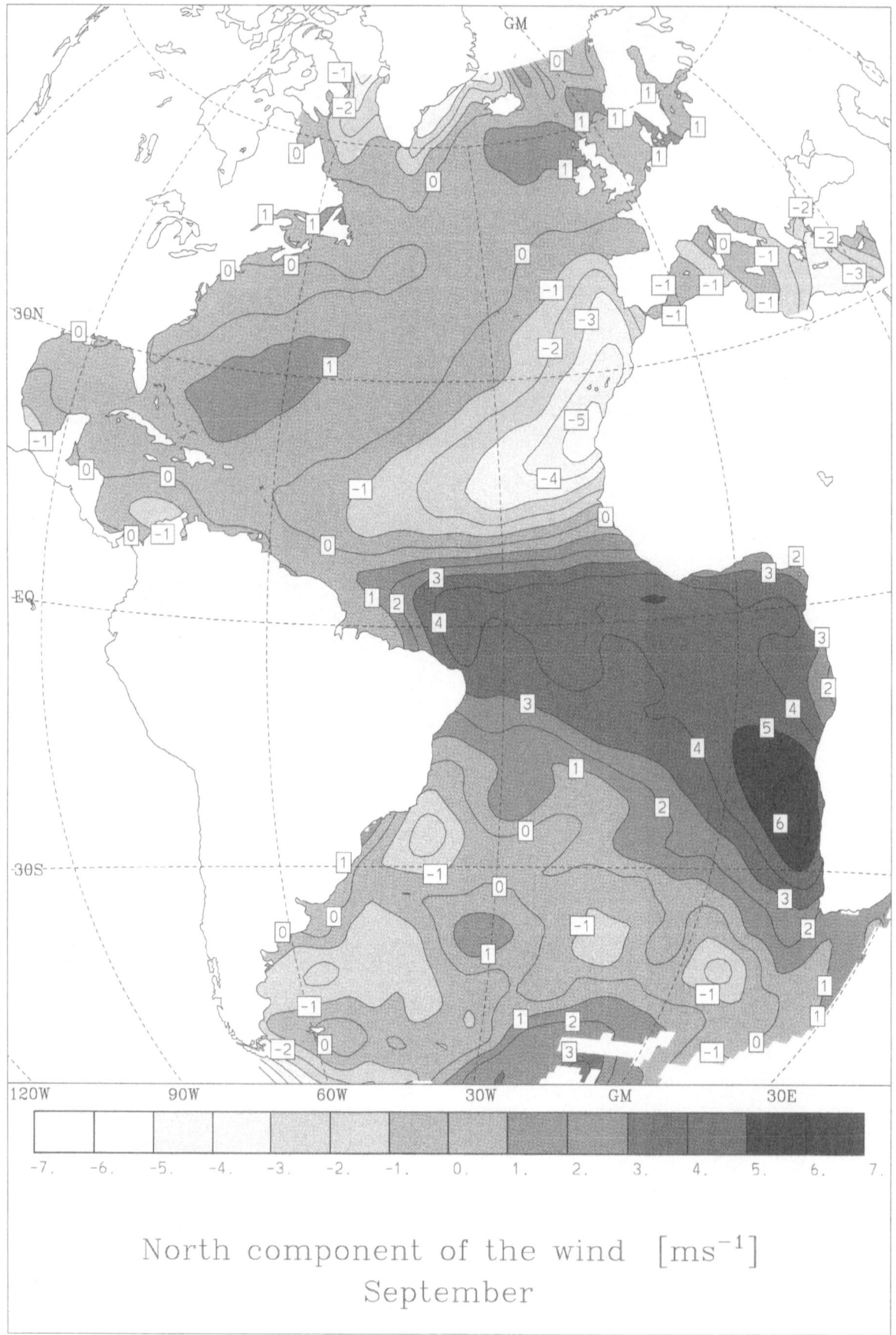

North component of the wind [ms⁻¹]
September

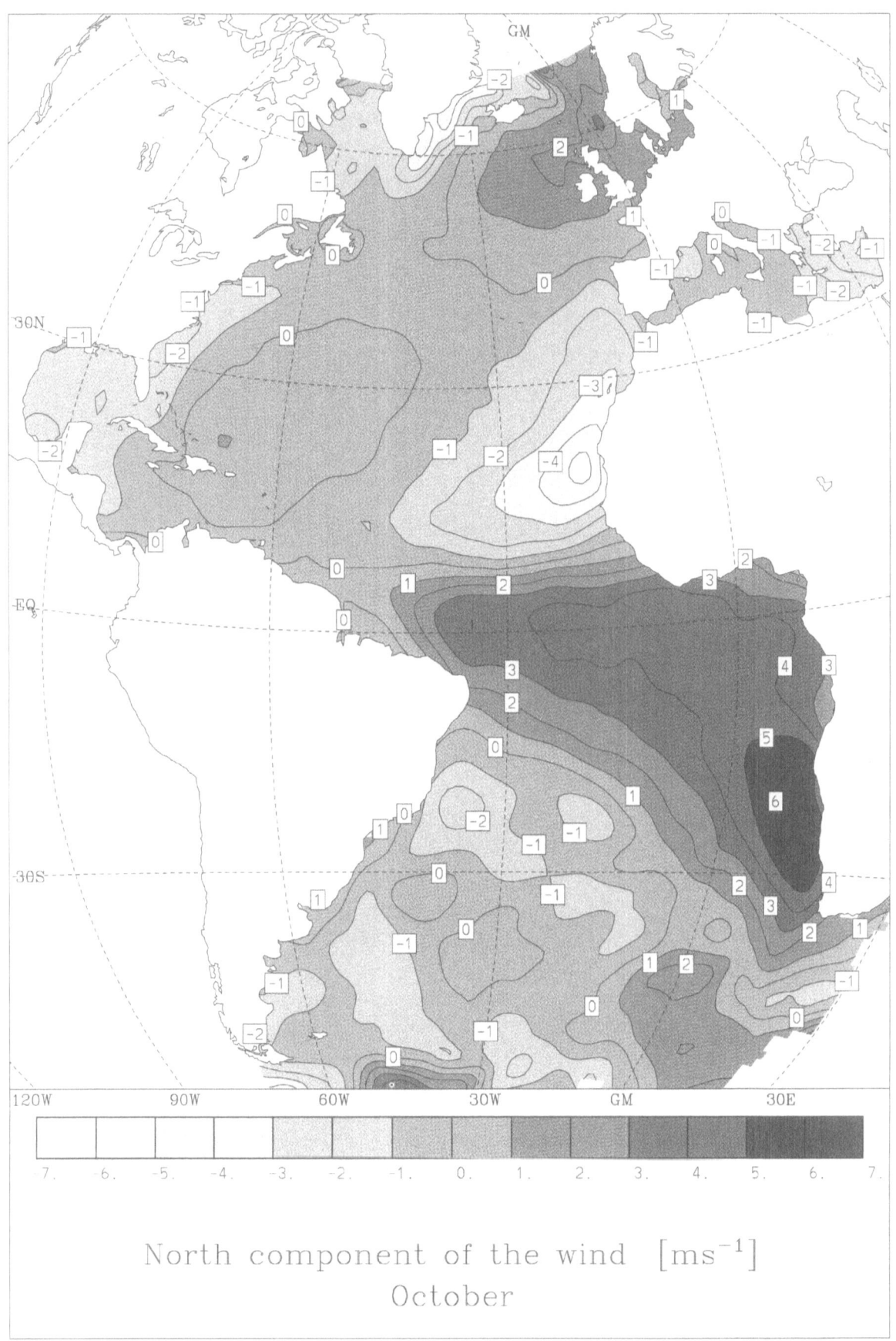

North component of the wind $[ms^{-1}]$
October

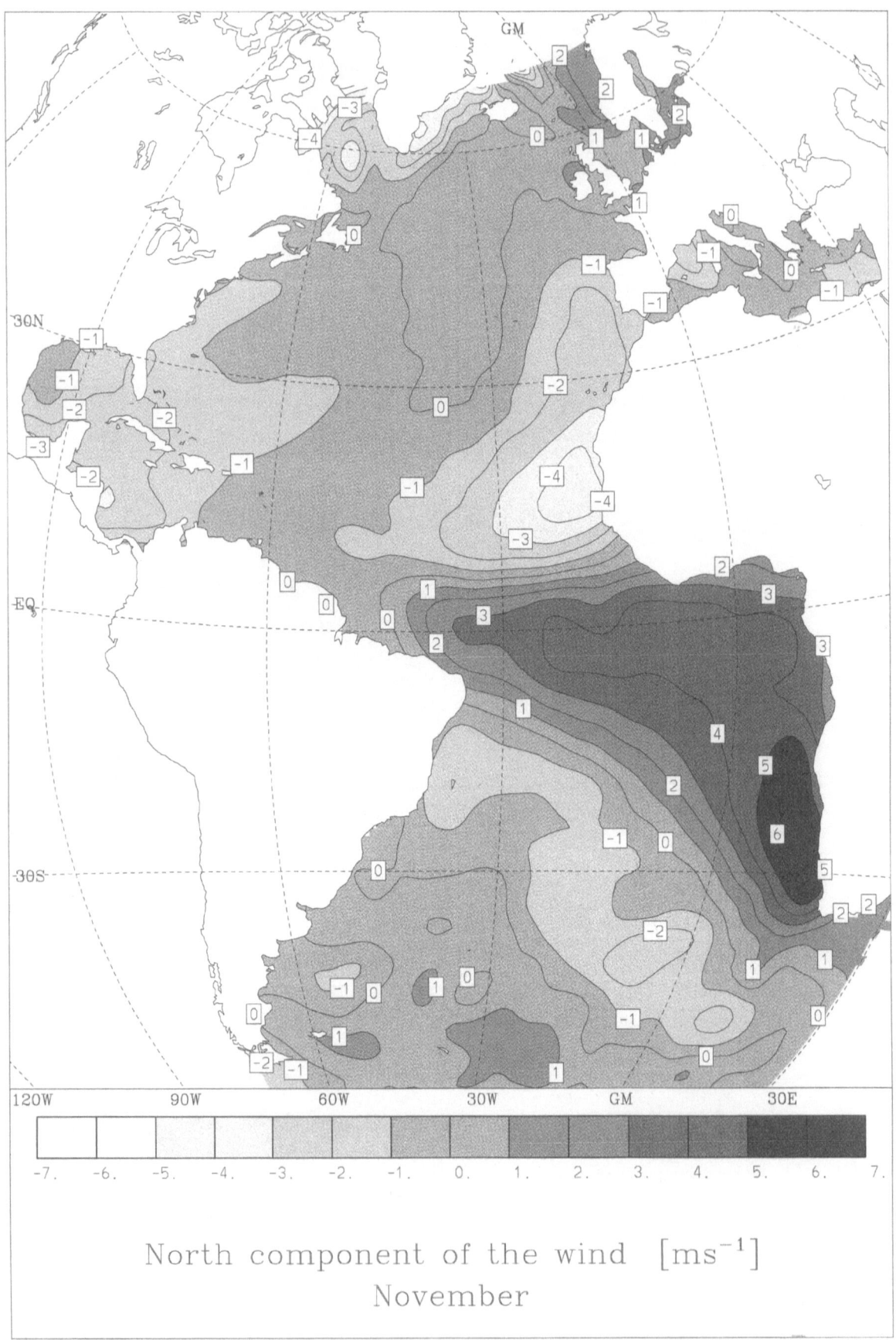

North component of the wind [ms⁻¹]
November

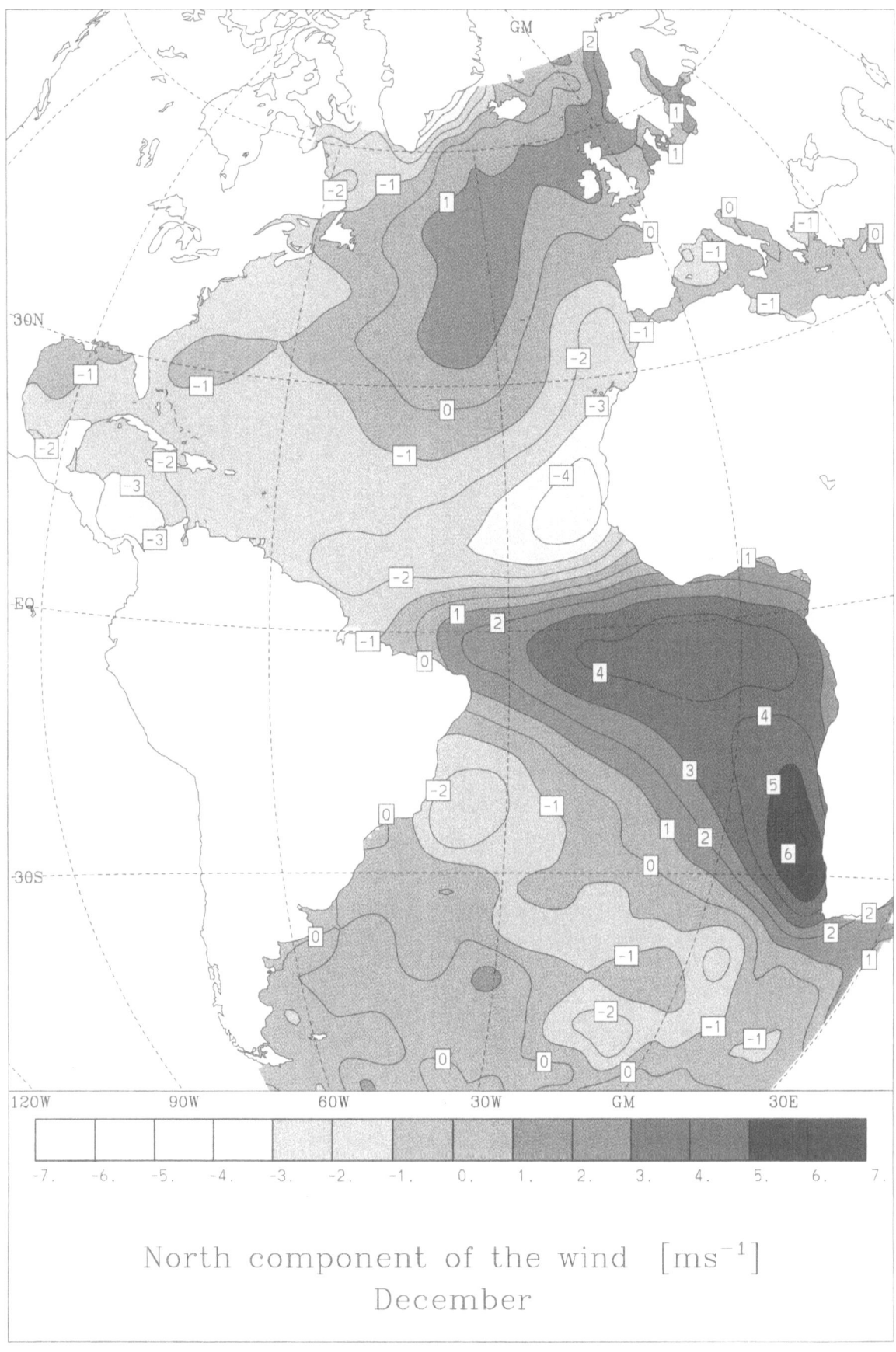

North component of the wind [ms^{-1}]
December

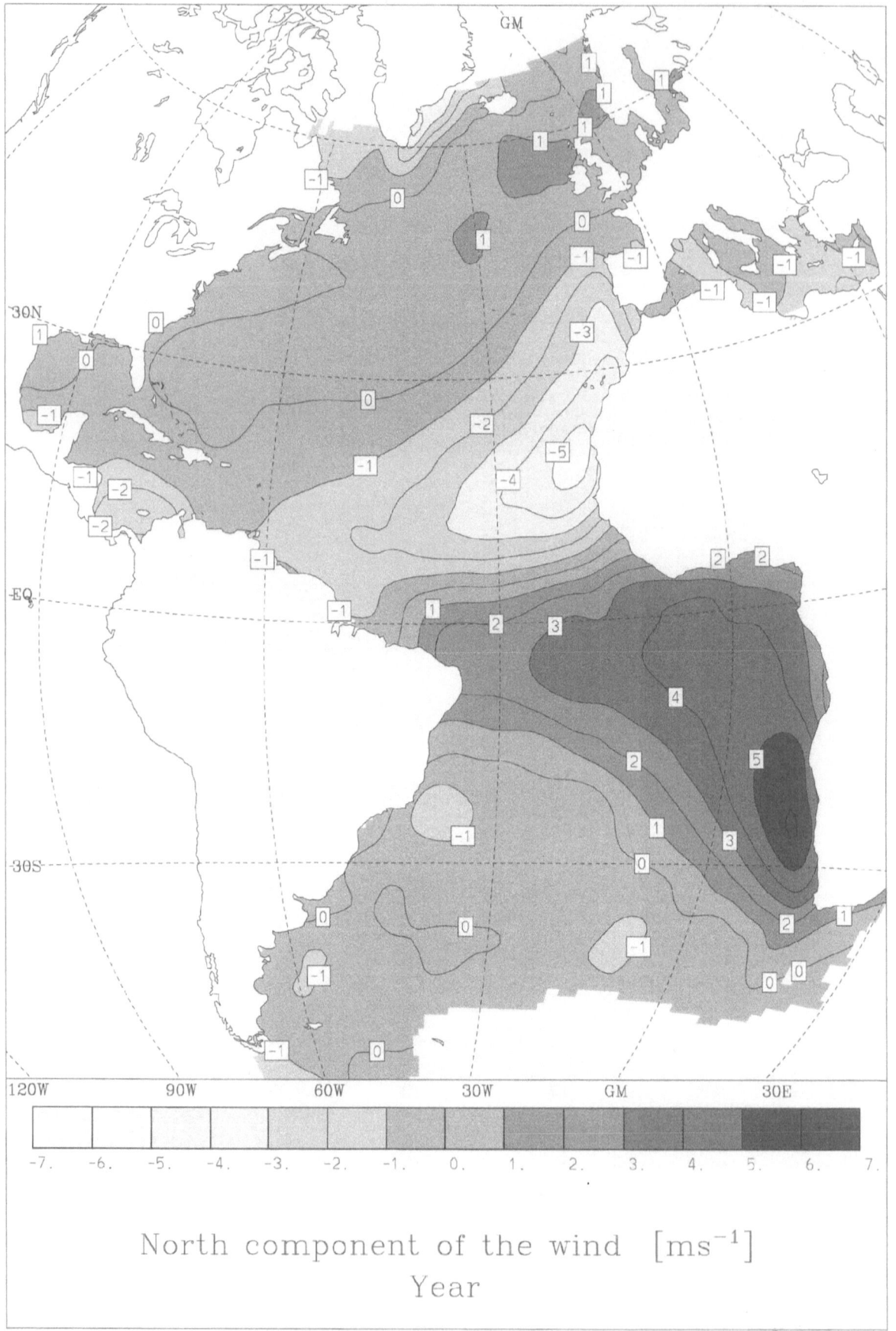

North component of the wind [ms⁻¹]
Year

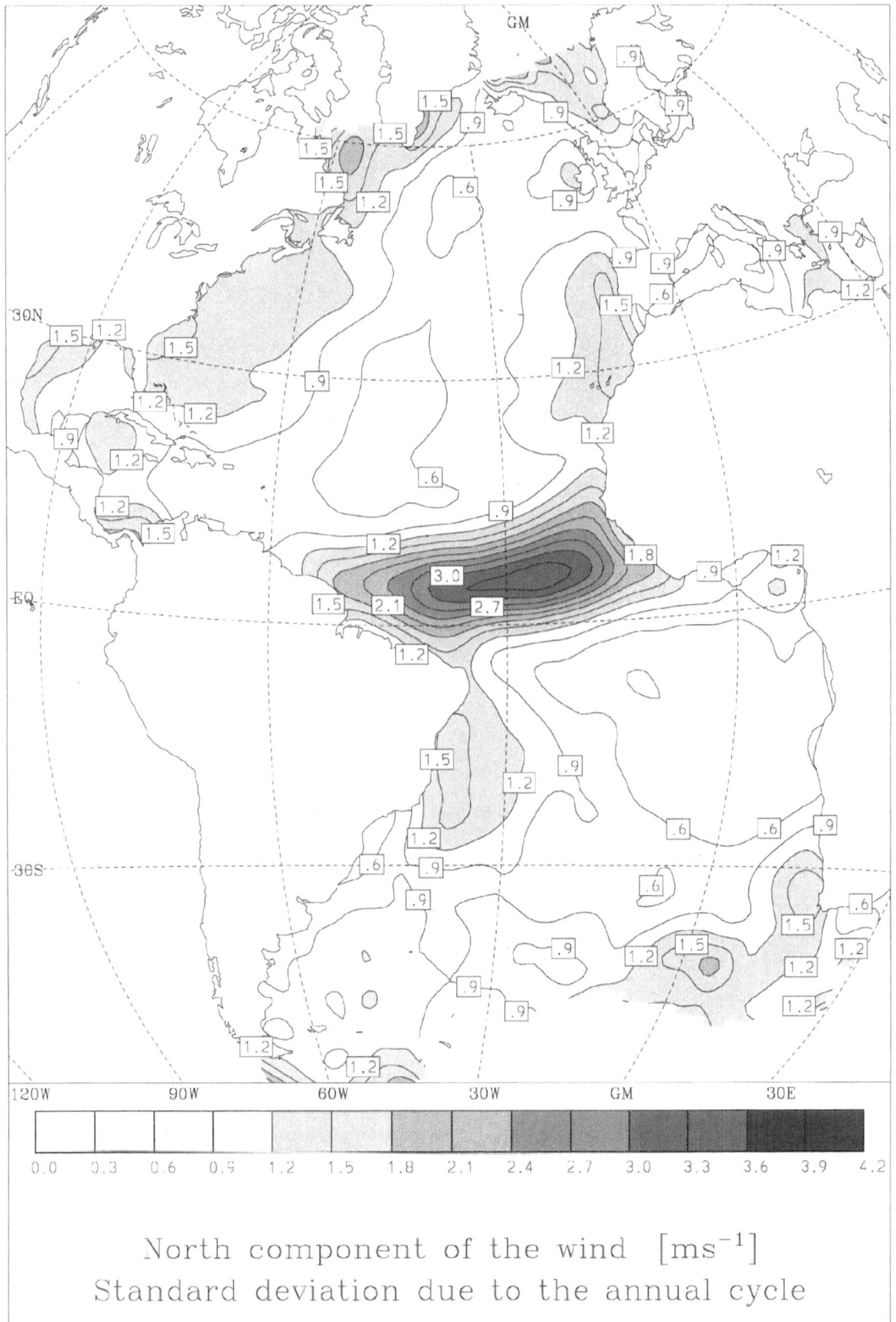

North component of the wind [ms⁻¹]
Standard deviation due to the annual cycle

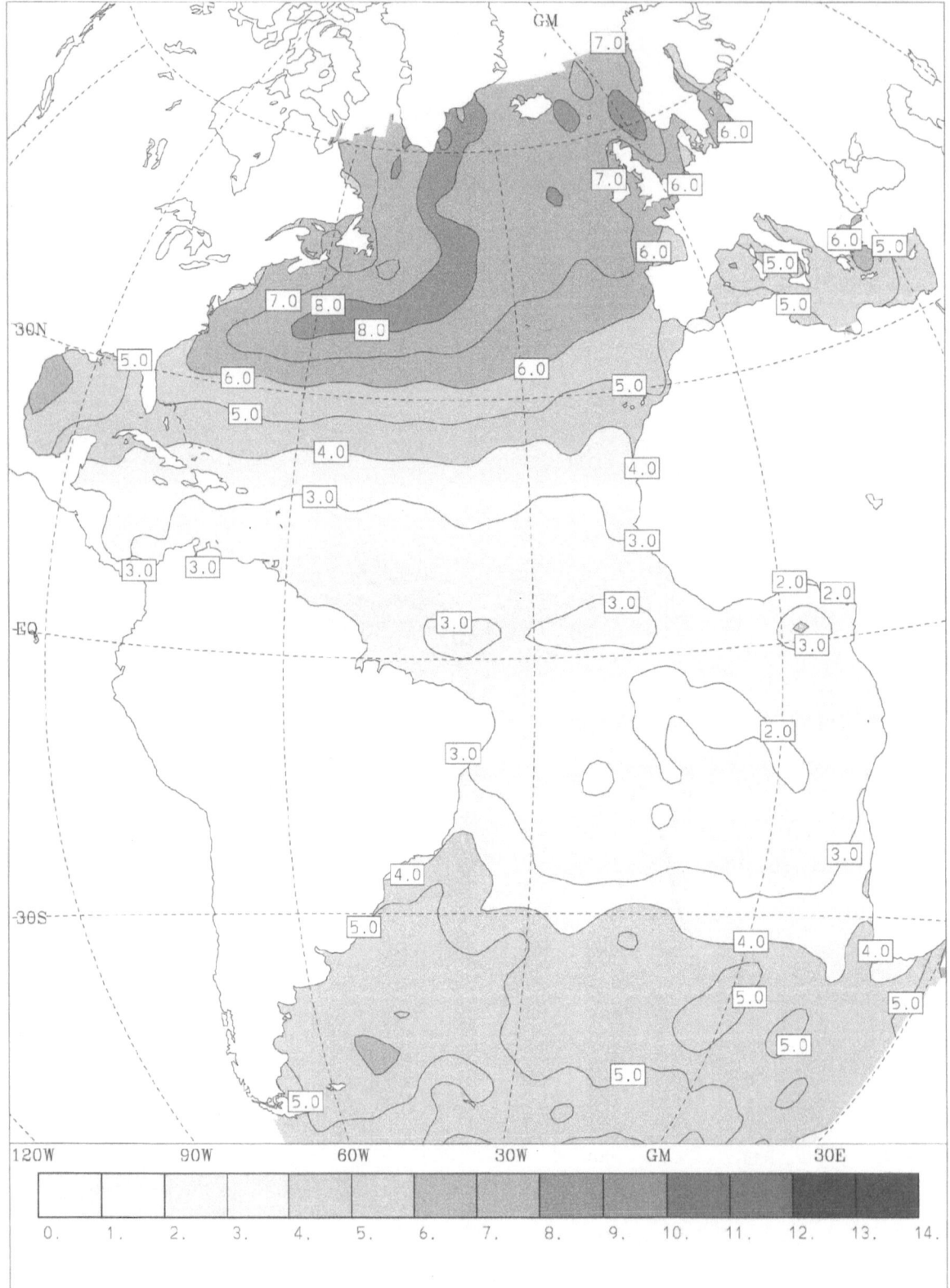

Standard deviation of north wind component [ms^{-1}]
January

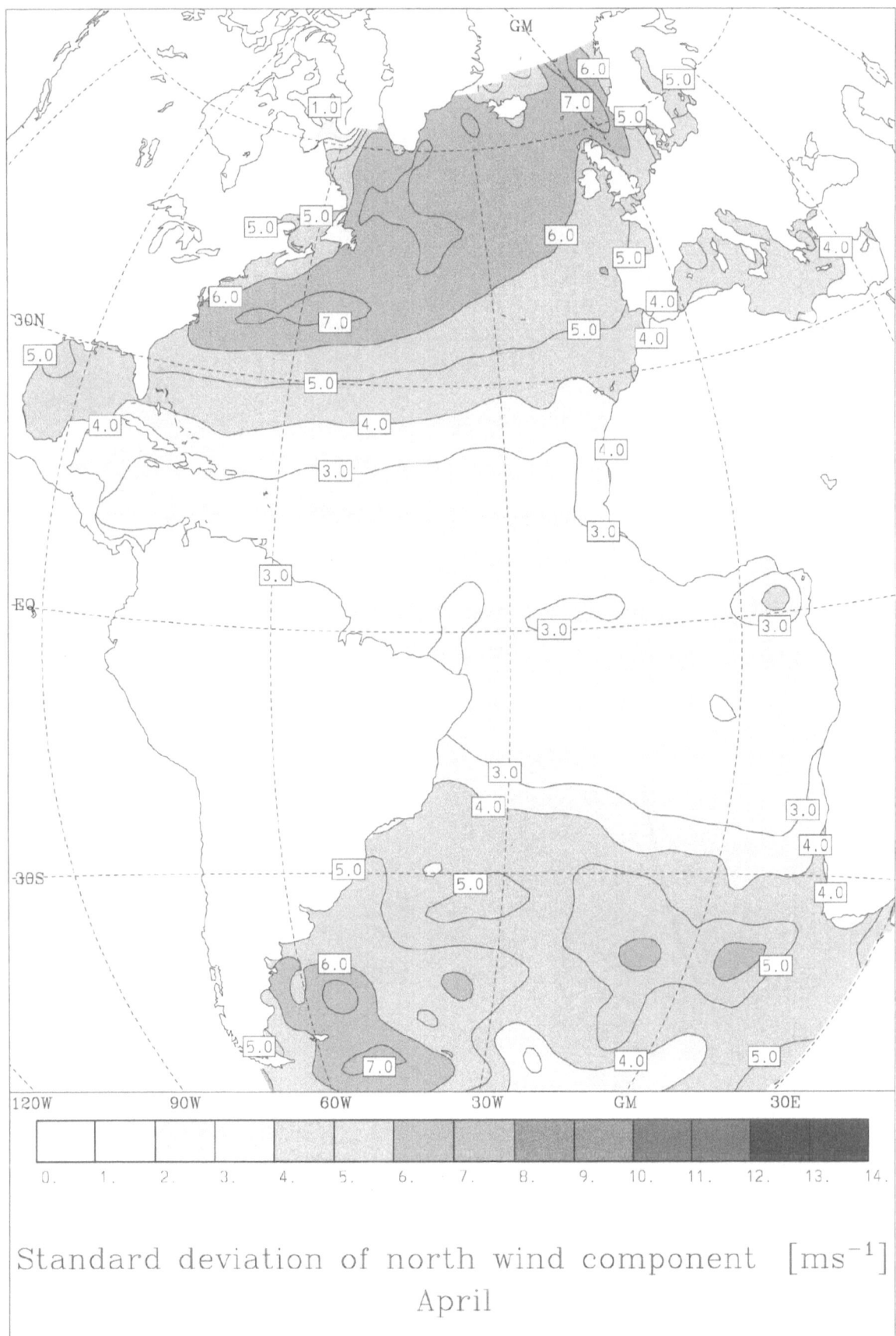

Standard deviation of north wind component [ms⁻¹]
April

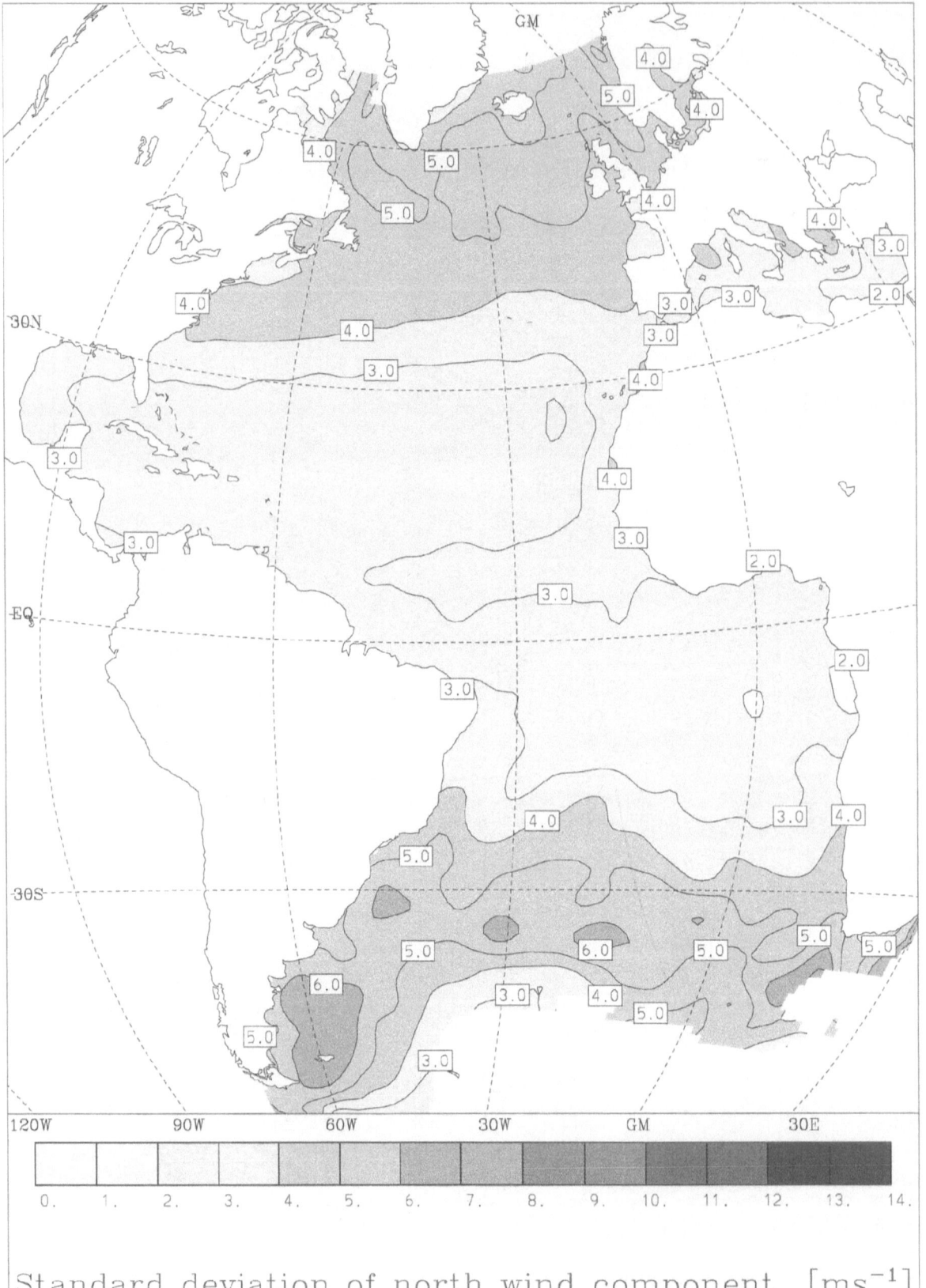

Standard deviation of north wind component [ms⁻¹]
July

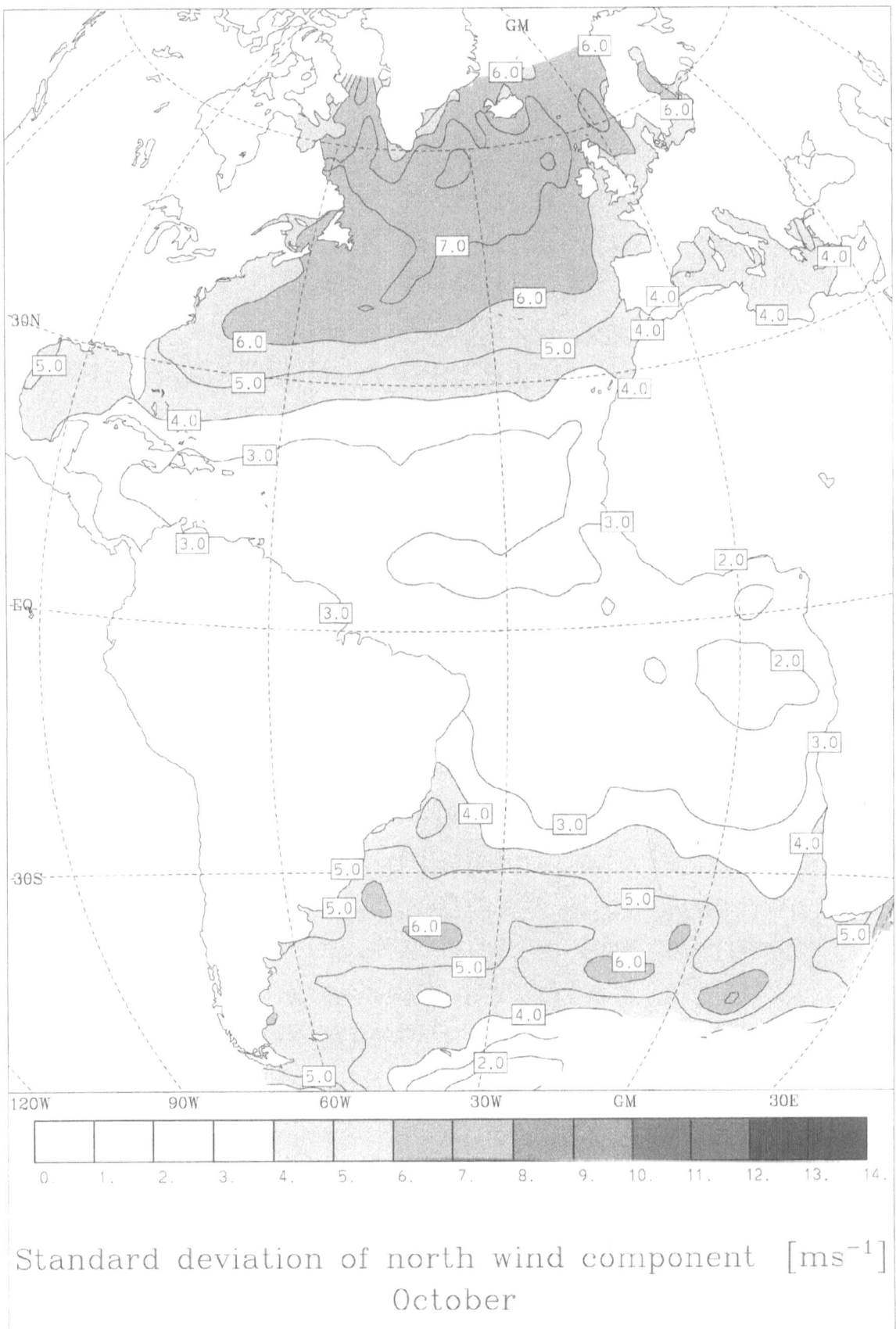

Standard deviation of north wind component [ms^{-1}]
October

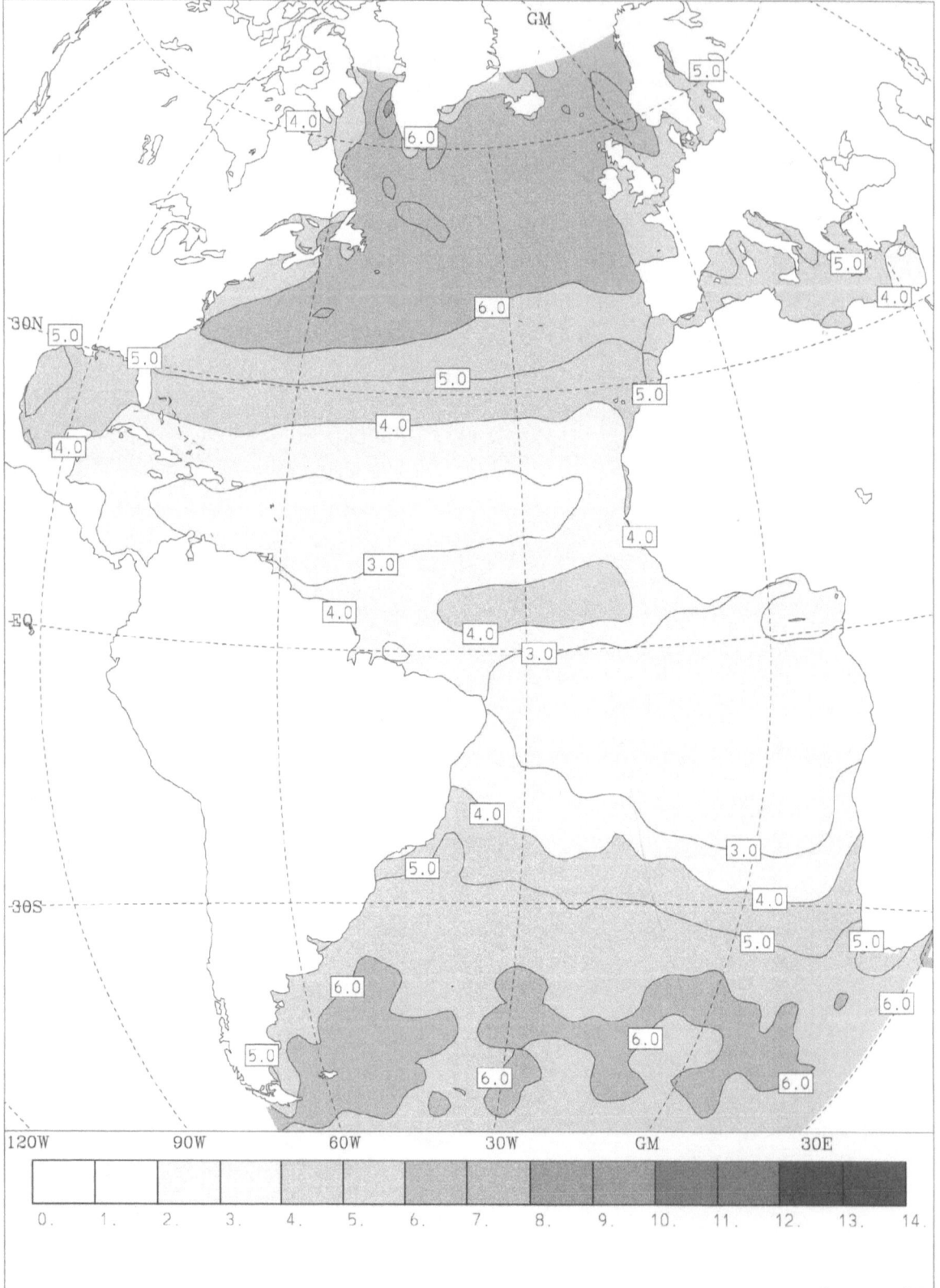

Standard deviation of north wind component $[\mathrm{ms}^{-1}]$
Year

Magnitude of Resultant Wind
plus Resultant Wind Vector

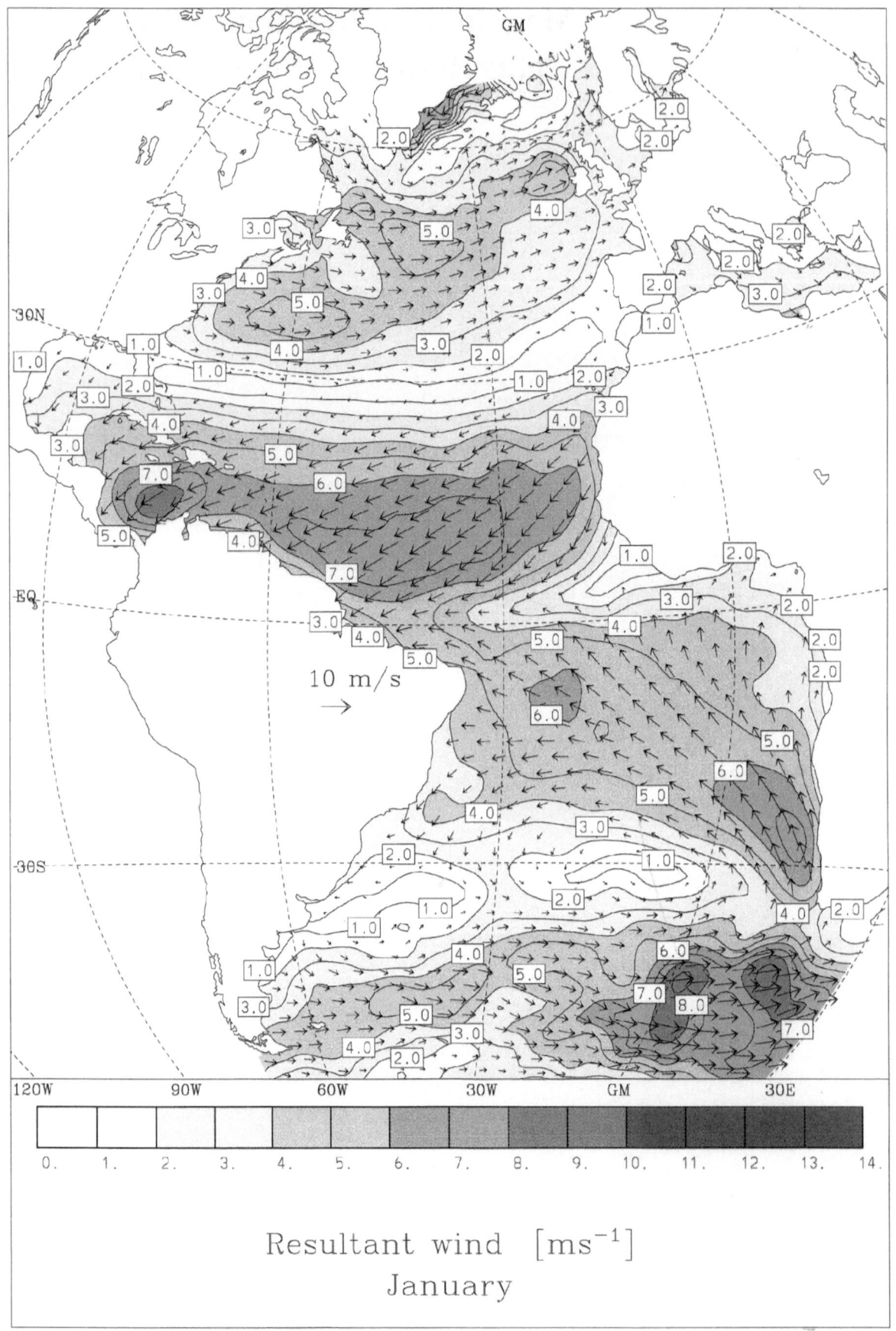

Resultant wind [ms^{-1}]
January

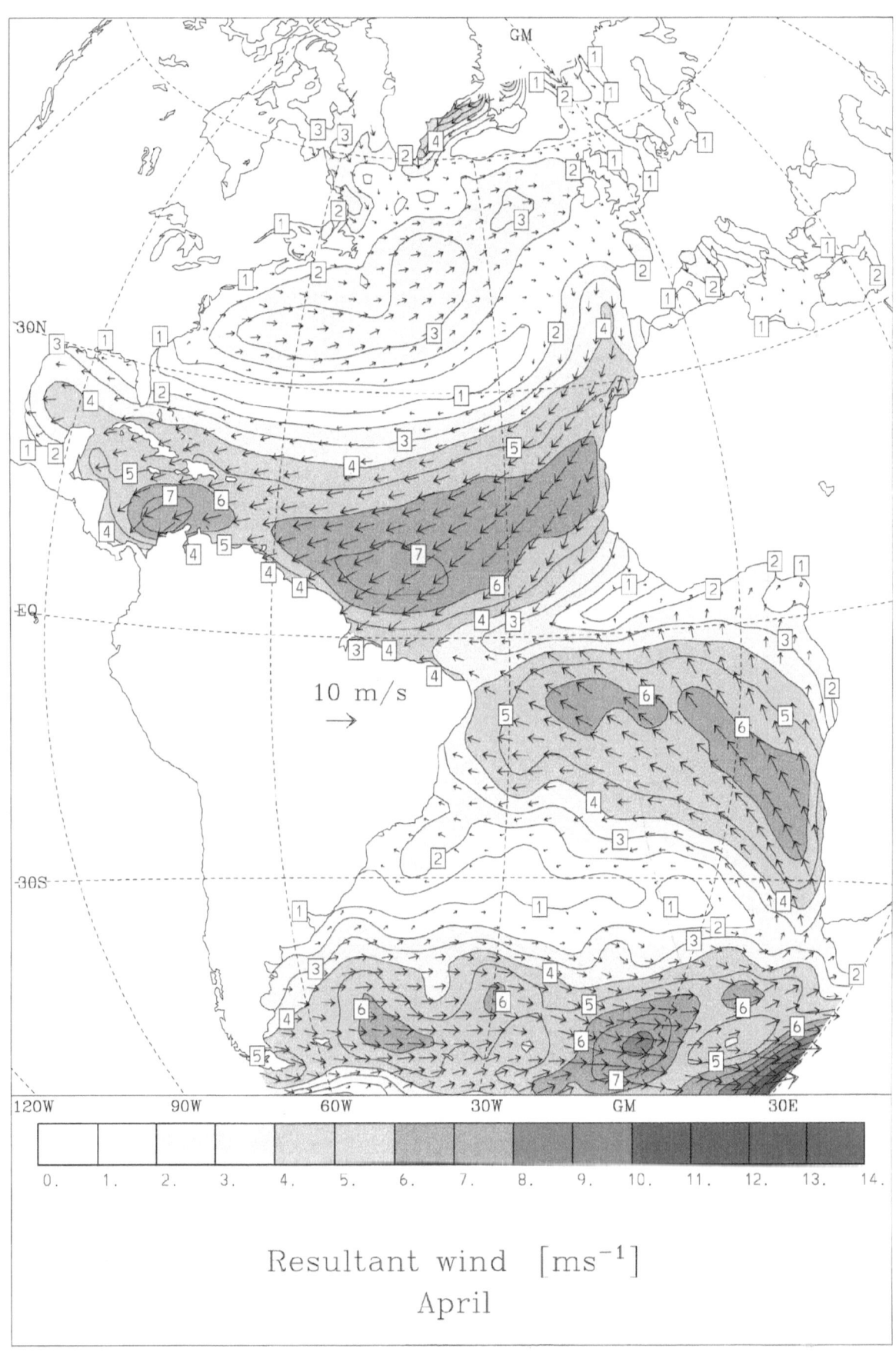

Resultant wind [ms^{-1}]
April

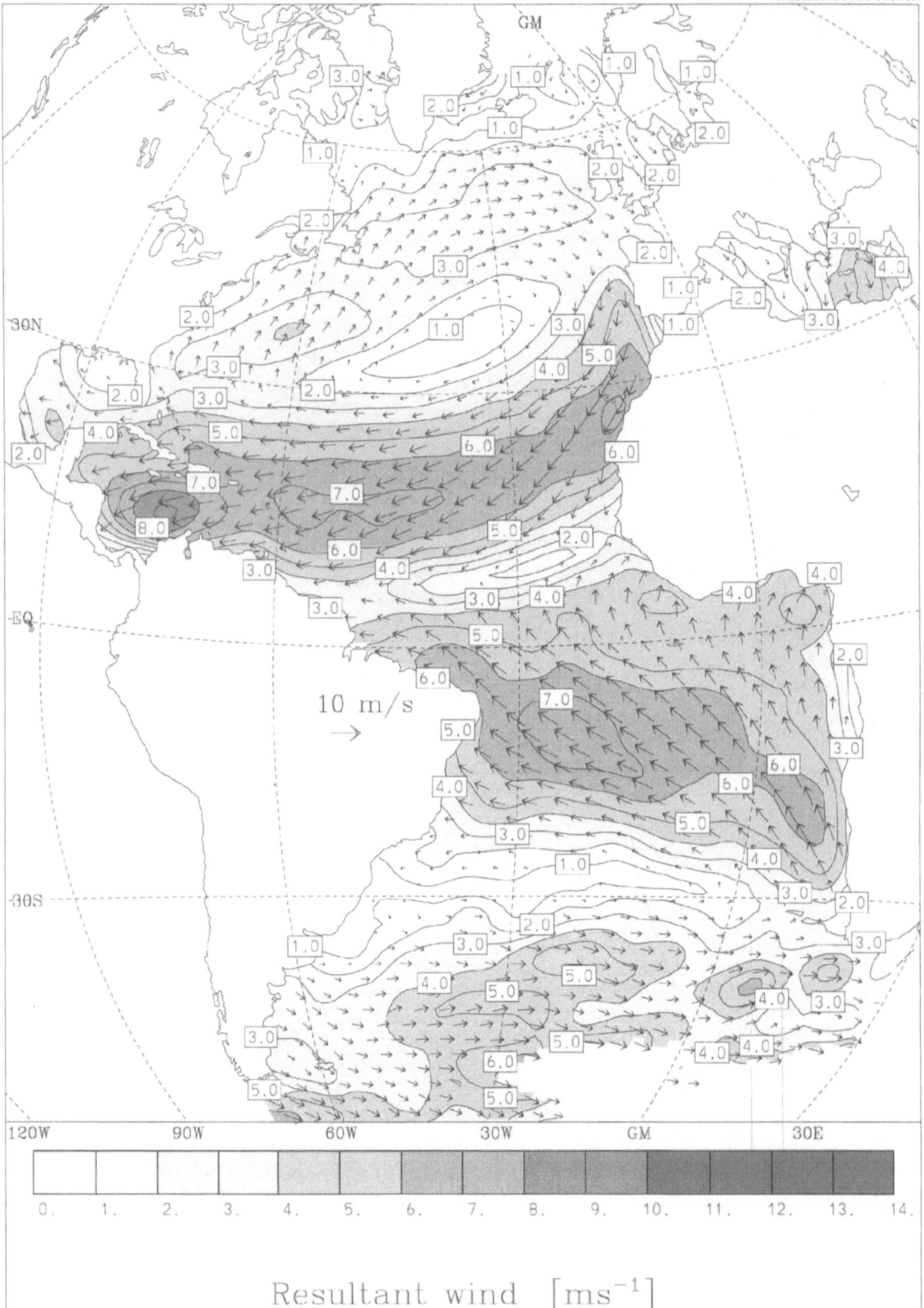

Resultant wind [ms^{-1}]

July

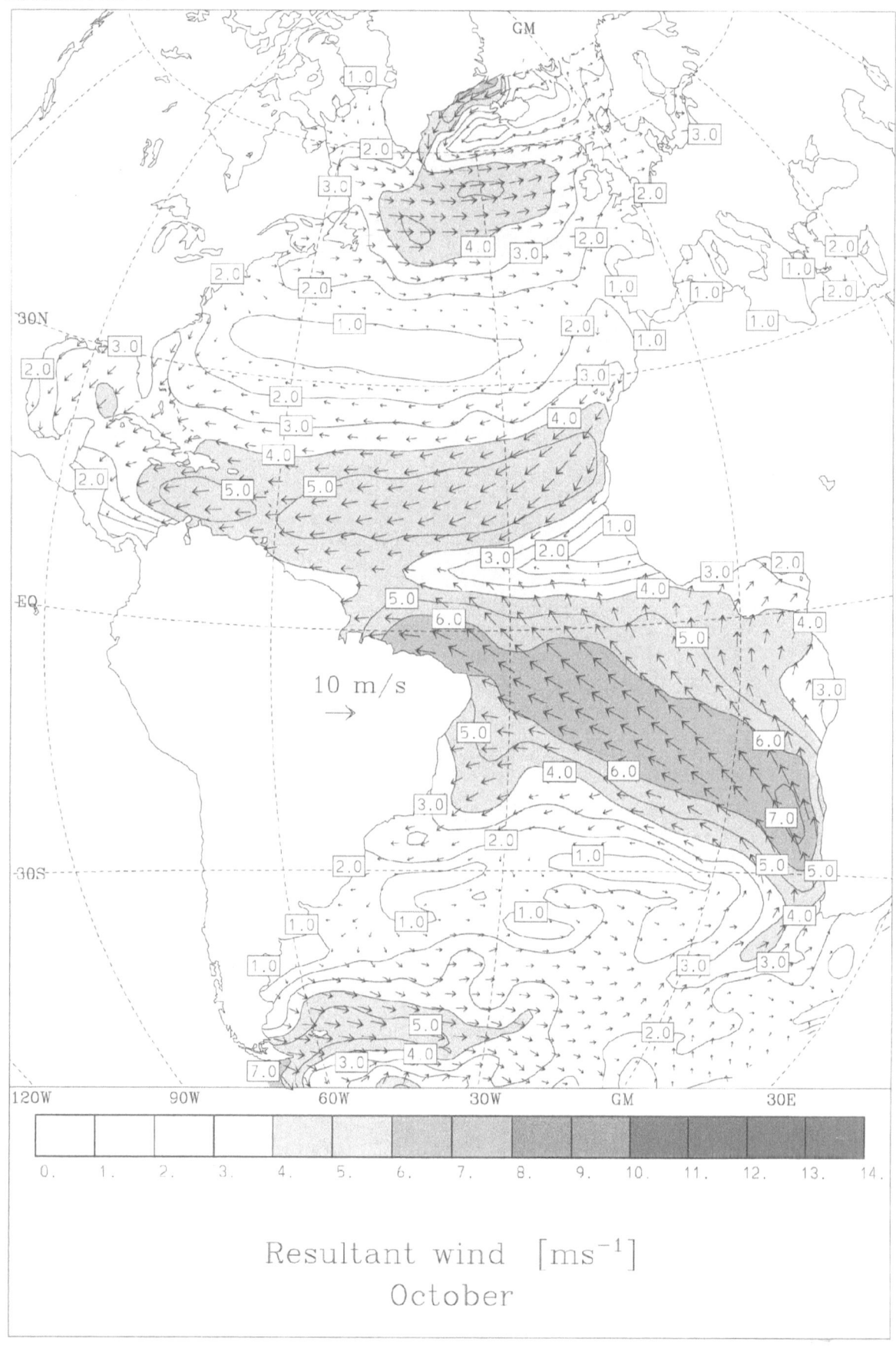

Resultant wind [ms^{-1}]

October

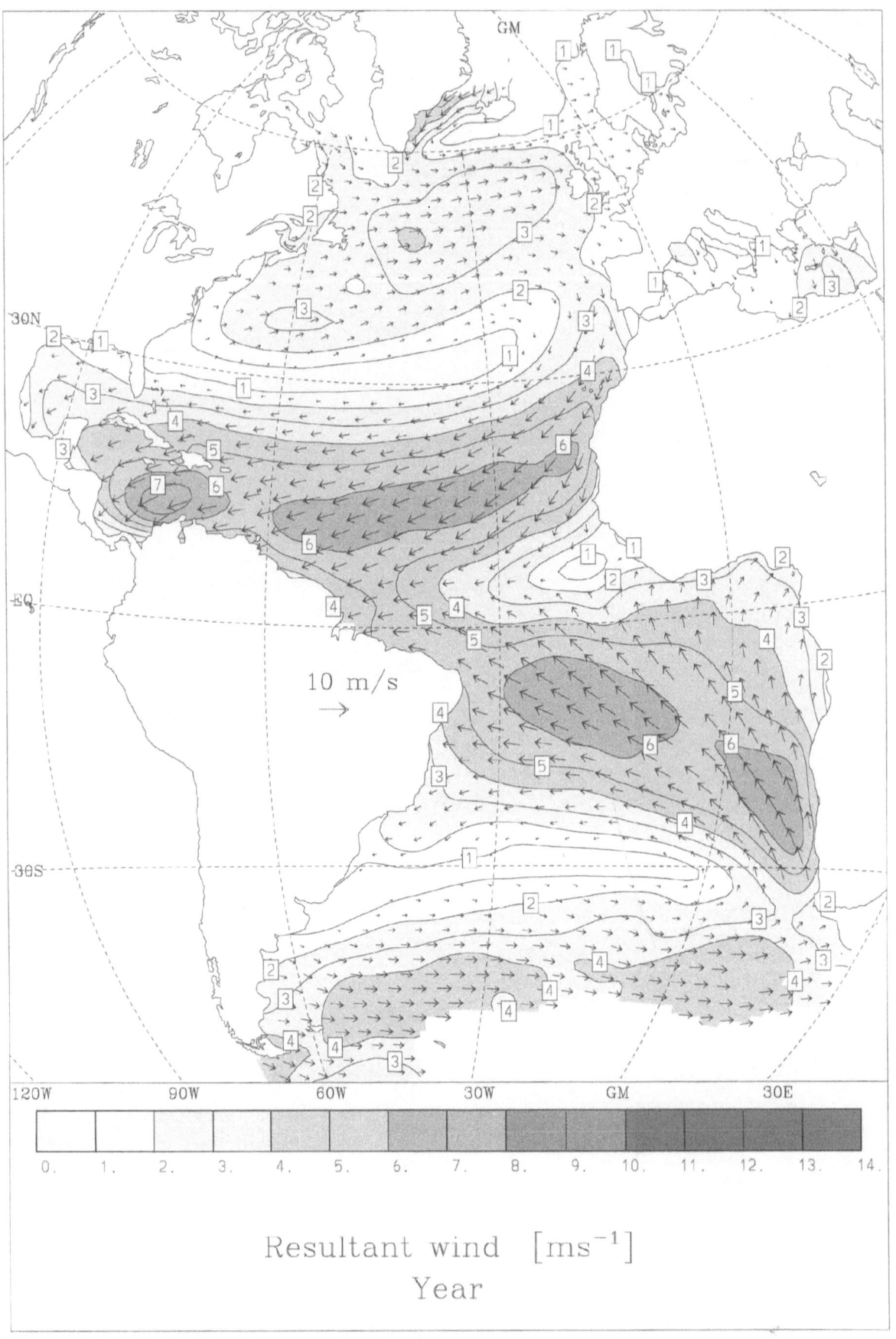

Resultant wind [ms^{-1}]
Year

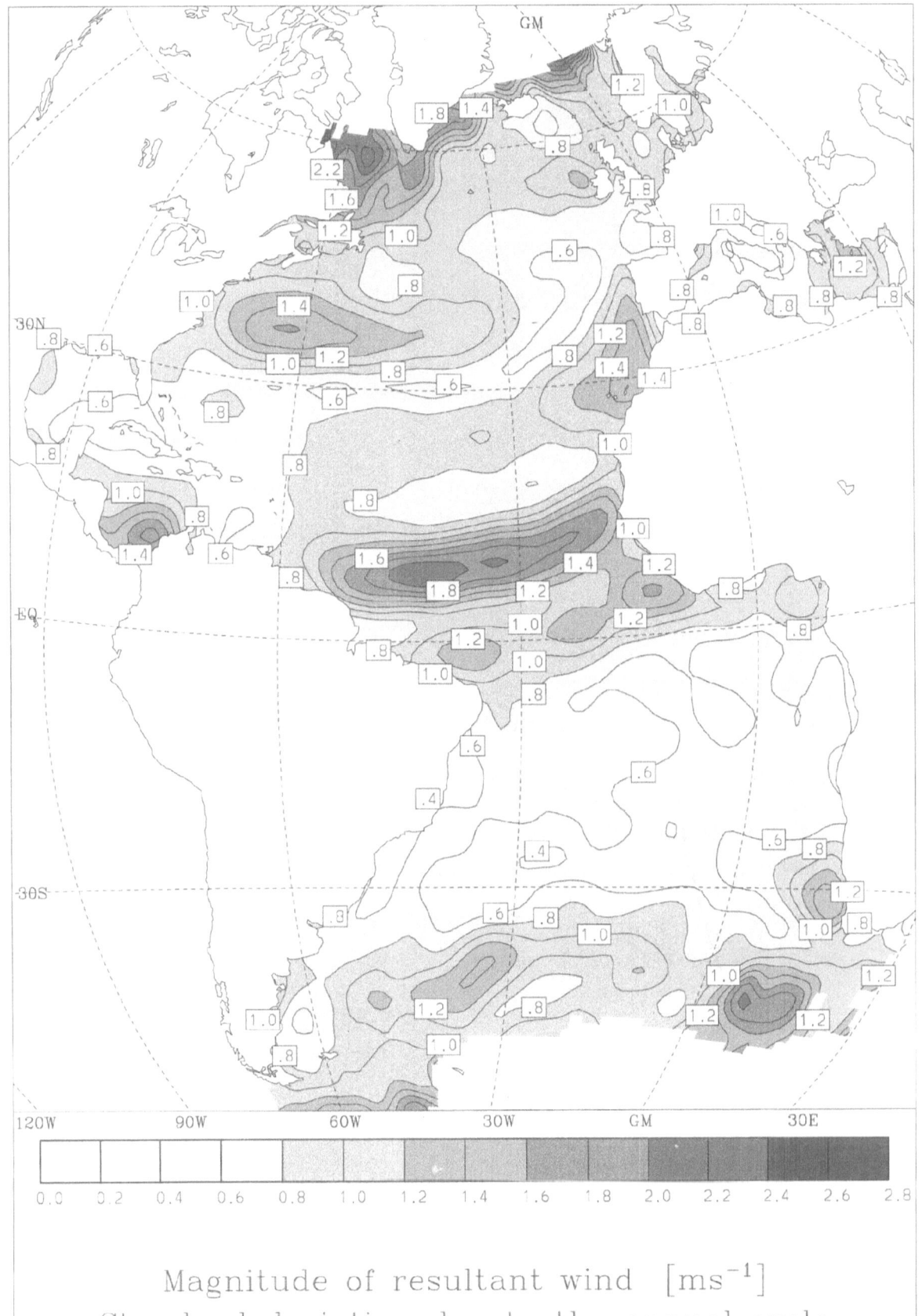

Magnitude of resultant wind [ms^{-1}]
Standard deviation due to the annual cycle

Direction of the Mean Wind Vector

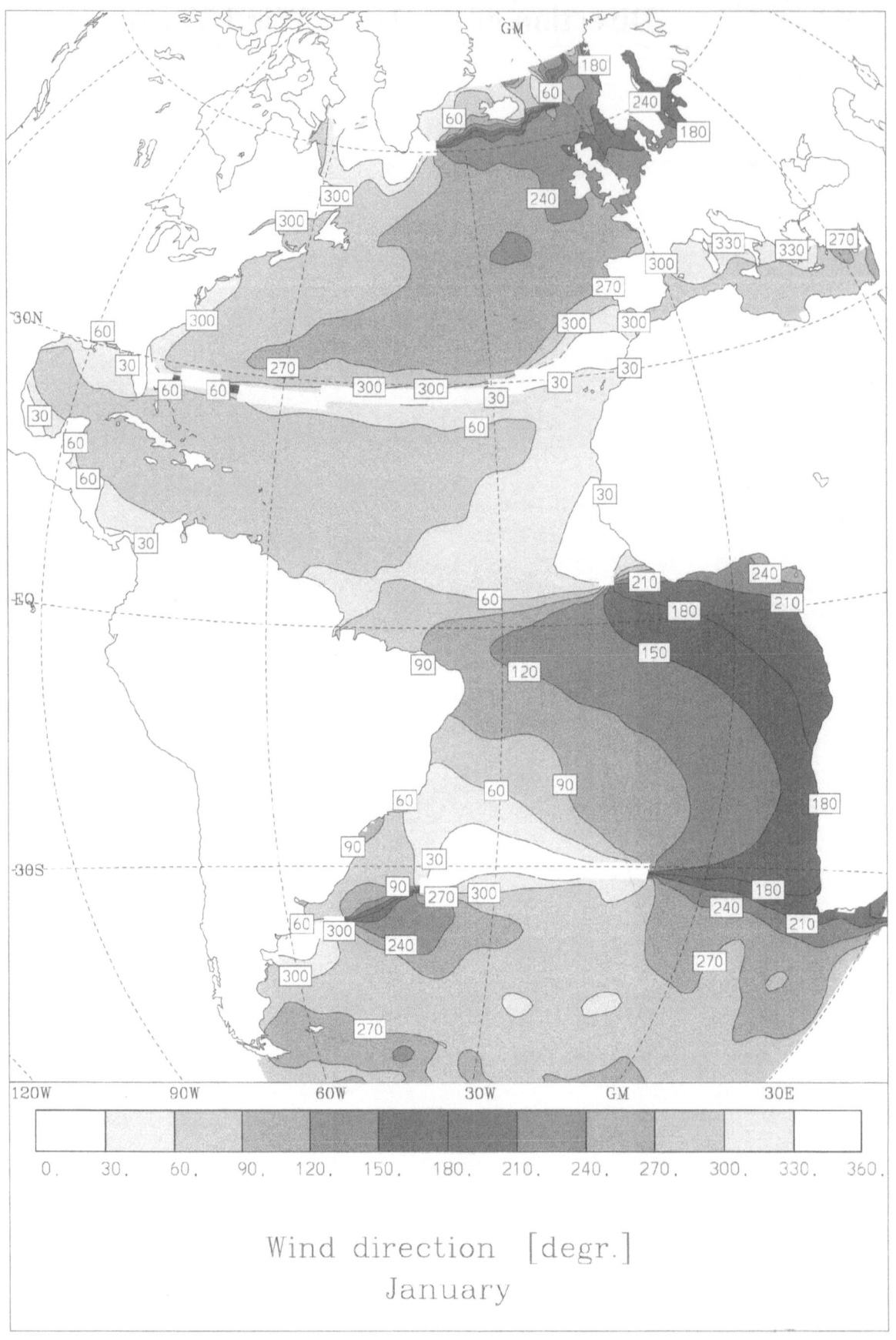

Wind direction [degr.]
January

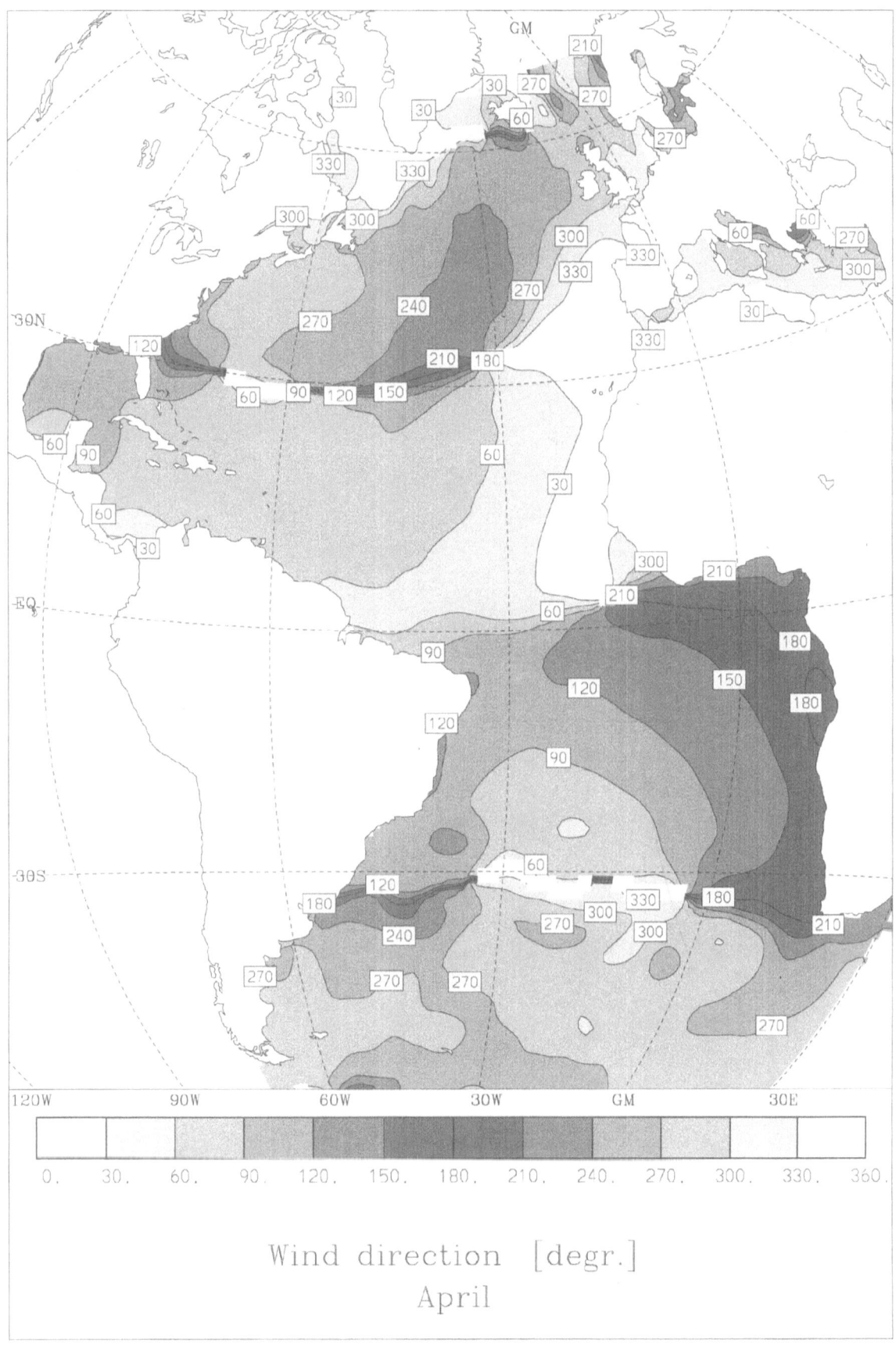

Wind direction [degr.]
April

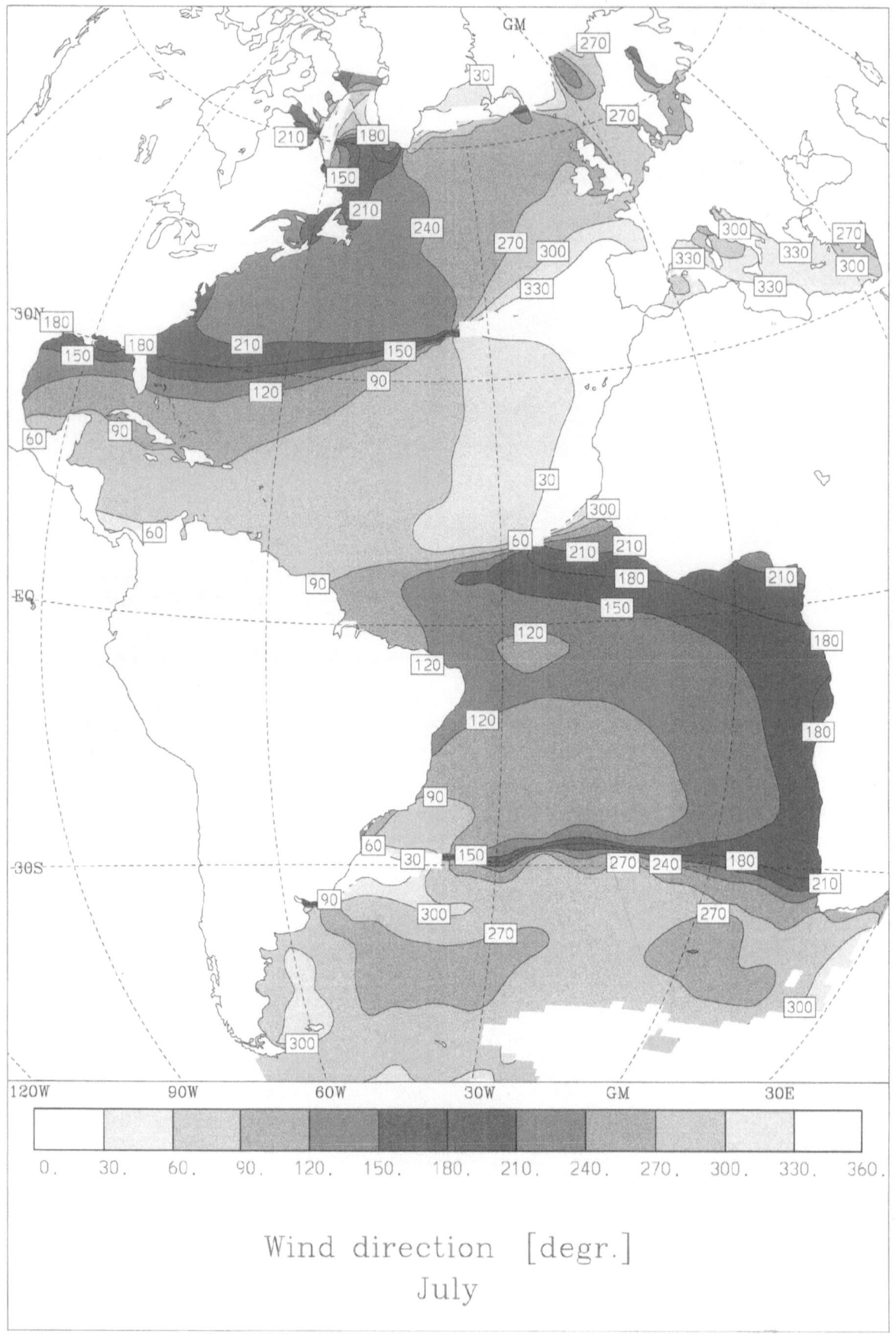

Wind direction [degr.]
July

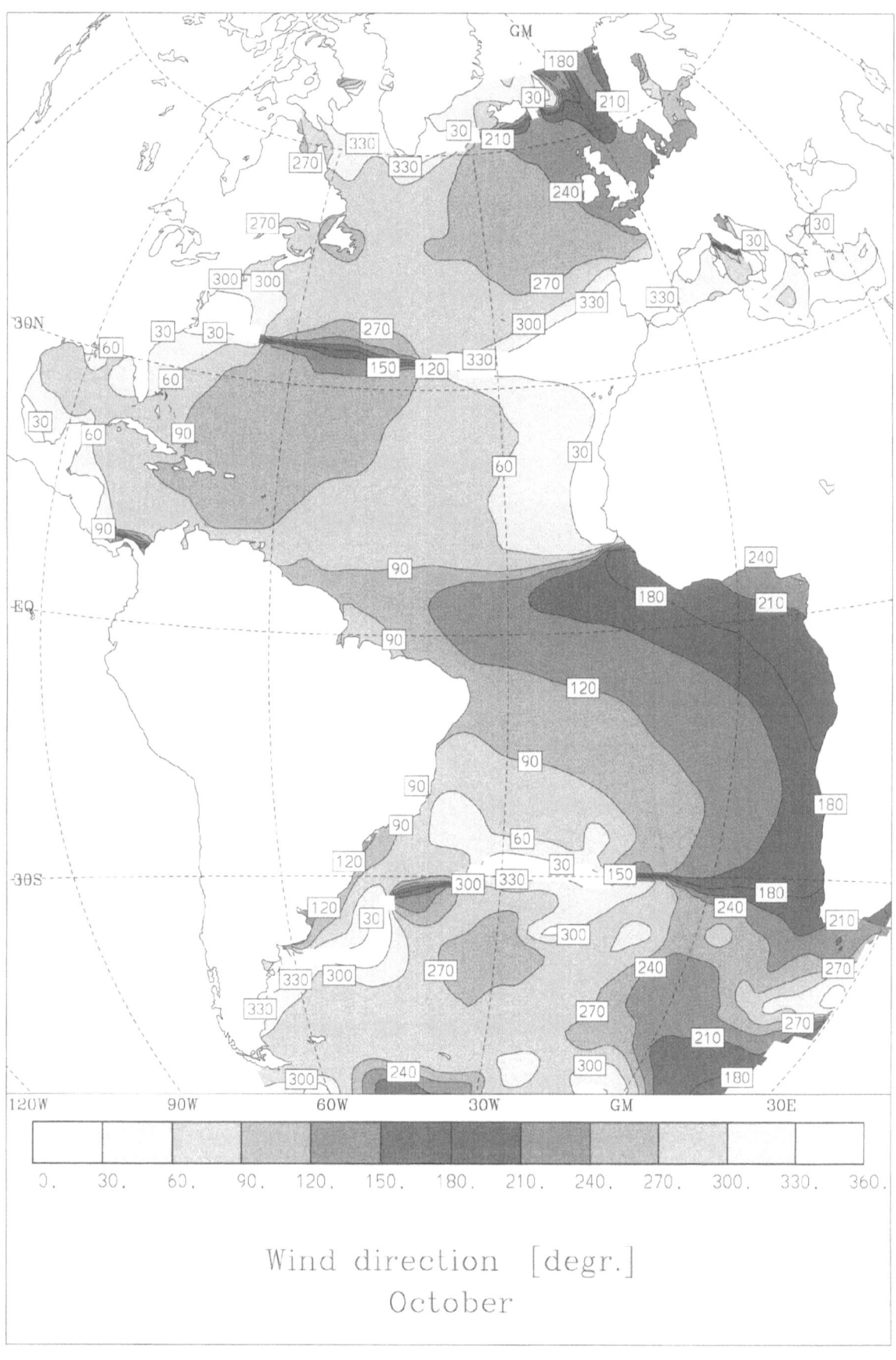

Wind direction [degr.]
October

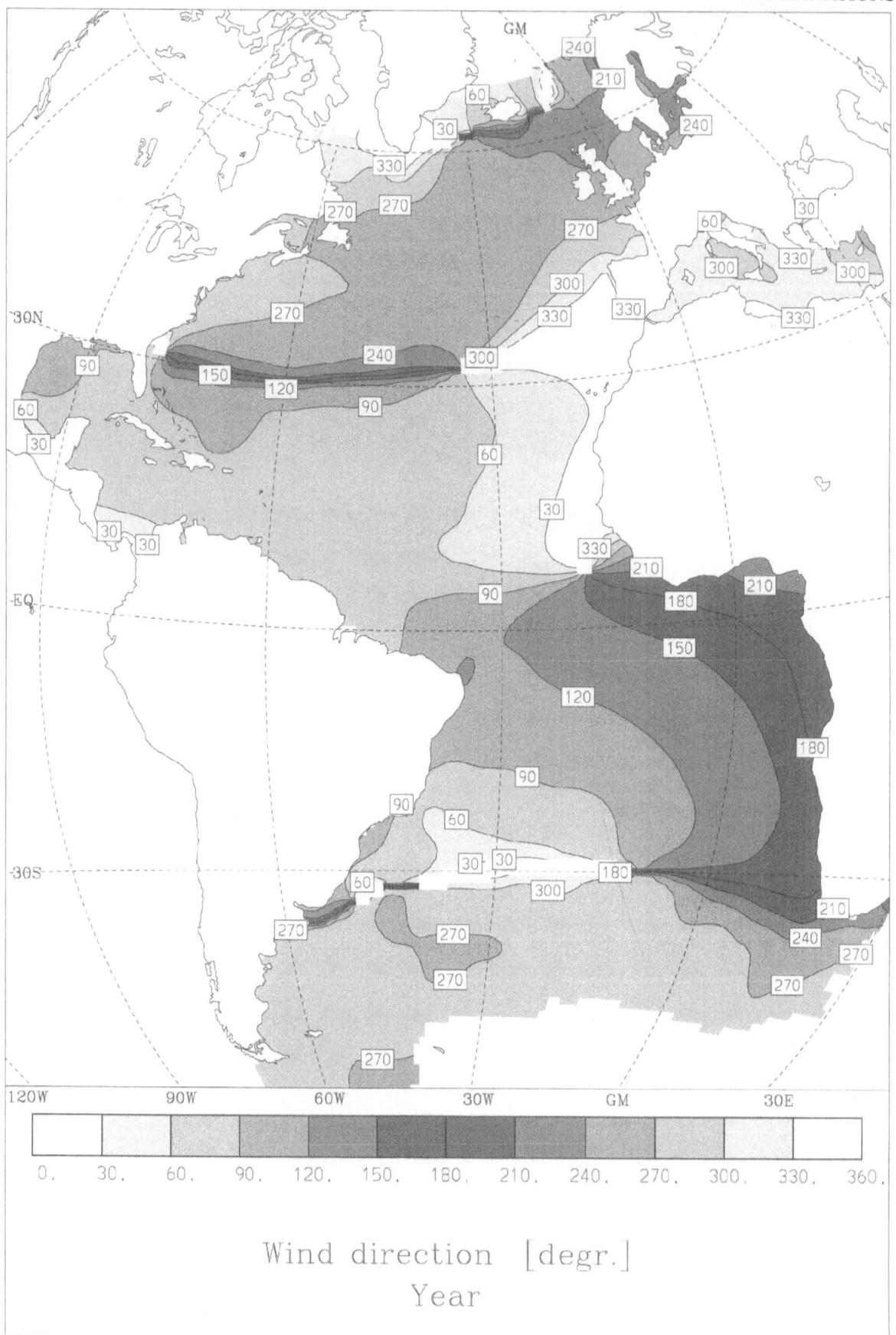

Wind direction [degr.]
Year

Directional Steadiness of the Wind

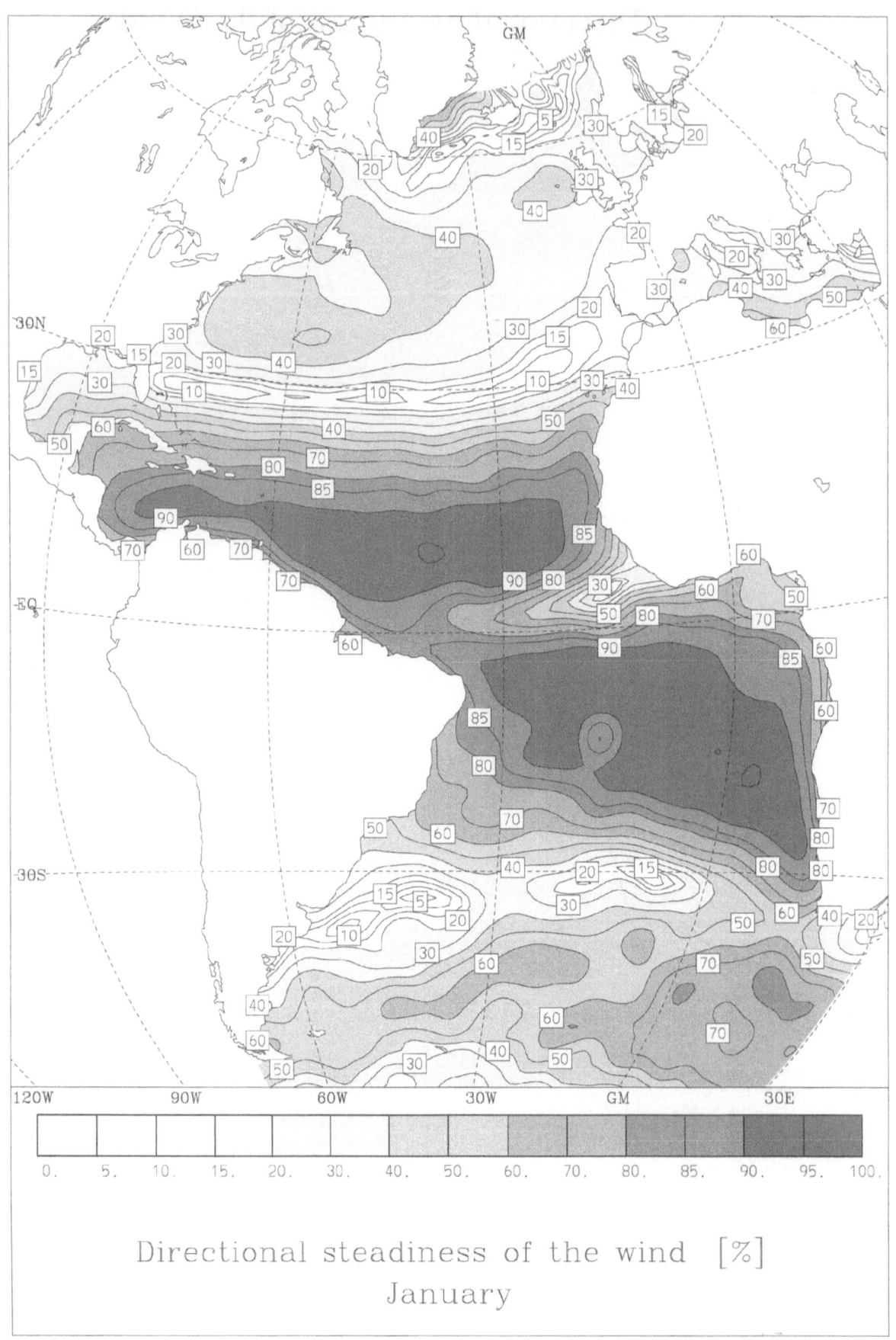

Directional steadiness of the wind [%]
January

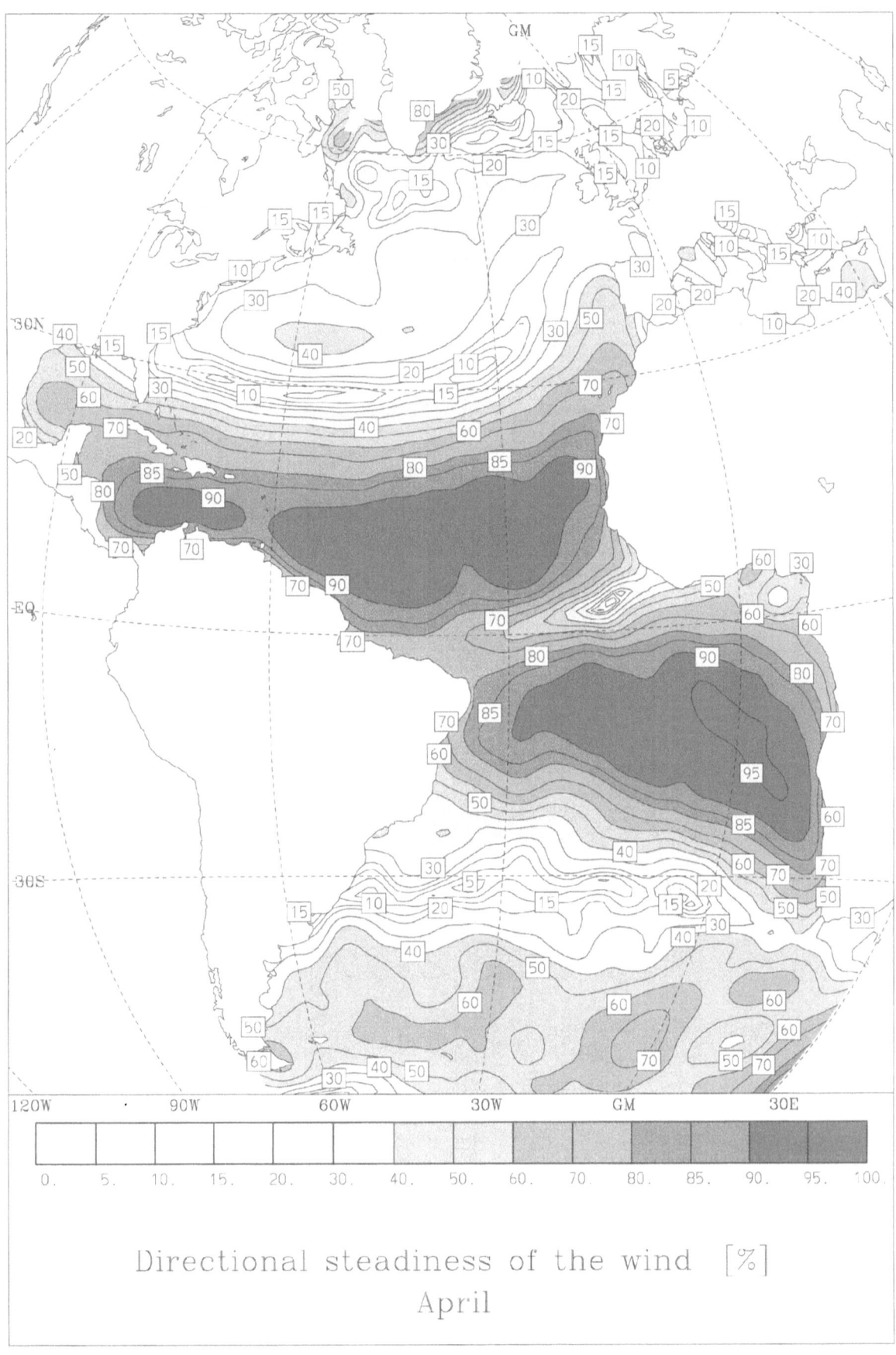

Directional steadiness of the wind [%]

April

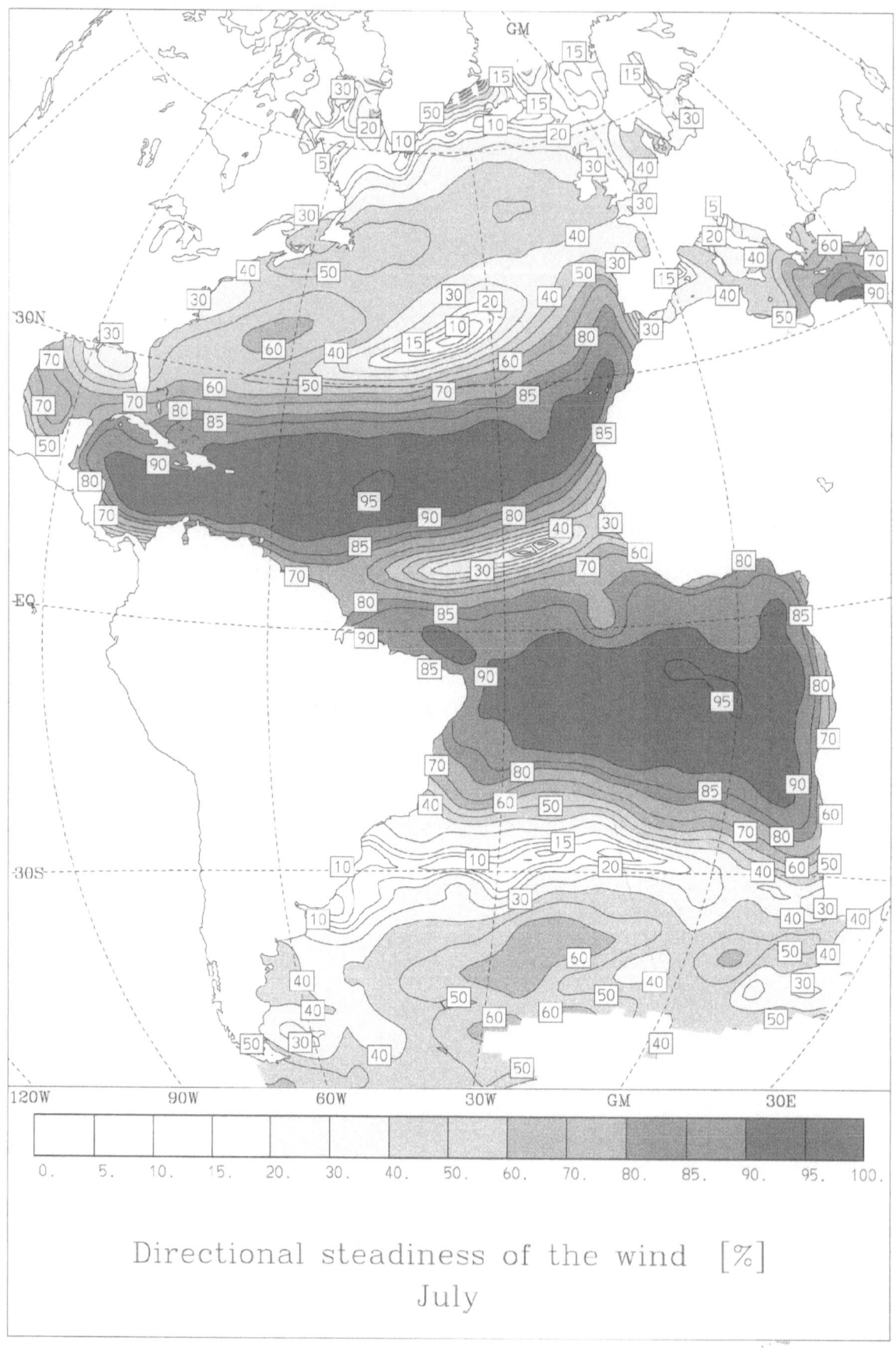

Directional steadiness of the wind [%]
July

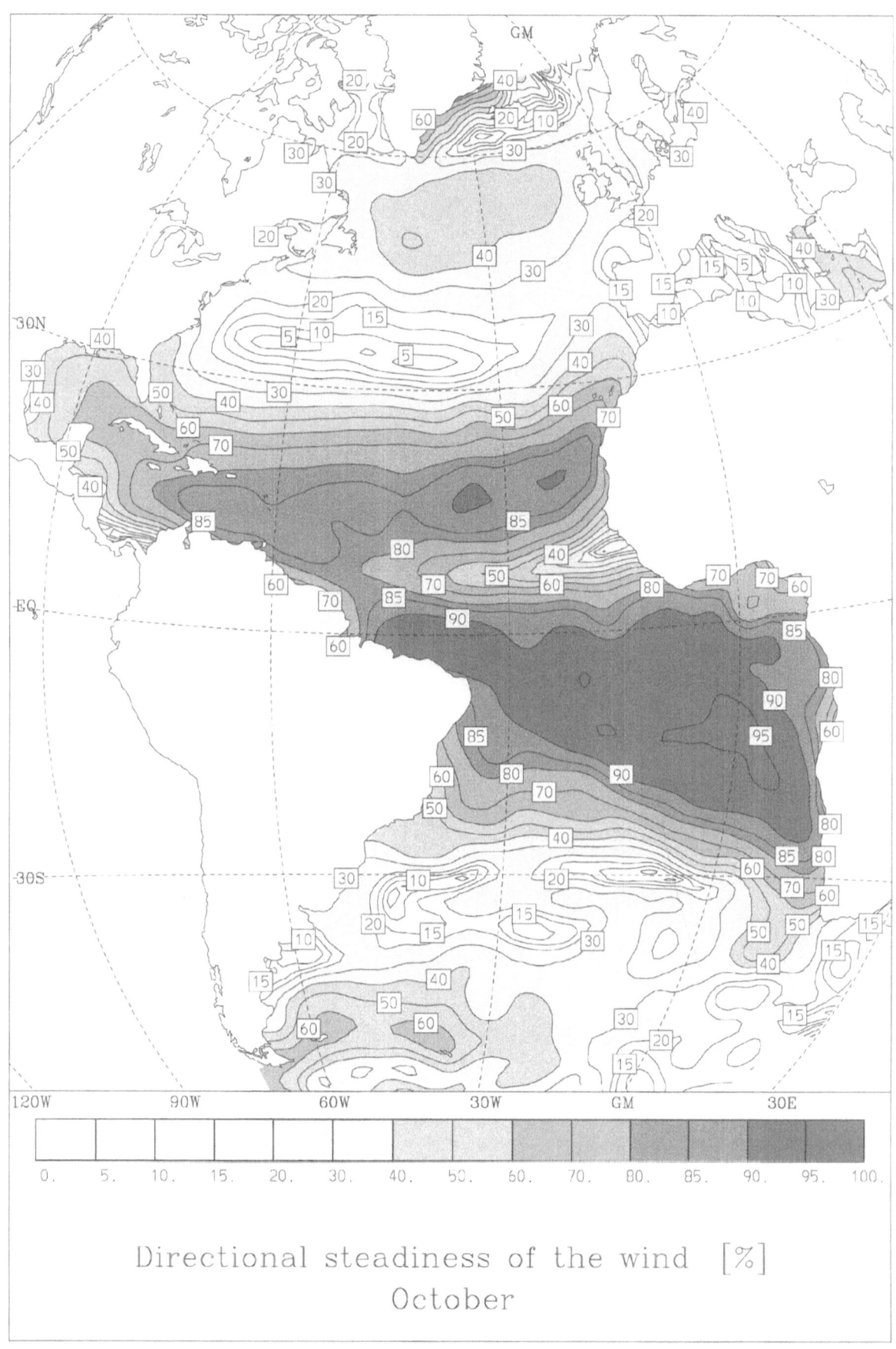

Directional steadiness of the wind [%]
October

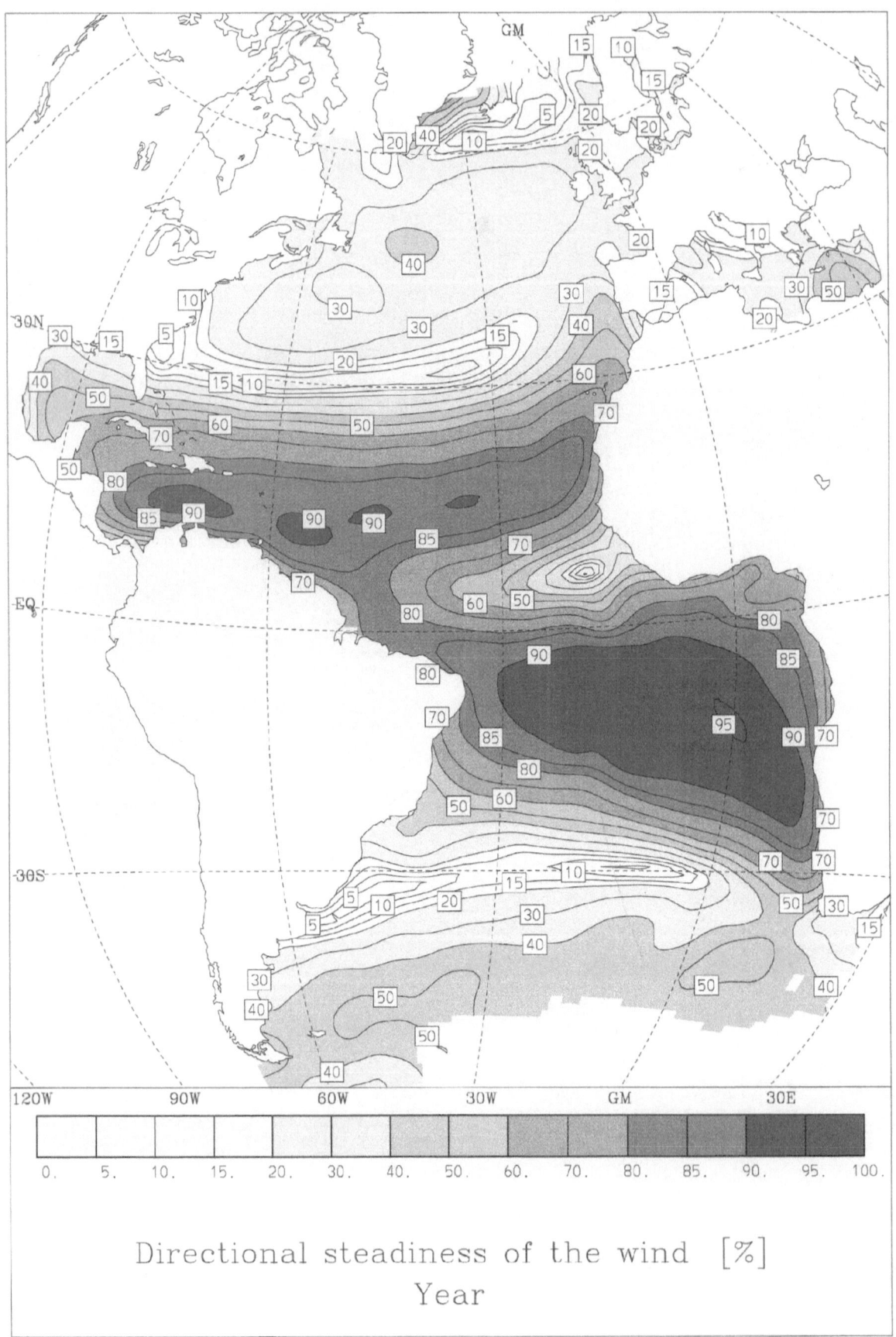

Directional steadiness of the wind [%]
Year

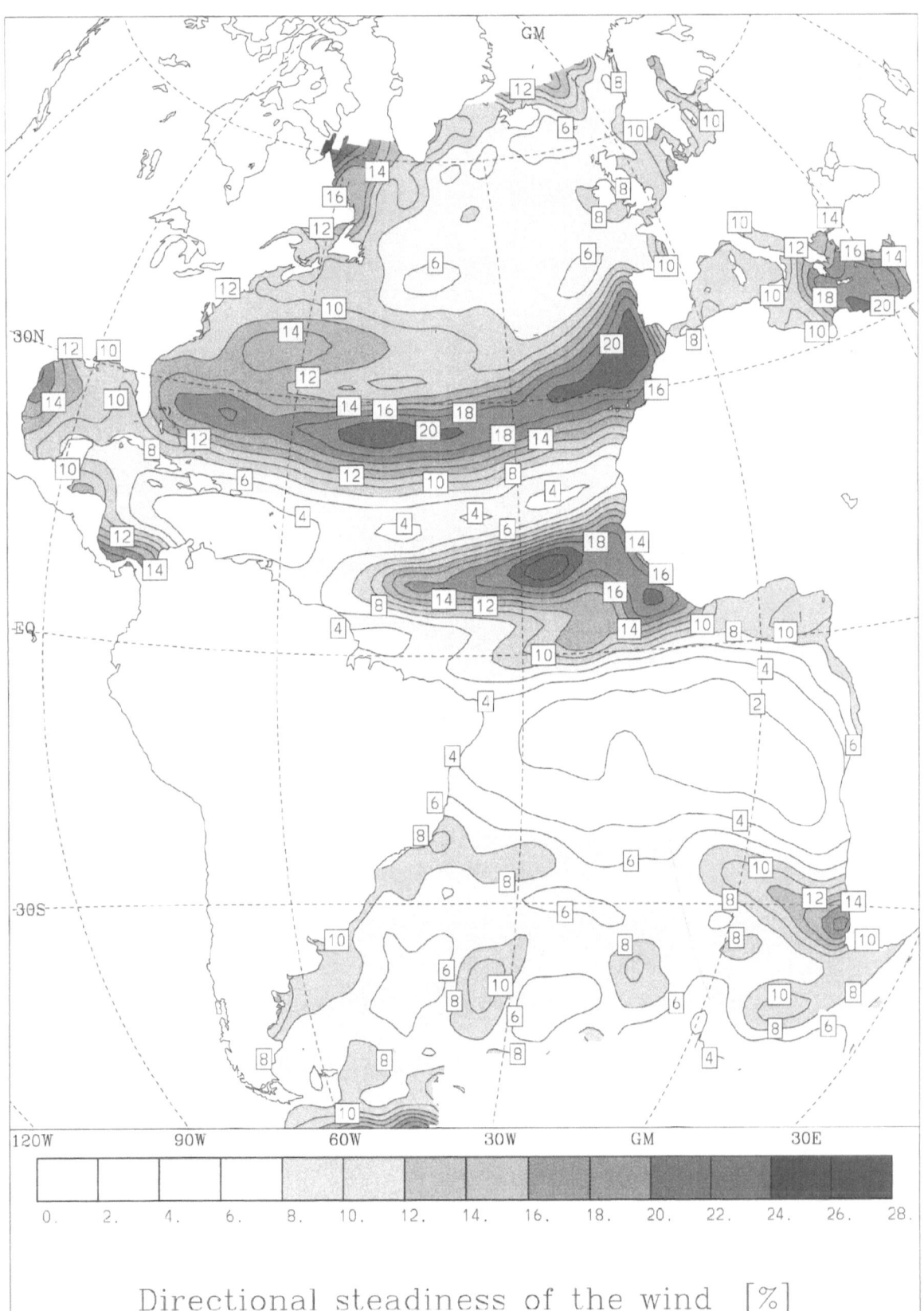

Directional steadiness of the wind [%]
Standard deviation due to the annual cycle

Divergence of the Wind

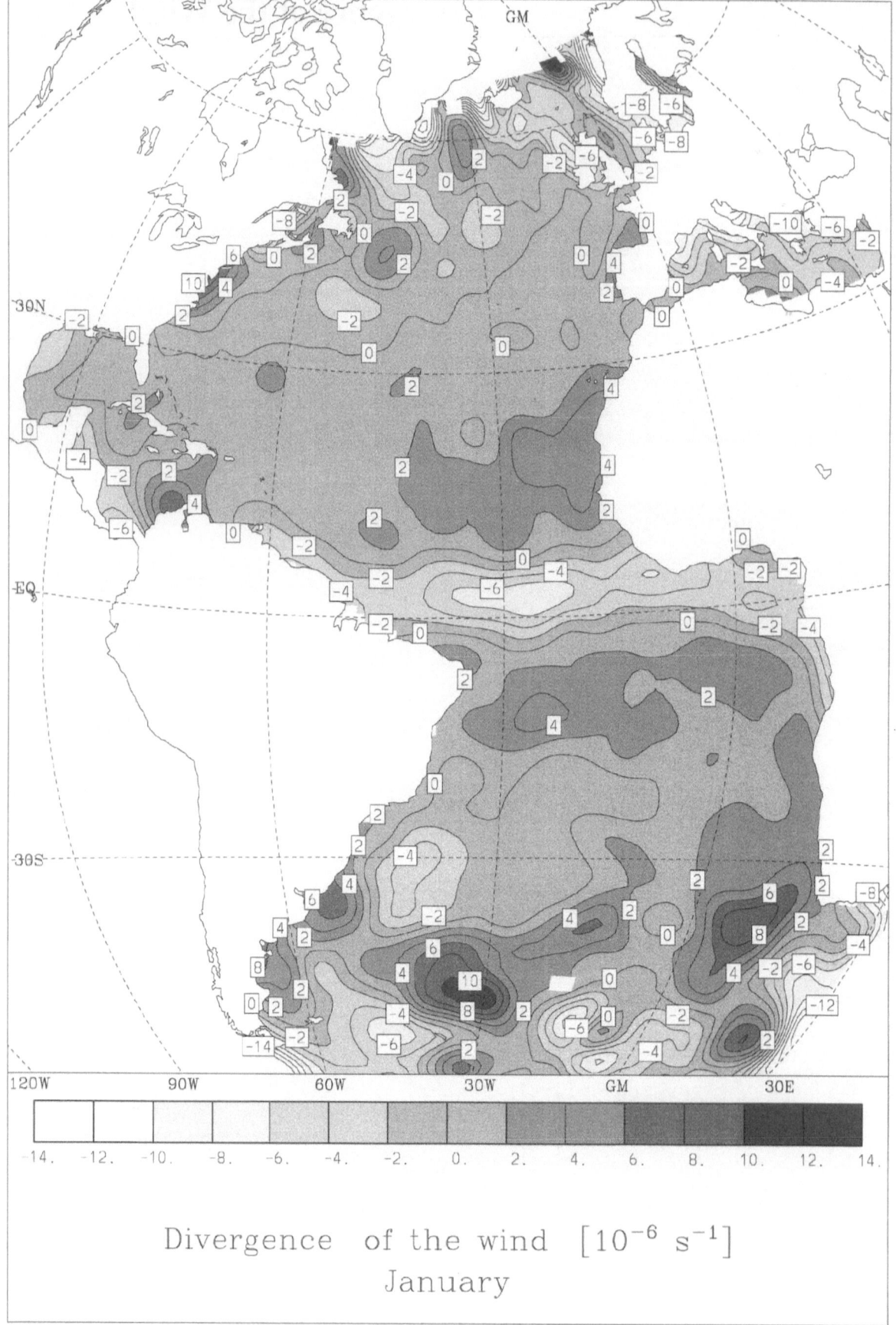

Divergence of the wind $[10^{-6}\ \mathrm{s}^{-1}]$
January

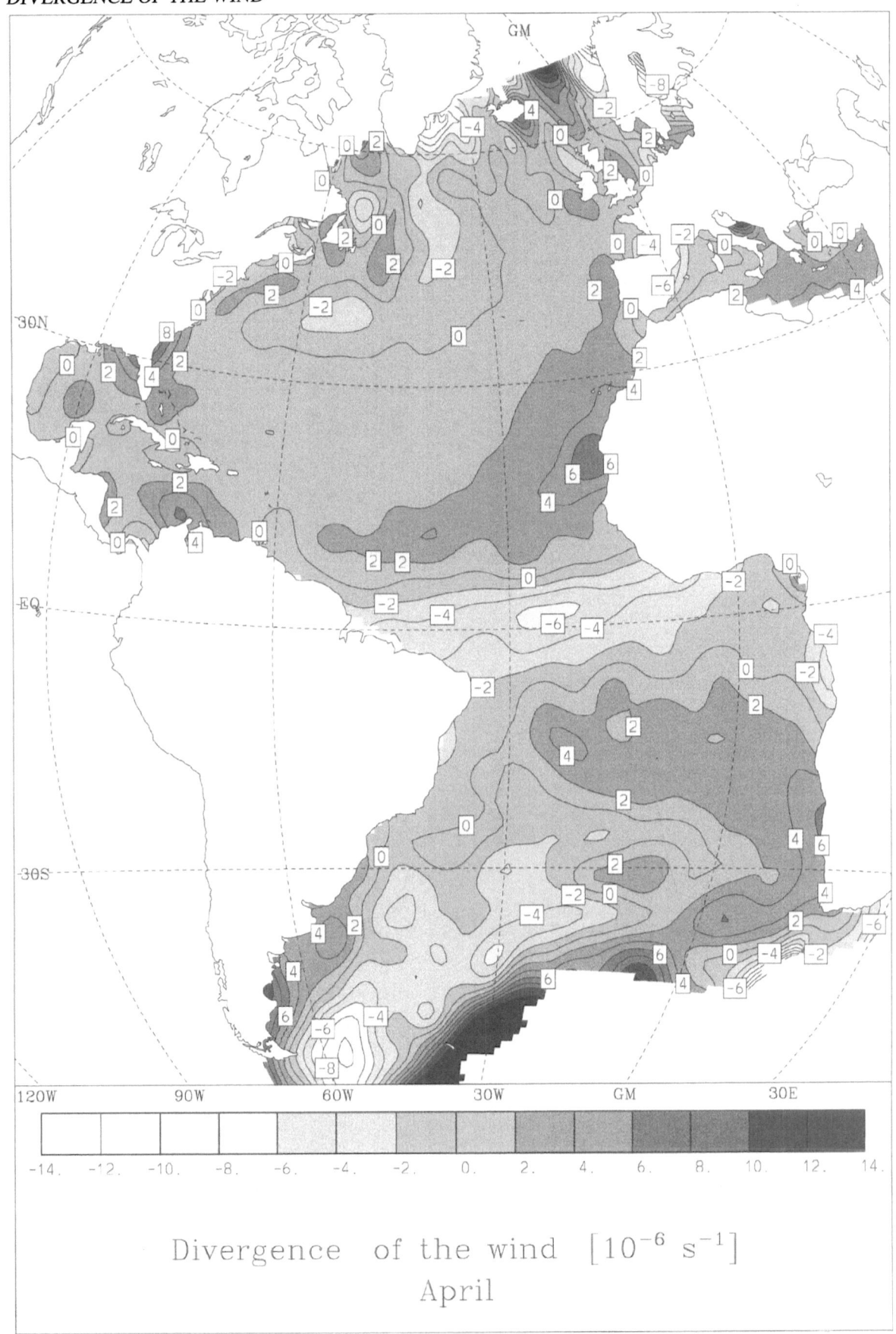

Divergence of the wind $[10^{-6} \text{ s}^{-1}]$
April

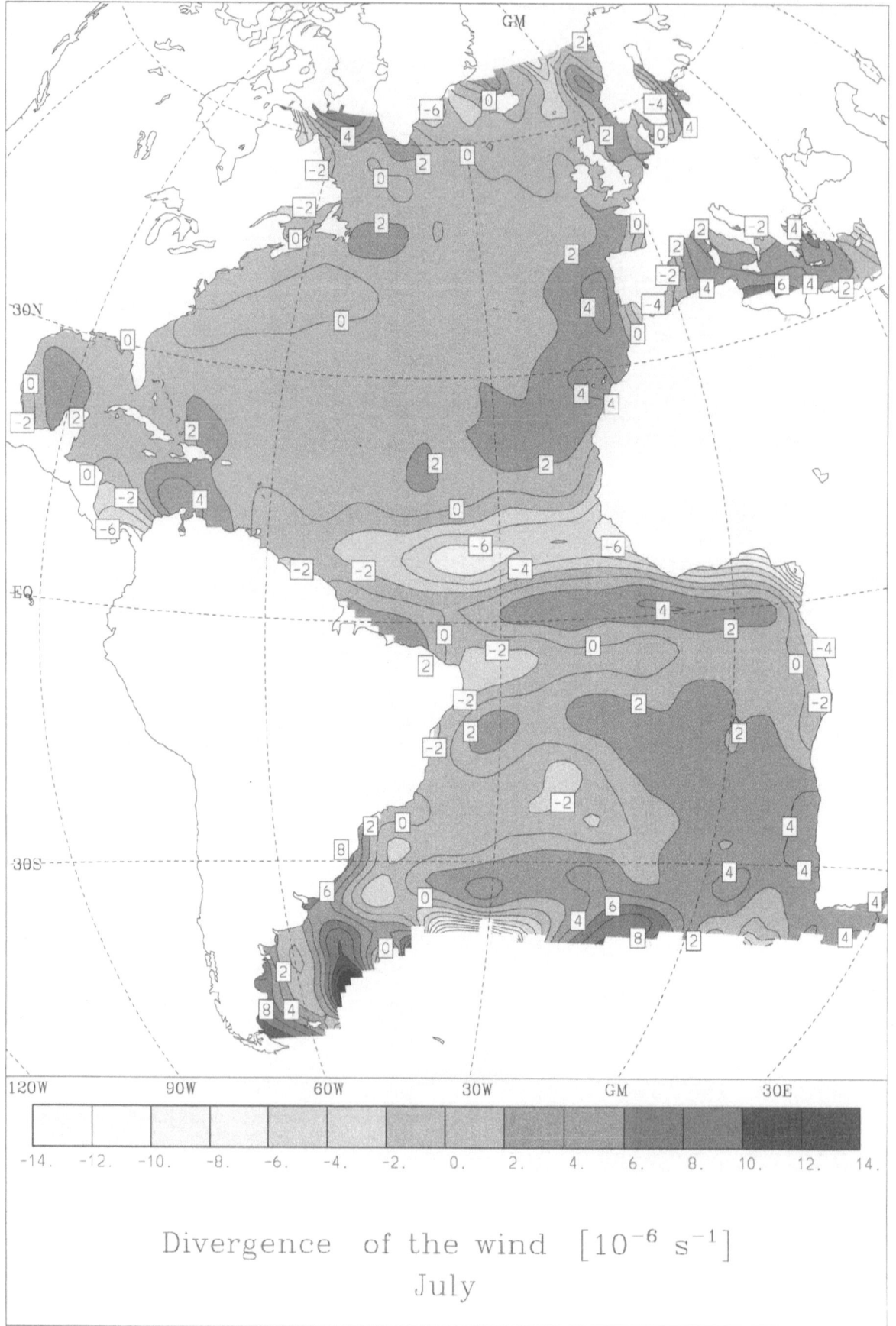

Divergence   of   the   wind $[10^{-6}\ \mathrm{s}^{-1}]$
July

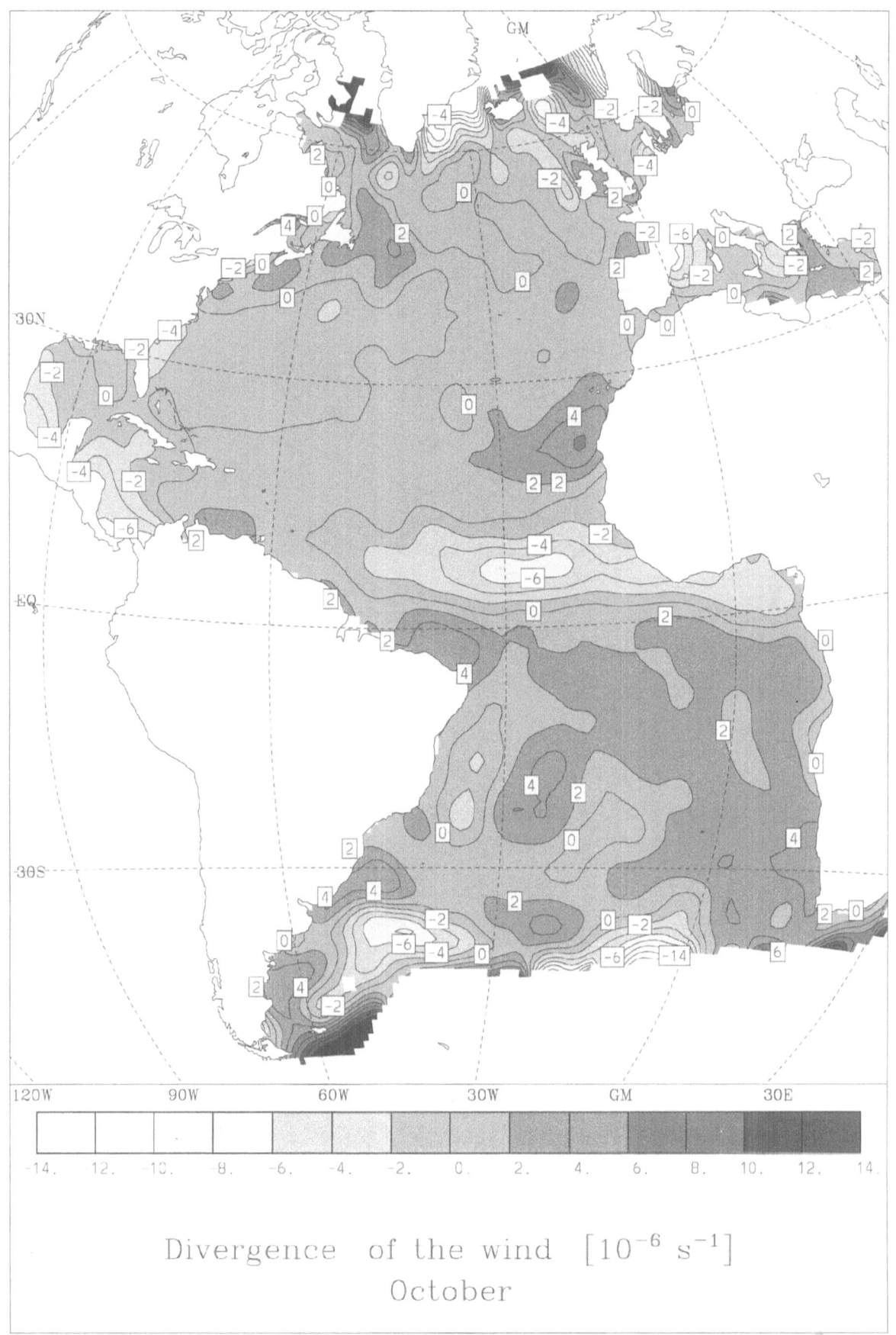

Divergence of the wind $[10^{-6}\ \mathrm{s}^{-1}]$
October

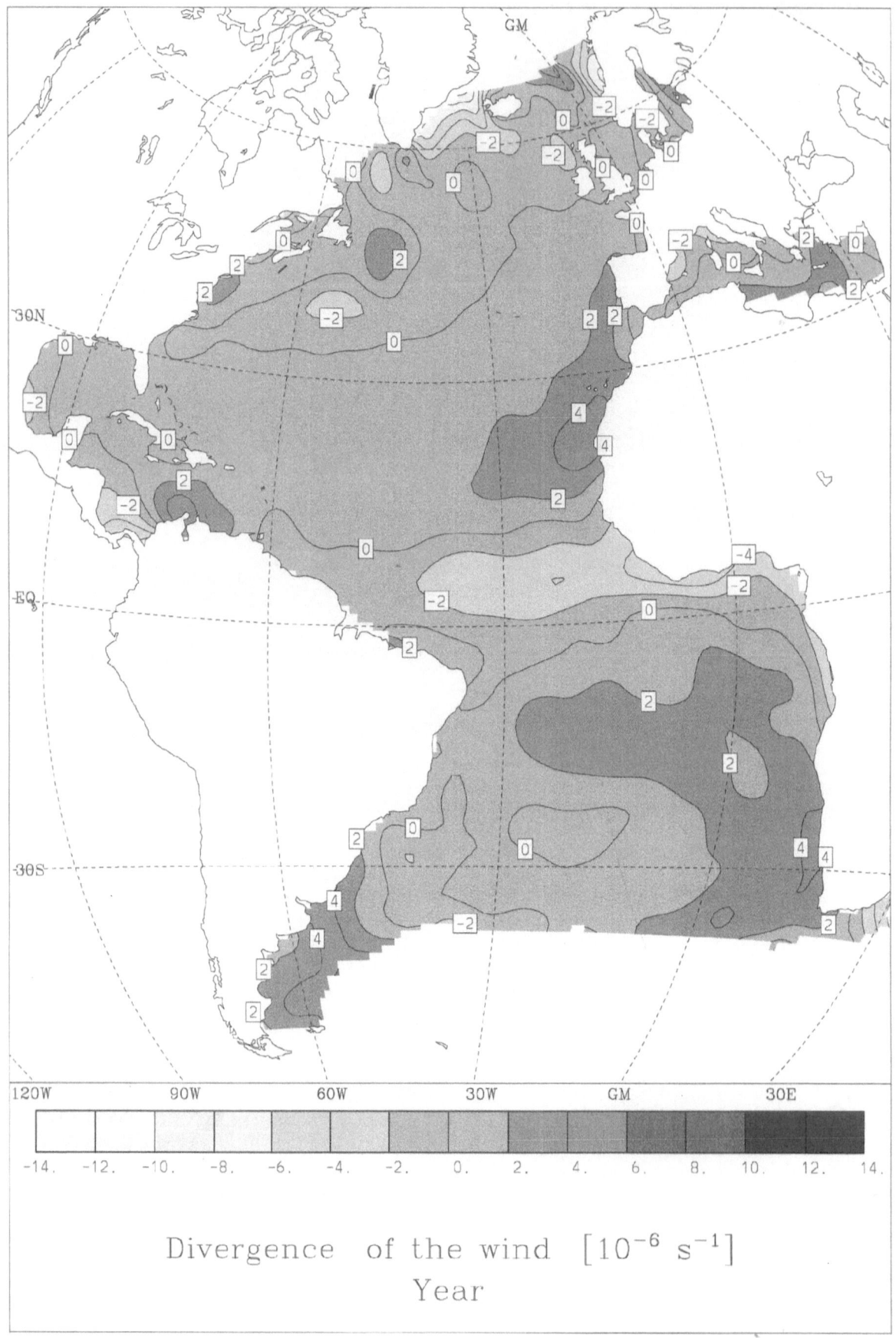

Divergence of the wind $[10^{-6} \ s^{-1}]$
Year

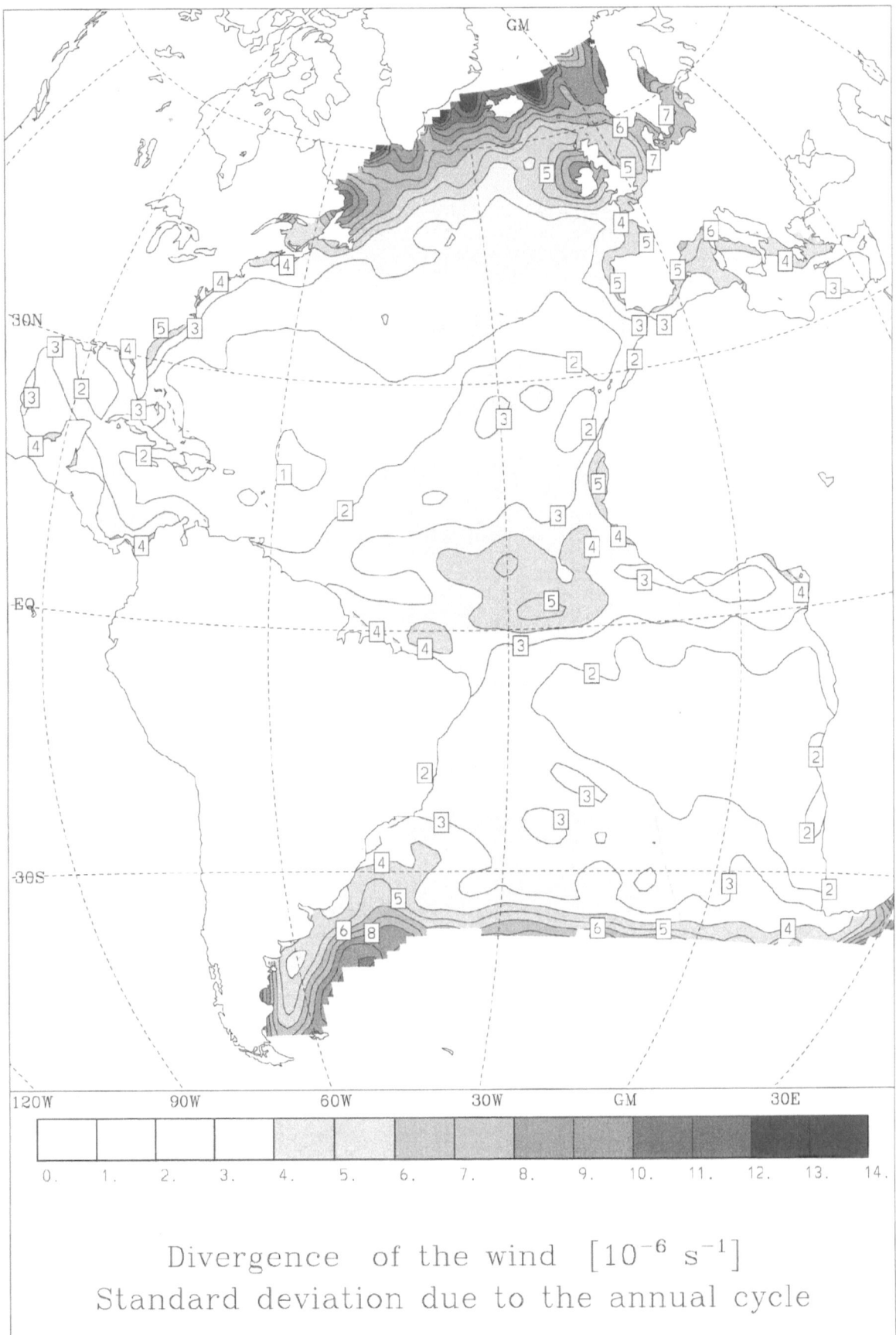

Divergence of the wind $[10^{-6} \text{ s}^{-1}]$
Standard deviation due to the annual cycle

Part II

Air-Sea Interactions

Part II

Air-Sea Interactions

Time-Latitude Diagrams

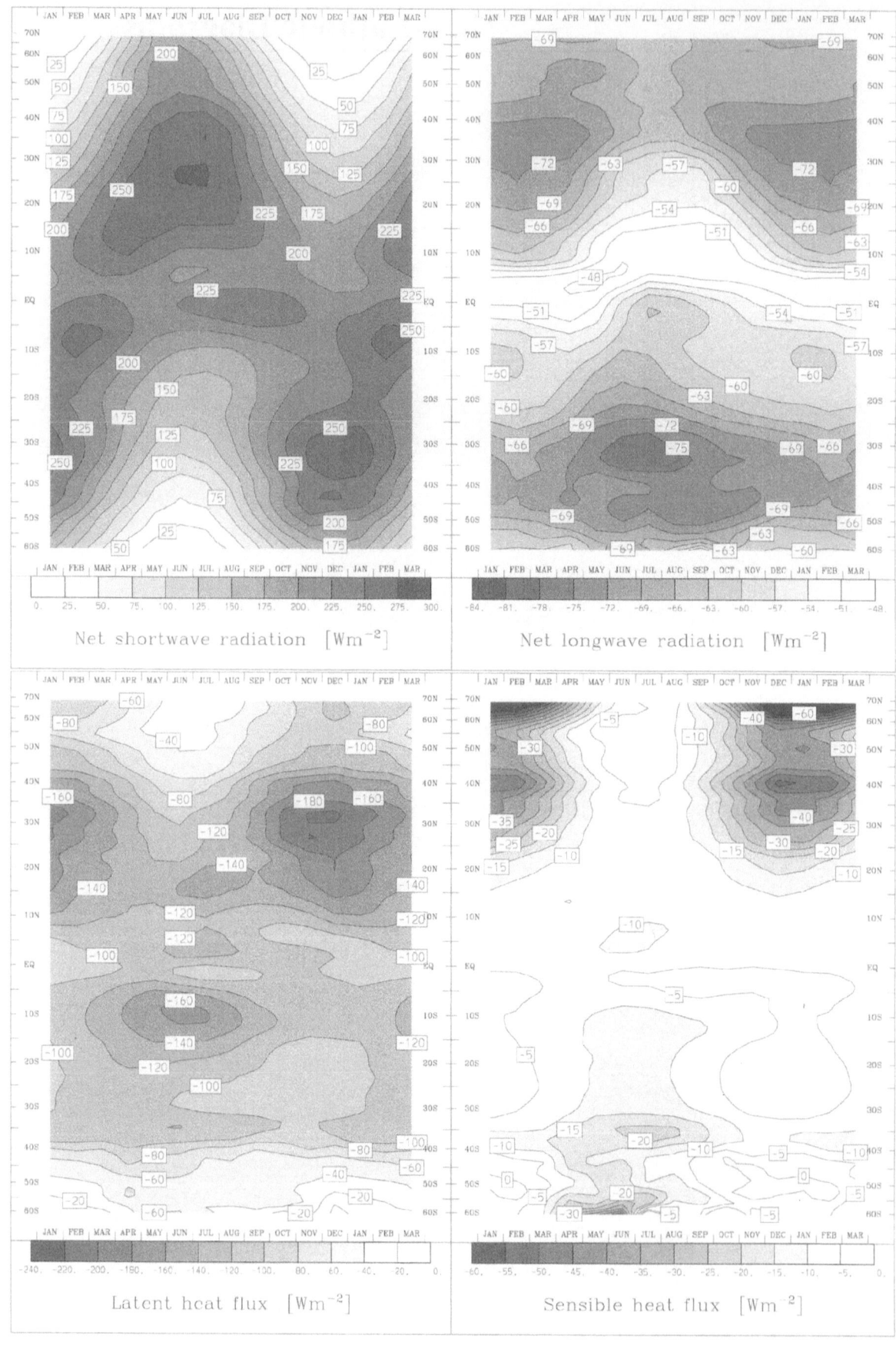

Net shortwave radiation [Wm^{-2}]

Net longwave radiation [Wm^{-2}]

Latent heat flux [Wm^{-2}]

Sensible heat flux [Wm^{-2}]

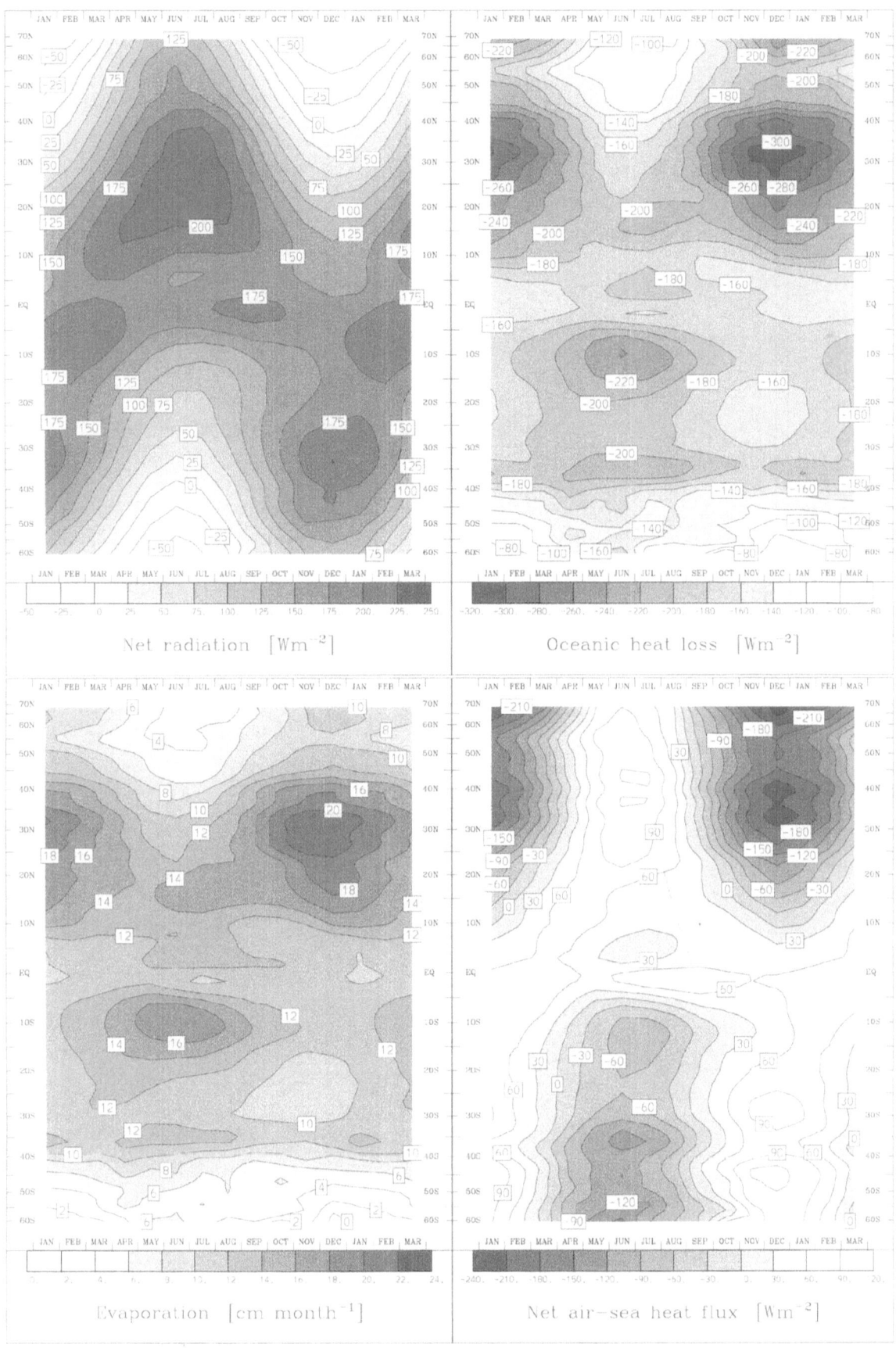

Net radiation $[Wm^{-2}]$

Oceanic heat loss $[Wm^{-2}]$

Evaporation $[cm\ month^{-1}]$

Net air-sea heat flux $[Wm^{-2}]$

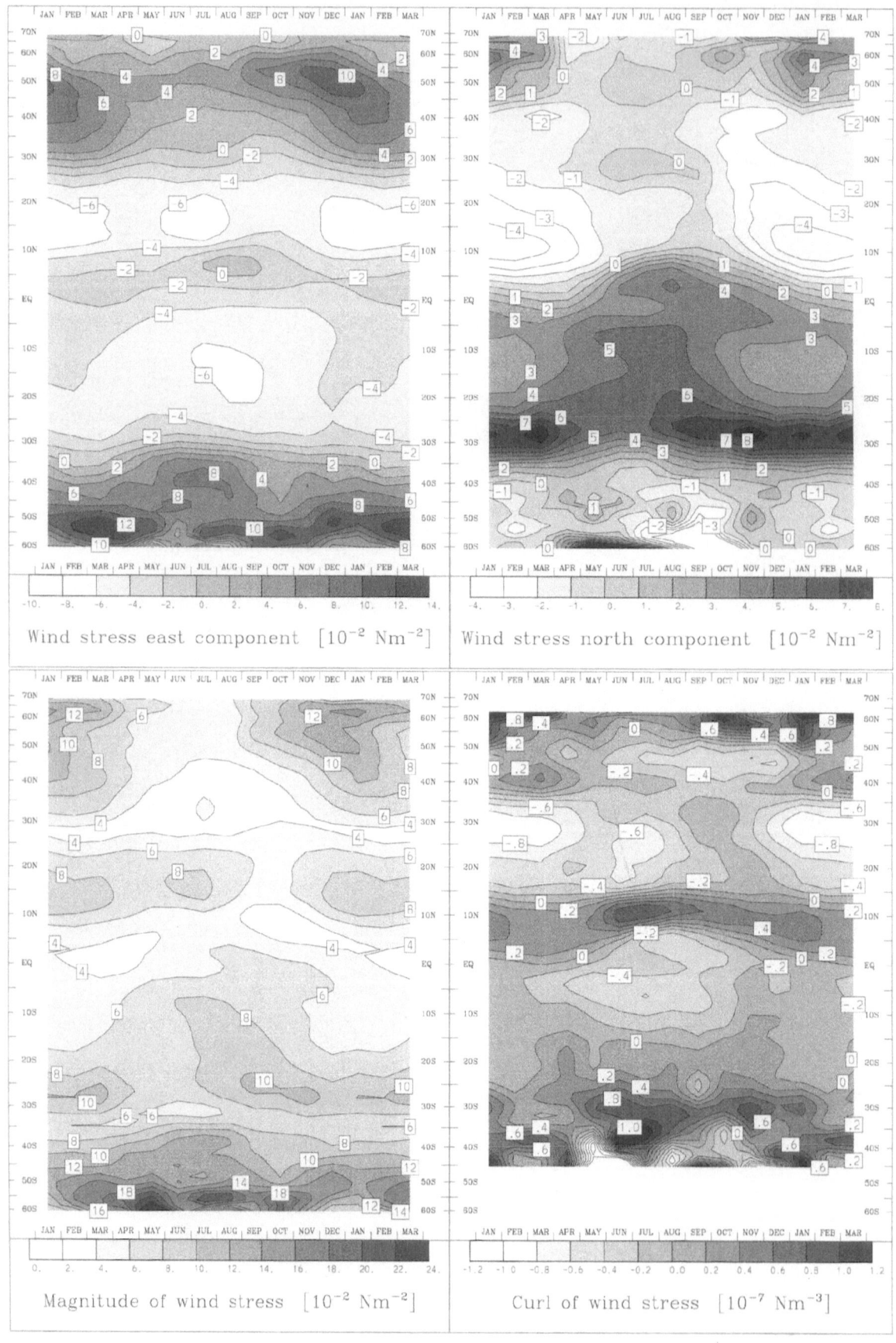

Wind stress east component [10^{-2} Nm^{-2}]

Wind stress north component [10^{-2} Nm^{-2}]

Magnitude of wind stress [10^{-2} Nm^{-2}]

Curl of wind stress [10^{-7} Nm^{-3}]

Annual Cycles at Selected Locations

Baltic Sea

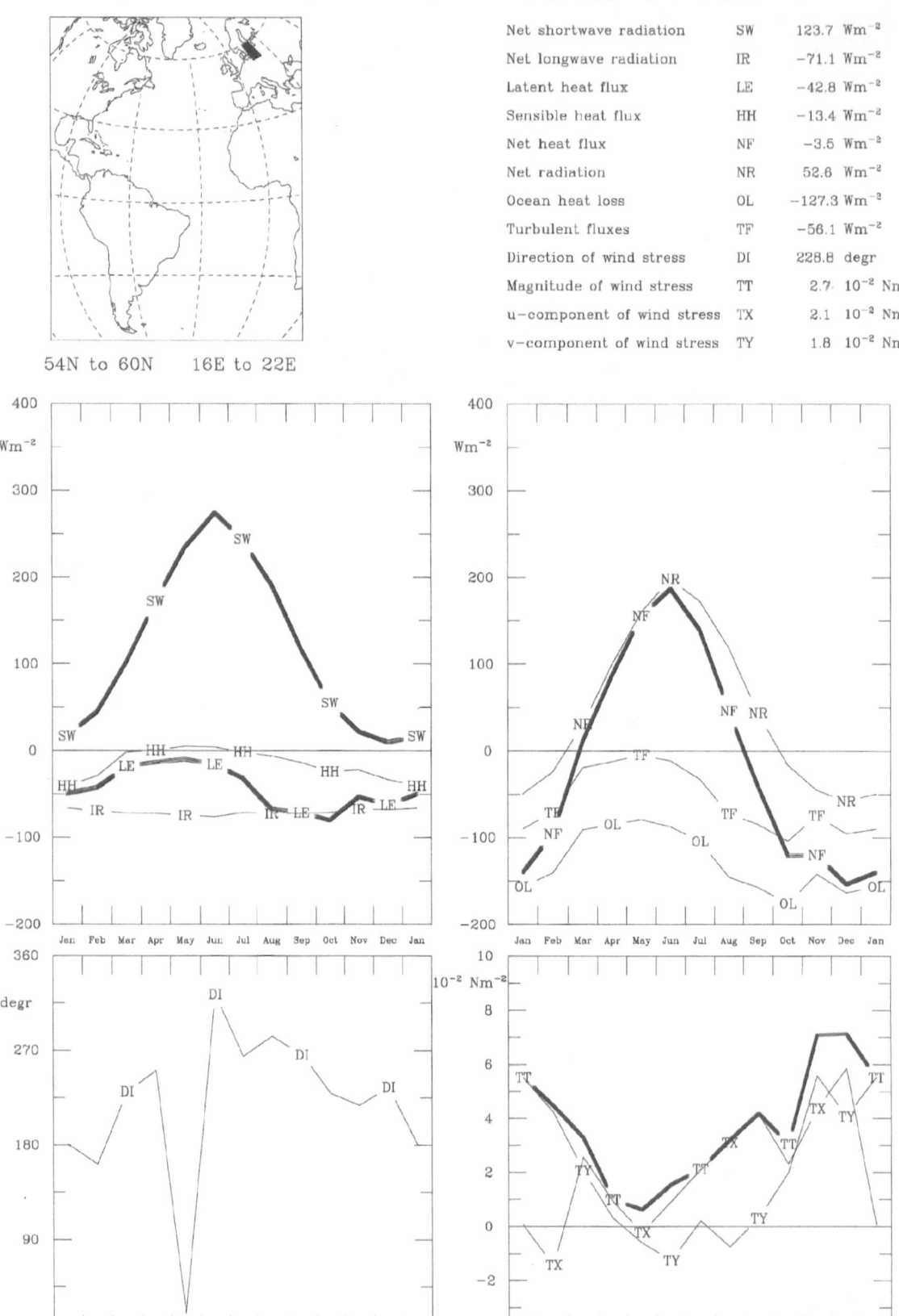

54N to 60N   16E to 22E

Net shortwave radiation	SW	123.7	Wm^{-2}
Net longwave radiation	IR	-71.1	Wm^{-2}
Latent heat flux	LE	-42.8	Wm^{-2}
Sensible heat flux	HH	-13.4	Wm^{-2}
Net heat flux	NF	-3.5	Wm^{-2}
Net radiation	NR	52.6	Wm^{-2}
Ocean heat loss	OL	-127.3	Wm^{-2}
Turbulent fluxes	TF	-56.1	Wm^{-2}
Direction of wind stress	DI	228.8	degr
Magnitude of wind stress	TT	2.7	10^{-2} Nm^{-2}
u-component of wind stress	TX	2.1	10^{-2} Nm^{-2}
v-component of wind stress	TY	1.8	10^{-2} Nm^{-2}

Gulf Stream

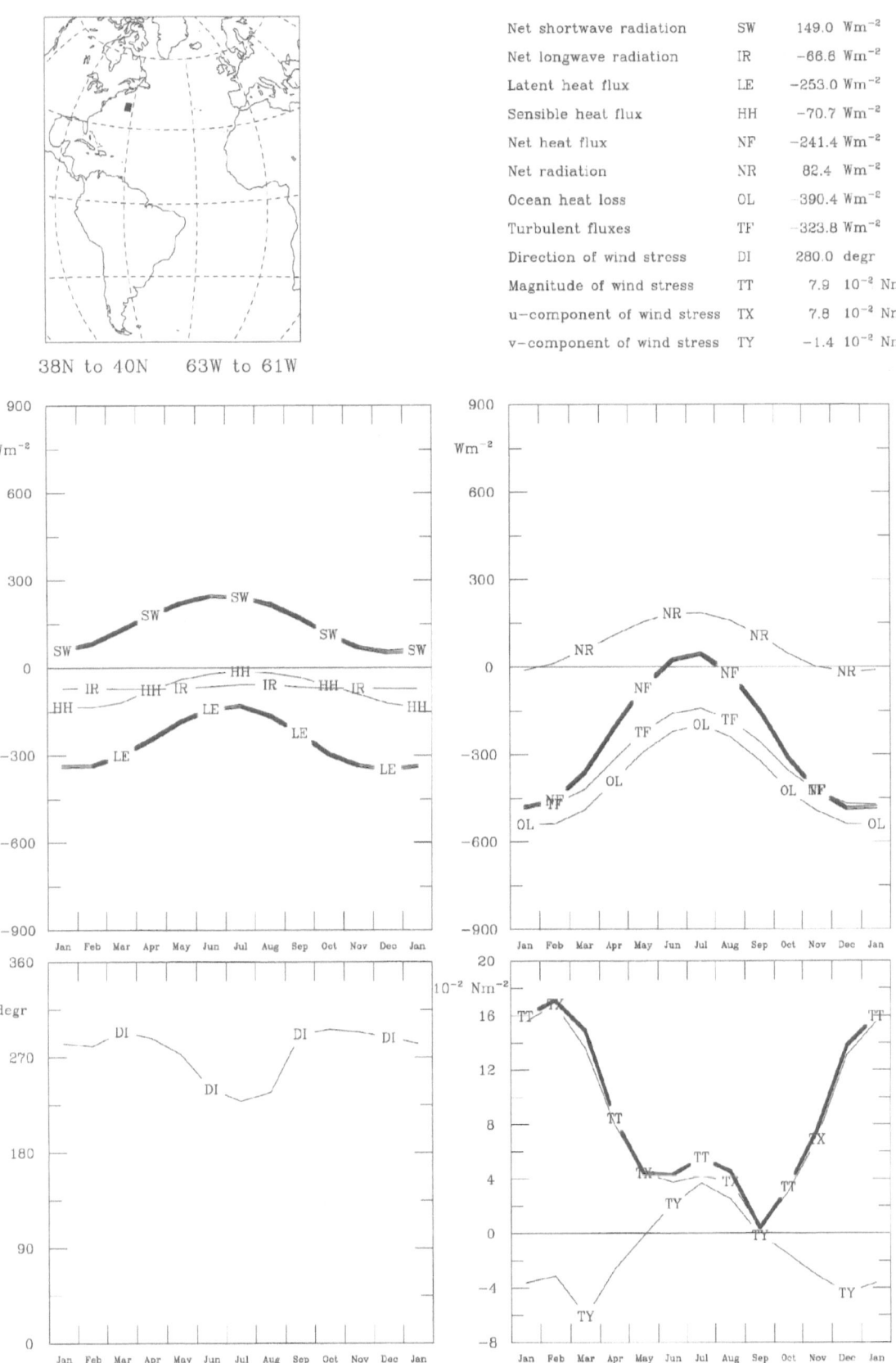

Net shortwave radiation	SW	149.0	Wm^{-2}
Net longwave radiation	IR	-66.6	Wm^{-2}
Latent heat flux	LE	-253.0	Wm^{-2}
Sensible heat flux	HH	-70.7	Wm^{-2}
Net heat flux	NF	-241.4	Wm^{-2}
Net radiation	NR	82.4	Wm^{-2}
Ocean heat loss	OL	-390.4	Wm^{-2}
Turbulent fluxes	TF	-323.8	Wm^{-2}
Direction of wind stress	DI	280.0	degr
Magnitude of wind stress	TT	7.9	10^{-2} Nm^{-2}
u-component of wind stress	TX	7.8	10^{-2} Nm^{-2}
v-component of wind stress	TY	-1.4	10^{-2} Nm^{-2}

38N to 40N   63W to 61W

Middle Atlantic Bight

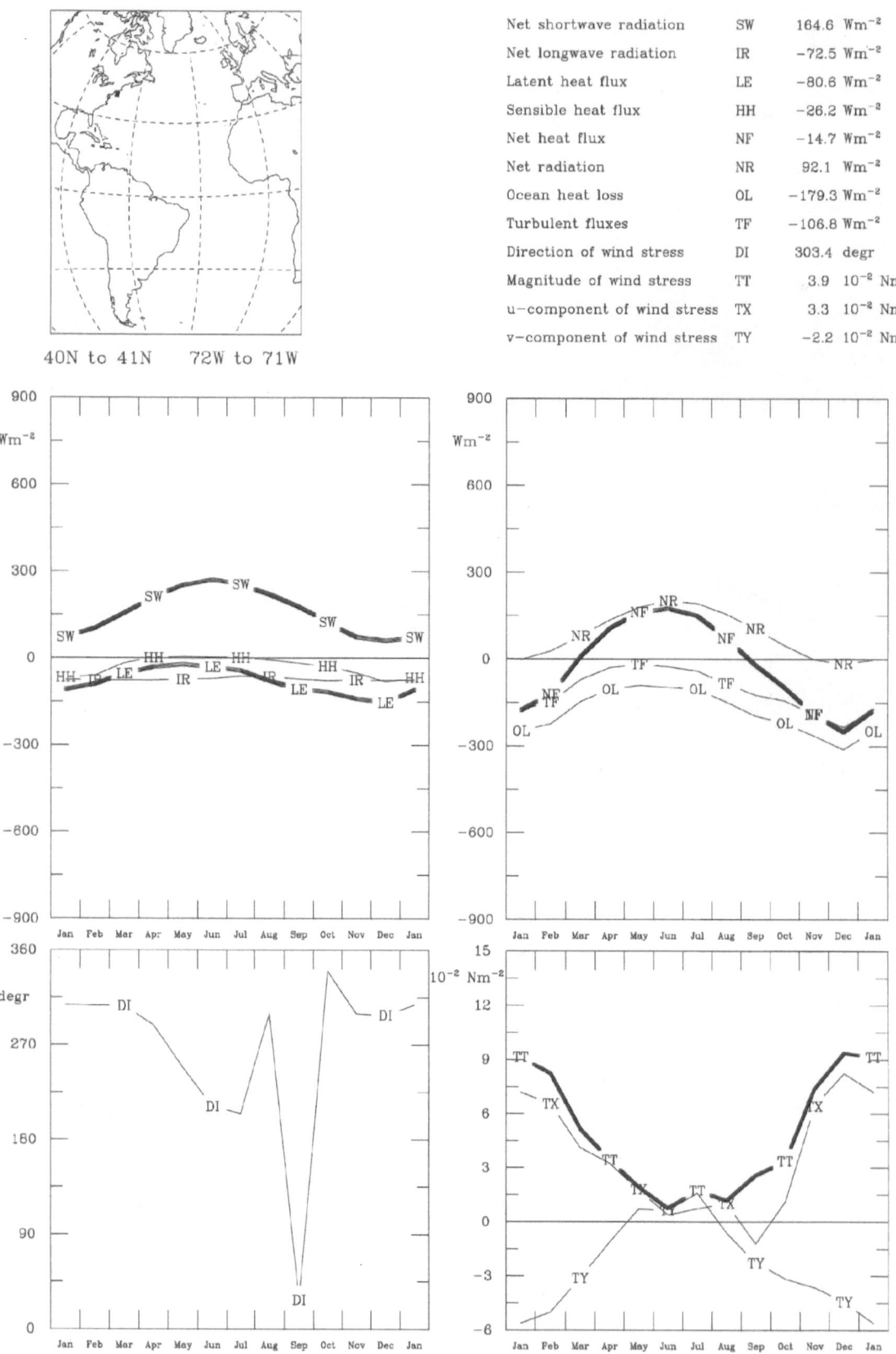

Net shortwave radiation	SW	164.6 Wm^{-2}
Net longwave radiation	IR	−72.5 Wm^{-2}
Latent heat flux	LE	−80.6 Wm^{-2}
Sensible heat flux	HH	−26.2 Wm^{-2}
Net heat flux	NF	−14.7 Wm^{-2}
Net radiation	NR	92.1 Wm^{-2}
Ocean heat loss	OL	−179.3 Wm^{-2}
Turbulent fluxes	TF	−106.8 Wm^{-2}
Direction of wind stress	DI	303.4 degr
Magnitude of wind stress	TT	3.9 10^{-2} Nm^{-2}
u−component of wind stress	TX	3.3 10^{-2} Nm^{-2}
v−component of wind stress	TY	−2.2 10^{-2} Nm^{-2}

40N to 41N   72W to 71W

Banks of Newfoundland

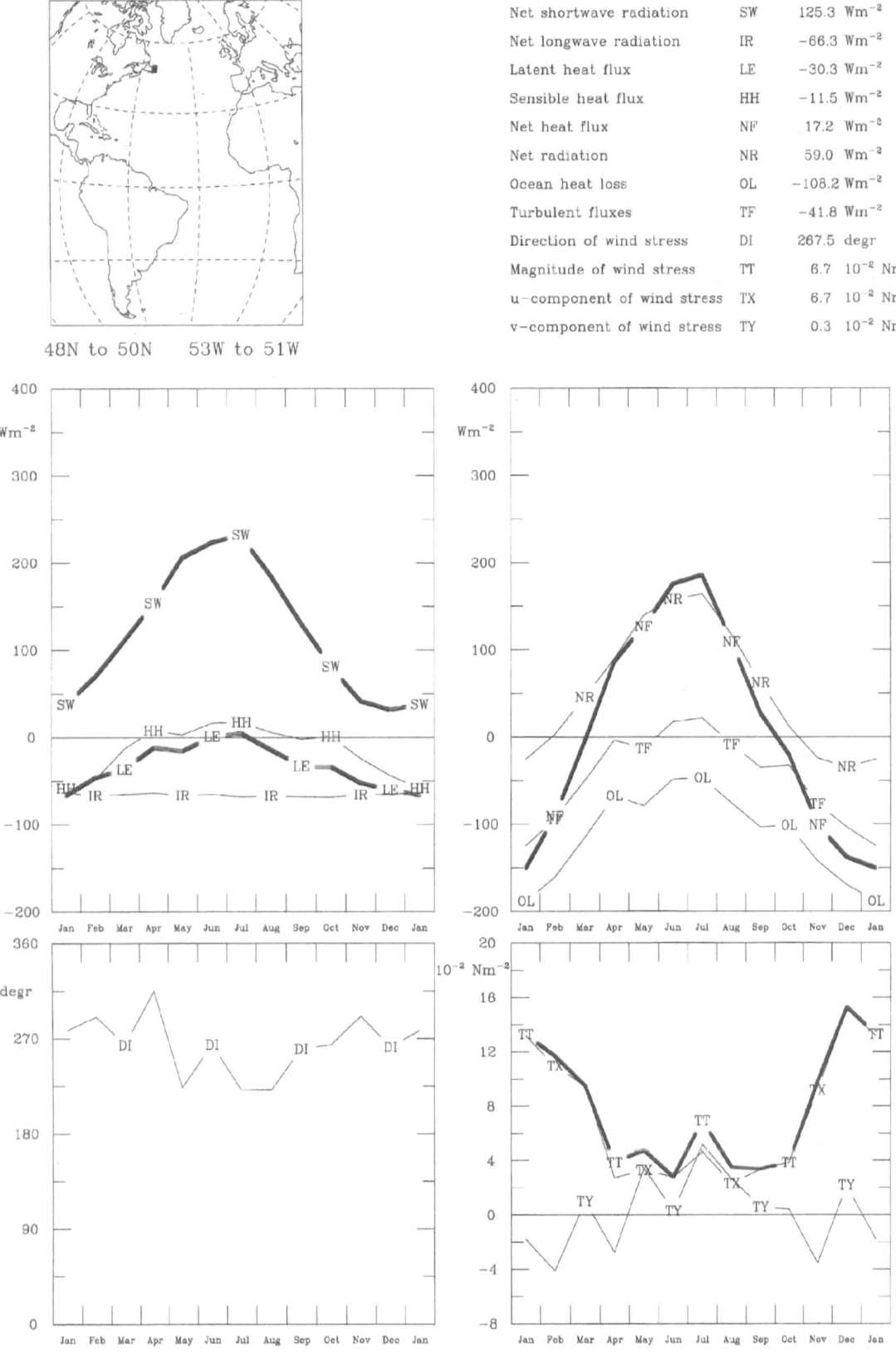

Net shortwave radiation	SW	125.3 Wm^{-2}
Net longwave radiation	IR	−66.3 Wm^{-2}
Latent heat flux	LE	−30.3 Wm^{-2}
Sensible heat flux	HH	−11.5 Wm^{-2}
Net heat flux	NF	17.2 Wm^{-2}
Net radiation	NR	59.0 Wm^{-2}
Ocean heat loss	OL	−108.2 Wm^{-2}
Turbulent fluxes	TF	−41.8 Wm^{-2}
Direction of wind stress	DI	267.5 degr
Magnitude of wind stress	TT	6.7 10^{-2} Nm^{-2}
u−component of wind stress	TX	6.7 10^{-2} Nm^{-2}
v−component of wind stress	TY	0.3 10^{-2} Nm^{-2}

48N to 50N   53W to 51W

Midocean westwind drift

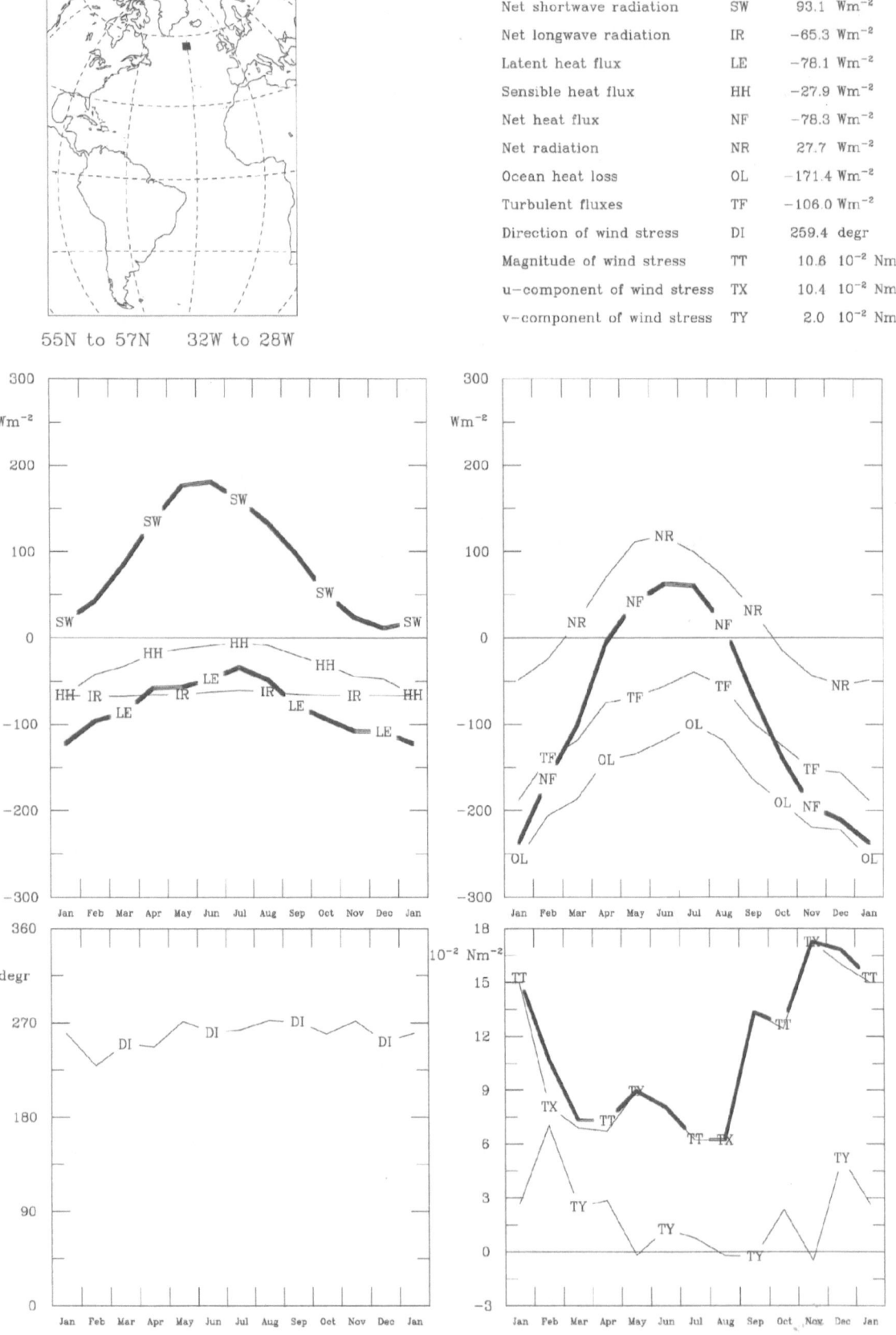

Net shortwave radiation	SW	93.1	Wm^{-2}
Net longwave radiation	IR	-65.3	Wm^{-2}
Latent heat flux	LE	-78.1	Wm^{-2}
Sensible heat flux	HH	-27.9	Wm^{-2}
Net heat flux	NF	-78.3	Wm^{-2}
Net radiation	NR	27.7	Wm^{-2}
Ocean heat loss	OL	-171.4	Wm^{-2}
Turbulent fluxes	TF	-106.0	Wm^{-2}
Direction of wind stress	DI	259.4	degr
Magnitude of wind stress	TT	10.6	10^{-2} Nm^{-2}
u-component of wind stress	TX	10.4	10^{-2} Nm^{-2}
v-component of wind stress	TY	2.0	10^{-2} Nm^{-2}

55N to 57N   32W to 28W

North Sea

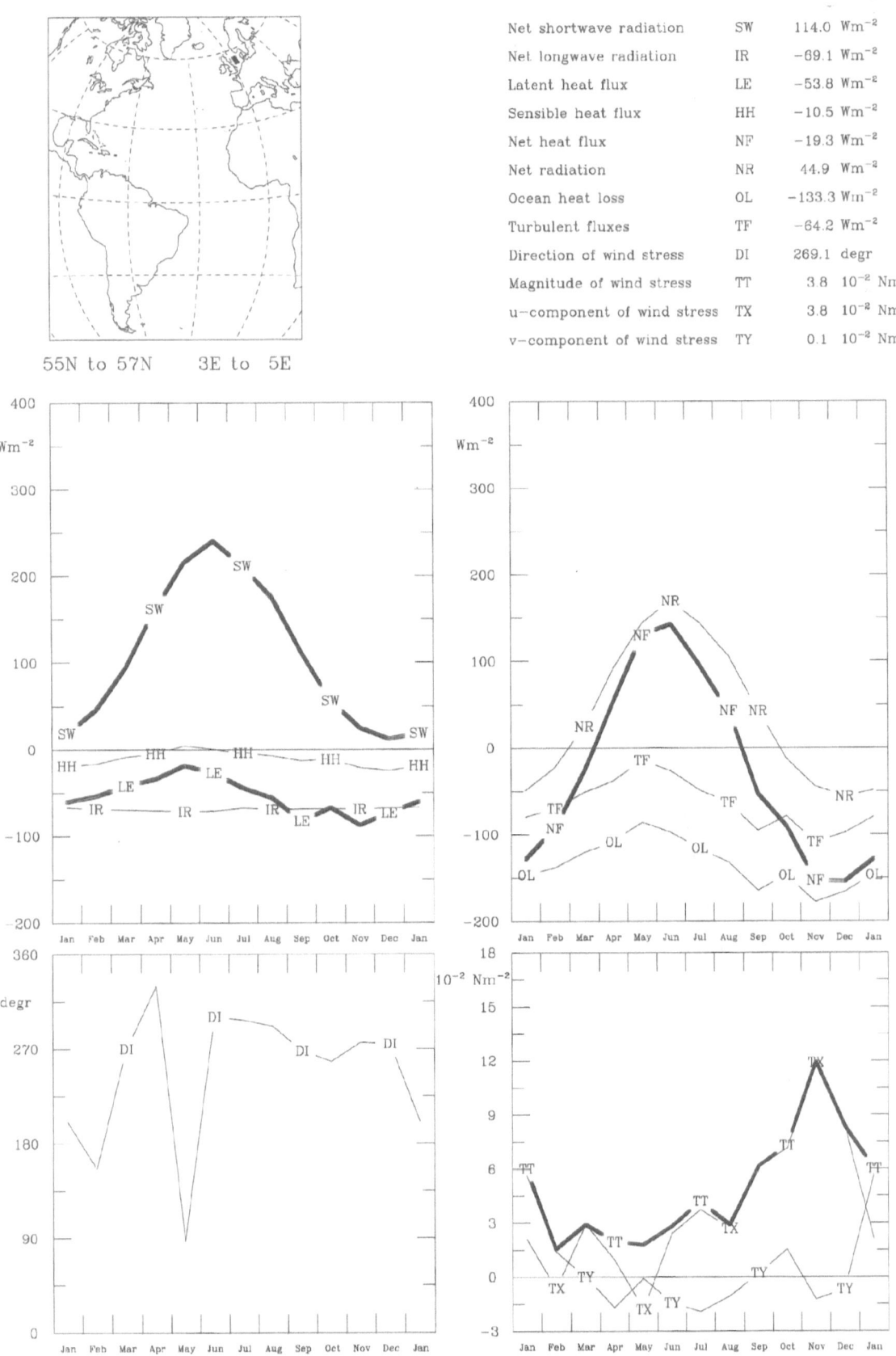

Net shortwave radiation	SW	114.0 Wm^{-2}
Net longwave radiation	IR	−69.1 Wm^{-2}
Latent heat flux	LE	−53.8 Wm^{-2}
Sensible heat flux	HH	−10.5 Wm^{-2}
Net heat flux	NF	−19.3 Wm^{-2}
Net radiation	NR	44.9 Wm^{-2}
Ocean heat loss	OL	−133.3 Wm^{-2}
Turbulent fluxes	TF	−64.2 Wm^{-2}
Direction of wind stress	DI	269.1 degr
Magnitude of wind stress	TT	3.8 10^{-2} Nm^{-2}
u−component of wind stress	TX	3.8 10^{-2} Nm^{-2}
v−component of wind stress	TY	0.1 10^{-2} Nm^{-2}

55N to 57N   3E to 5E

Subtropical convergence

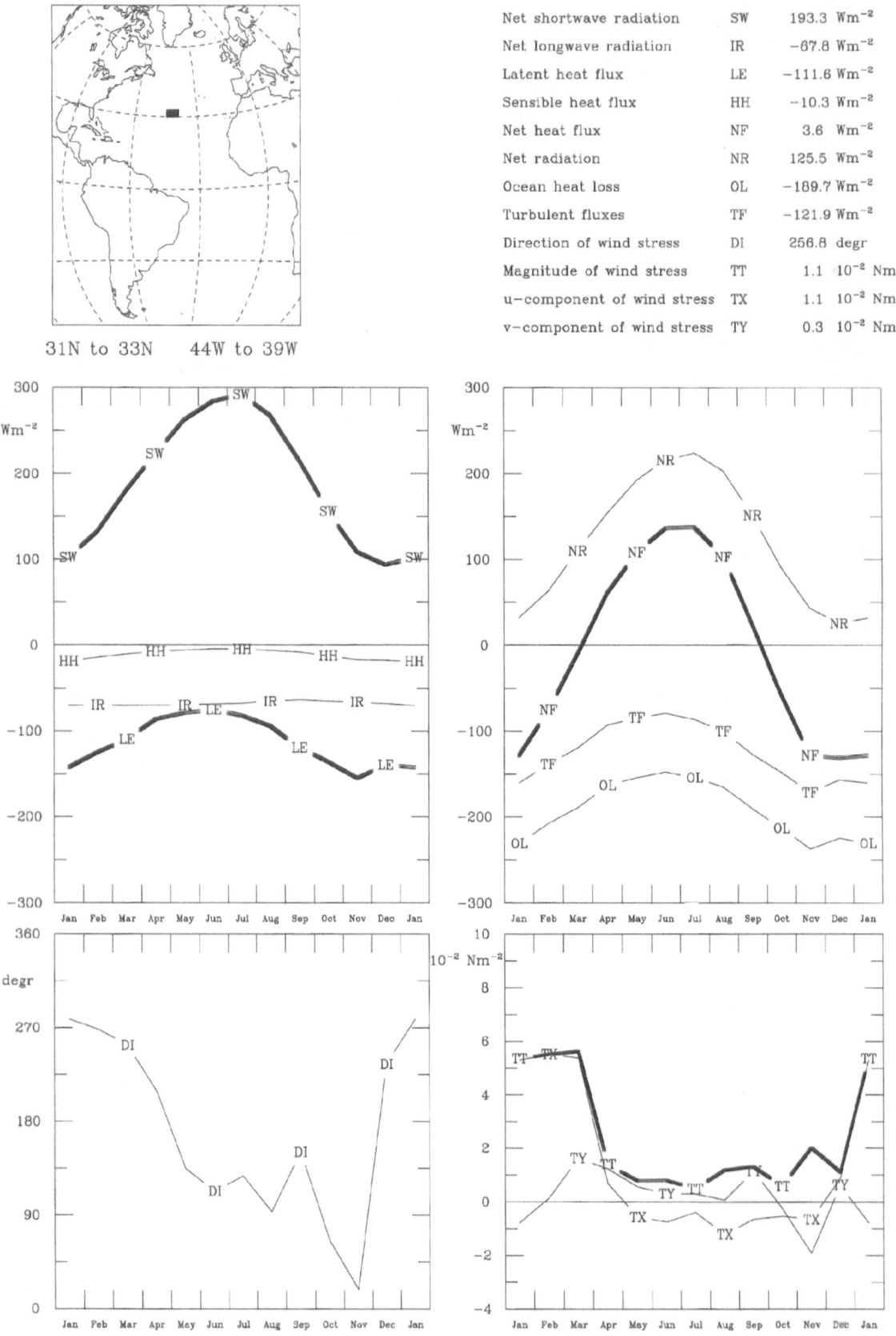

Net shortwave radiation	SW	193.3 Wm^{-2}
Net longwave radiation	IR	−67.8 Wm^{-2}
Latent heat flux	LE	−111.6 Wm^{-2}
Sensible heat flux	HH	−10.3 Wm^{-2}
Net heat flux	NF	3.6 Wm^{-2}
Net radiation	NR	125.5 Wm^{-2}
Ocean heat loss	OL	−189.7 Wm^{-2}
Turbulent fluxes	TF	−121.9 Wm^{-2}
Direction of wind stress	DI	256.8 degr
Magnitude of wind stress	TT	1.1 10^{-2} Nm^{-2}
u−component of wind stress	TX	1.1 10^{-2} Nm^{-2}
v−component of wind stress	TY	0.3 10^{-2} Nm^{-2}

31N to 33N   44W to 39W

Coast of Senegal

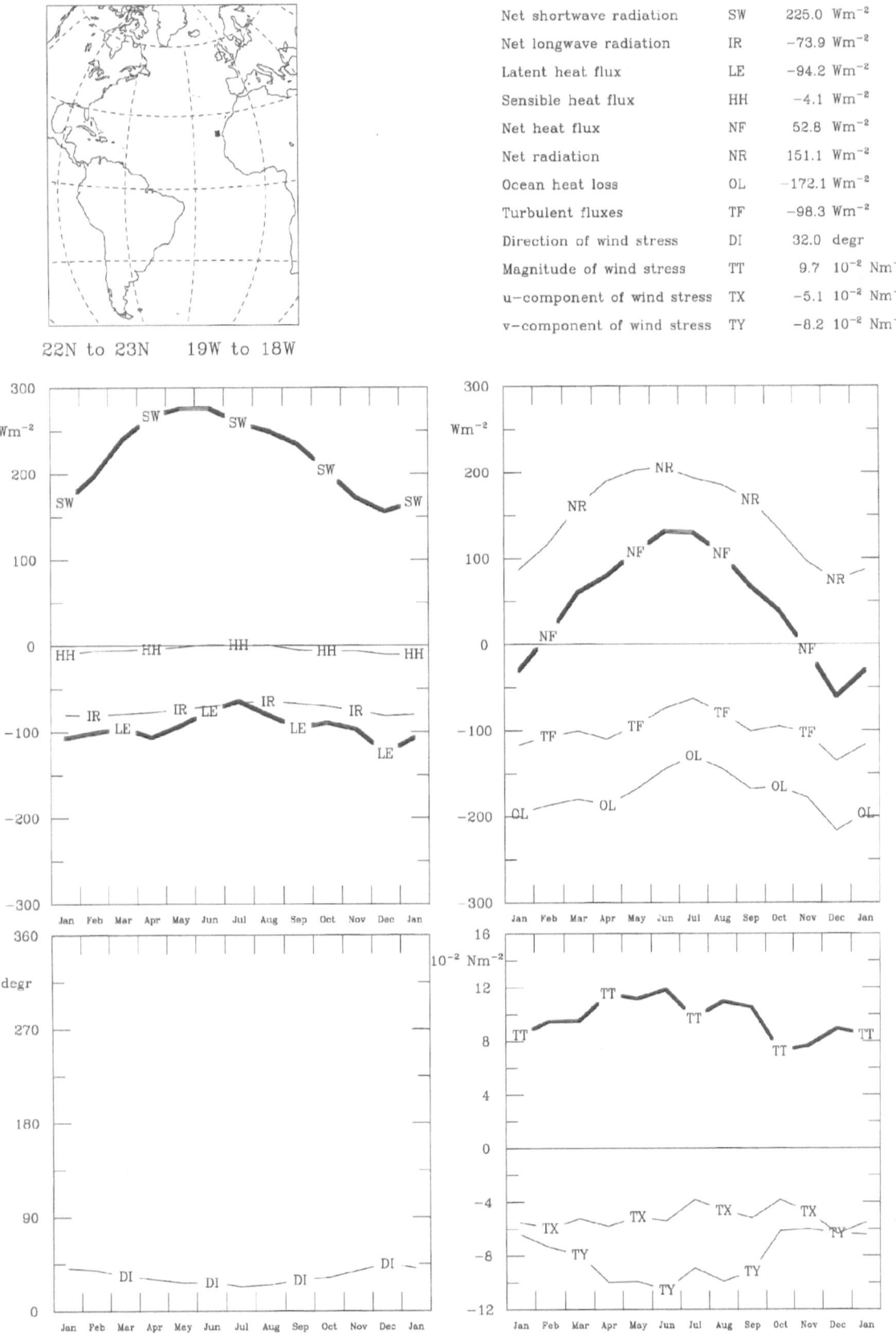

Net shortwave radiation	SW	225.0	Wm⁻²
Net longwave radiation	IR	−73.9	Wm⁻²
Latent heat flux	LE	−94.2	Wm⁻²
Sensible heat flux	HH	−4.1	Wm⁻²
Net heat flux	NF	52.8	Wm⁻²
Net radiation	NR	151.1	Wm⁻²
Ocean heat loss	OL	−172.1	Wm⁻²
Turbulent fluxes	TF	−98.3	Wm⁻²
Direction of wind stress	DI	32.0	degr
Magnitude of wind stress	TT	9.7	10⁻² Nm⁻²
u-component of wind stress	TX	−5.1	10⁻² Nm⁻²
v-component of wind stress	TY	−8.2	10⁻² Nm⁻²

22N to 23N   19W to 18W

Trade wind region

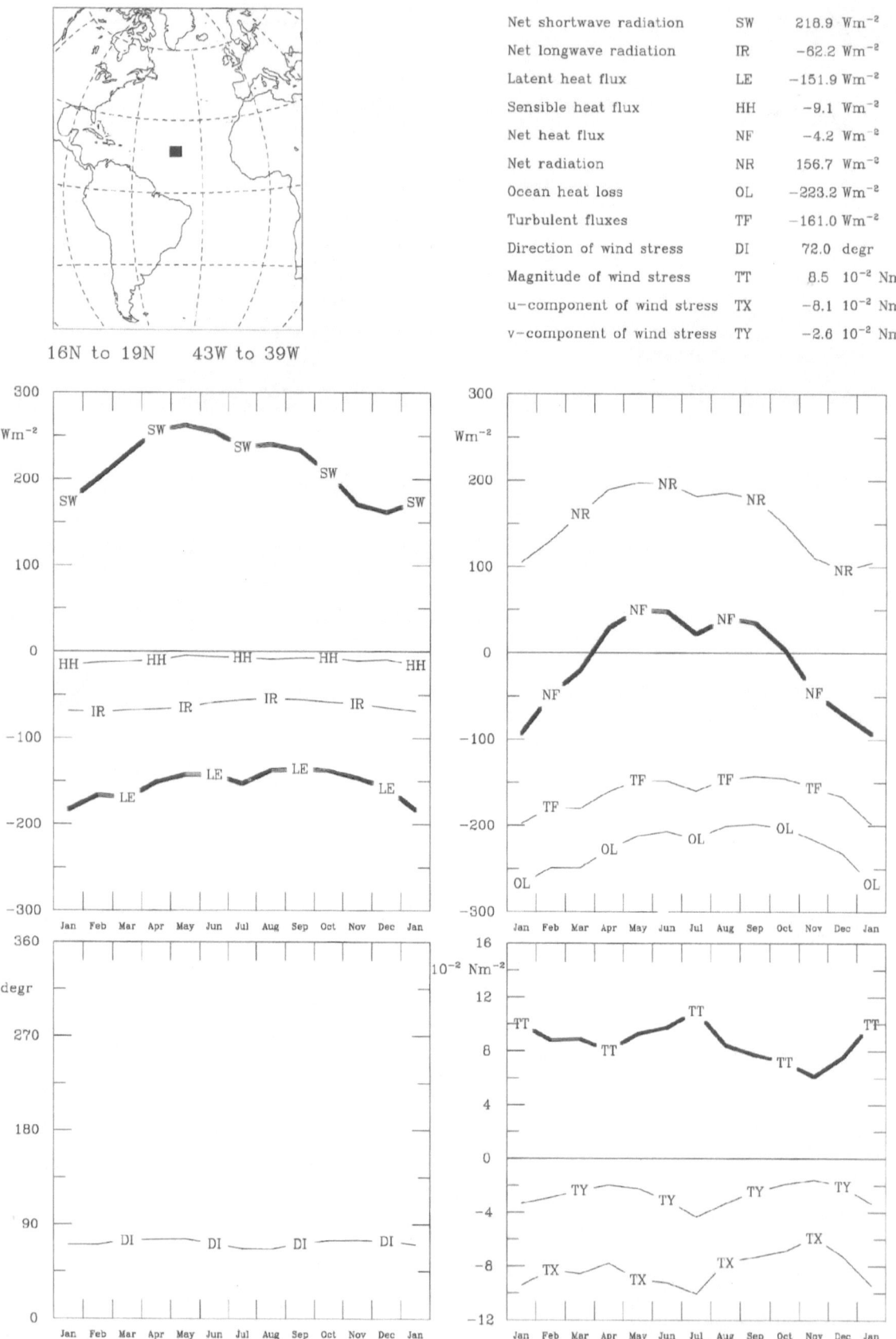

16N to 19N   43W to 39W

Net shortwave radiation	SW	218.9 Wm^{-2}
Net longwave radiation	IR	−62.2 Wm^{-2}
Latent heat flux	LE	−151.9 Wm^{-2}
Sensible heat flux	HH	−9.1 Wm^{-2}
Net heat flux	NF	−4.2 Wm^{-2}
Net radiation	NR	156.7 Wm^{-2}
Ocean heat loss	OL	−223.2 Wm^{-2}
Turbulent fluxes	TF	−161.0 Wm^{-2}
Direction of wind stress	DI	72.0 degr
Magnitude of wind stress	TT	8.5 10^{-2} Nm^{-2}
u−component of wind stress	TX	−8.1 10^{-2} Nm^{-2}
v−component of wind stress	TY	−2.6 10^{-2} Nm^{-2}

Caribbean Sea

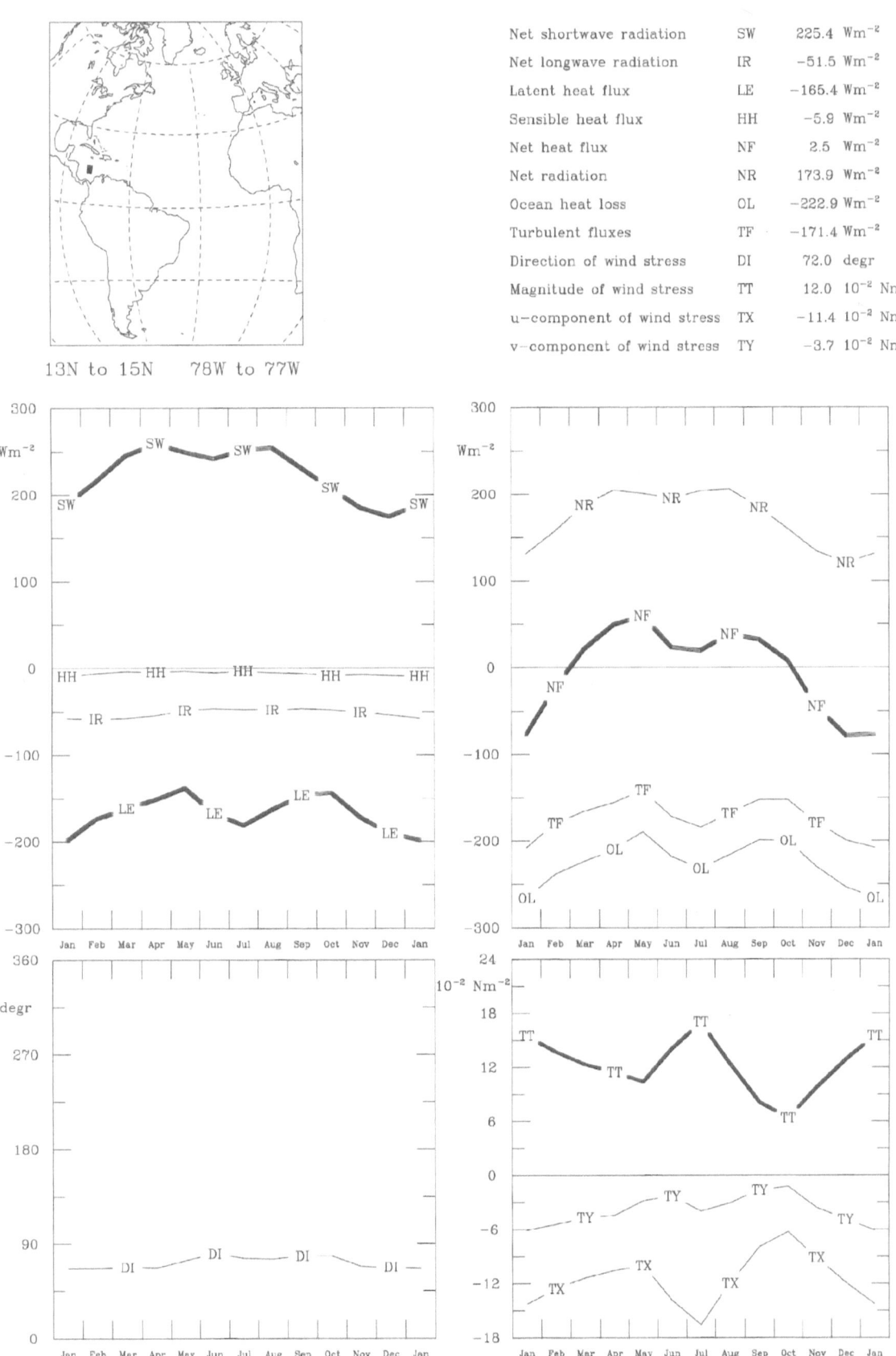

13N to 15N   78W to 77W

Net shortwave radiation	SW	225.4 Wm^{-2}
Net longwave radiation	IR	−51.5 Wm^{-2}
Latent heat flux	LE	−165.4 Wm^{-2}
Sensible heat flux	HH	−5.9 Wm^{-2}
Net heat flux	NF	2.5 Wm^{-2}
Net radiation	NR	173.9 Wm^{-2}
Ocean heat loss	OL	−222.9 Wm^{-2}
Turbulent fluxes	TF	−171.4 Wm^{-2}
Direction of wind stress	DI	72.0 degr
Magnitude of wind stress	TT	12.0 $10^{-2} Nm^{-2}$
u−component of wind stress	TX	−11.4 $10^{-2} Nm^{-2}$
v−component of wind stress	TY	−3.7 $10^{-2} Nm^{-2}$

Cape Hoorn

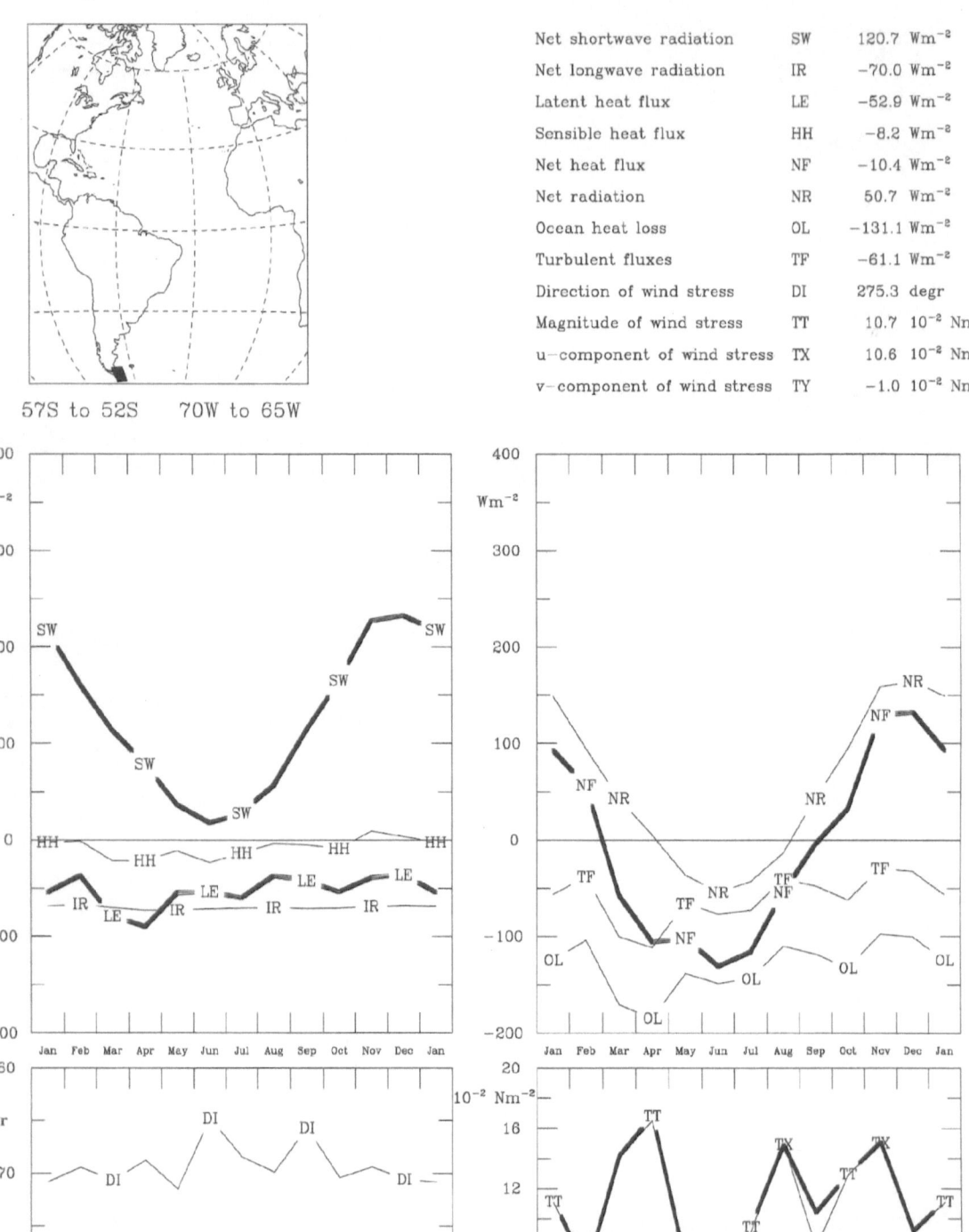

Net shortwave radiation	SW	120.7	Wm^{-2}
Net longwave radiation	IR	−70.0	Wm^{-2}
Latent heat flux	LE	−52.9	Wm^{-2}
Sensible heat flux	HH	−8.2	Wm^{-2}
Net heat flux	NF	−10.4	Wm^{-2}
Net radiation	NR	50.7	Wm^{-2}
Ocean heat loss	OL	−131.1	Wm^{-2}
Turbulent fluxes	TF	−61.1	Wm^{-2}
Direction of wind stress	DI	275.3	degr
Magnitude of wind stress	TT	10.7	10^{-2} Nm^{-2}
u−component of wind stress	TX	10.6	10^{-2} Nm^{-2}
v−component of wind stress	TY	−1.0	10^{-2} Nm^{-2}

57S to 52S   70W to 65W

South Georgia

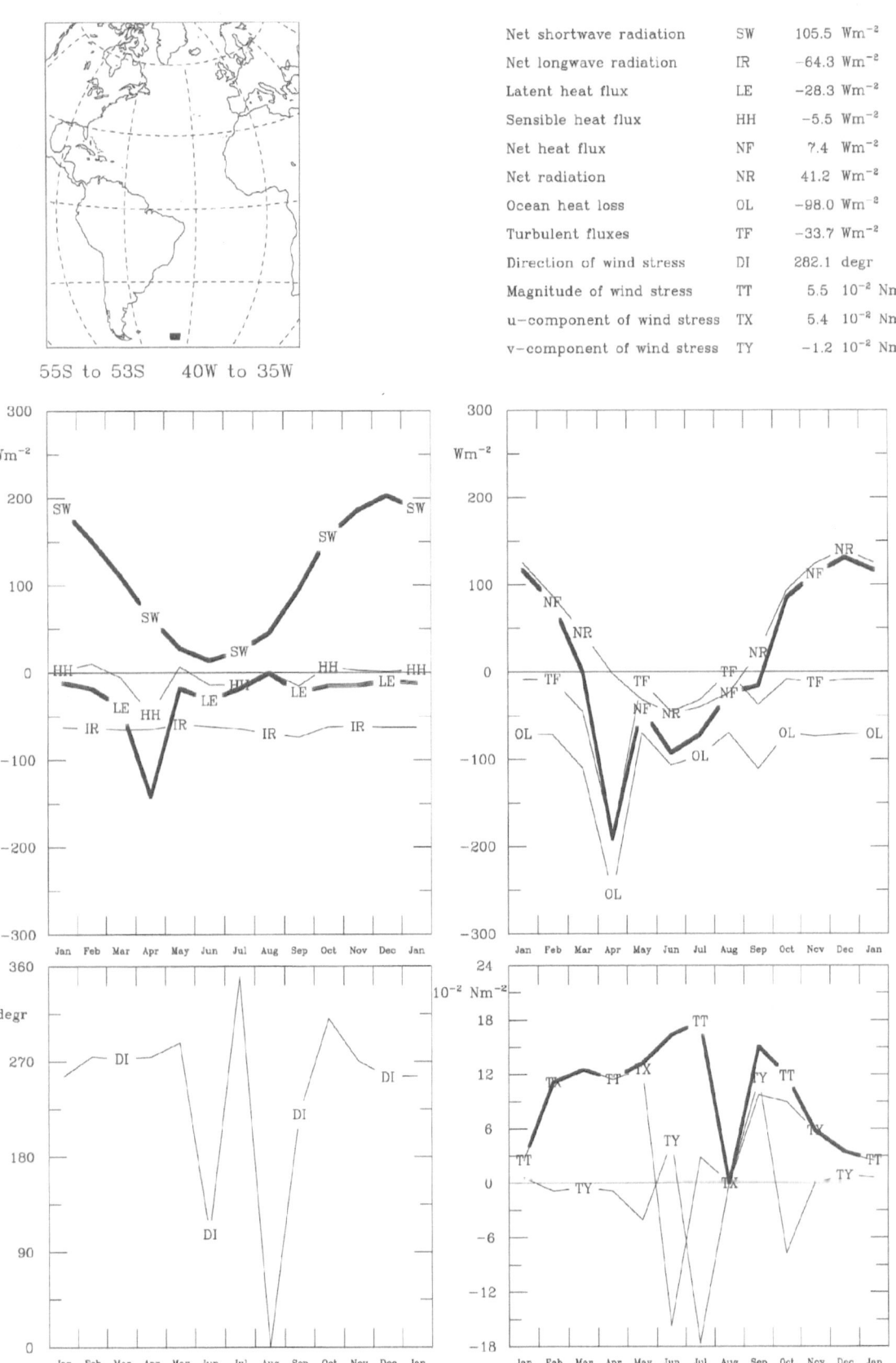

Net shortwave radiation	SW	105.5 Wm^{-2}
Net longwave radiation	IR	−64.3 Wm^{-2}
Latent heat flux	LE	−28.3 Wm^{-2}
Sensible heat flux	HH	−5.5 Wm^{-2}
Net heat flux	NF	7.4 Wm^{-2}
Net radiation	NR	41.2 Wm^{-2}
Ocean heat loss	OL	−98.0 Wm^{-2}
Turbulent fluxes	TF	−33.7 Wm^{-2}
Direction of wind stress	DI	282.1 degr
Magnitude of wind stress	TT	5.5 10^{-2} Nm^{-2}
u−component of wind stress	TX	5.4 10^{-2} Nm^{-2}
v−component of wind stress	TY	−1.2 10^{-2} Nm^{-2}

55S to 53S   40W to 35W

Labrador Sea

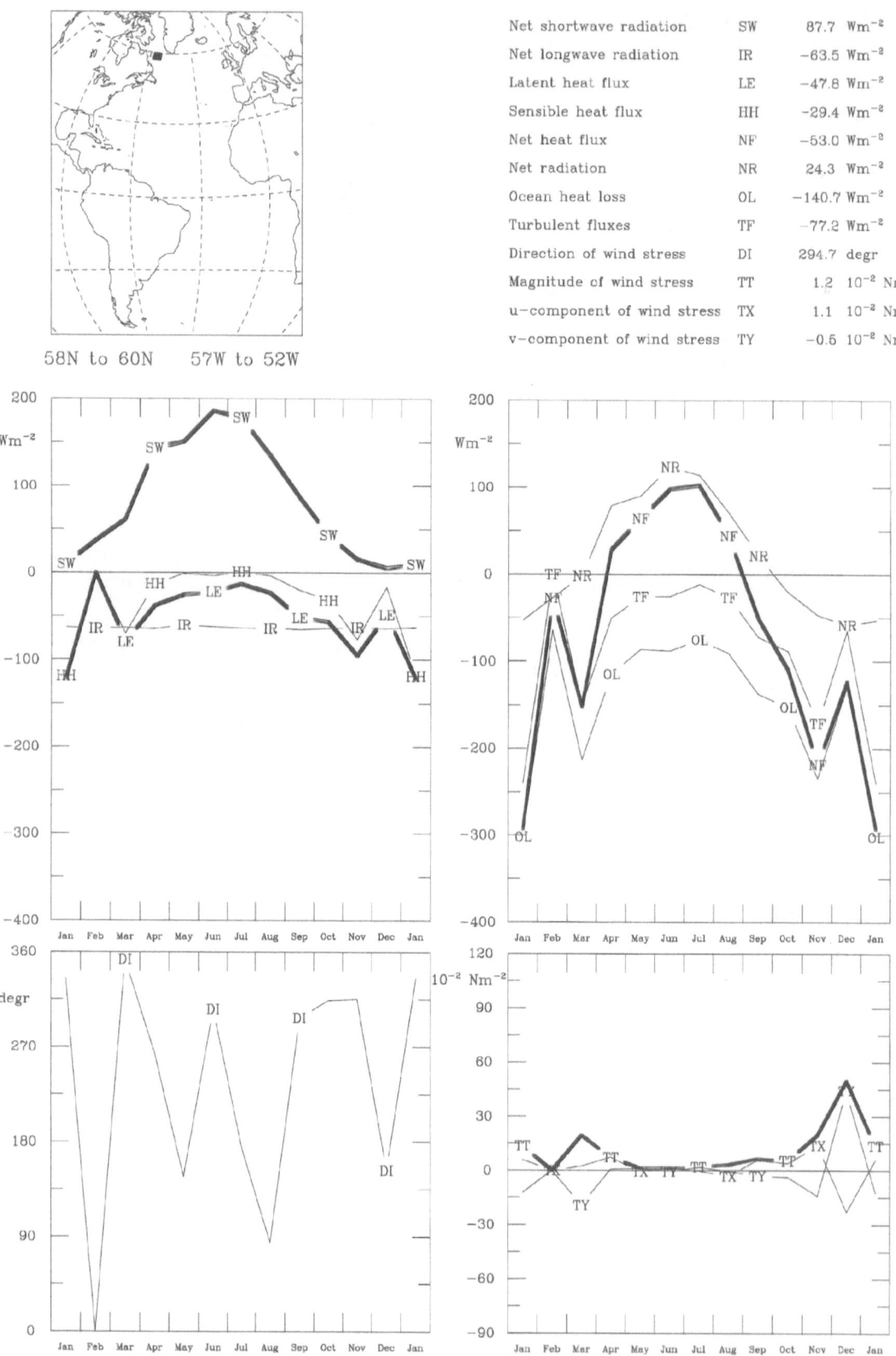

58N to 60N   57W to 52W

Net shortwave radiation	SW	87.7	Wm^{-2}
Net longwave radiation	IR	−63.5	Wm^{-2}
Latent heat flux	LE	−47.8	Wm^{-2}
Sensible heat flux	HH	−29.4	Wm^{-2}
Net heat flux	NF	−53.0	Wm^{-2}
Net radiation	NR	24.3	Wm^{-2}
Ocean heat loss	OL	−140.7	Wm^{-2}
Turbulent fluxes	TF	−77.2	Wm^{-2}
Direction of wind stress	DI	294.7	degr
Magnitude of wind stress	TT	1.2	$10^{-2}\ Nm^{-2}$
u−component of wind stress	TX	1.1	$10^{-2}\ Nm^{-2}$
v−component of wind stress	TY	−0.5	$10^{-2}\ Nm^{-2}$

Irminger Sea

59N to 61N   38W to 33W

Net shortwave radiation	SW	86.8	Wm^{-2}
Net longwave radiation	IR	−65.1	Wm^{-2}
Latent heat flux	LE	−64.2	Wm^{-2}
Sensible heat flux	HH	−26.9	Wm^{-2}
Net heat flux	NF	−69.3	Wm^{-2}
Net radiation	NR	21.8	Wm^{-2}
Ocean heat loss	OL	−156.1	Wm^{-2}
Turbulent fluxes	TF	−91.1	Wm^{-2}
Direction of wind stress	DI	257.5	degr
Magnitude of wind stress	TT	6.4	10^{-2} Nm^{-2}
u−component of wind stress	TX	6.2	10^{-2} Nm^{-2}
v−component of wind stress	TY	1.4	10^{-2} Nm^{-2}

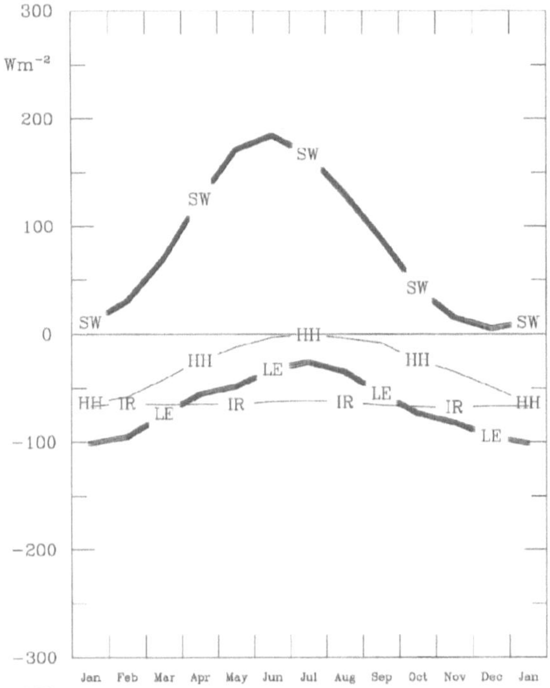

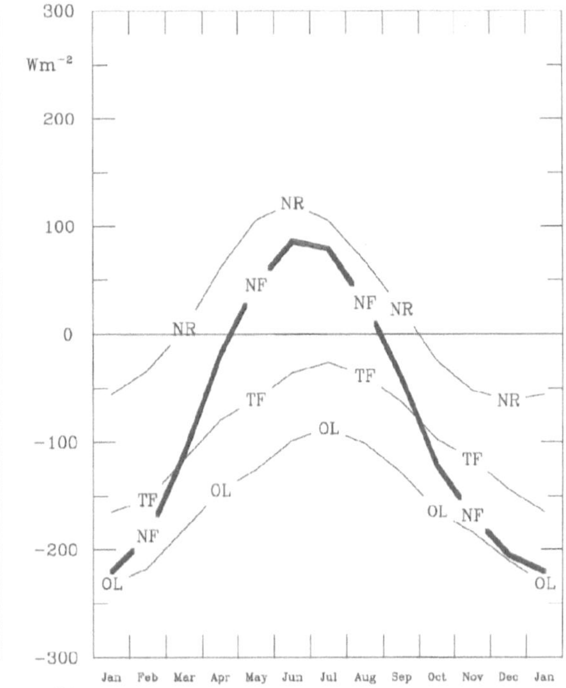

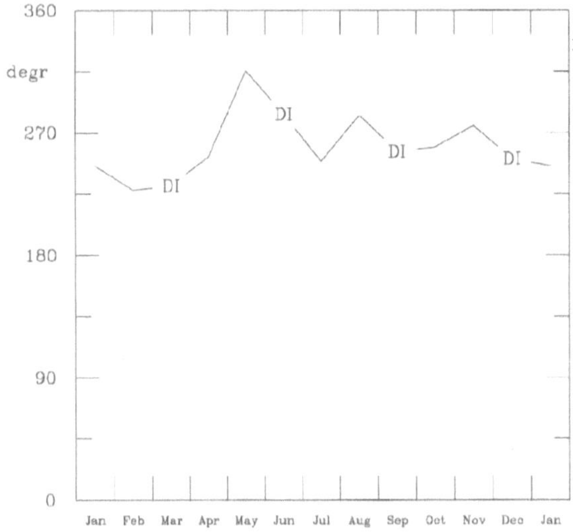

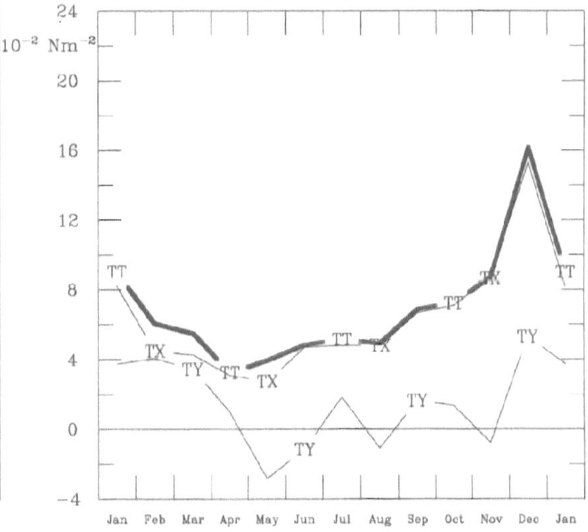

Agulhas Current

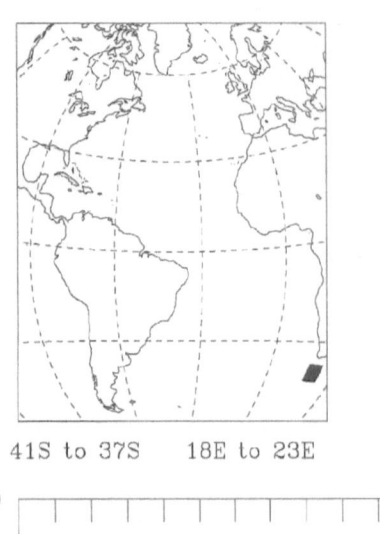

41S to 37S   18E to 23E

Net shortwave radiation	SW	162.2	Wm^{-2}
Net longwave radiation	IR	−72.2	Wm^{-2}
Latent heat flux	LE	−223.3	Wm^{-2}
Sensible heat flux	HH	−47.3	Wm^{-2}
Net heat flux	NF	−180.6	Wm^{-2}
Net radiation	NR	90.1	Wm^{-2}
Ocean heat loss	OL	−342.8	Wm^{-2}
Turbulent fluxes	TF	−270.6	Wm^{-2}
Direction of wind stress	DI	261.5	degr
Magnitude of wind stress	TT	10.5	10^{-2} Nm^{-2}
u−component of wind stress	TX	10.3	10^{-2} Nm^{-2}
v−component of wind stress	TY	1.6	10^{-2} Nm^{-2}

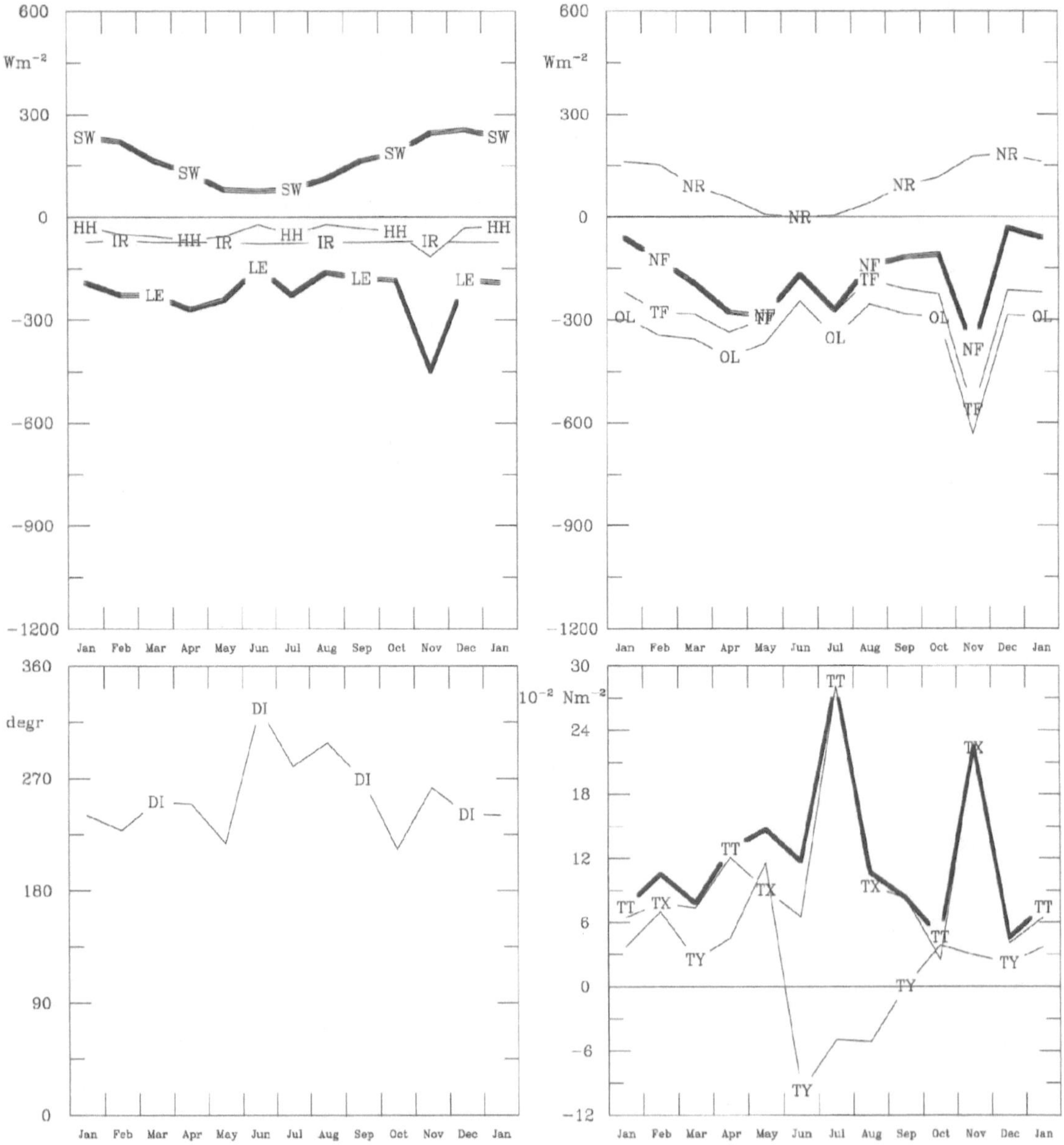

Subtropic convergence

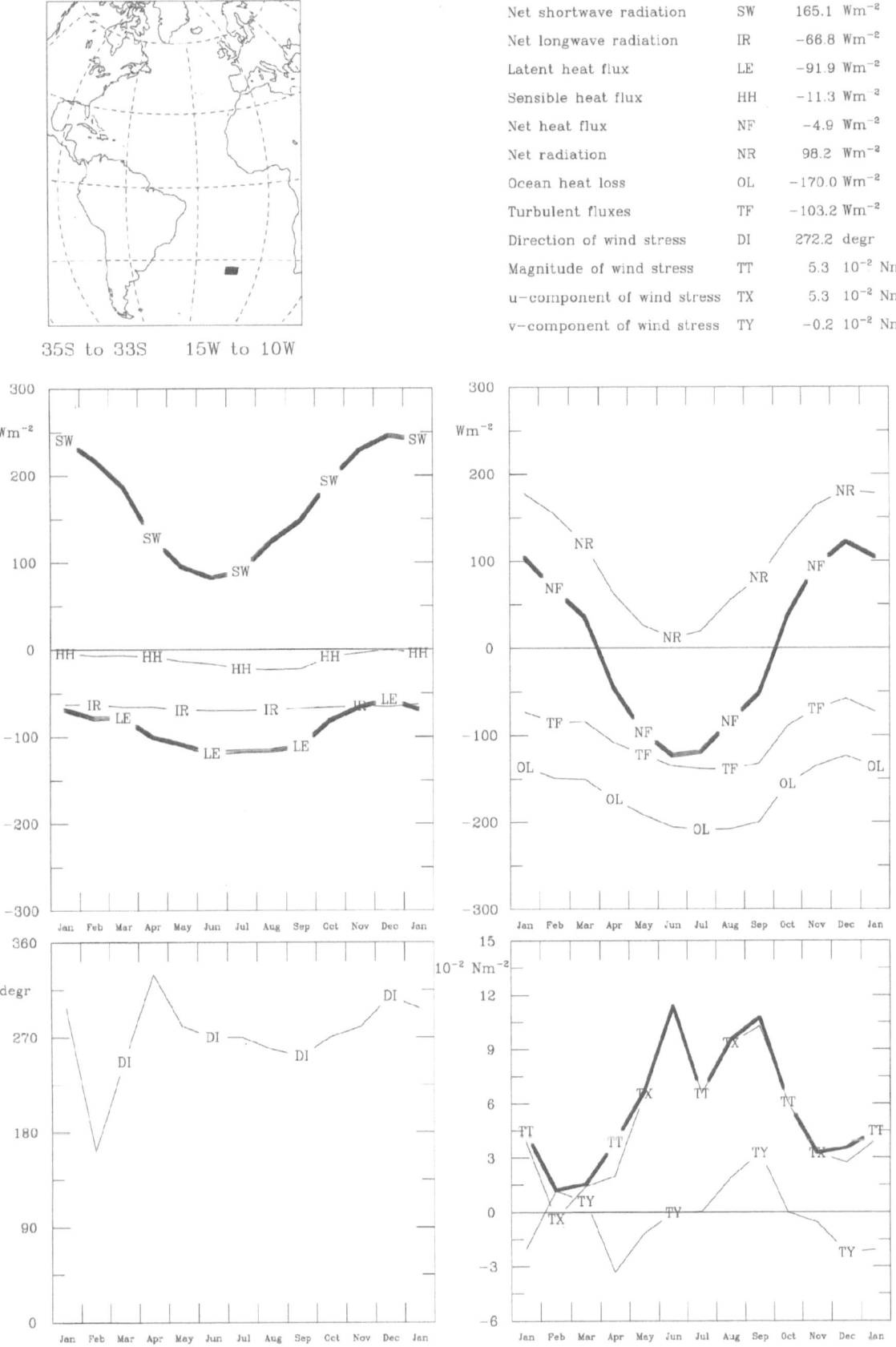

Net shortwave radiation	SW	165.1	Wm^{-2}
Net longwave radiation	IR	−66.8	Wm^{-2}
Latent heat flux	LE	−91.9	Wm^{-2}
Sensible heat flux	HH	−11.3	Wm^{-2}
Net heat flux	NF	−4.9	Wm^{-2}
Net radiation	NR	98.2	Wm^{-2}
Ocean heat loss	OL	−170.0	Wm^{-2}
Turbulent fluxes	TF	−103.2	Wm^{-2}
Direction of wind stress	DI	272.2	degr
Magnitude of wind stress	TT	5.3	10^{-2} Nm^{-2}
u−component of wind stress	TX	5.3	10^{-2} Nm^{-2}
v−component of wind stress	TY	−0.2	10^{-2} Nm^{-2}

35S to 33S   15W to 10W

Benguela Current

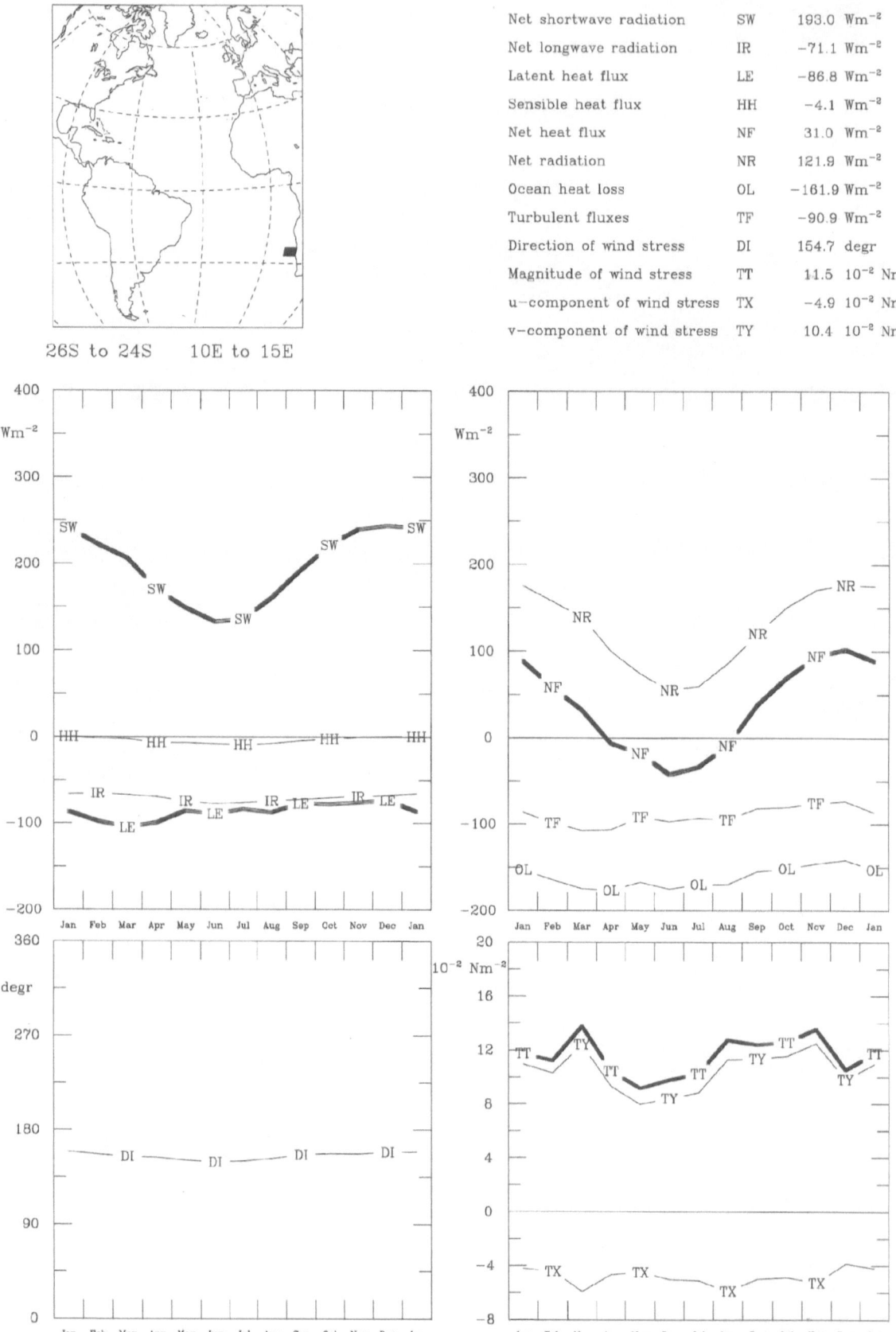

Net shortwave radiation	SW	193.0 Wm^{-2}
Net longwave radiation	IR	−71.1 Wm^{-2}
Latent heat flux	LE	−86.8 Wm^{-2}
Sensible heat flux	HH	−4.1 Wm^{-2}
Net heat flux	NF	31.0 Wm^{-2}
Net radiation	NR	121.9 Wm^{-2}
Ocean heat loss	OL	−161.9 Wm^{-2}
Turbulent fluxes	TF	−90.9 Wm^{-2}
Direction of wind stress	DI	154.7 degr
Magnitude of wind stress	TT	11.5 10^{-2} Nm^{-2}
u−component of wind stress	TX	−4.9 10^{-2} Nm^{-2}
v−component of wind stress	TY	10.4 10^{-2} Nm^{-2}

26S to 24S   10E to 15E

South East Trade Wind

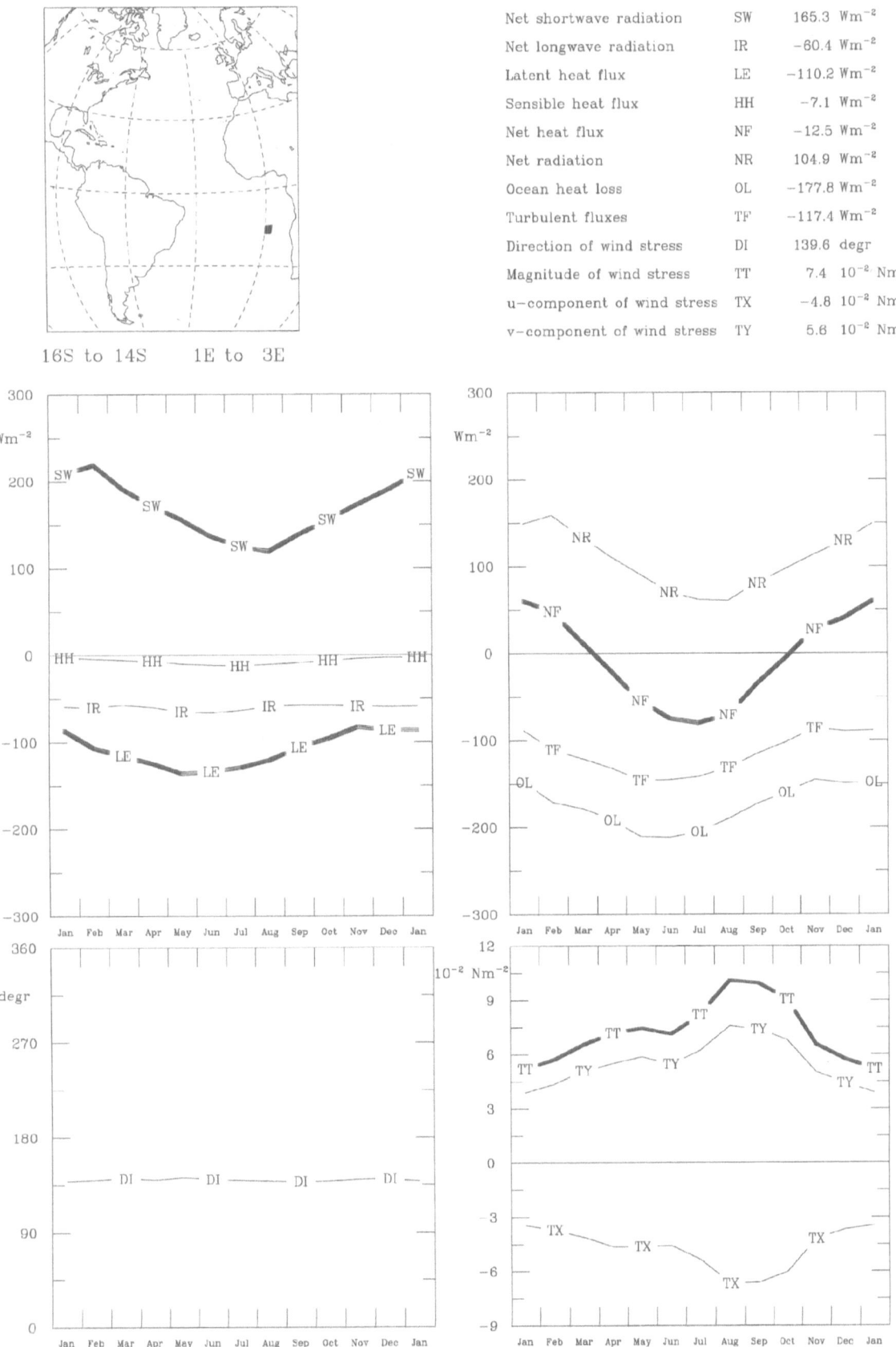

Net shortwave radiation	SW	165.3	Wm^{-2}
Net longwave radiation	IR	−60.4	Wm^{-2}
Latent heat flux	LE	−110.2	Wm^{-2}
Sensible heat flux	HH	−7.1	Wm^{-2}
Net heat flux	NF	−12.5	Wm^{-2}
Net radiation	NR	104.9	Wm^{-2}
Ocean heat loss	OL	−177.8	Wm^{-2}
Turbulent fluxes	TF	−117.4	Wm^{-2}
Direction of wind stress	DI	139.6	degr
Magnitude of wind stress	TT	7.4	10^{-2} Nm^{-2}
u−component of wind stress	TX	−4.8	10^{-2} Nm^{-2}
v−component of wind stress	TY	5.6	10^{-2} Nm^{-2}

16S to 14S   1E to 3E

Mediterranean Sea

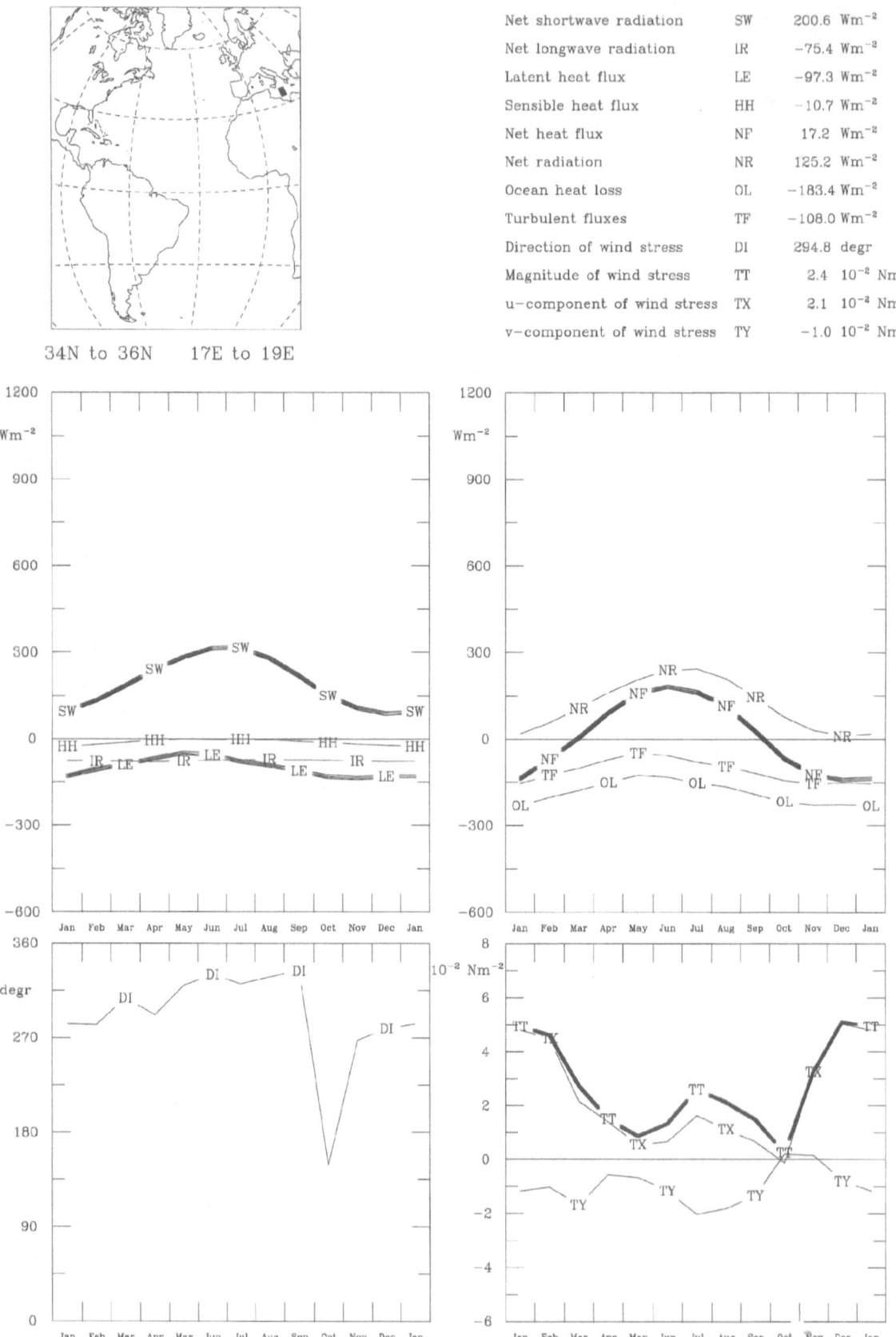

34N to 36N   17E to 19E

Net shortwave radiation	SW	200.6	Wm^{-2}
Net longwave radiation	IR	−75.4	Wm^{-2}
Latent heat flux	LE	−97.3	Wm^{-2}
Sensible heat flux	HH	−10.7	Wm^{-2}
Net heat flux	NF	17.2	Wm^{-2}
Net radiation	NR	125.2	Wm^{-2}
Ocean heat loss	OL	−183.4	Wm^{-2}
Turbulent fluxes	TF	−108.0	Wm^{-2}
Direction of wind stress	DI	294.8	degr
Magnitude of wind stress	TT	2.4	10^{-2} Nm^{-2}
u−component of wind stress	TX	2.1	10^{-2} Nm^{-2}
v−component of wind stress	TY	−1.0	10^{-2} Nm^{-2}

Tropical convergence

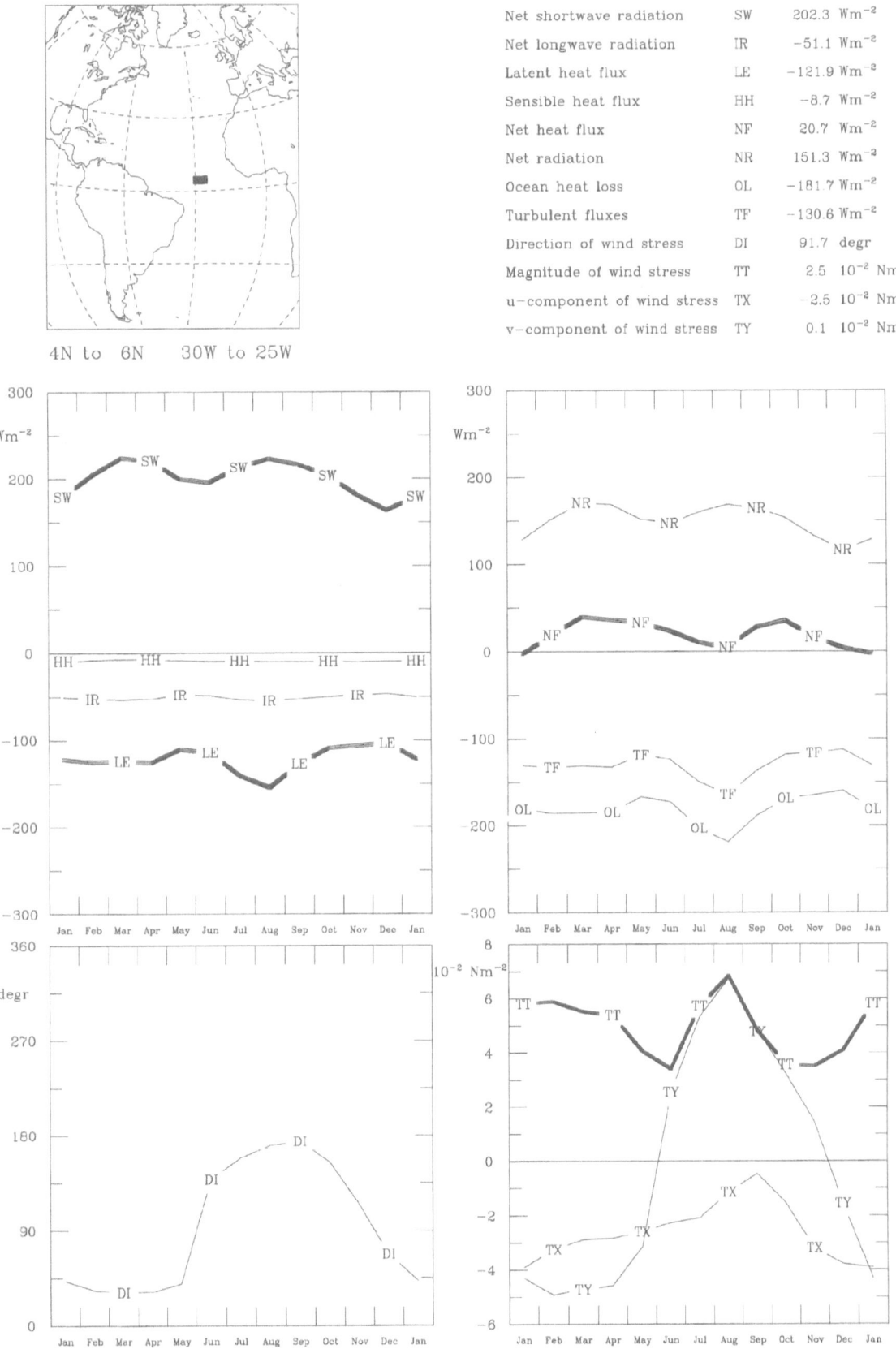

4N to 6N   30W to 25W

Net shortwave radiation	SW	202.3 Wm^{-2}
Net longwave radiation	IR	−51.1 Wm^{-2}
Latent heat flux	LE	−121.9 Wm^{-2}
Sensible heat flux	HH	−8.7 Wm^{-2}
Net heat flux	NF	20.7 Wm^{-2}
Net radiation	NR	151.3 Wm^{-2}
Ocean heat loss	OL	−181.7 Wm^{-2}
Turbulent fluxes	TF	−130.6 Wm^{-2}
Direction of wind stress	DI	91.7 degr
Magnitude of wind stress	TT	2.5 10^{-2} Nm^{-2}
u−component of wind stress	TX	−2.5 10^{-2} Nm^{-2}
v−component of wind stress	TY	0.1 10^{-2} Nm^{-2}

Net Shortwave Radiation

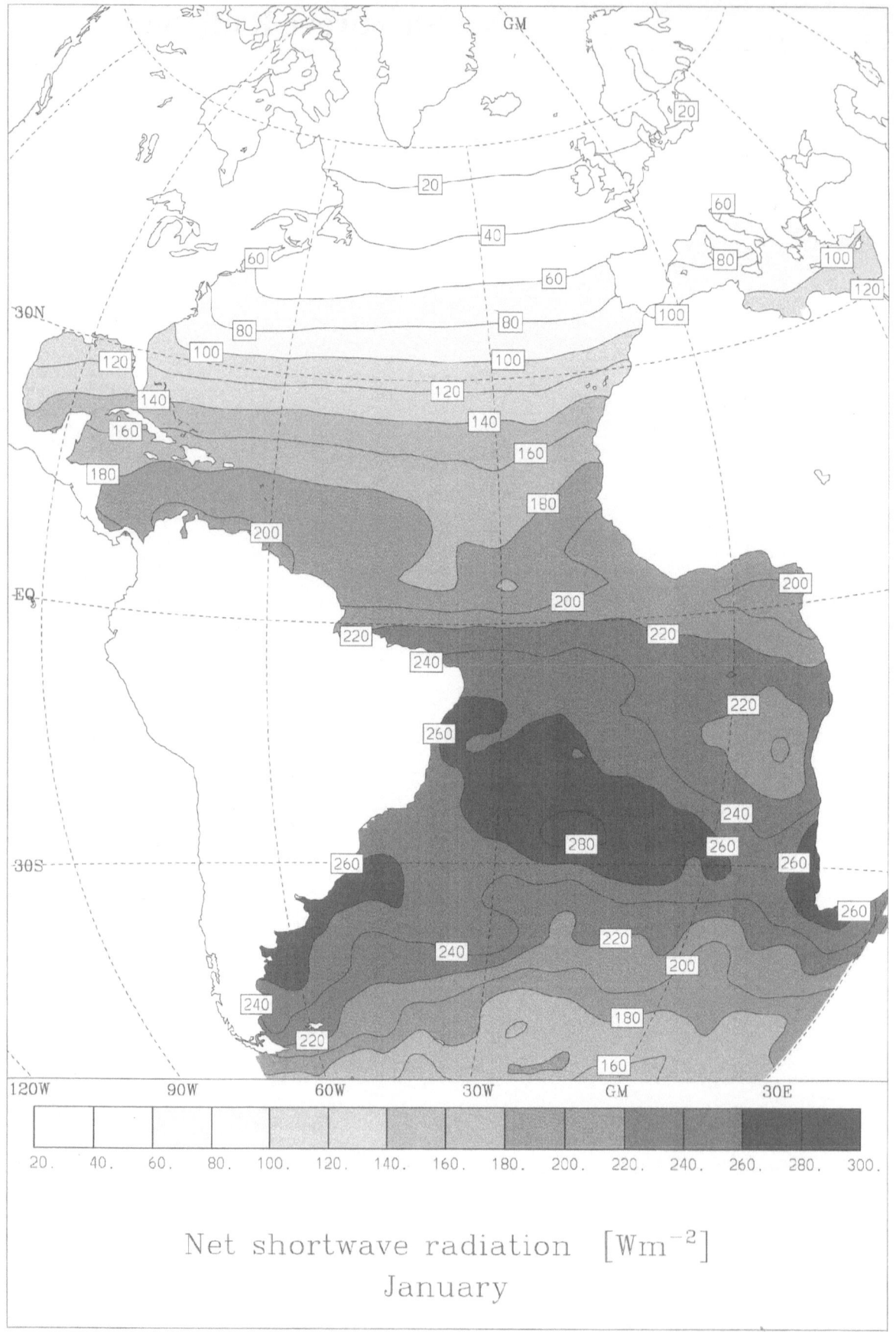

Net shortwave radiation [Wm^{-2}]
January

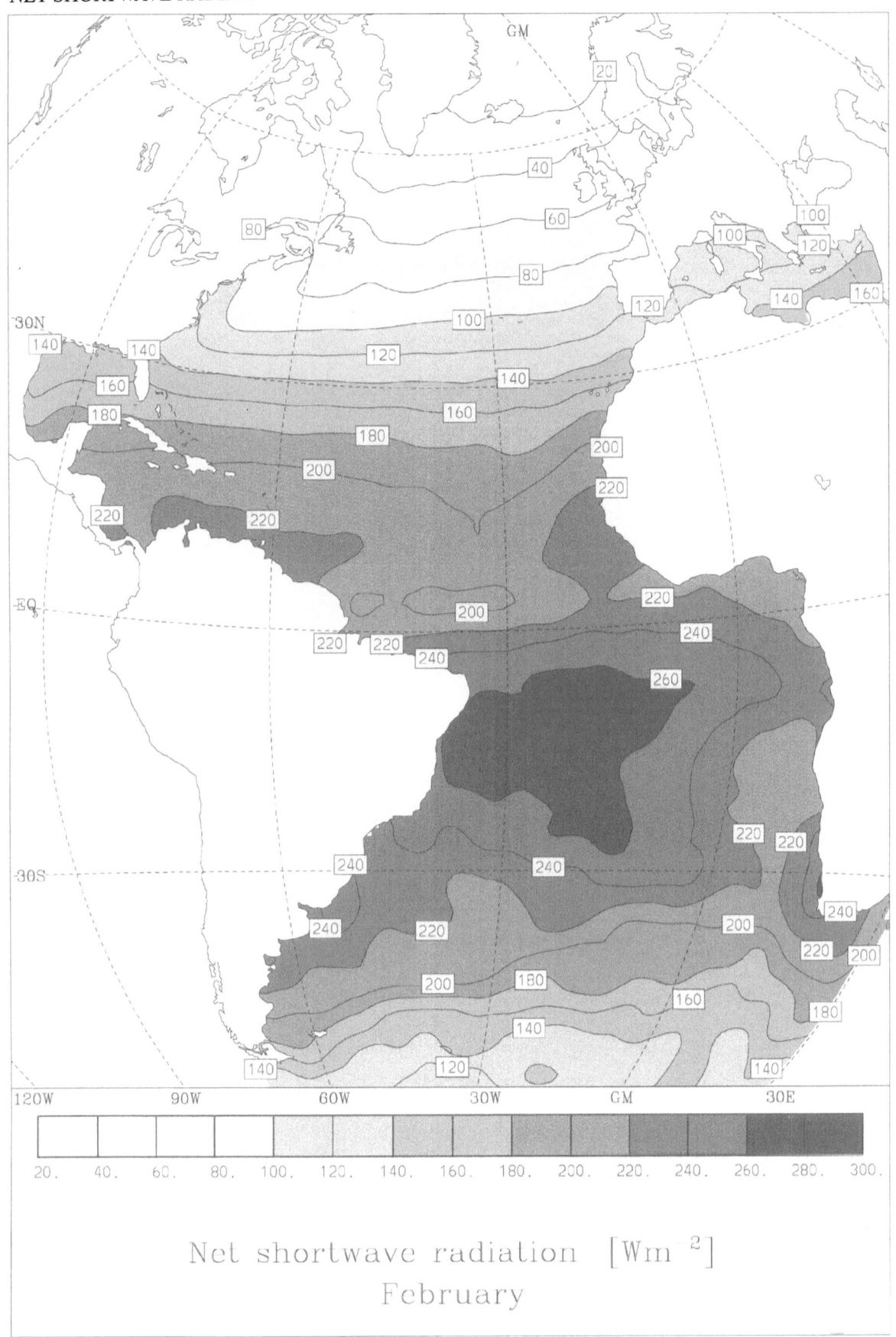

Net shortwave radiation $[Wm^{-2}]$

February

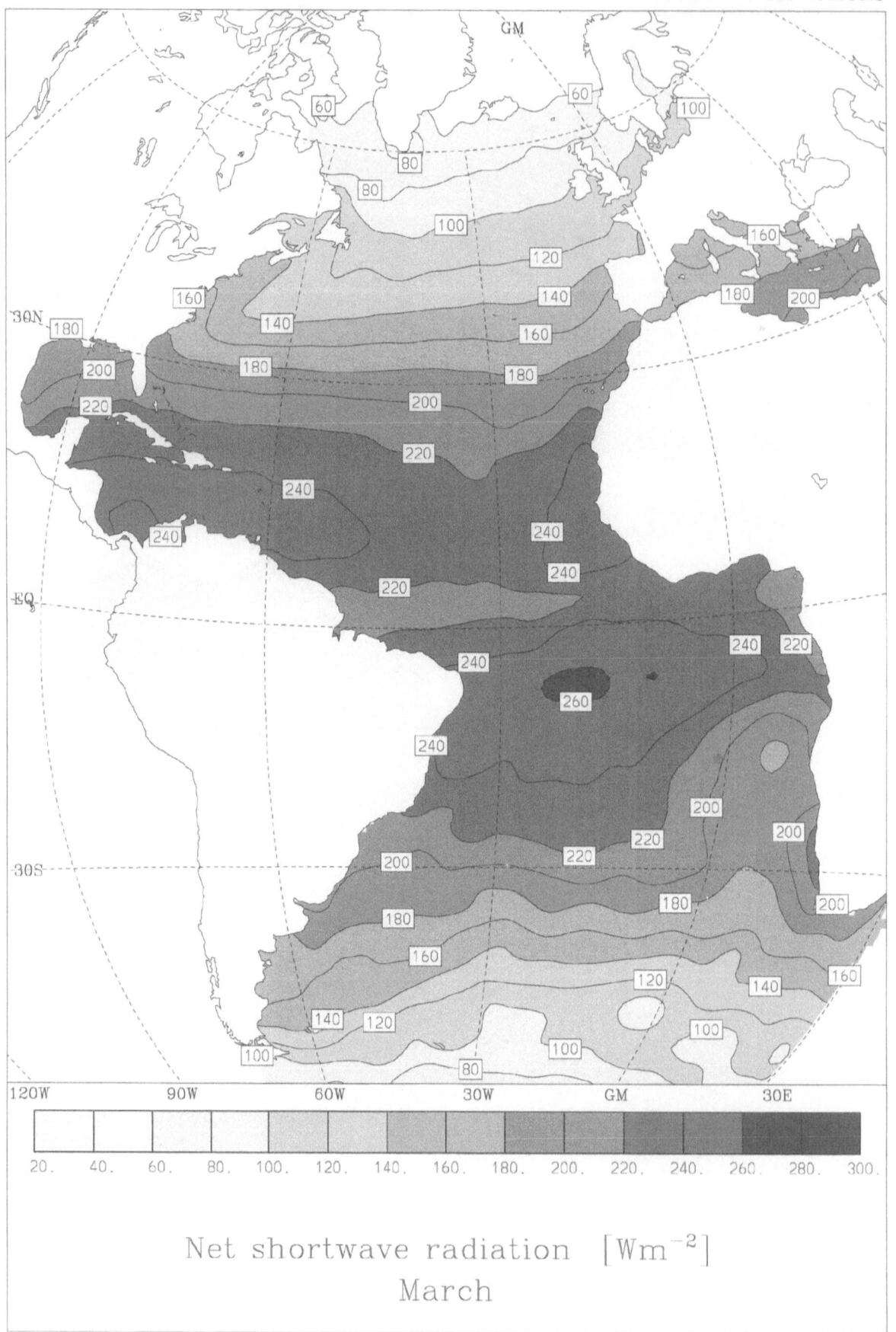

Net shortwave radiation [Wm^{-2}]
March

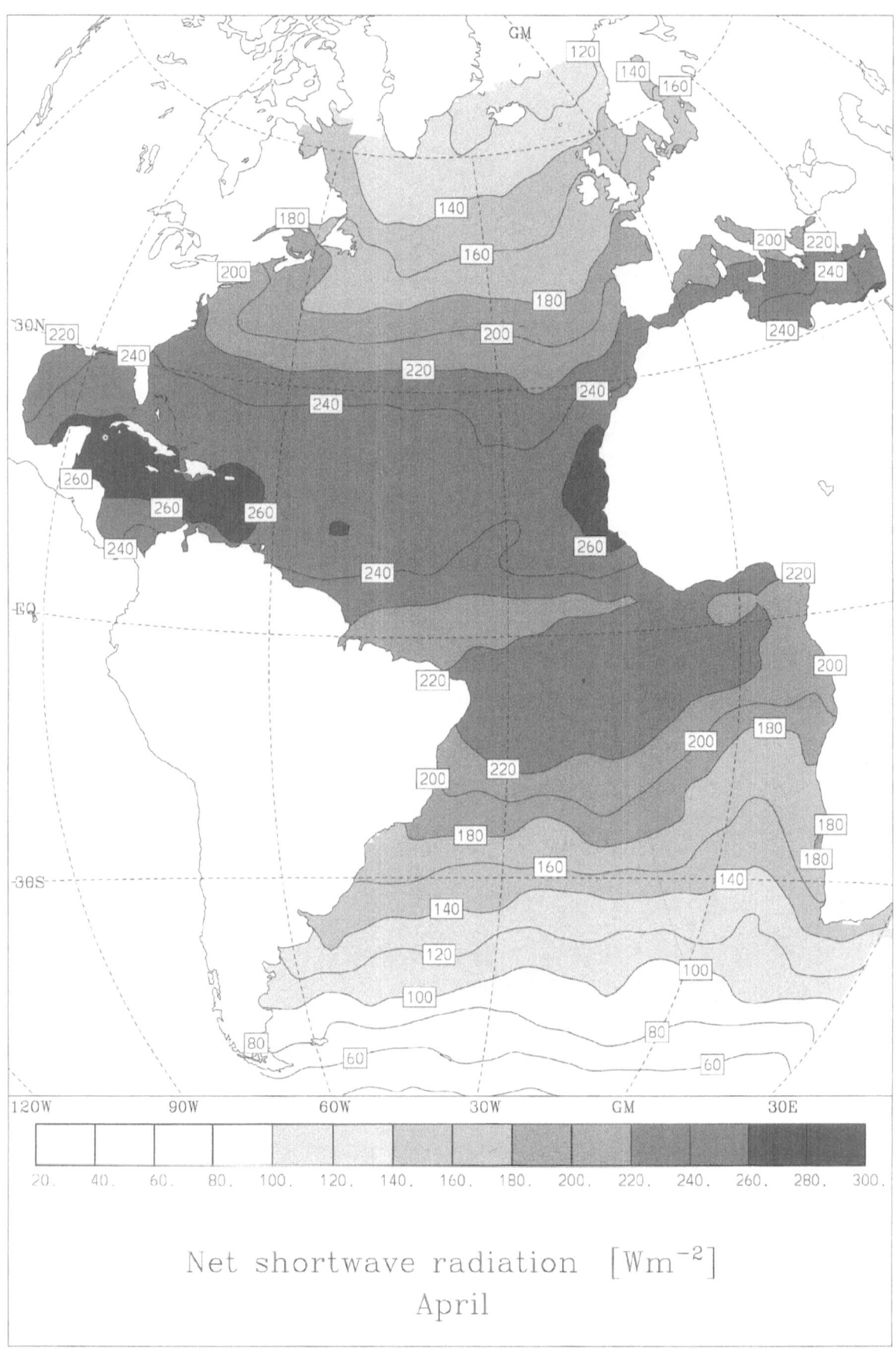

Net shortwave radiation [Wm^{-2}]
April

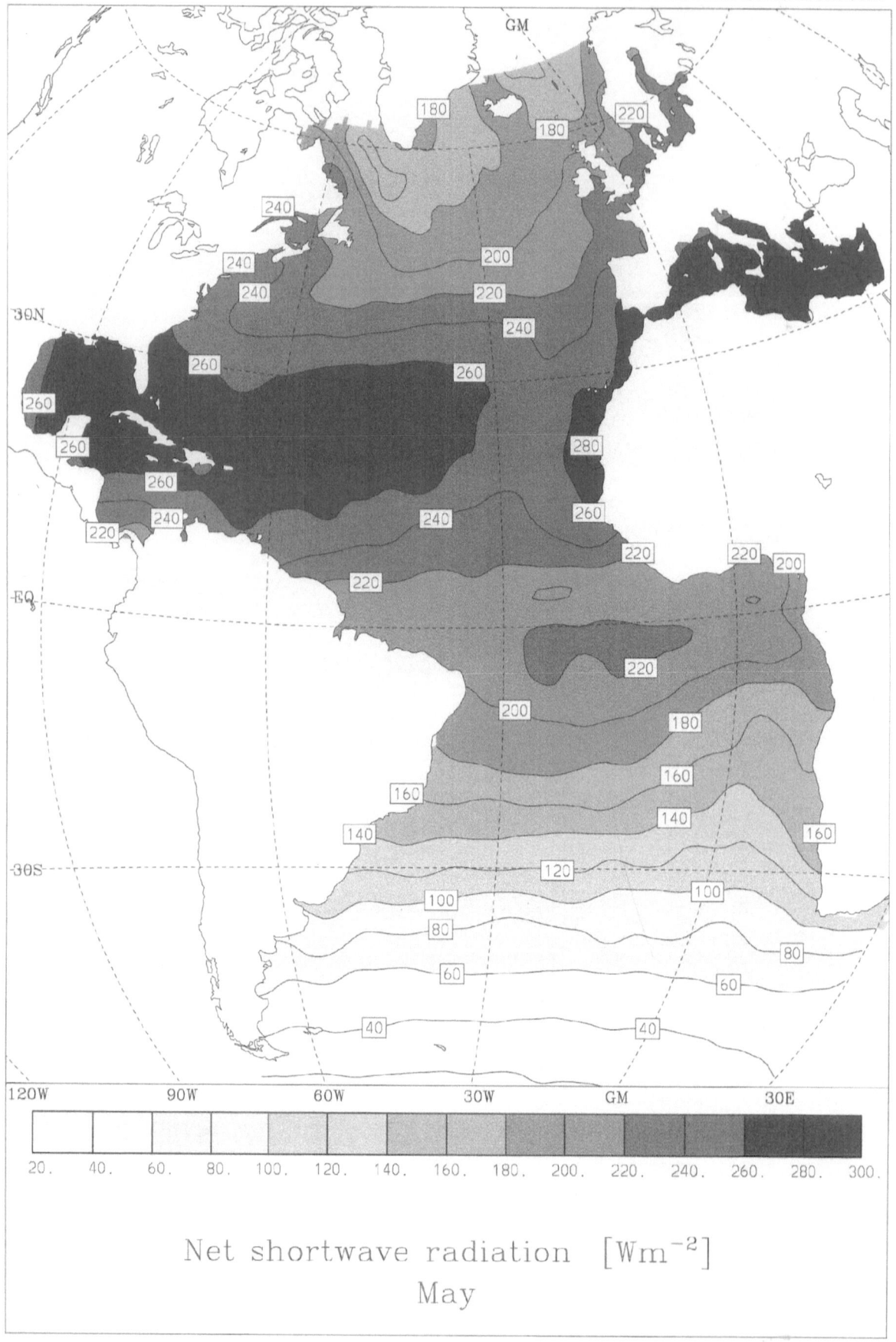

Net shortwave radiation $[Wm^{-2}]$
May

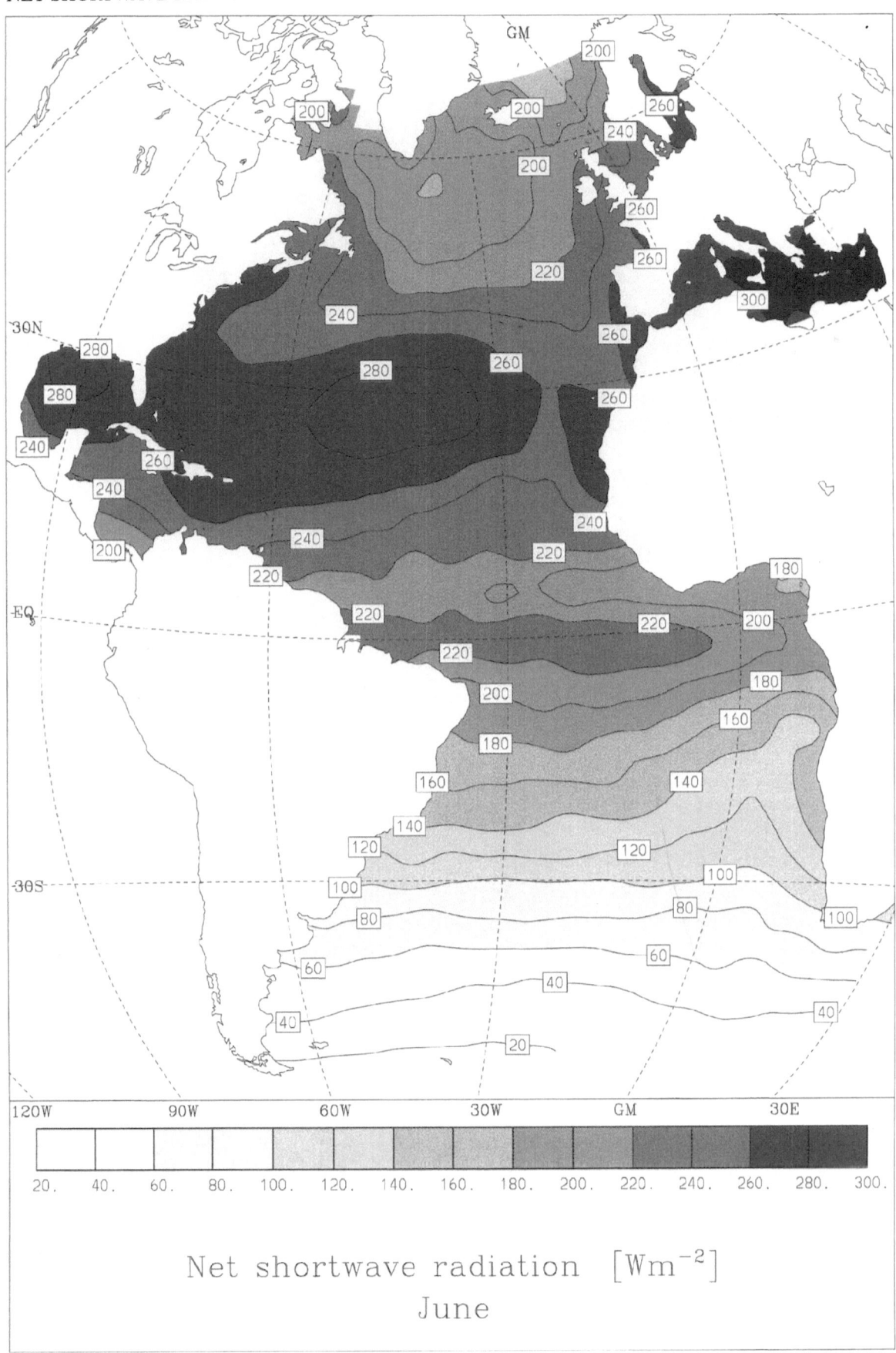

Net shortwave radiation [Wm^{-2}]
June

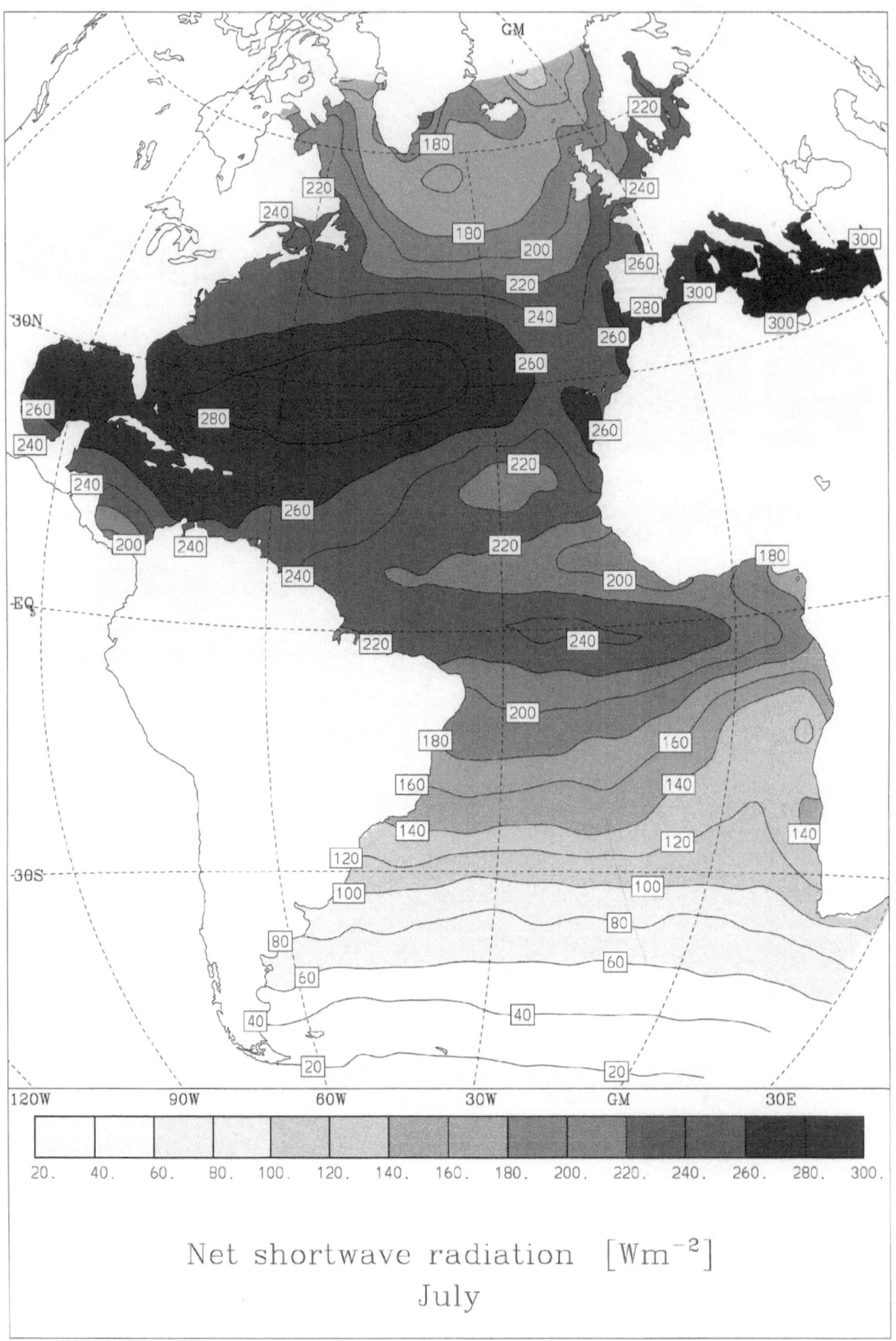

Net shortwave radiation [Wm^{-2}]
July

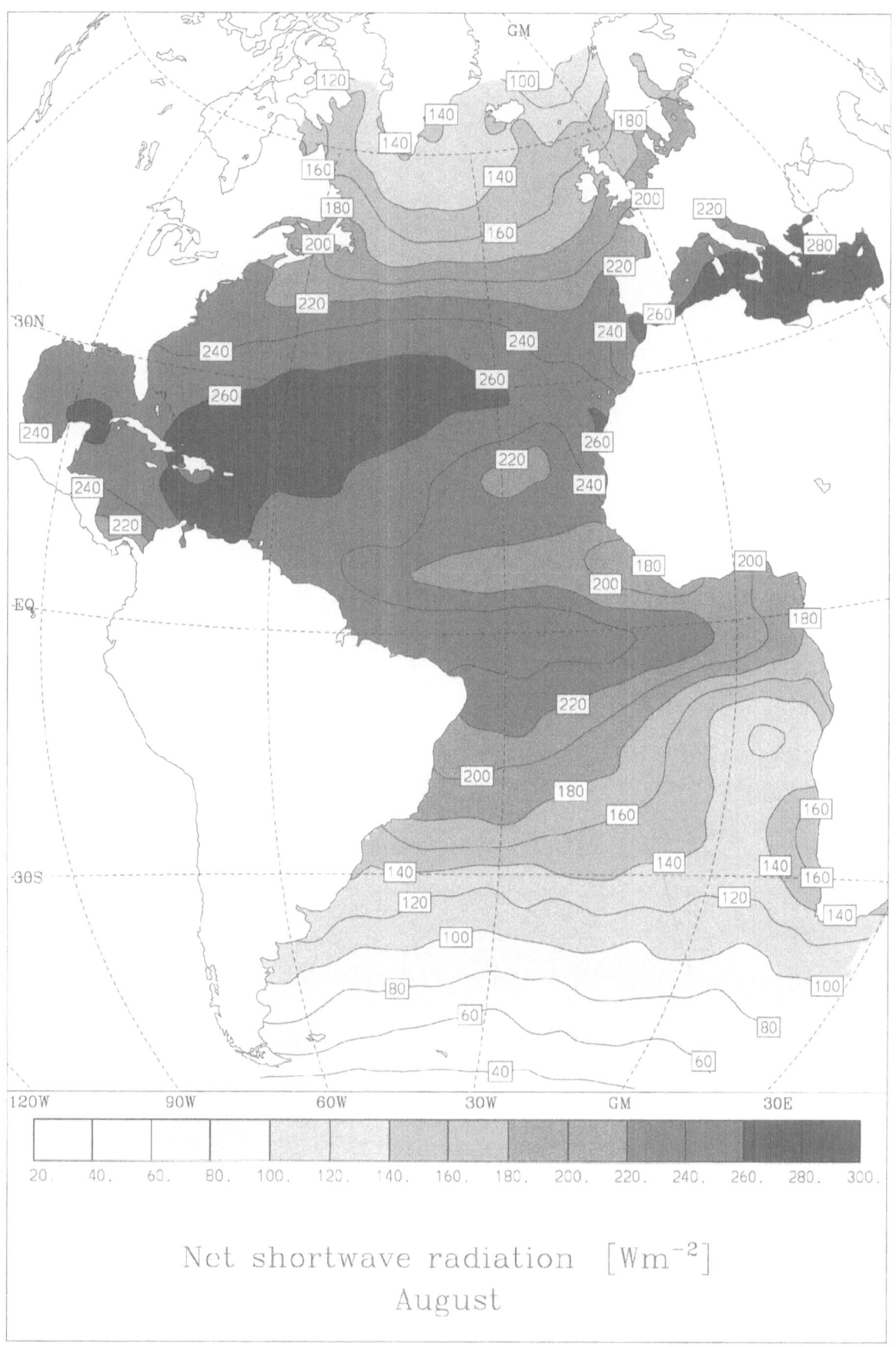

Net shortwave radiation [Wm^{-2}]
August

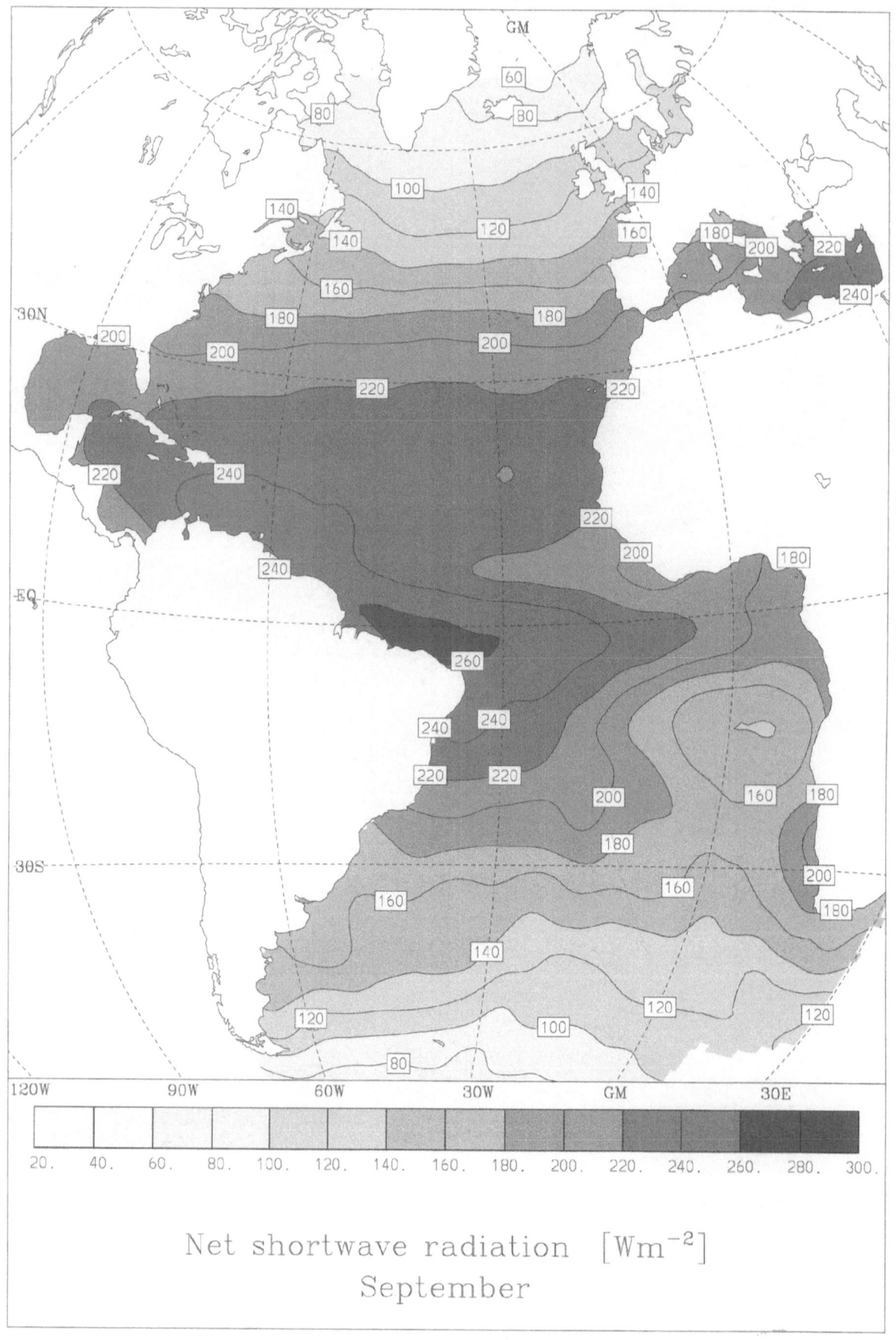

Net shortwave radiation [Wm^{-2}]
September

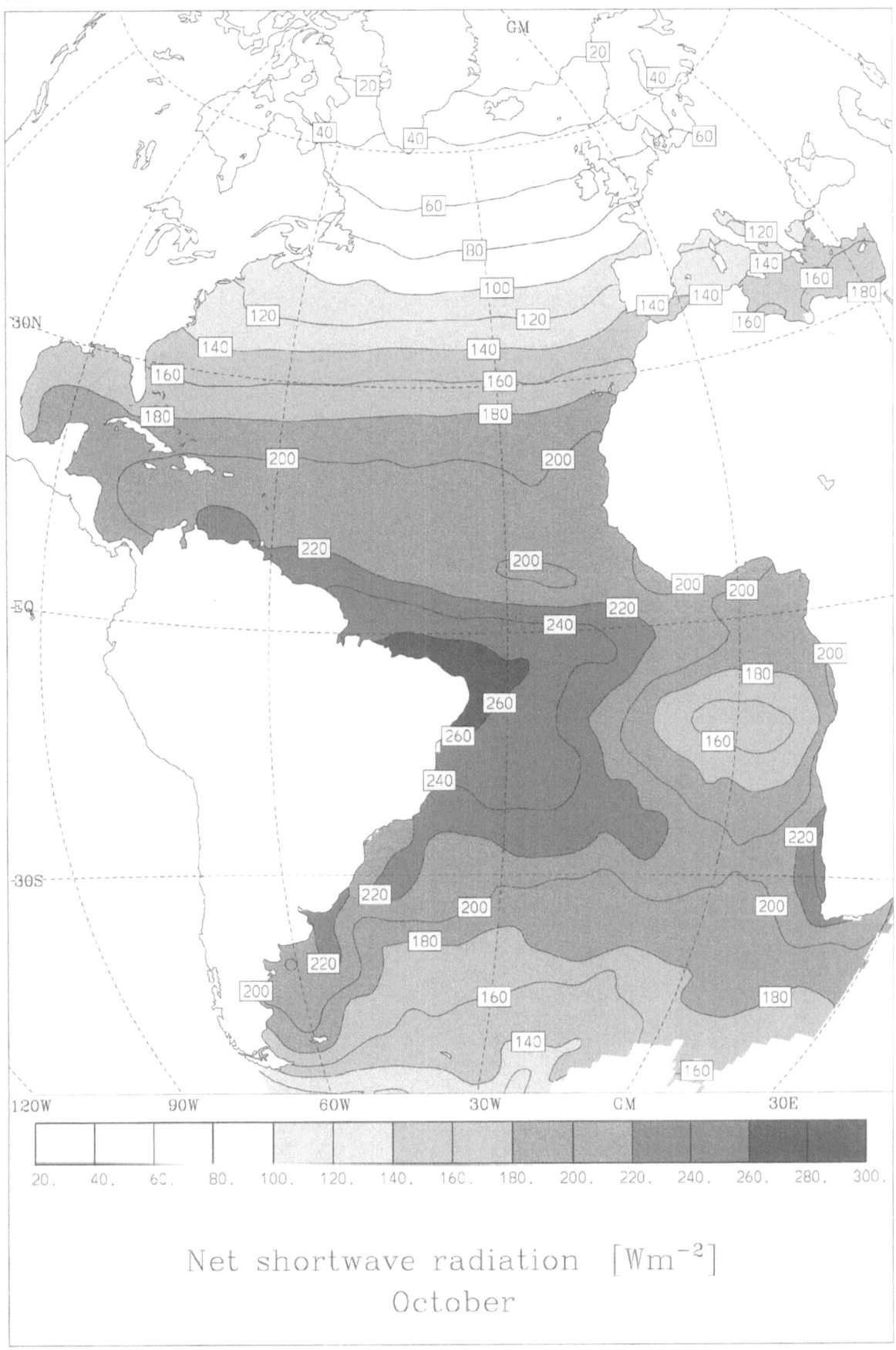

Net shortwave radiation $[Wm^{-2}]$

October

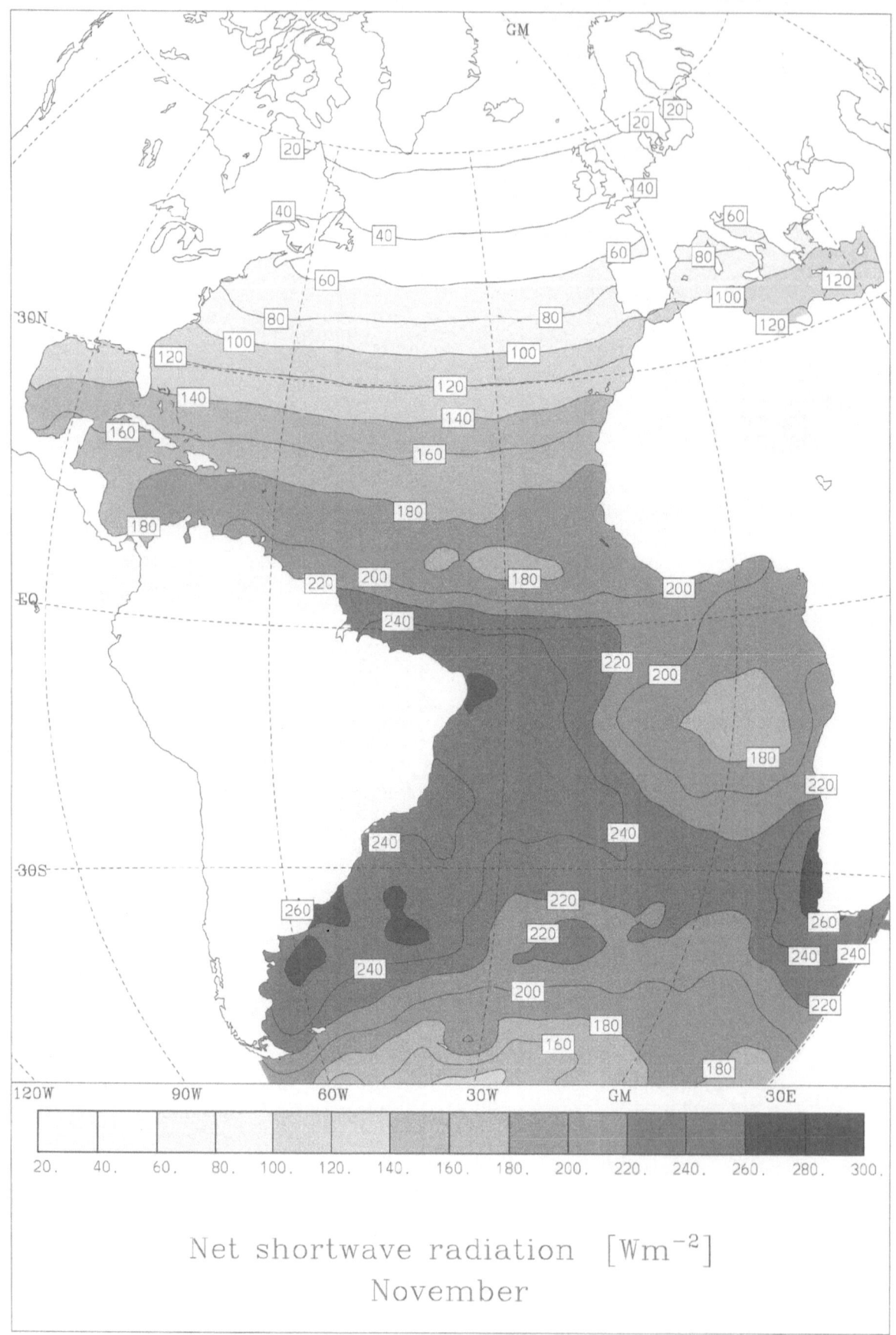

Net shortwave radiation [Wm^{-2}]
November

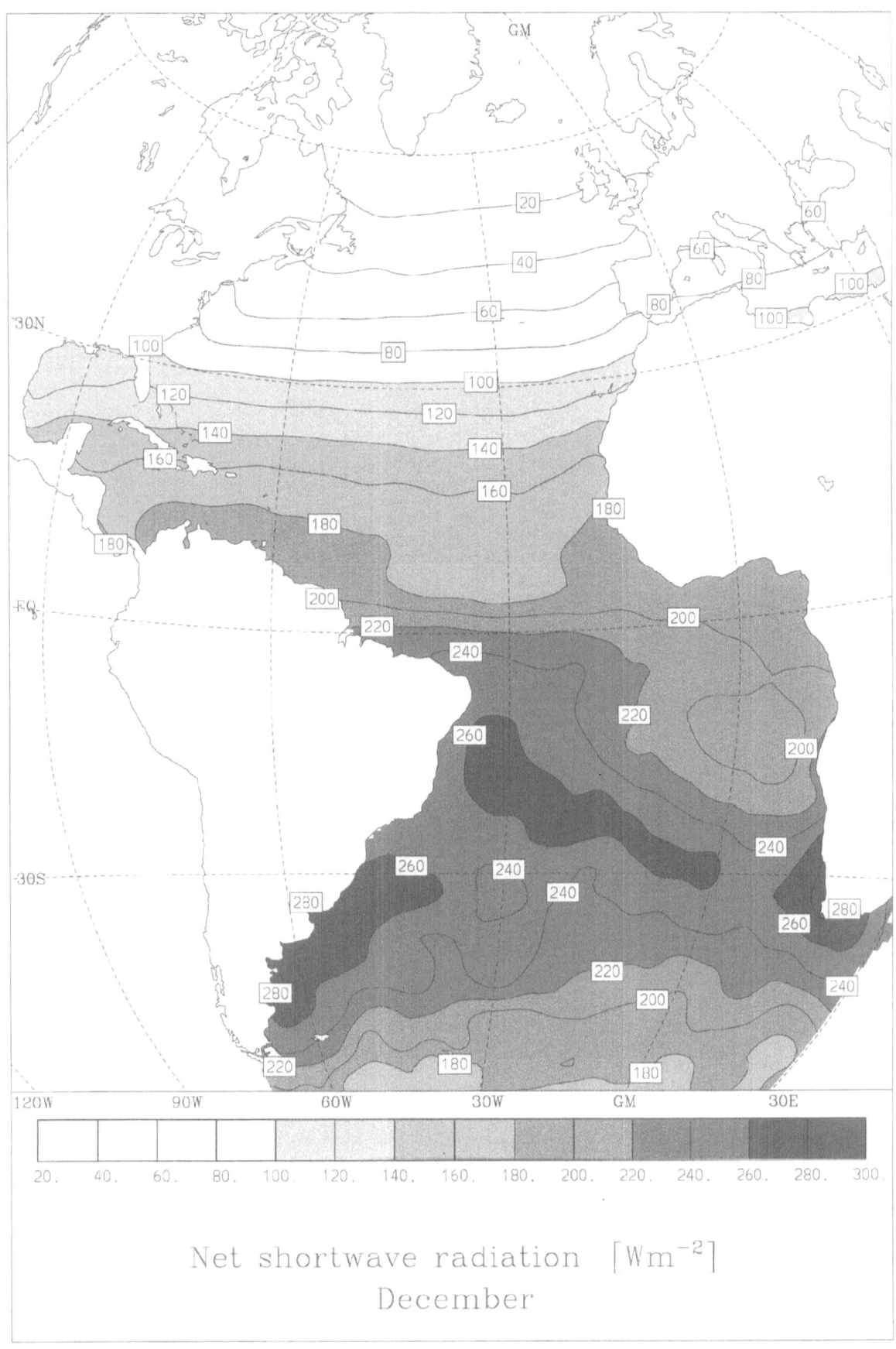

Net shortwave radiation $[Wm^{-2}]$

December

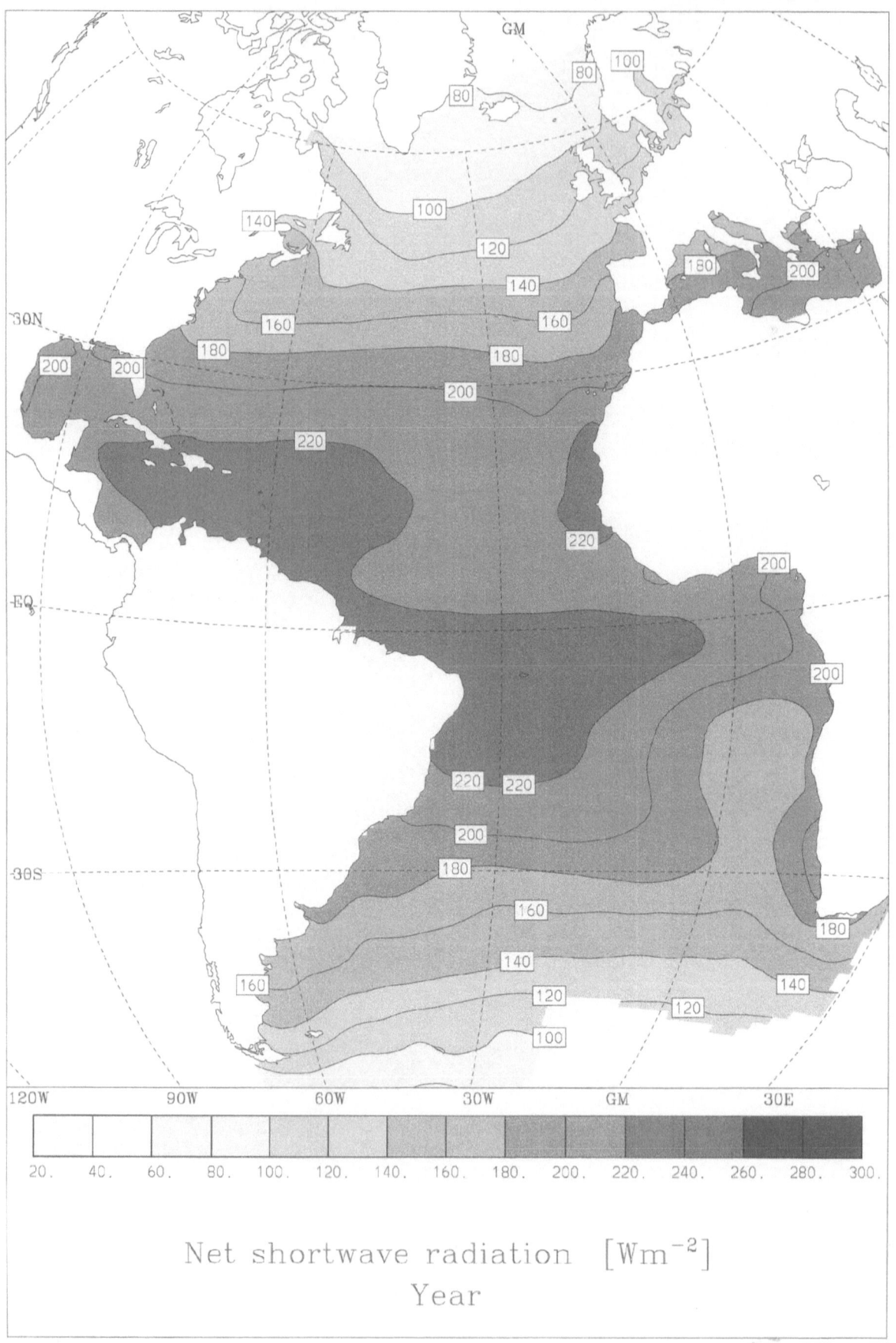

Net shortwave radiation [Wm^{-2}]
Year

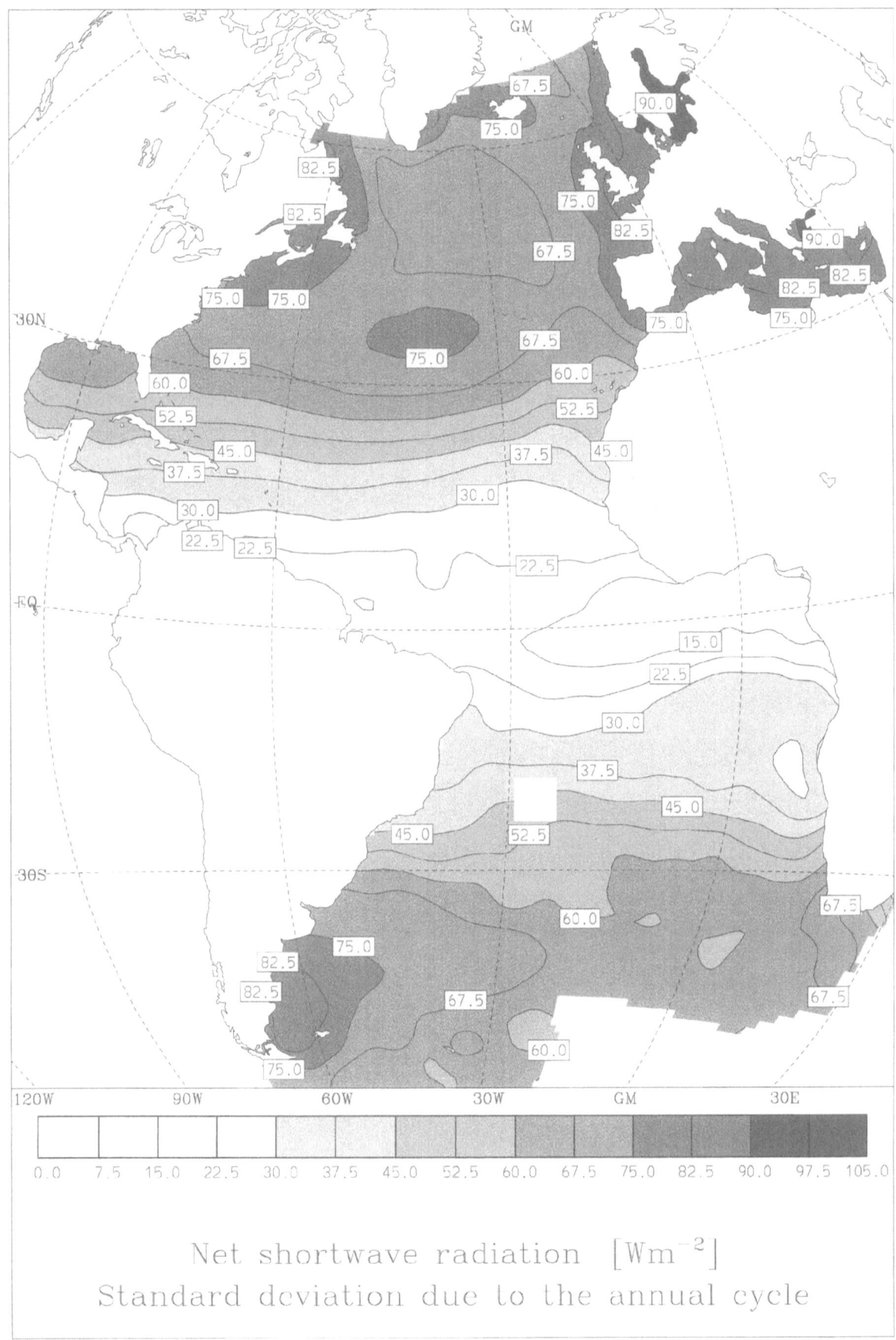

Net shortwave radiation [Wm⁻²]
Standard deviation due to the annual cycle

Net Longwave Radiation

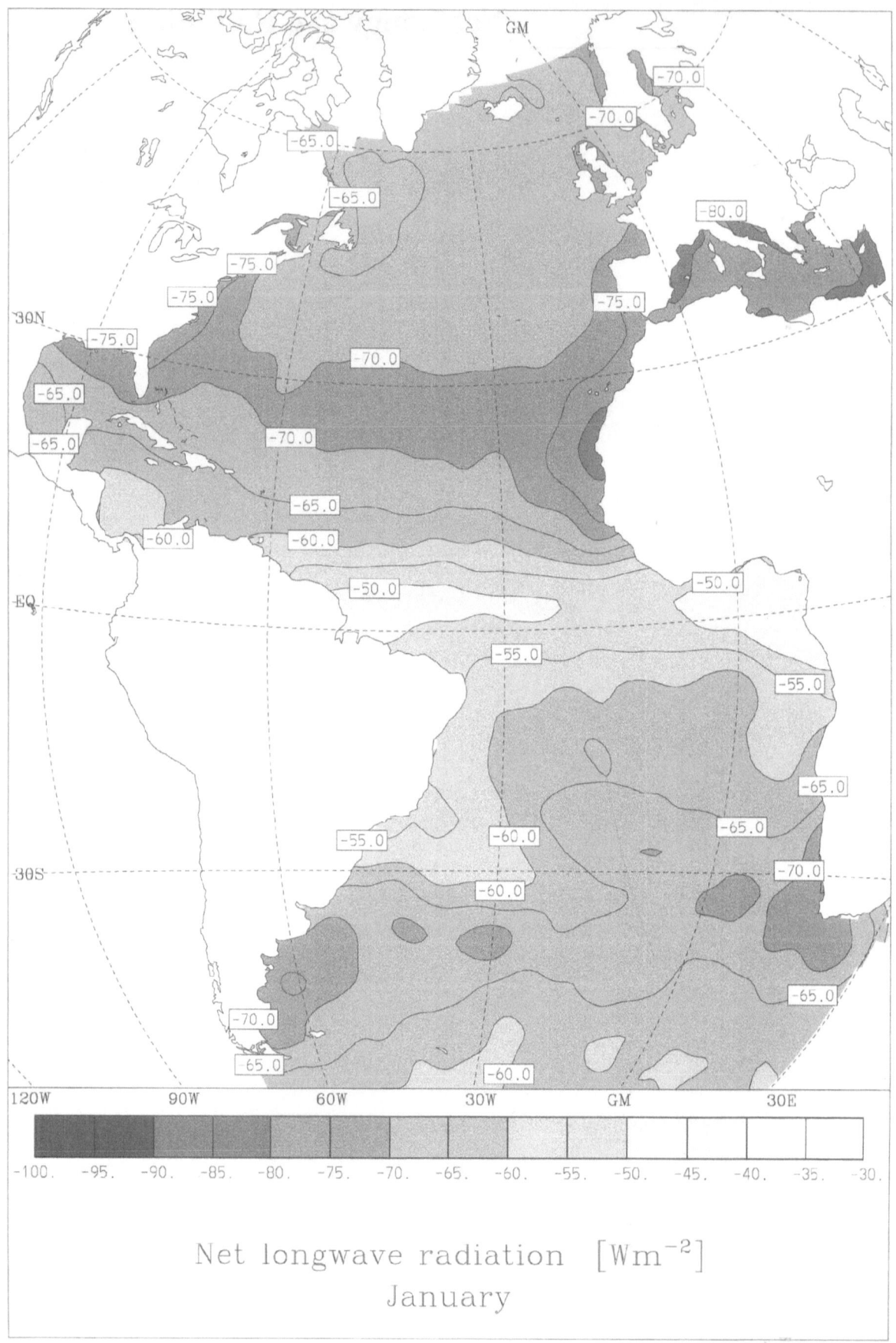

Net longwave radiation [Wm^{-2}]
January

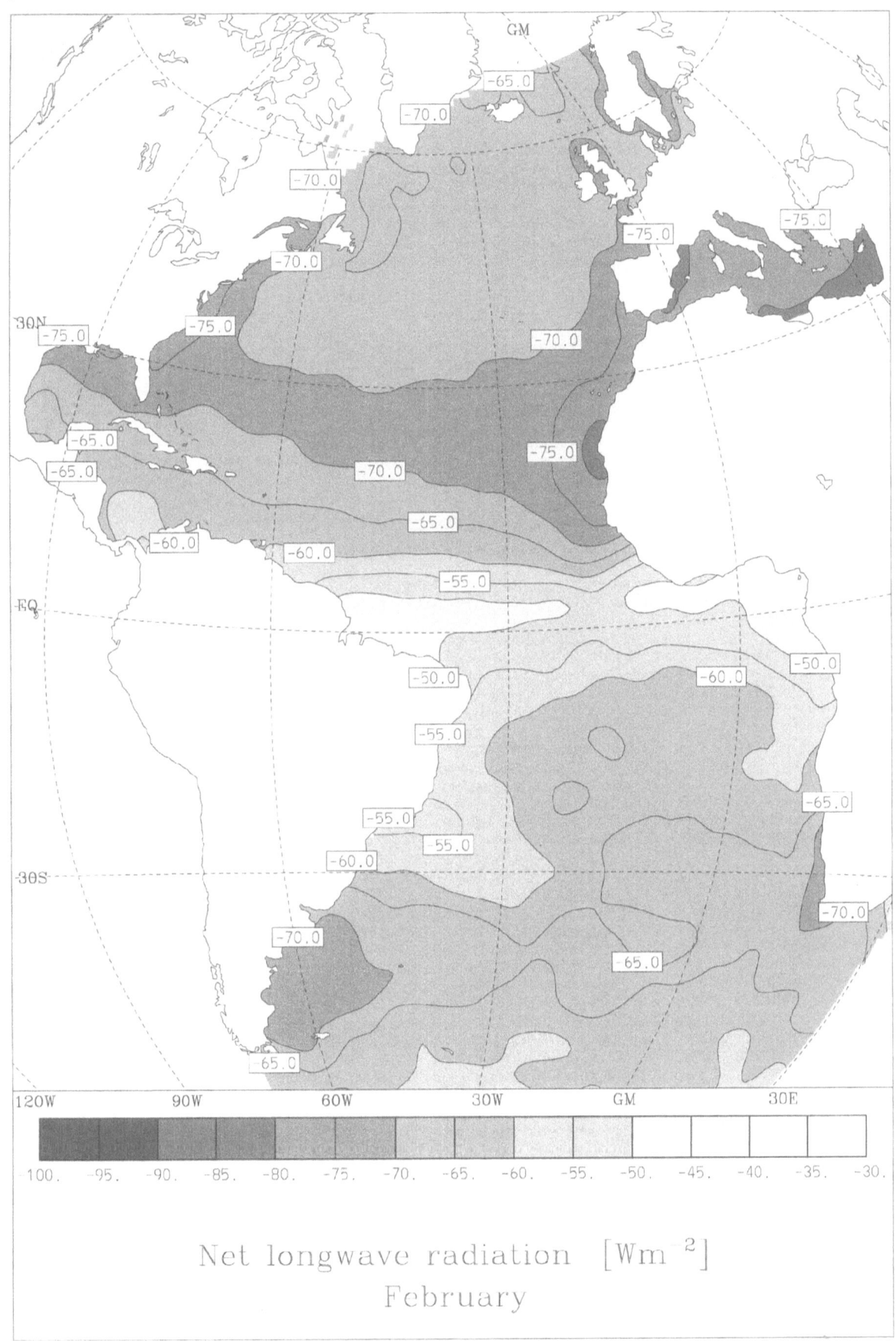

Net longwave radiation [Wm^{-2}]
February

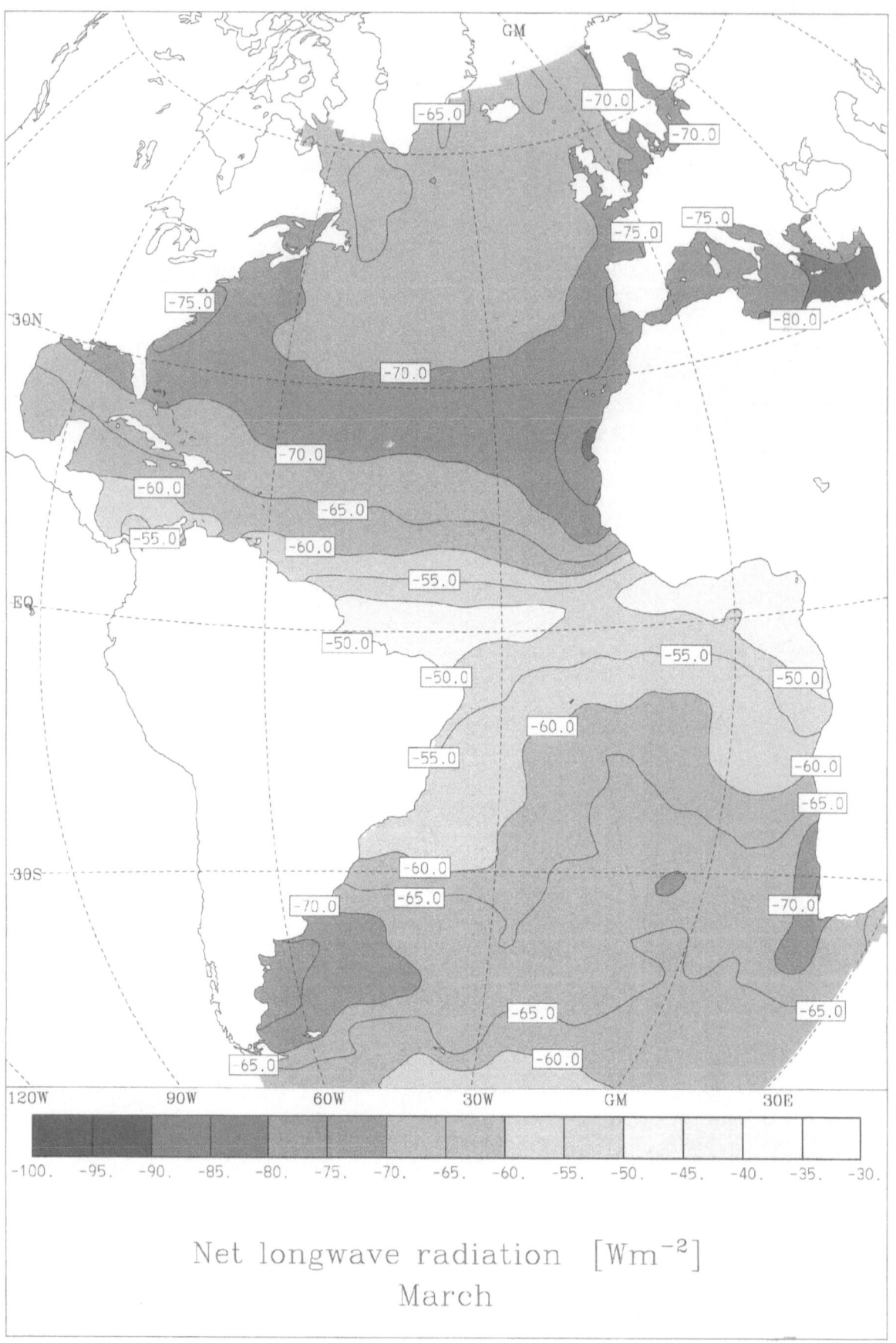

Net longwave radiation [Wm^{-2}]
March

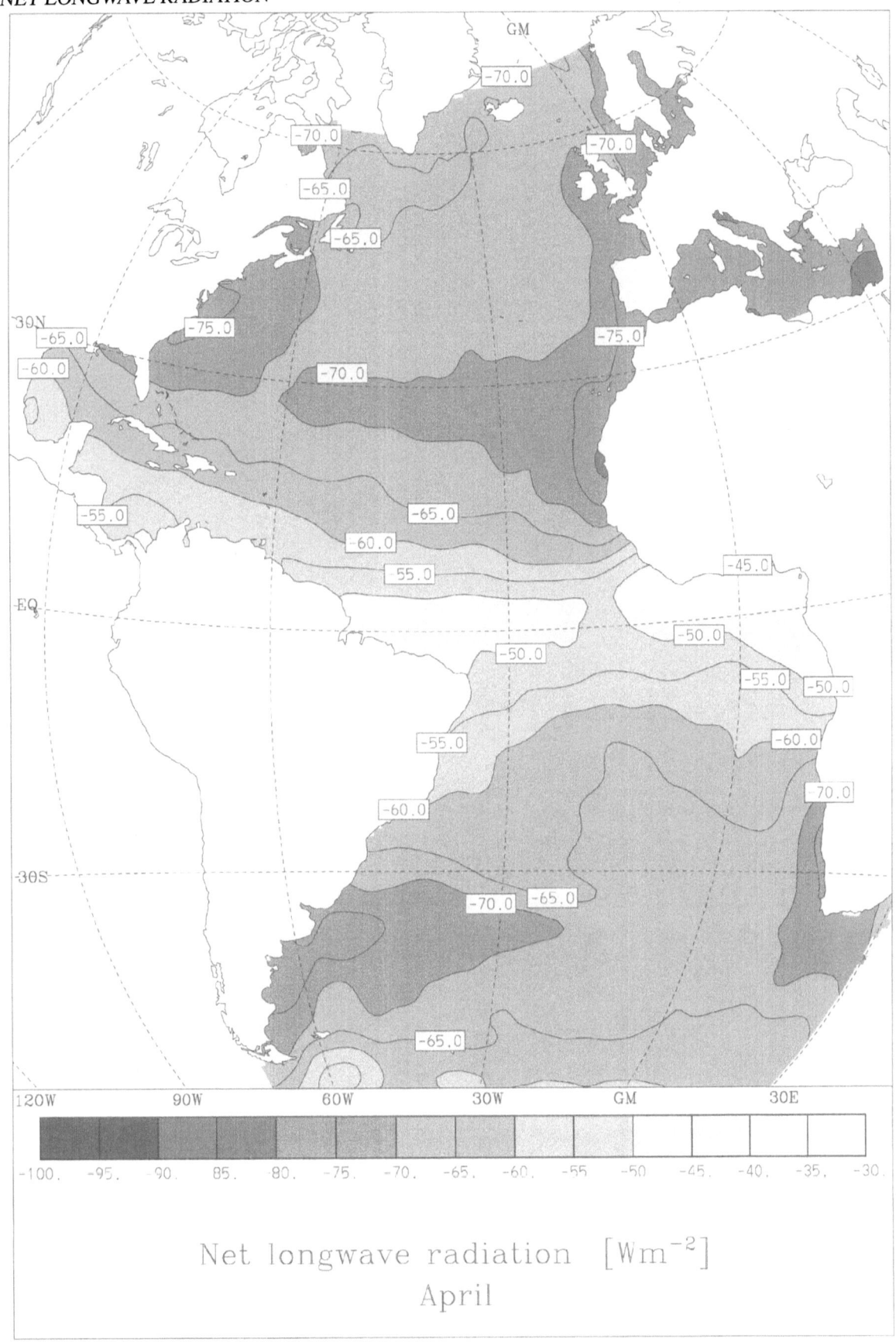

Net longwave radiation [Wm^{-2}]
April

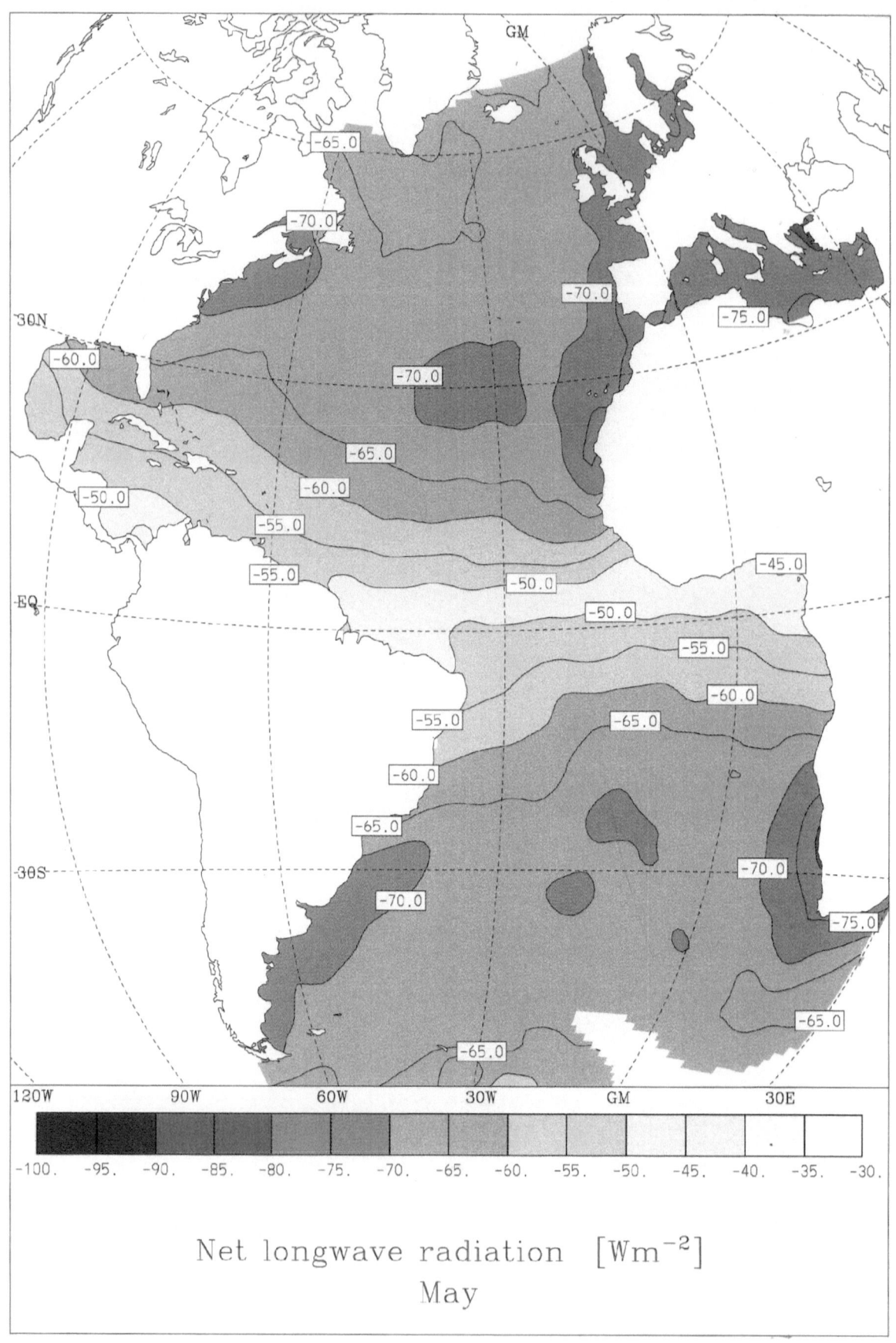

Net longwave radiation $[\text{Wm}^{-2}]$
May

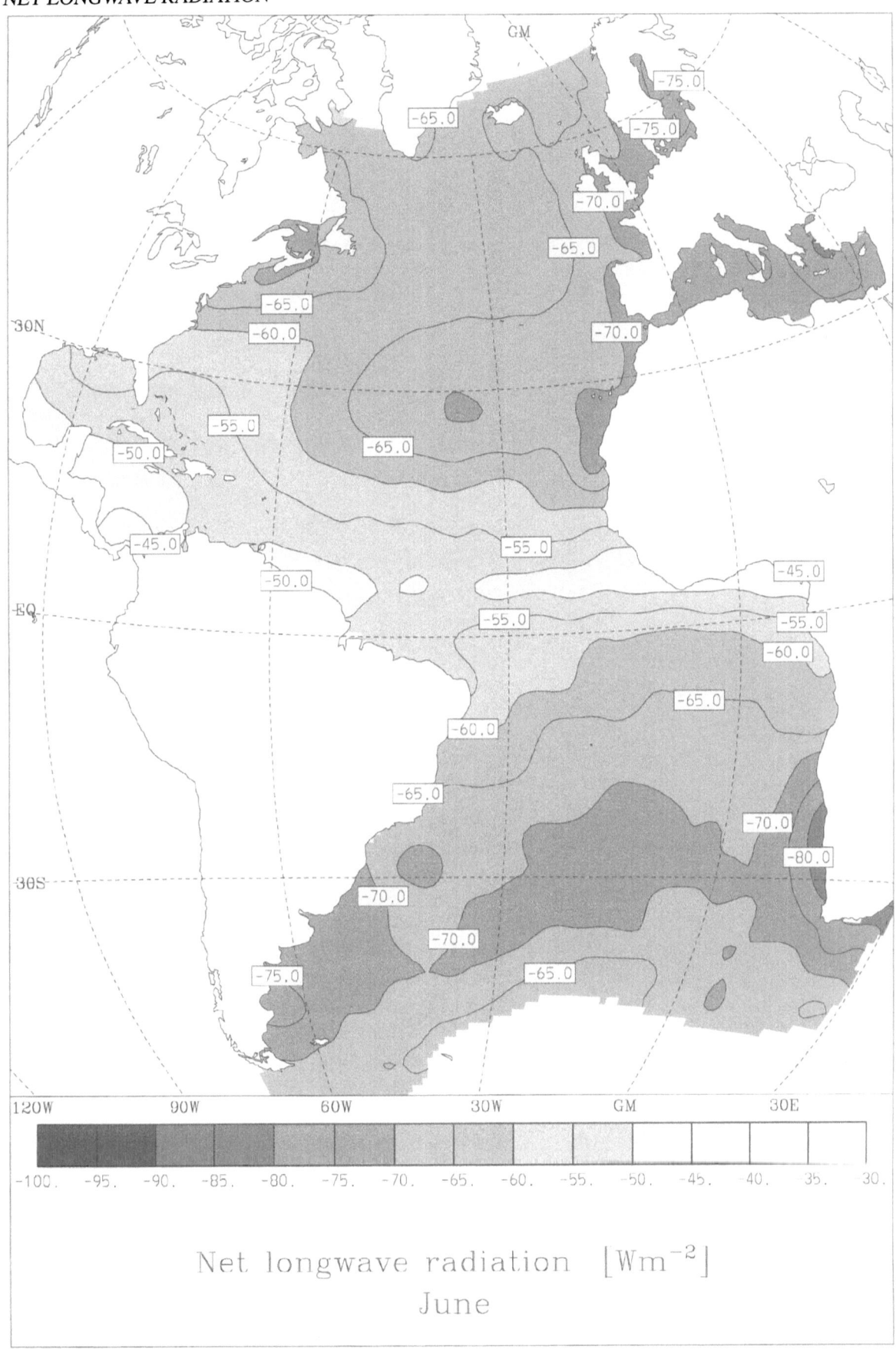

Net longwave radiation $[Wm^{-2}]$

June

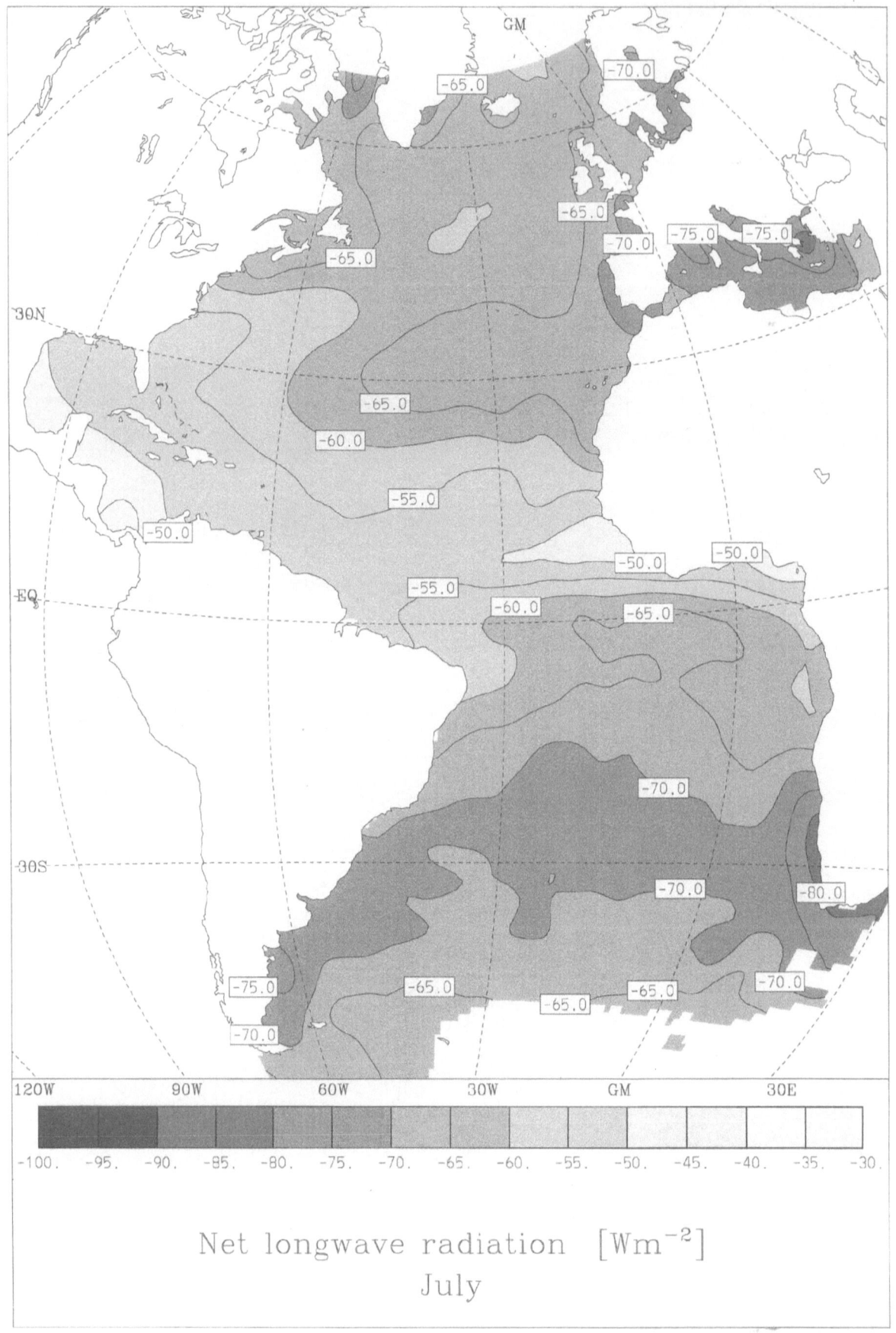

Net longwave radiation [Wm^{-2}]
July

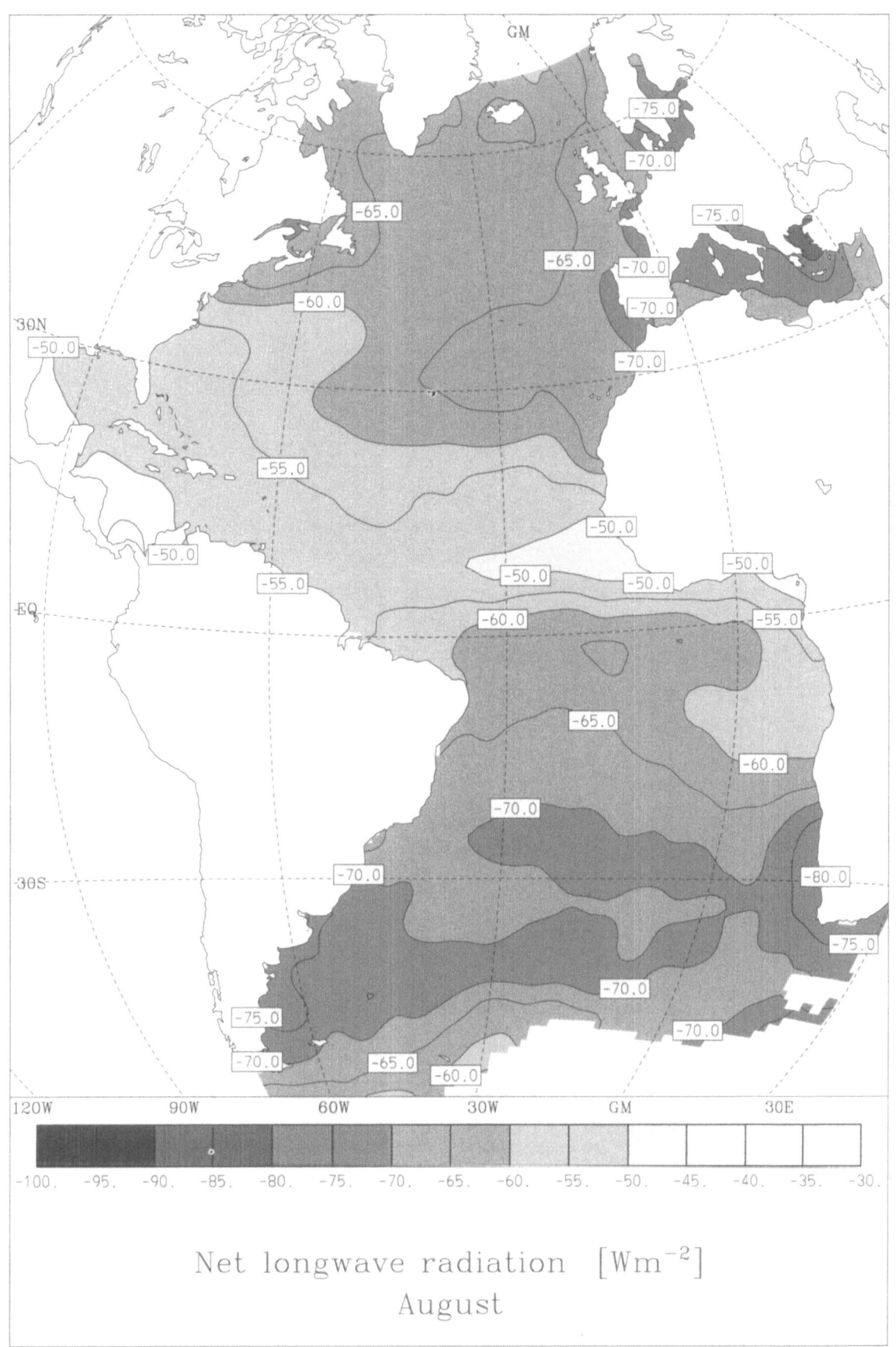

Net longwave radiation [Wm^{-2}]
August

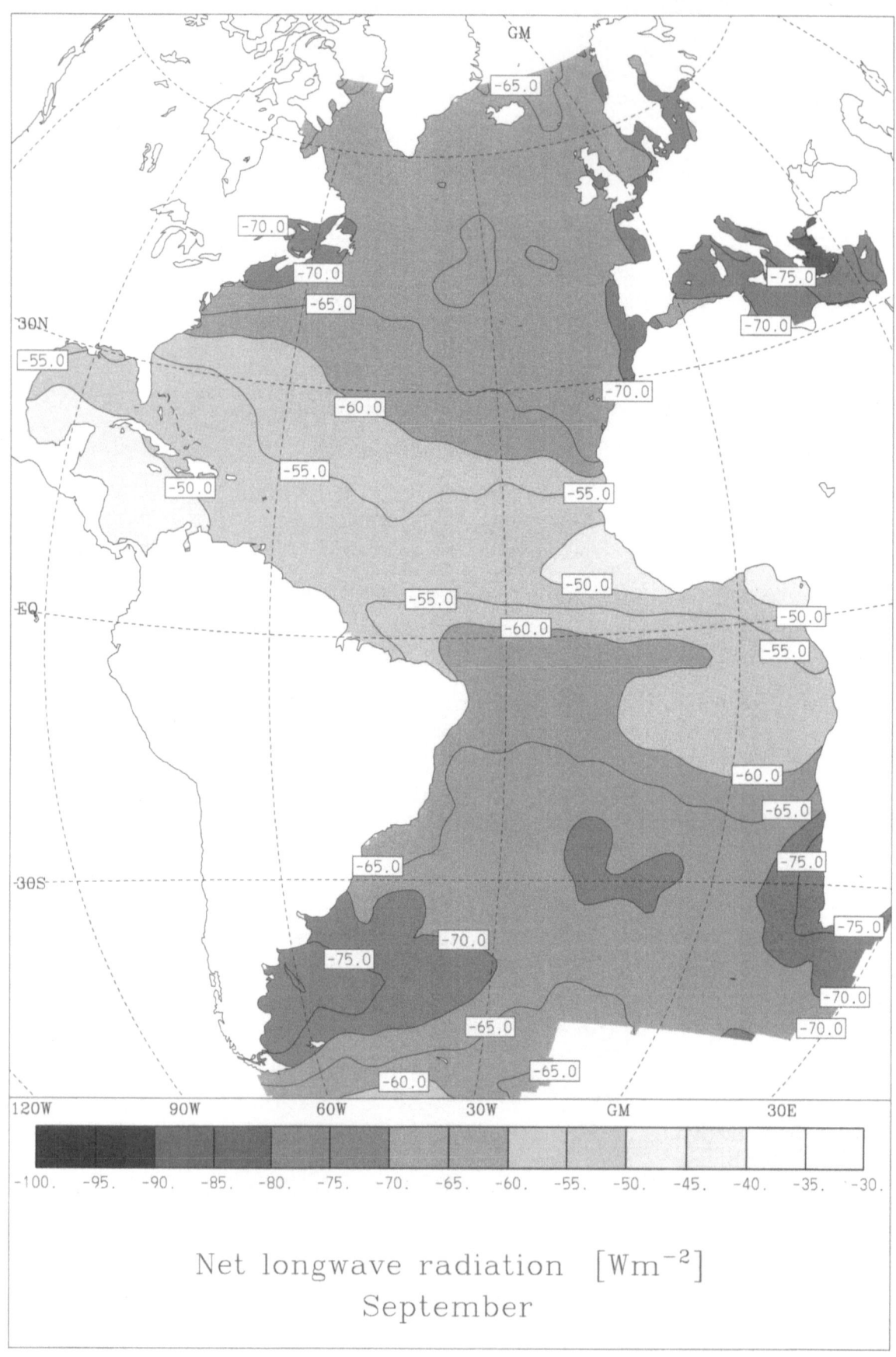

Net longwave radiation [Wm^{-2}]
September

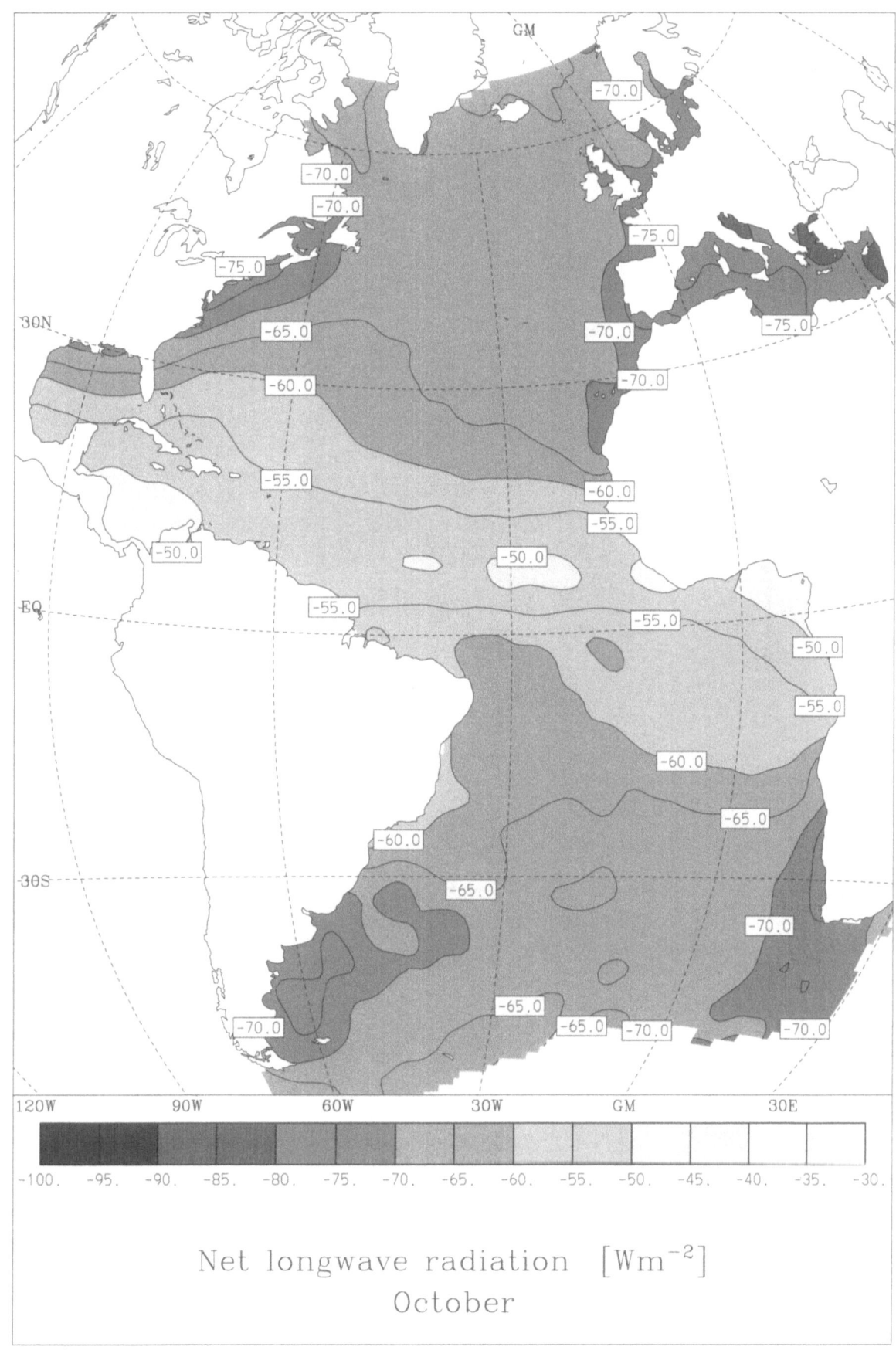

Net longwave radiation [Wm^{-2}]
October

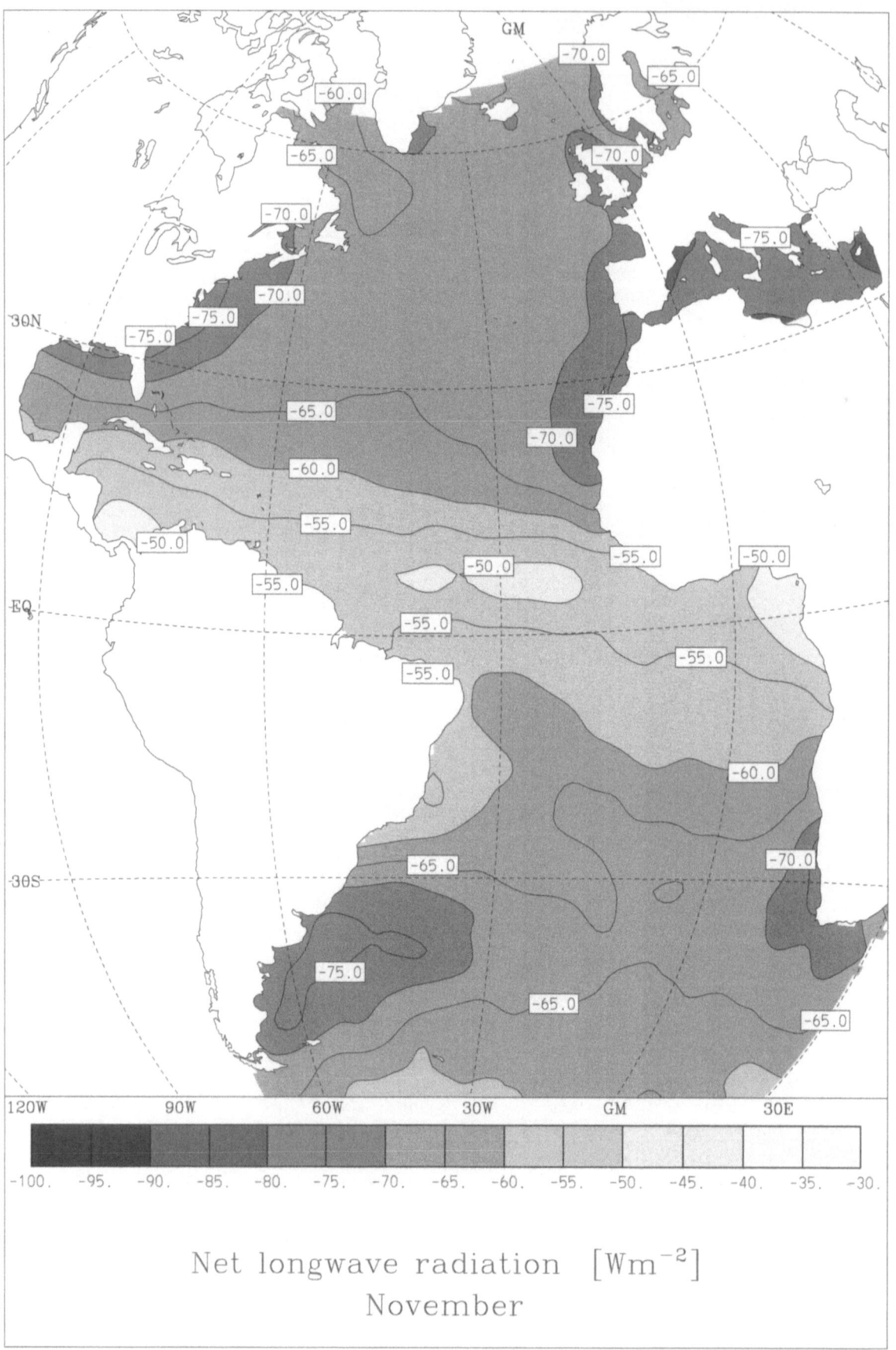

Net longwave radiation $[\mathrm{Wm}^{-2}]$
November

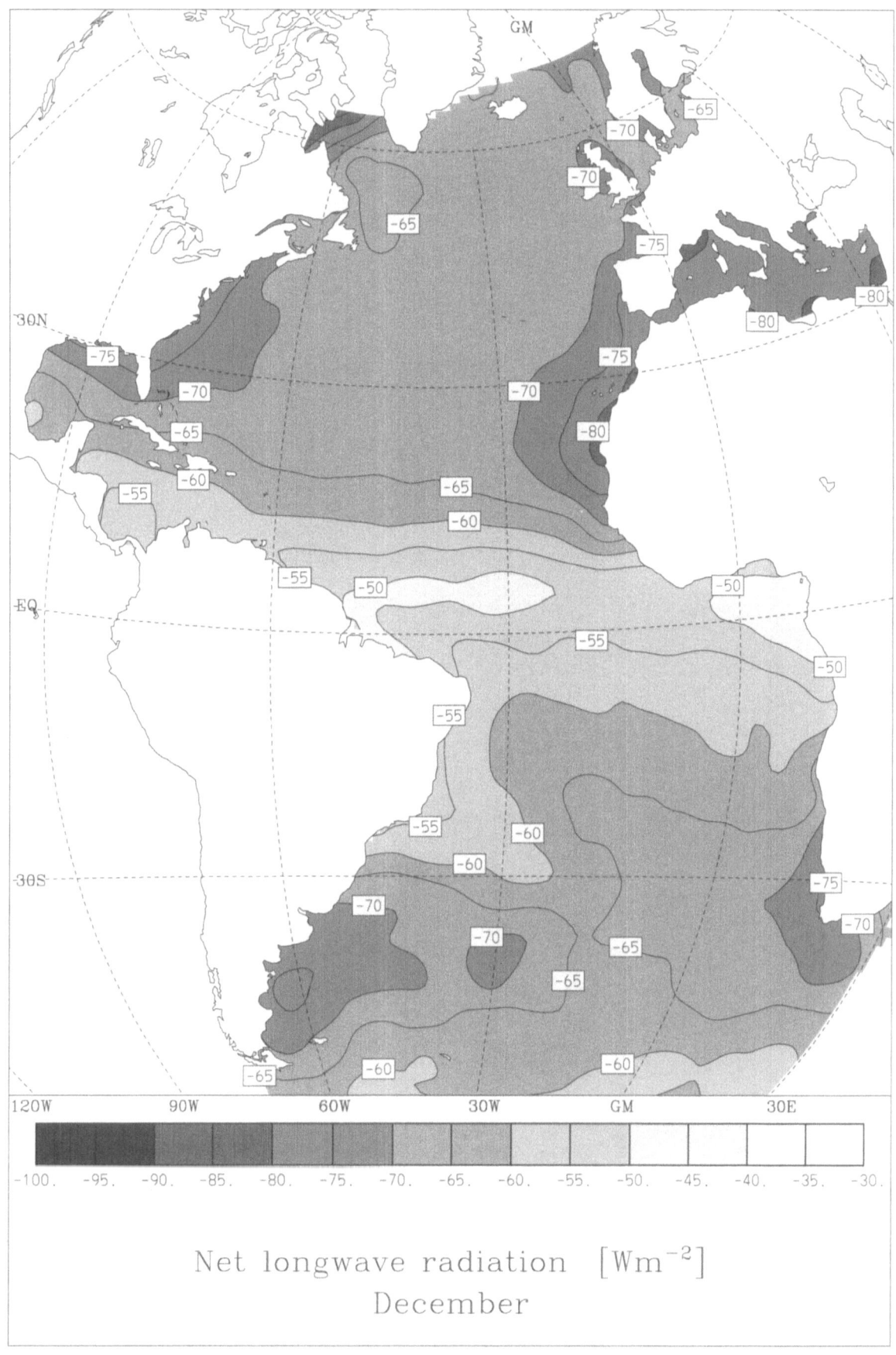

Net longwave radiation [Wm^{-2}]
December

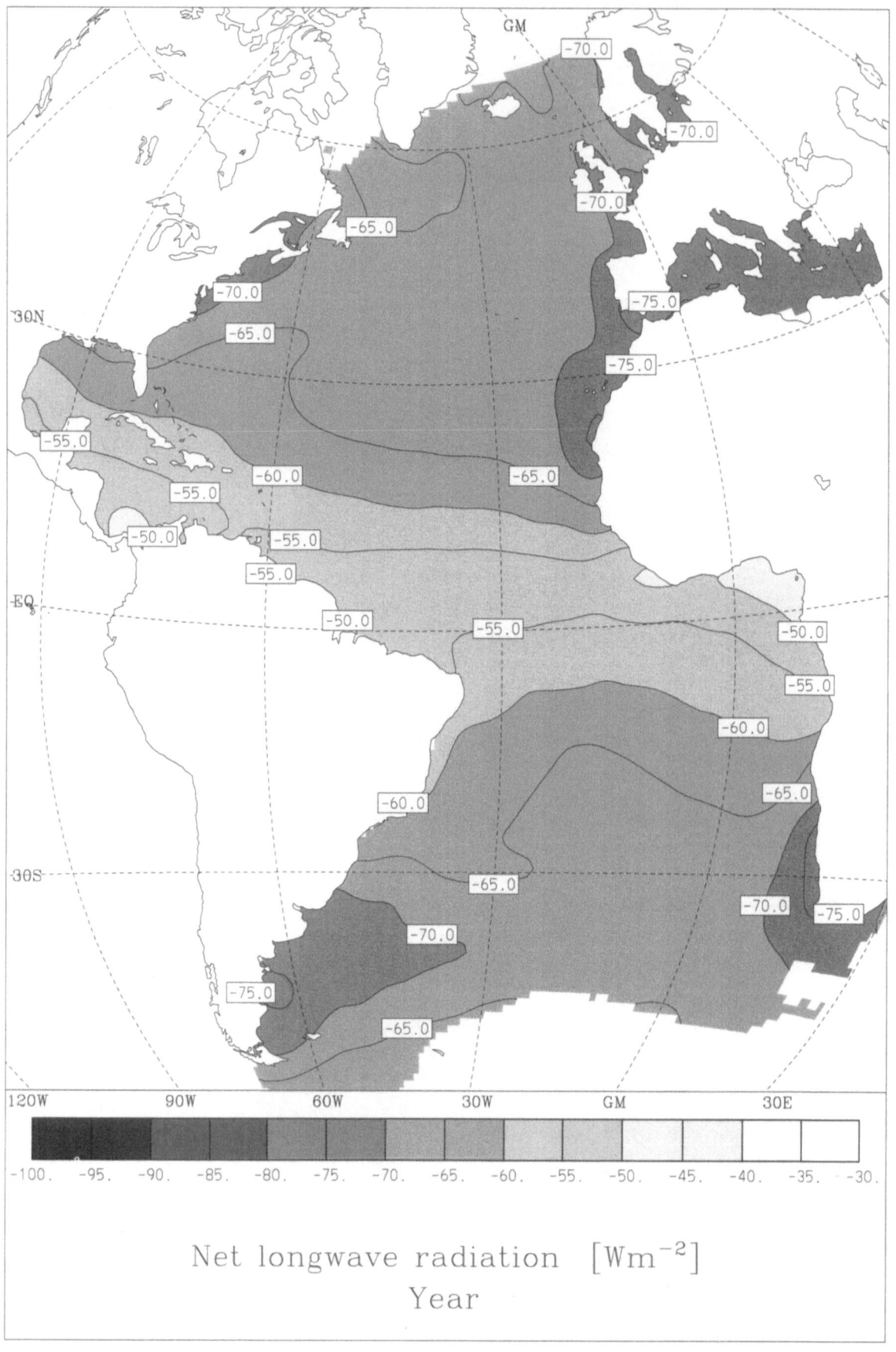

Net longwave radiation $[\mathrm{Wm}^{-2}]$
Year

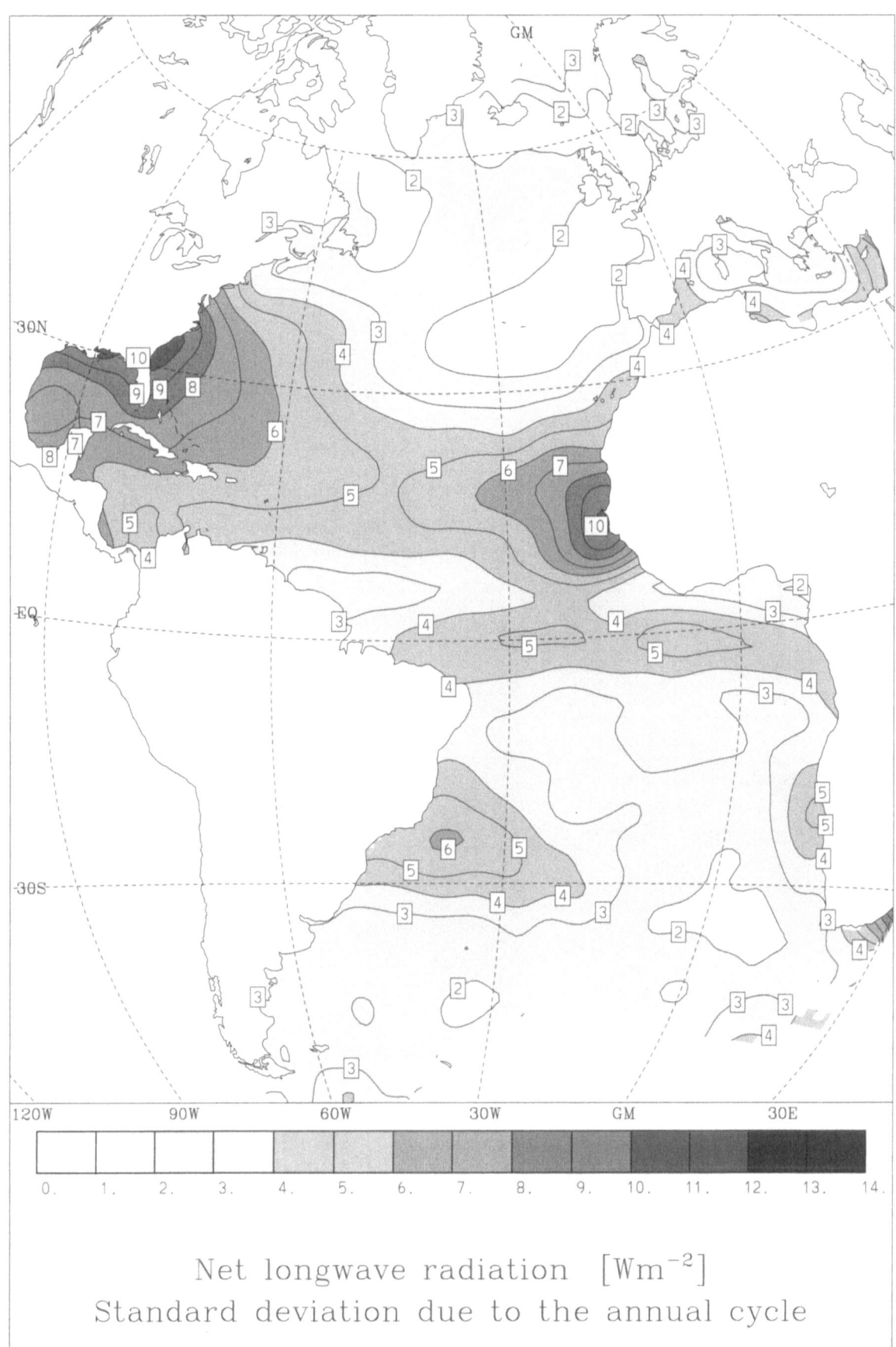

Net longwave radiation [Wm^{-2}]
Standard deviation due to the annual cycle

Net Radiation

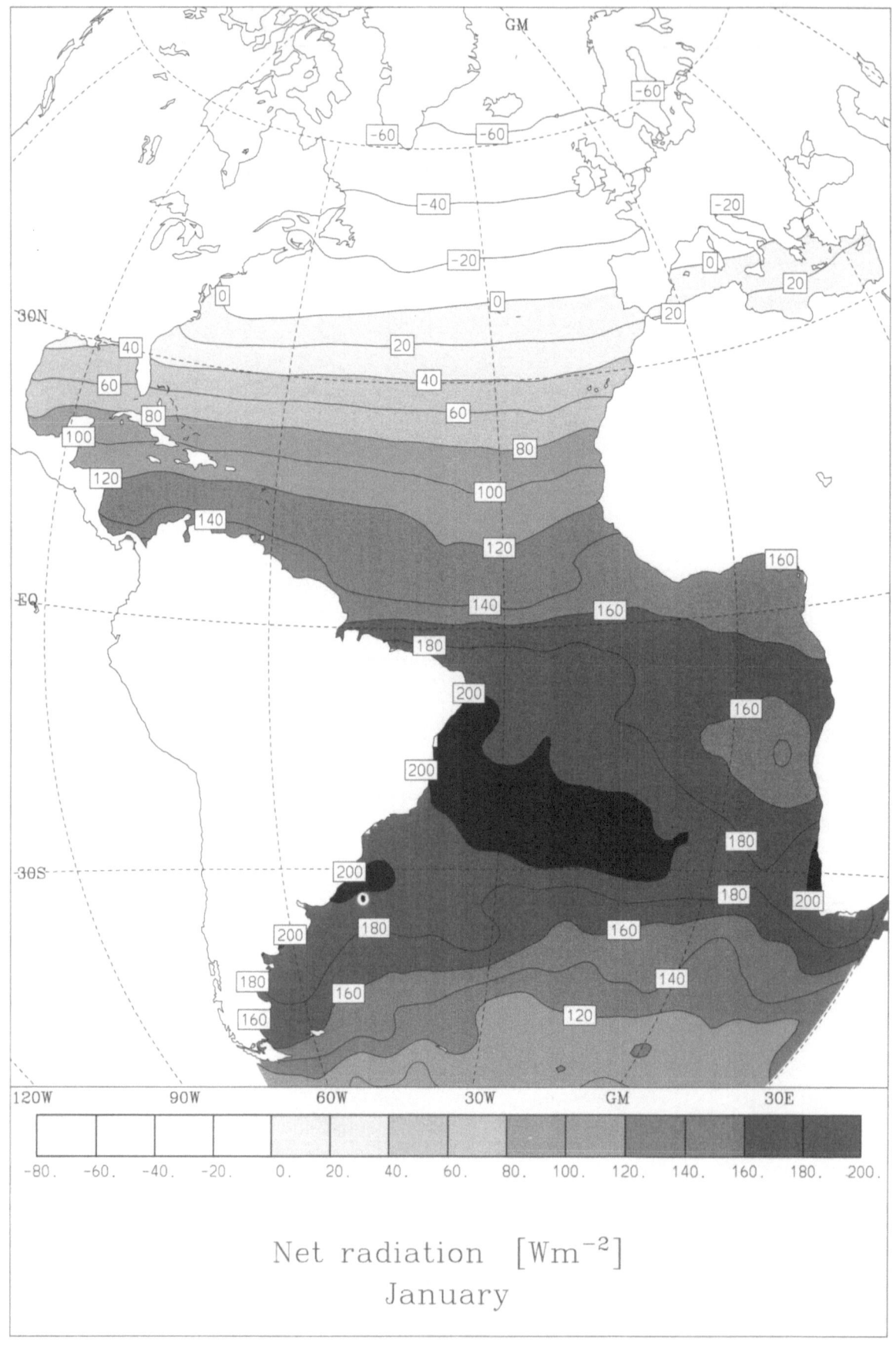

Net radiation [Wm^{-2}]
January

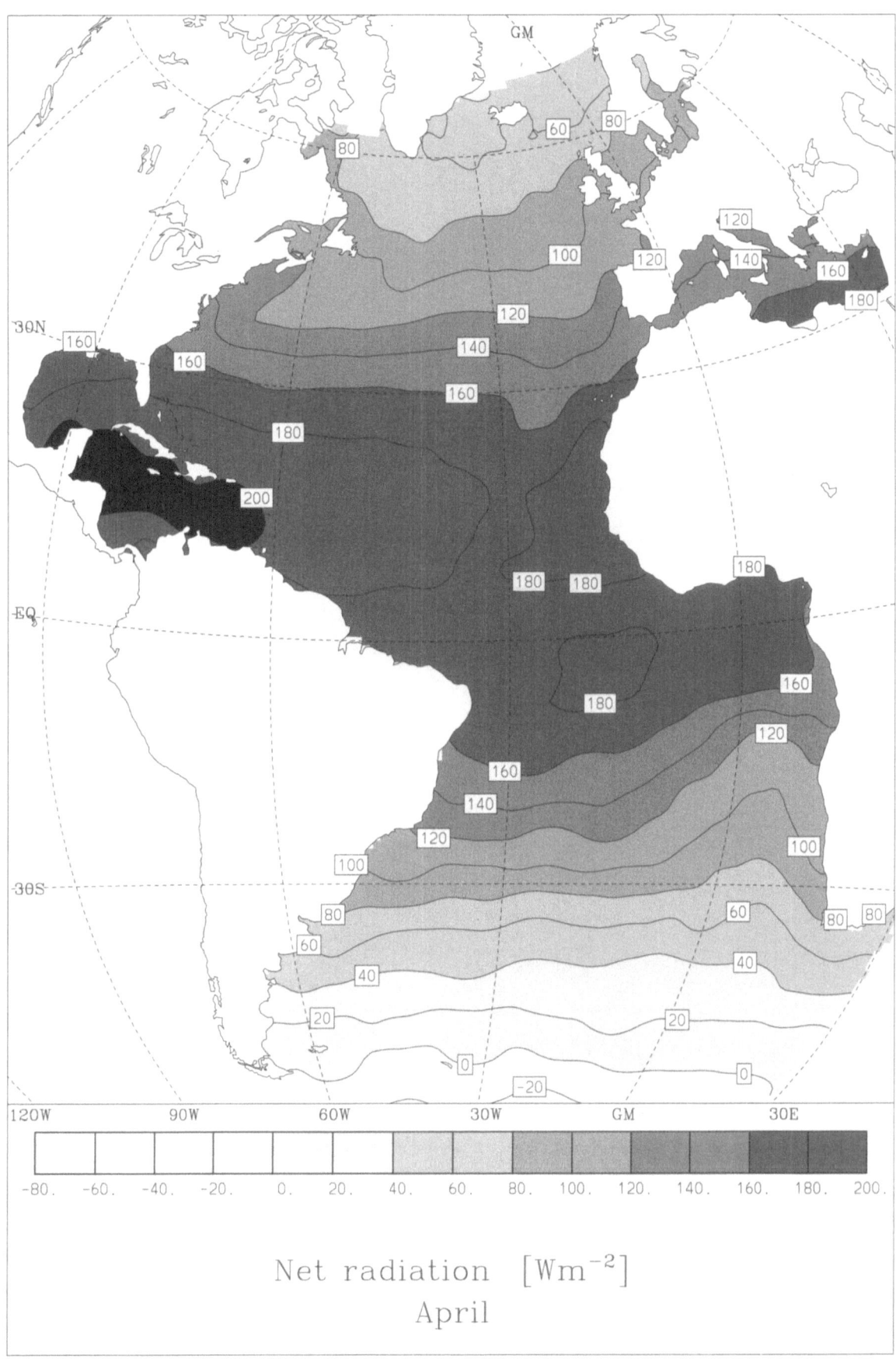

Net radiation [Wm^{-2}]
April

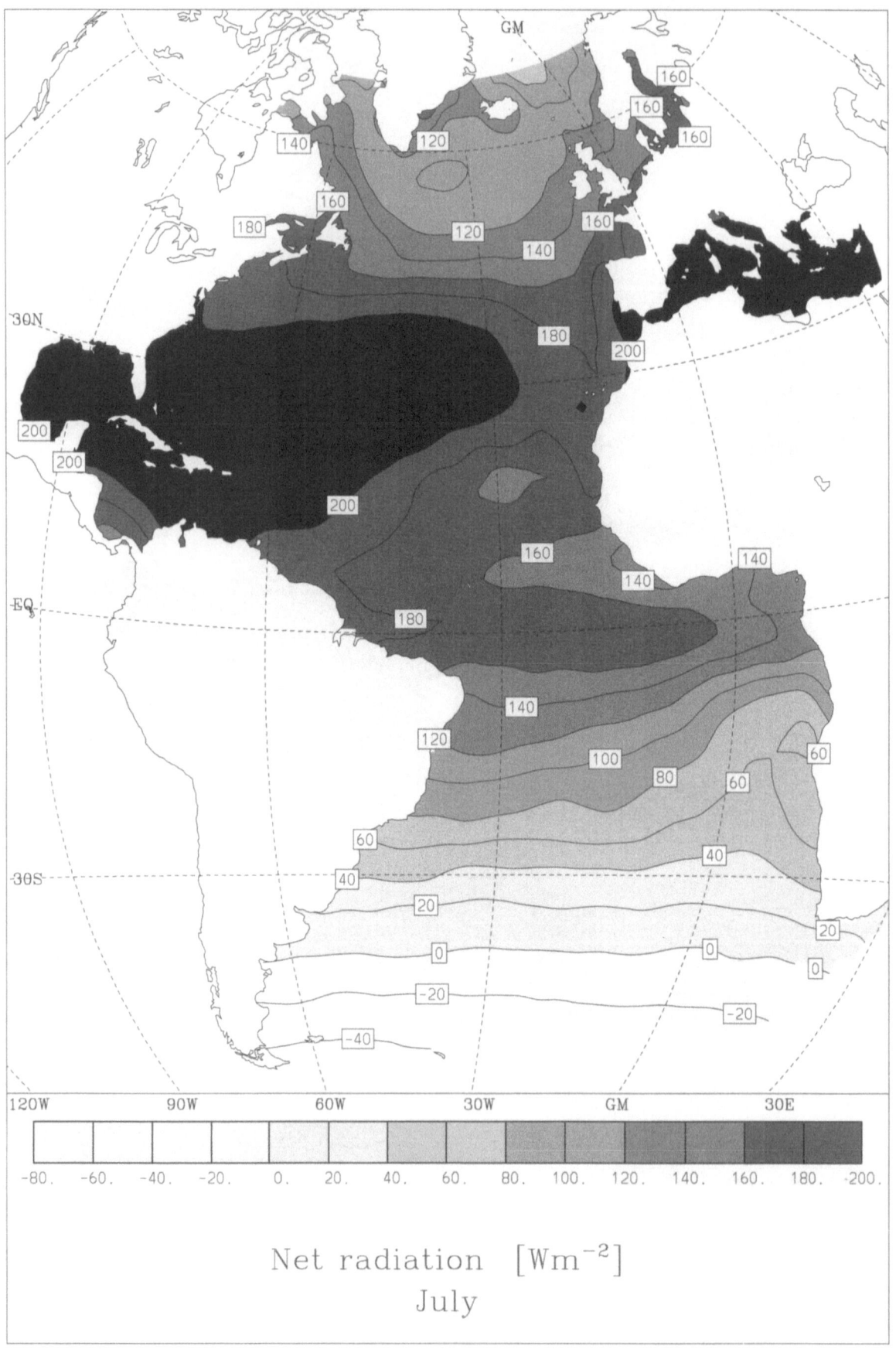

Net radiation [Wm^{-2}]
July

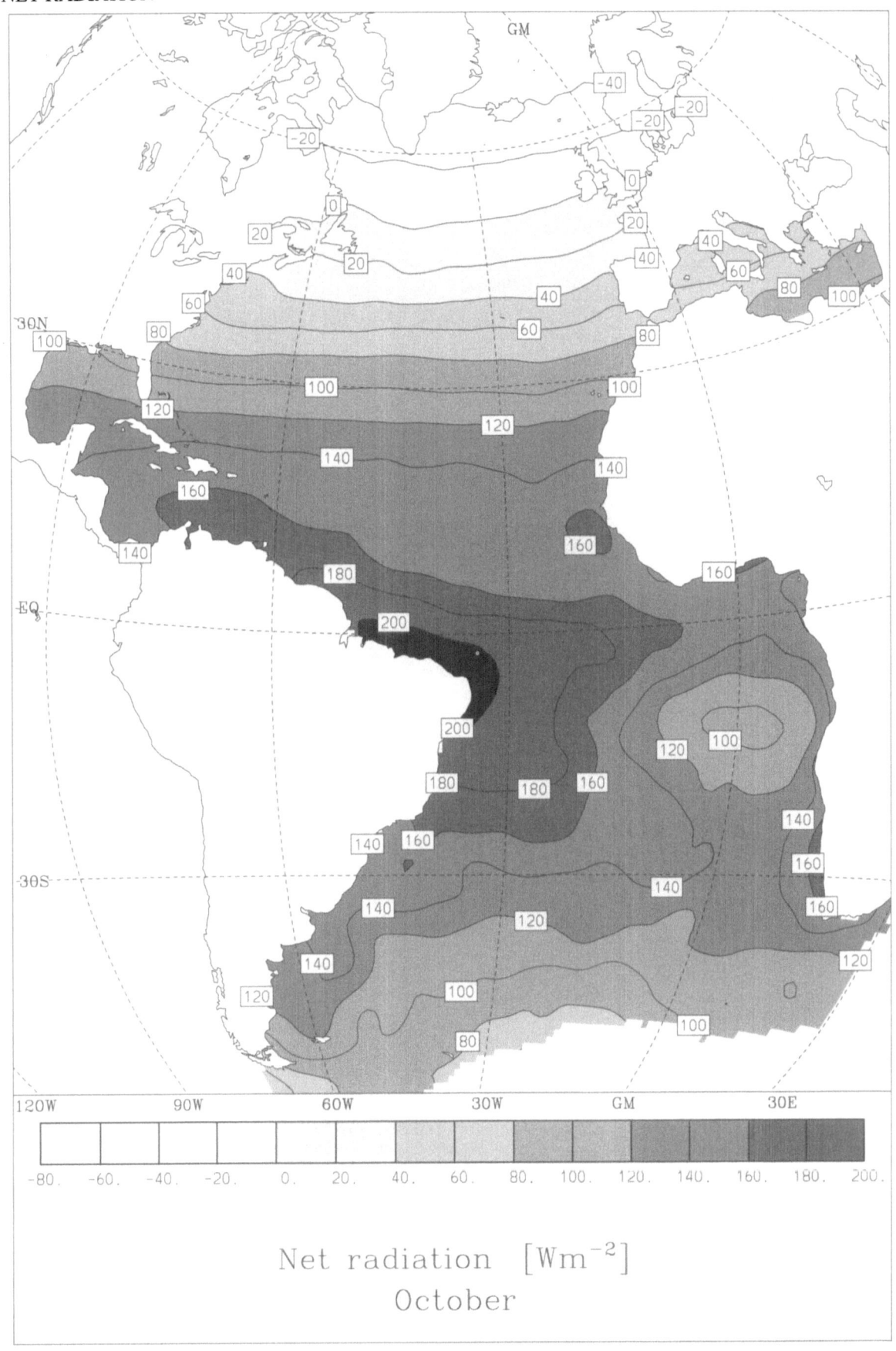

Net radiation [Wm^{-2}]
October

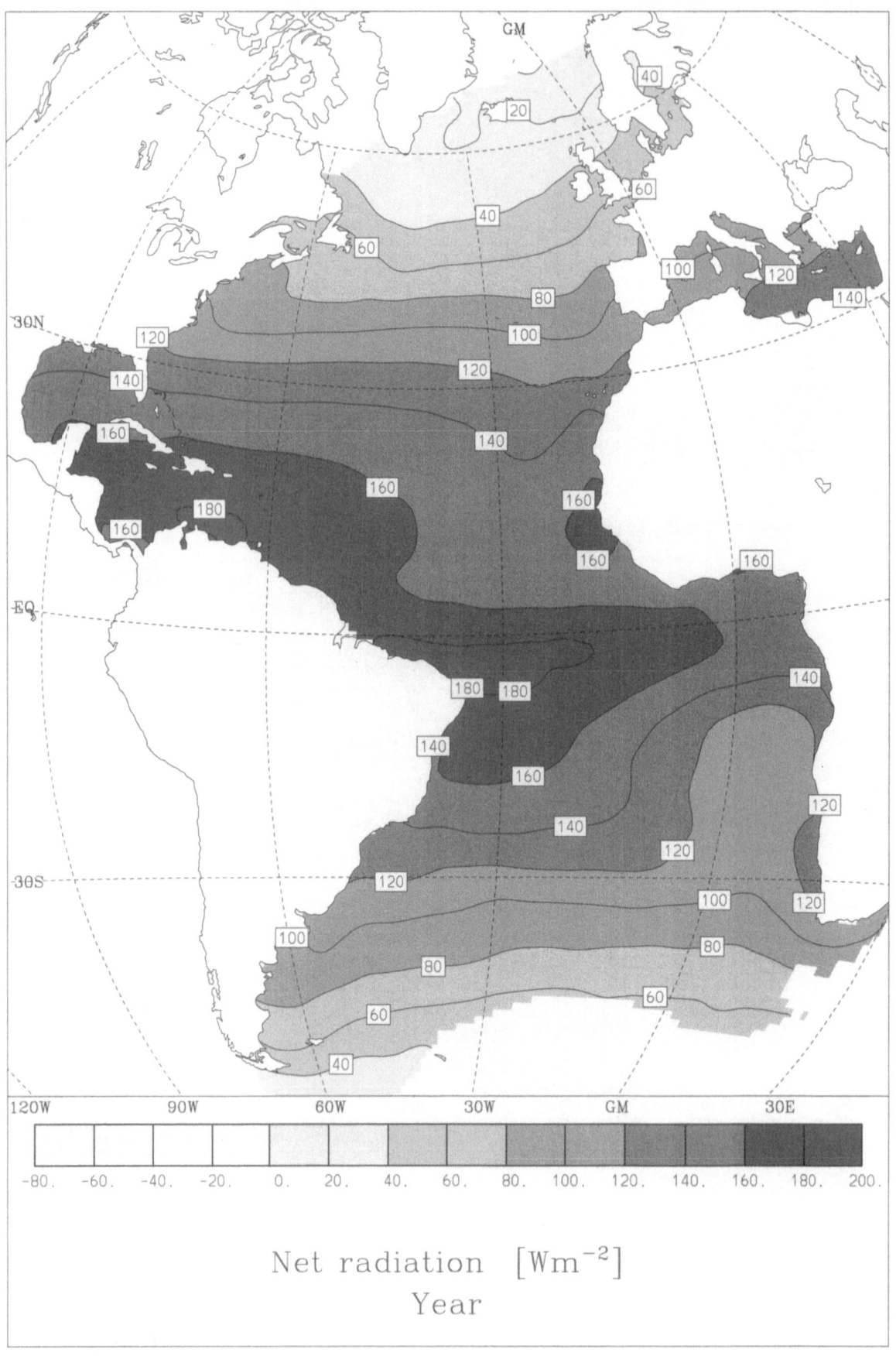

Net radiation [Wm^{-2}]
Year

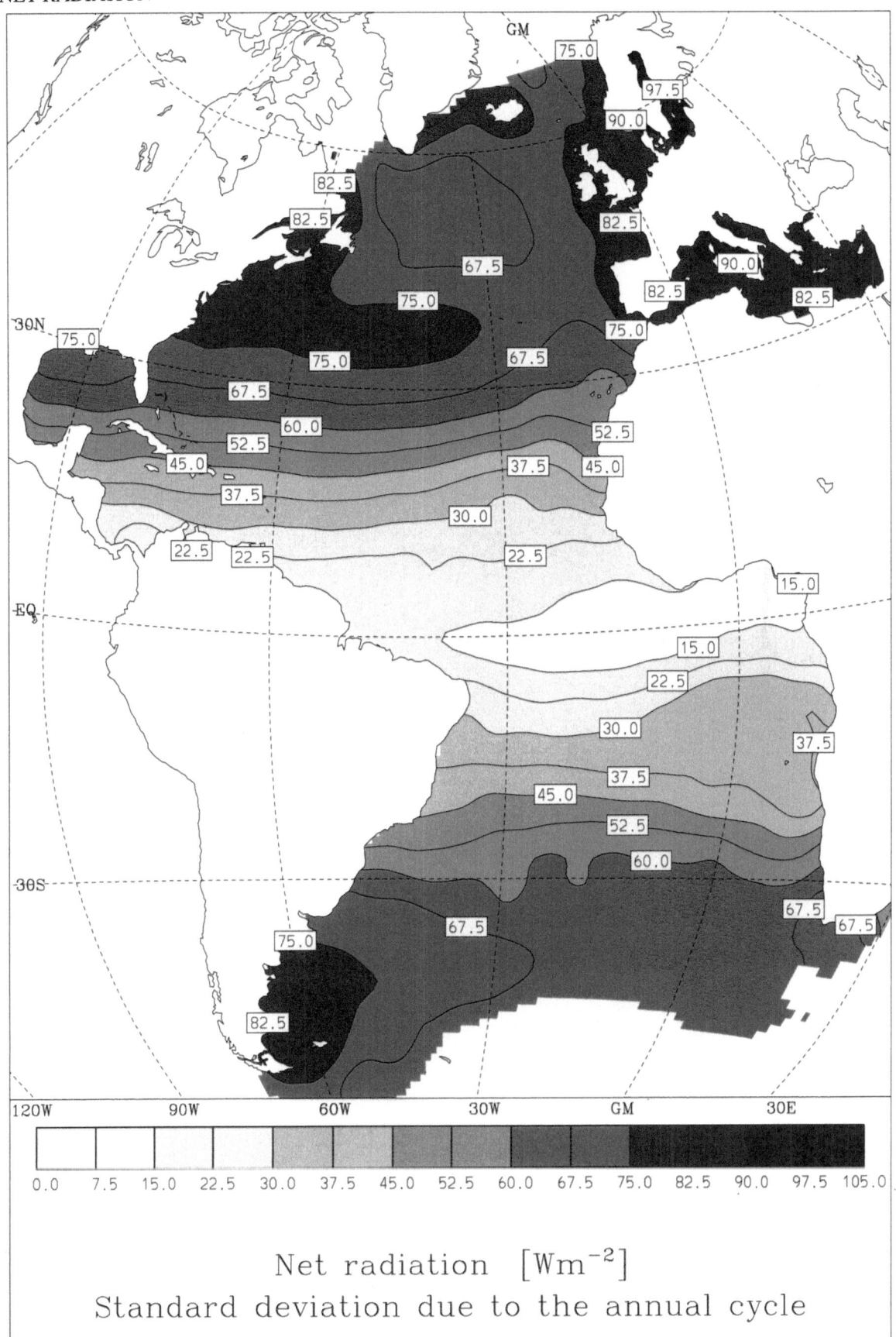

Net radiation [Wm^{-2}]
Standard deviation due to the annual cycle

Latent Heat Flux

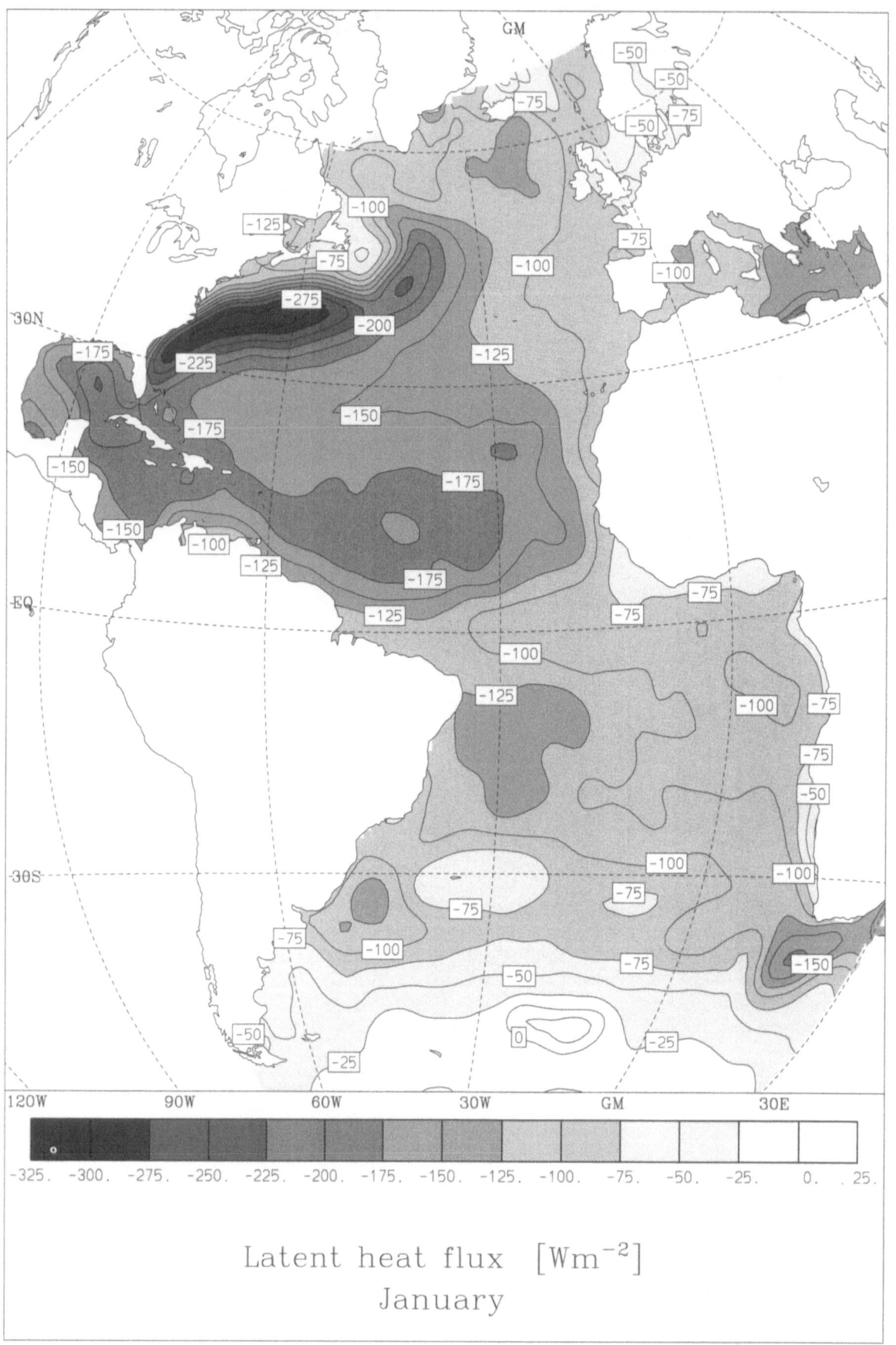

Latent heat flux [Wm^{-2}]
January

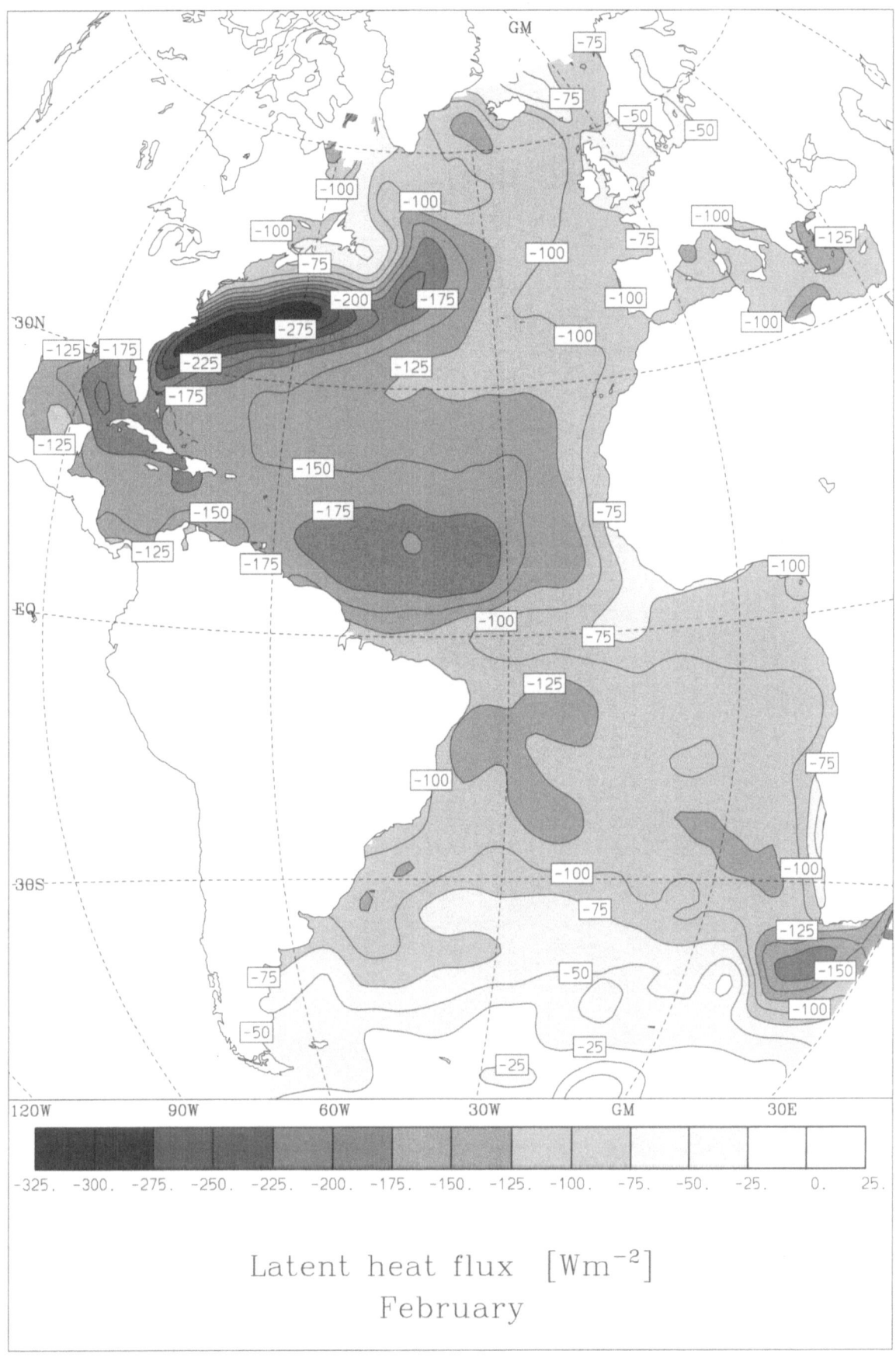

Latent heat flux [Wm^{-2}]
February

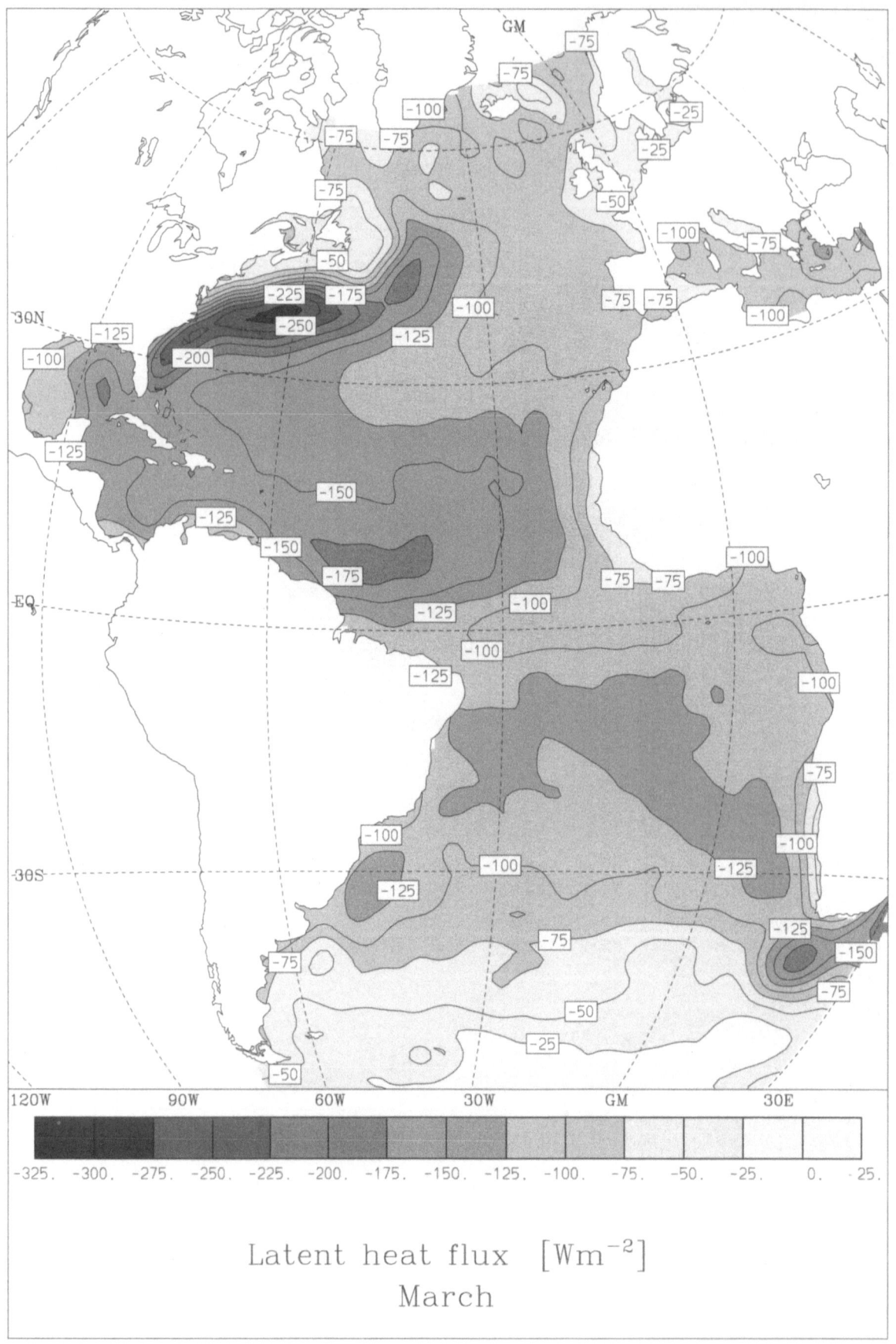

Latent heat flux [Wm^{-2}]
March

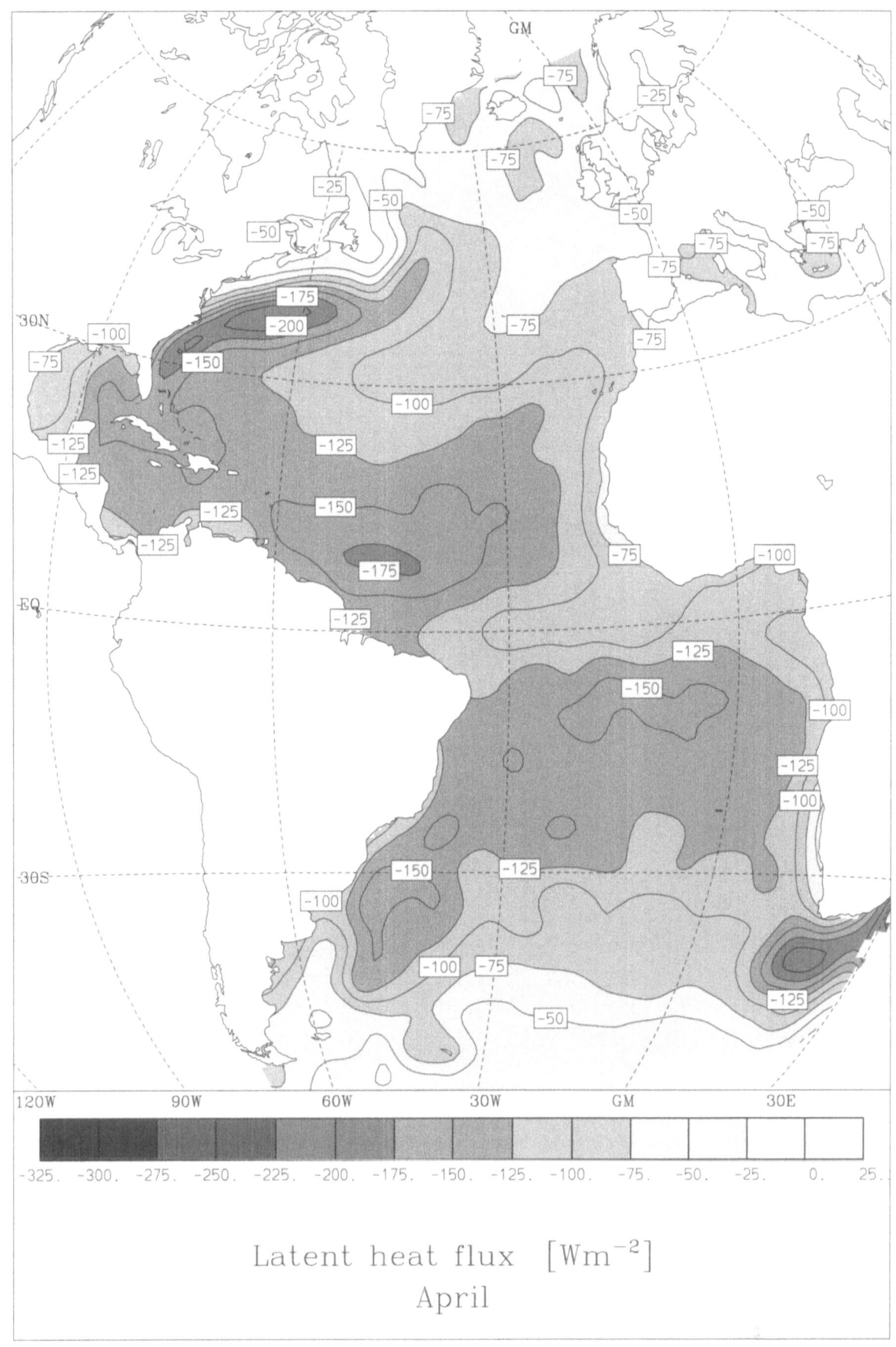

Latent heat flux [Wm^{-2}]
April

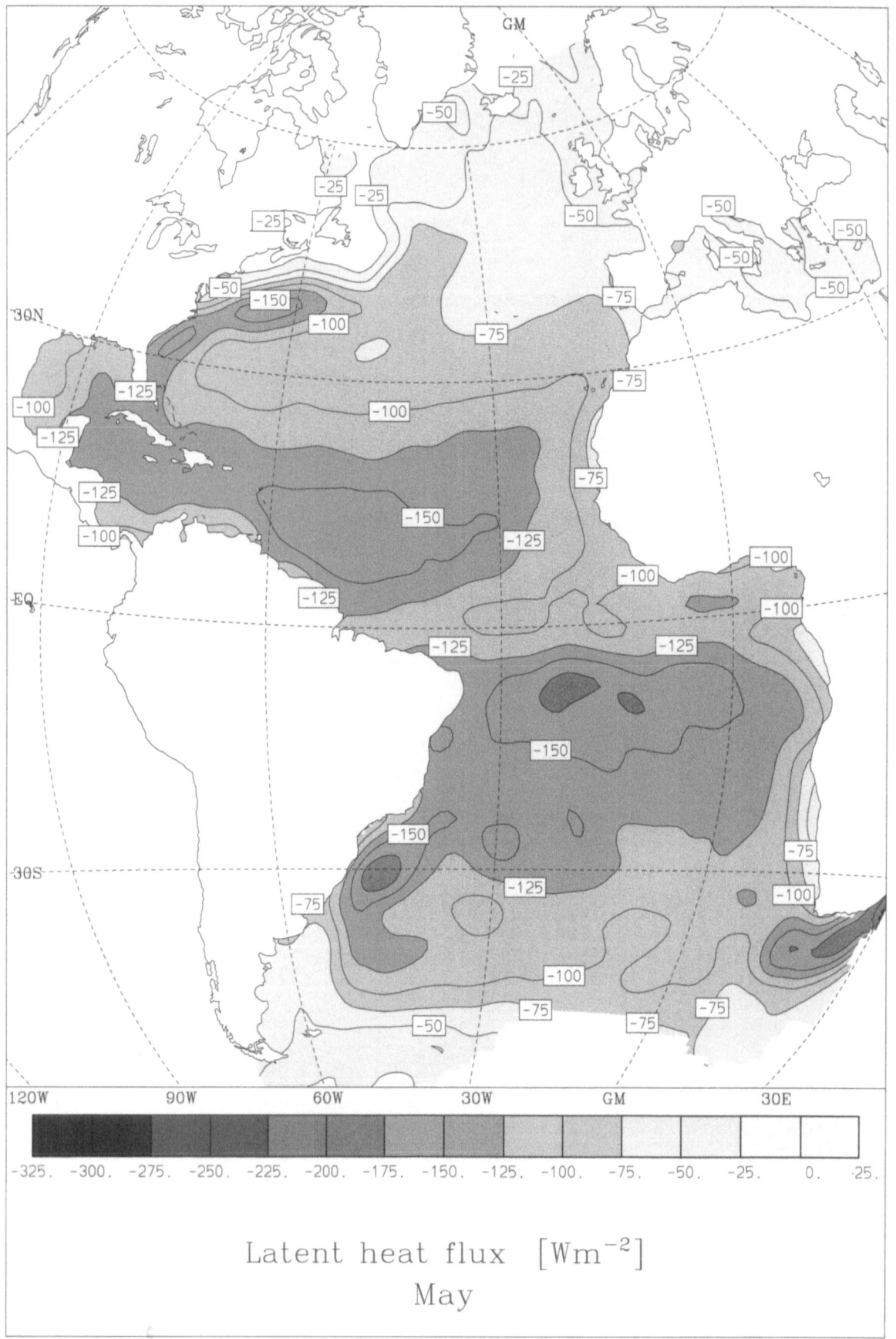

Latent heat flux $[Wm^{-2}]$
May

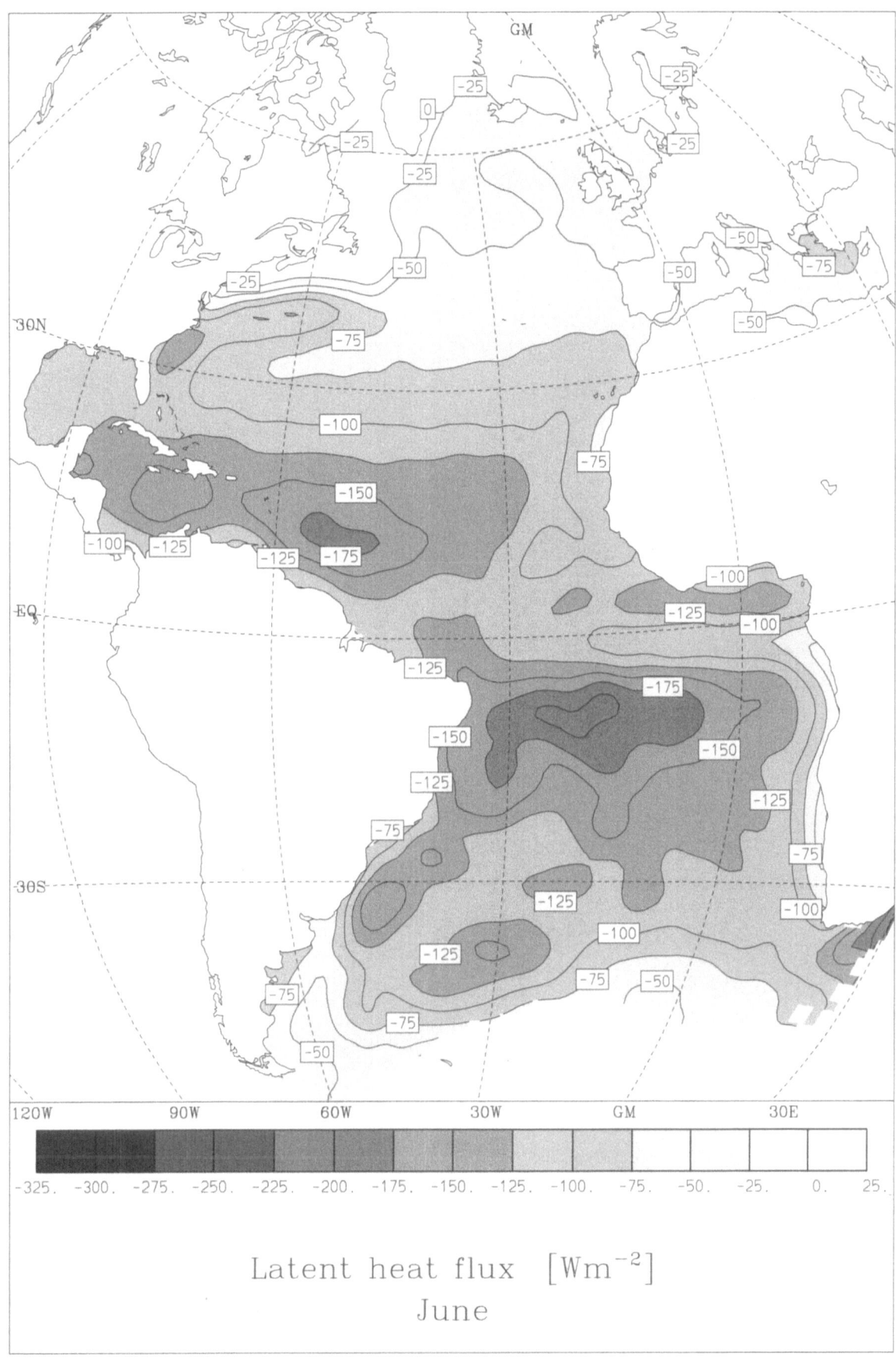

Latent heat flux [Wm^{-2}]
June

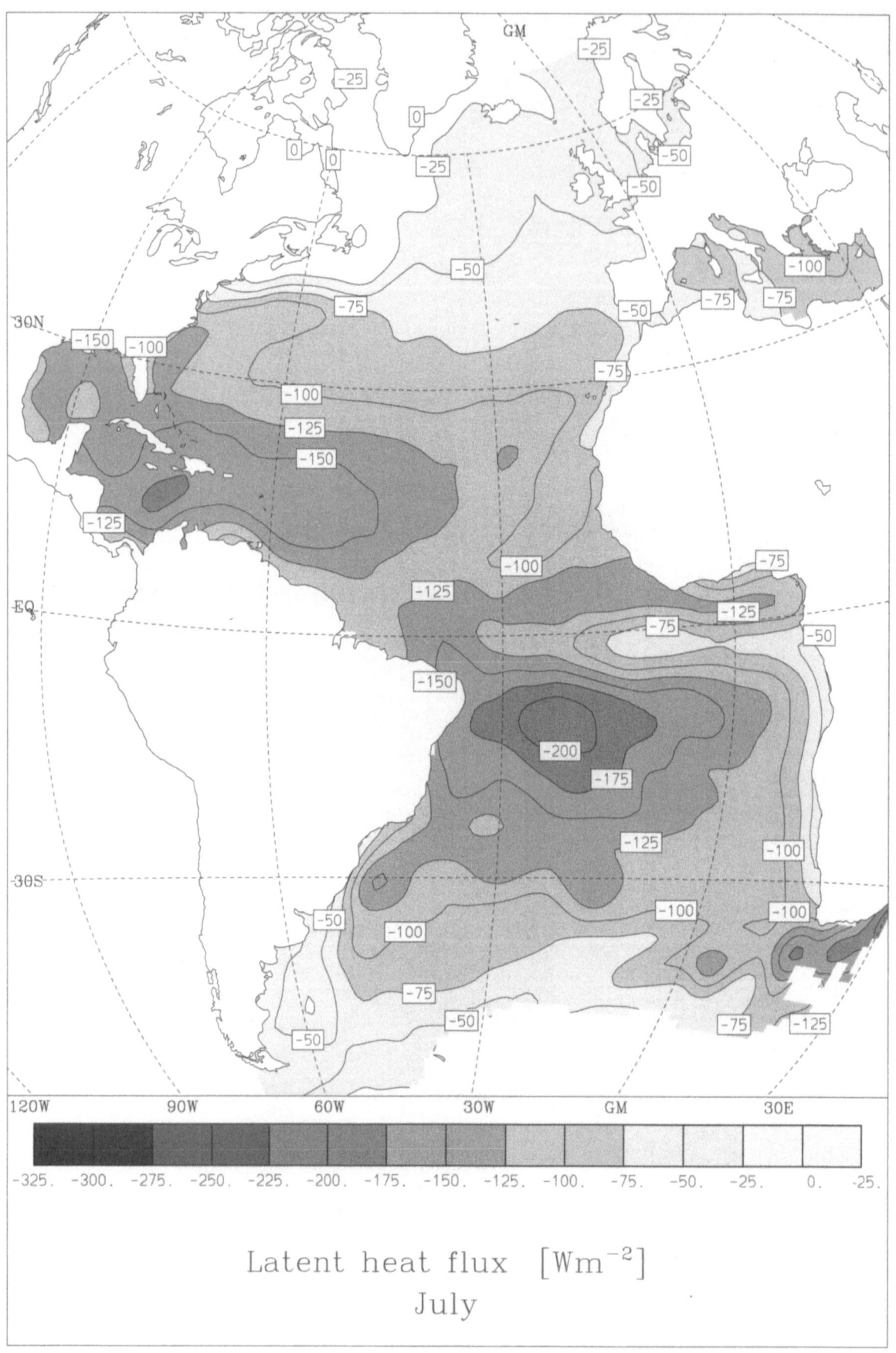

Latent heat flux [Wm^{-2}]
July

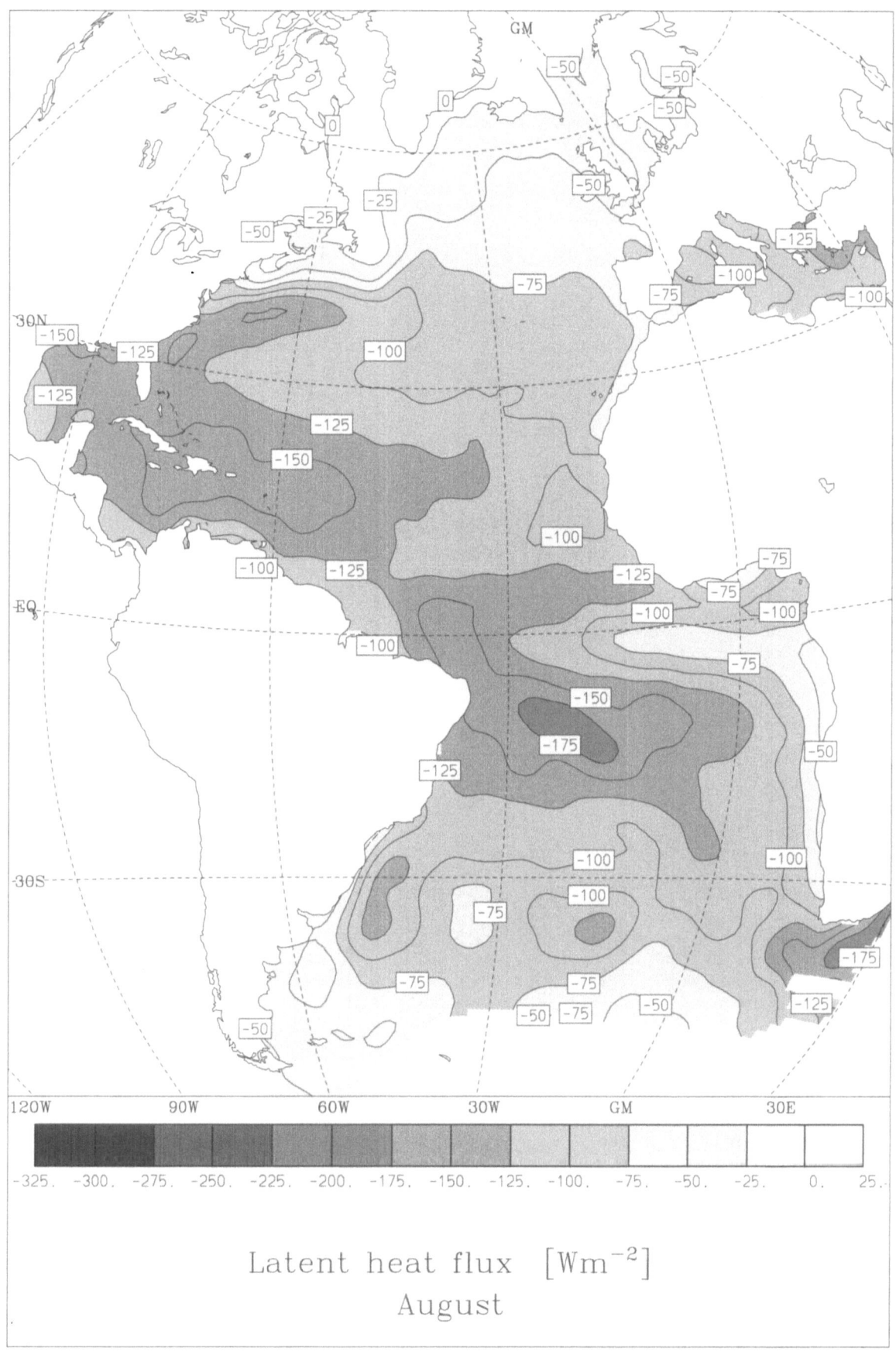

Latent heat flux $[Wm^{-2}]$

August

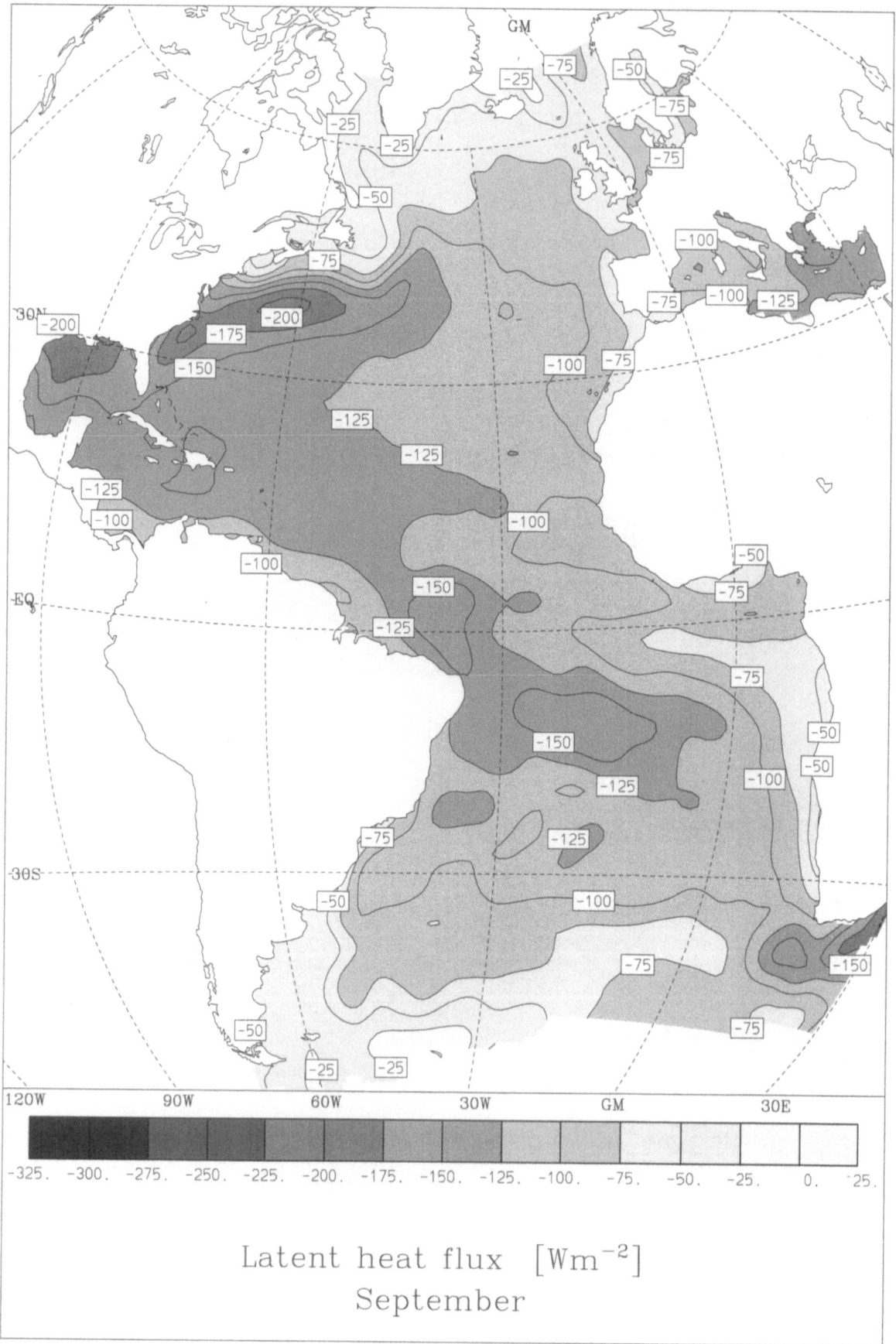

Latent heat flux [Wm^{-2}]
September

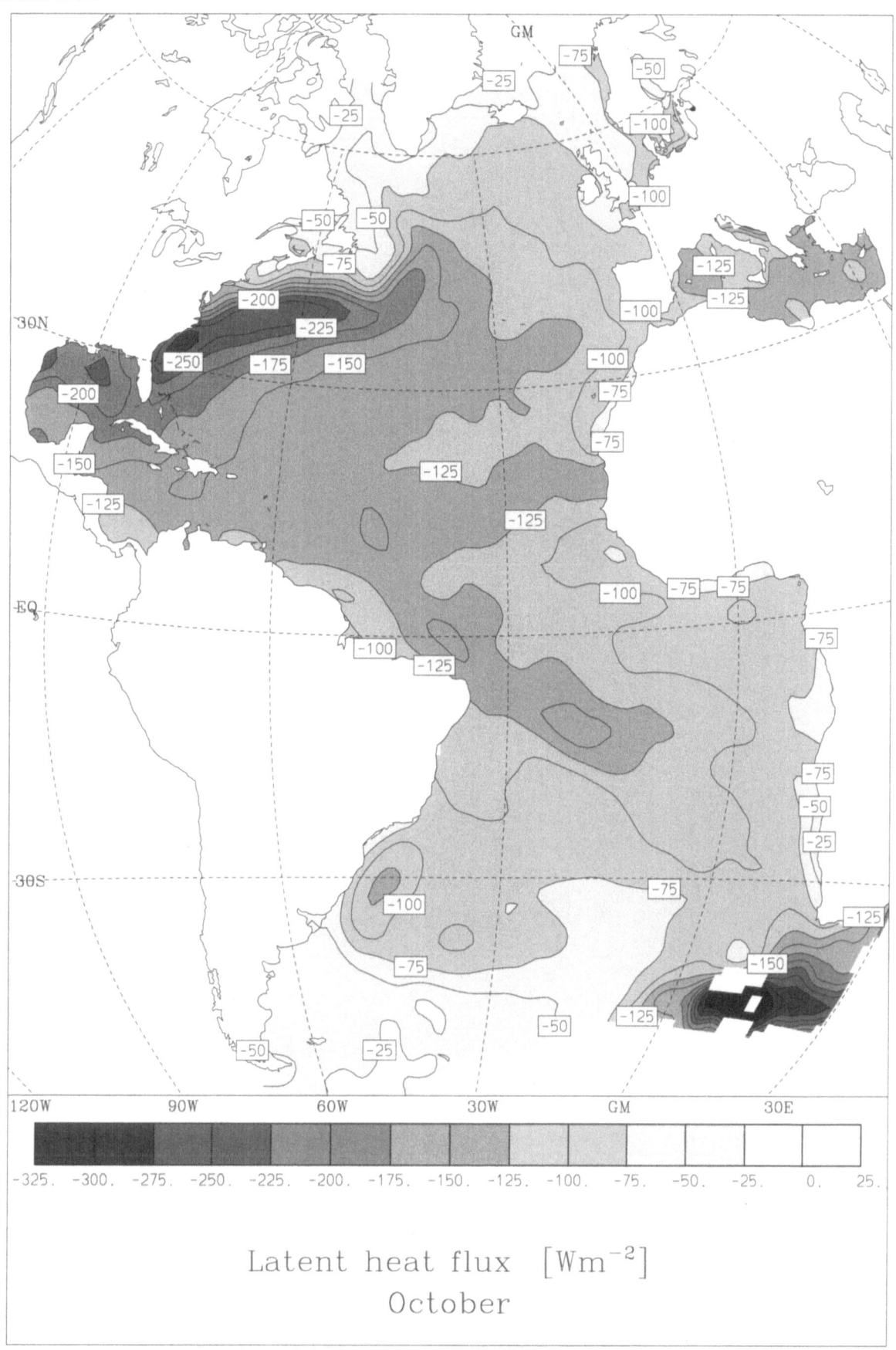

Latent heat flux [Wm^{-2}]
October

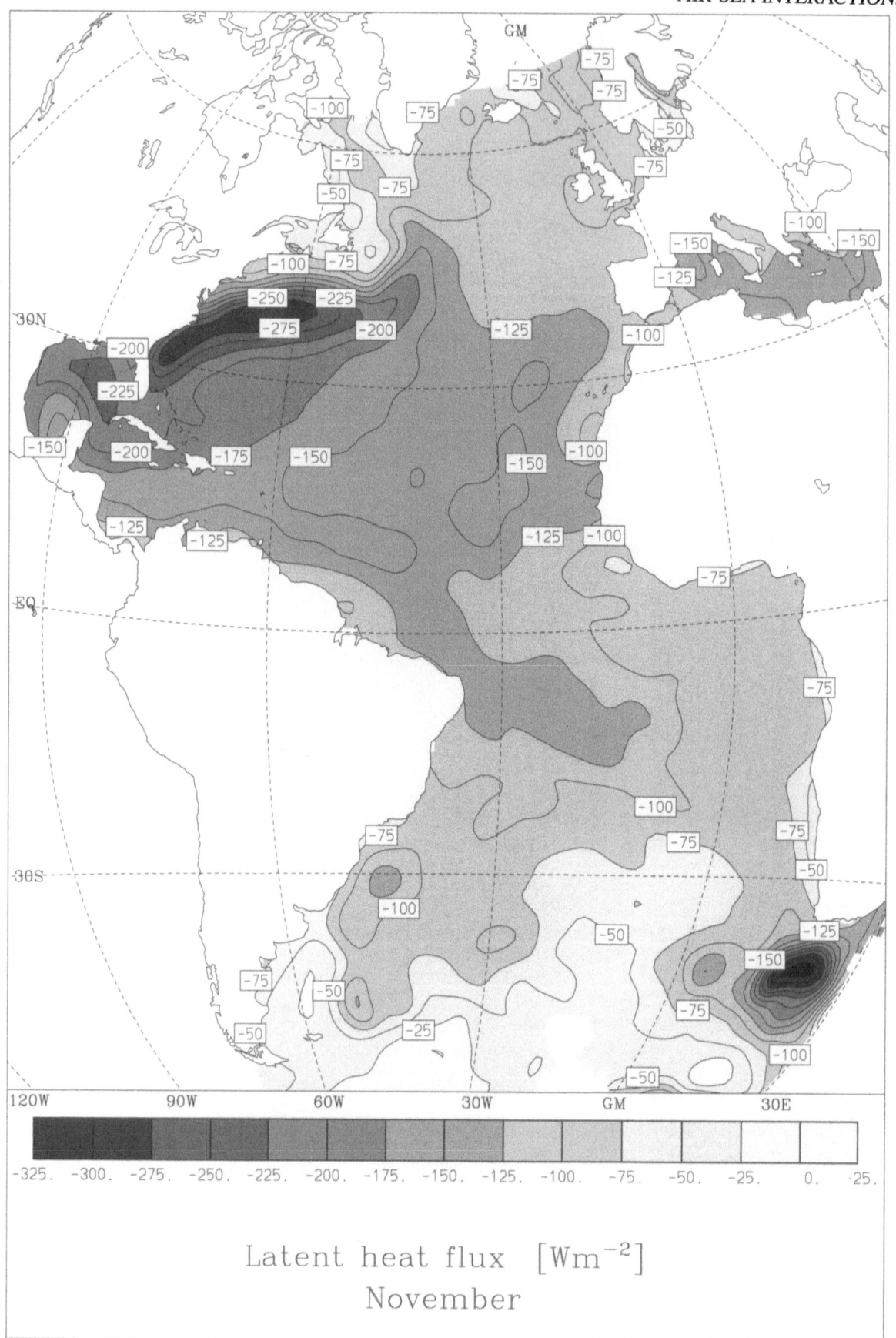

Latent heat flux [Wm^{-2}]
November

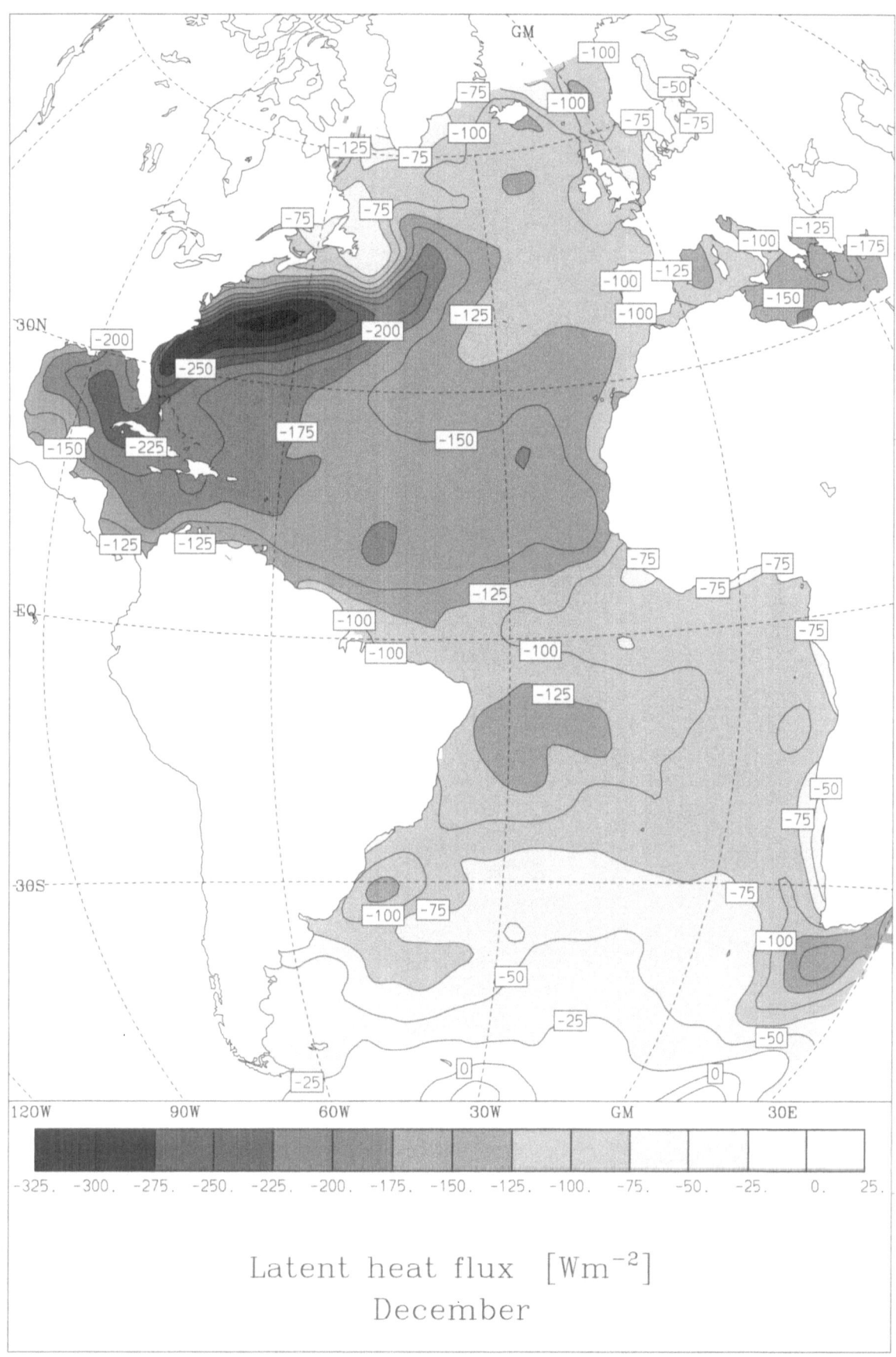

Latent heat flux [Wm^{-2}]
December

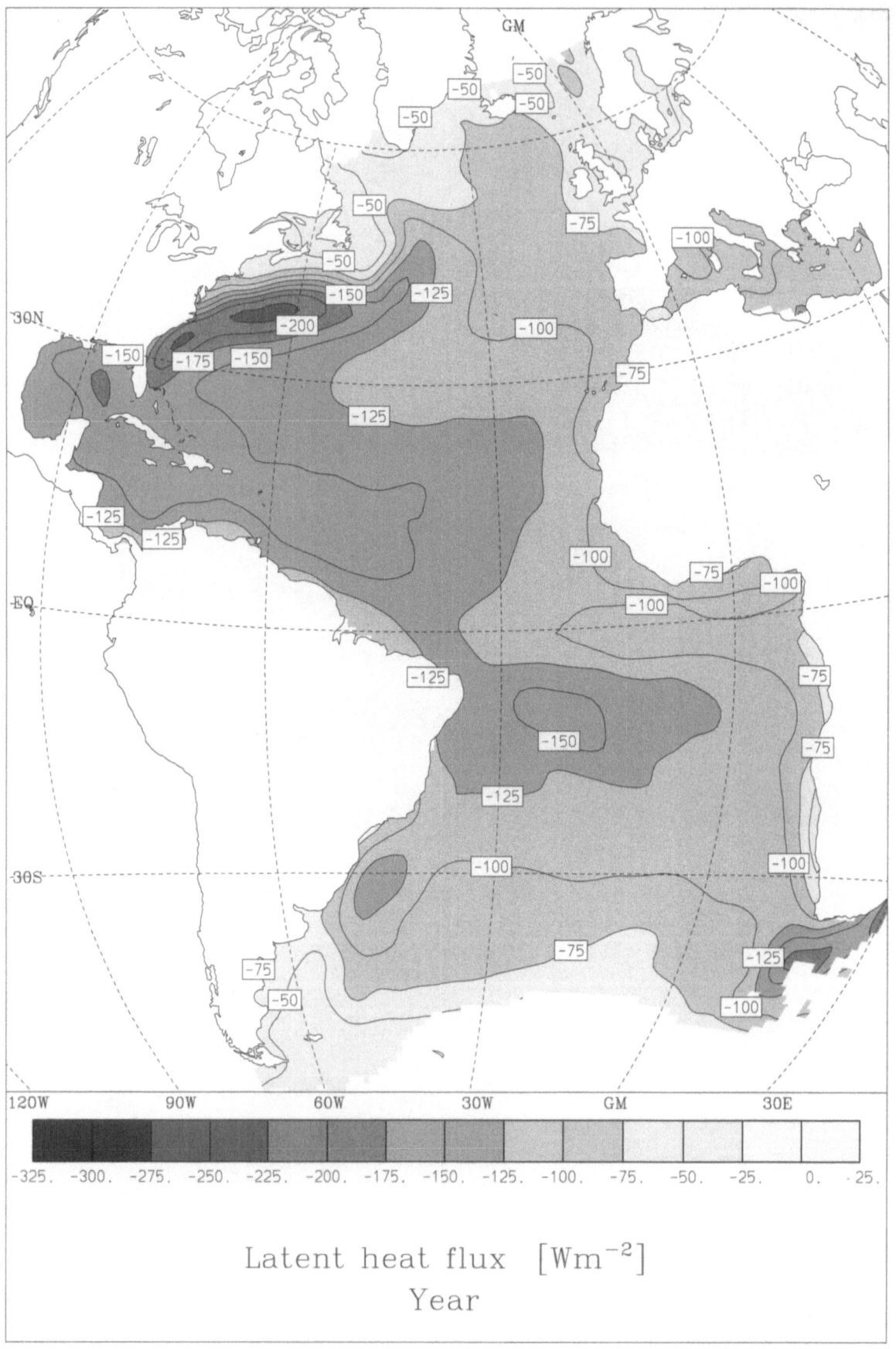

Latent heat flux [Wm^{-2}]
Year

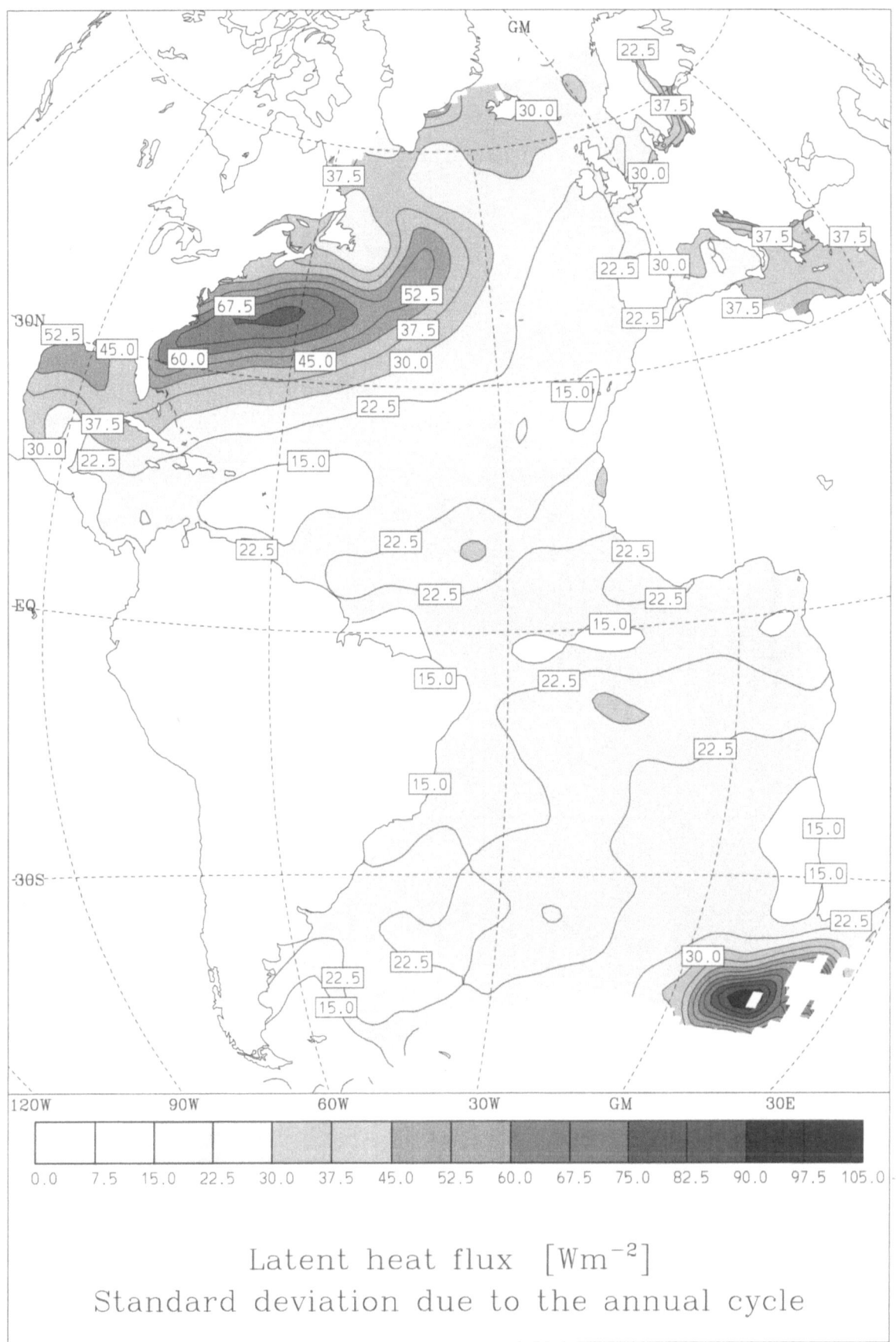

Latent heat flux [Wm^{-2}]
Standard deviation due to the annual cycle

Evaporation

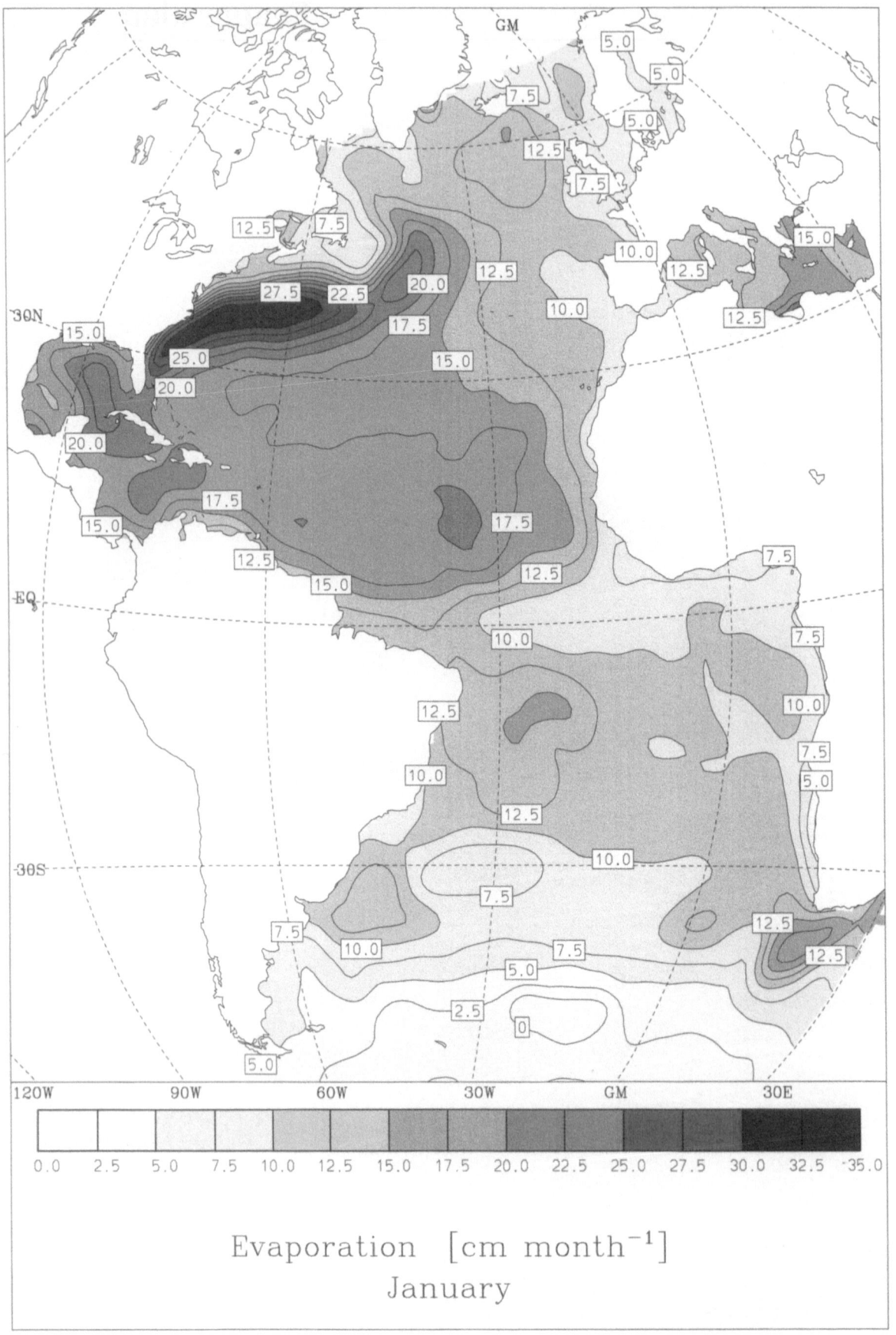

Evaporation [cm month^{-1}]
January

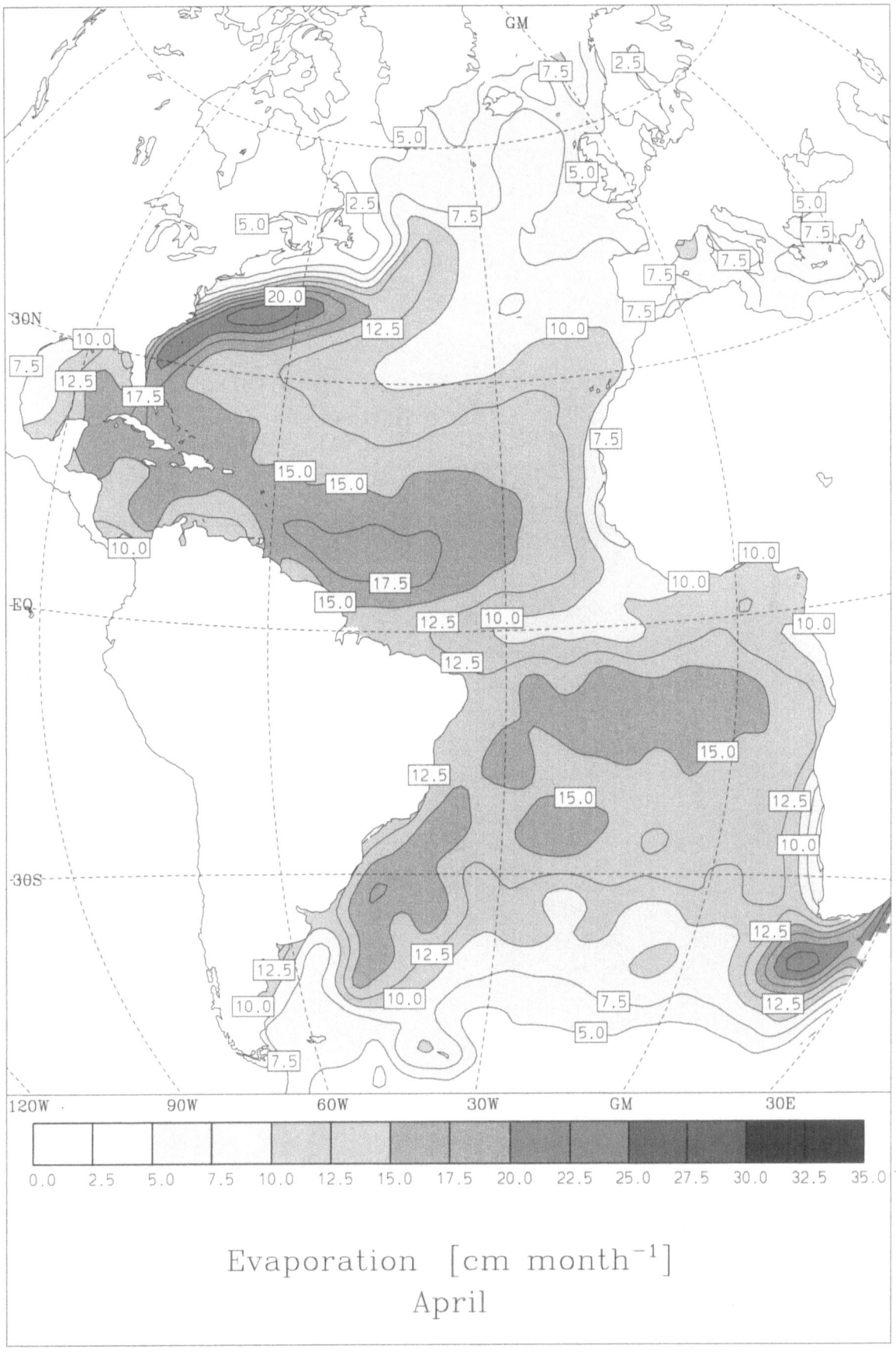

Evaporation [cm month^{-1}]
April

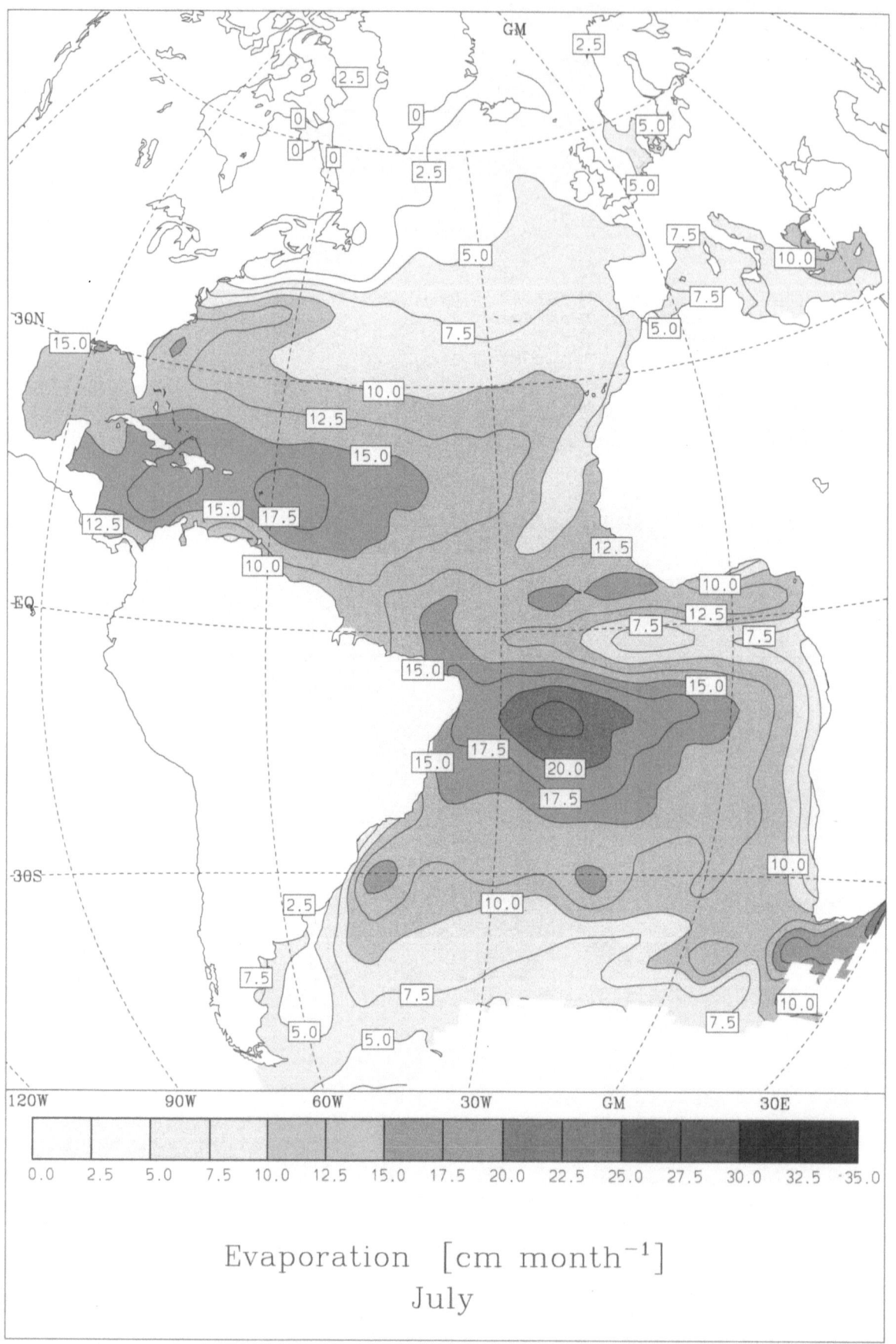

Evaporation [cm month^{-1}]
July

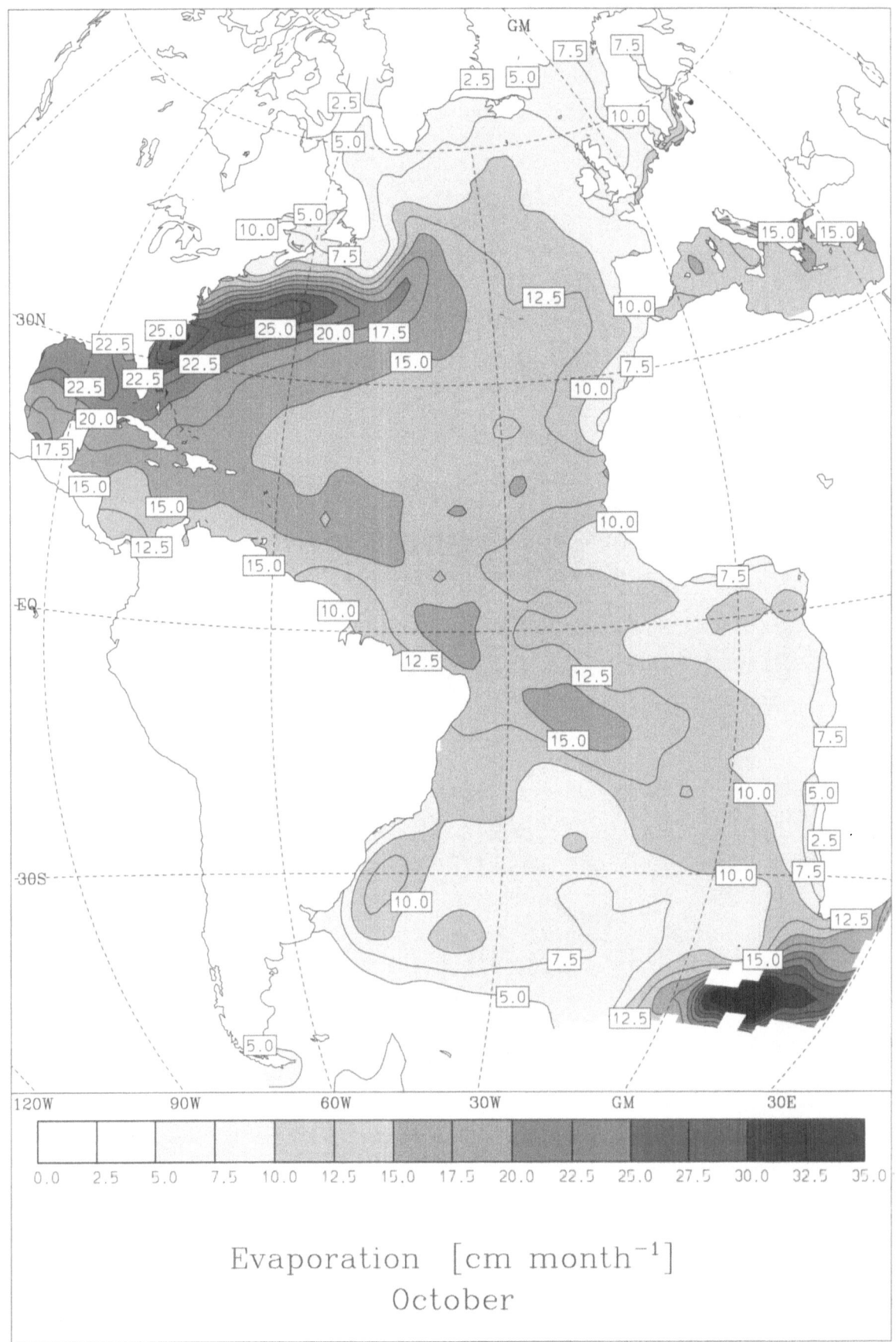

Evaporation [cm month^{-1}]
October

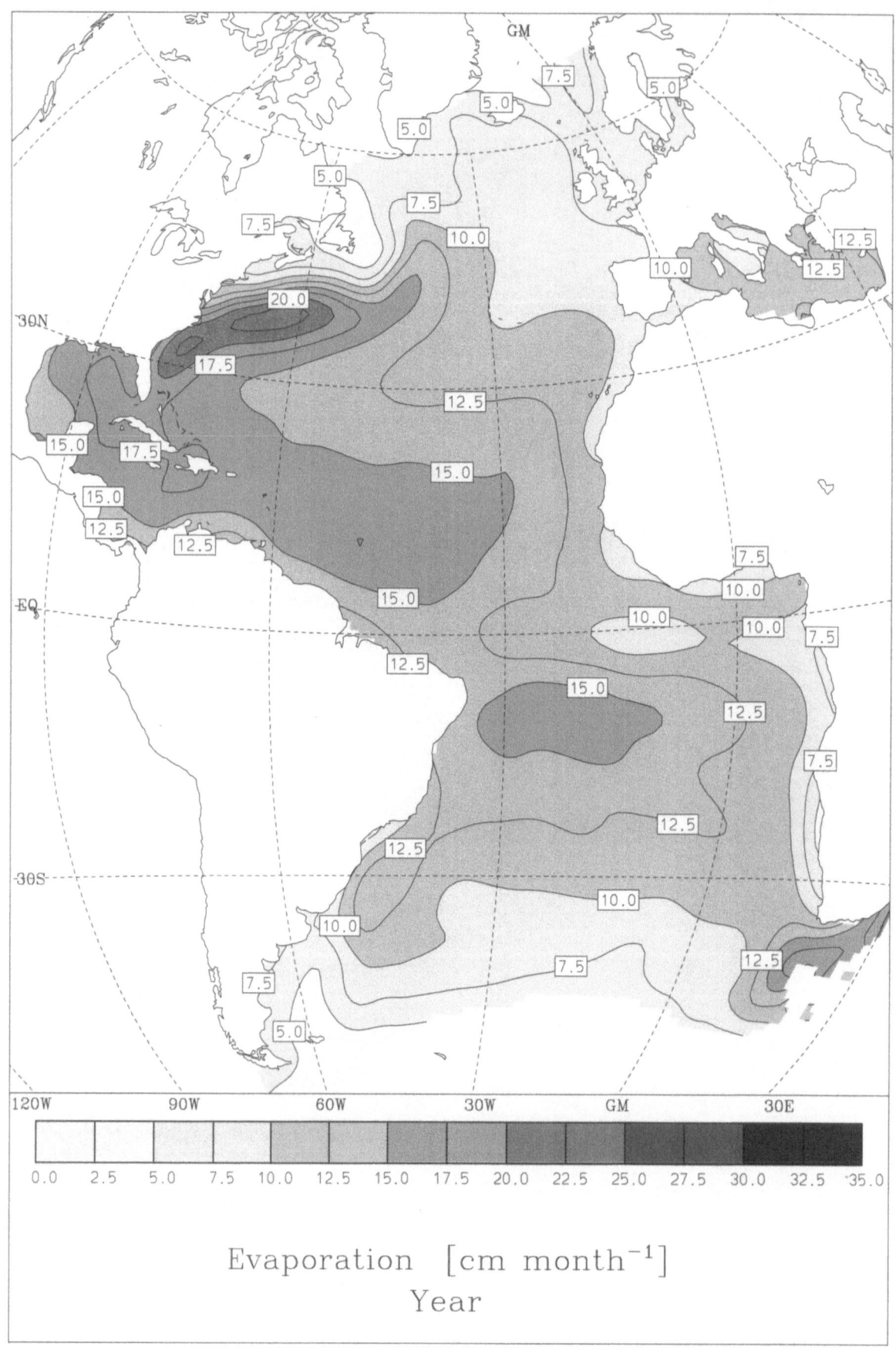

Evaporation [cm month^{-1}]
Year

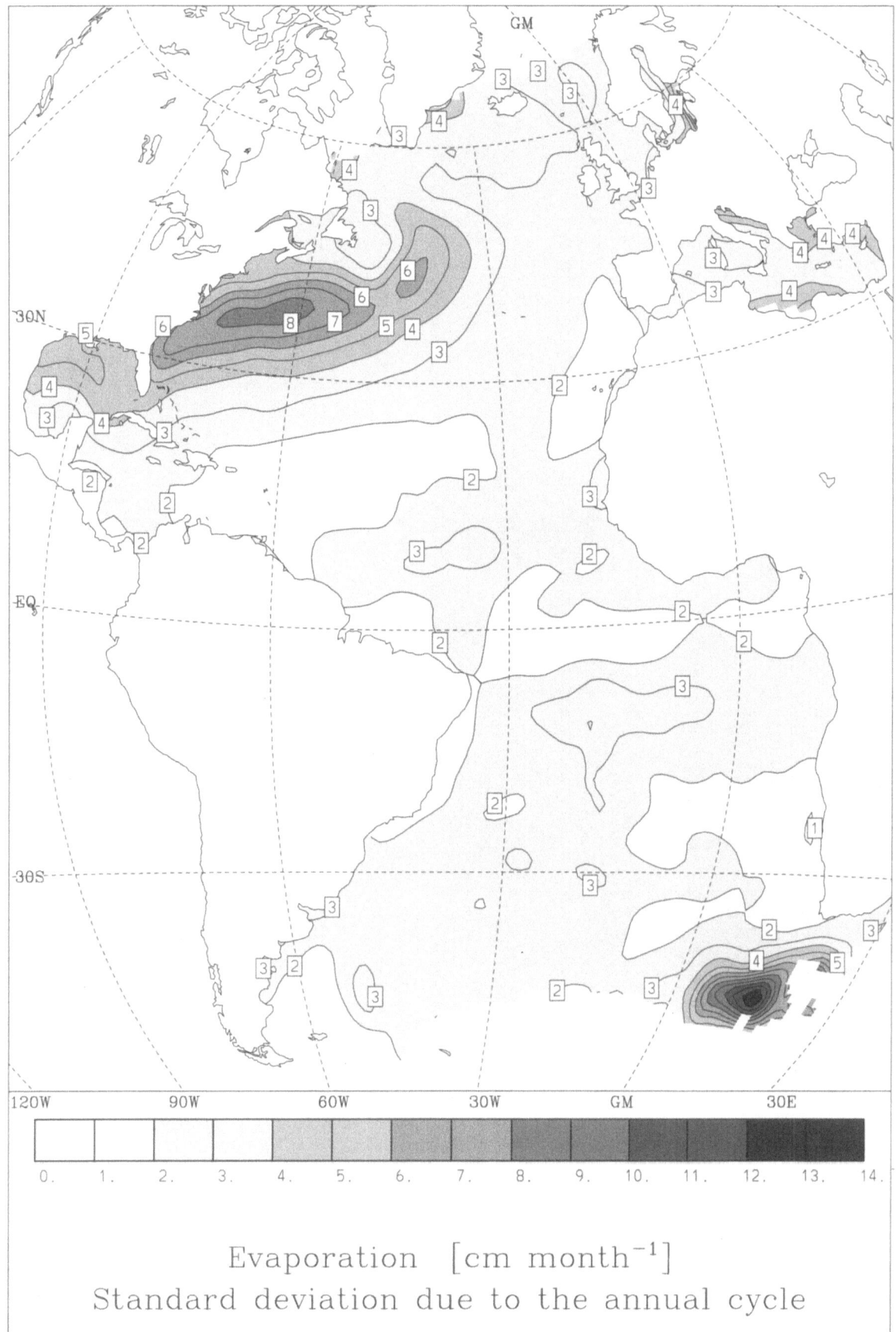

Evaporation [cm month^{-1}]
Standard deviation due to the annual cycle

Sensible Heat Flux

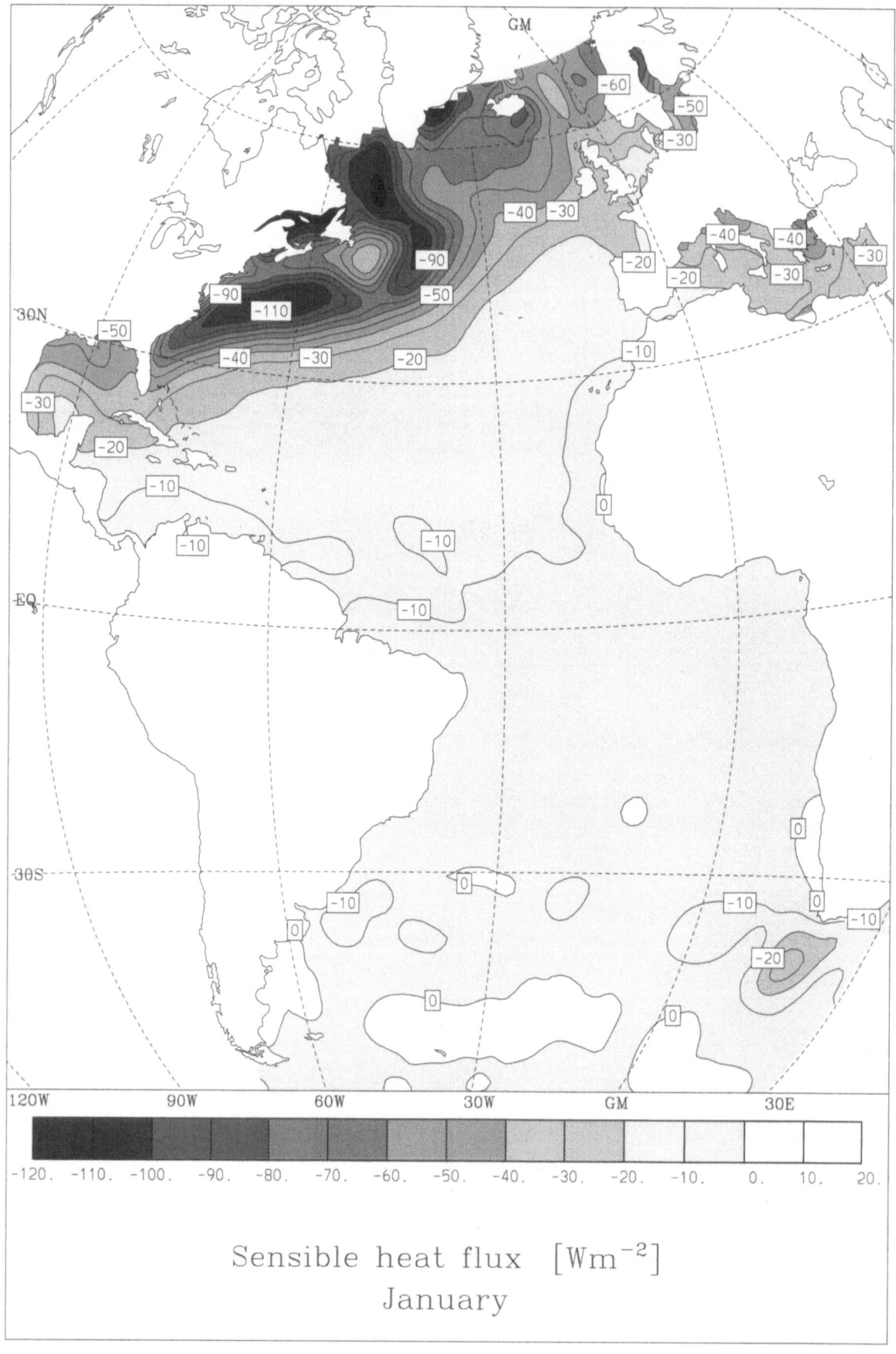

Sensible heat flux [Wm^{-2}]
January

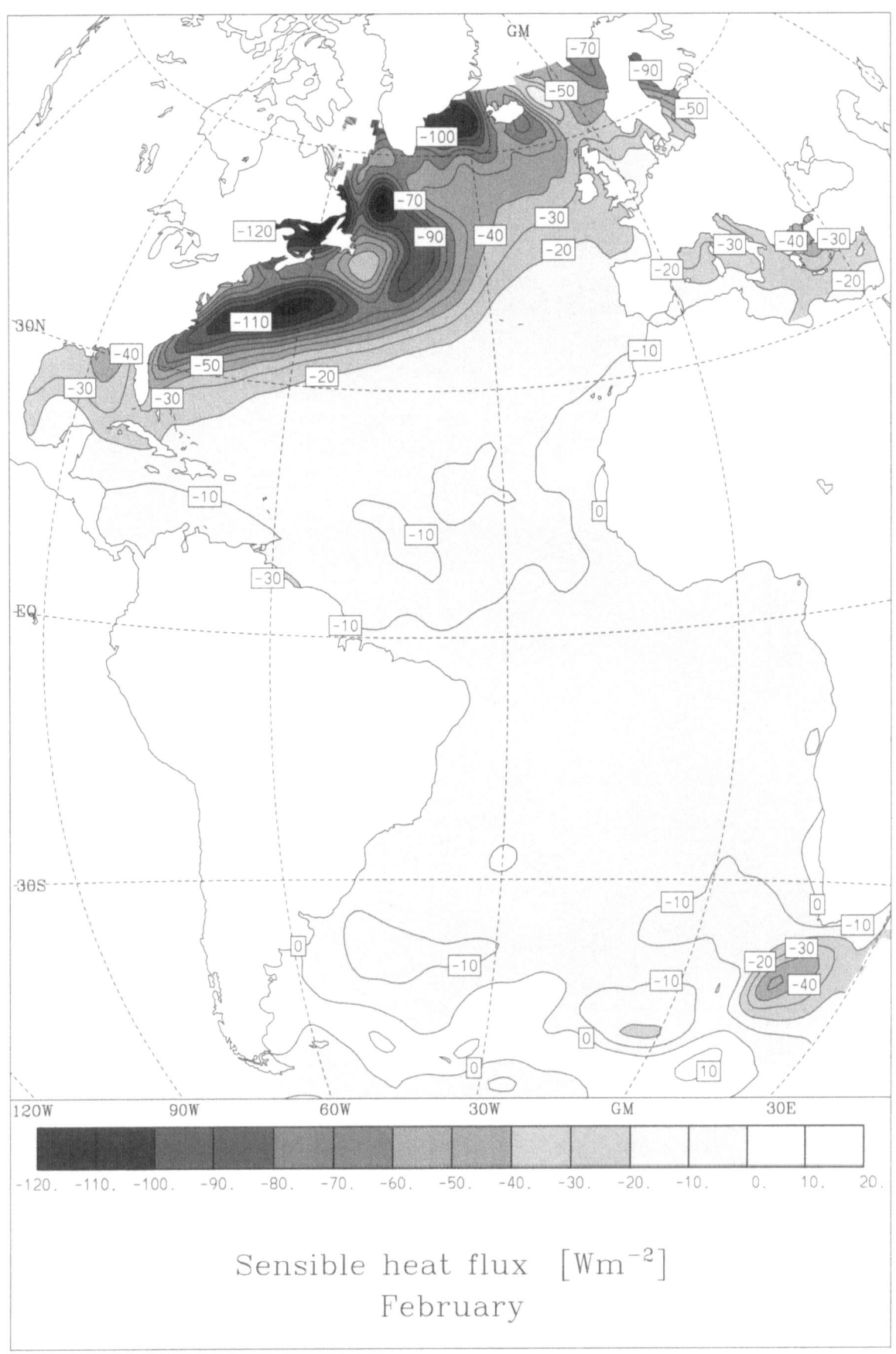

Sensible heat flux $[Wm^{-2}]$
February

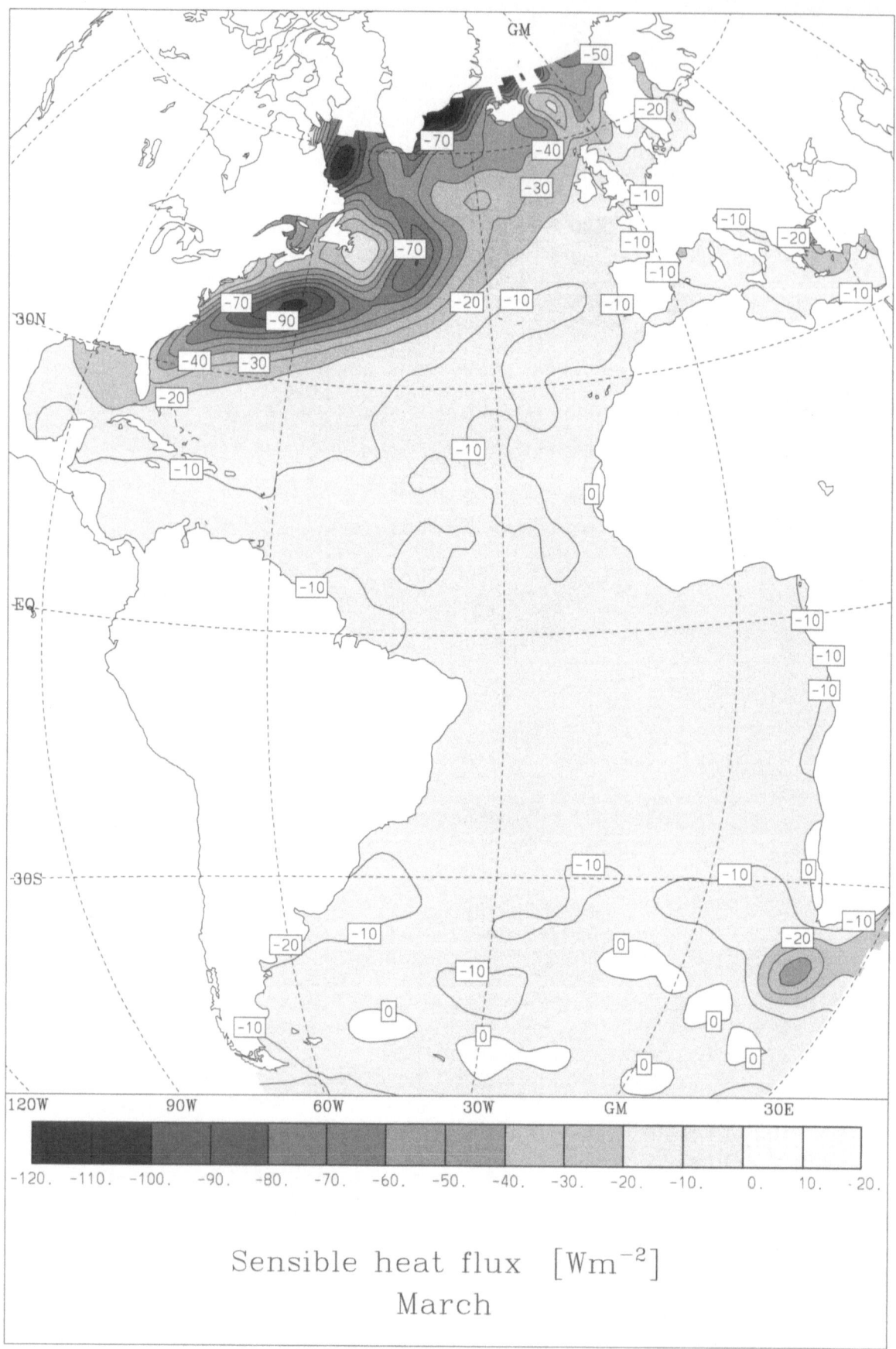

Sensible heat flux [Wm^{-2}]
March

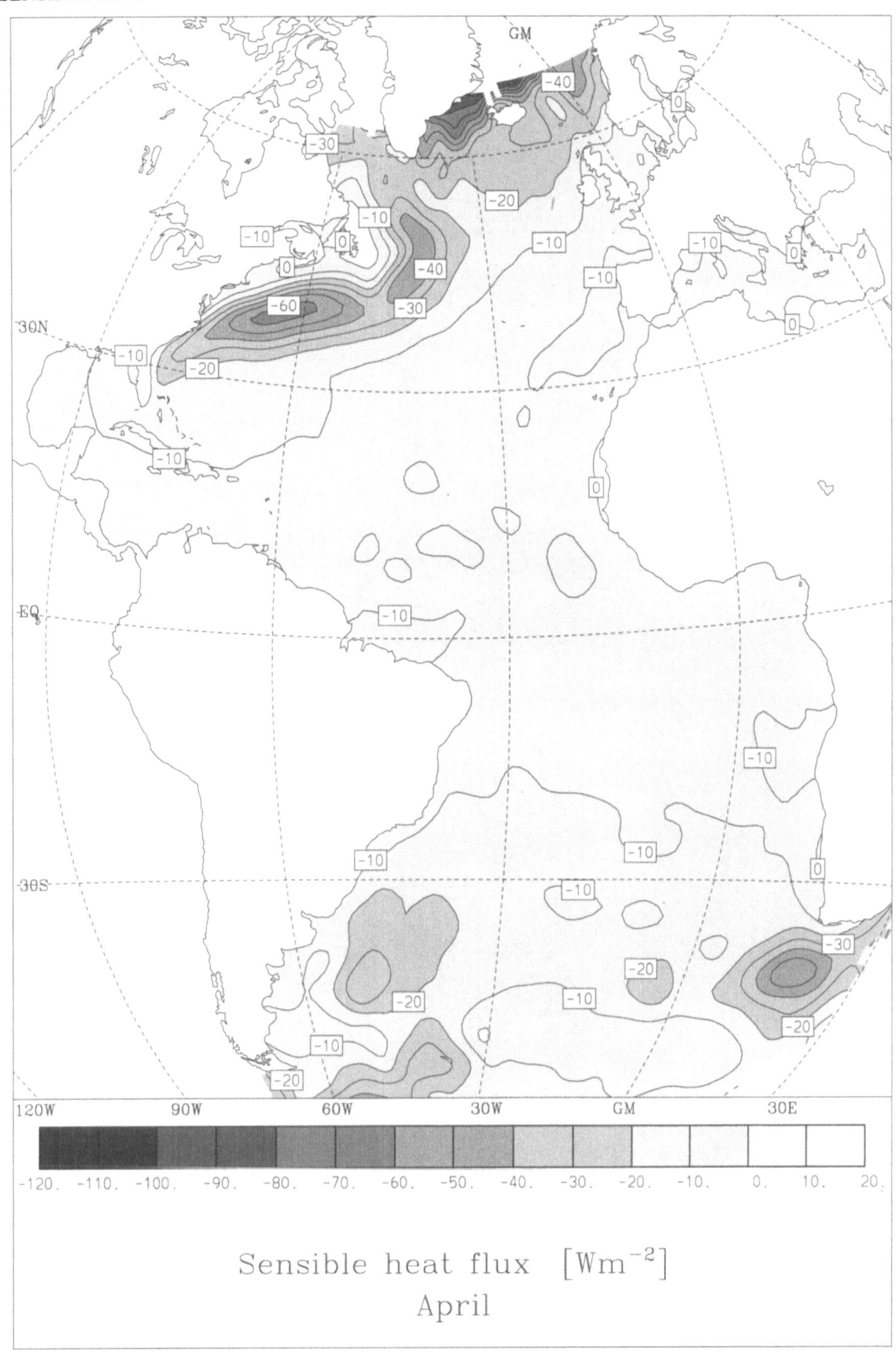

Sensible heat flux $[Wm^{-2}]$

April

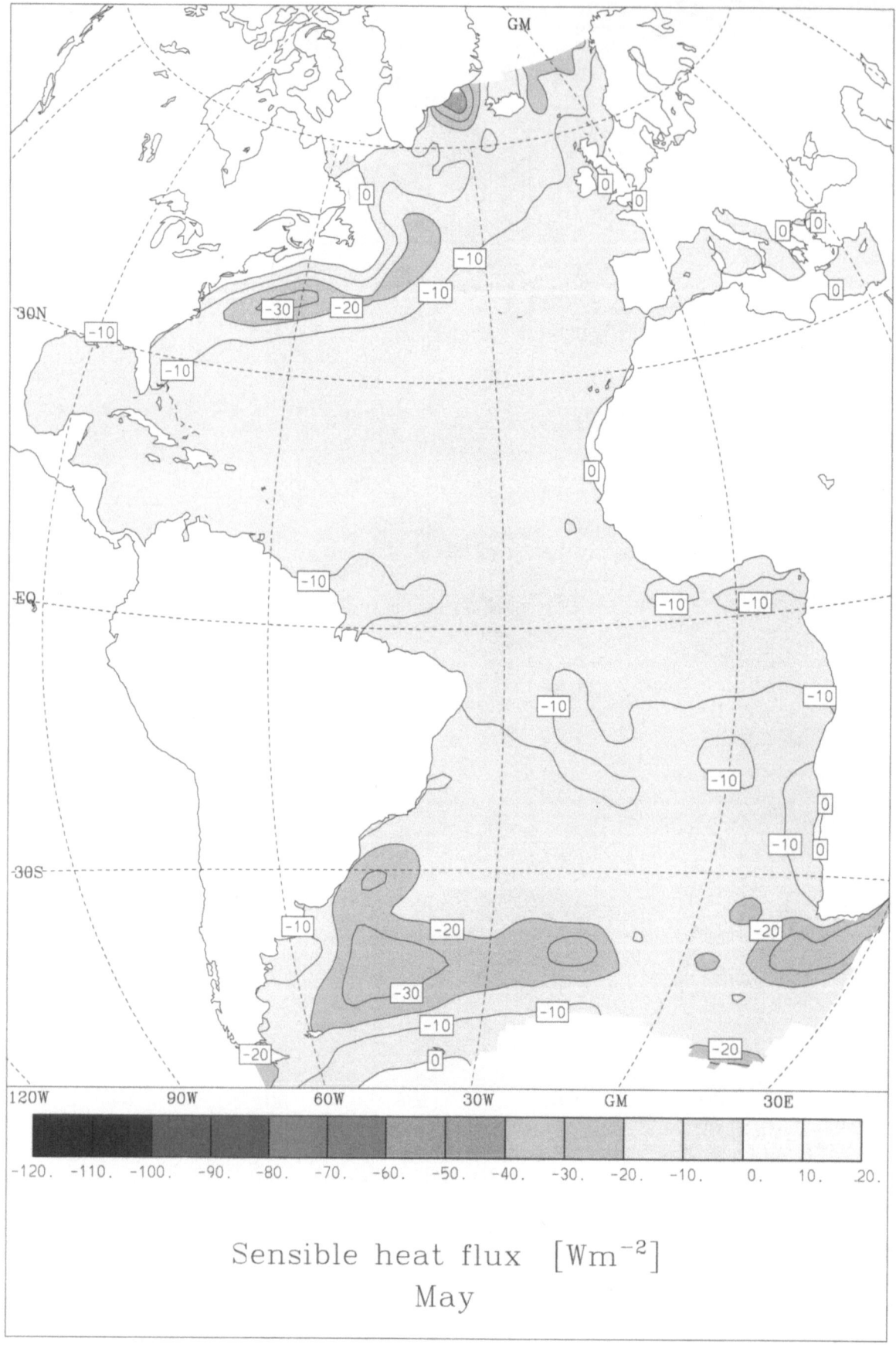

Sensible heat flux [Wm^{-2}]
May

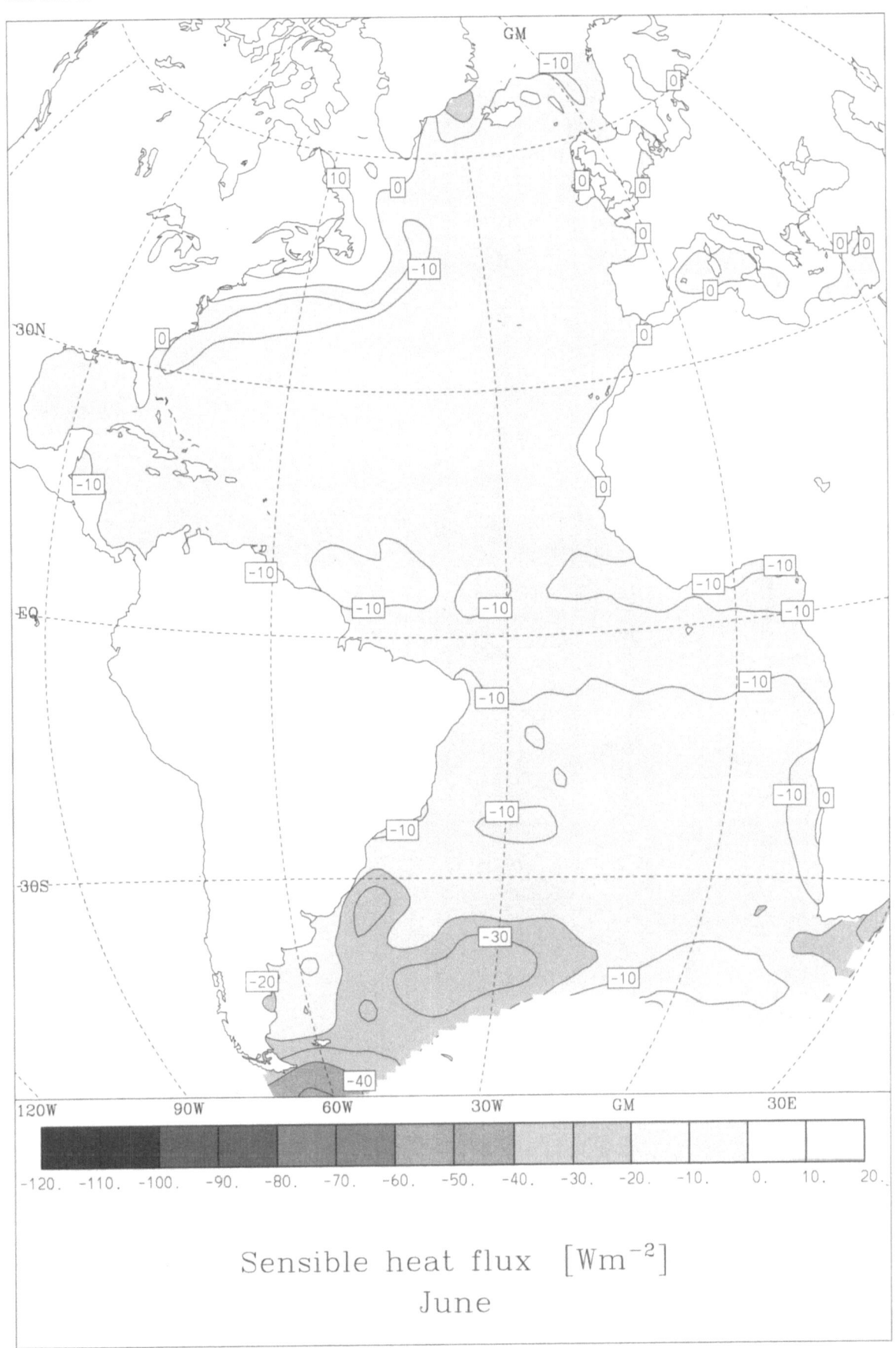

Sensible heat flux [Wm^{-2}]
June

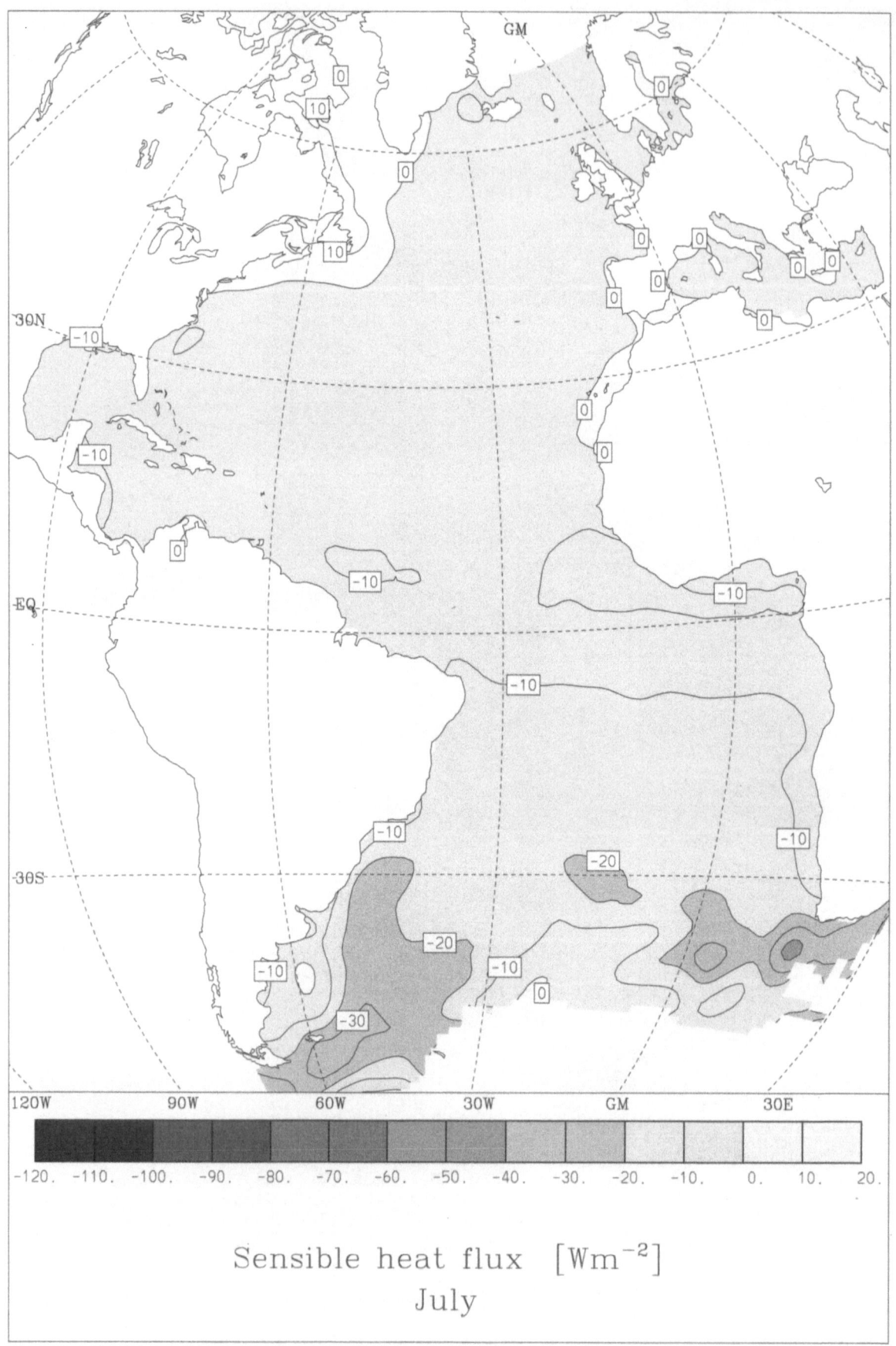

Sensible heat flux [Wm^{-2}]
July

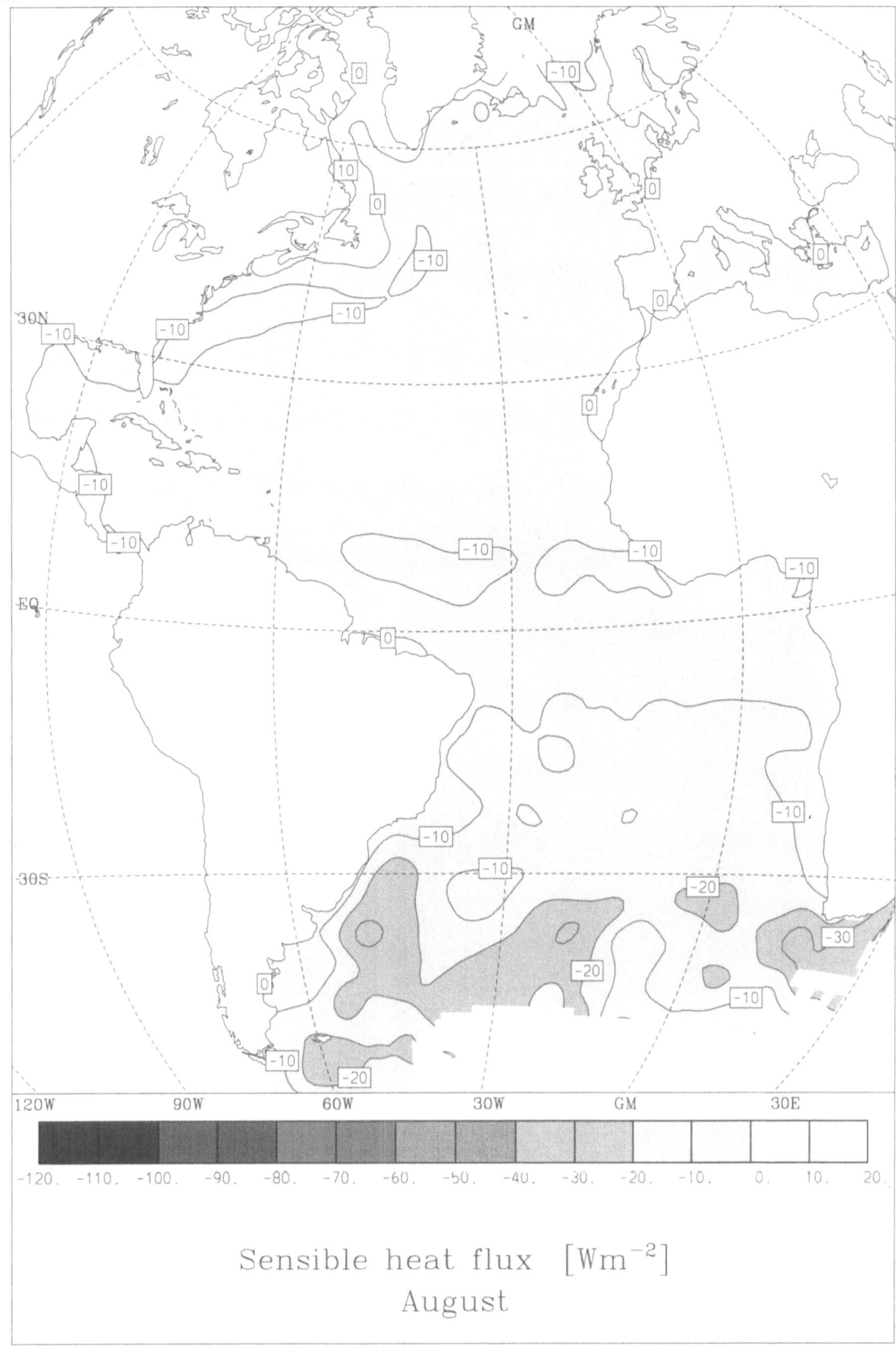

Sensible heat flux [Wm^{-2}]
August

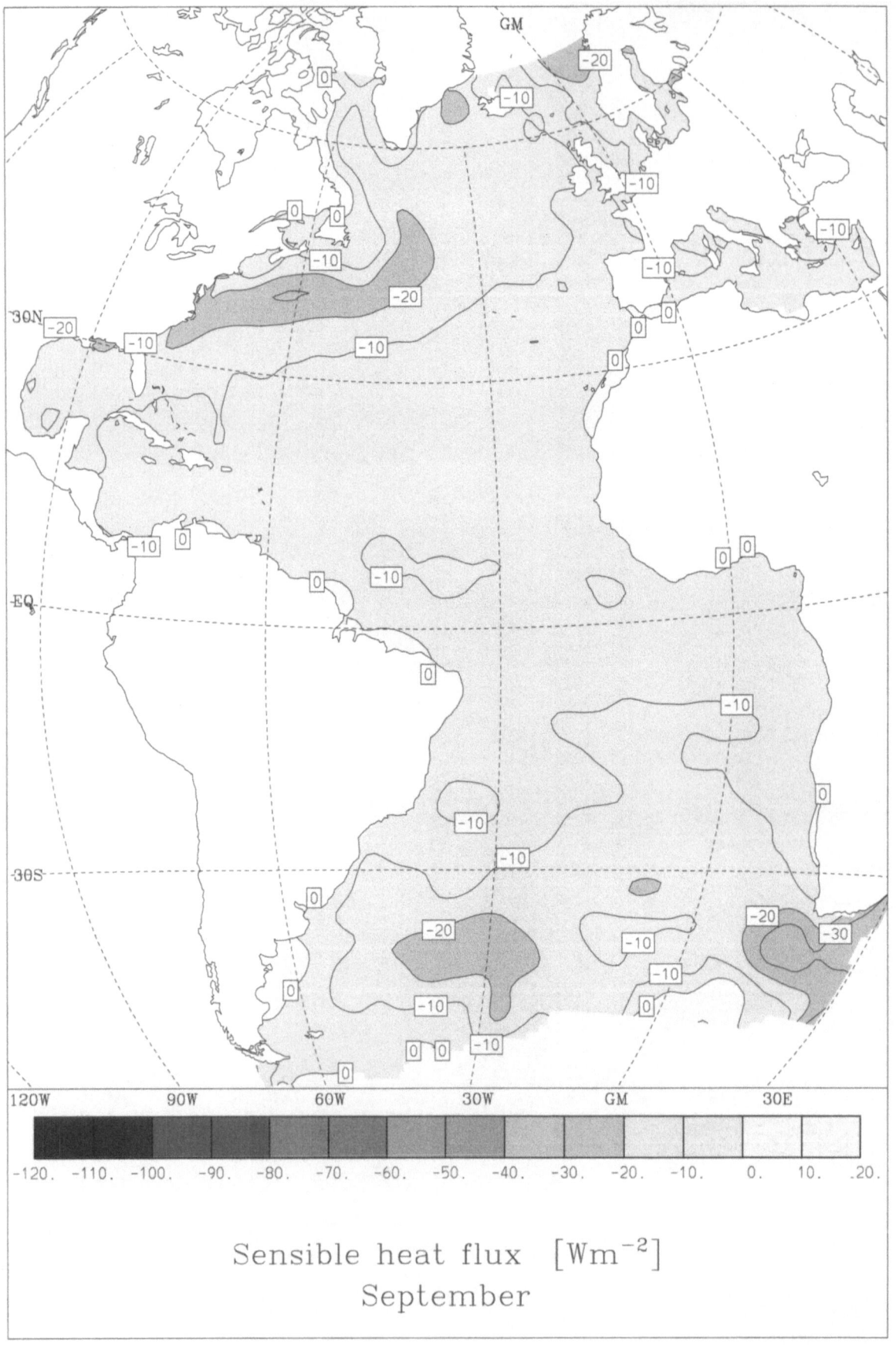

Sensible heat flux $[Wm^{-2}]$
September

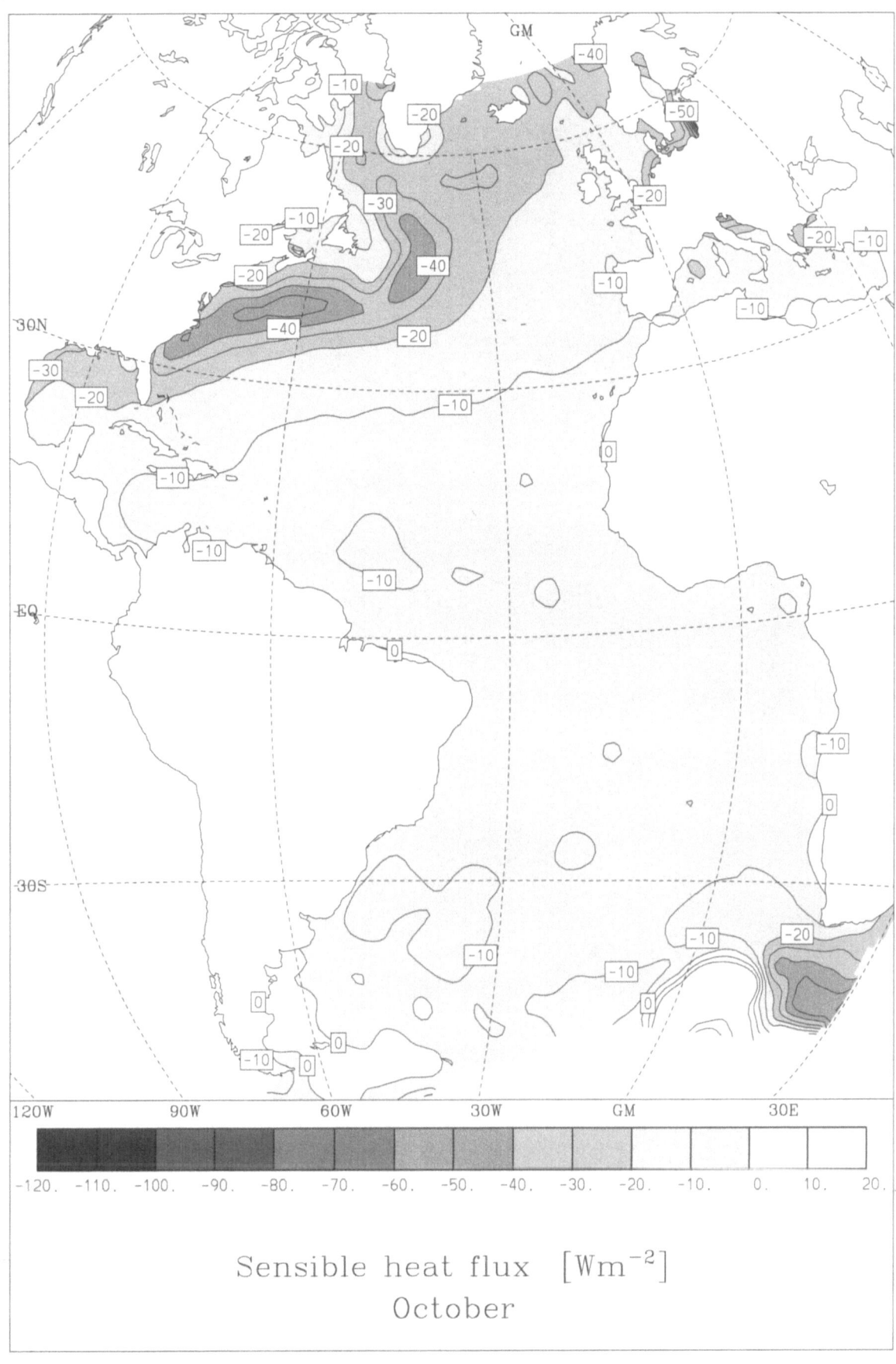

Sensible heat flux $[Wm^{-2}]$

October

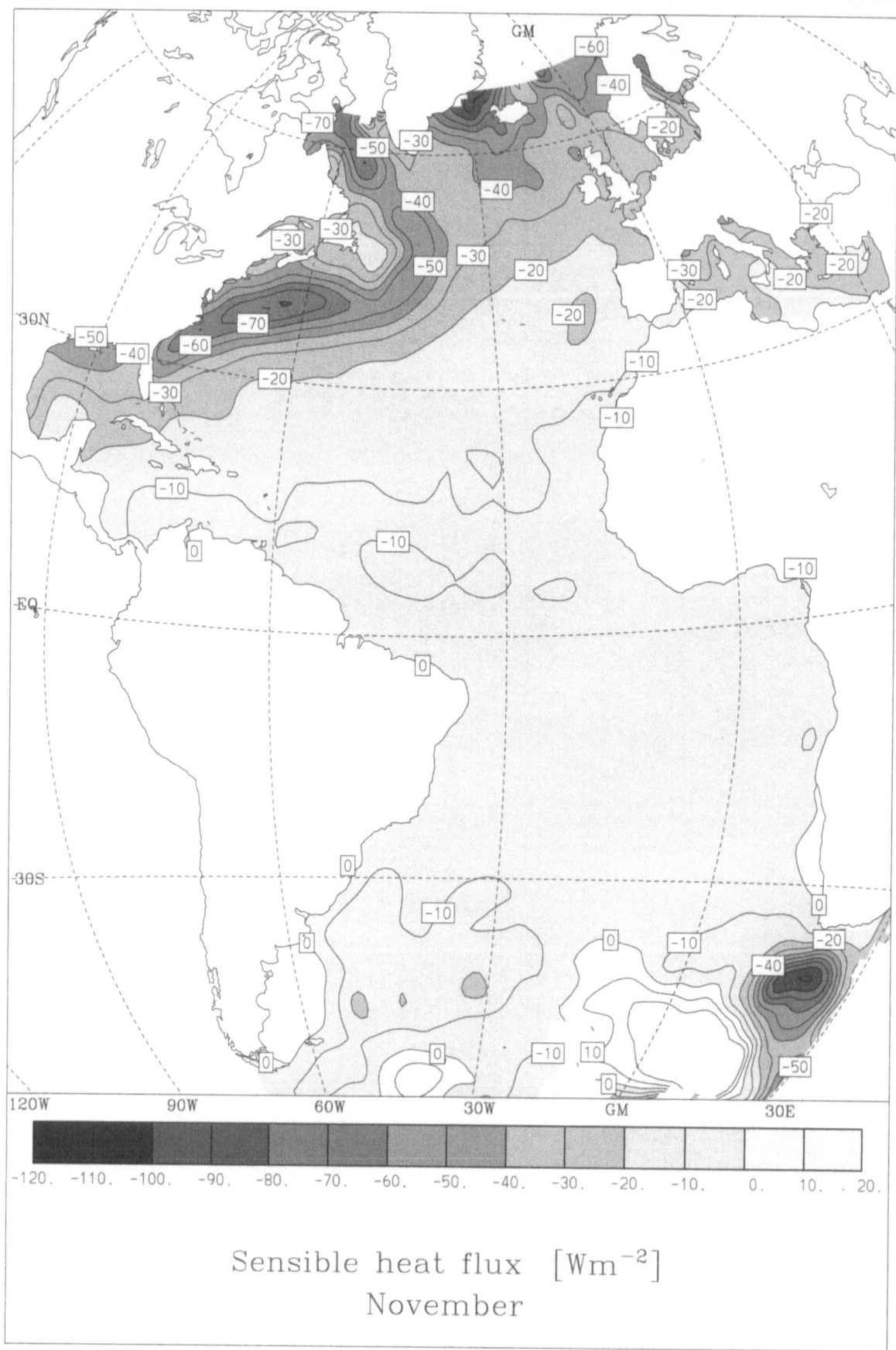

Sensible heat flux [Wm^{-2}]
November

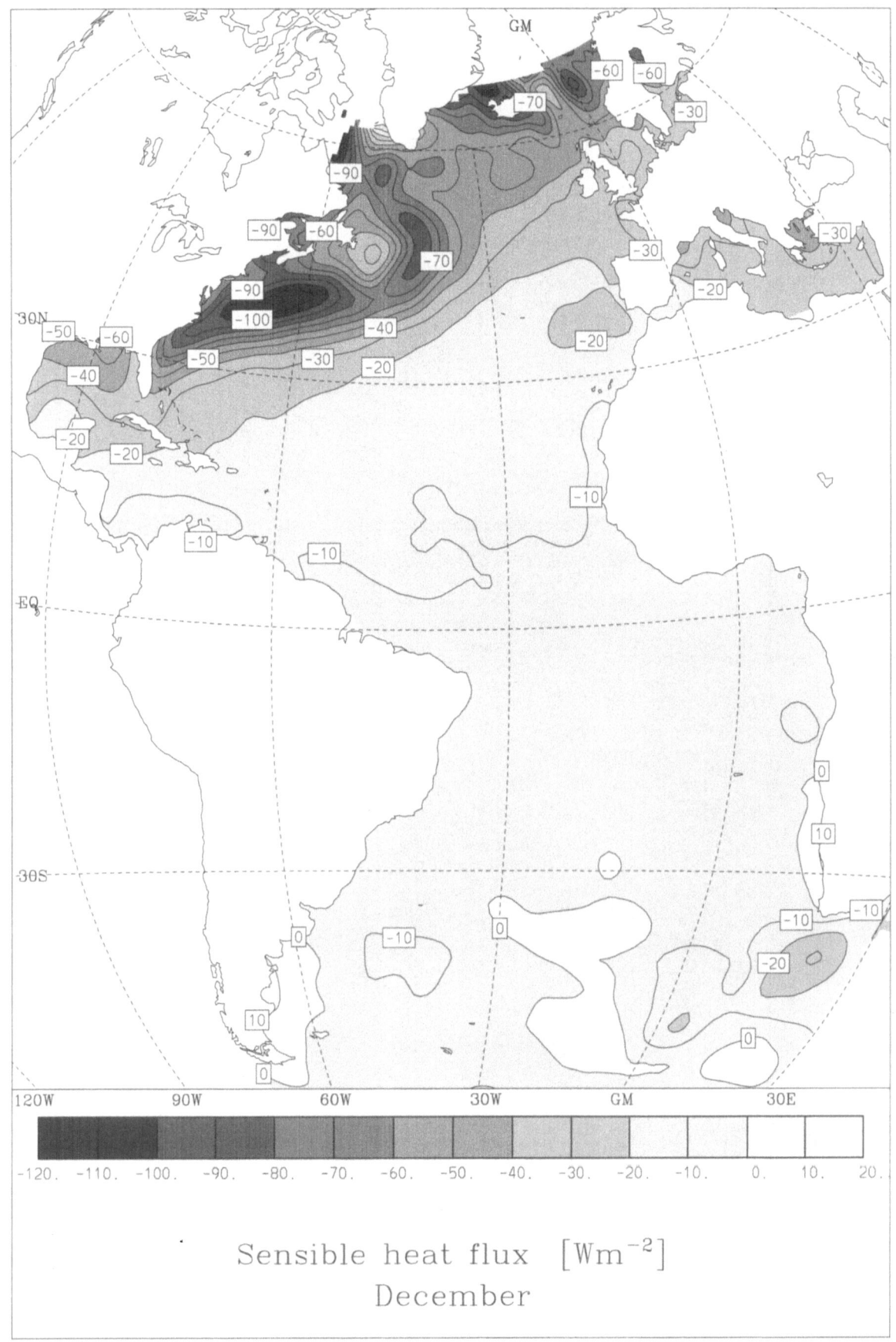

Sensible heat flux [Wm^{-2}]
December

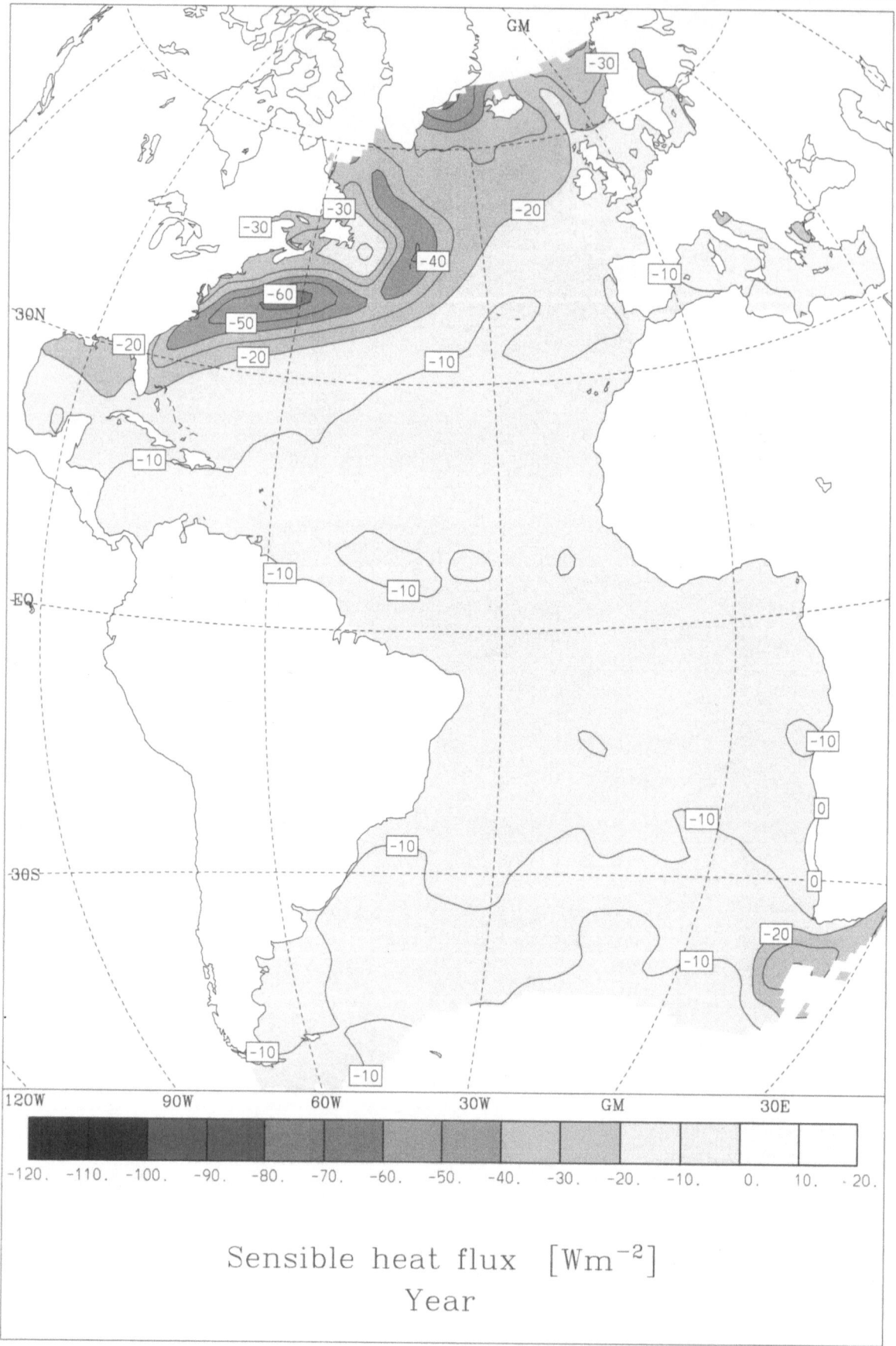

Sensible heat flux [Wm^{-2}]
Year

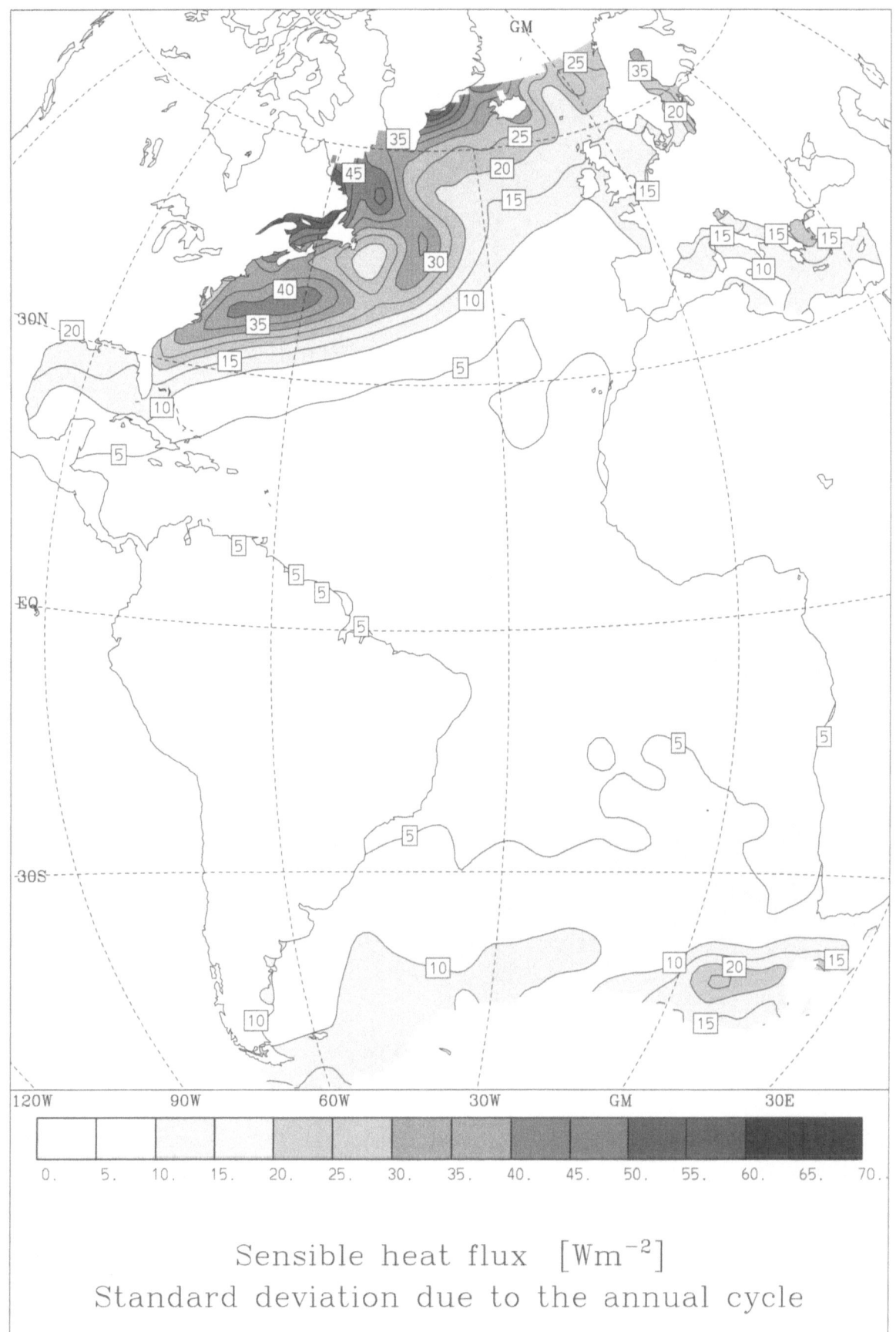

Sensible heat flux [Wm^{-2}]
Standard deviation due to the annual cycle

Total Net Air-Sea Heat Flux

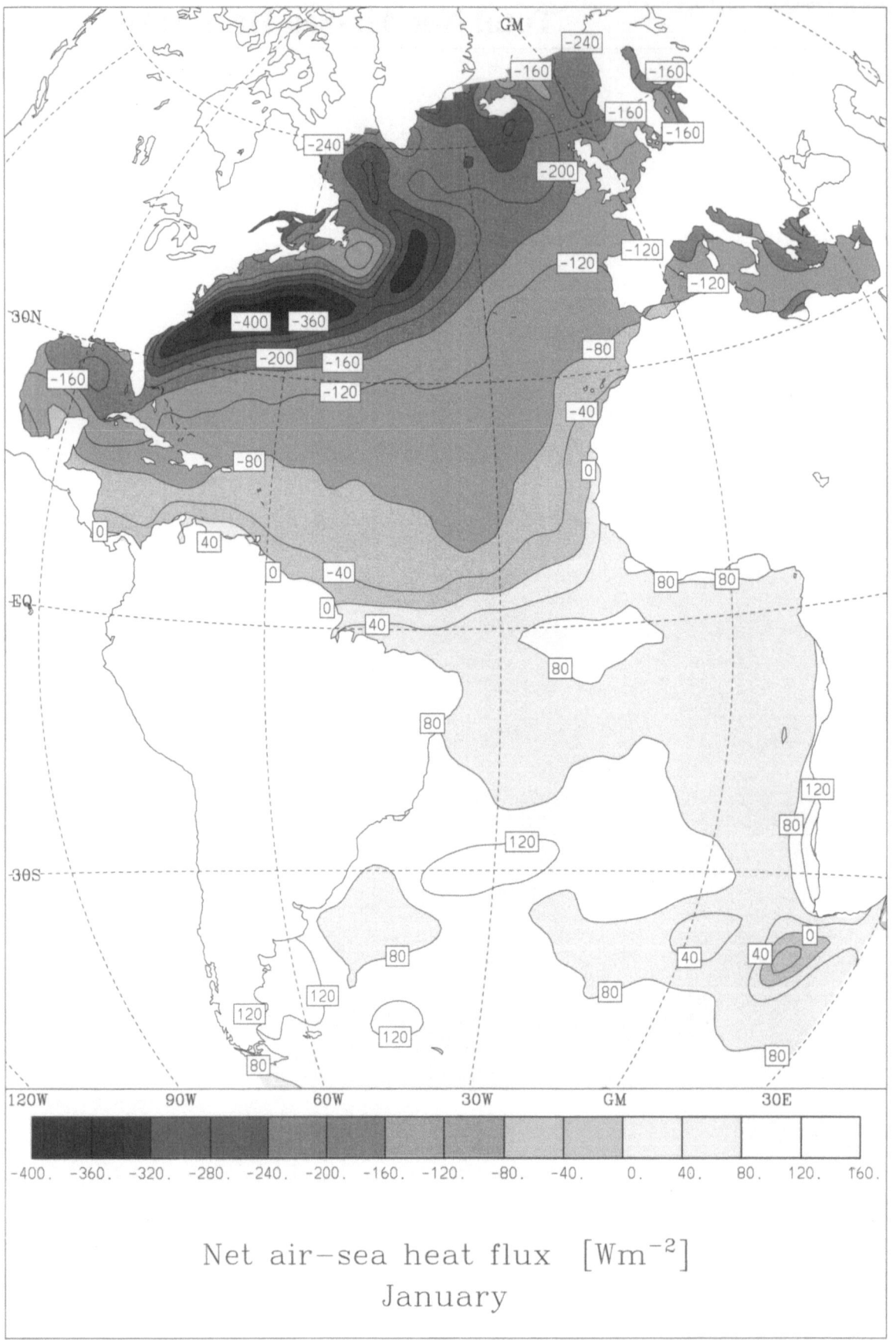

Net air-sea heat flux [Wm^{-2}]
January

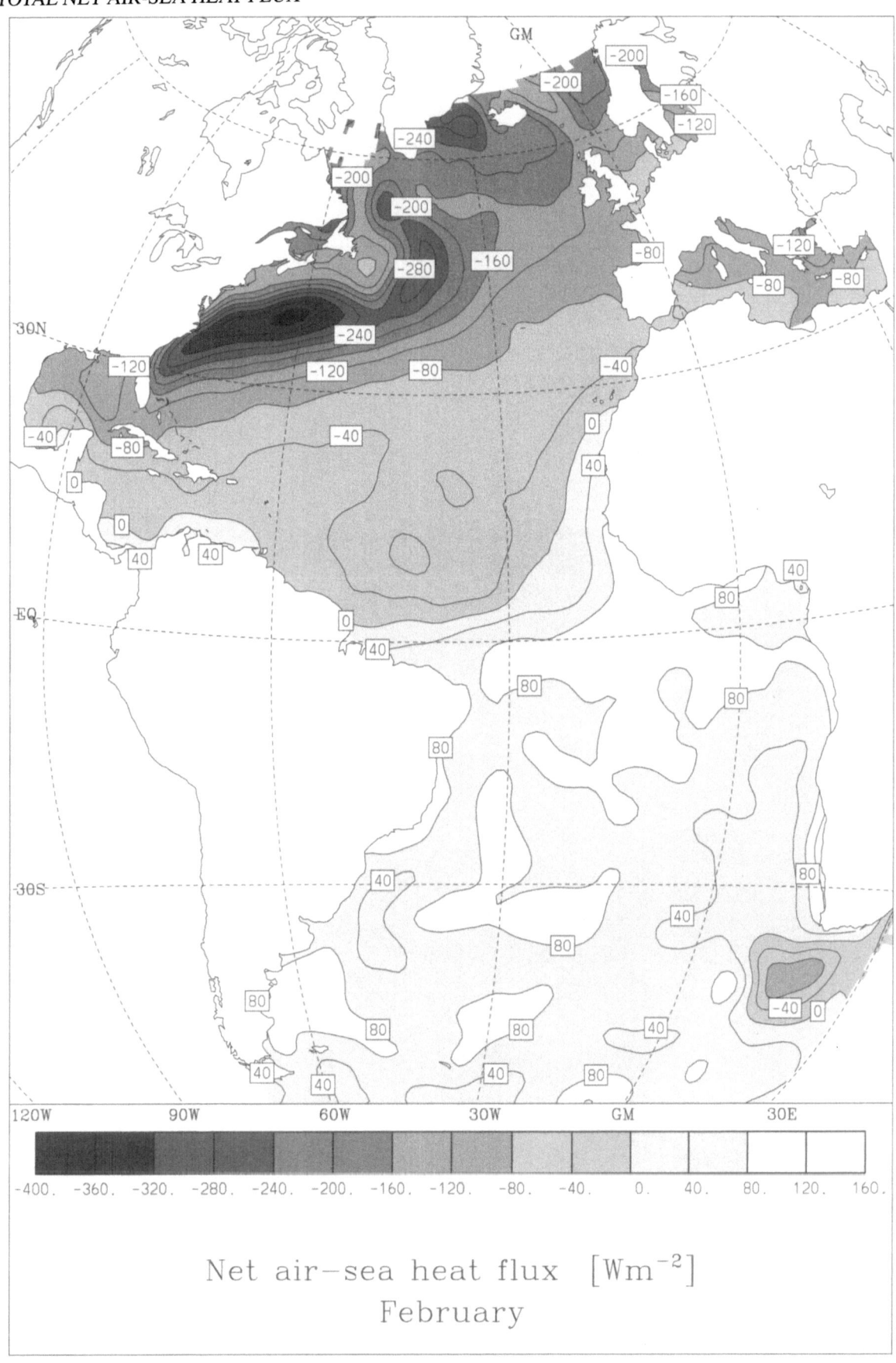

Net air-sea heat flux [Wm^{-2}]
February

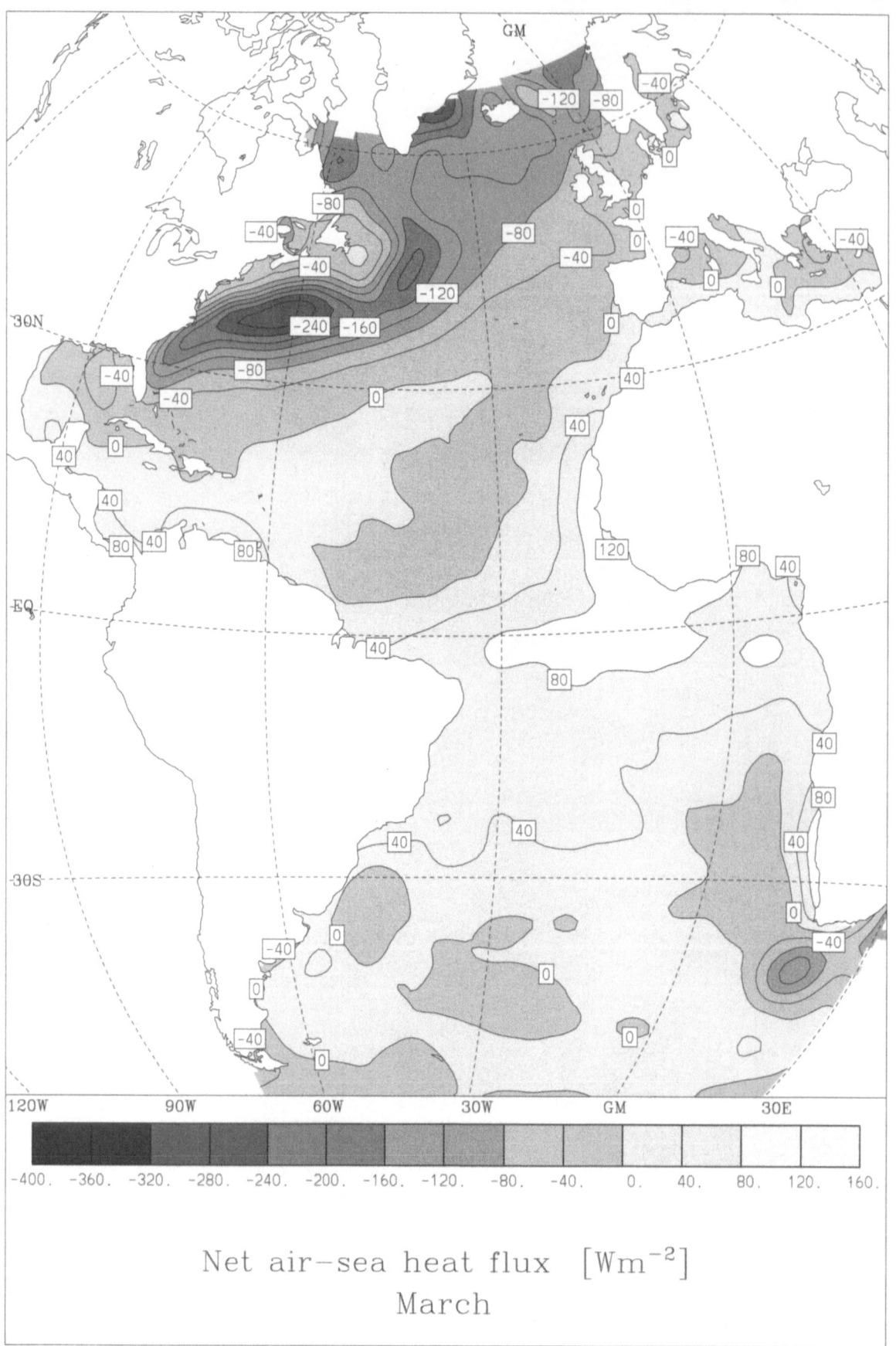

Net air–sea heat flux [Wm^{-2}]
March

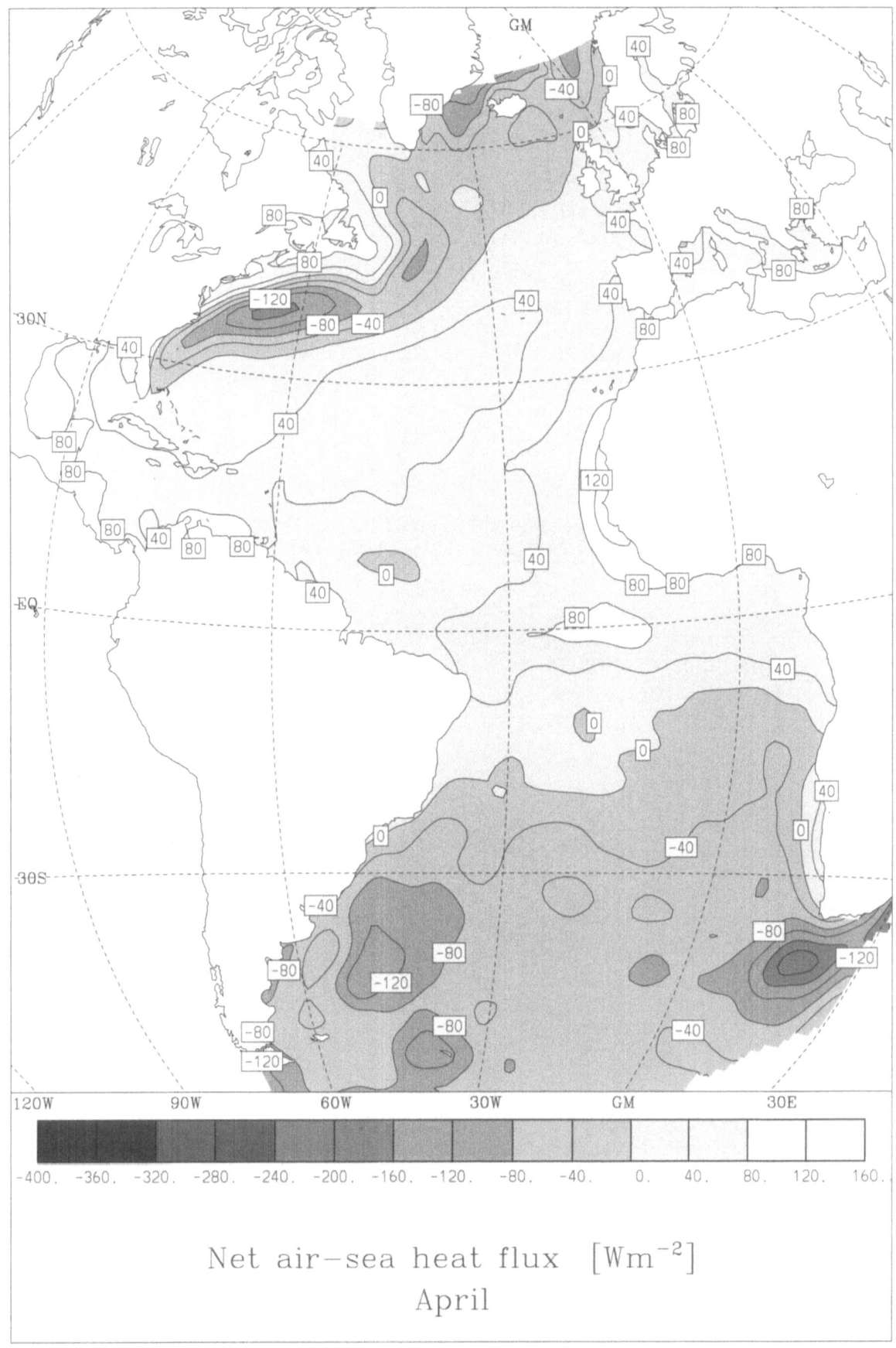

Net air−sea heat flux [Wm⁻²]
April

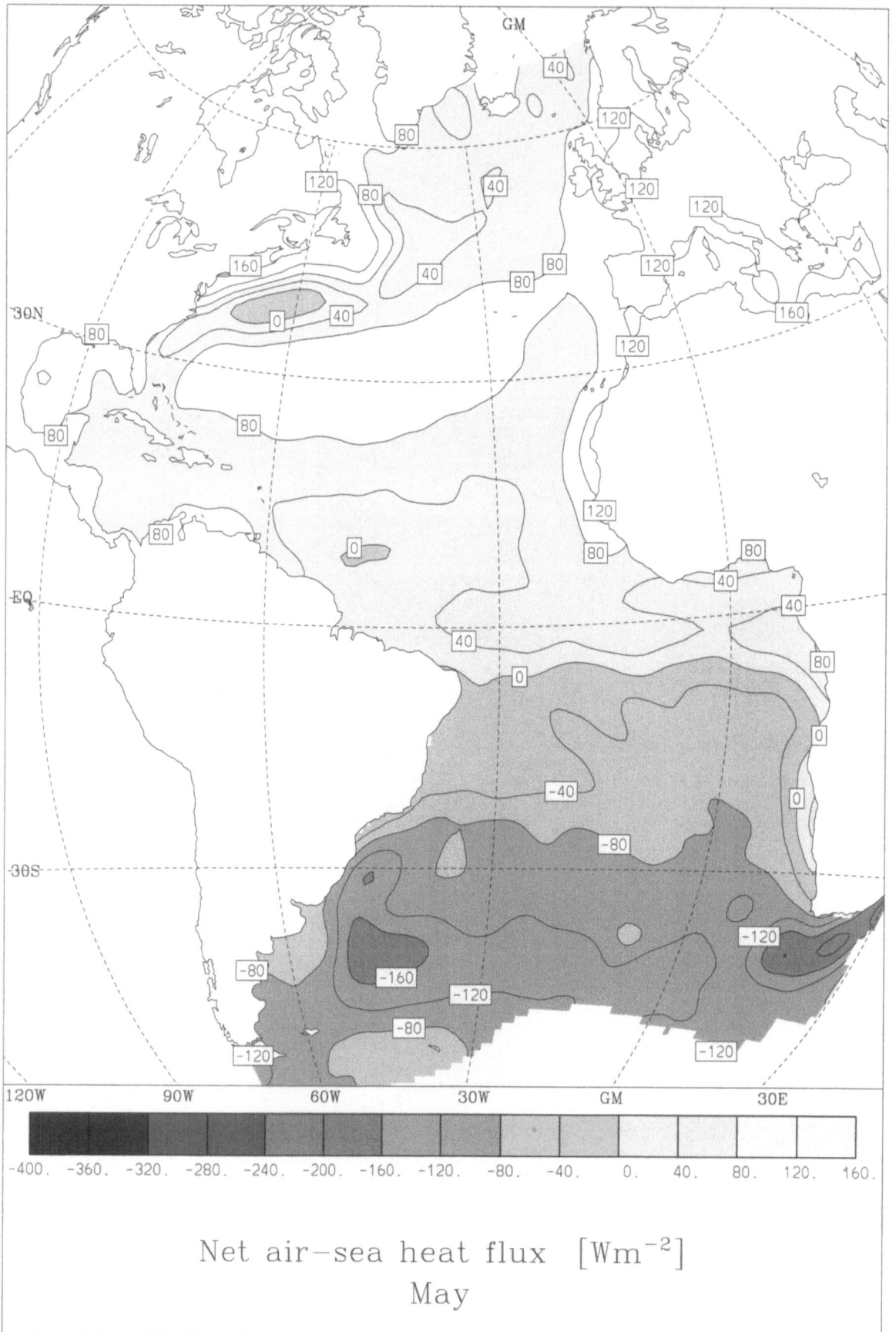

Net air−sea heat flux [Wm⁻²]
May

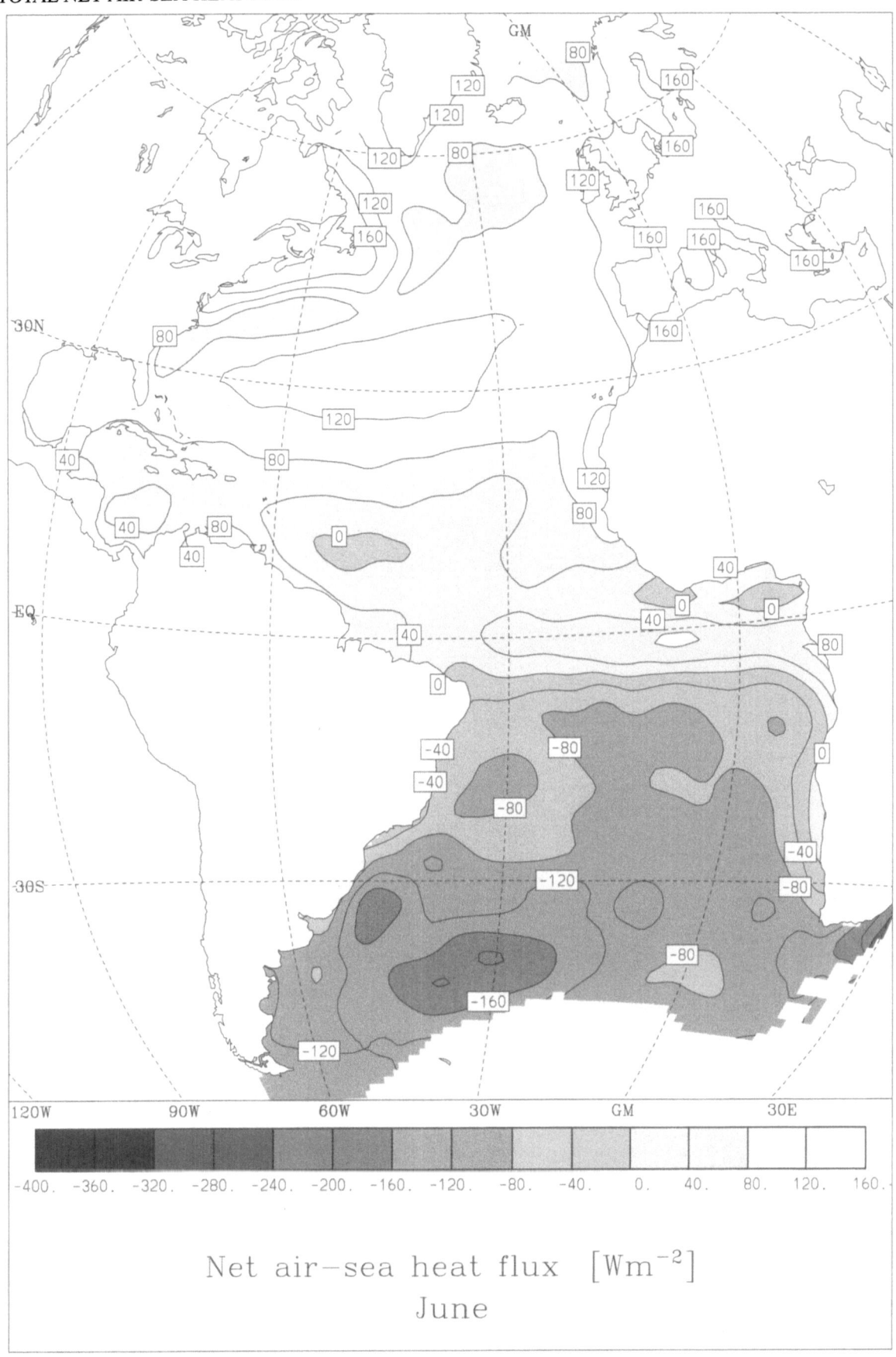

Net air−sea heat flux [Wm⁻²]

June

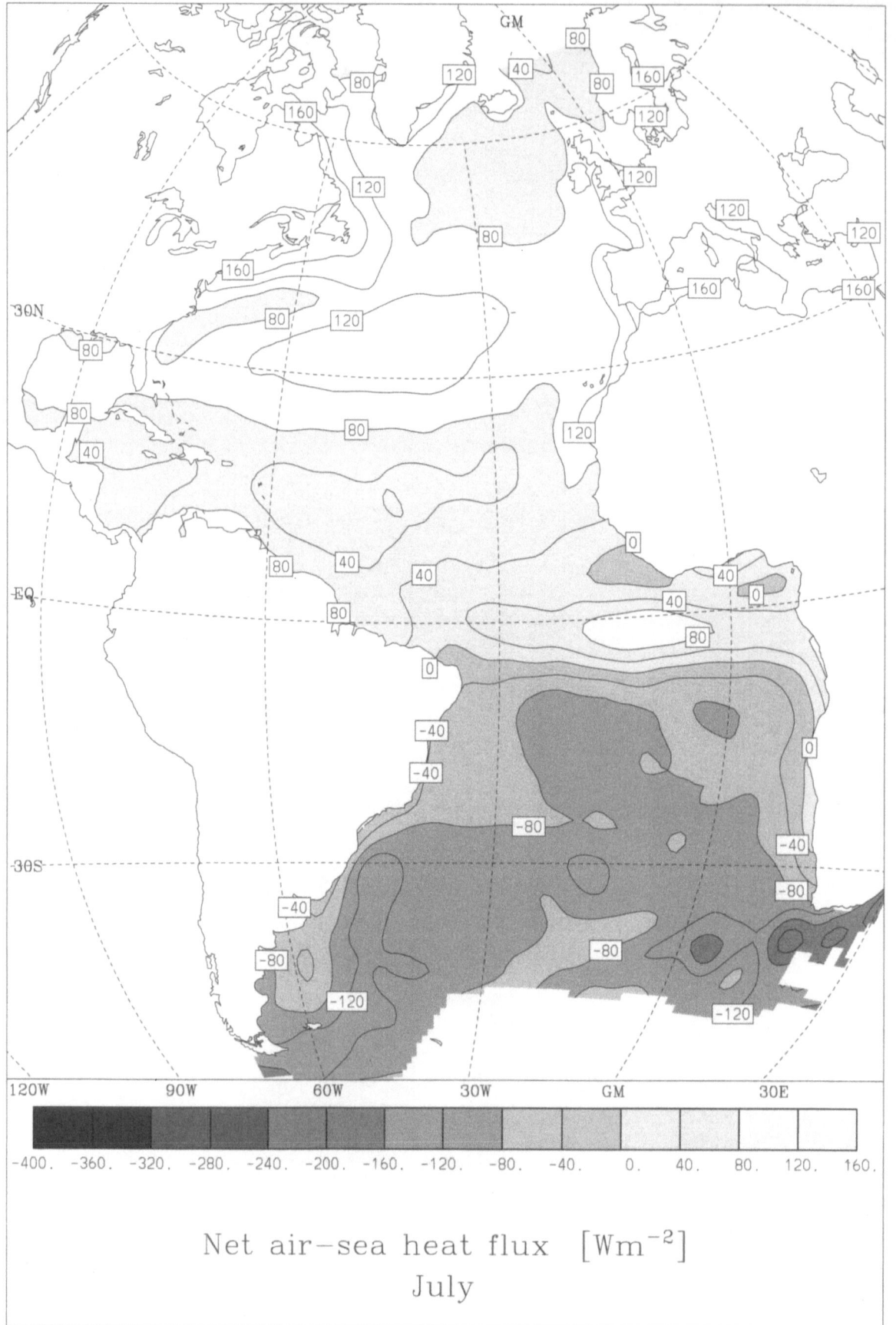

Net air-sea heat flux $[Wm^{-2}]$
July

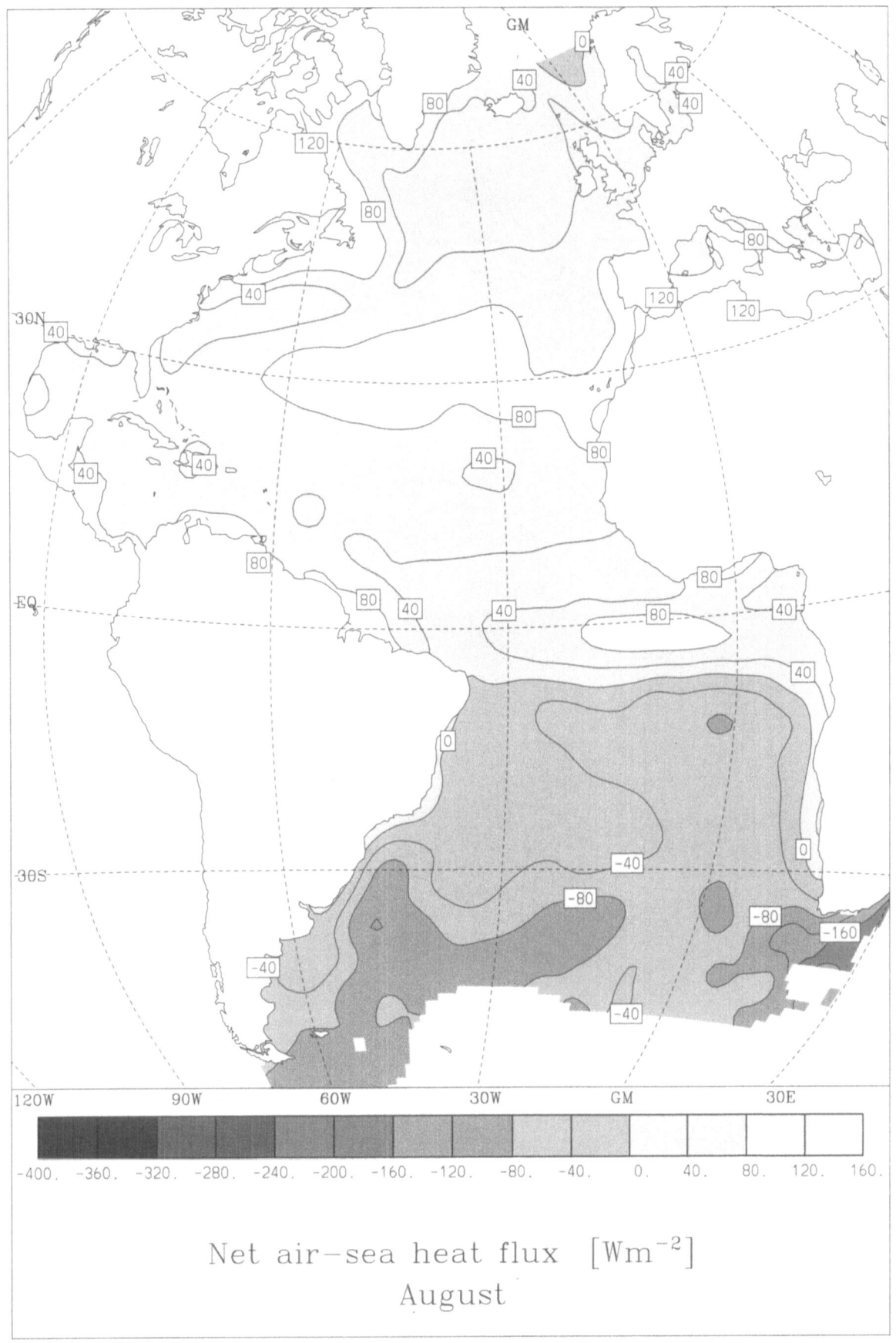

Net air-sea heat flux [Wm^{-2}]
August

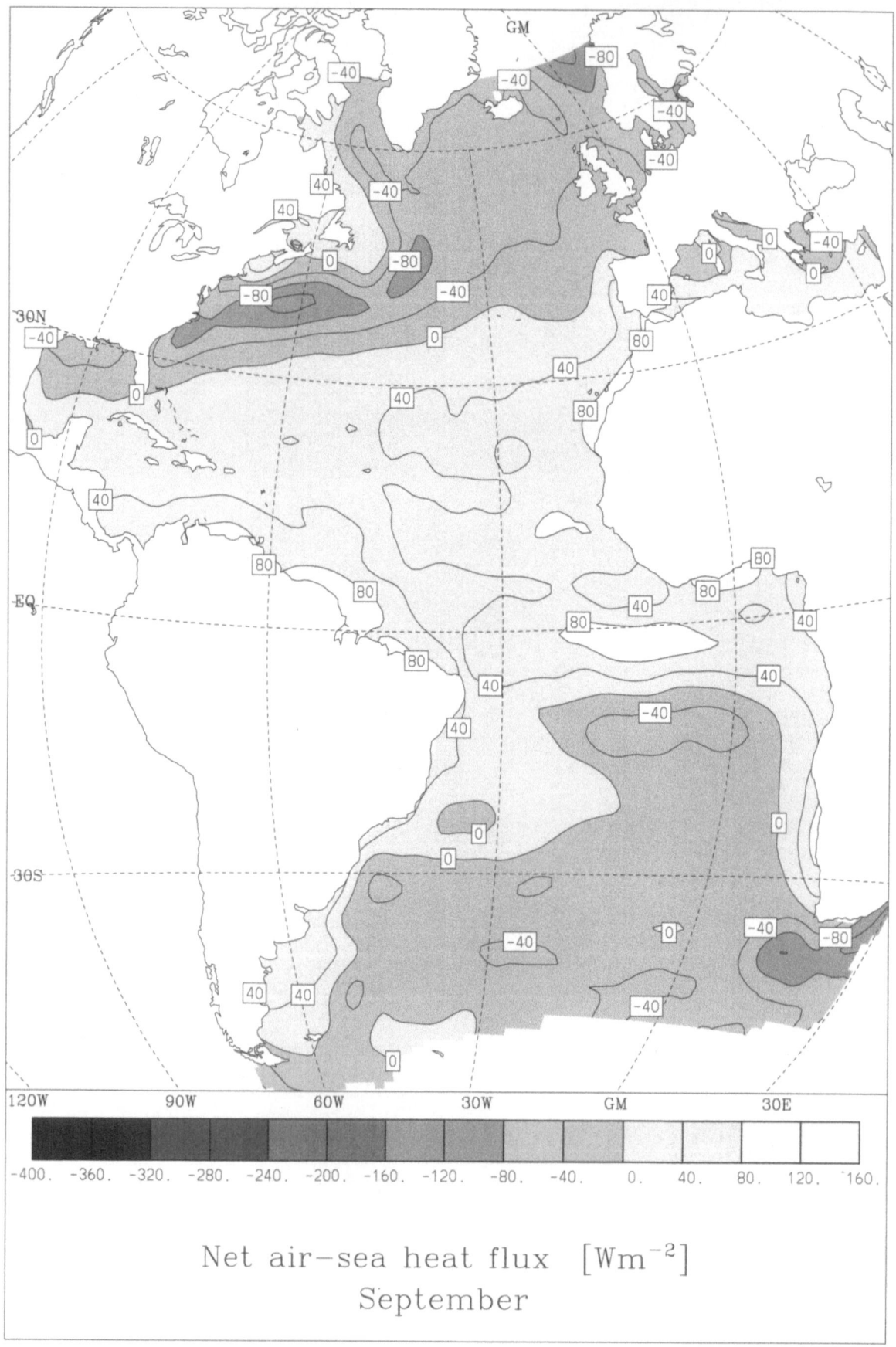

Net air−sea heat flux [Wm^{-2}]
September

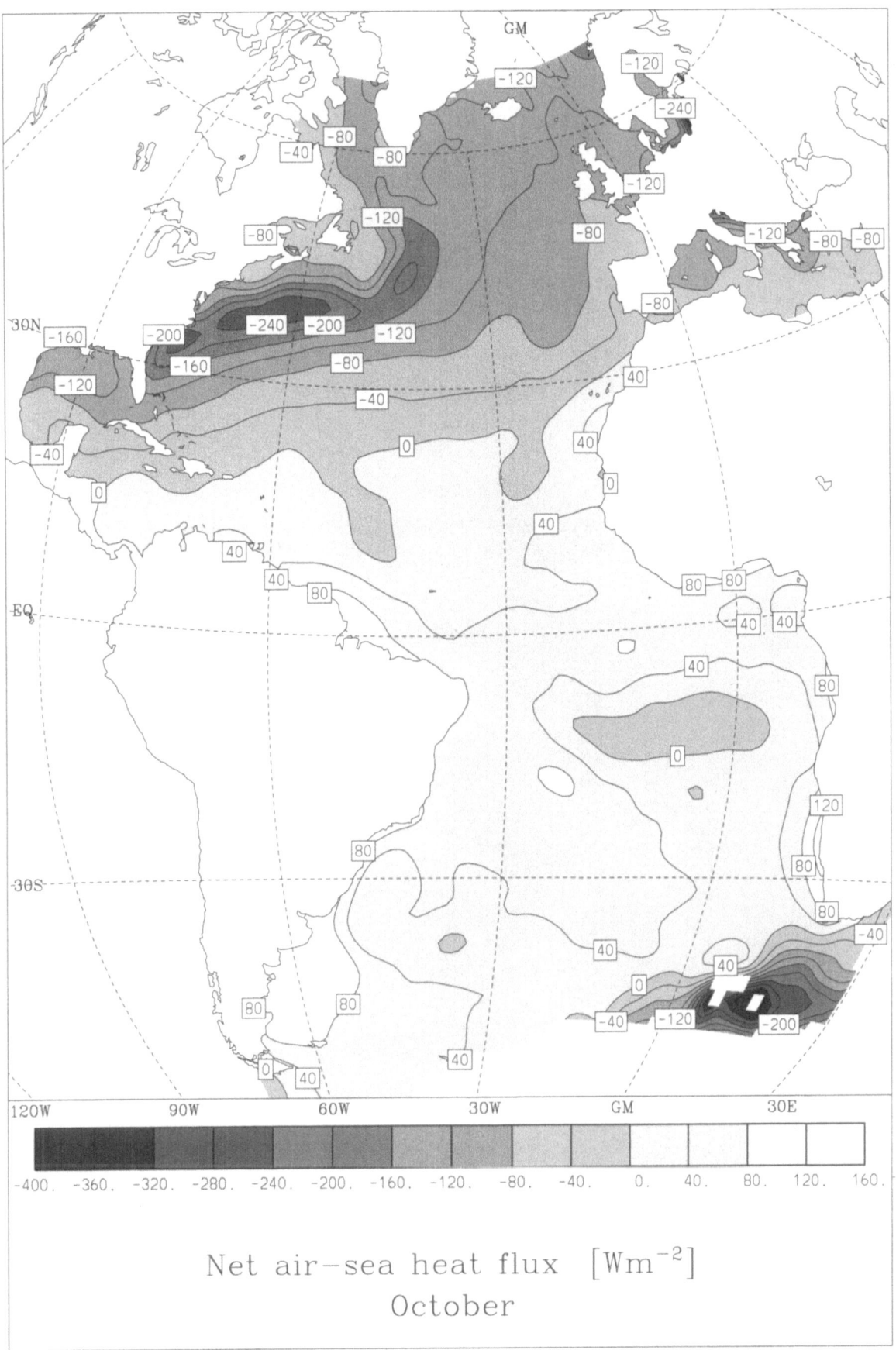

Net air−sea heat flux [Wm^{-2}]
October

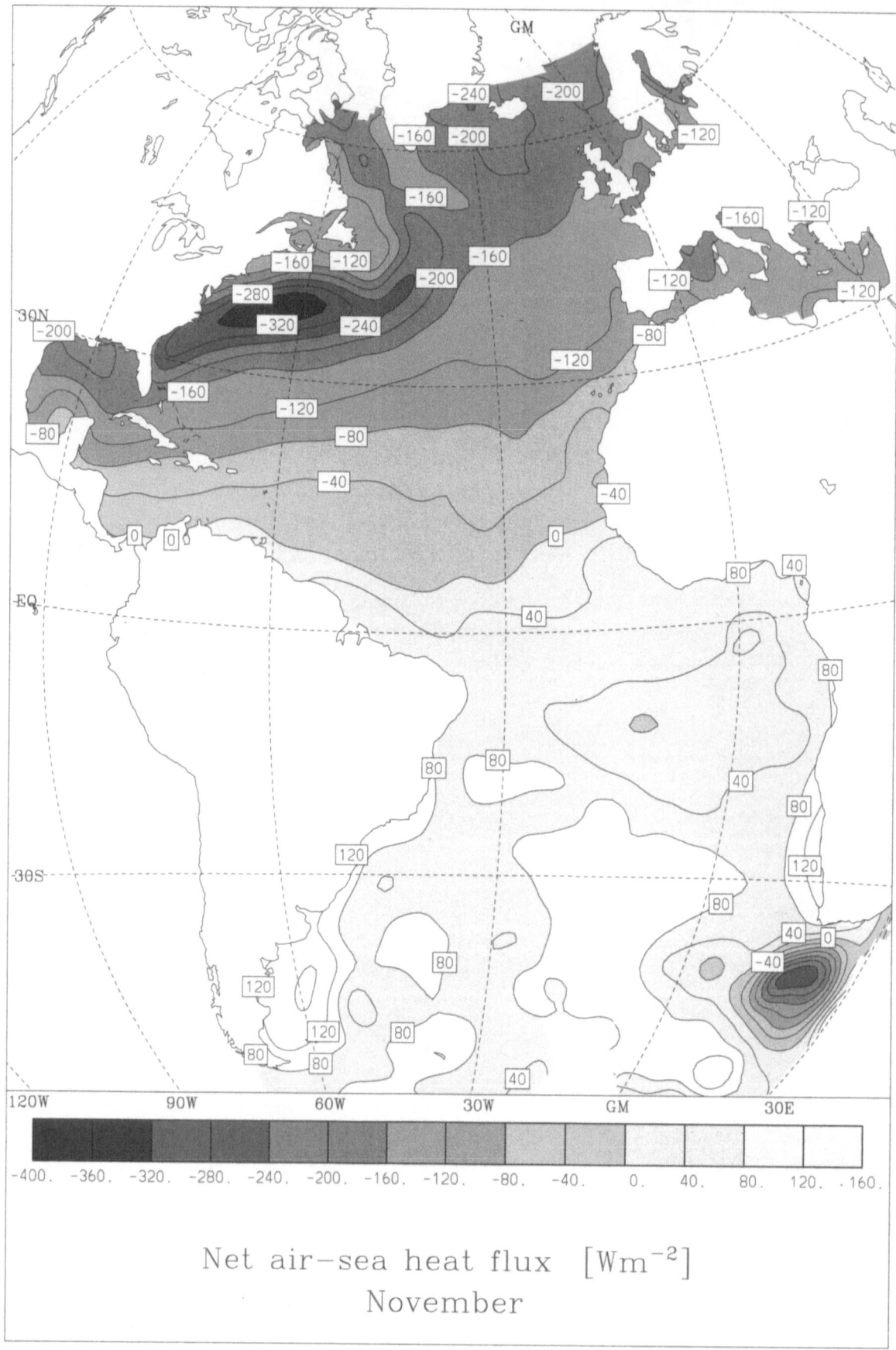

Net air-sea heat flux [Wm^{-2}]
November

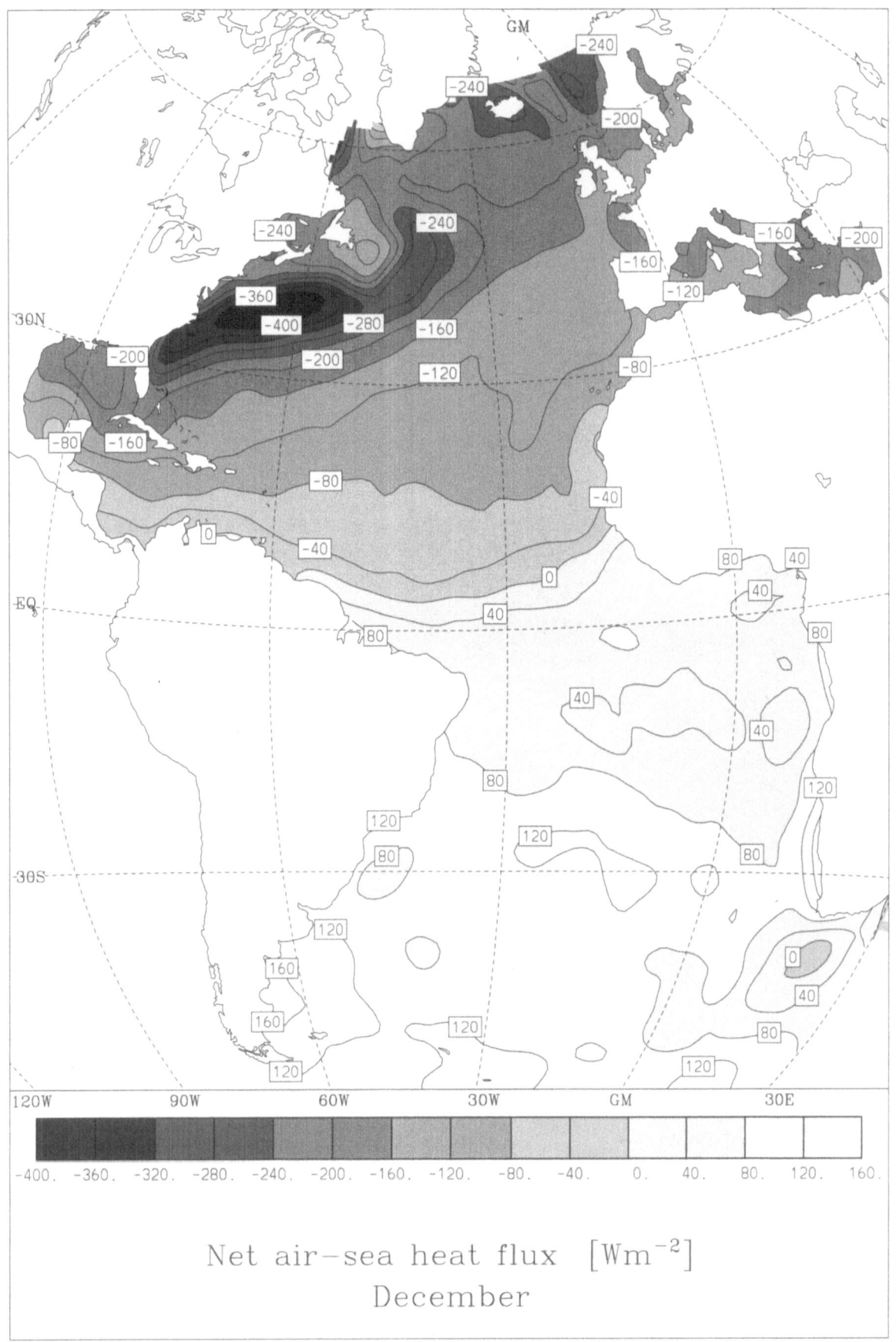

Net air−sea heat flux [Wm^{-2}]
December

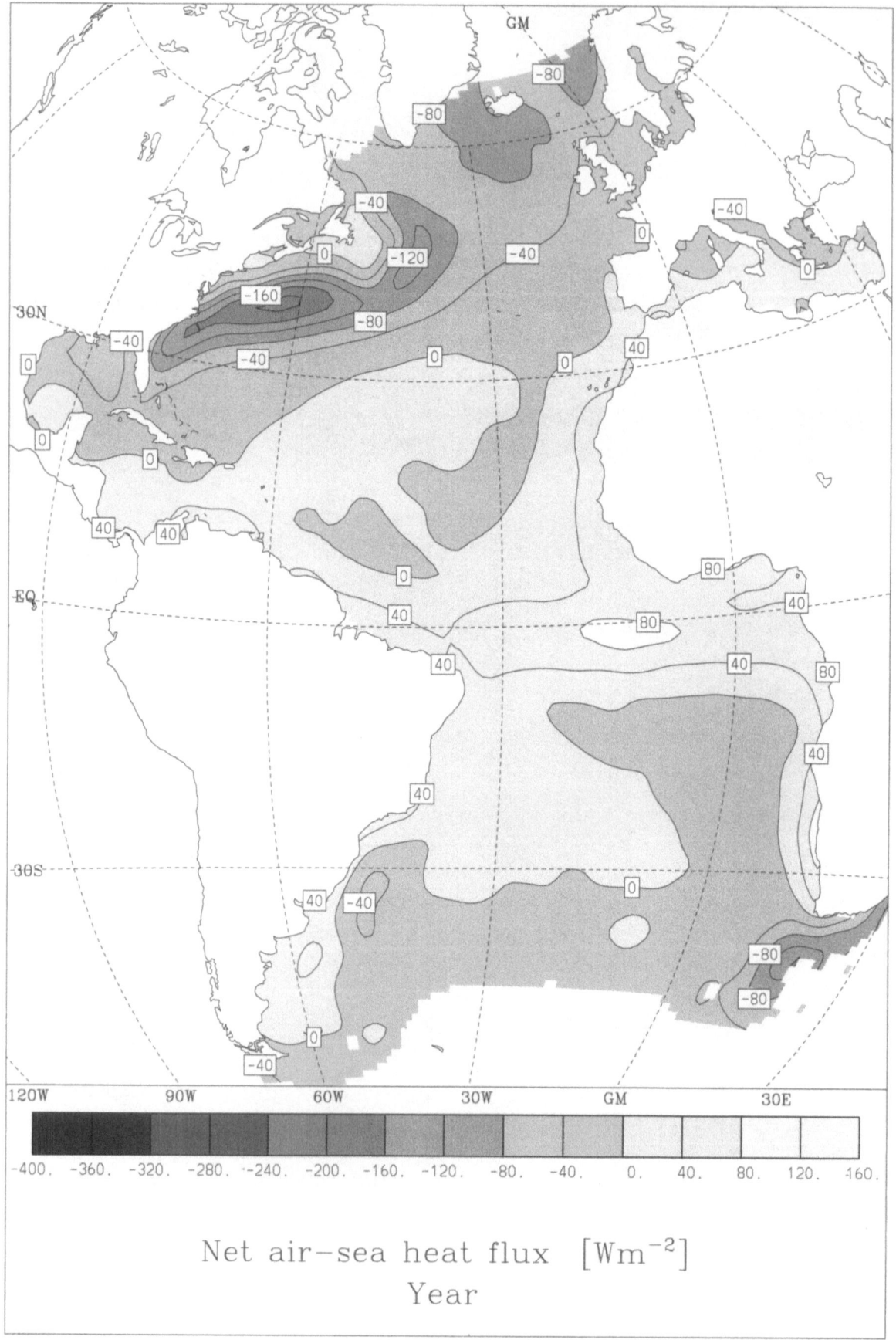

Net air−sea heat flux [Wm^{-2}]
Year

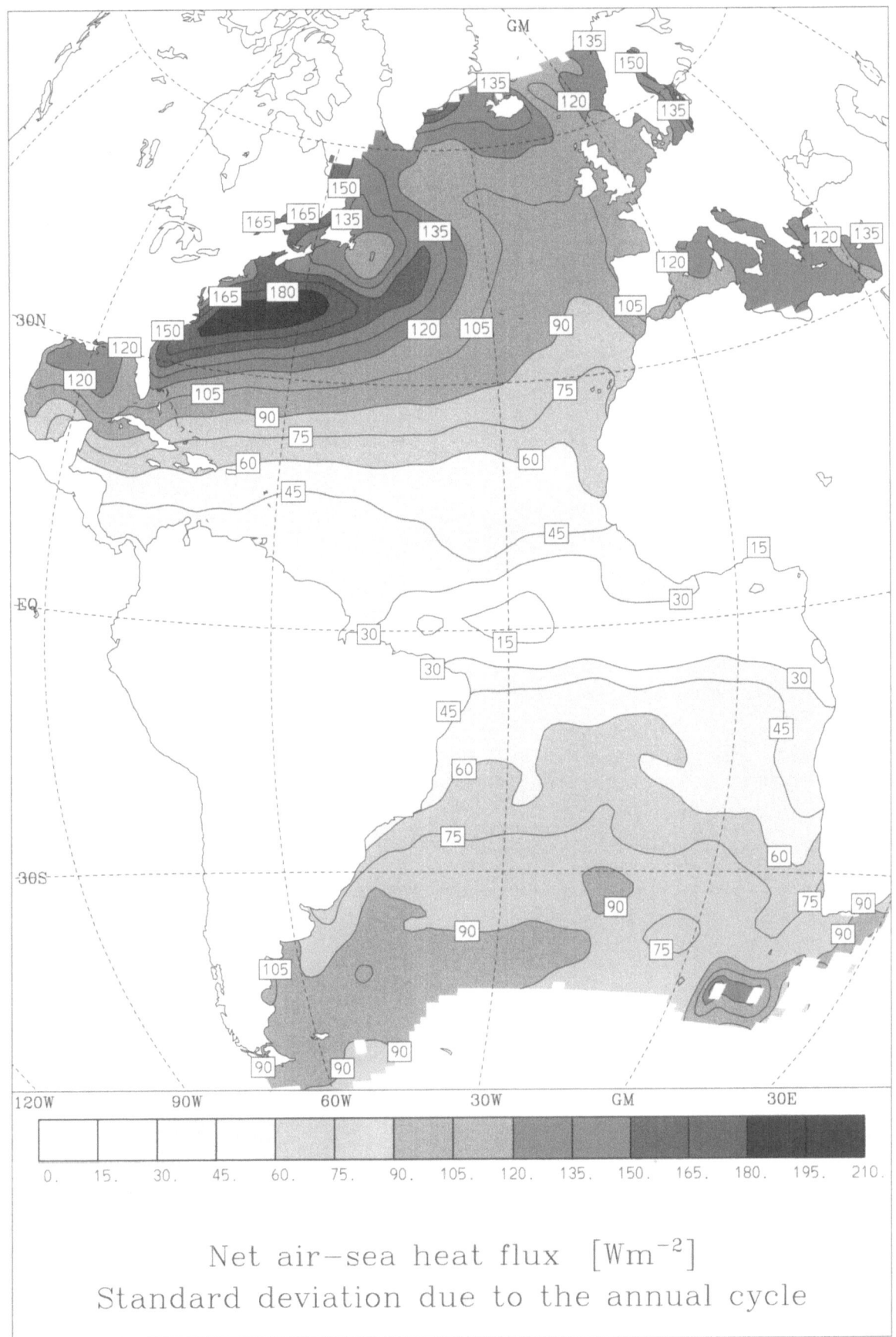

Net air−sea heat flux [Wm⁻²]
Standard deviation due to the annual cycle

Wind Stress

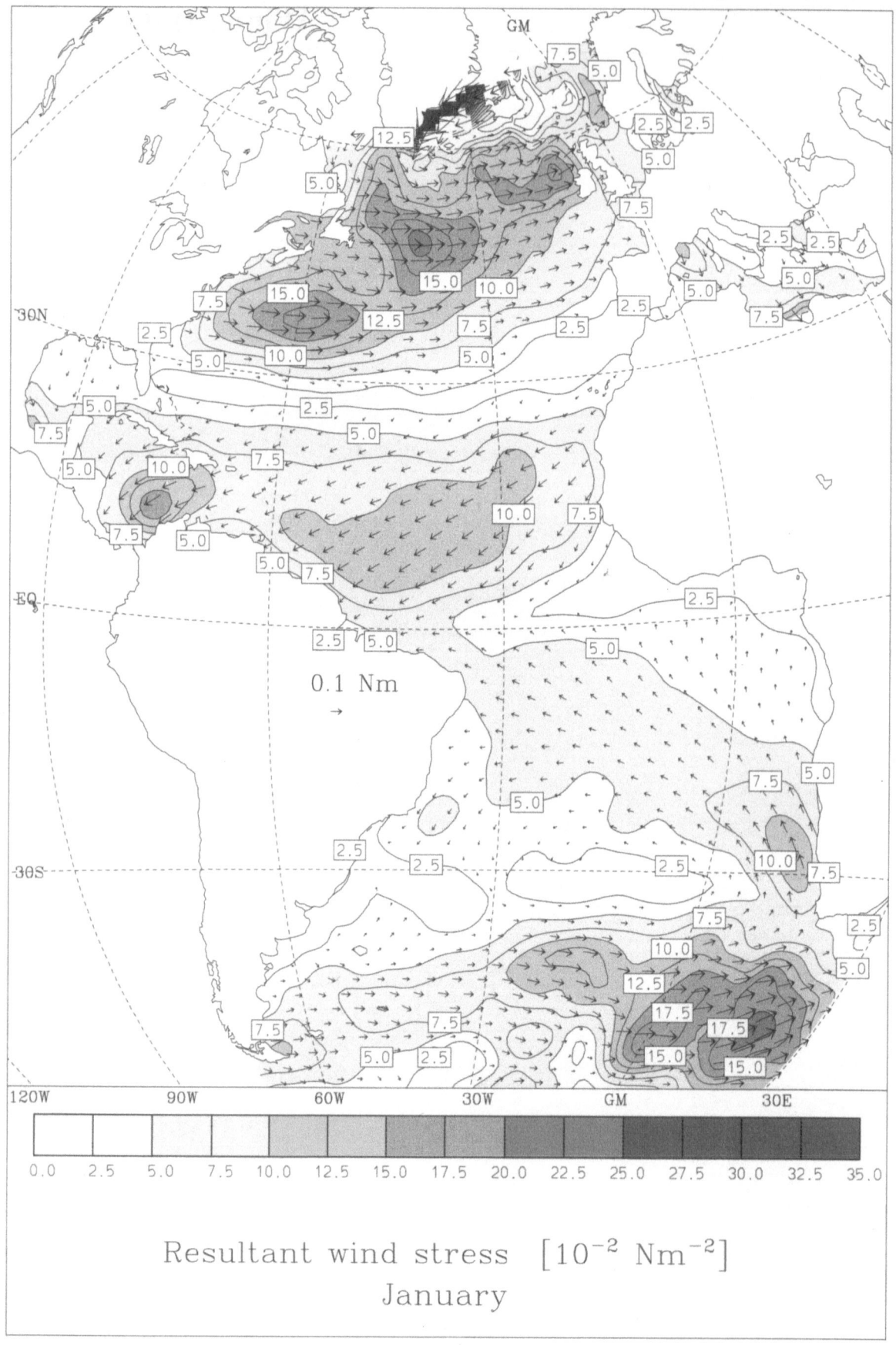

0.1 Nm

Resultant wind stress $[10^{-2} \, \mathrm{Nm}^{-2}]$
January

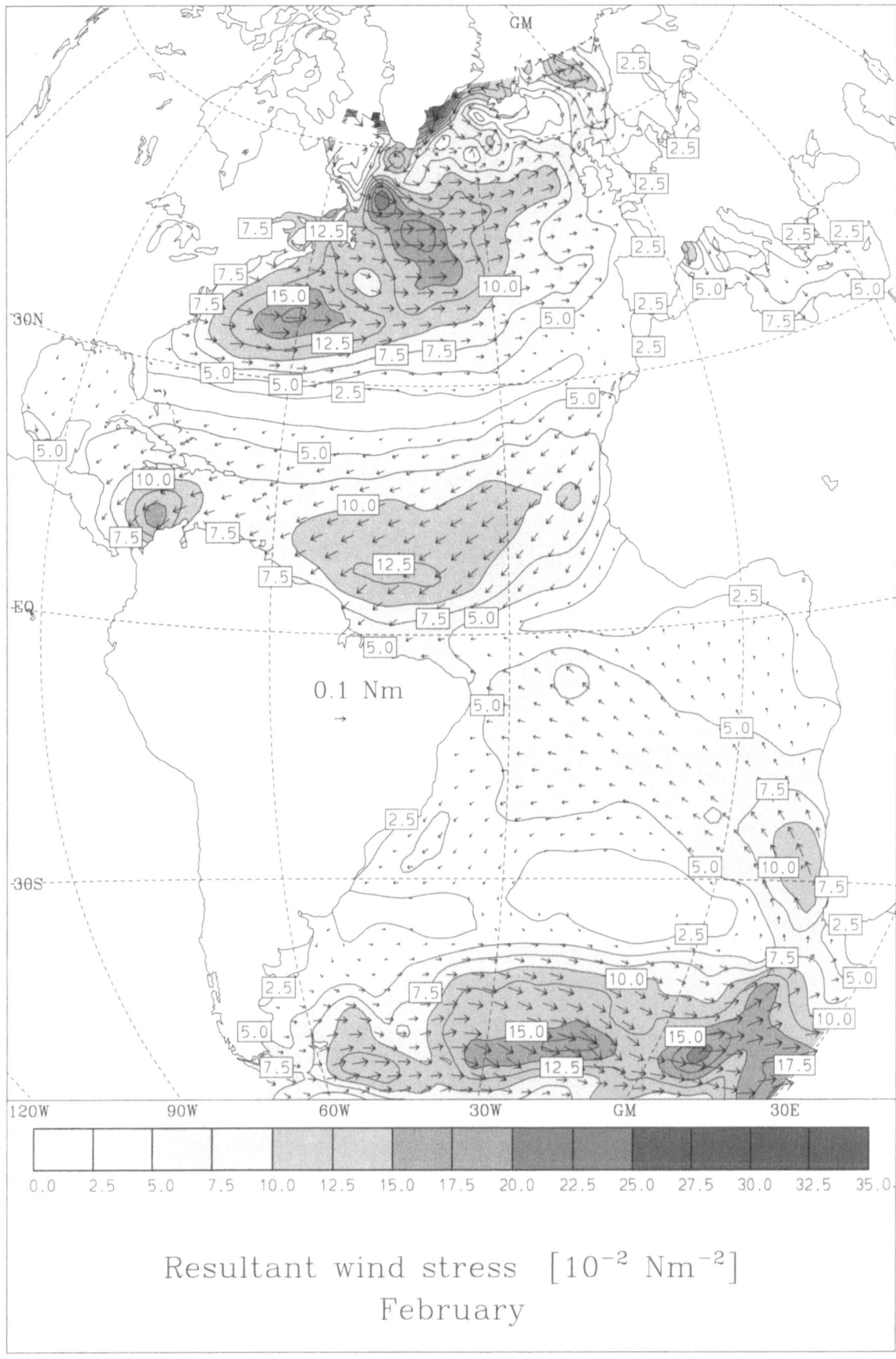

Resultant wind stress $[10^{-2} \ Nm^{-2}]$
February

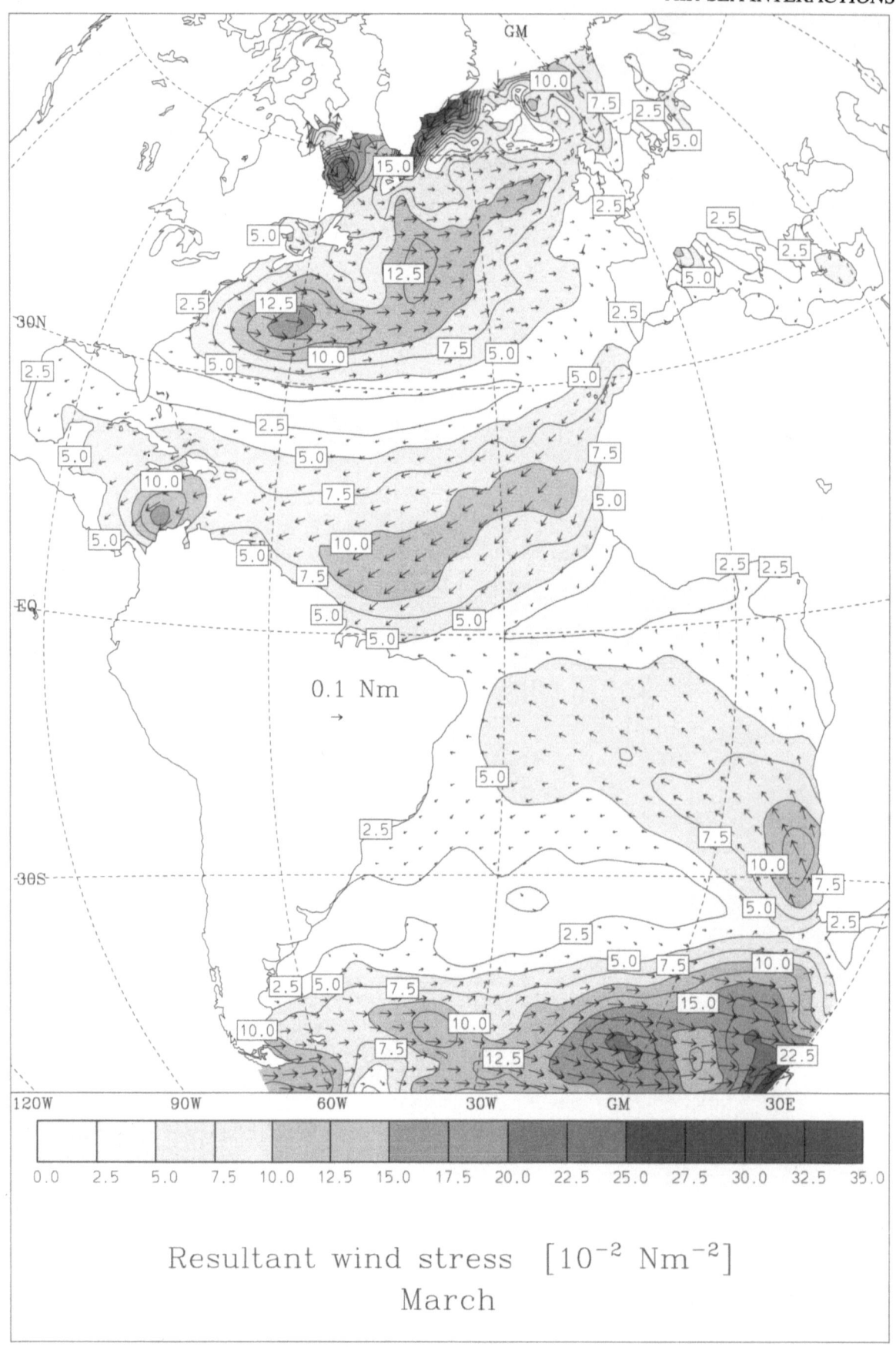

Resultant wind stress [10^{-2} Nm^{-2}]
March

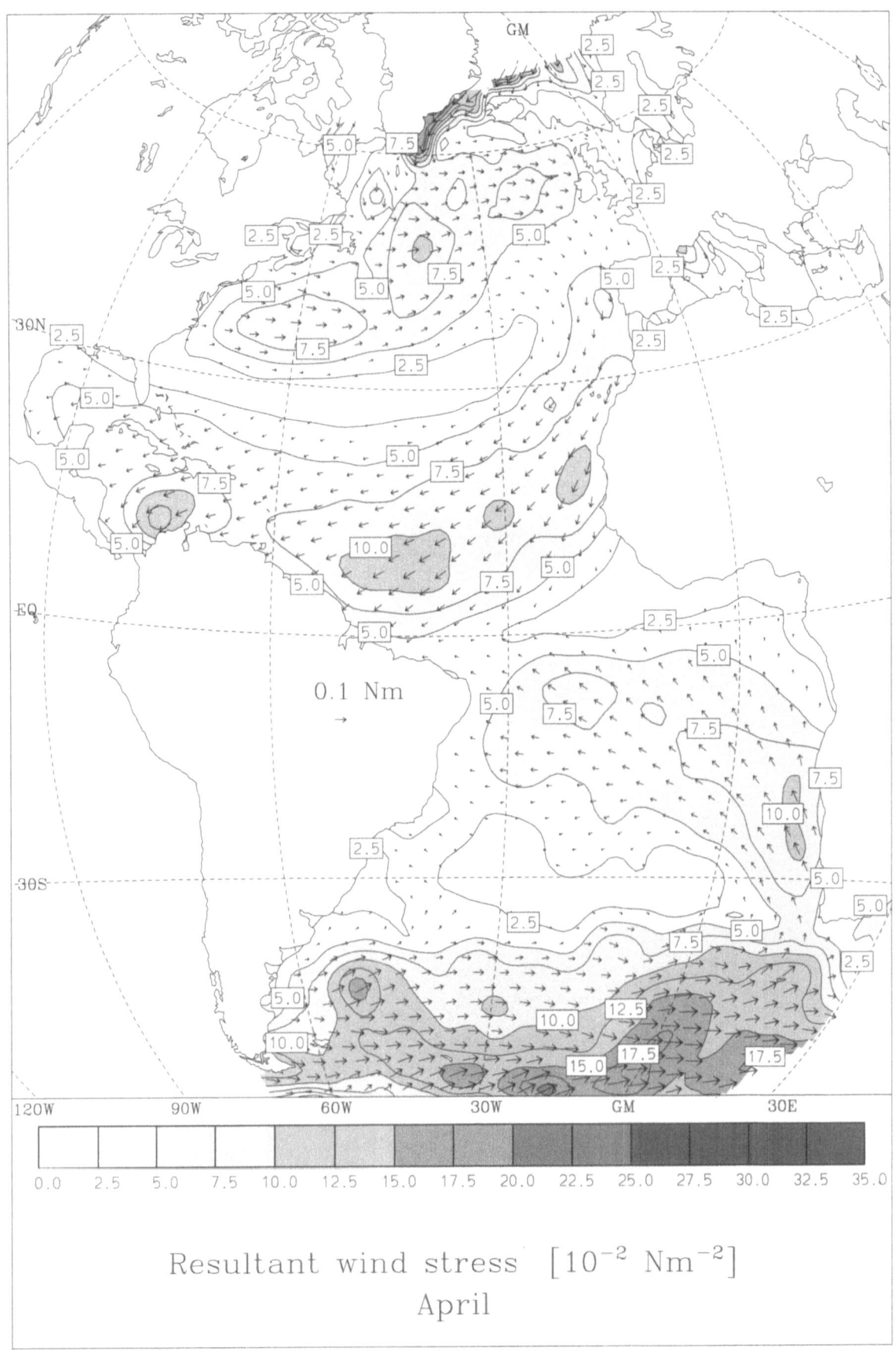

Resultant wind stress $[10^{-2} \ \mathrm{Nm^{-2}}]$
April

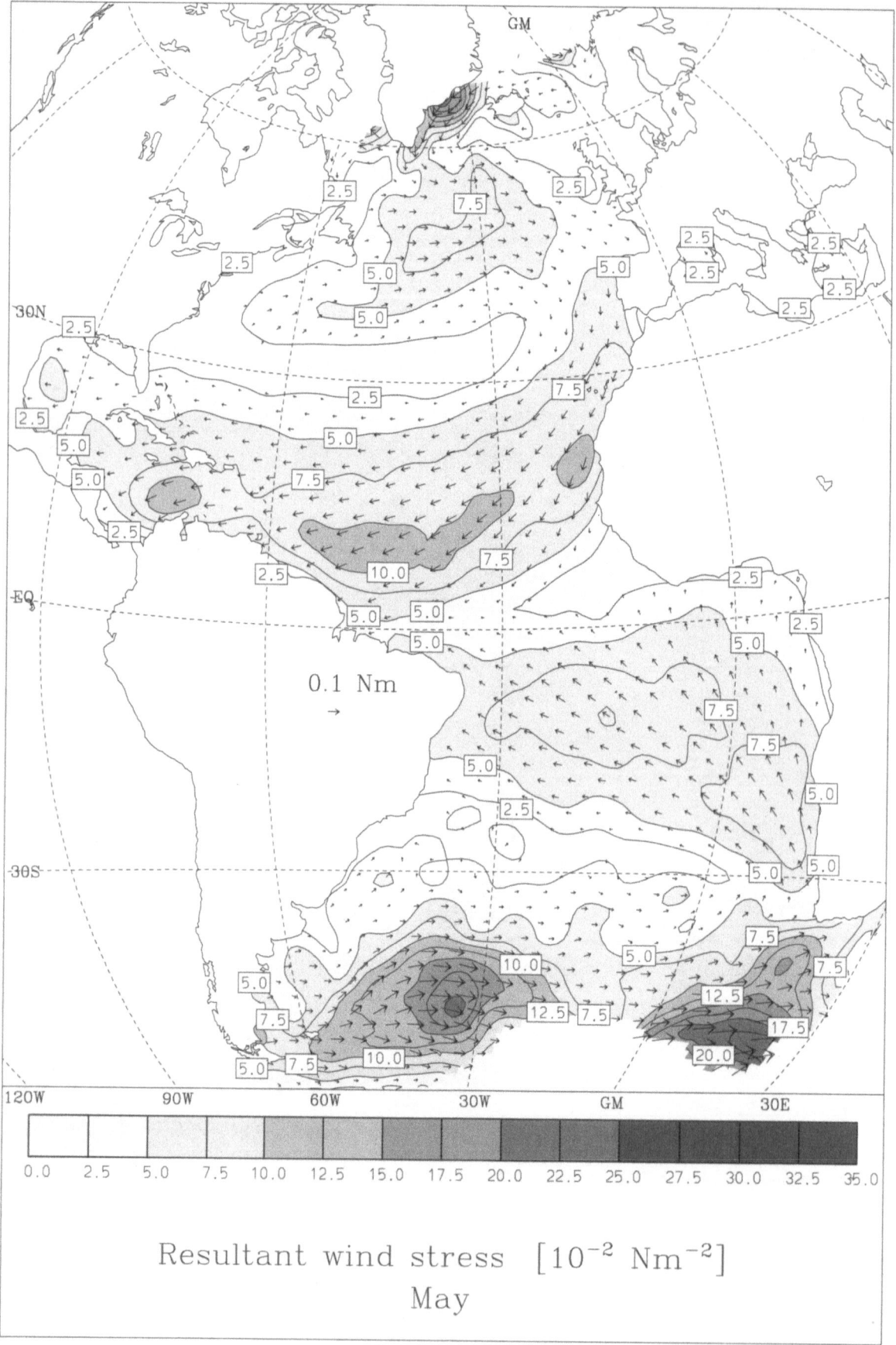

Resultant wind stress $[10^{-2} \text{ Nm}^{-2}]$

May

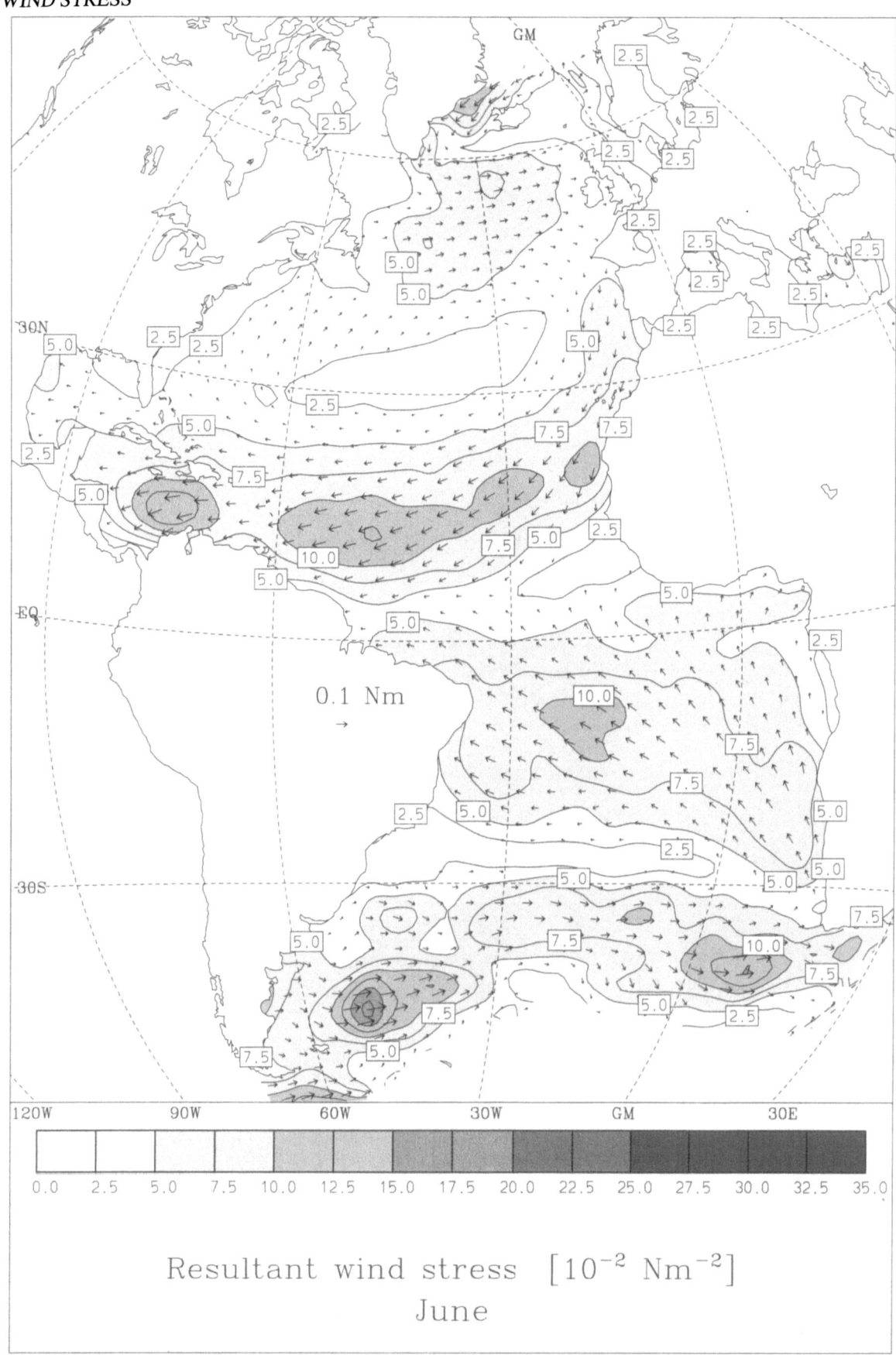

0.1 Nm
→

Resultant wind stress $[10^{-2}\ Nm^{-2}]$
June

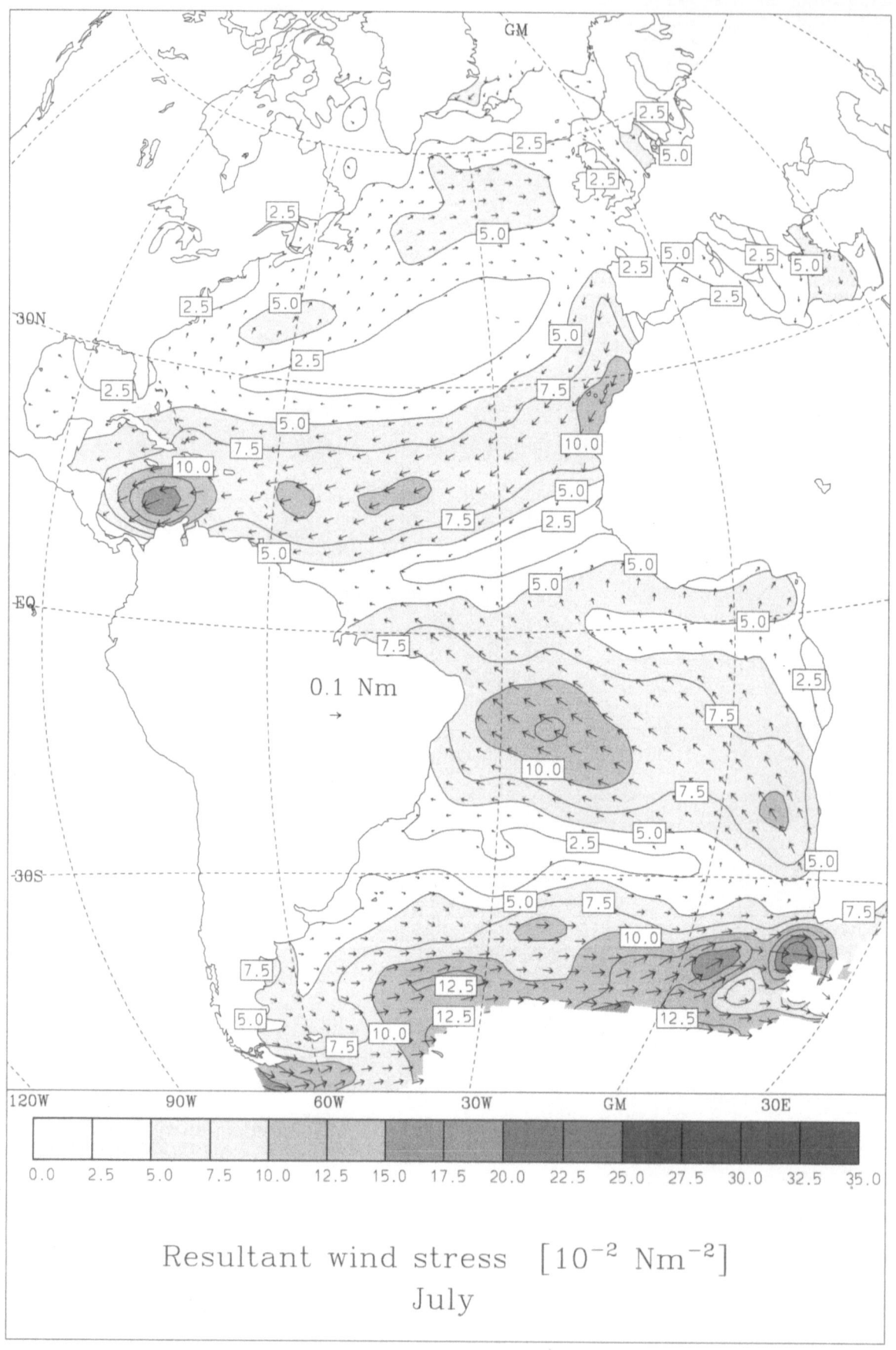

Resultant wind stress $[10^{-2}\ Nm^{-2}]$
July

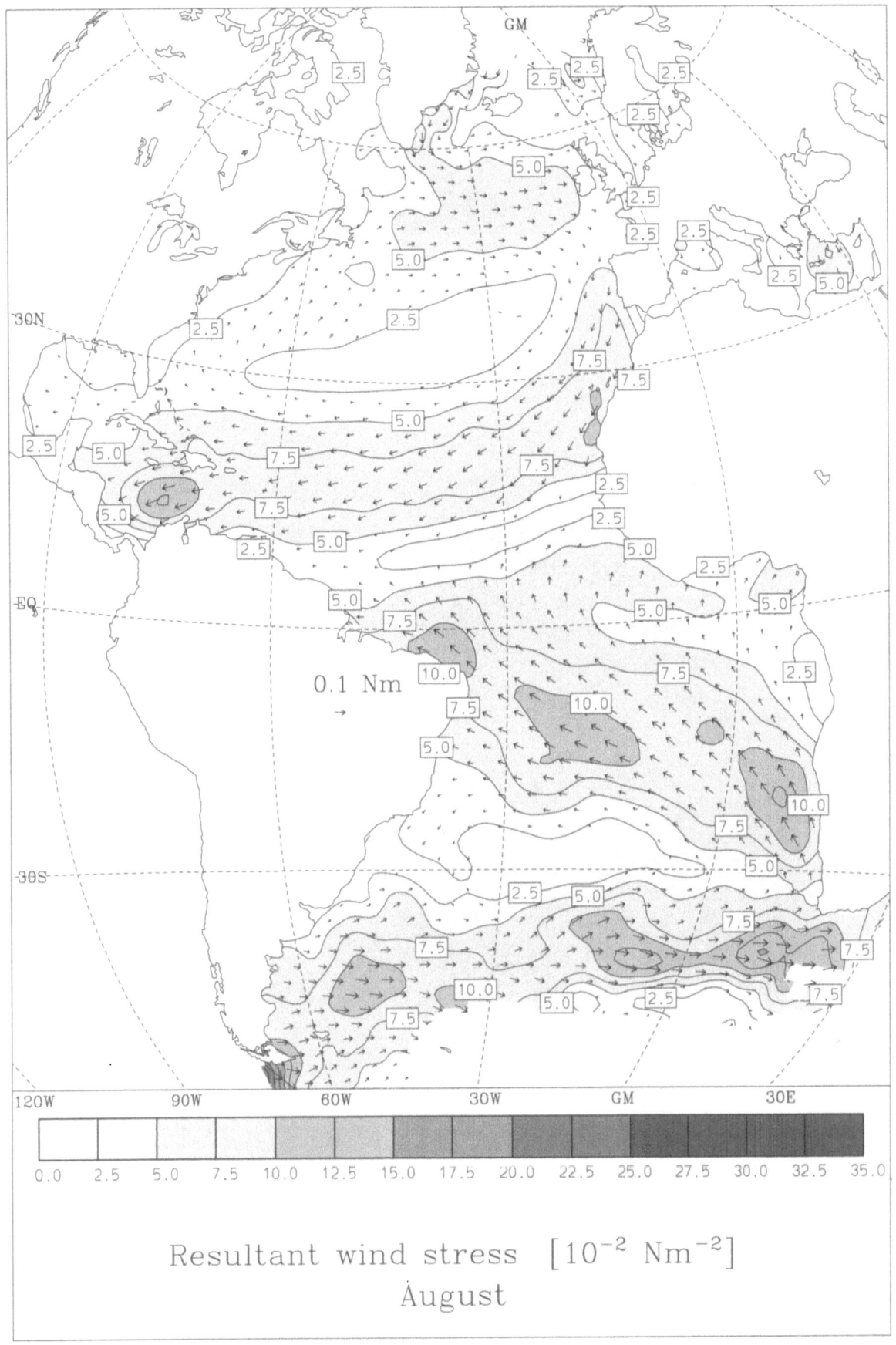

Resultant wind stress $[10^{-2} \text{ Nm}^{-2}]$
August

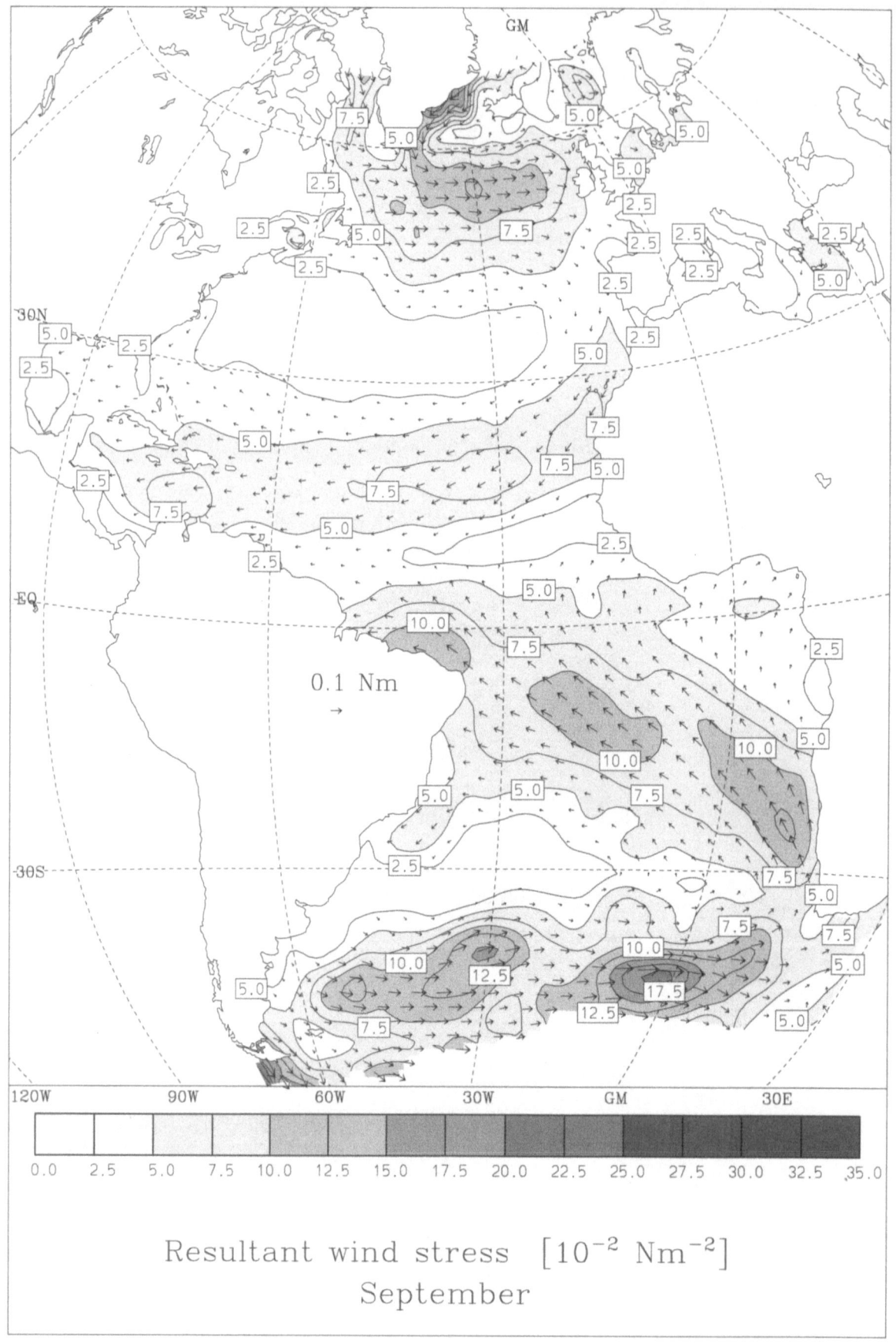

Resultant wind stress $[10^{-2}\ \mathrm{Nm^{-2}}]$
September

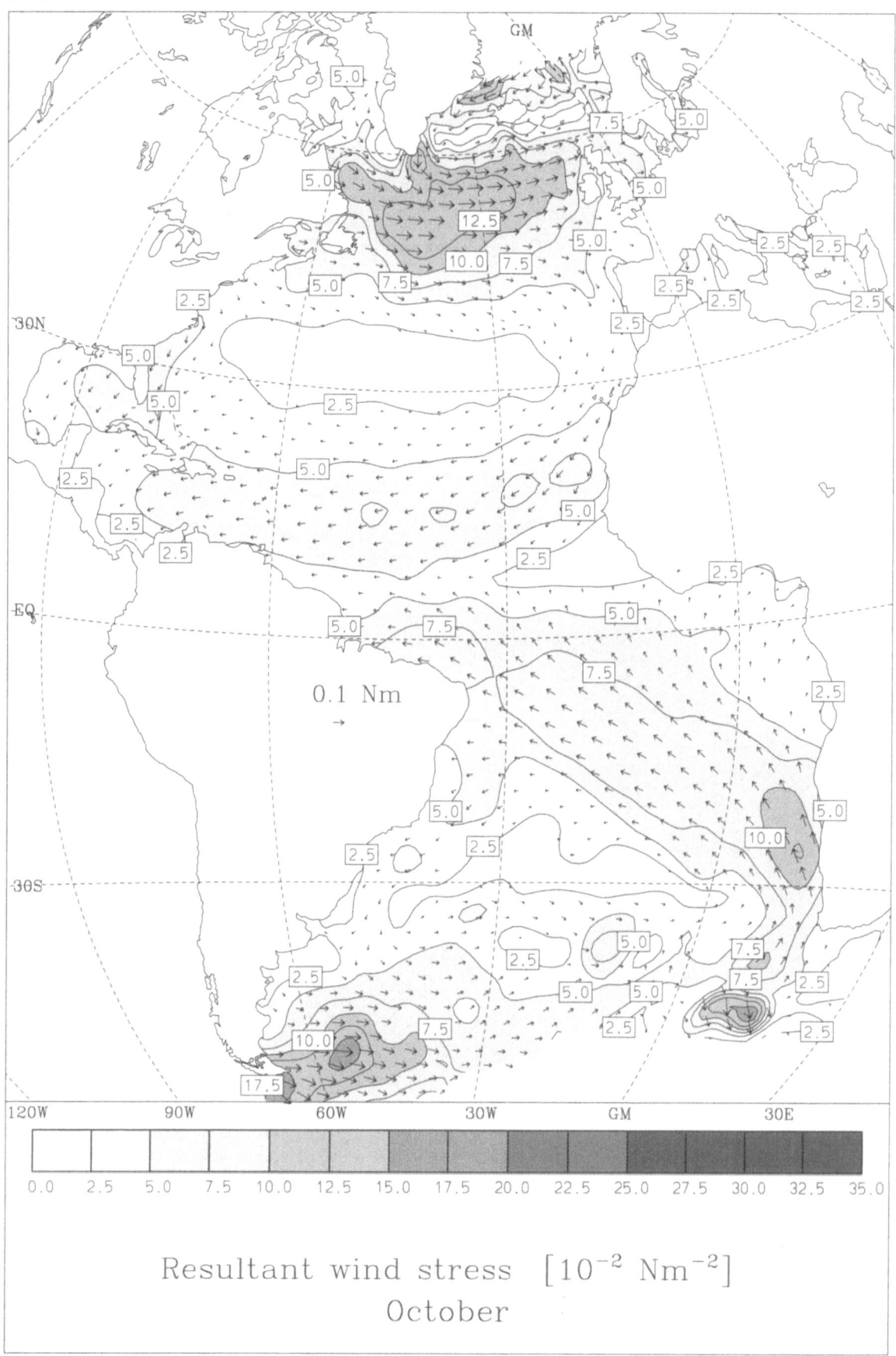

Resultant wind stress $[10^{-2}\ \mathrm{Nm}^{-2}]$
October

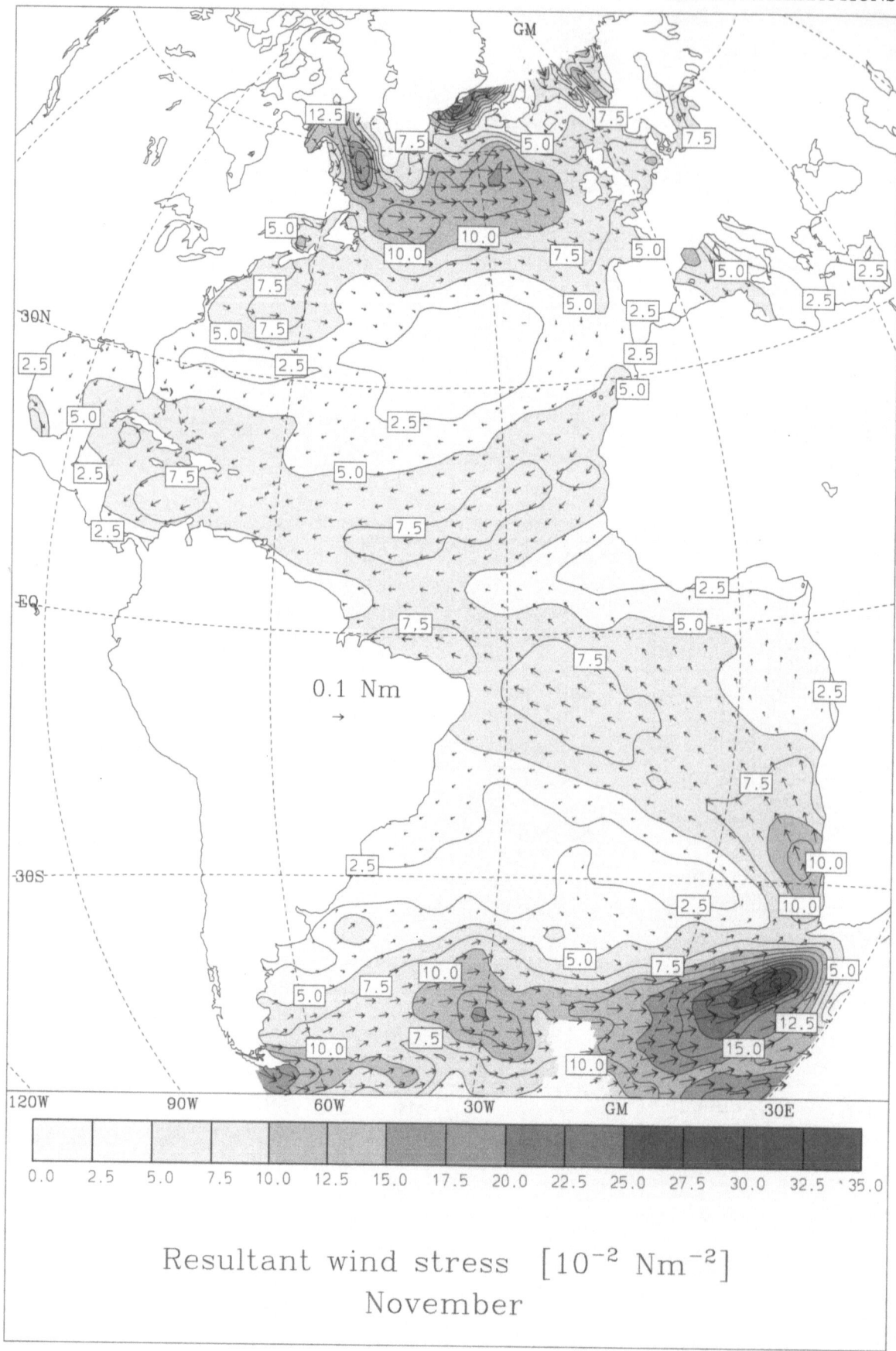

Resultant wind stress [10^{-2} Nm^{-2}]
November

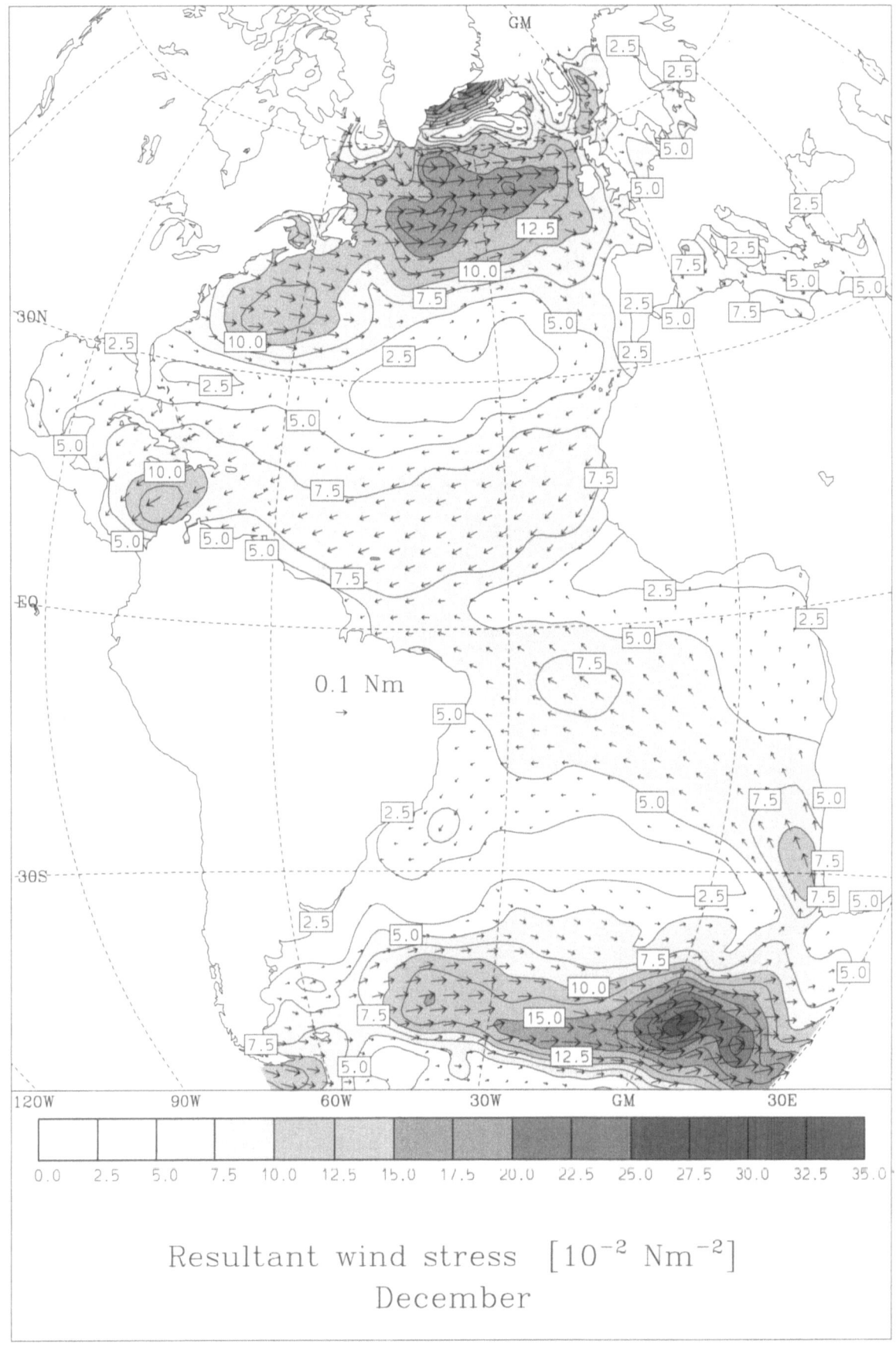

Resultant wind stress $[10^{-2} \text{ Nm}^{-2}]$
December

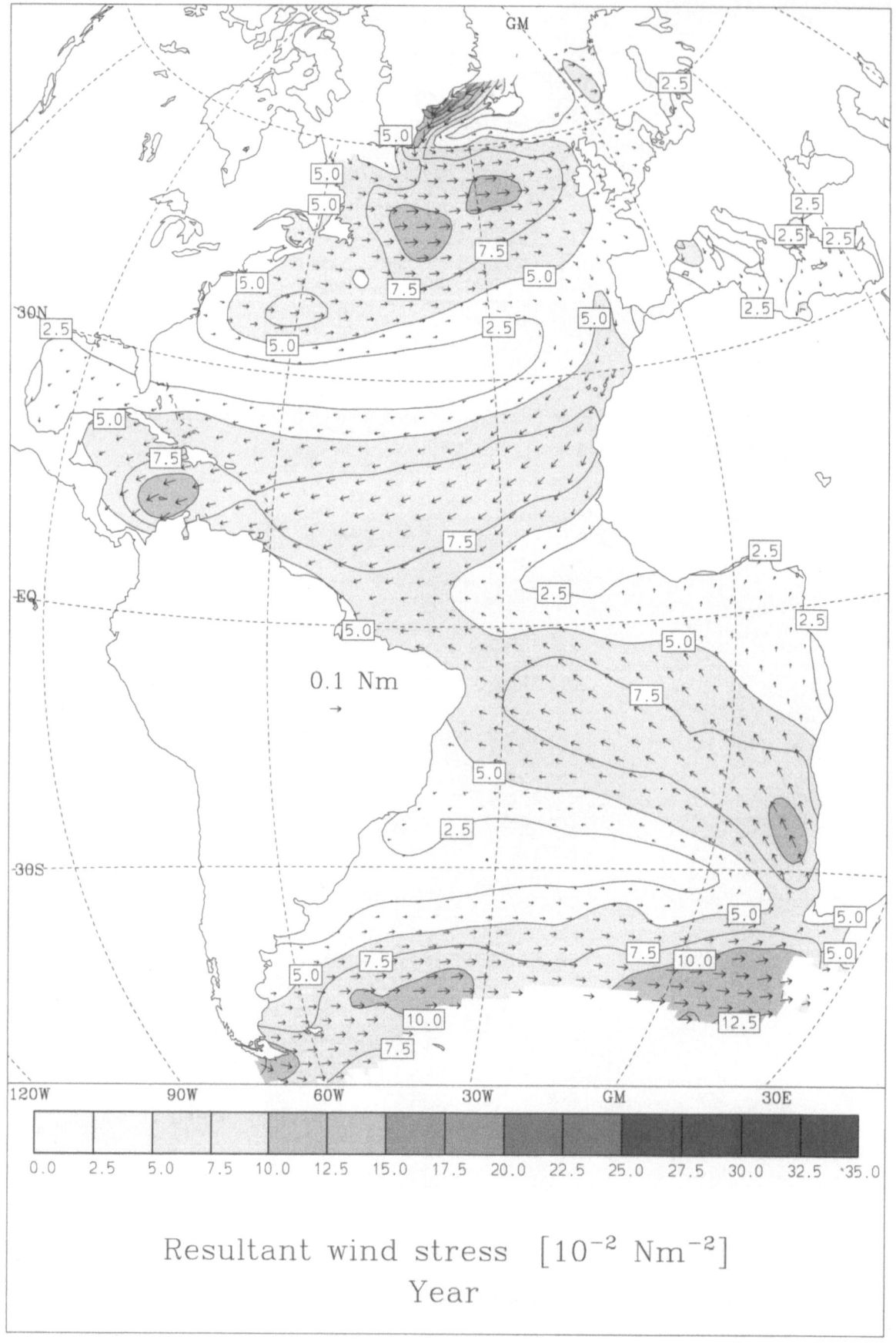

Resultant wind stress $[10^{-2} \ Nm^{-2}]$
Year

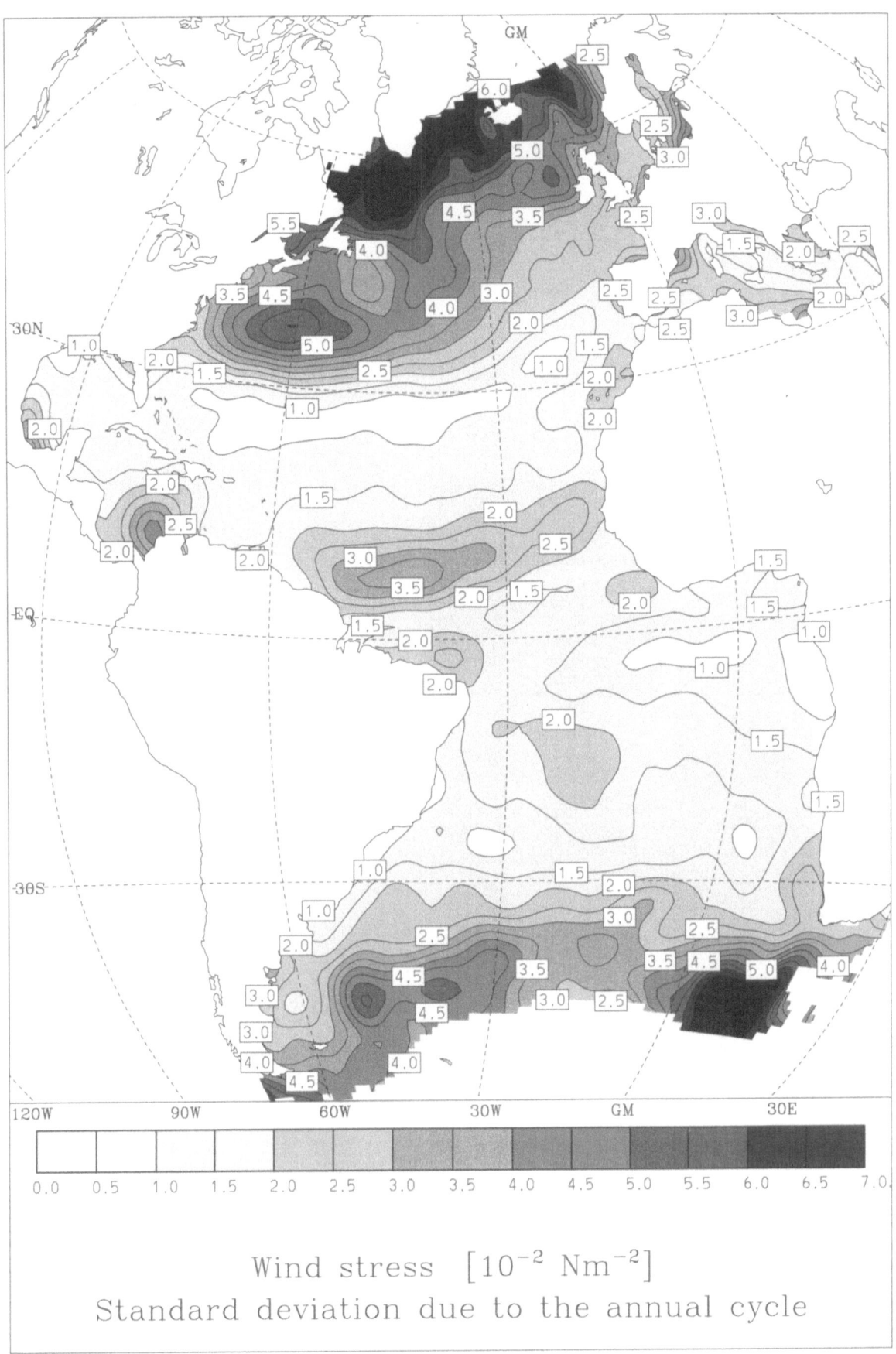

Wind stress $[10^{-2} \ Nm^{-2}]$
Standard deviation due to the annual cycle

East Component of the Wind Stress

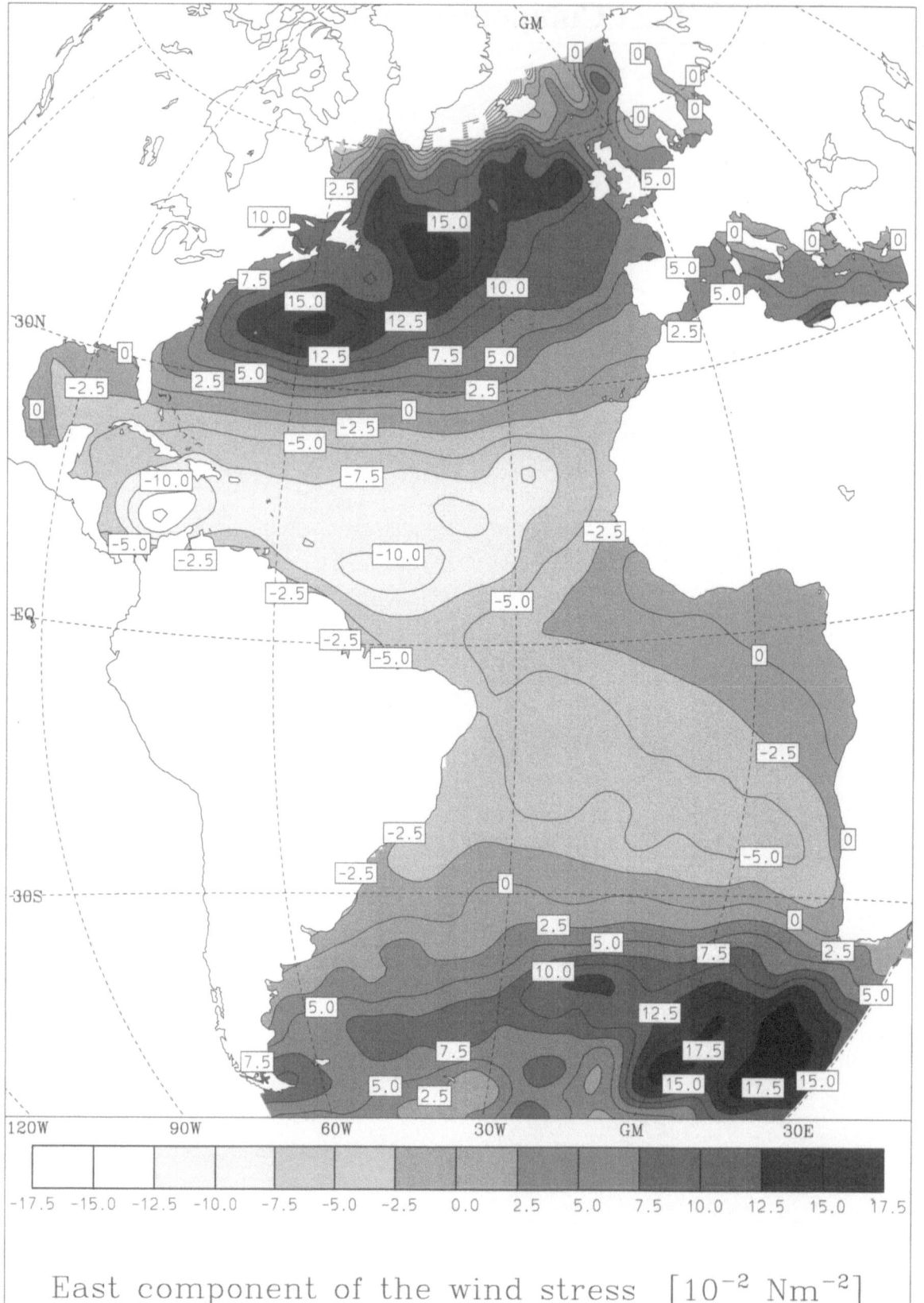

East component of the wind stress $[10^{-2}\ \mathrm{Nm}^{-2}]$

January

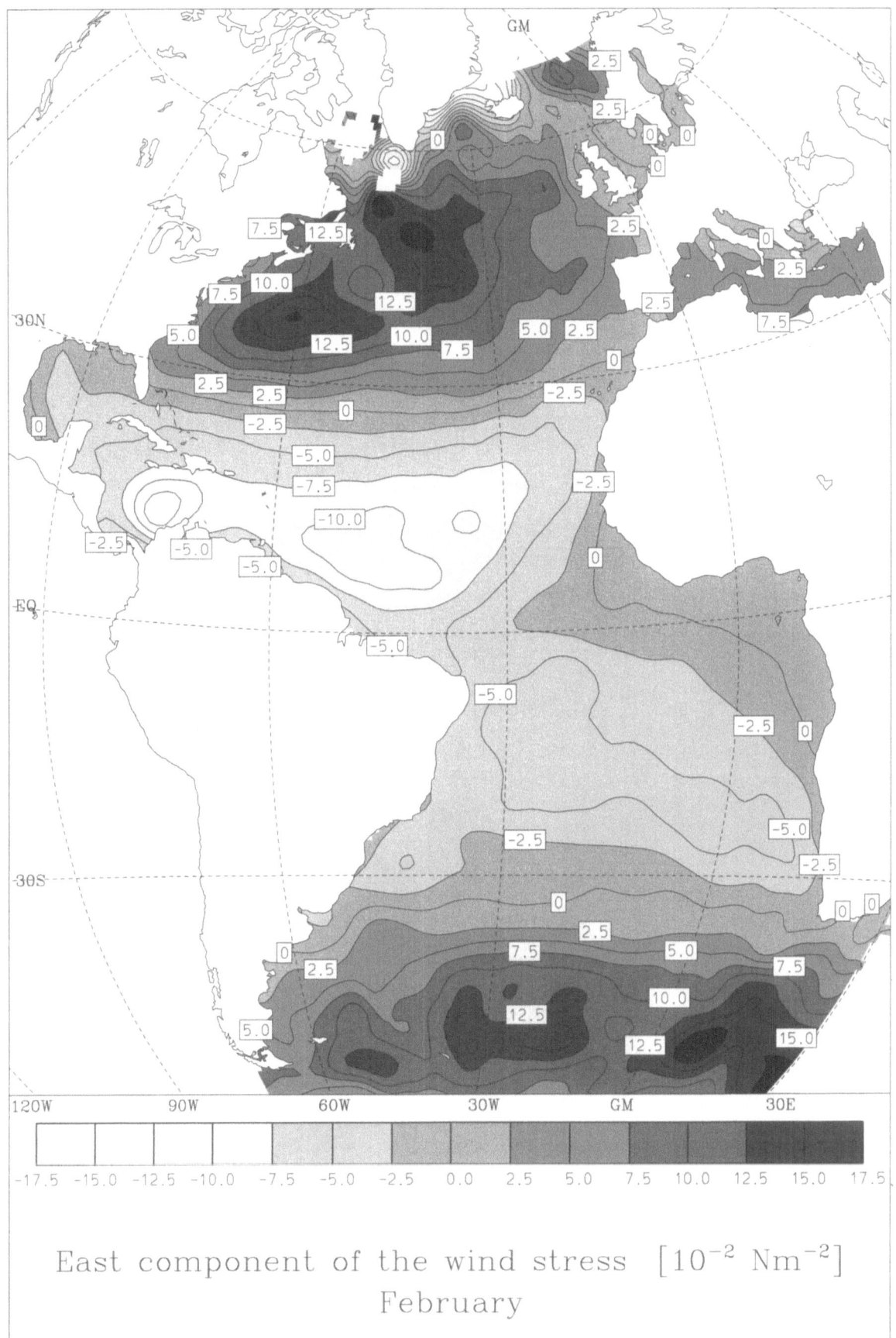

East component of the wind stress $[10^{-2}\ \mathrm{Nm}^{-2}]$
February

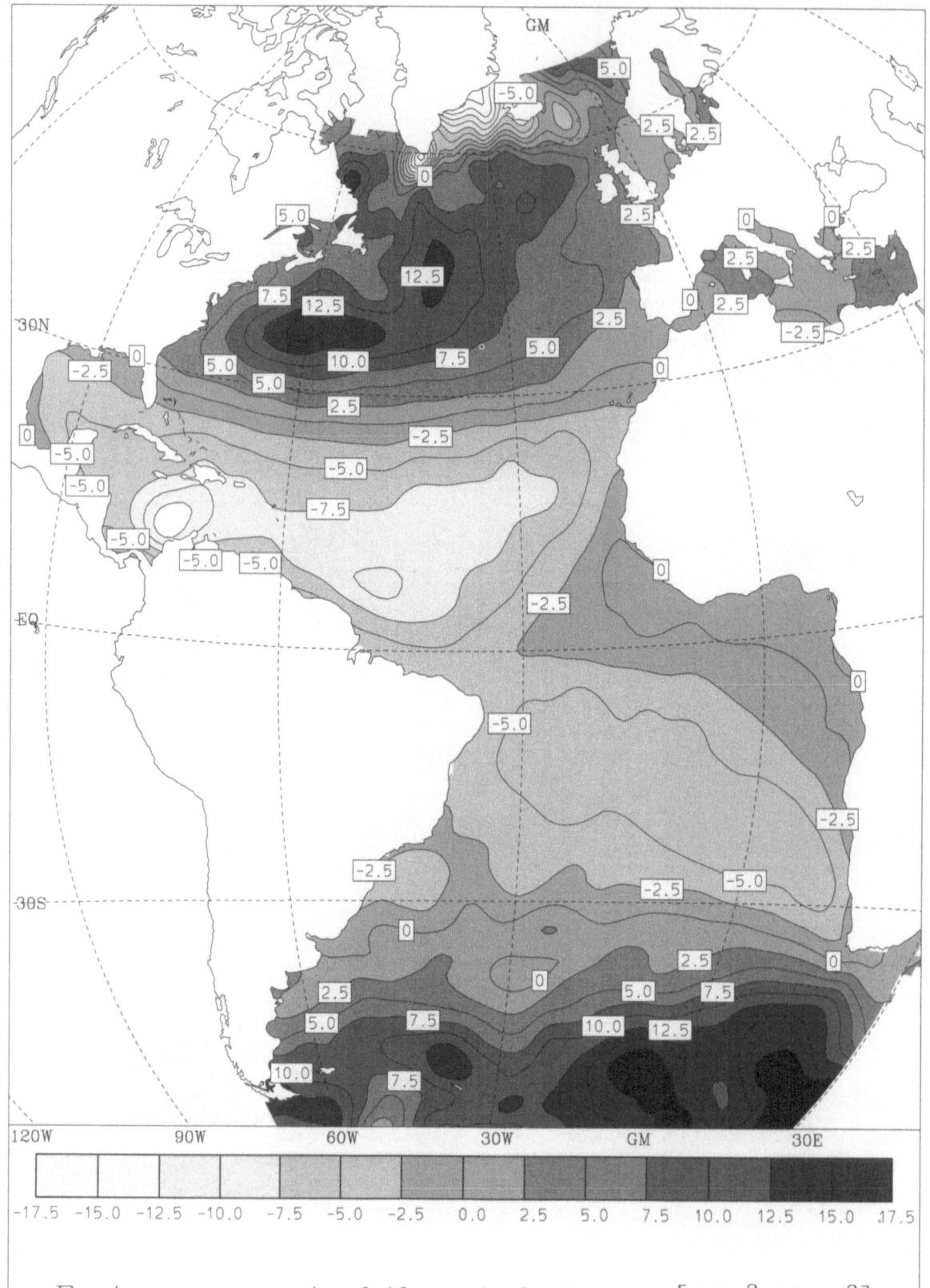

East component of the wind stress [10^{-2} Nm^{-2}]
March

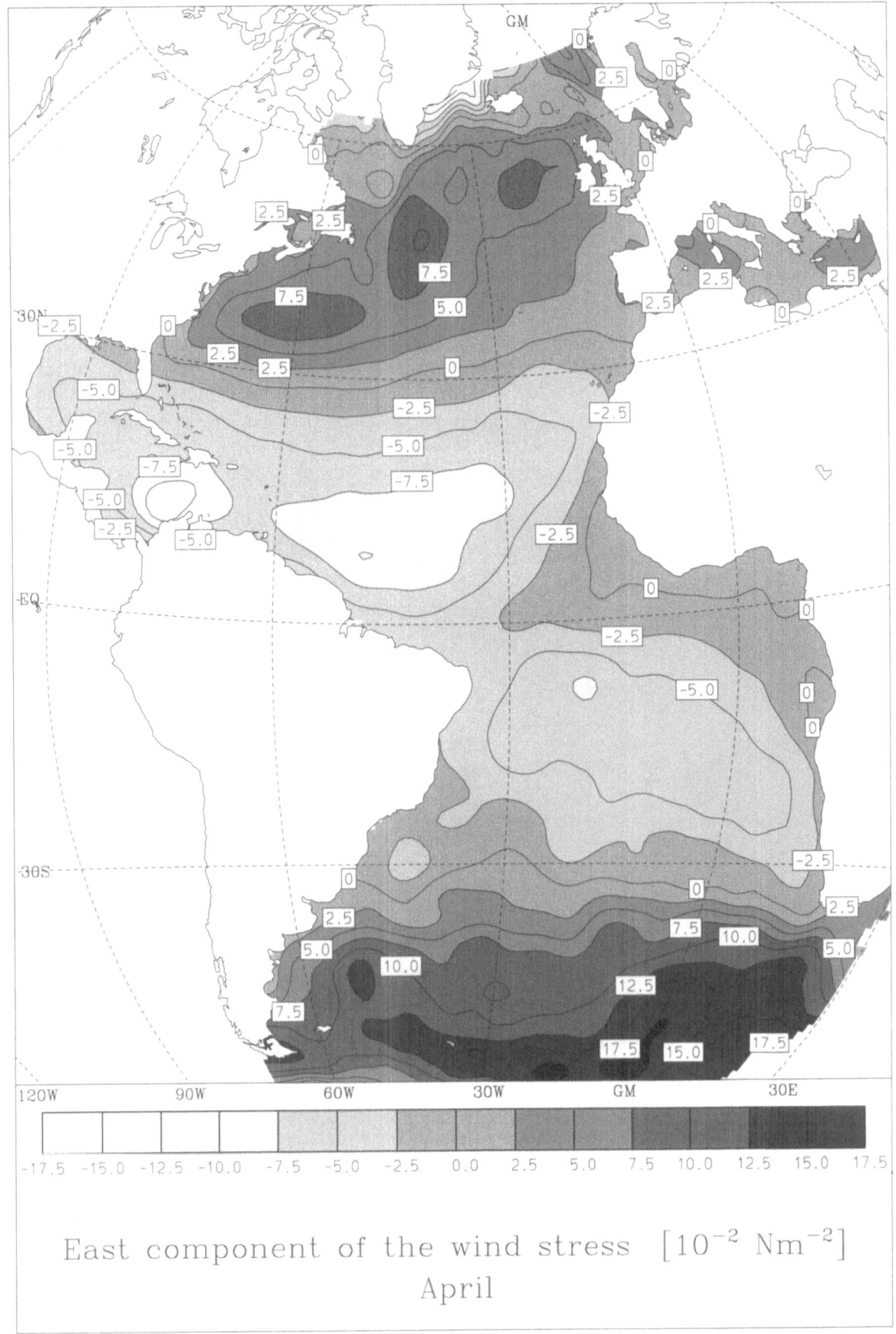

East component of the wind stress $[10^{-2}\ \mathrm{Nm}^{-2}]$
April

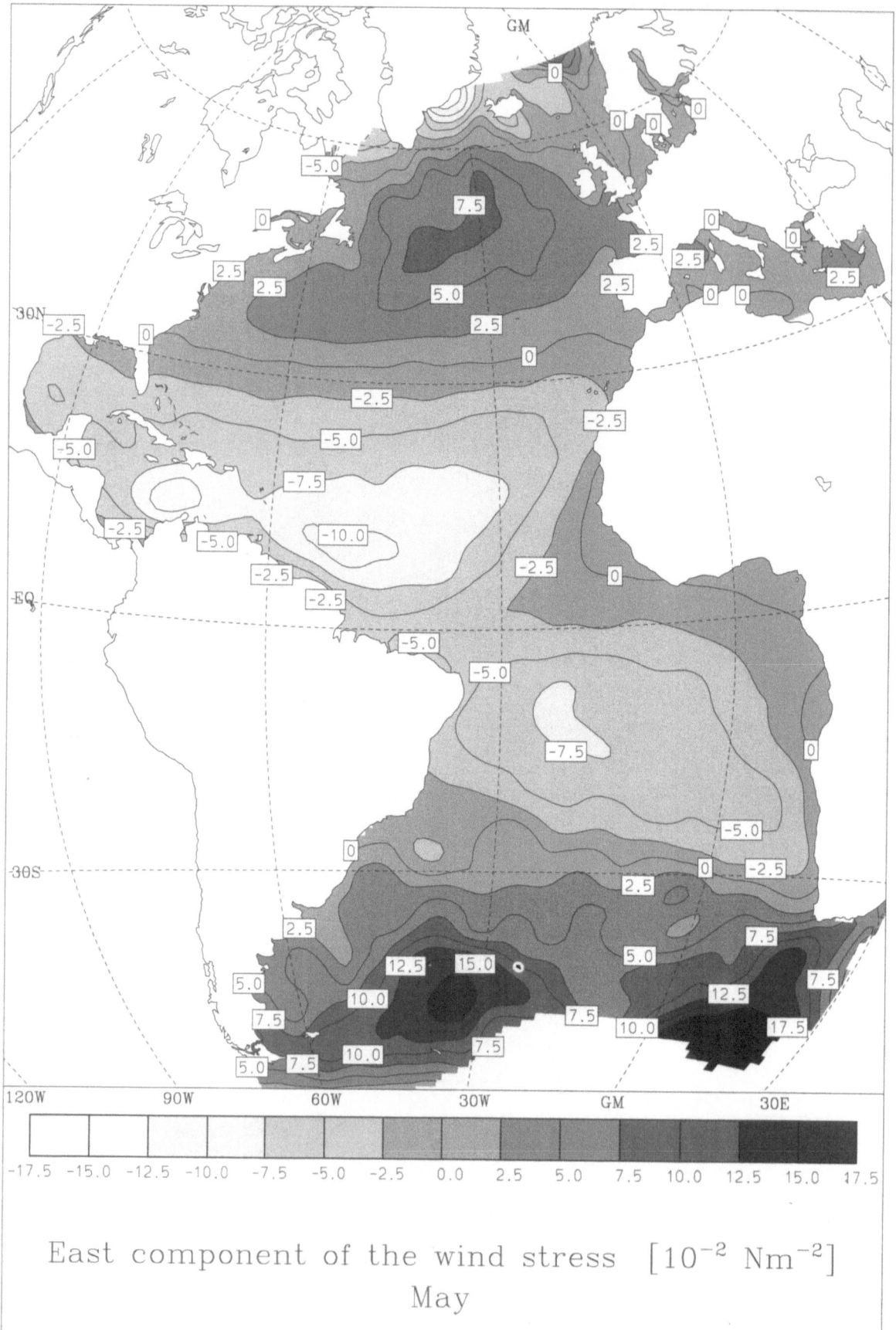

East component of the wind stress [10^{-2} Nm^{-2}]
May

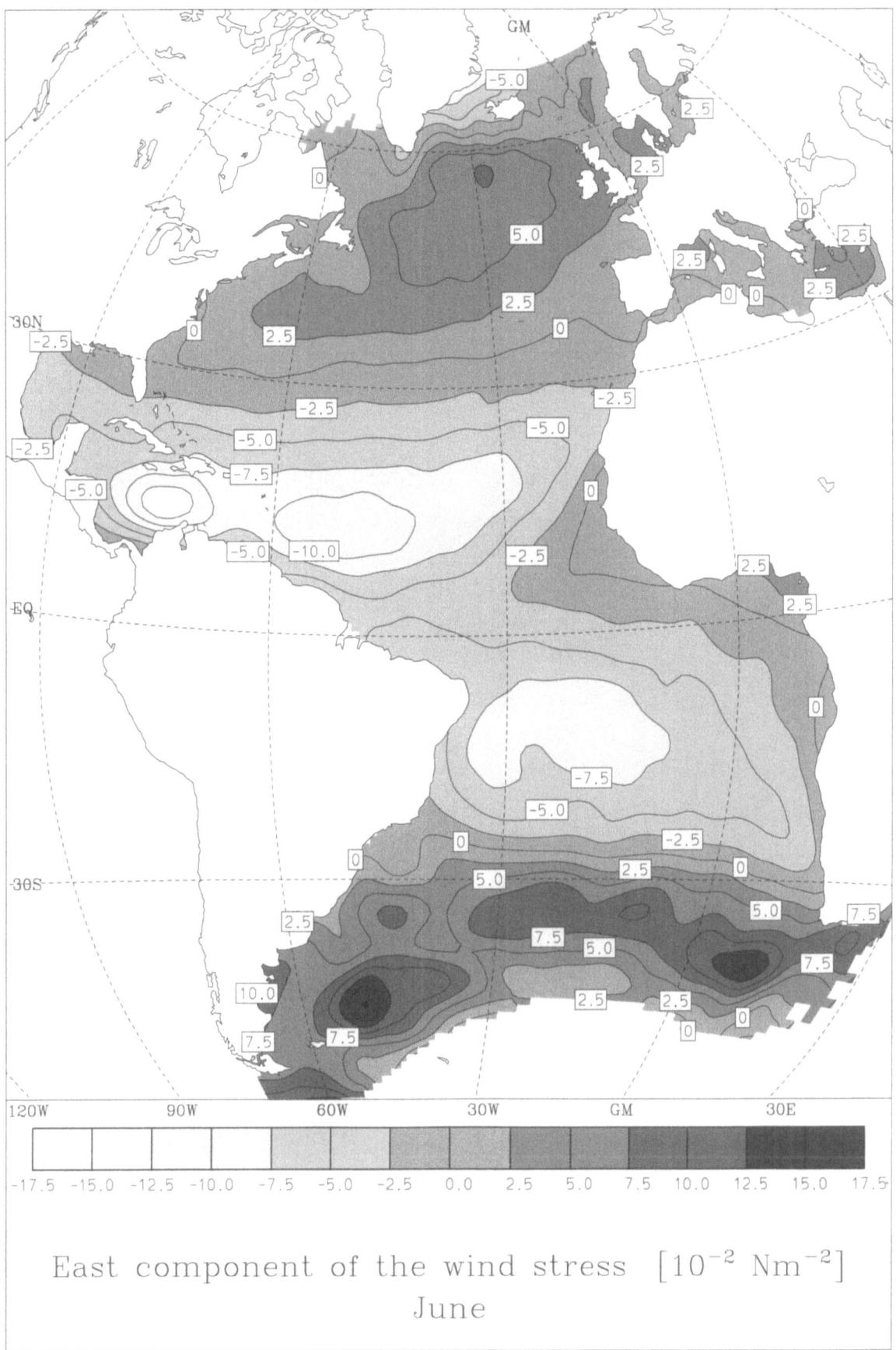

East component of the wind stress $[10^{-2} \ \mathrm{N m^{-2}}]$
June

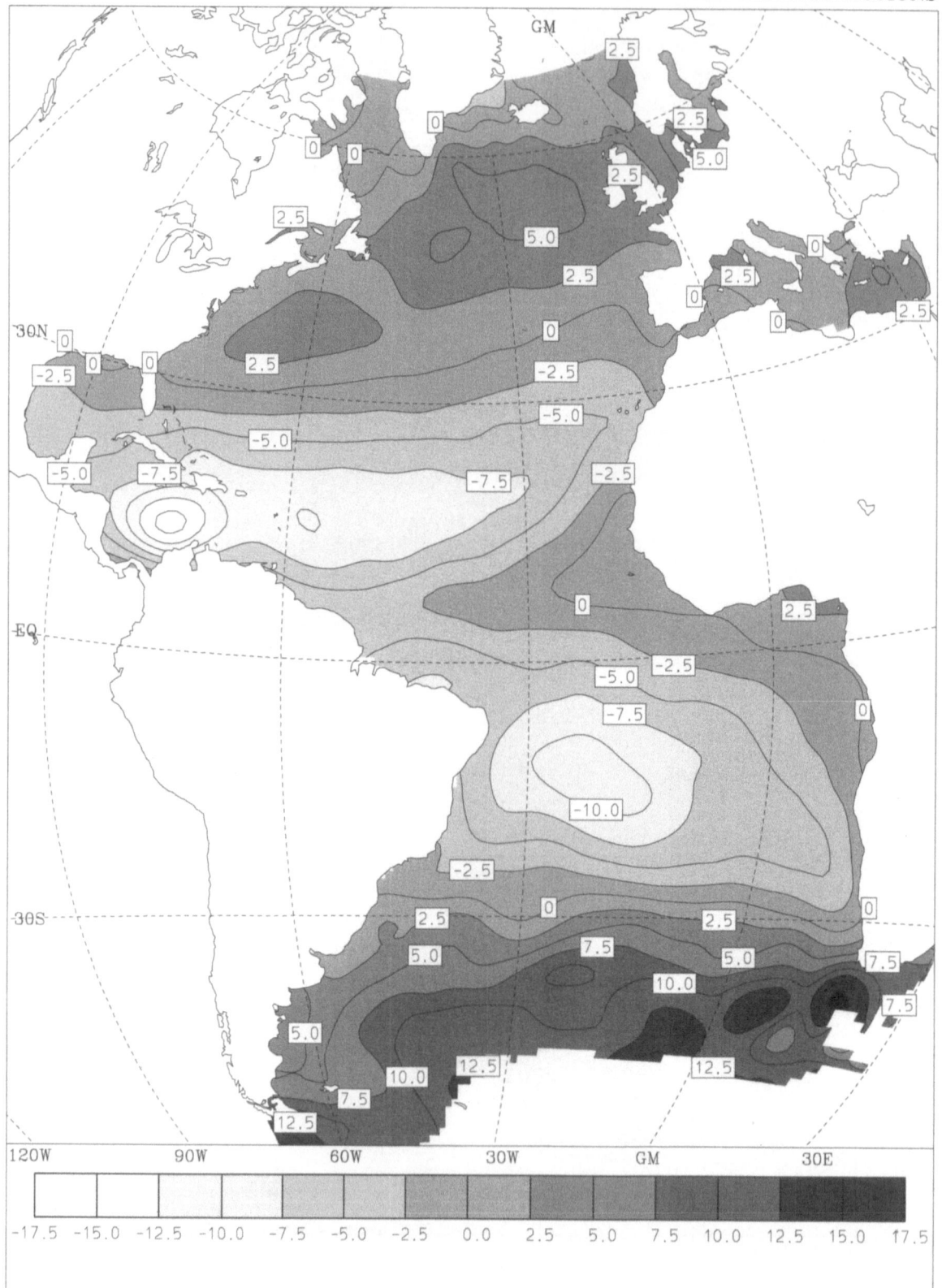

East component of the wind stress $[10^{-2}\ Nm^{-2}]$

July

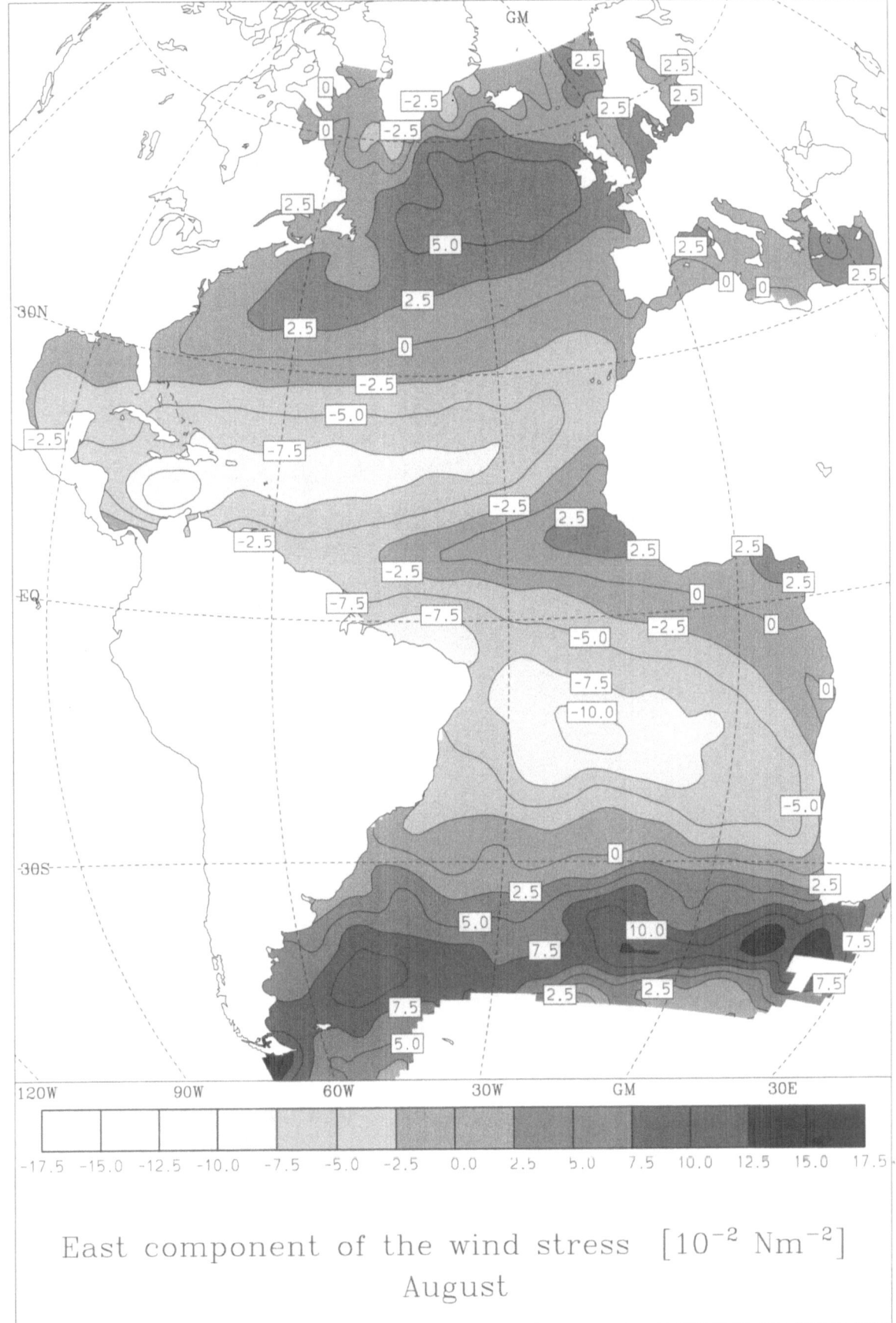

East component of the wind stress $[10^{-2} \ \mathrm{Nm}^{-2}]$
August

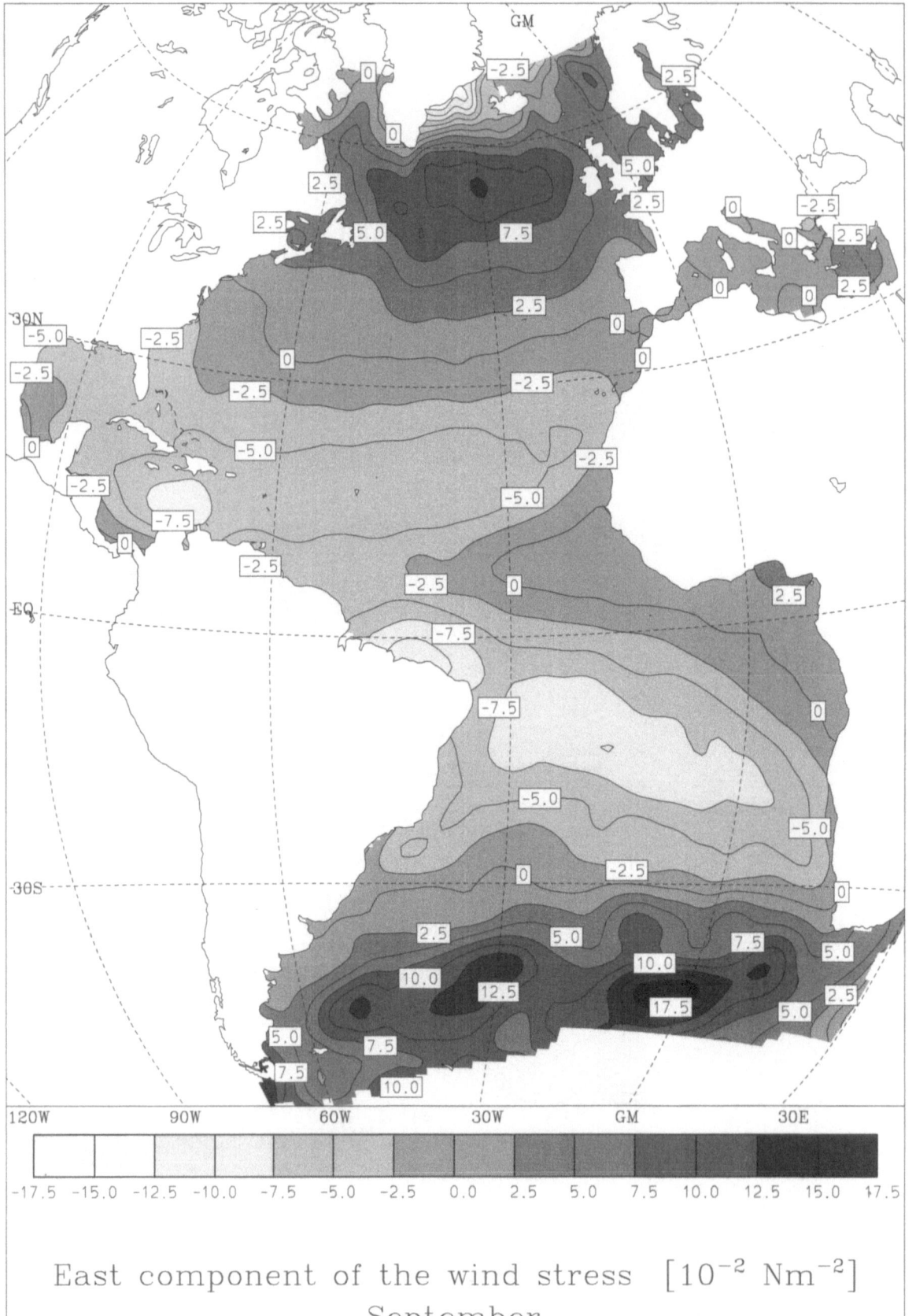

East component of the wind stress $[10^{-2}\ \mathrm{Nm}^{-2}]$
September

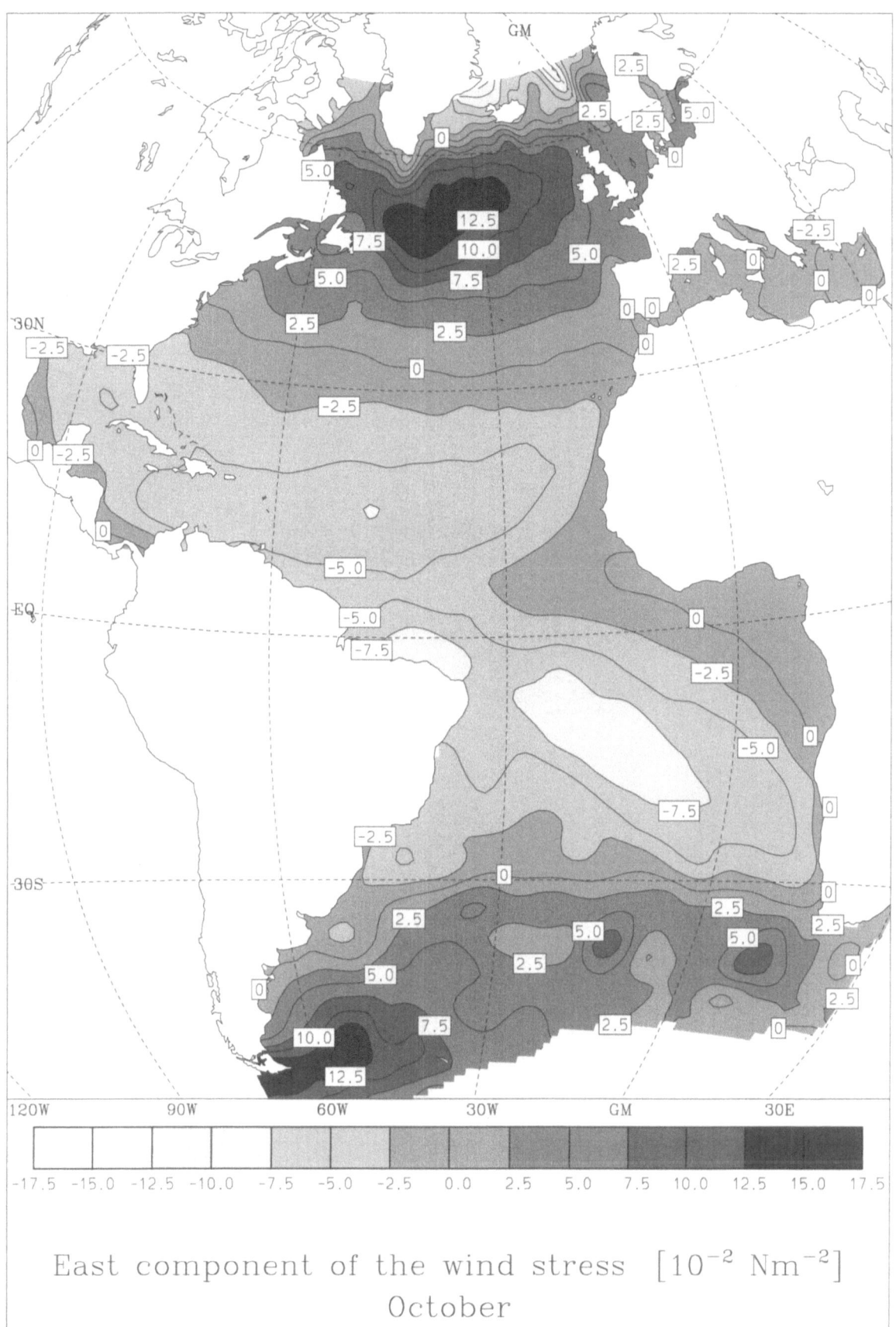

East component of the wind stress $[10^{-2}\ \mathrm{Nm}^{-2}]$
October

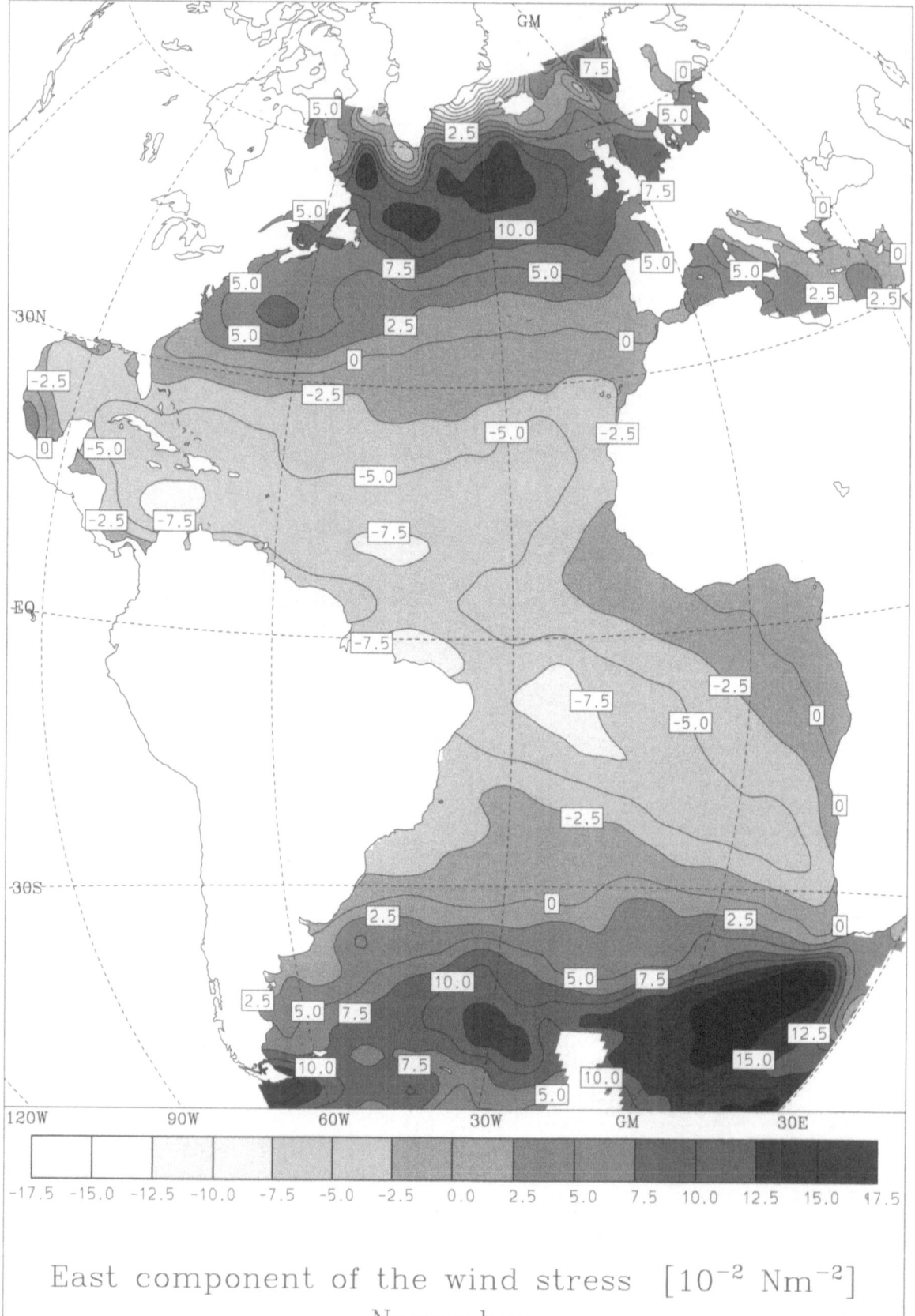

East component of the wind stress $[10^{-2}\ \mathrm{Nm}^{-2}]$
November

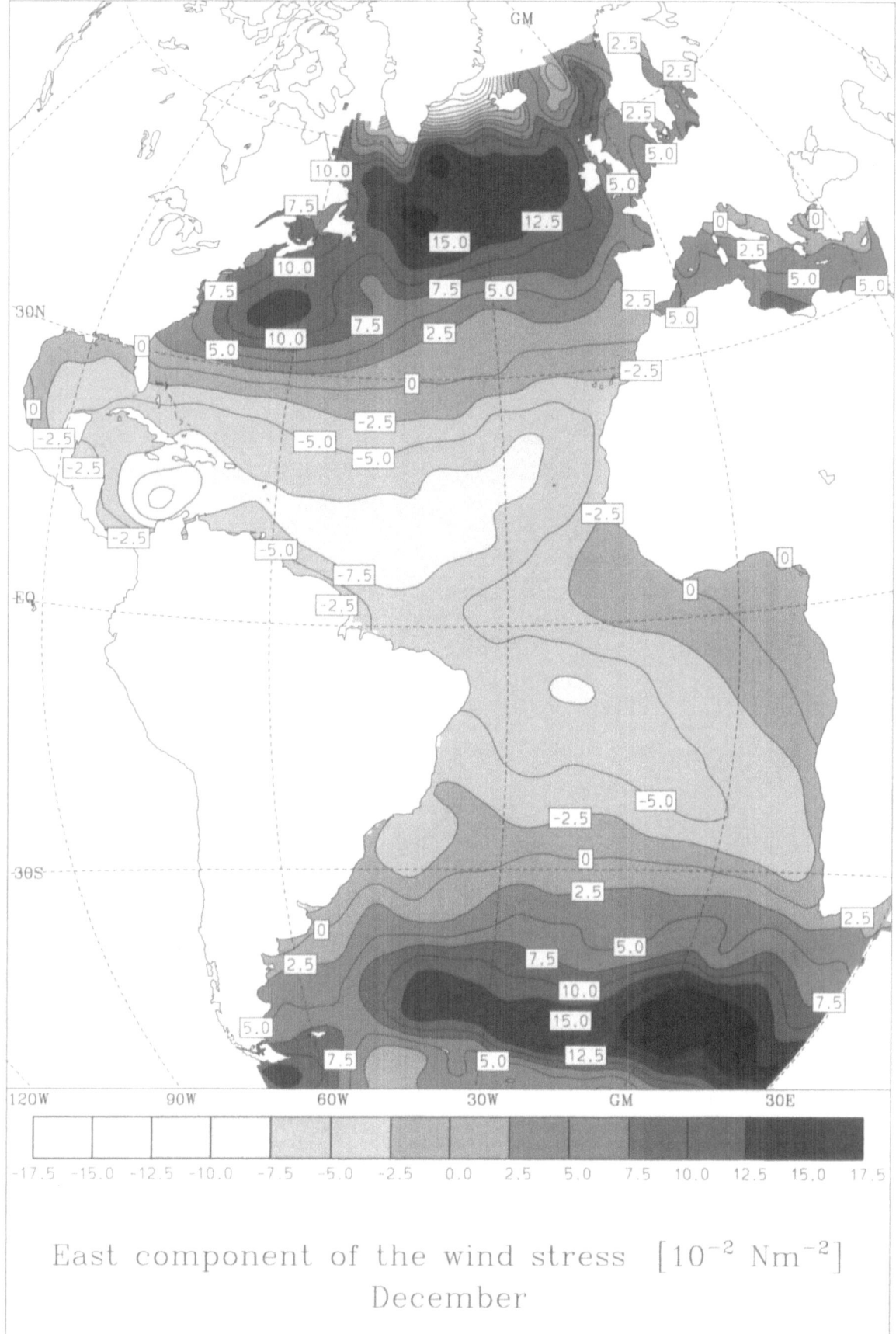

East component of the wind stress $[10^{-2} \, \text{Nm}^{-2}]$
December

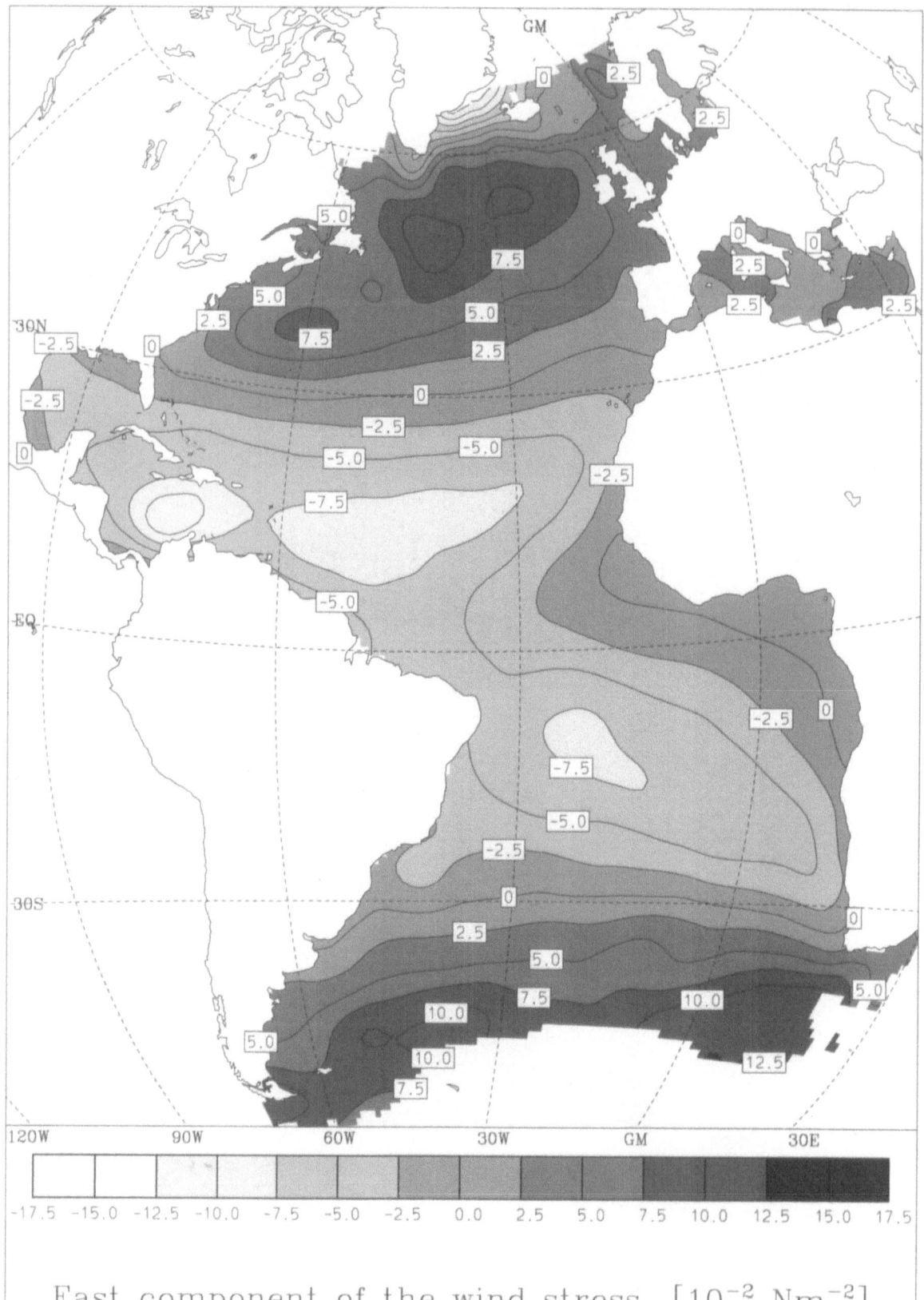

East component of the wind stress $[10^{-2} \text{ Nm}^{-2}]$
Year

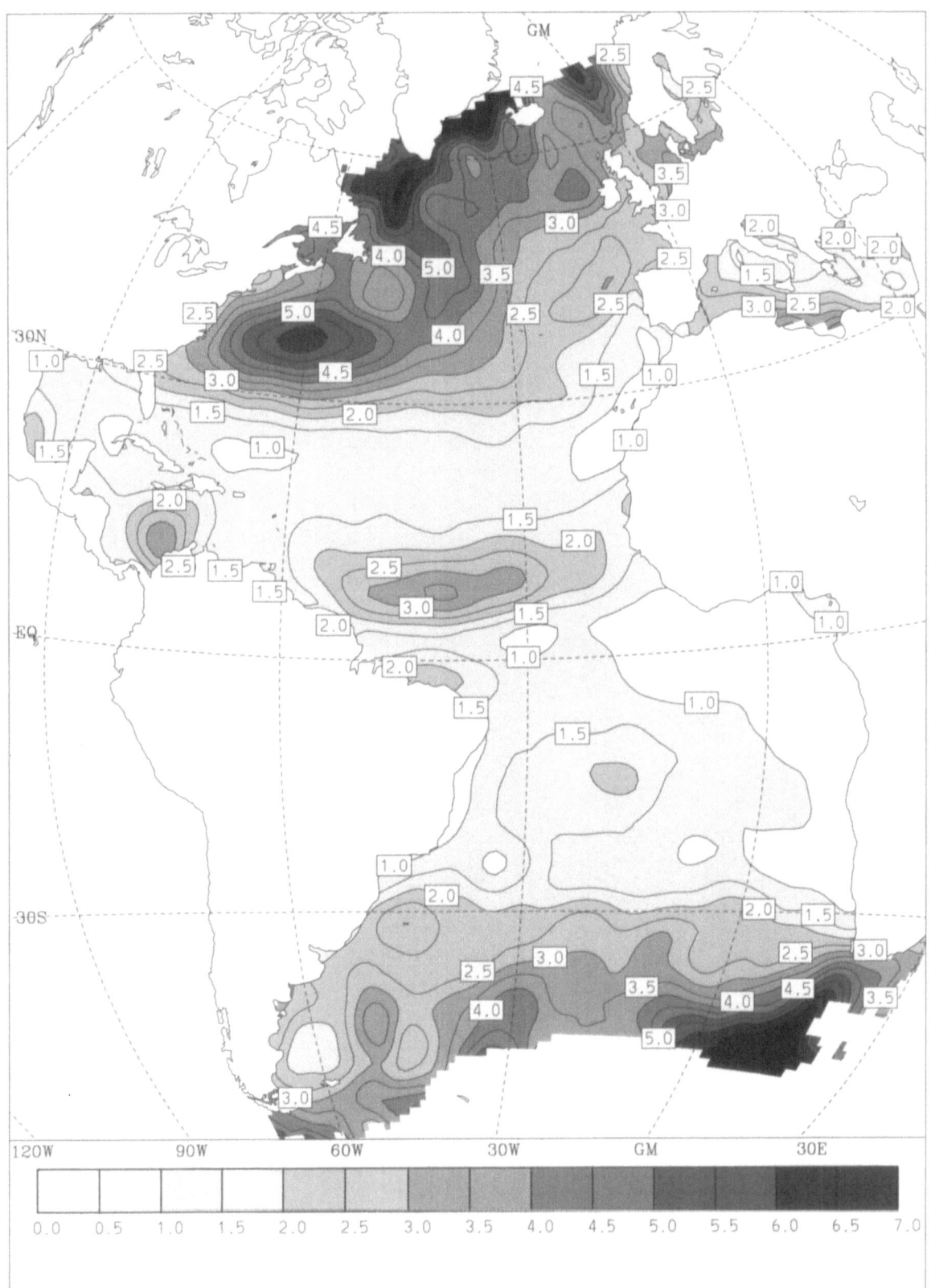

East component of the wind stress $[10^{-2} \ \mathrm{Nm}^{-2}]$
Standard deviation due to the annual cycle

North Component of the Wind Stress

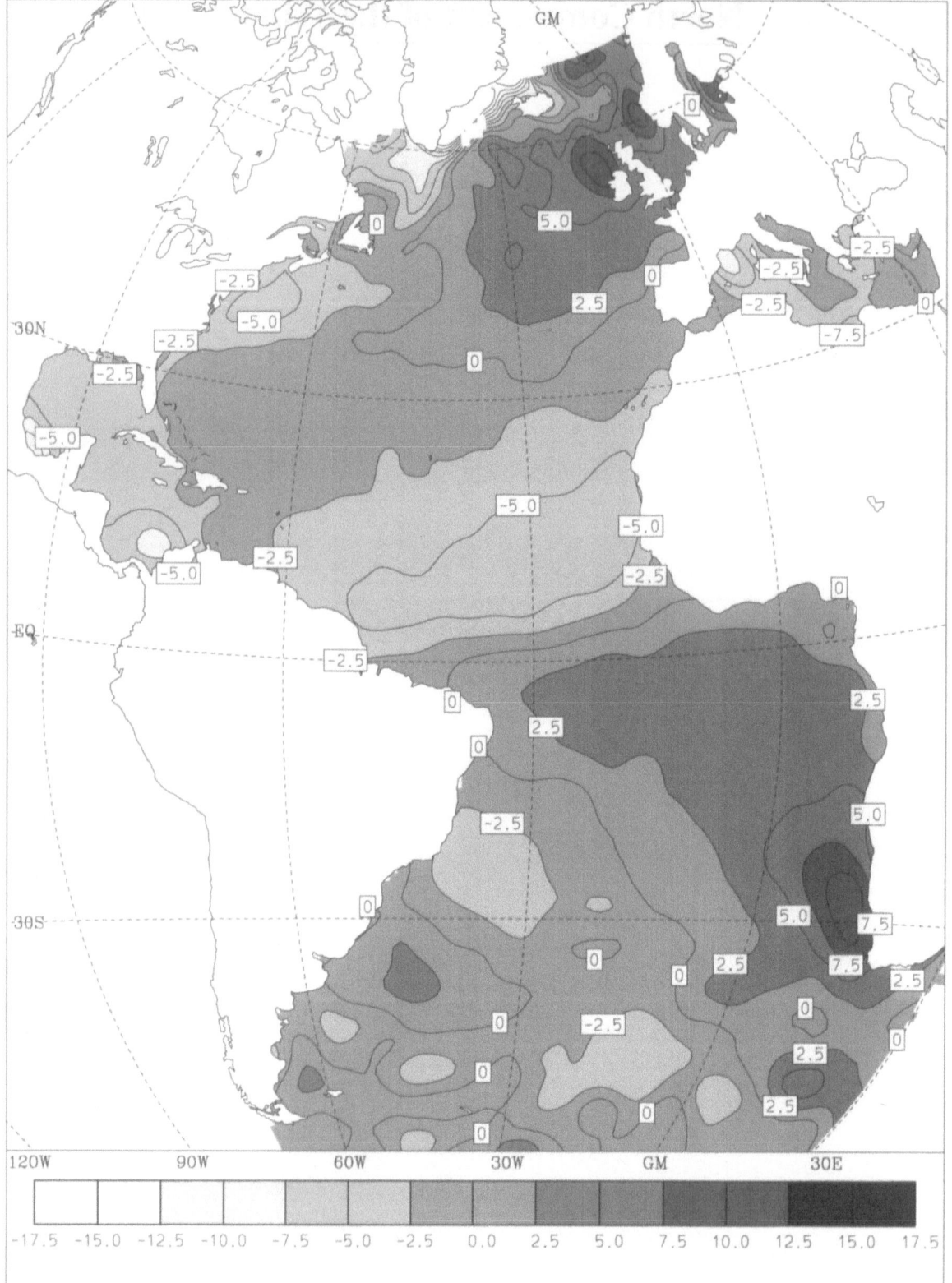

North component of the wind stress [10⁻² Nm⁻²]
January

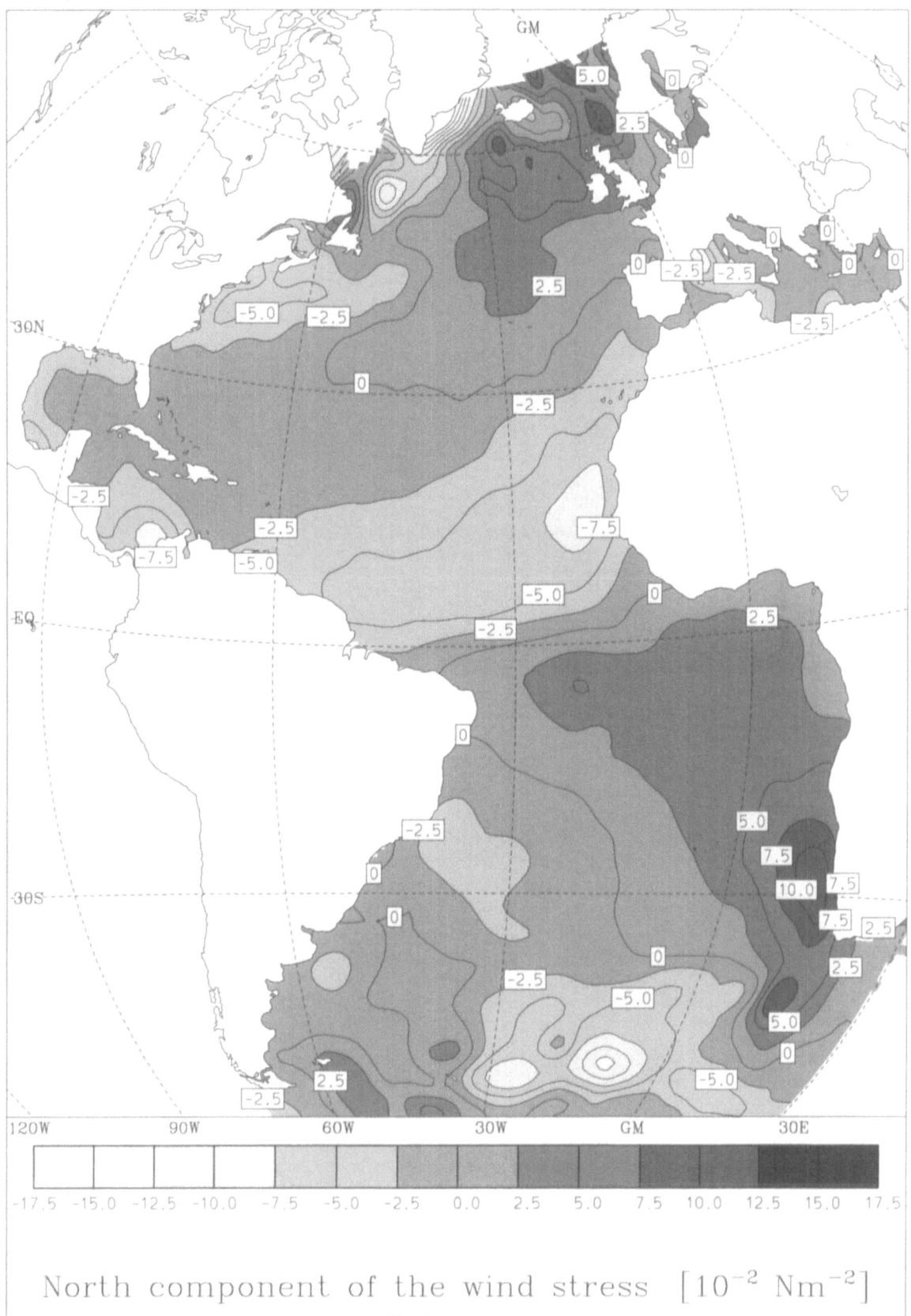

North component of the wind stress $[10^{-2}\ \mathrm{Nm^{-2}}]$
February

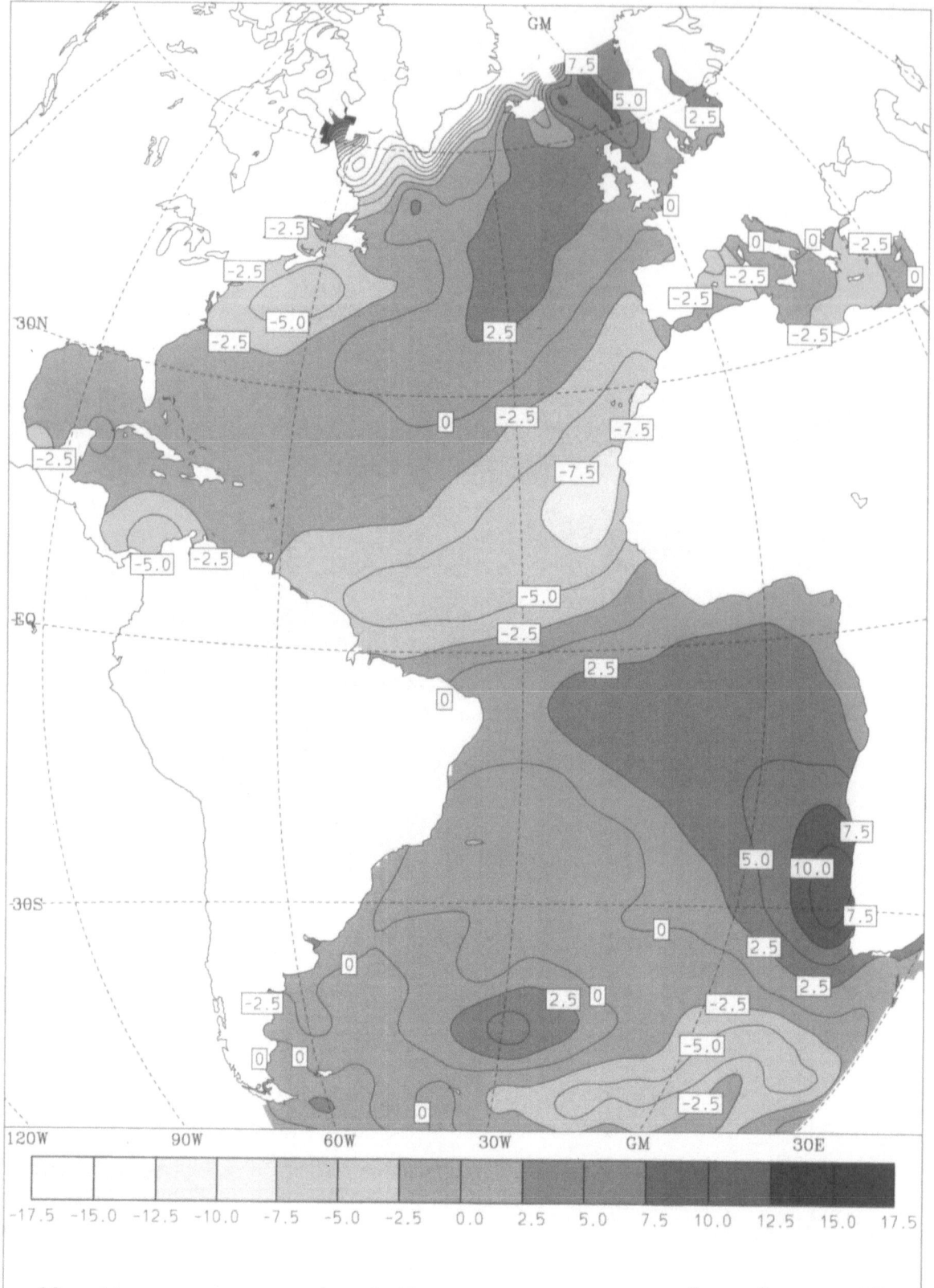

North component of the wind stress $[10^{-2}\ \mathrm{Nm^{-2}}]$
March

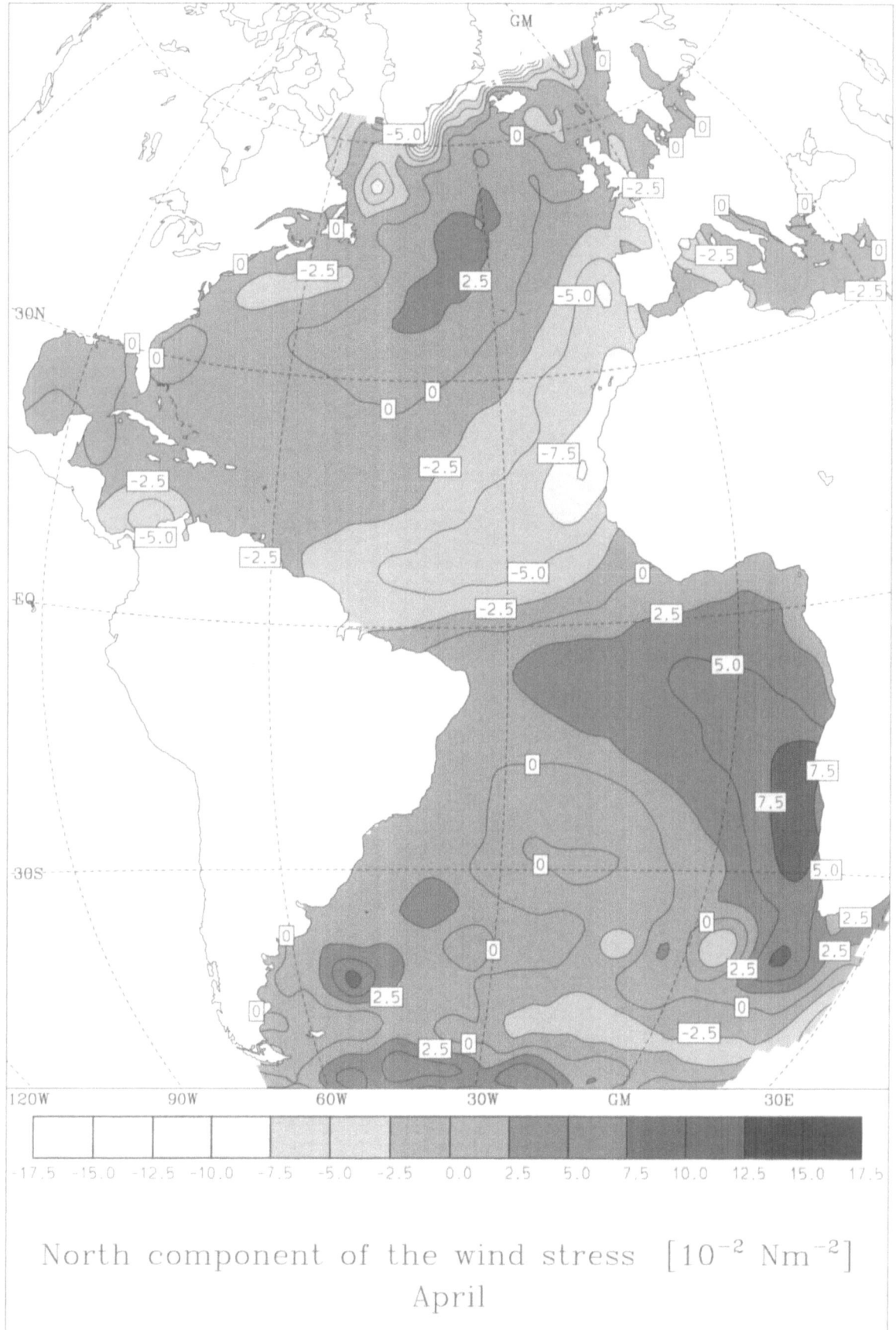

North component of the wind stress $[10^{-2}\ Nm^{-2}]$
April

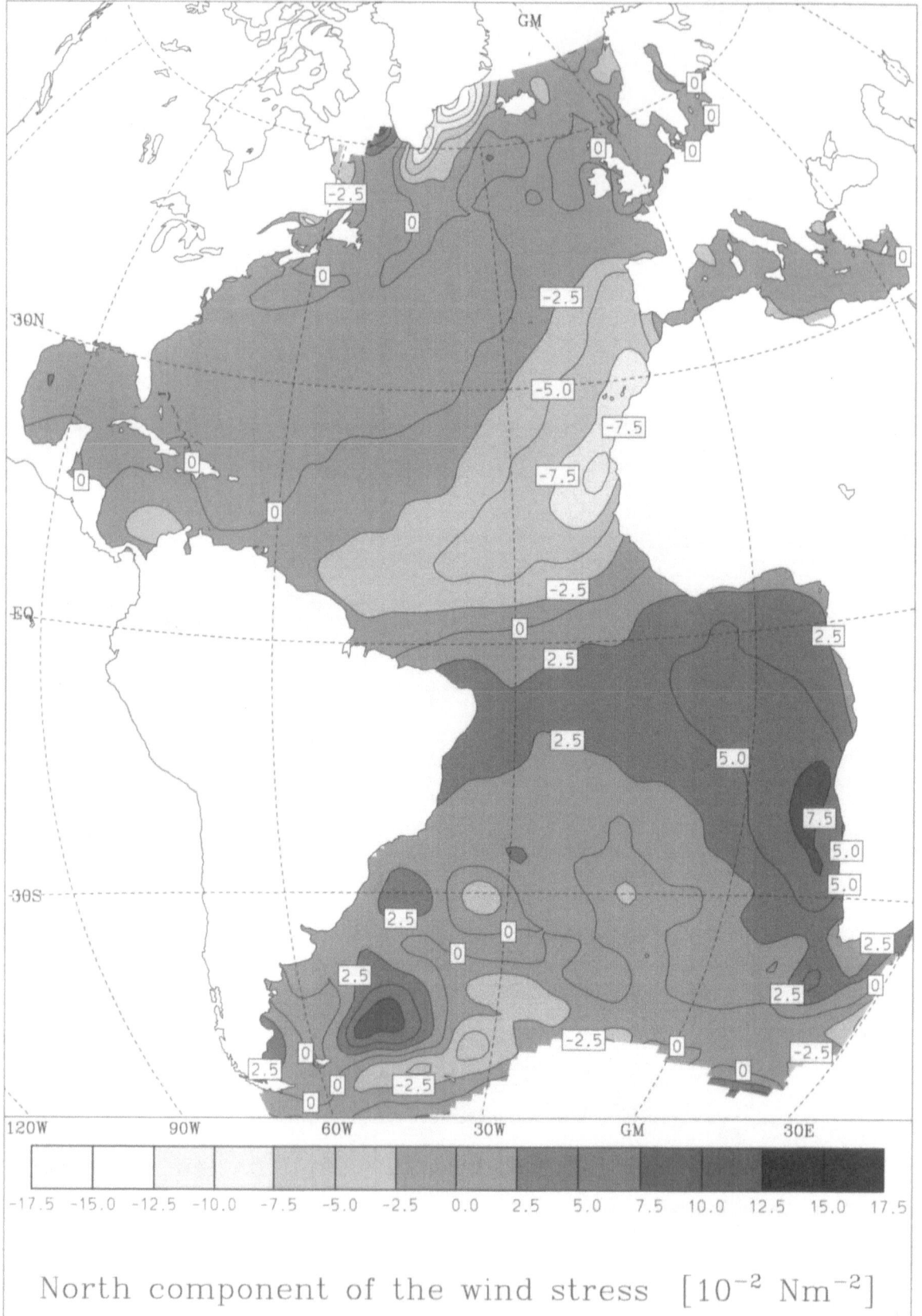

North component of the wind stress $[10^{-2} \text{ Nm}^{-2}]$
May

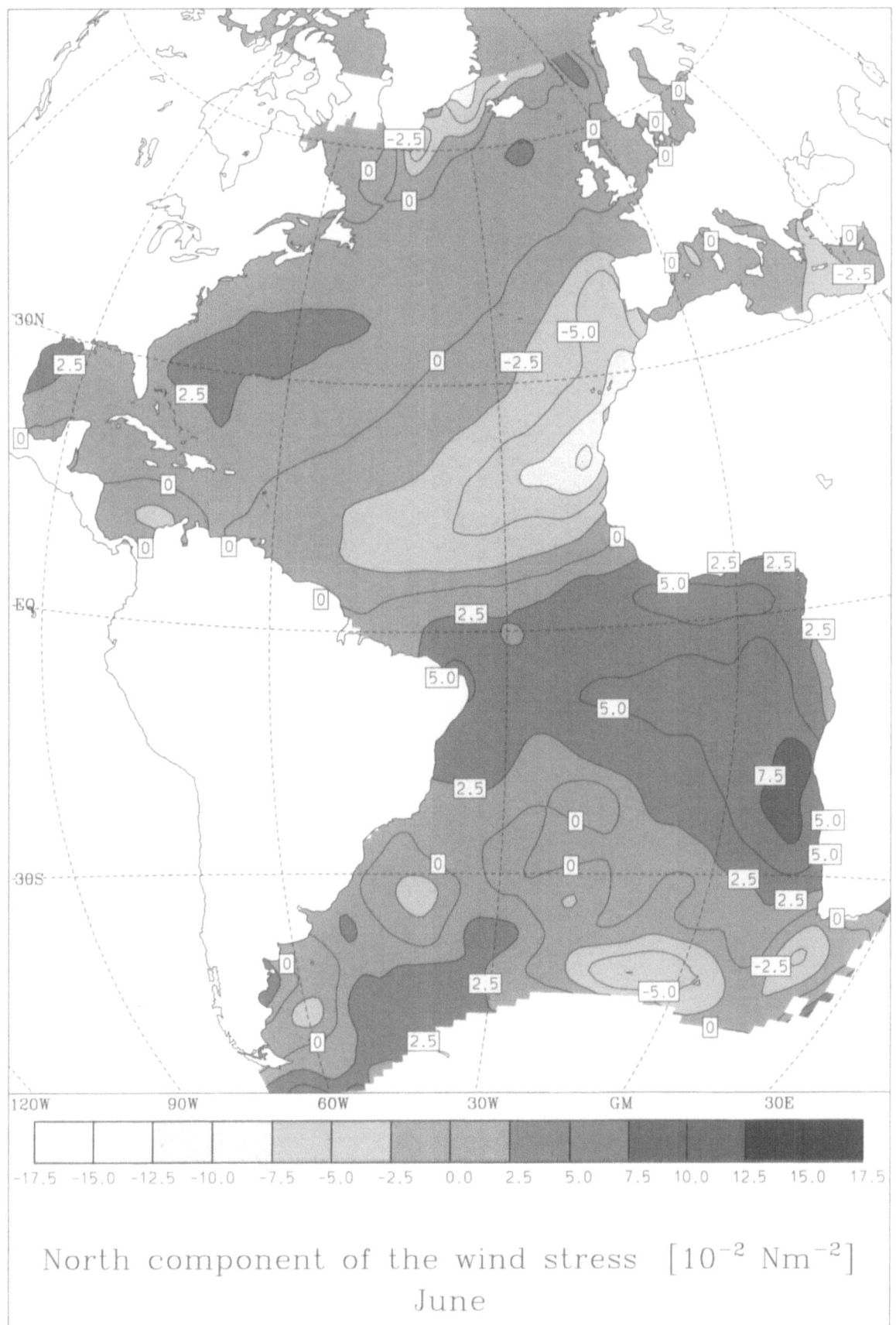

North component of the wind stress $[10^{-2}\ \mathrm{Nm}^{-2}]$

June

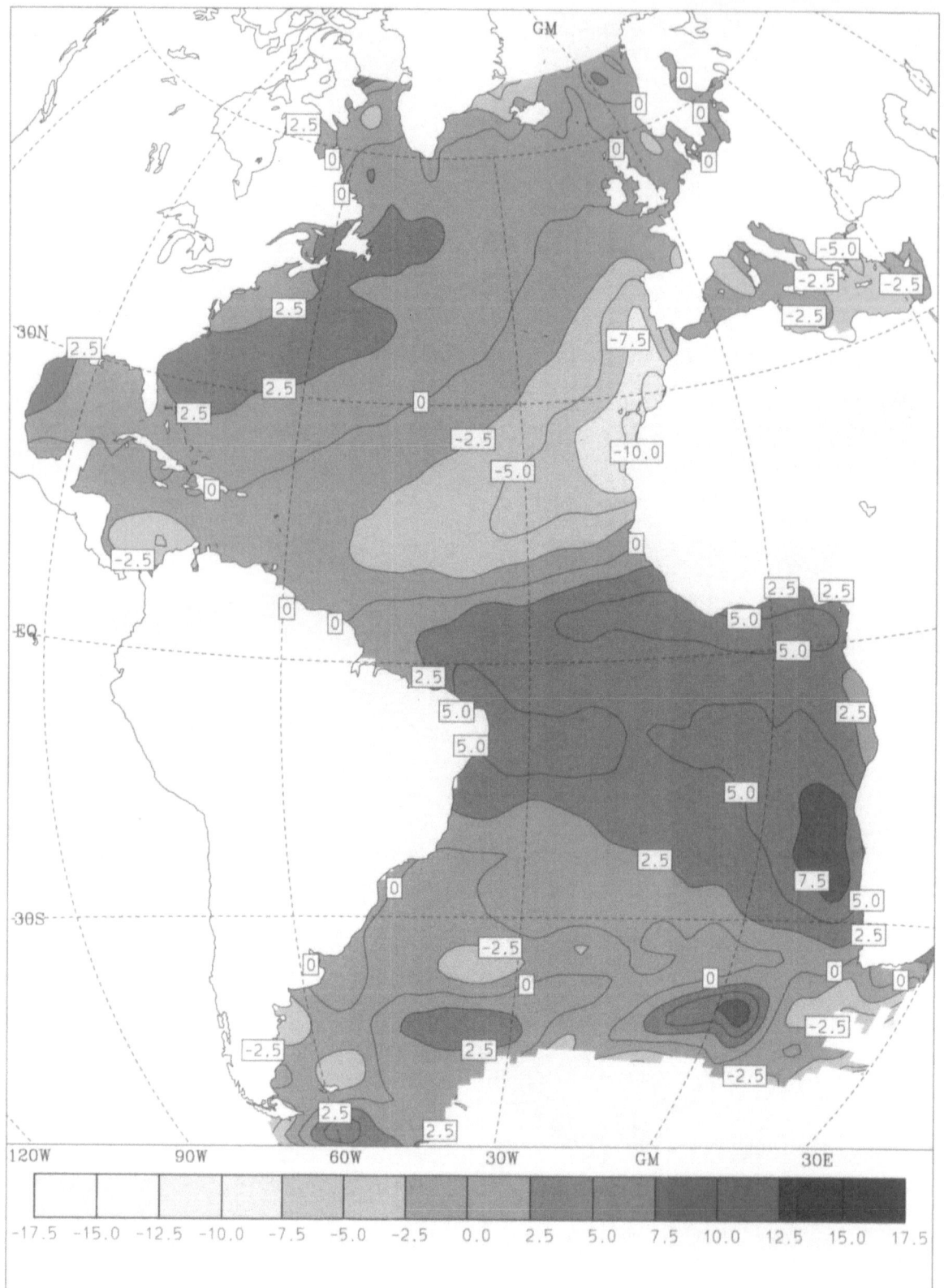

North component of the wind stress $[10^{-2}\ Nm^{-2}]$
July

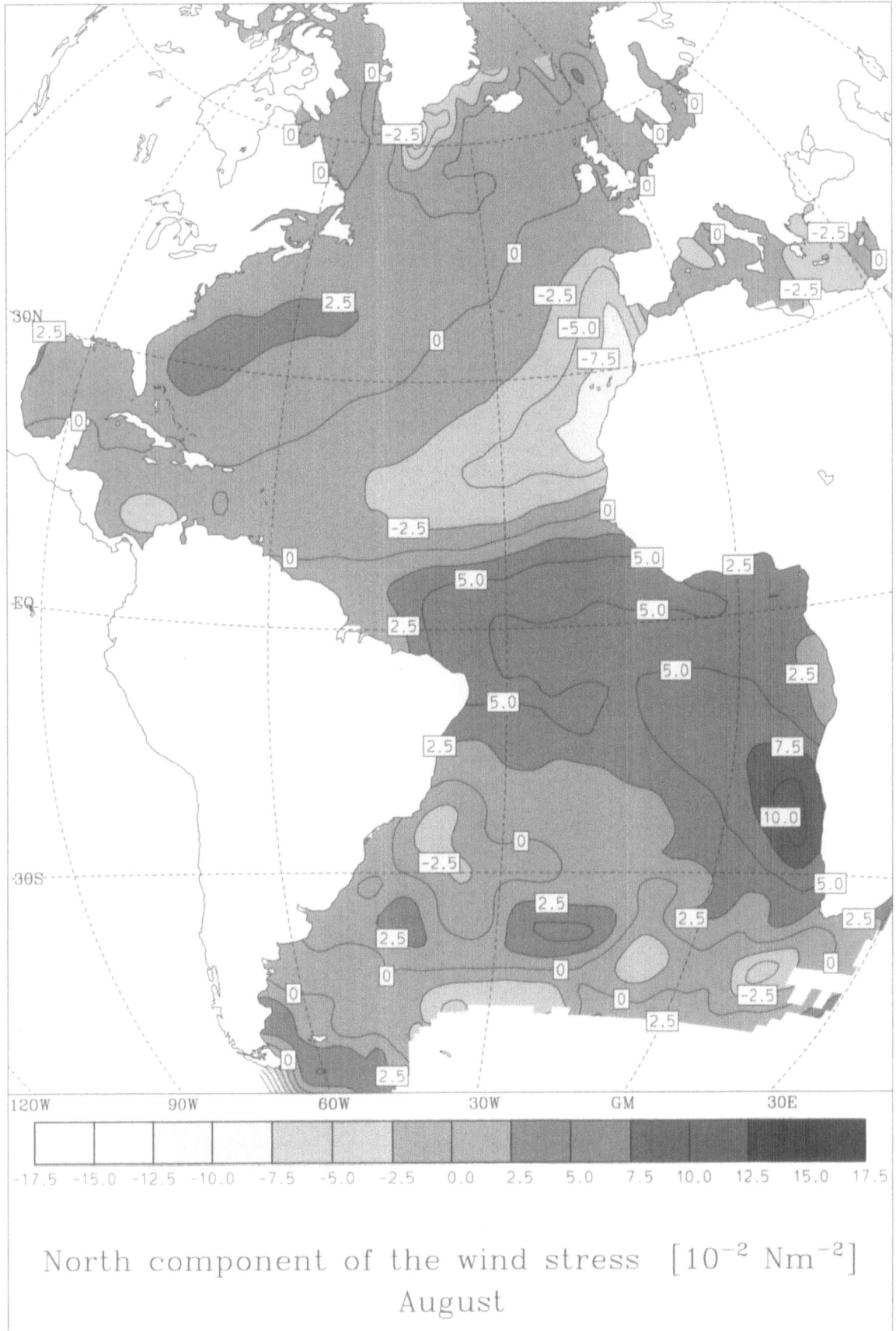

North component of the wind stress $[10^{-2} \ Nm^{-2}]$
August

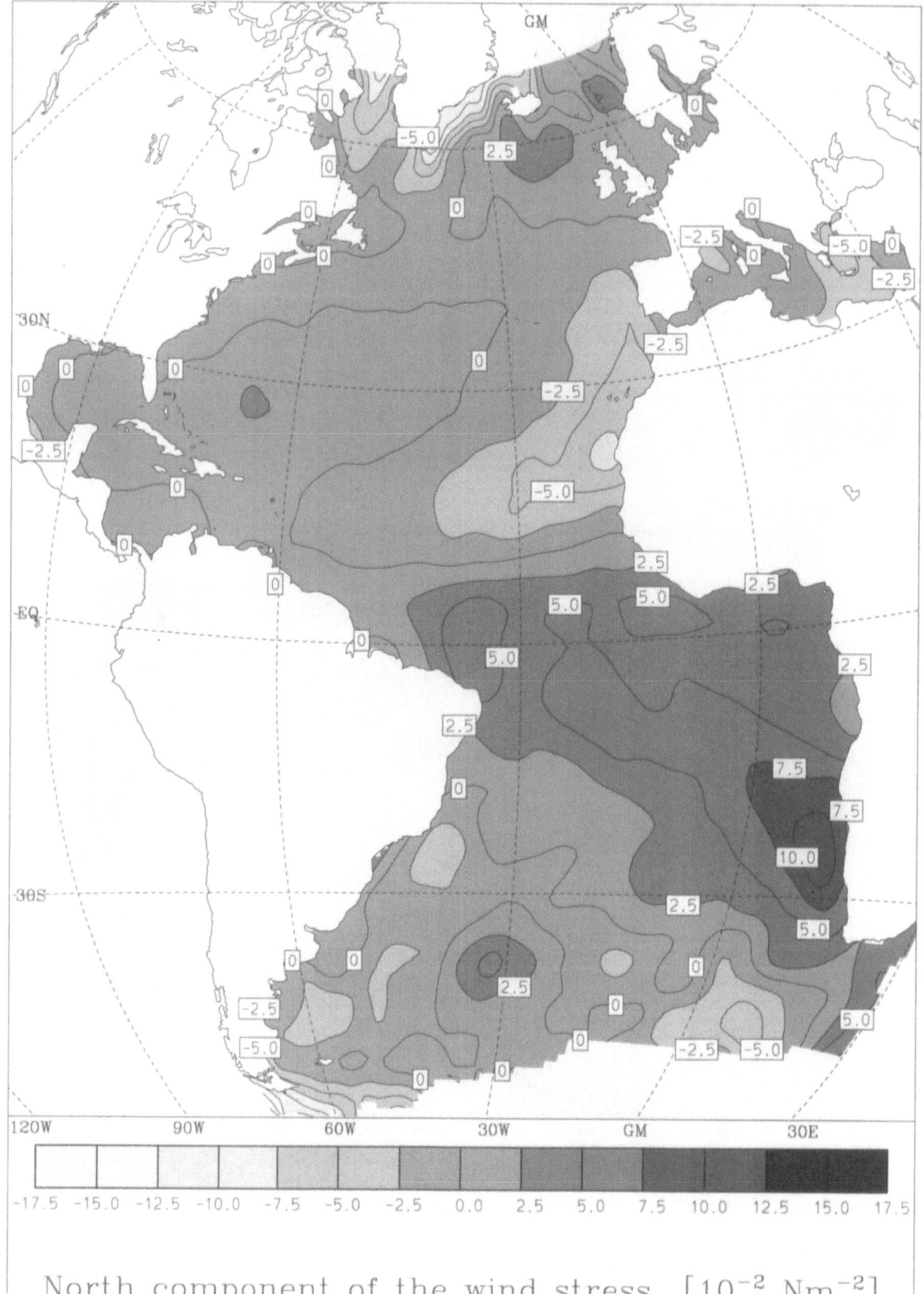

North component of the wind stress $[10^{-2} \text{ Nm}^{-2}]$
September

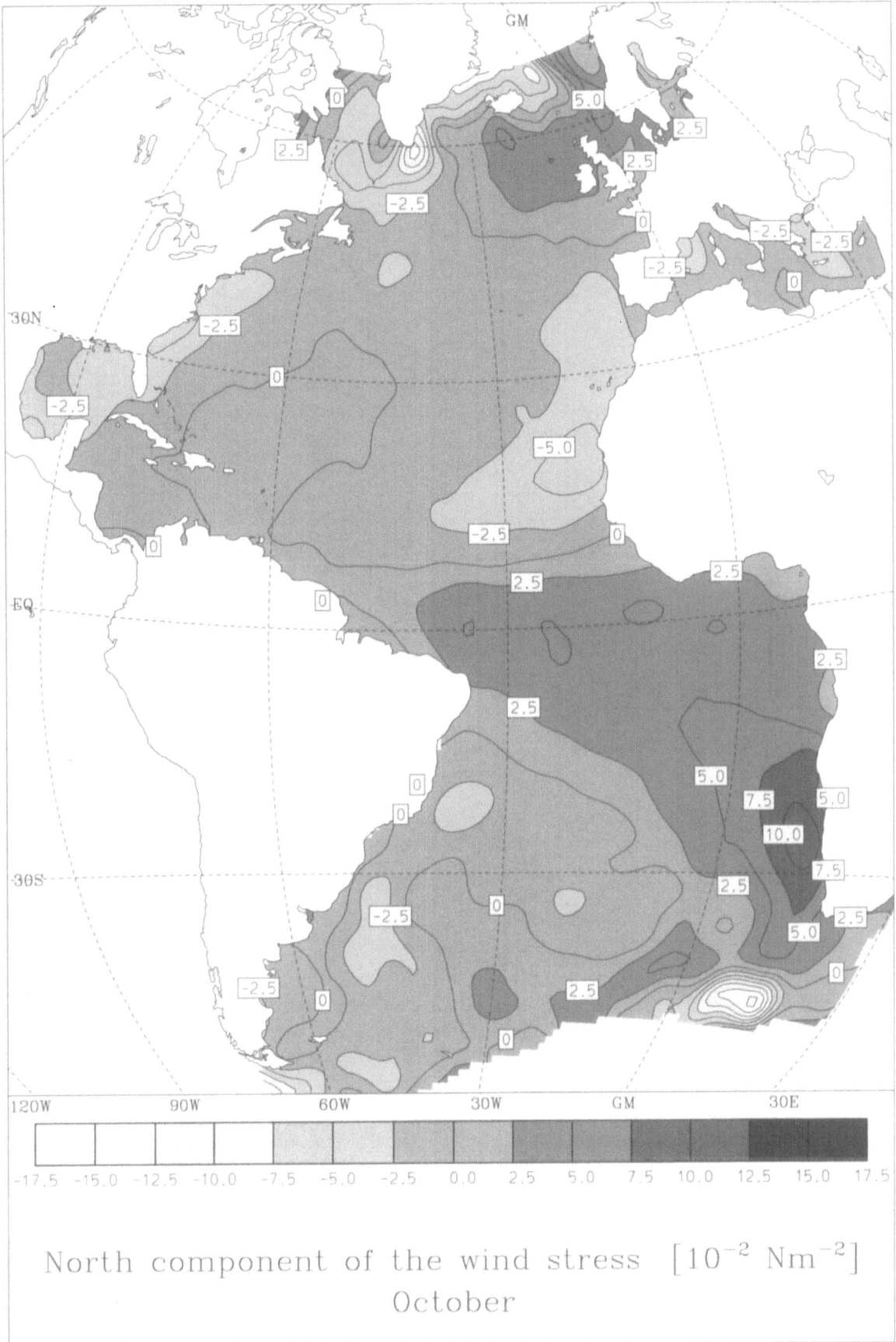

North component of the wind stress $[10^{-2}\ Nm^{-2}]$
October

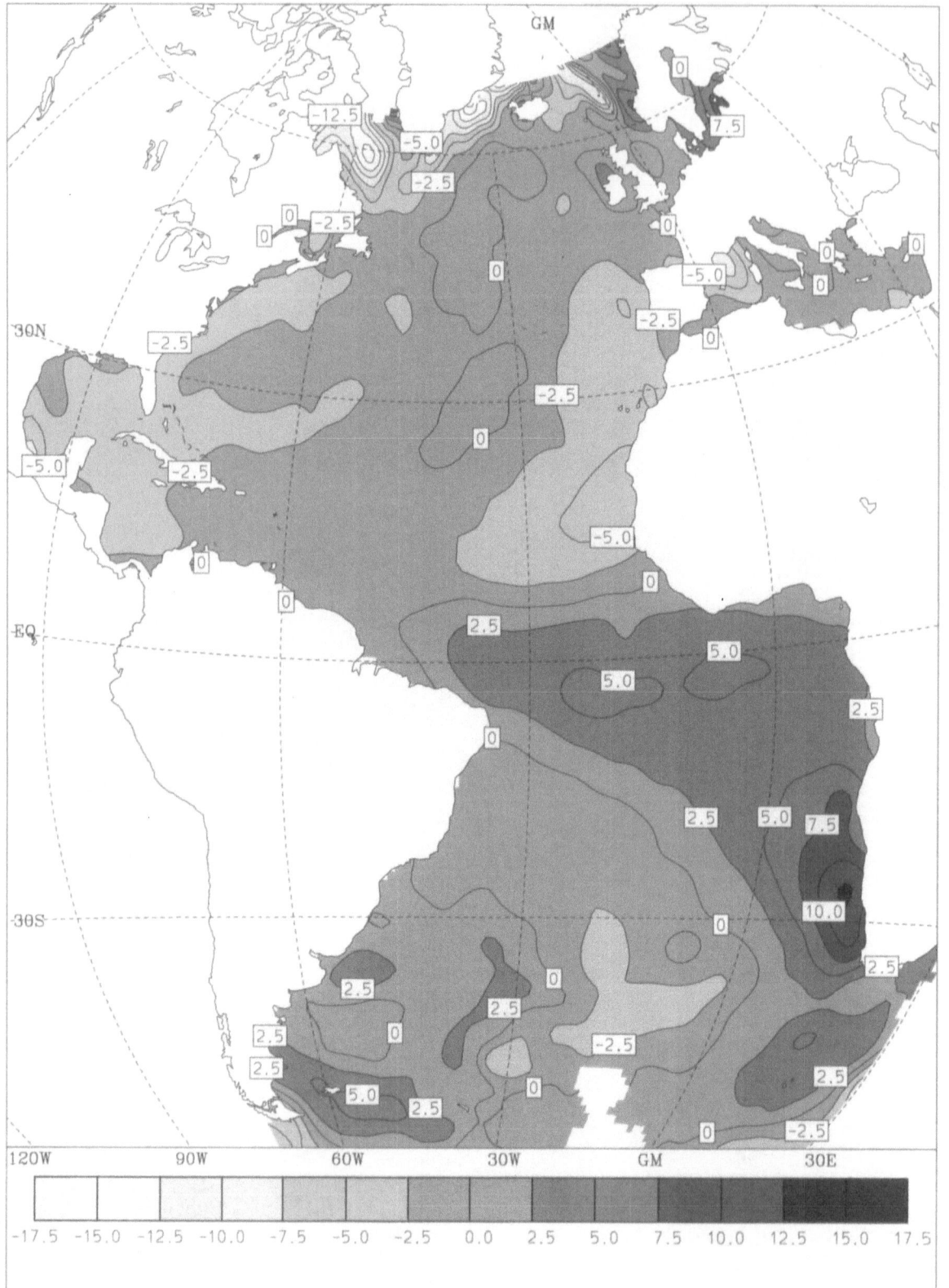

North component of the wind stress $[10^{-2} \ Nm^{-2}]$
November

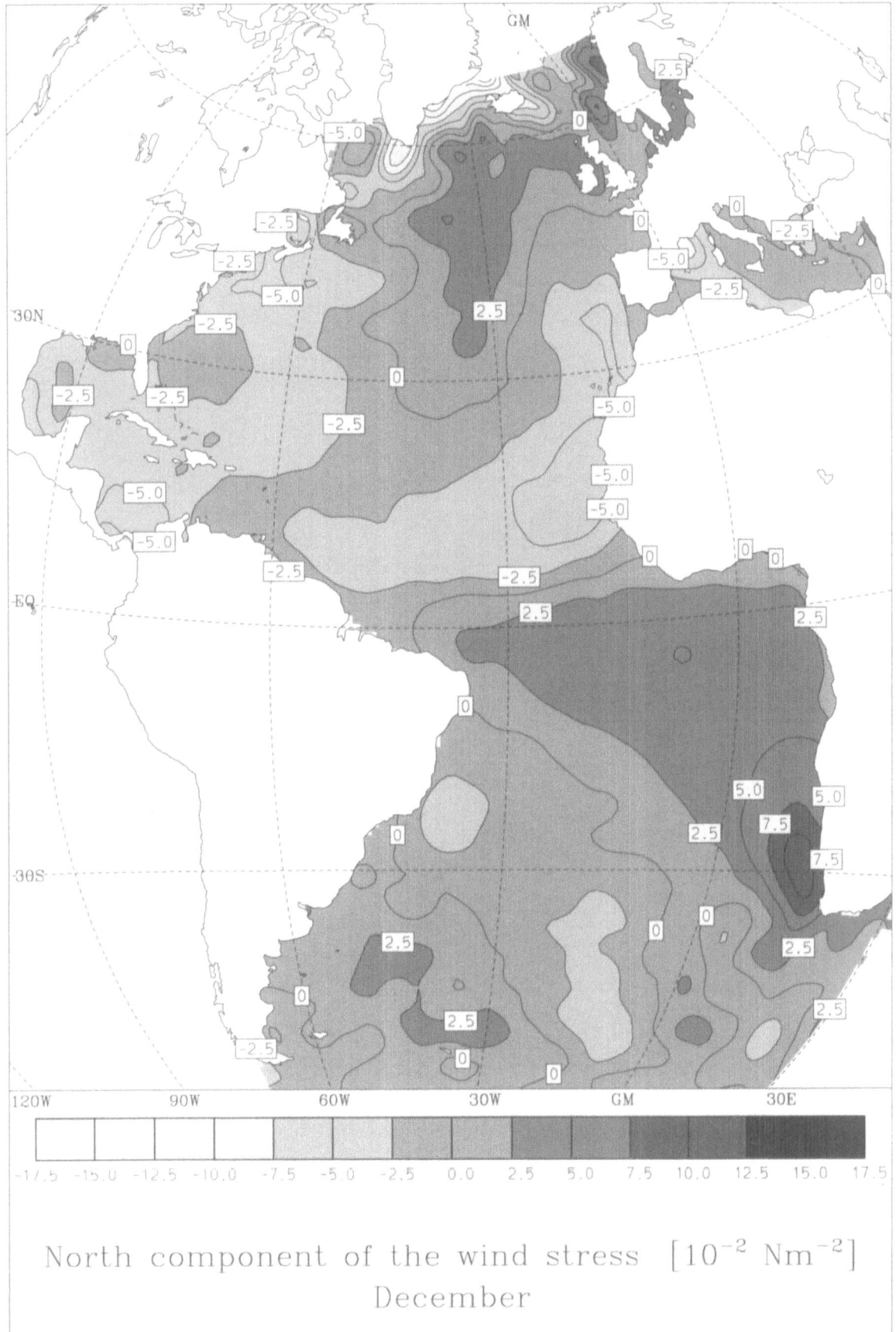

North component of the wind stress $[10^{-2}\ Nm^{-2}]$
December

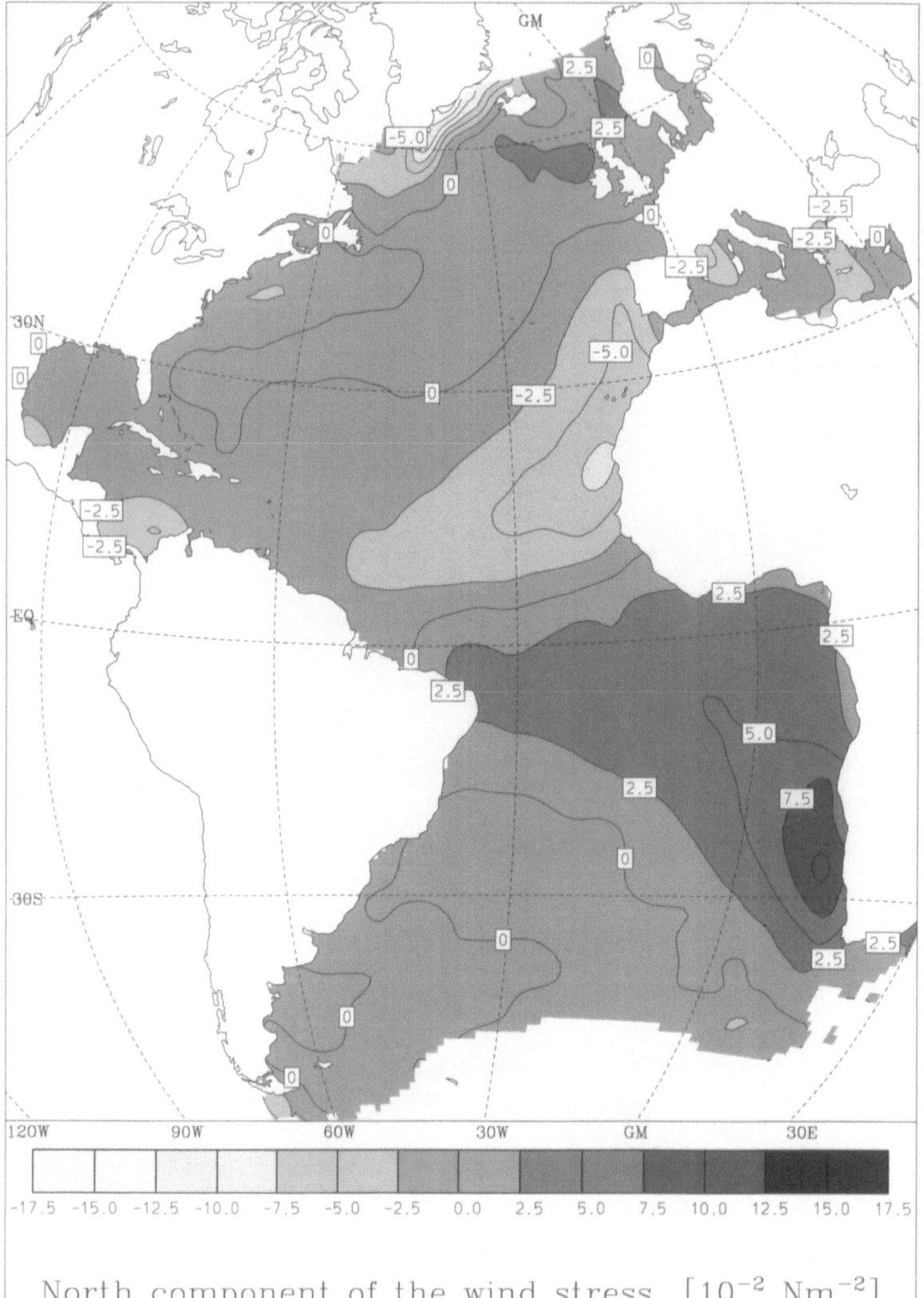

North component of the wind stress $[10^{-2} \text{ Nm}^{-2}]$
Year

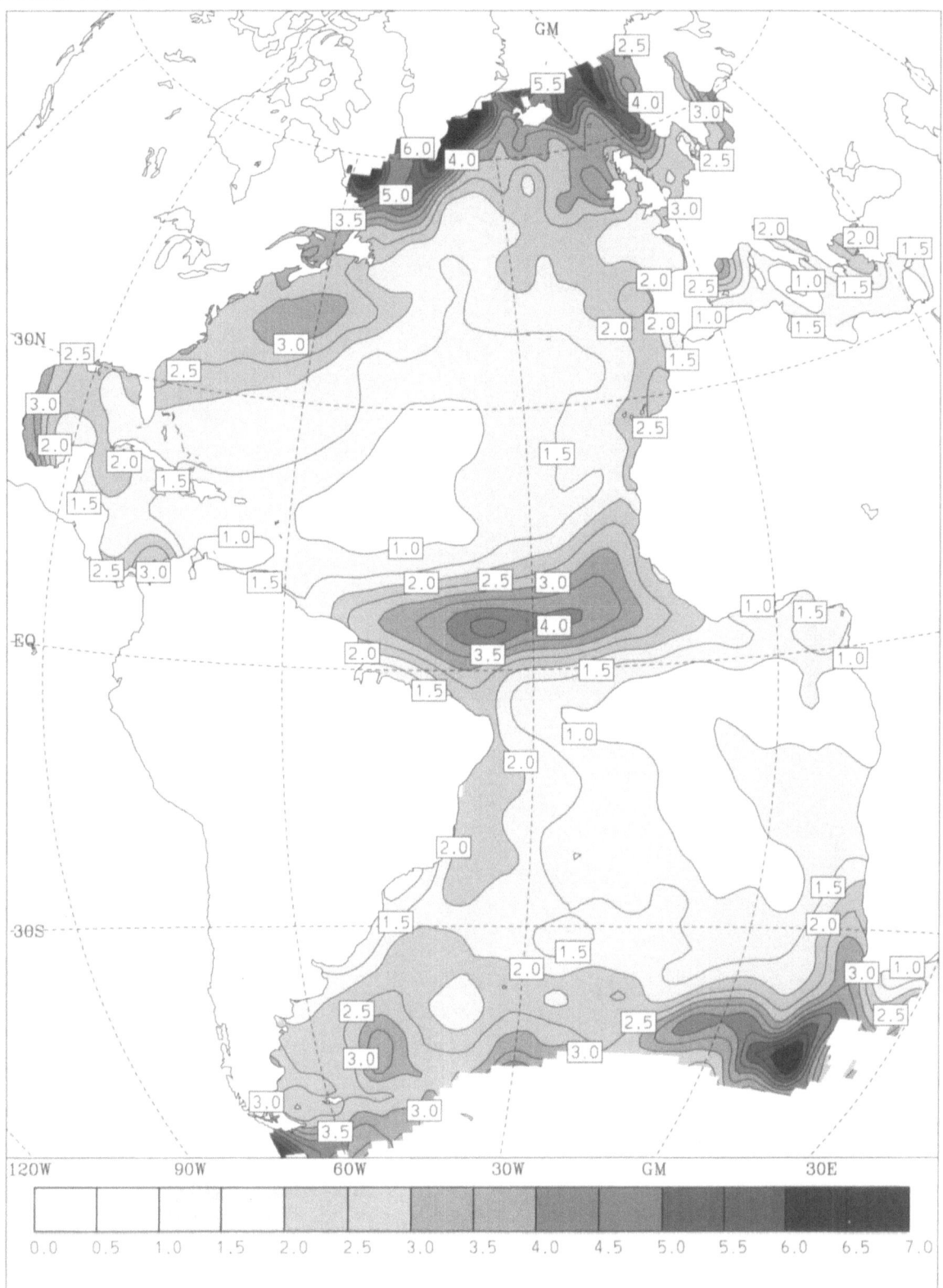

North component of the wind stress $[10^{-2} \text{ Nm}^{-2}]$
Standard deviation due to the annual cycle

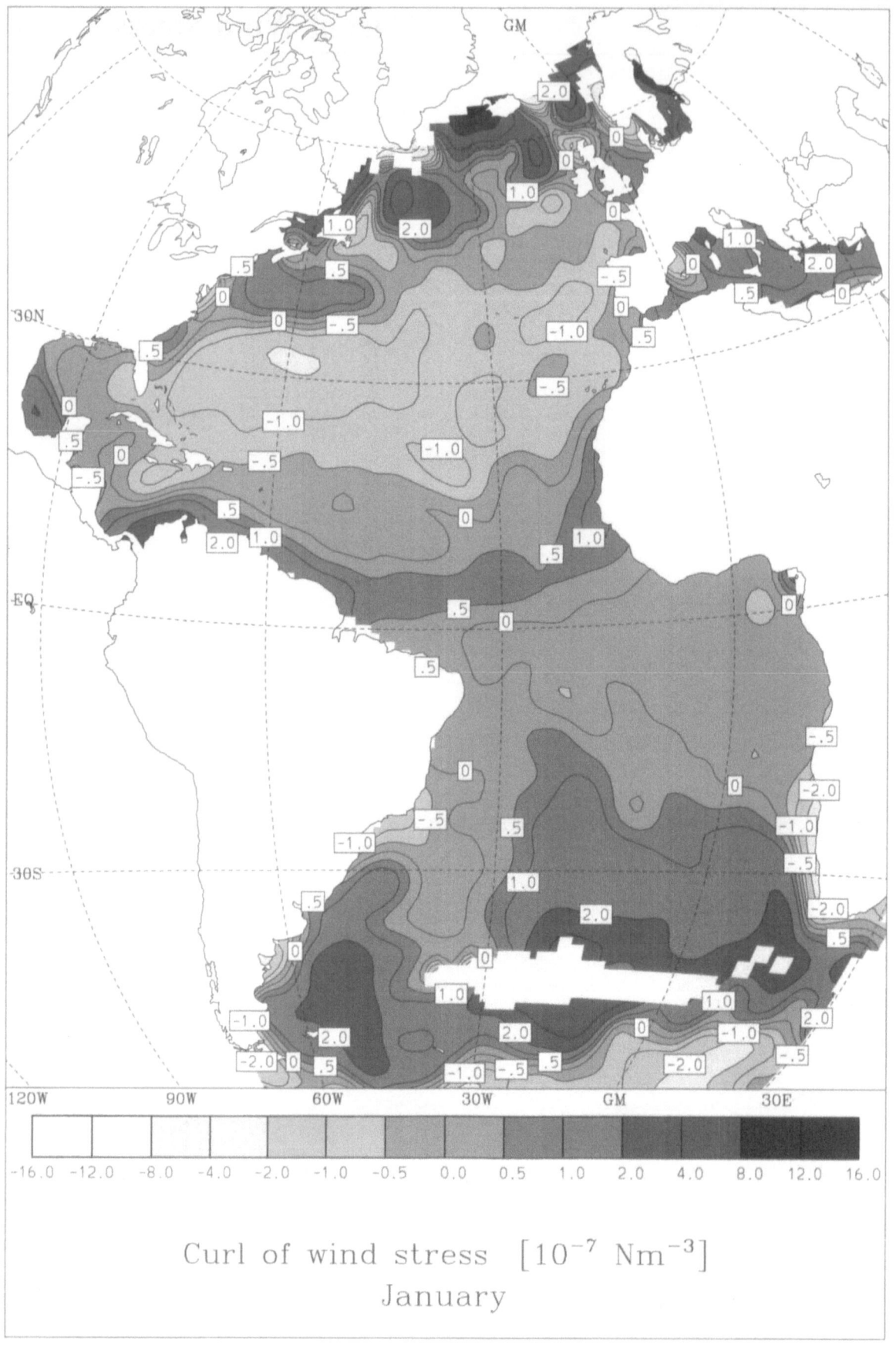

Curl of wind stress $[10^{-7} \ Nm^{-3}]$
January

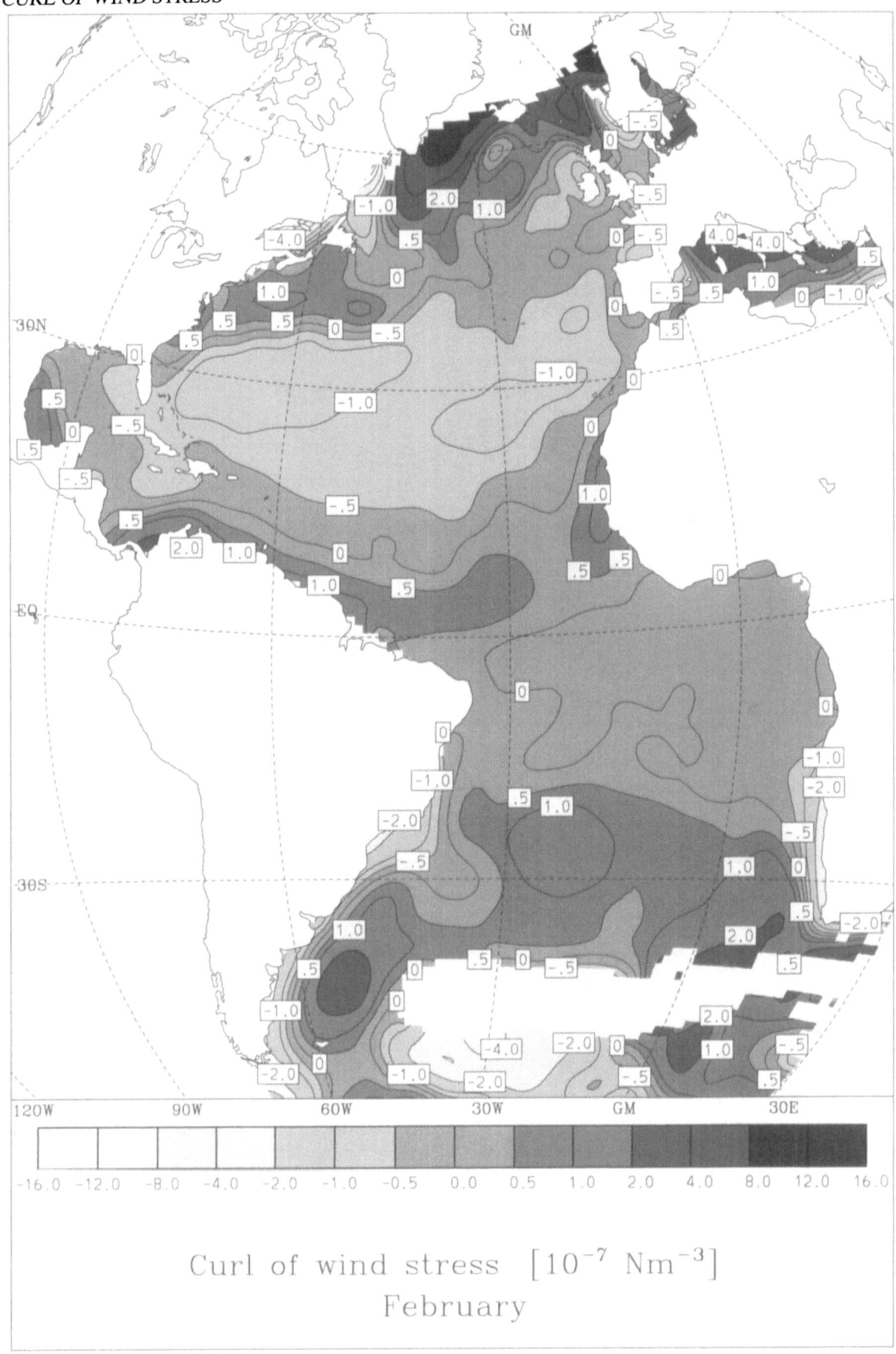

Curl of wind stress $[10^{-7} \ Nm^{-3}]$
February

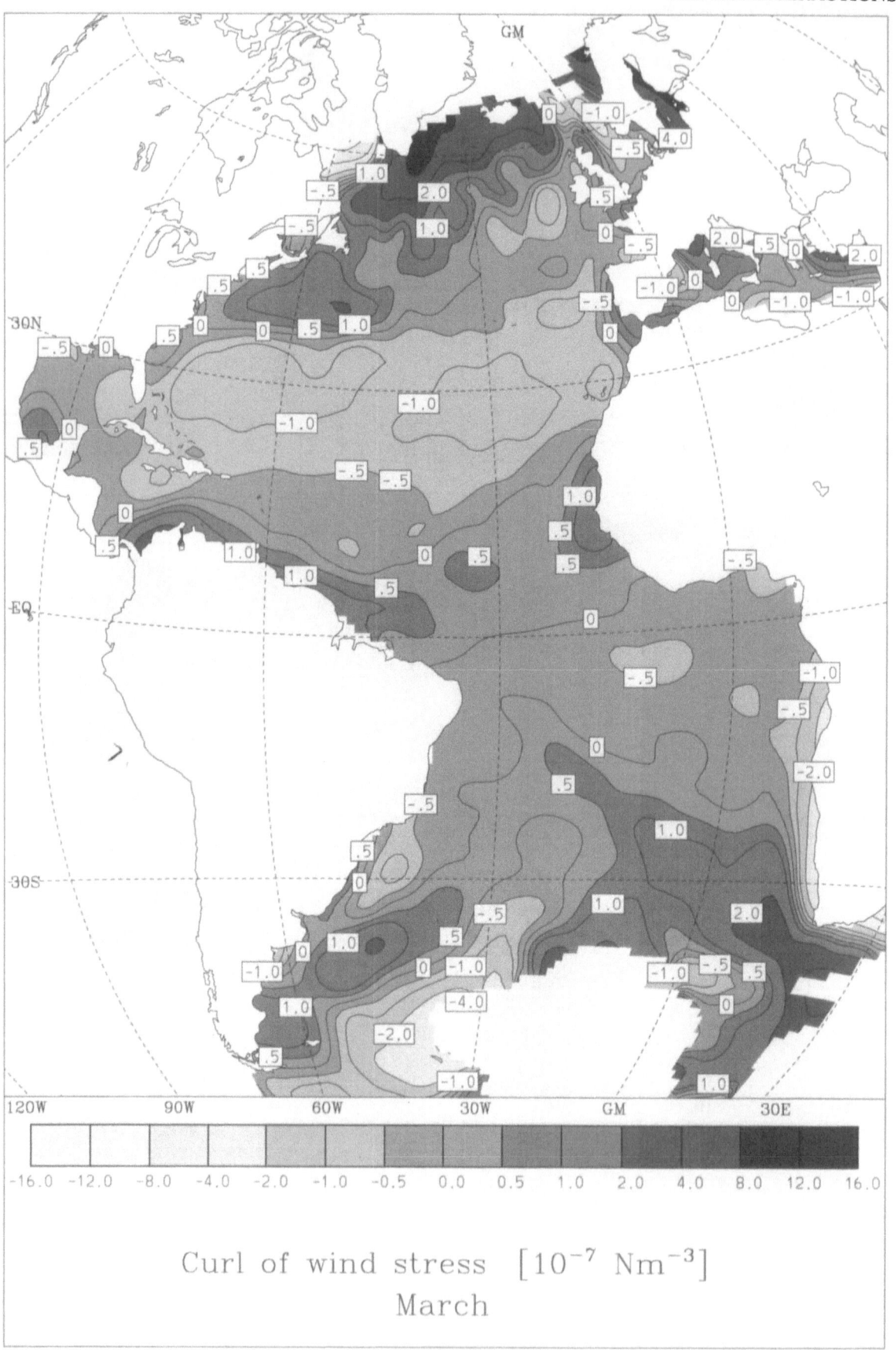

Curl of wind stress $[10^{-7} \ Nm^{-3}]$

March

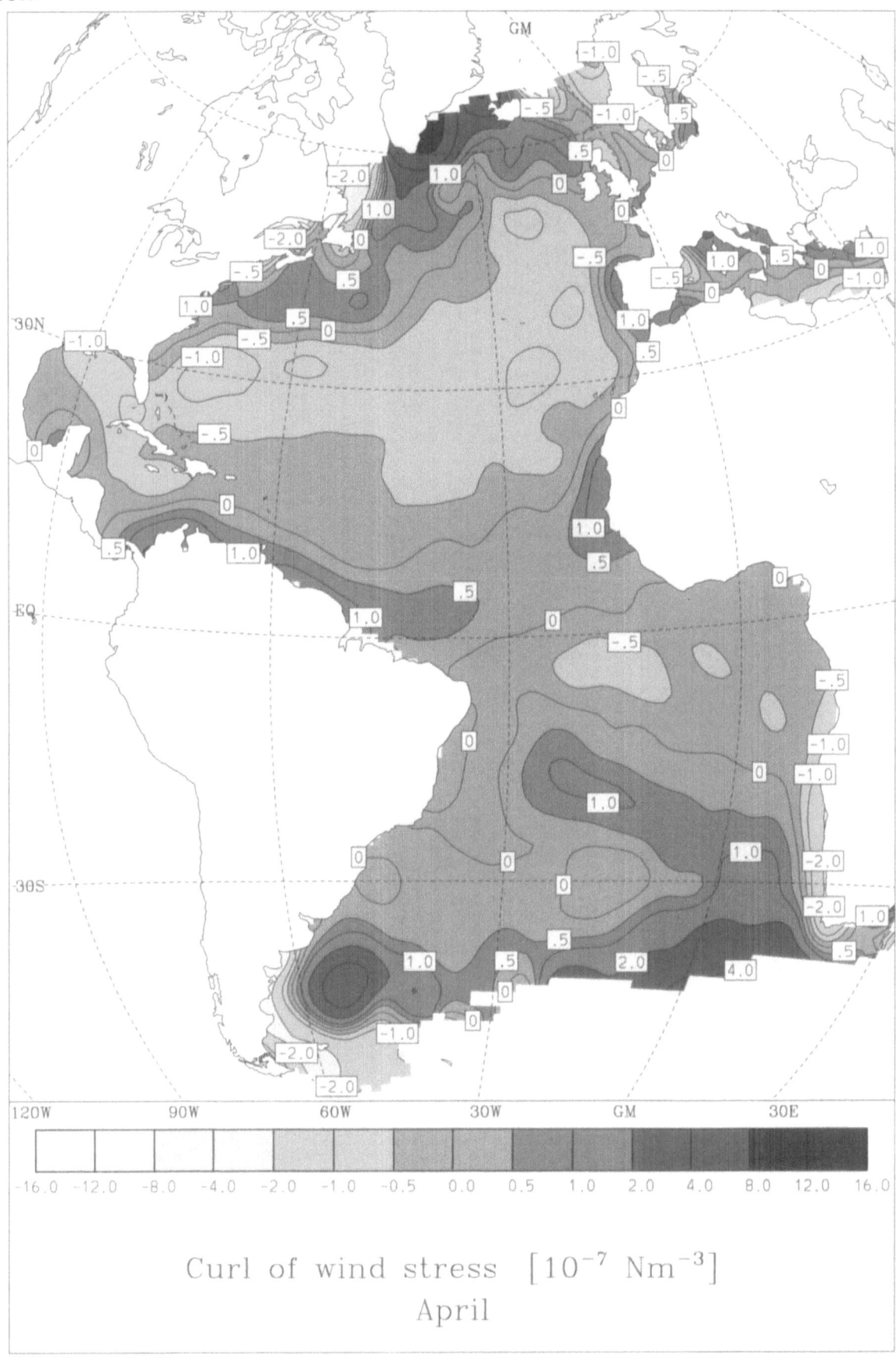

Curl of wind stress $[10^{-7} \text{ Nm}^{-3}]$
April

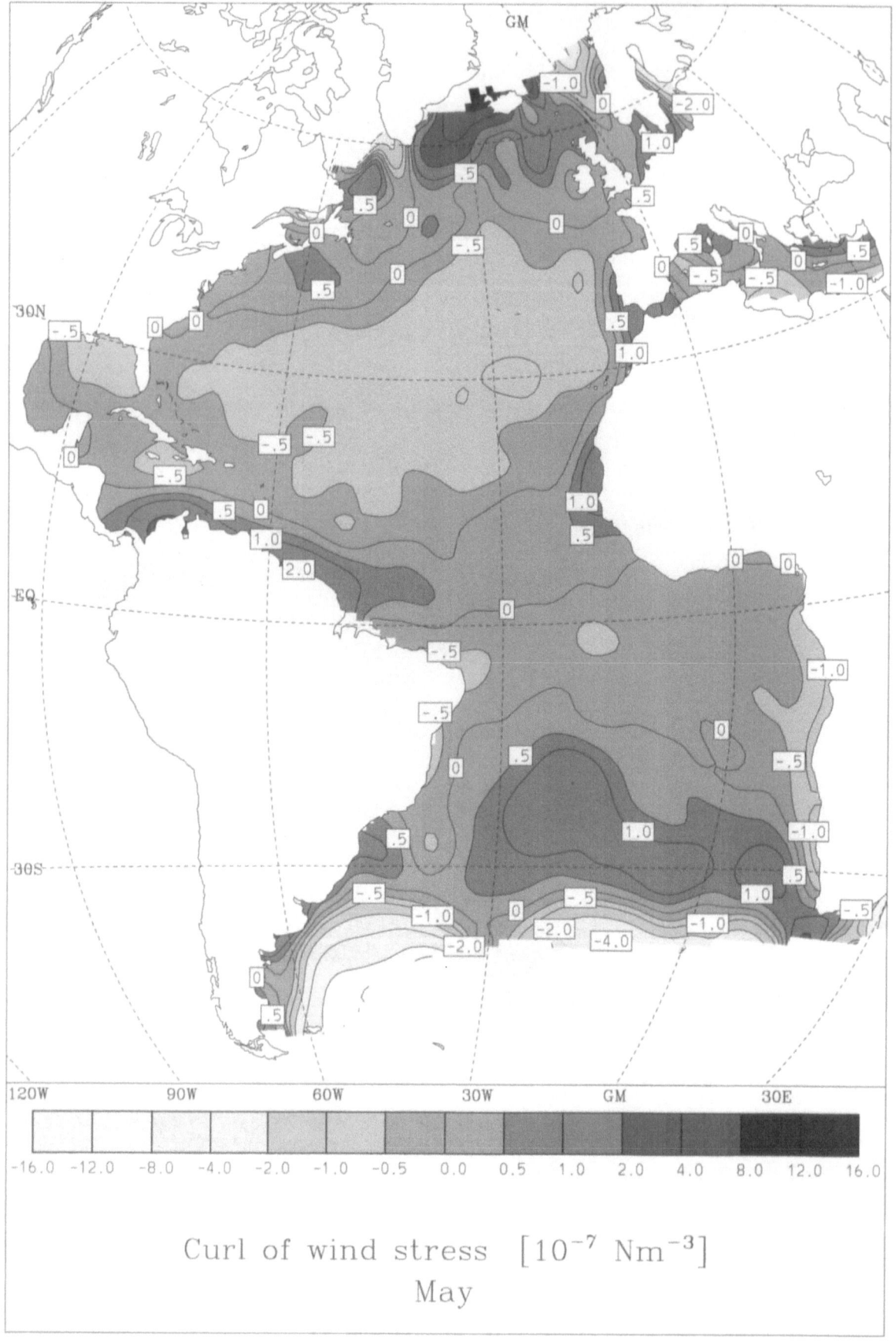

Curl of wind stress $[10^{-7} \text{ Nm}^{-3}]$
May

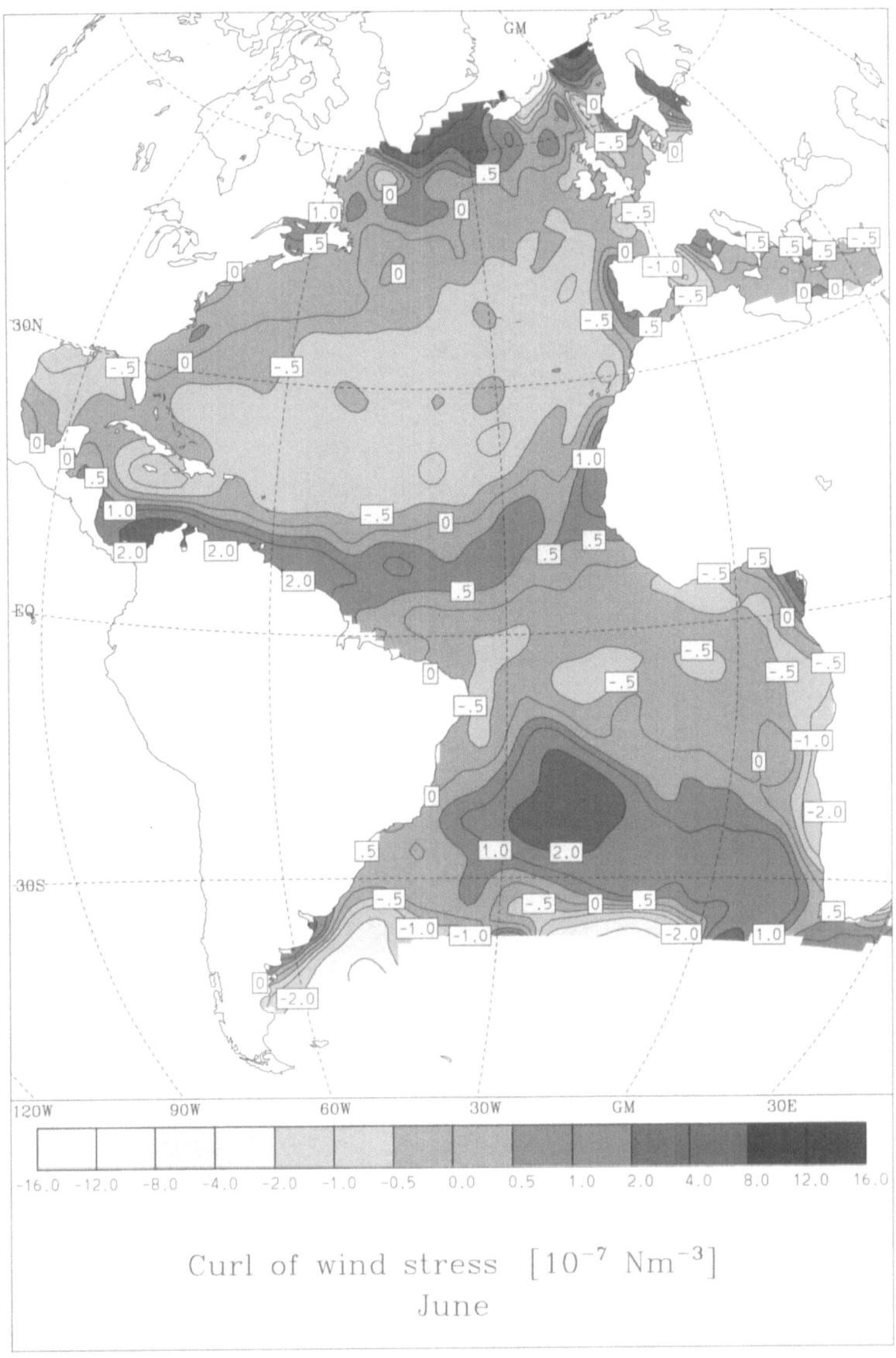

Curl of wind stress $[10^{-7} \text{ Nm}^{-3}]$
June

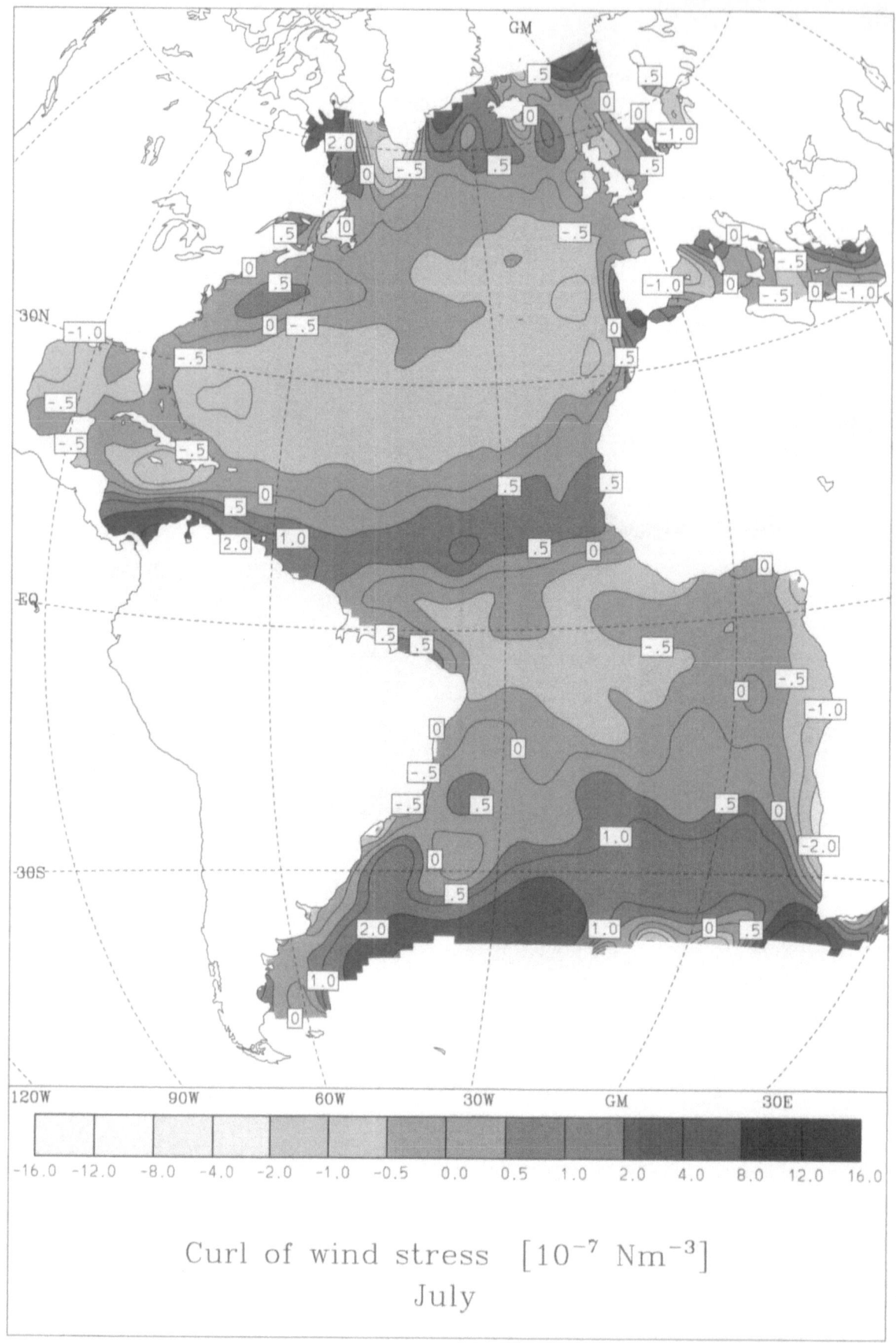

Curl of wind stress $[10^{-7} \; Nm^{-3}]$

July

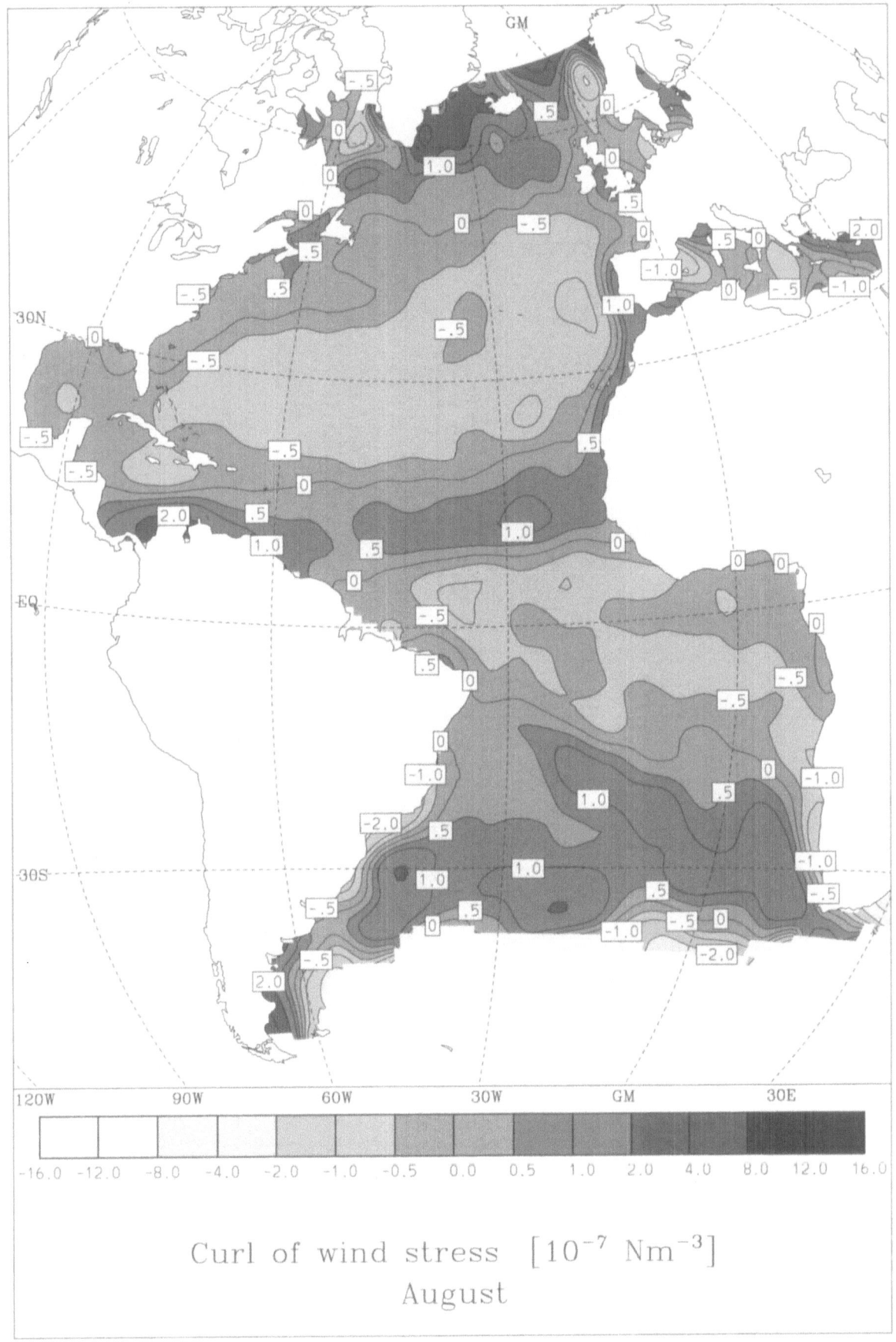

Curl of wind stress $[10^{-7} \ \mathrm{Nm^{-3}}]$
August

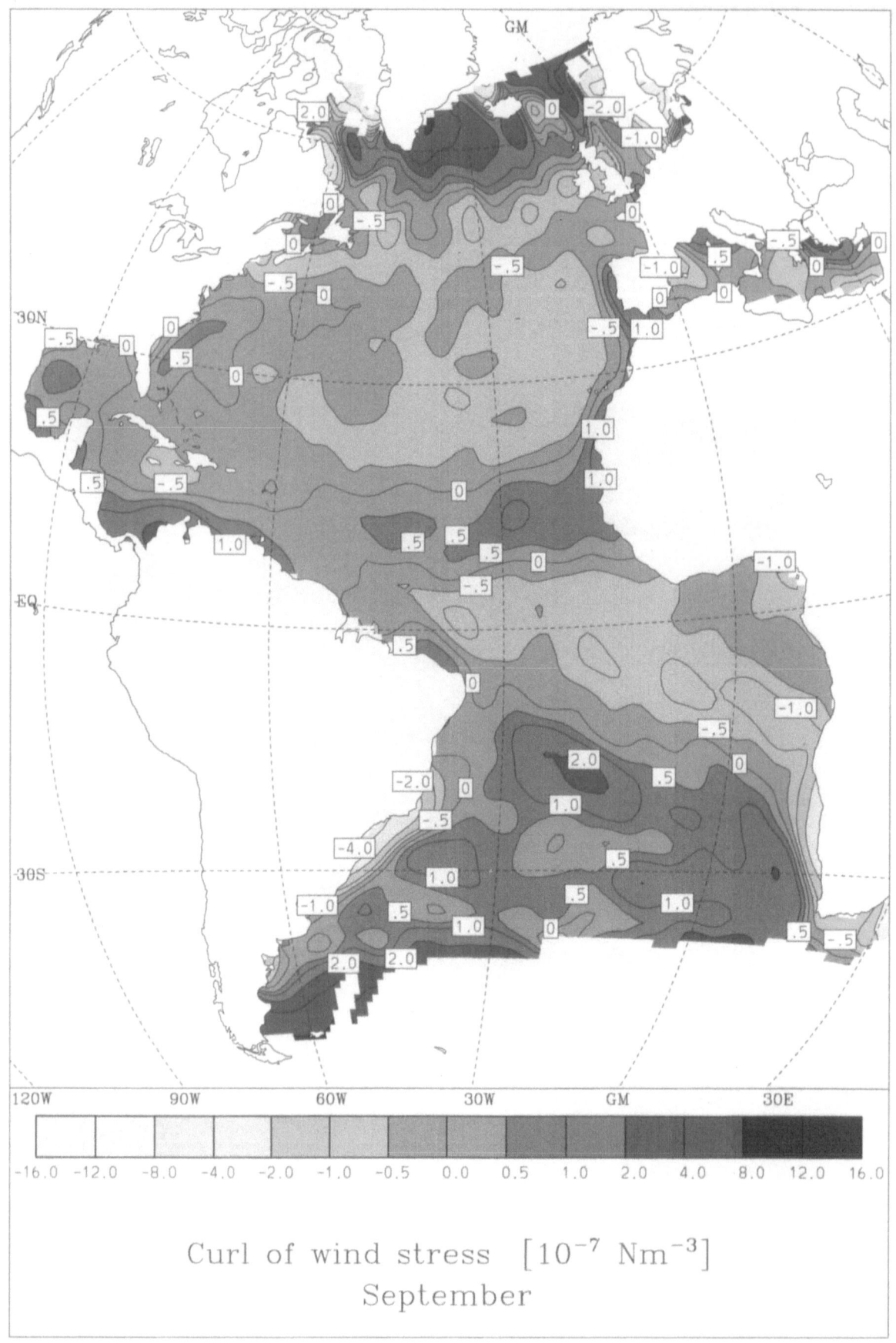

Curl of wind stress $[10^{-7} \ Nm^{-3}]$
September

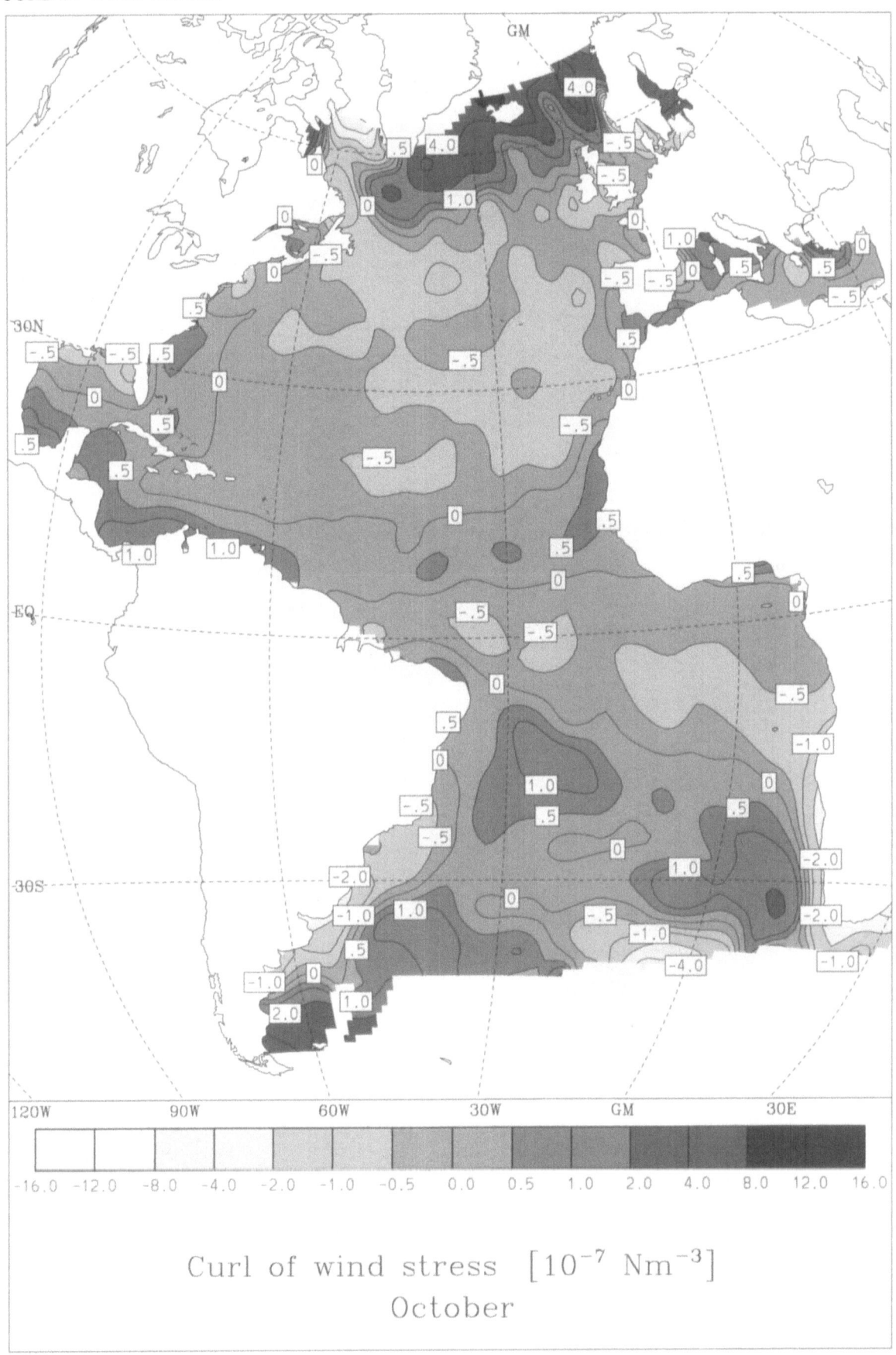

Curl of wind stress $[10^{-7} \text{ Nm}^{-3}]$
October

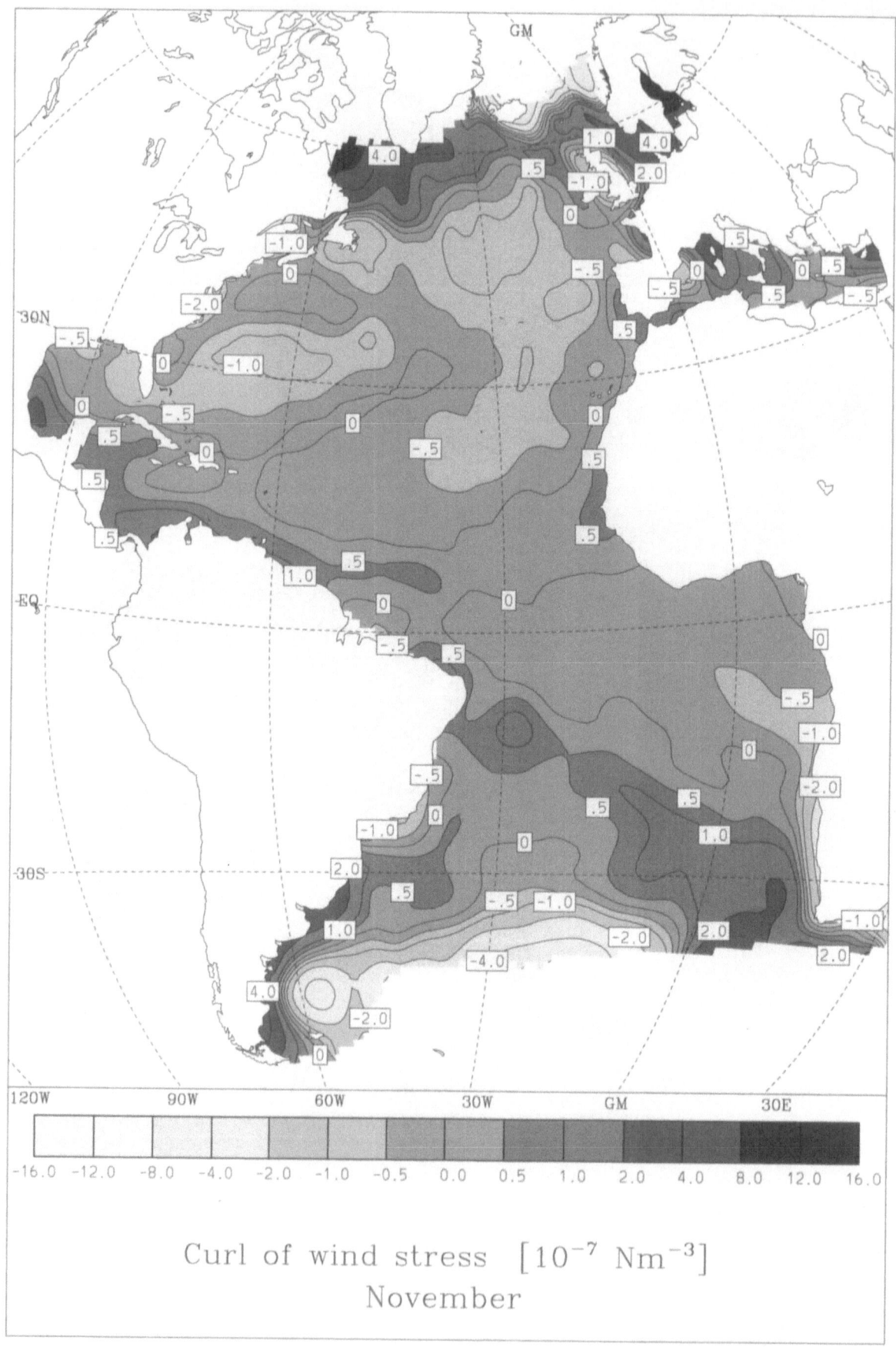

Curl of wind stress $[10^{-7}\ \mathrm{Nm^{-3}}]$
November

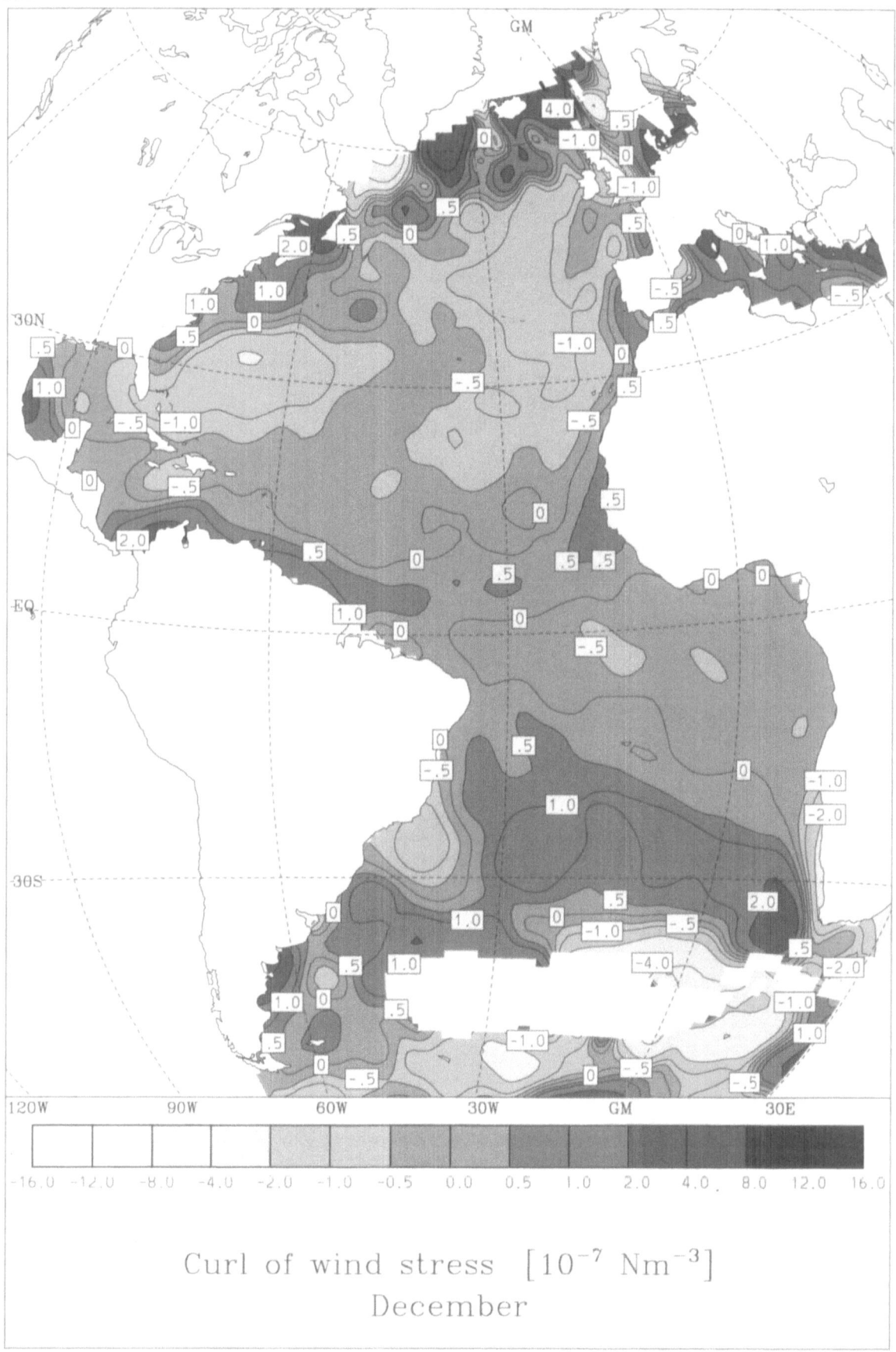

Curl of wind stress $[10^{-7} \text{ Nm}^{-3}]$
December

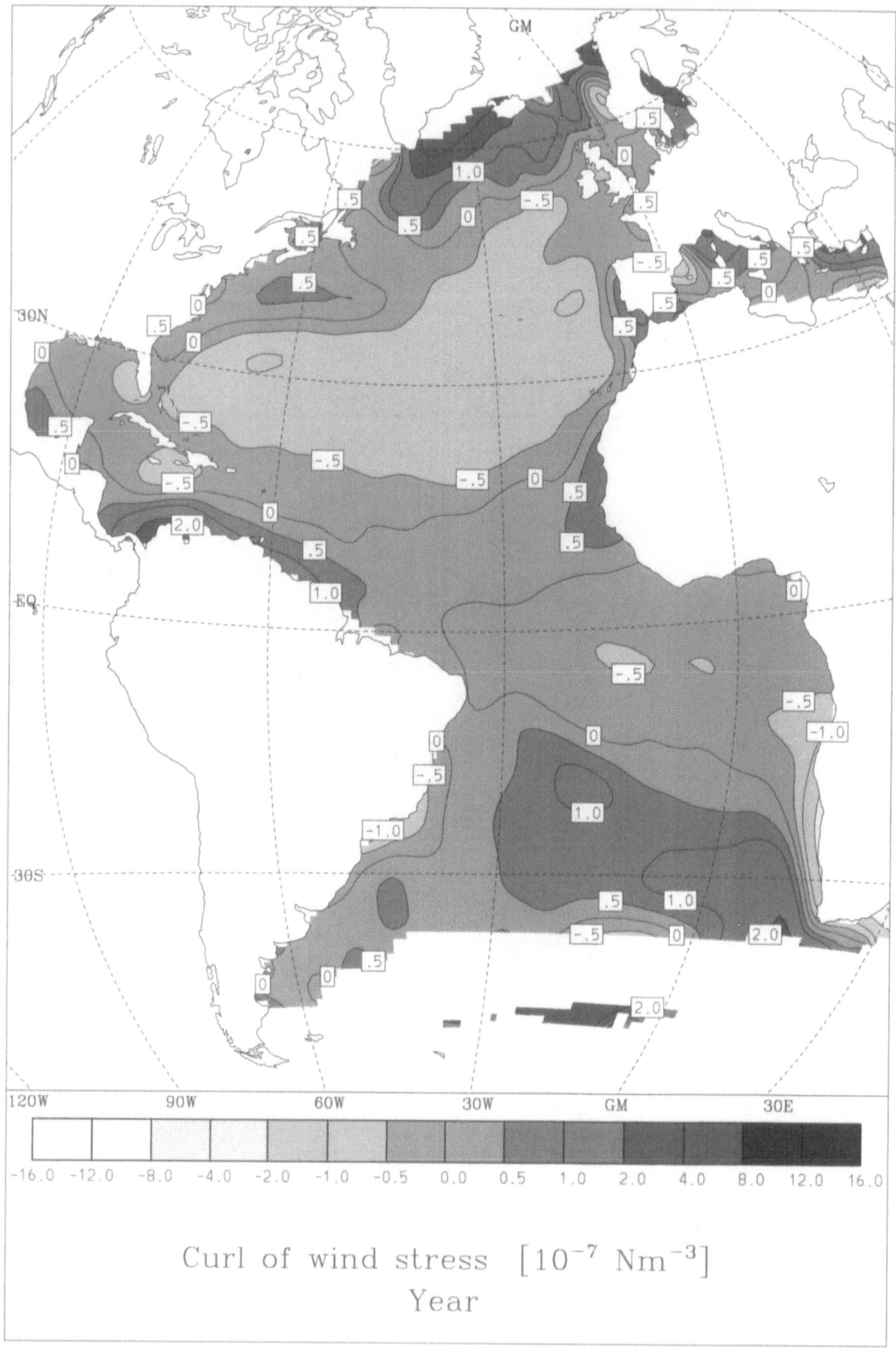

Curl of wind stress $[10^{-7}\ Nm^{-3}]$
Year

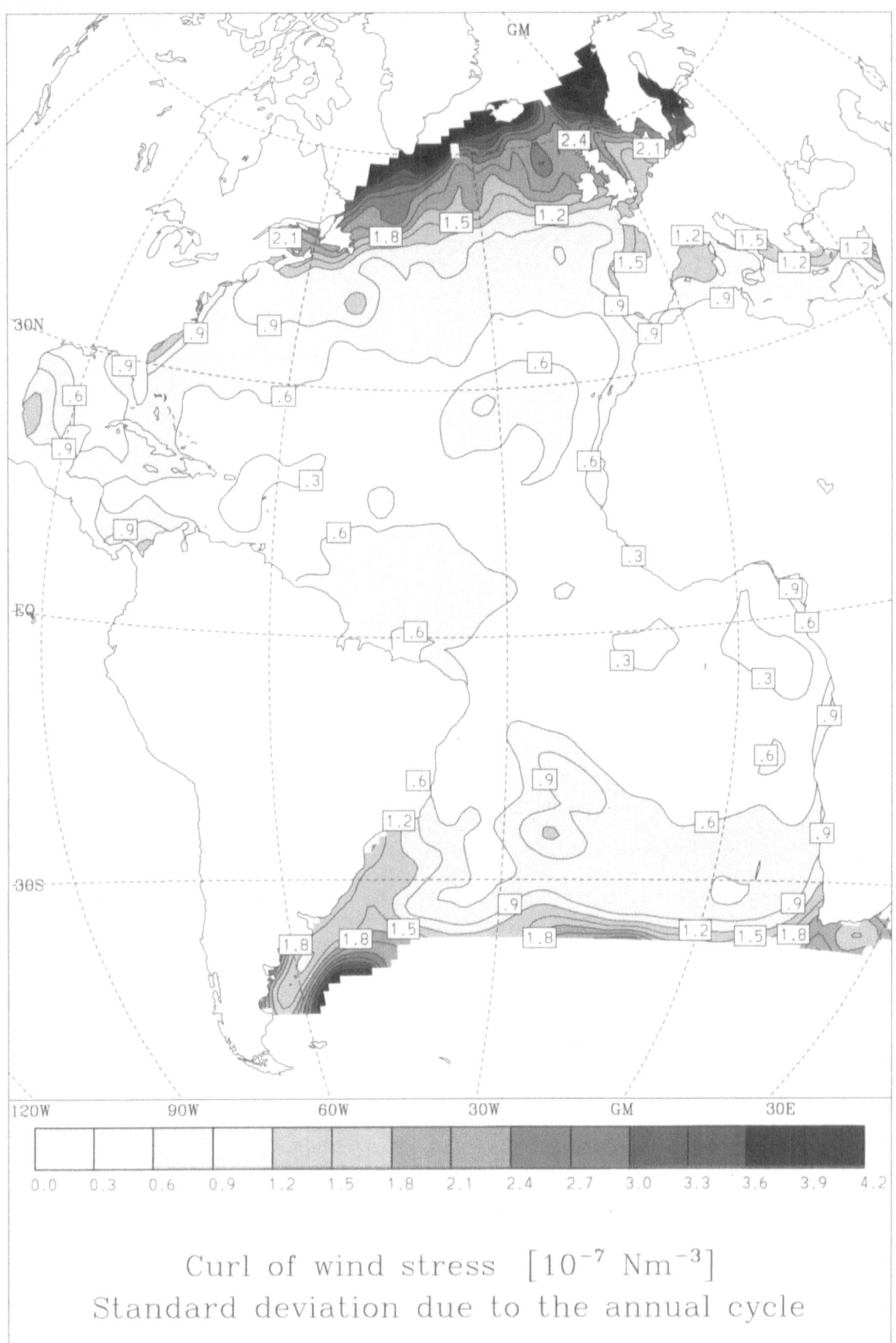

Curl of wind stress $[10^{-7} \ \mathrm{Nm^{-3}}]$
Standard deviation due to the annual cycle

Ekman Volume Transport

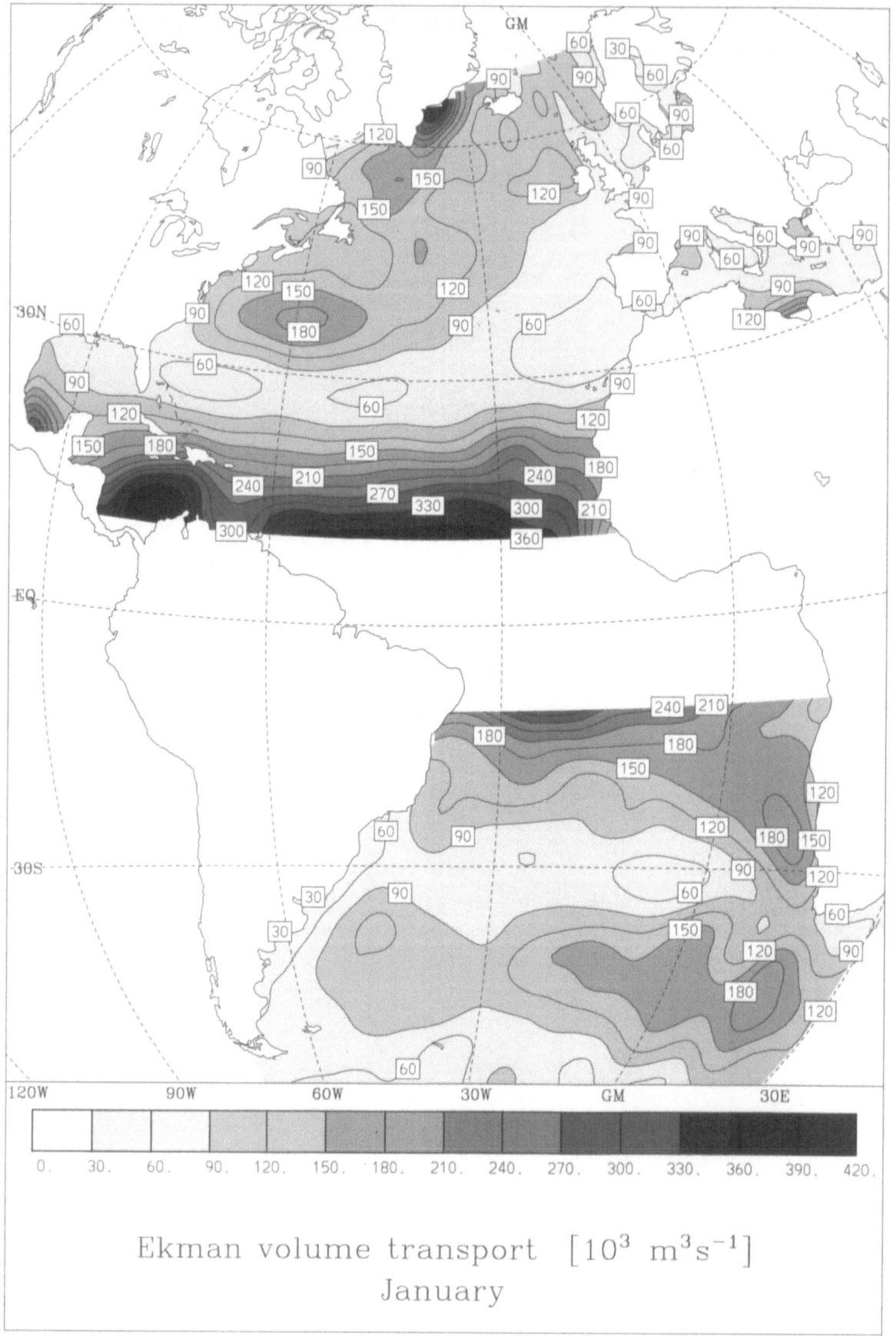

Ekman volume transport $[10^3 \ m^3 s^{-1}]$
January

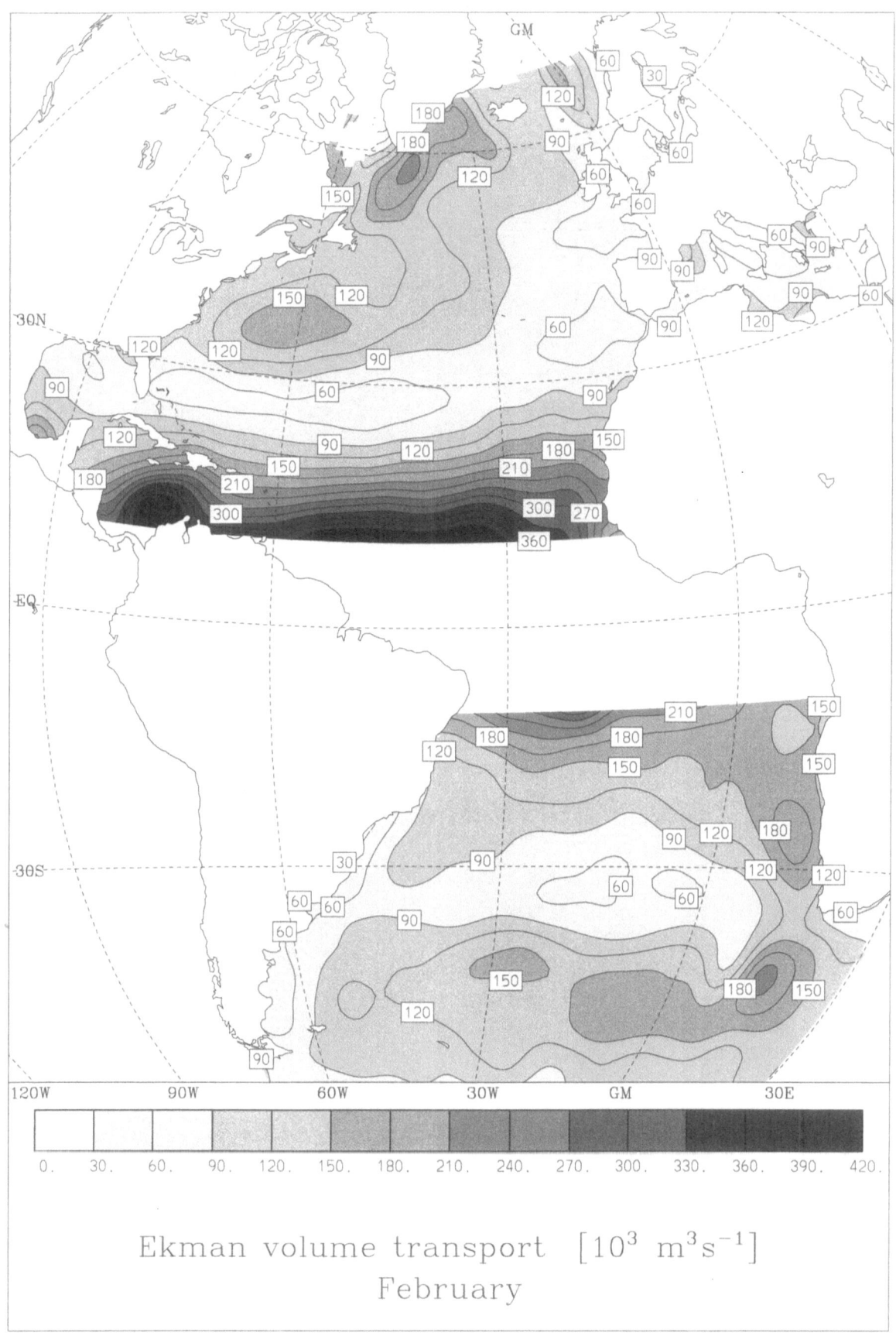

Ekman volume transport $[10^3 \ m^3 s^{-1}]$
February

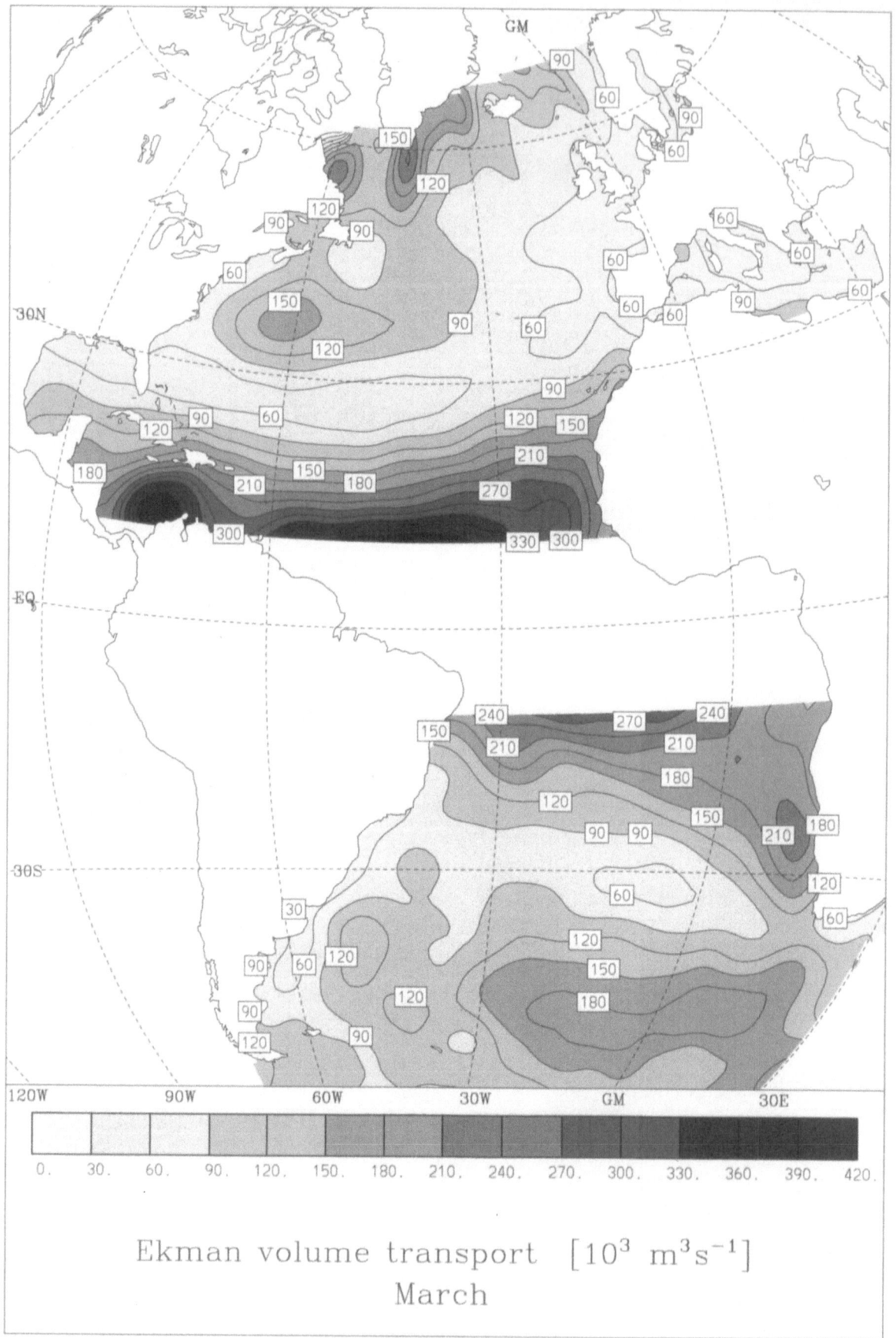

Ekman volume transport $[10^3 \ m^3 s^{-1}]$
March

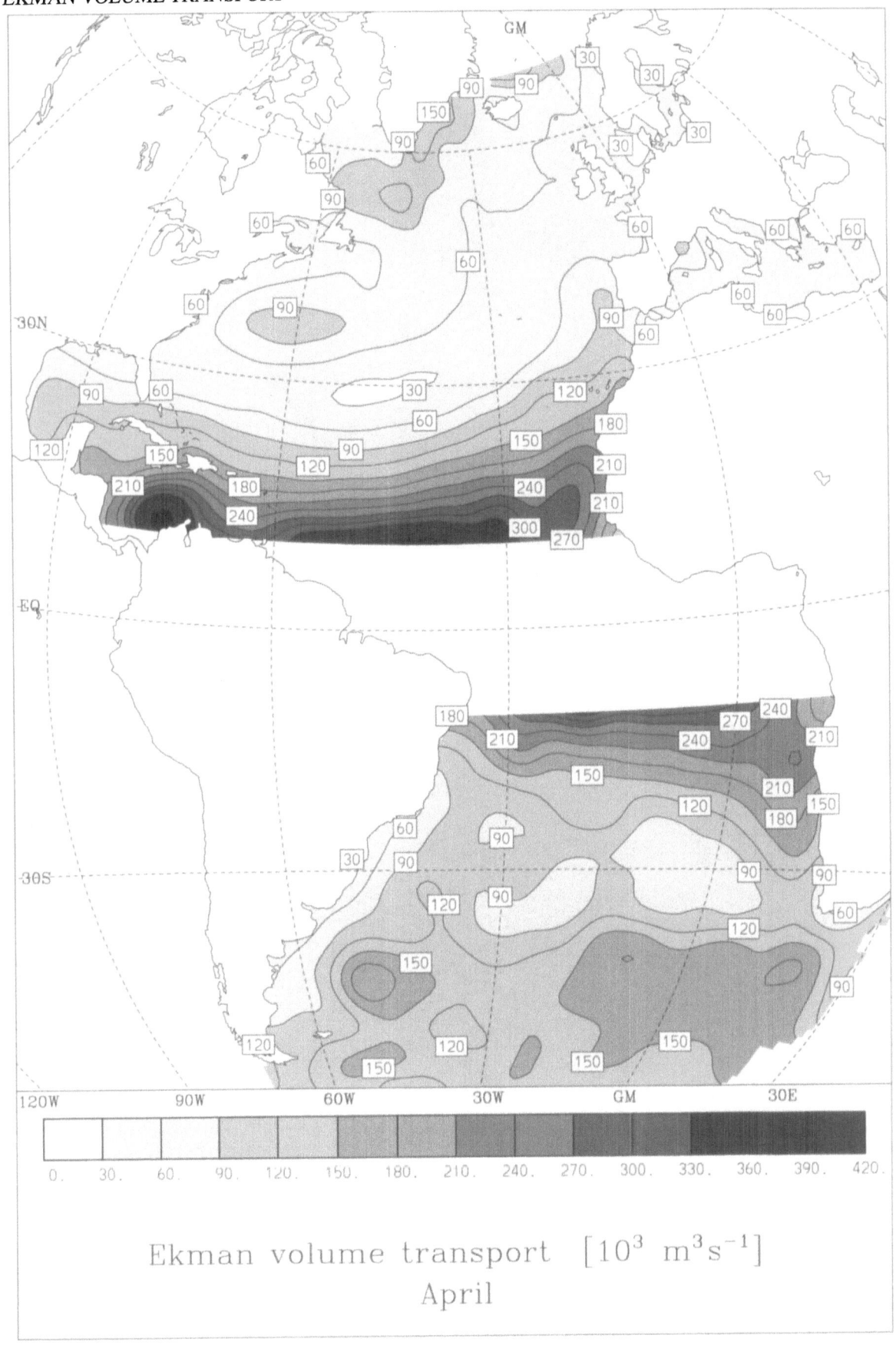

Ekman volume transport $[10^3 \ m^3 s^{-1}]$
April

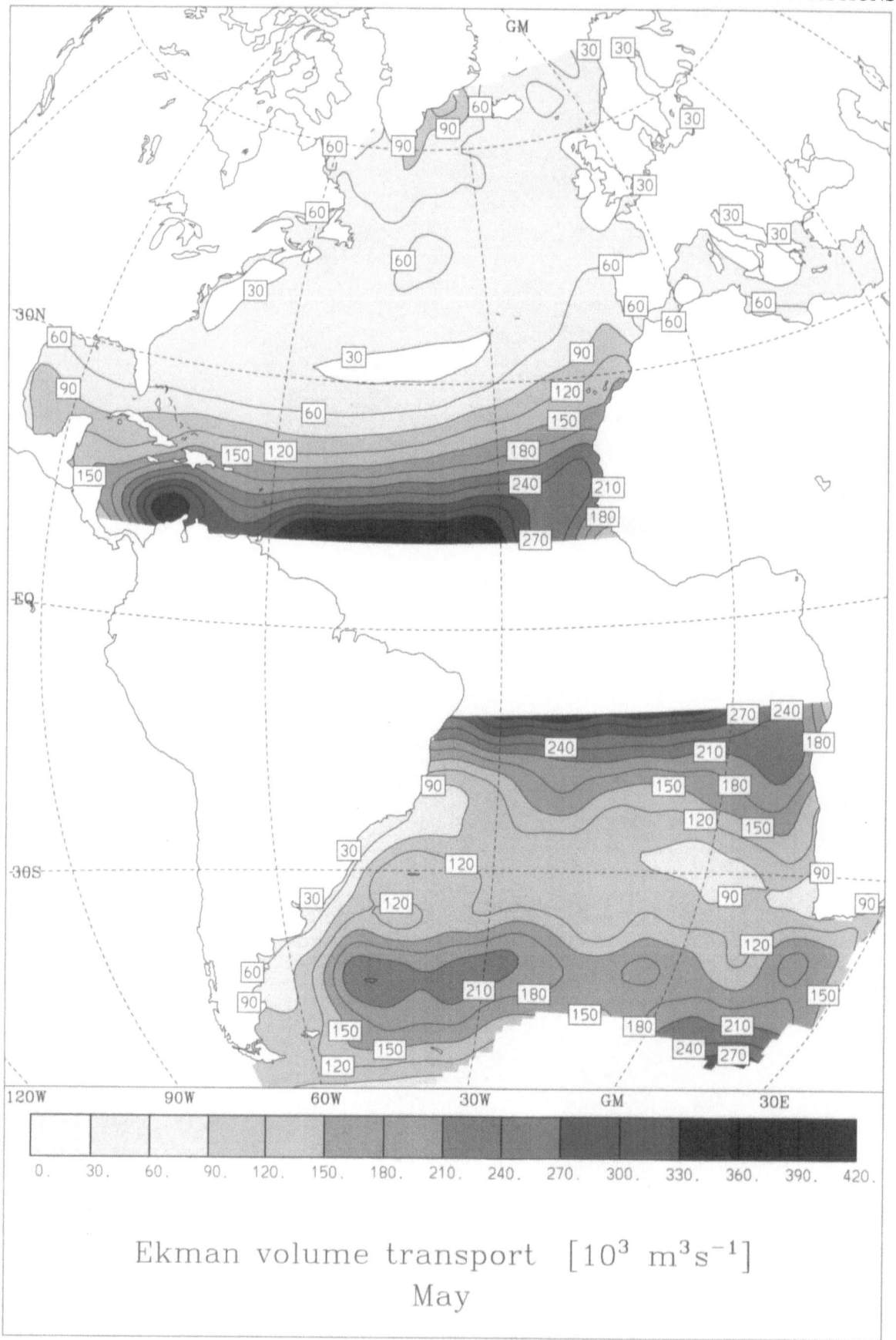

Ekman volume transport $[10^3 \ \mathrm{m}^3\mathrm{s}^{-1}]$
May

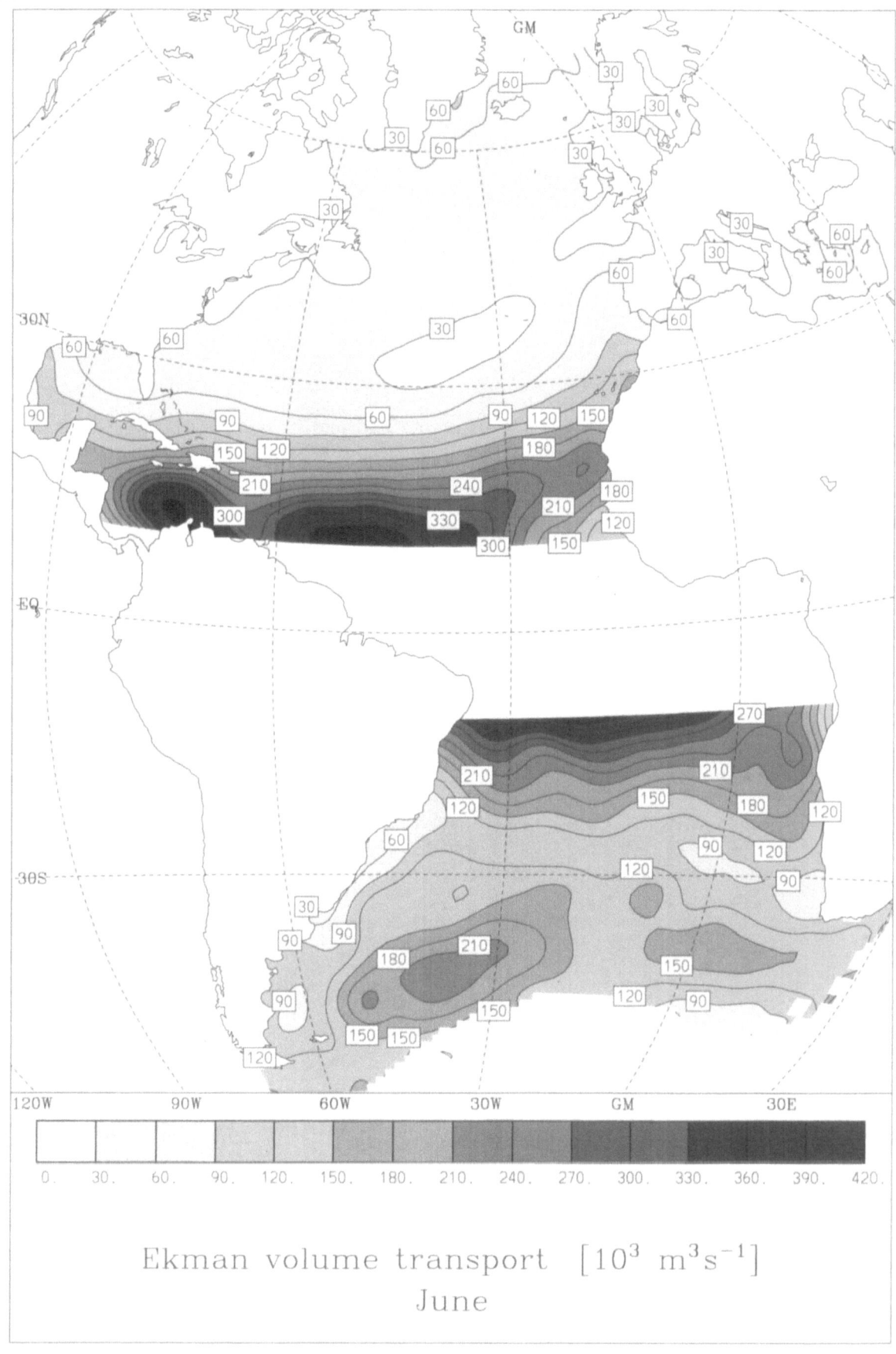

Ekman volume transport $[10^3 \ m^3 s^{-1}]$

June

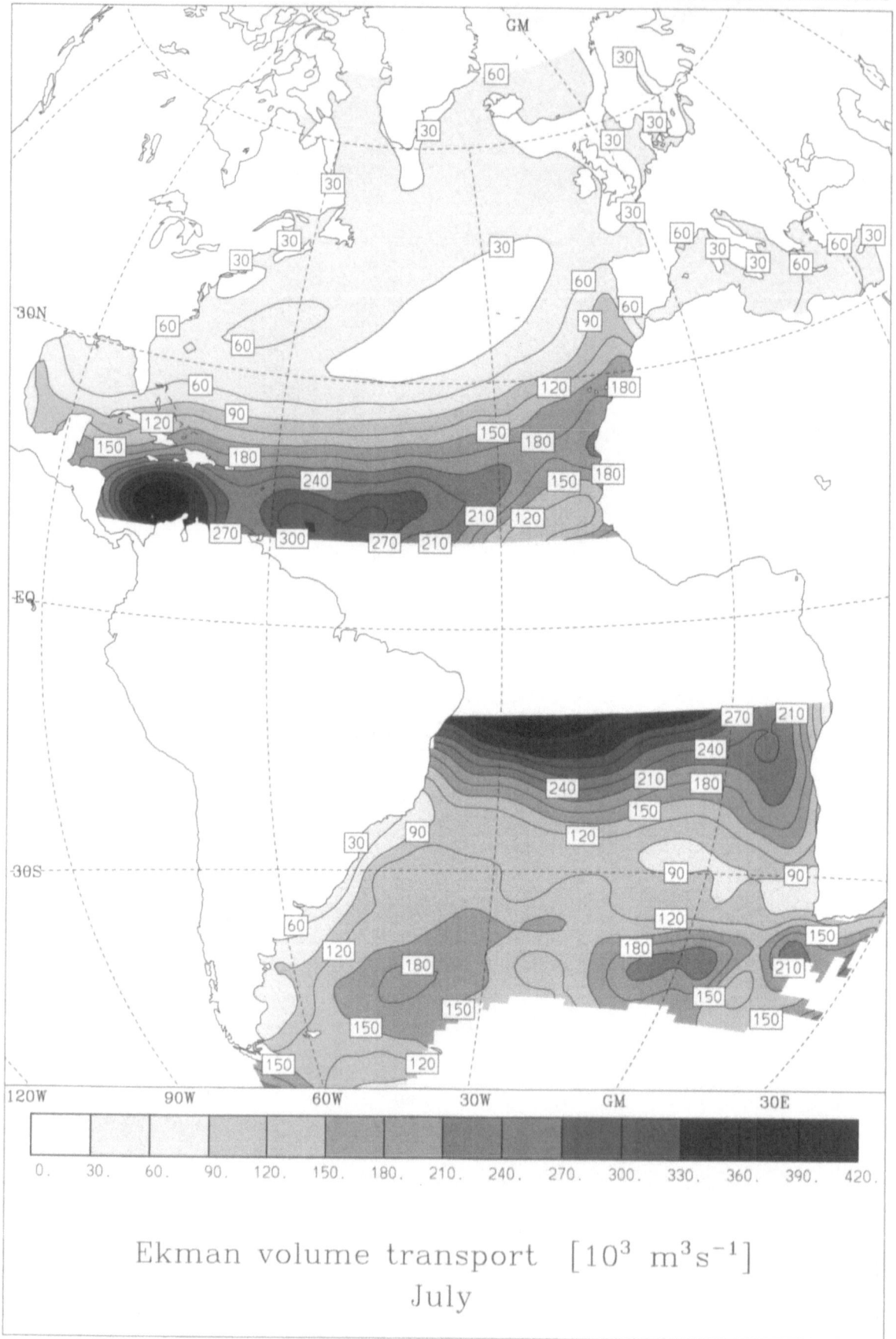

Ekman volume transport [10^3 m^{3}s^{-1}]
July

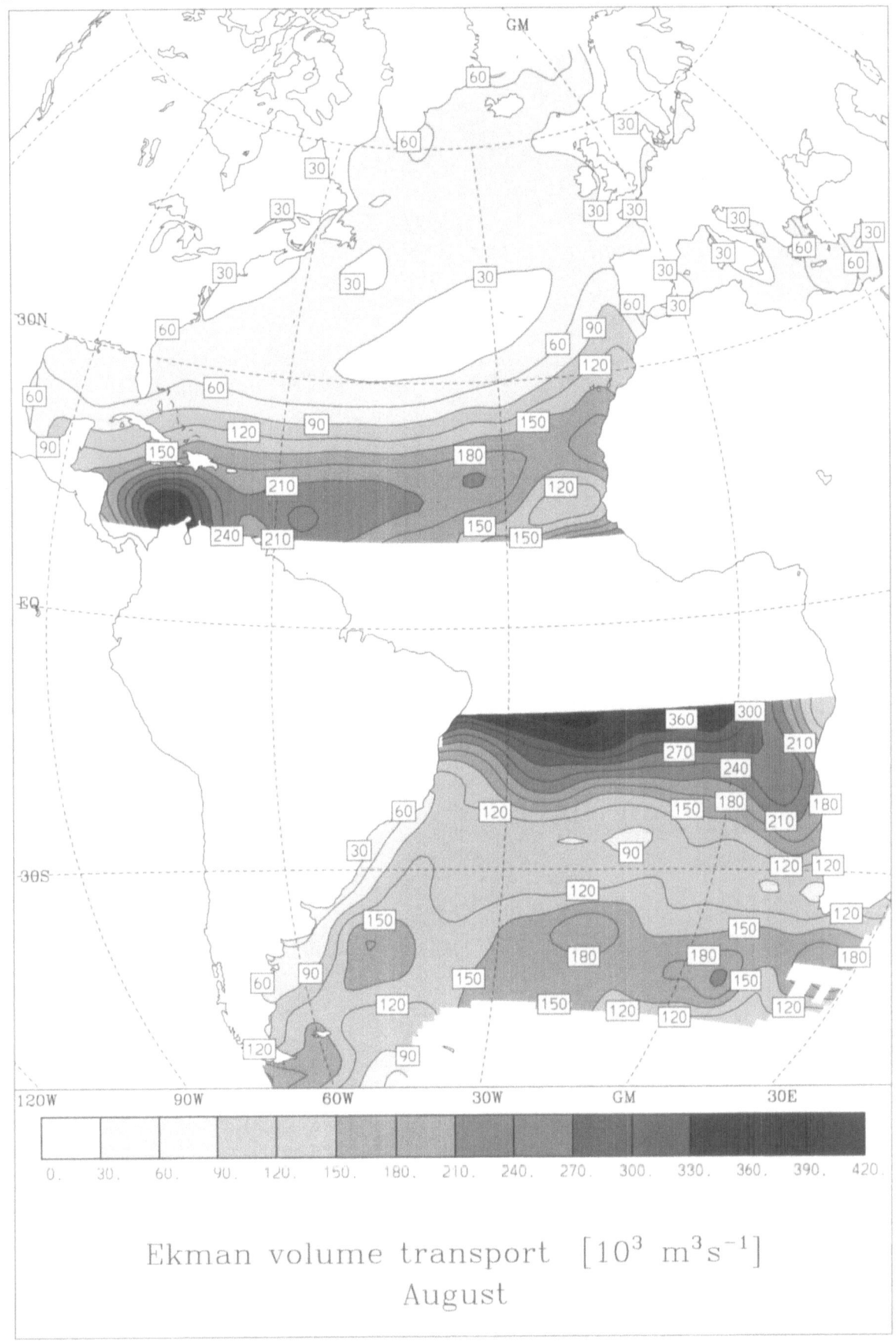

Ekman volume transport $[10^3 \ m^3 s^{-1}]$
August

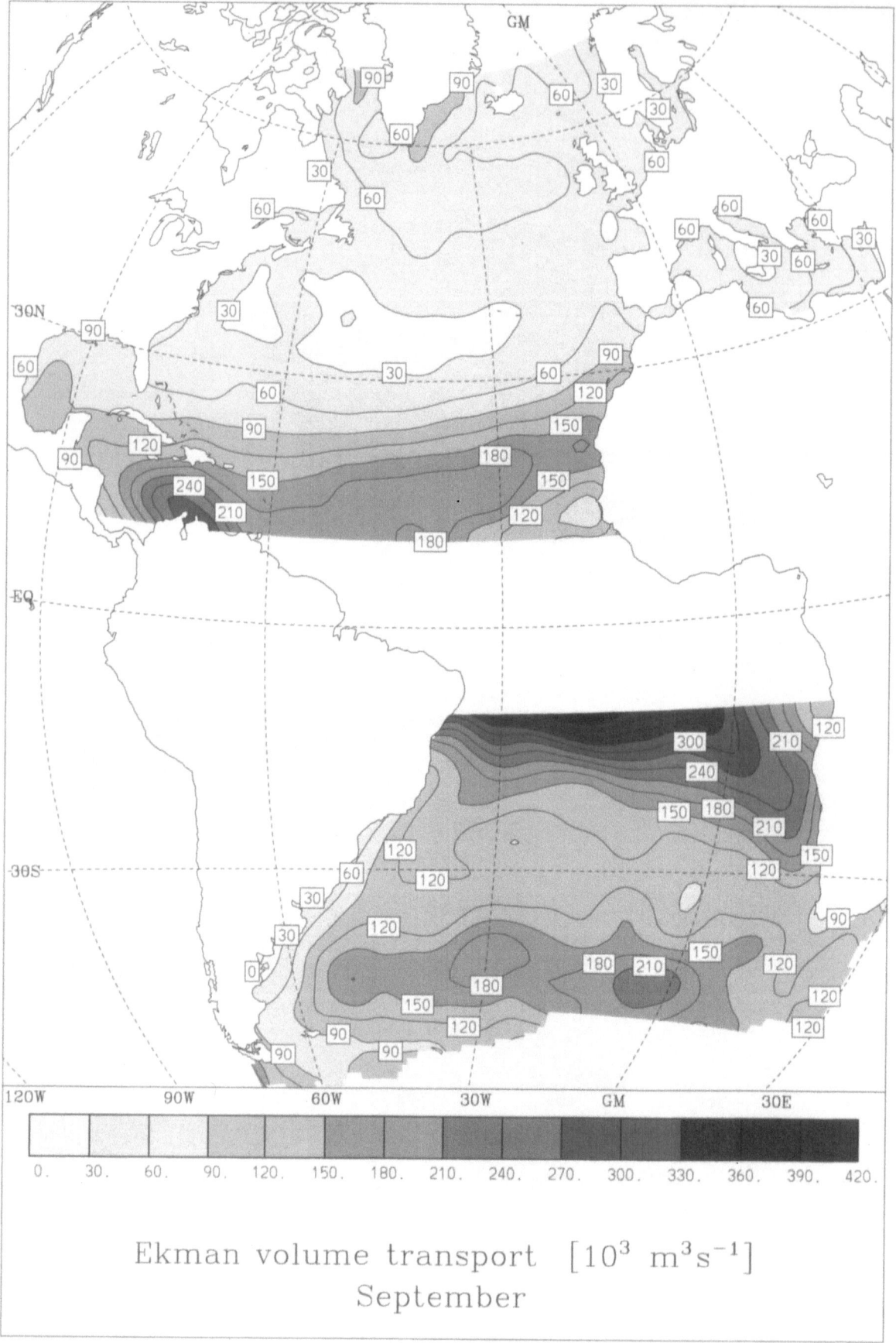

Ekman volume transport $[10^3 \ m^3 s^{-1}]$
September

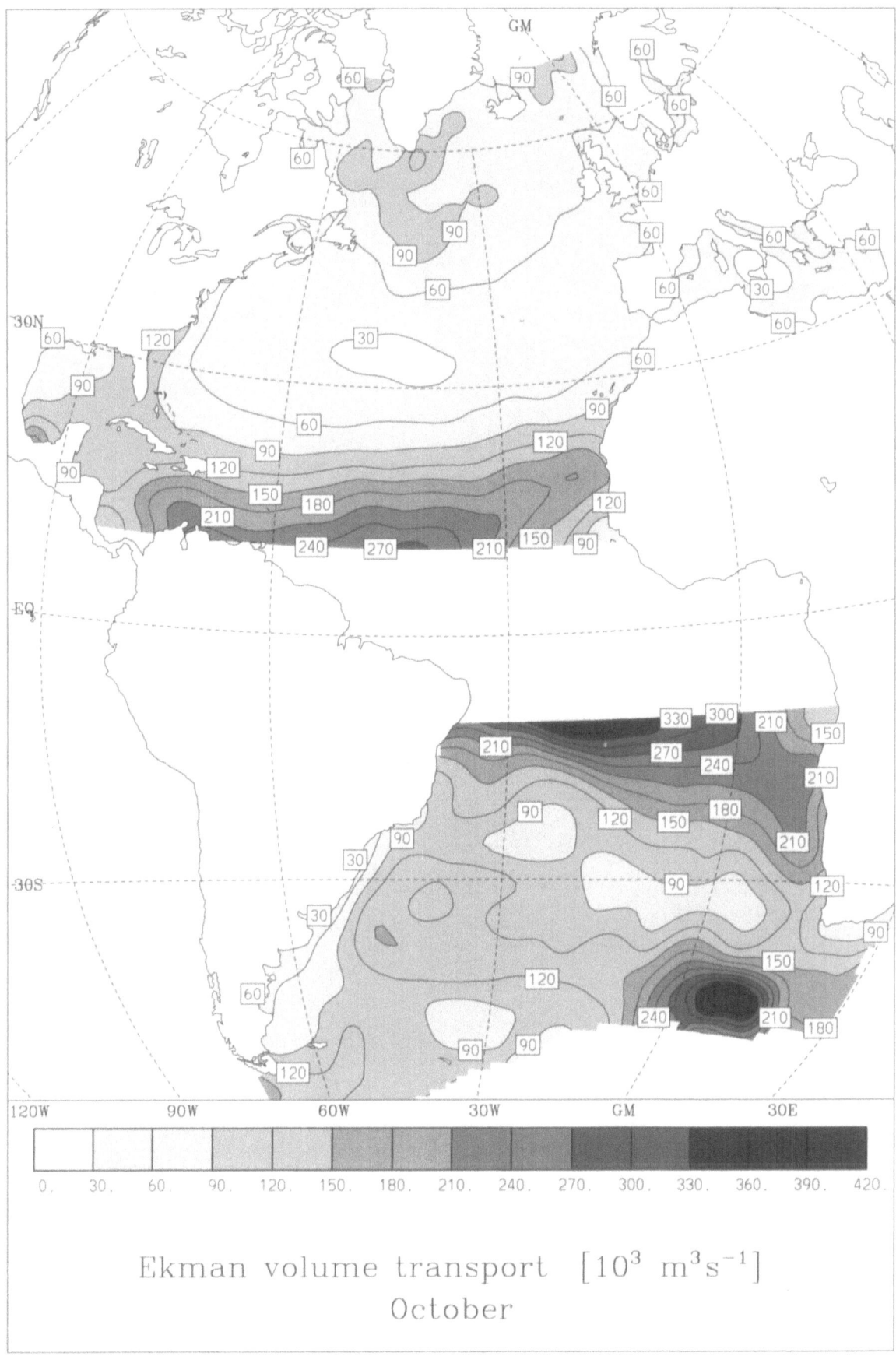

Ekman volume transport $[10^3 \ m^3 s^{-1}]$
October

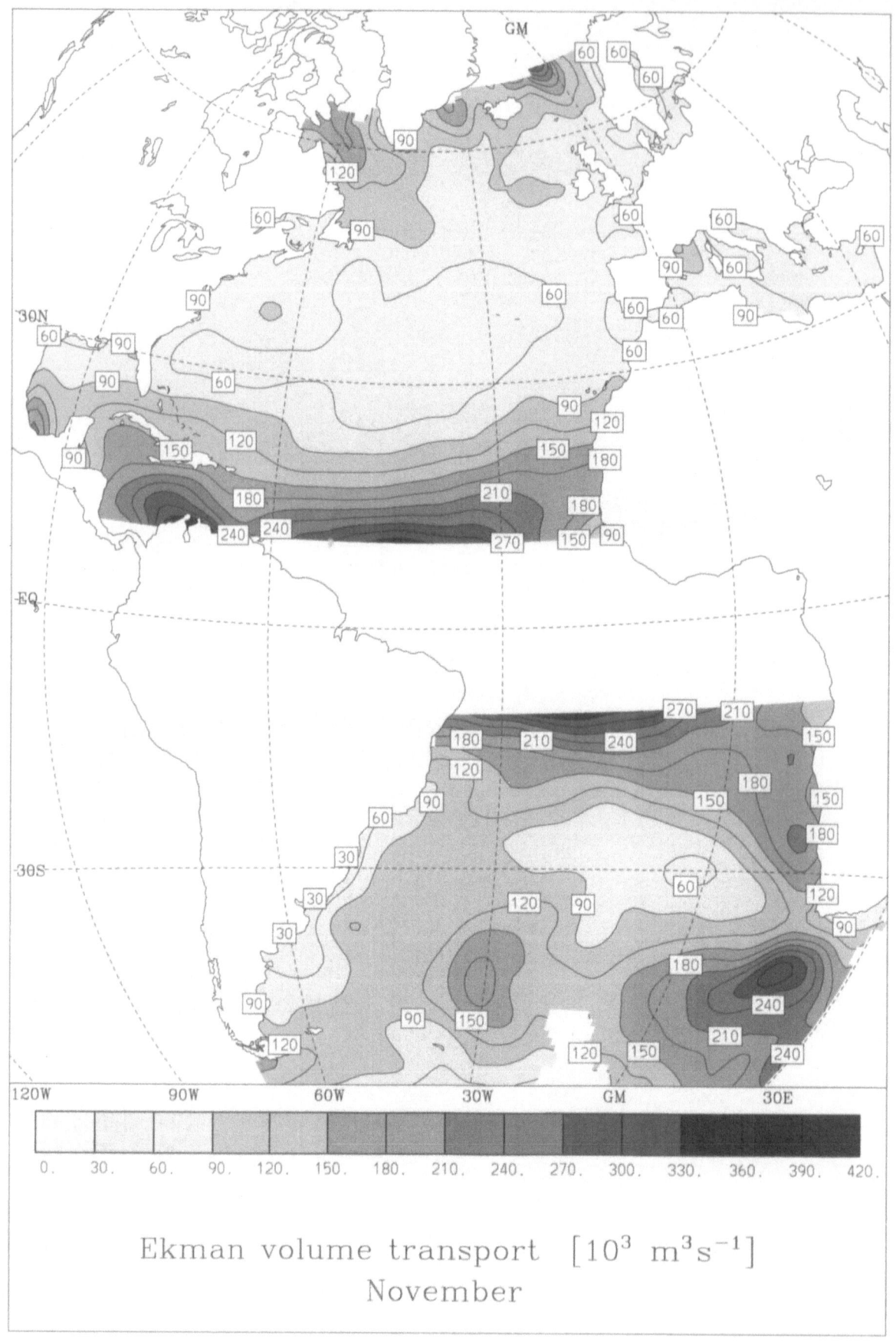

Ekman volume transport $[10^3 \ m^3 s^{-1}]$
November

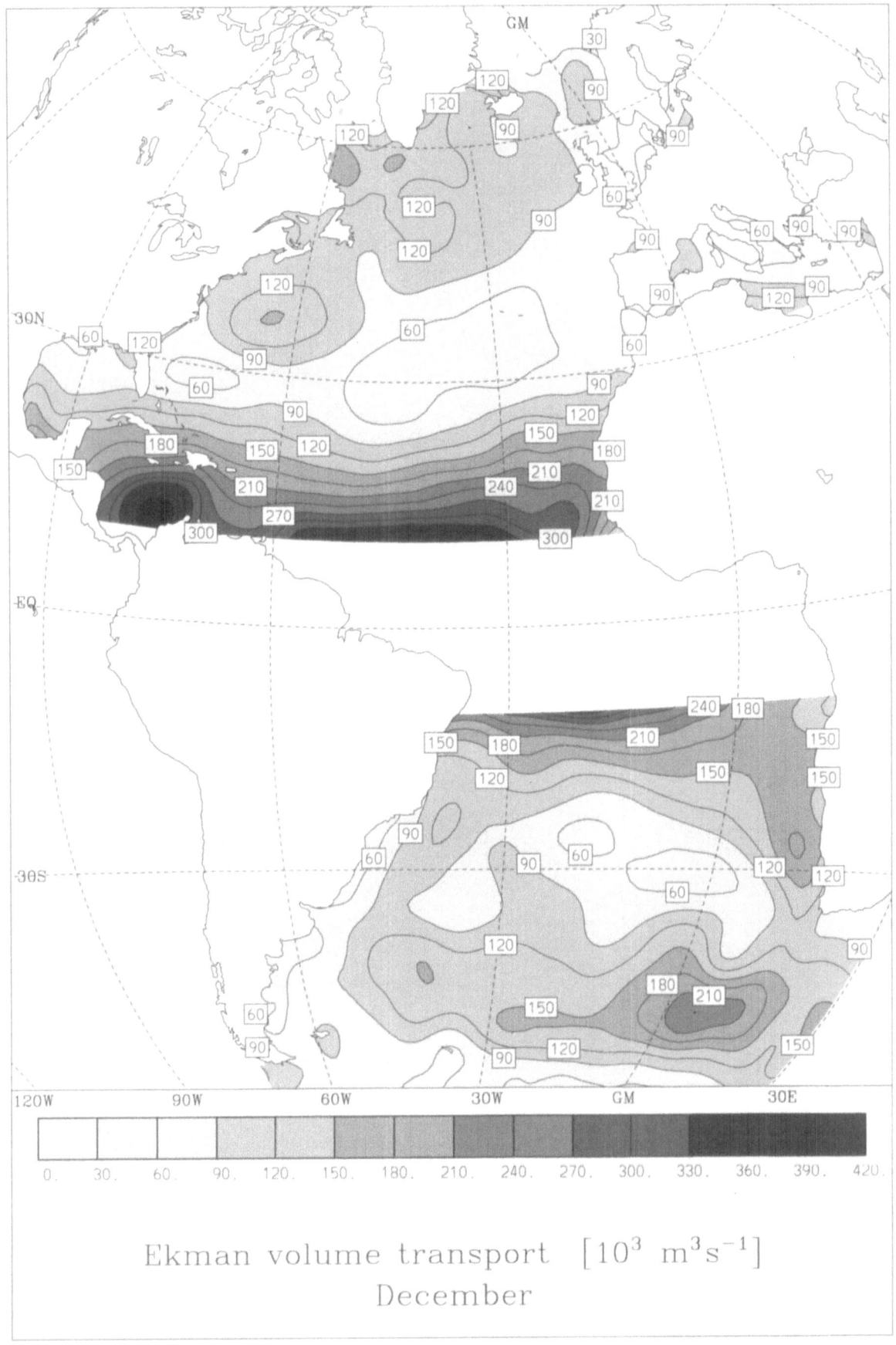

Ekman volume transport $[10^3 \text{ m}^3\text{s}^{-1}]$
December

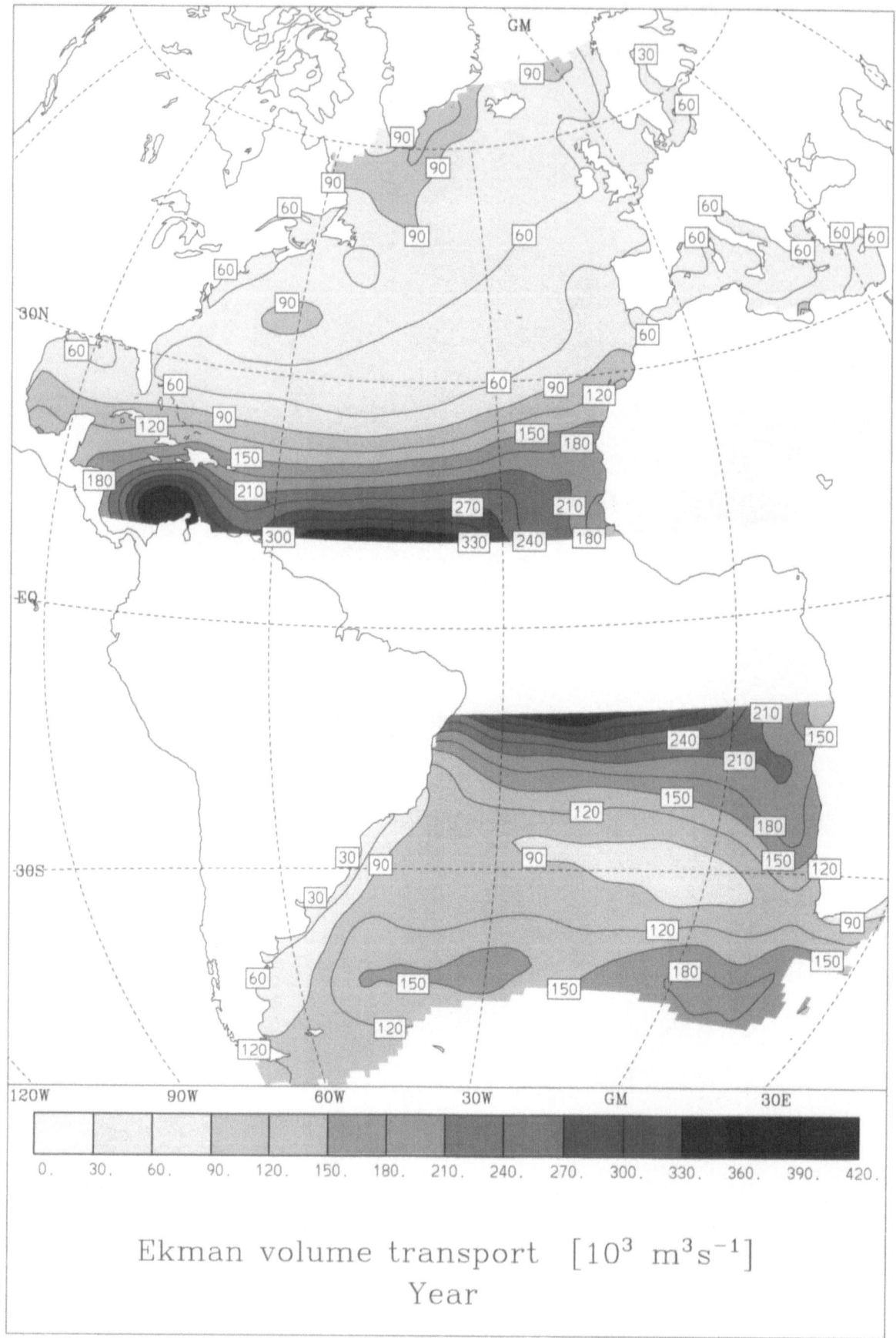

Ekman volume transport $[10^3 \ \mathrm{m^3 s^{-1}}]$
Year

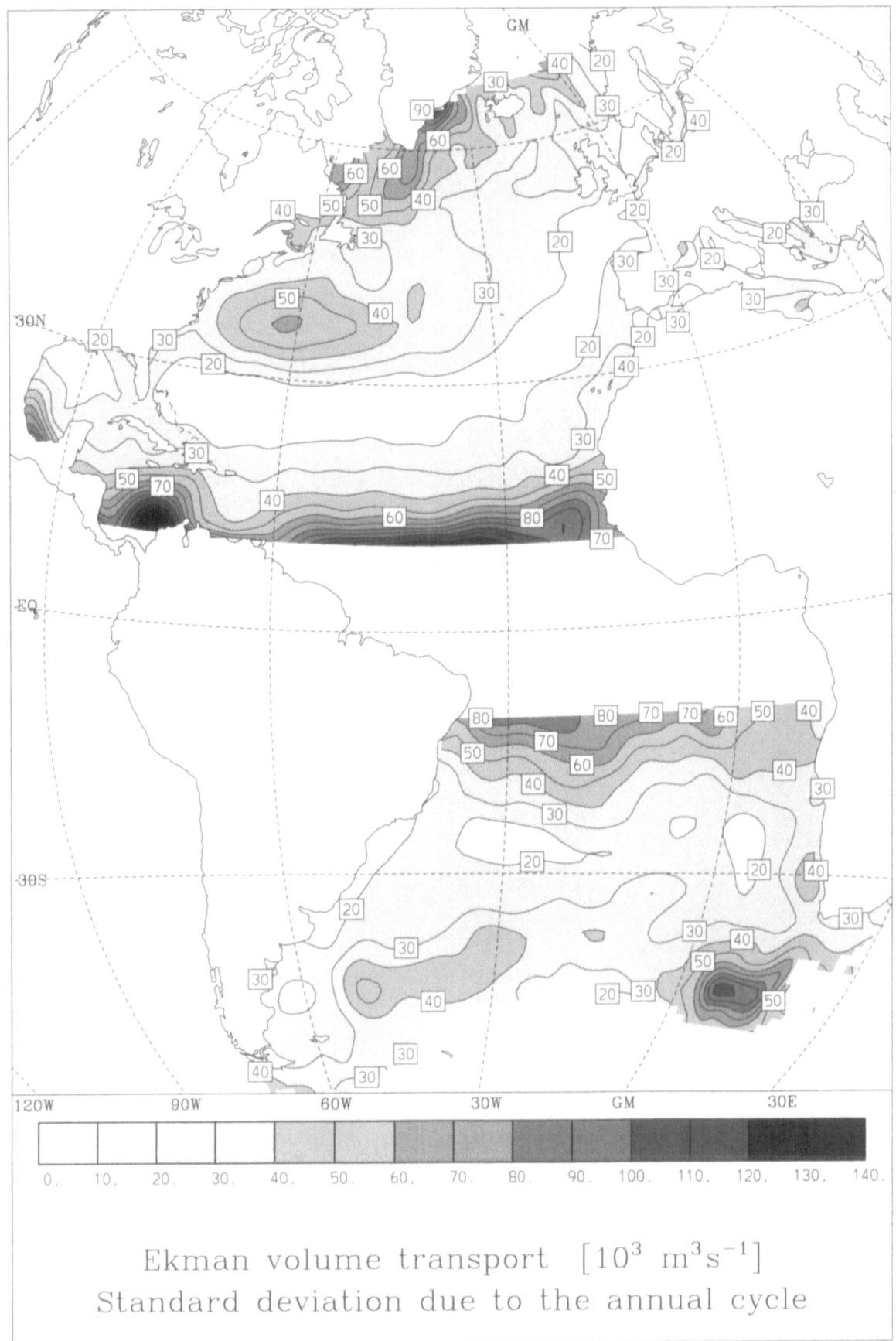

Ekman volume transport $[10^3 \mathrm{\ m^3 s^{-1}}]$
Standard deviation due to the annual cycle

Vertikal Ekman Velocity

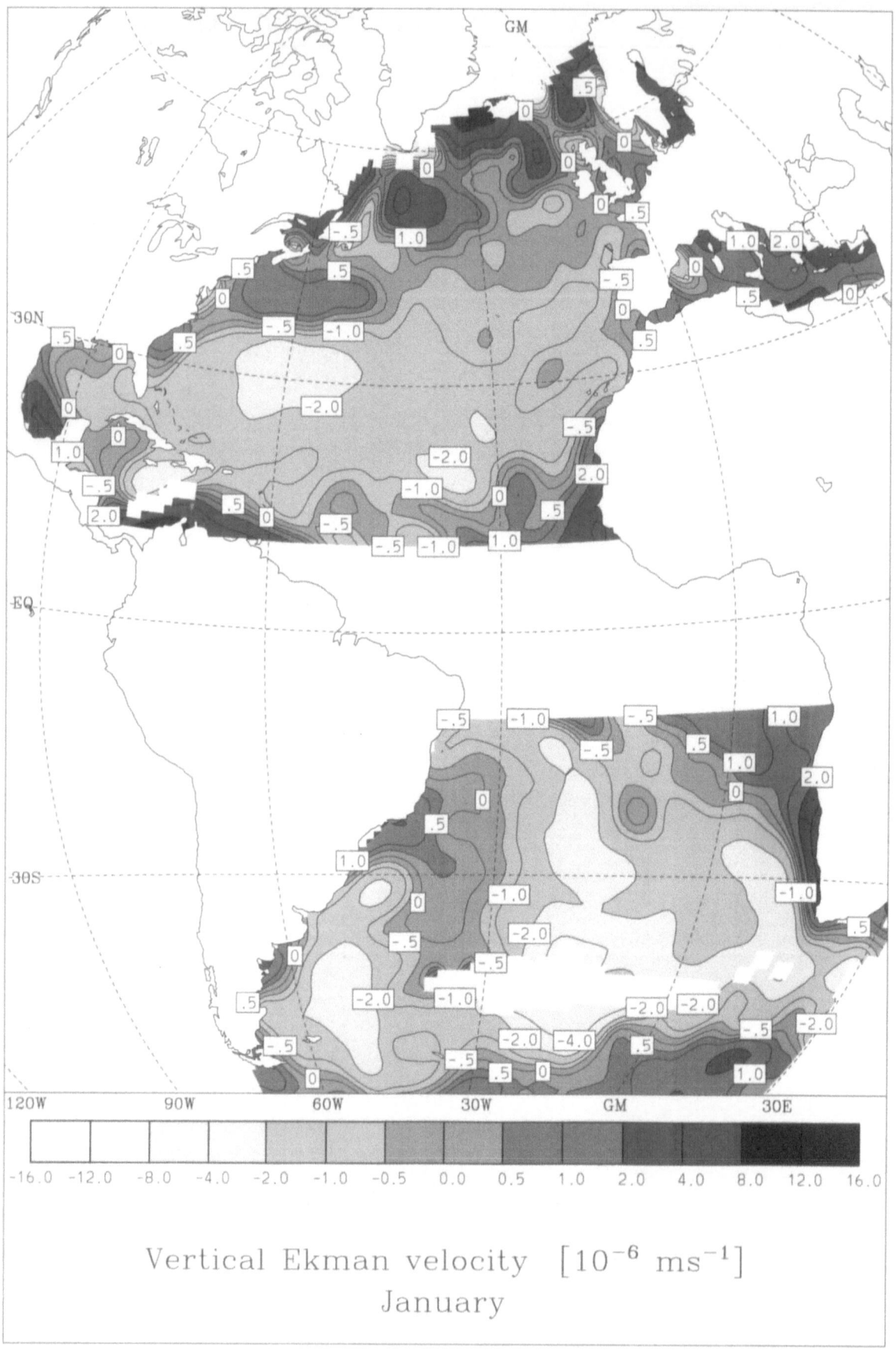

Vertical Ekman velocity $[10^{-6} \text{ ms}^{-1}]$
January

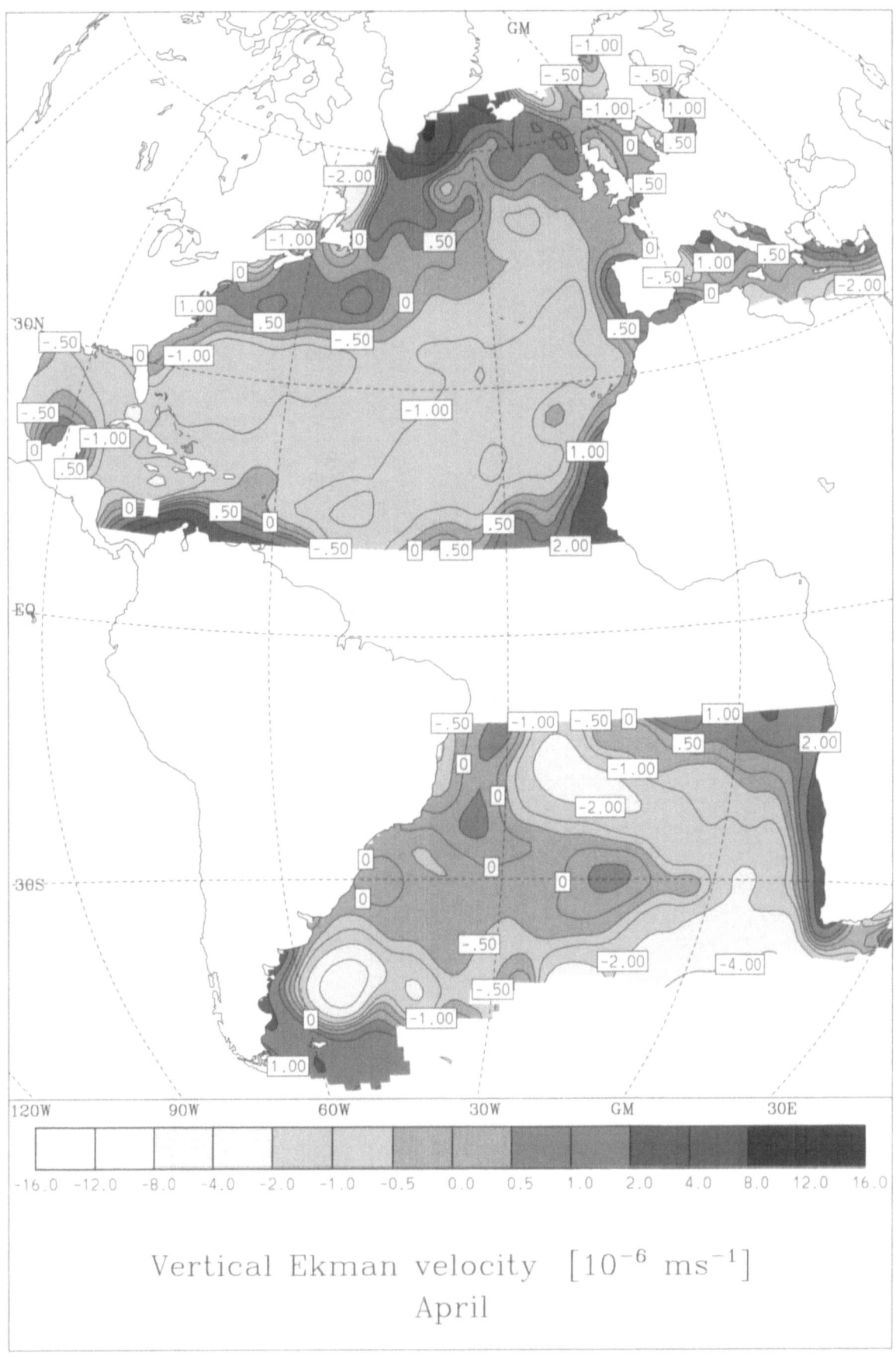

Vertical Ekman velocity $[10^{-6}\ \mathrm{ms}^{-1}]$
April

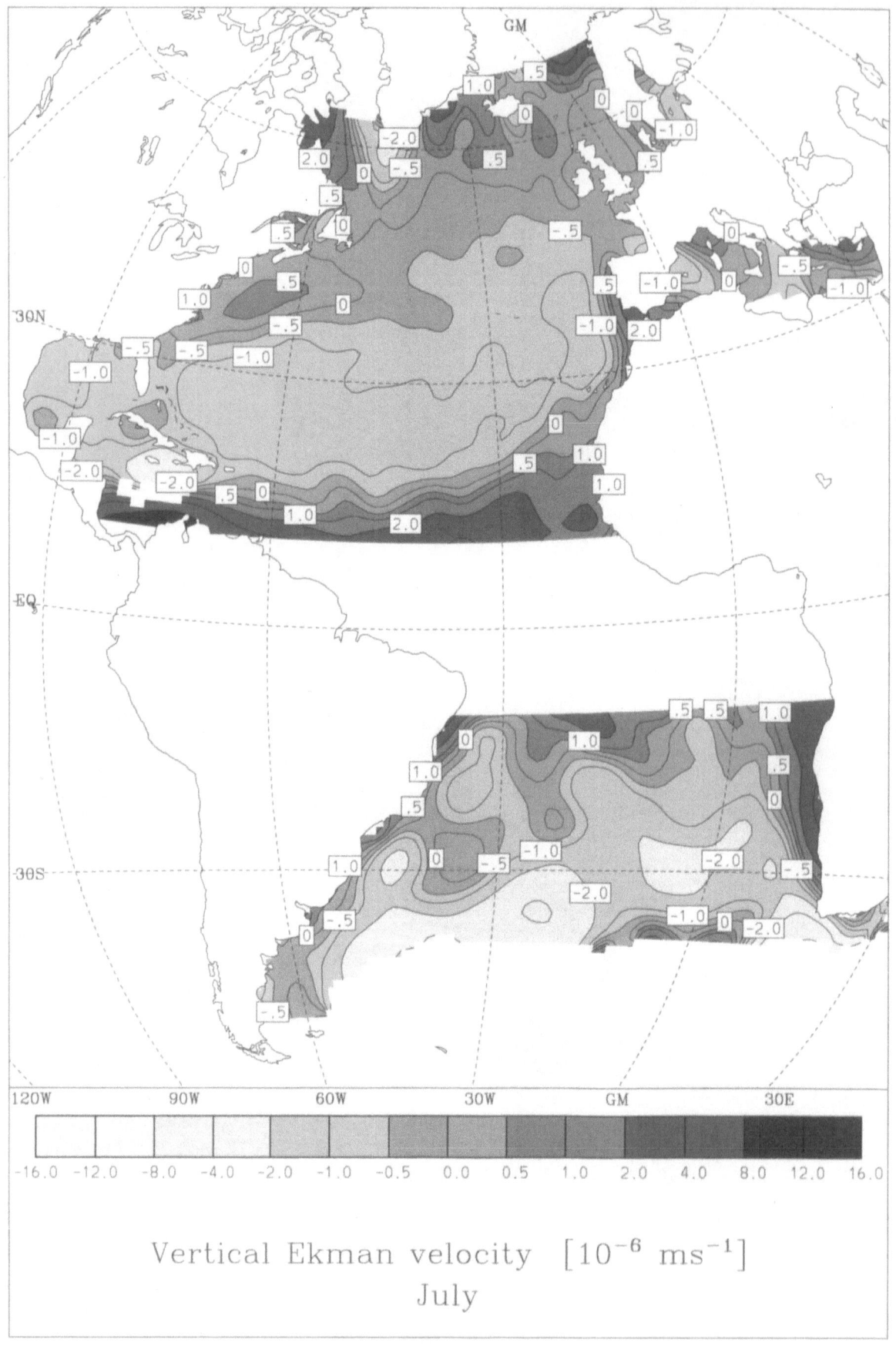

Vertical Ekman velocity $[10^{-6}\ \mathrm{ms}^{-1}]$
July

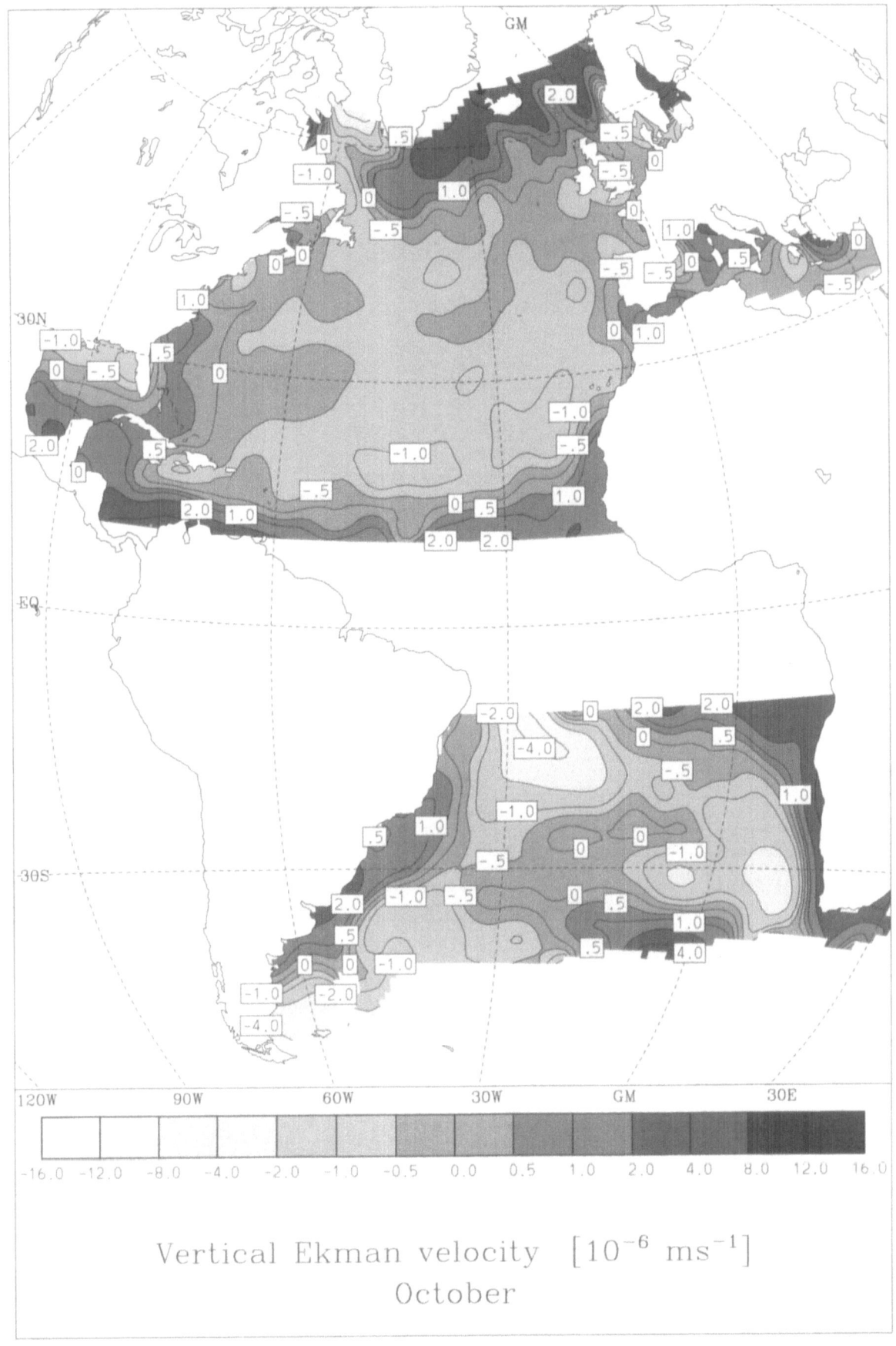

Vertical Ekman velocity $[10^{-6} \text{ ms}^{-1}]$
October

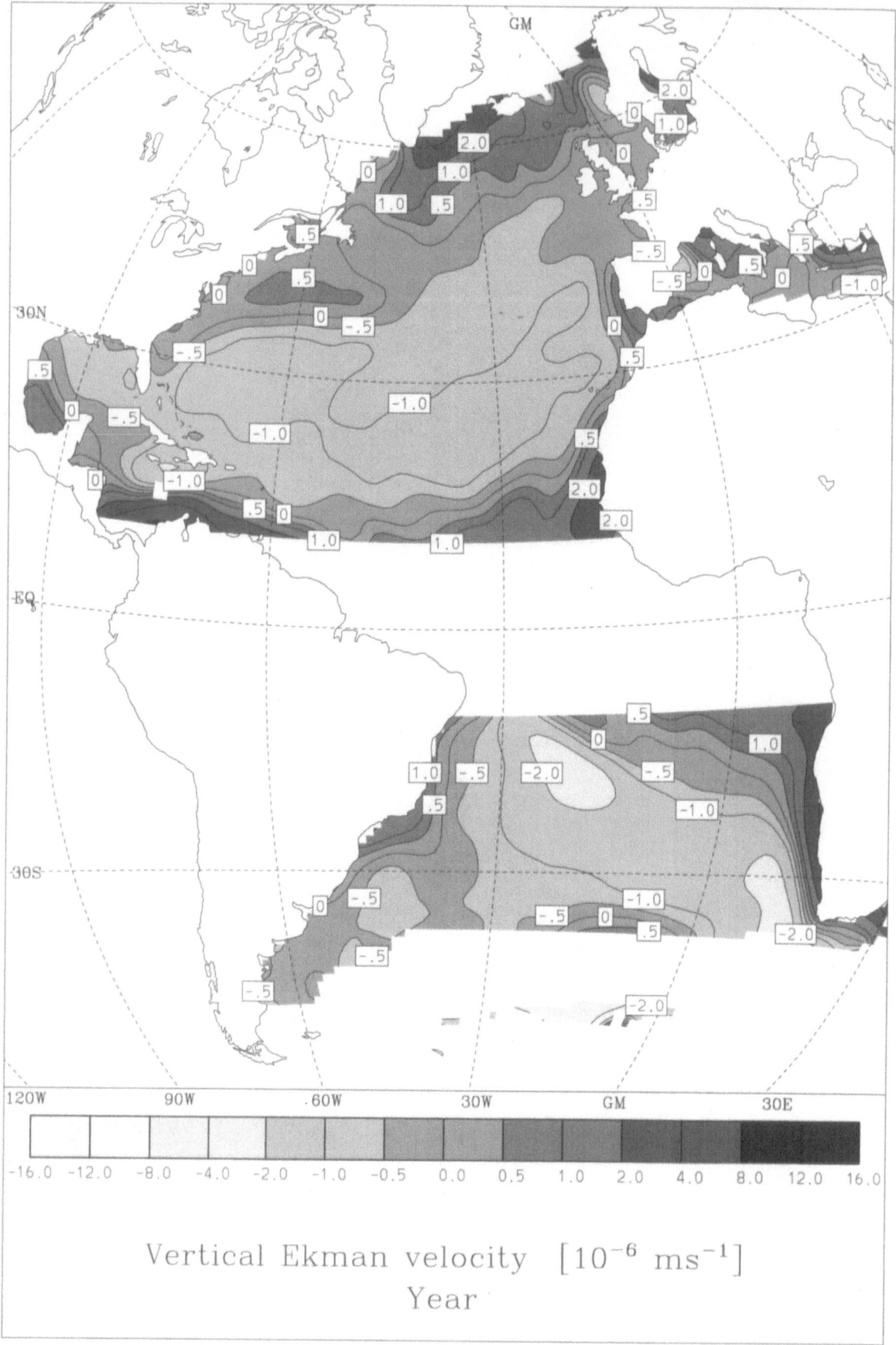

Vertical Ekman velocity $[10^{-6}\ ms^{-1}]$
Year

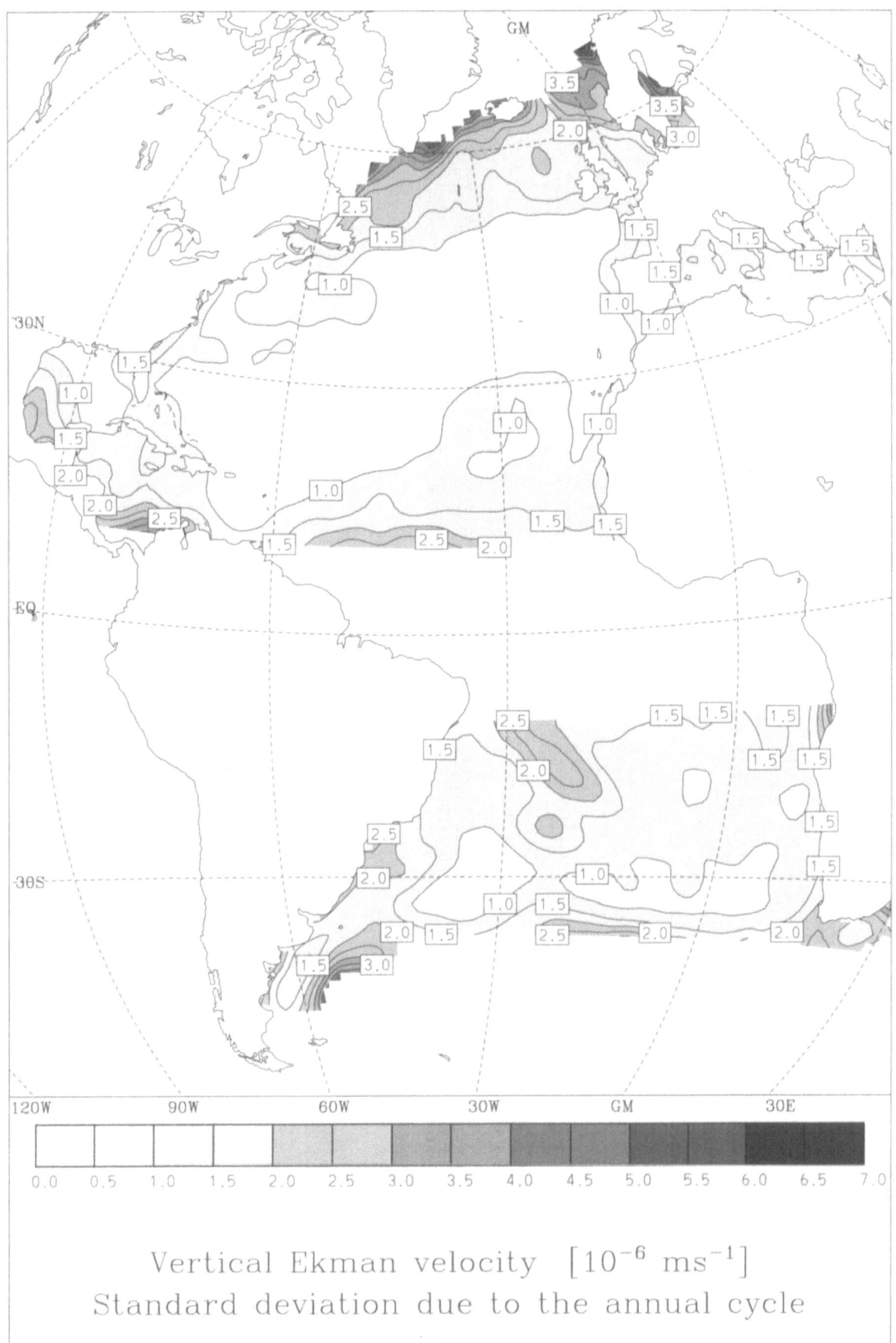

Vertical Ekman velocity $[10^{-6}\ ms^{-1}]$
Standard deviation due to the annual cycle

Sverdrup Transport

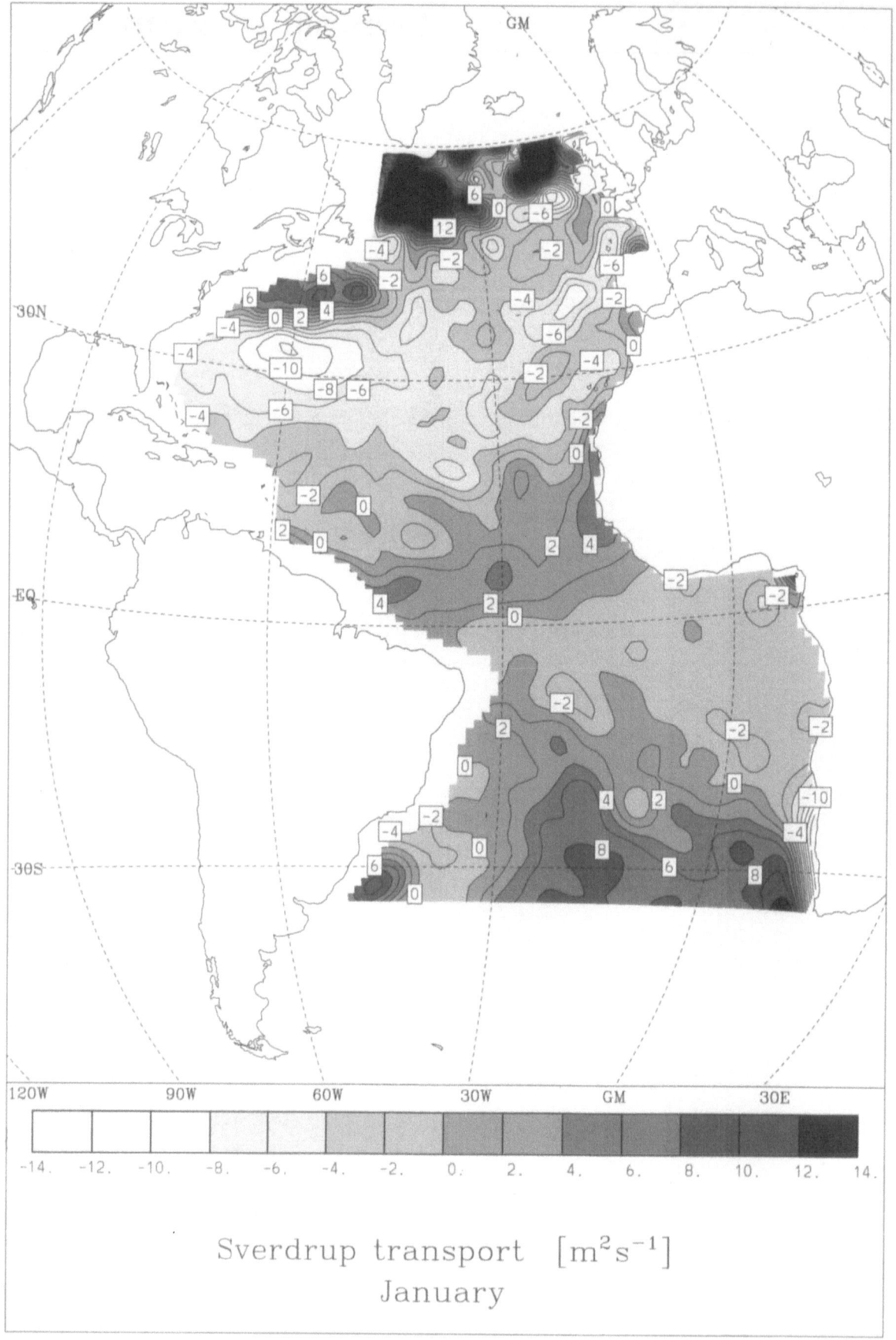

Sverdrup transport $[\mathrm{m^2 s^{-1}}]$
January

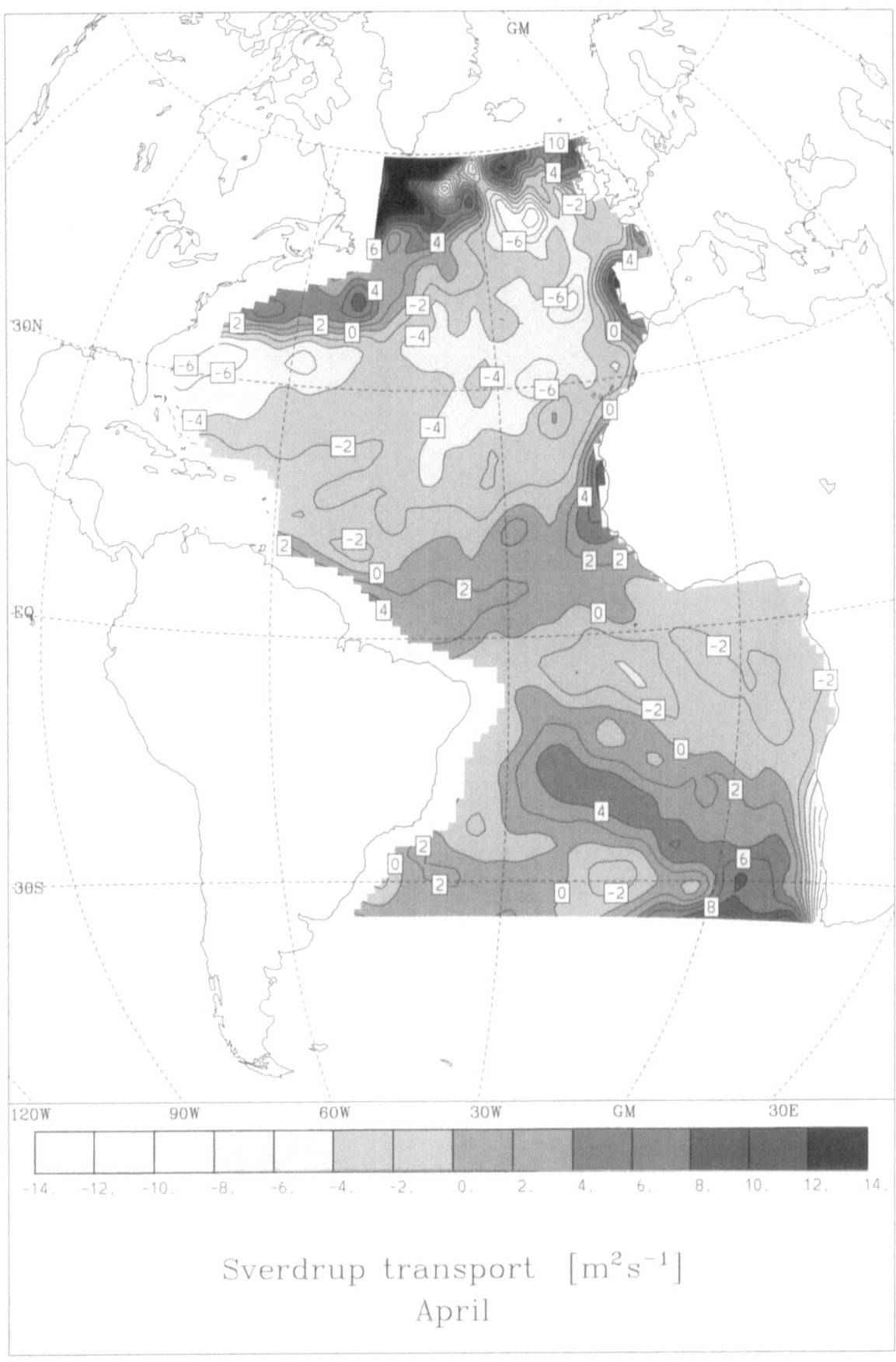

Sverdrup transport $[\mathrm{m^2 s^{-1}}]$

April

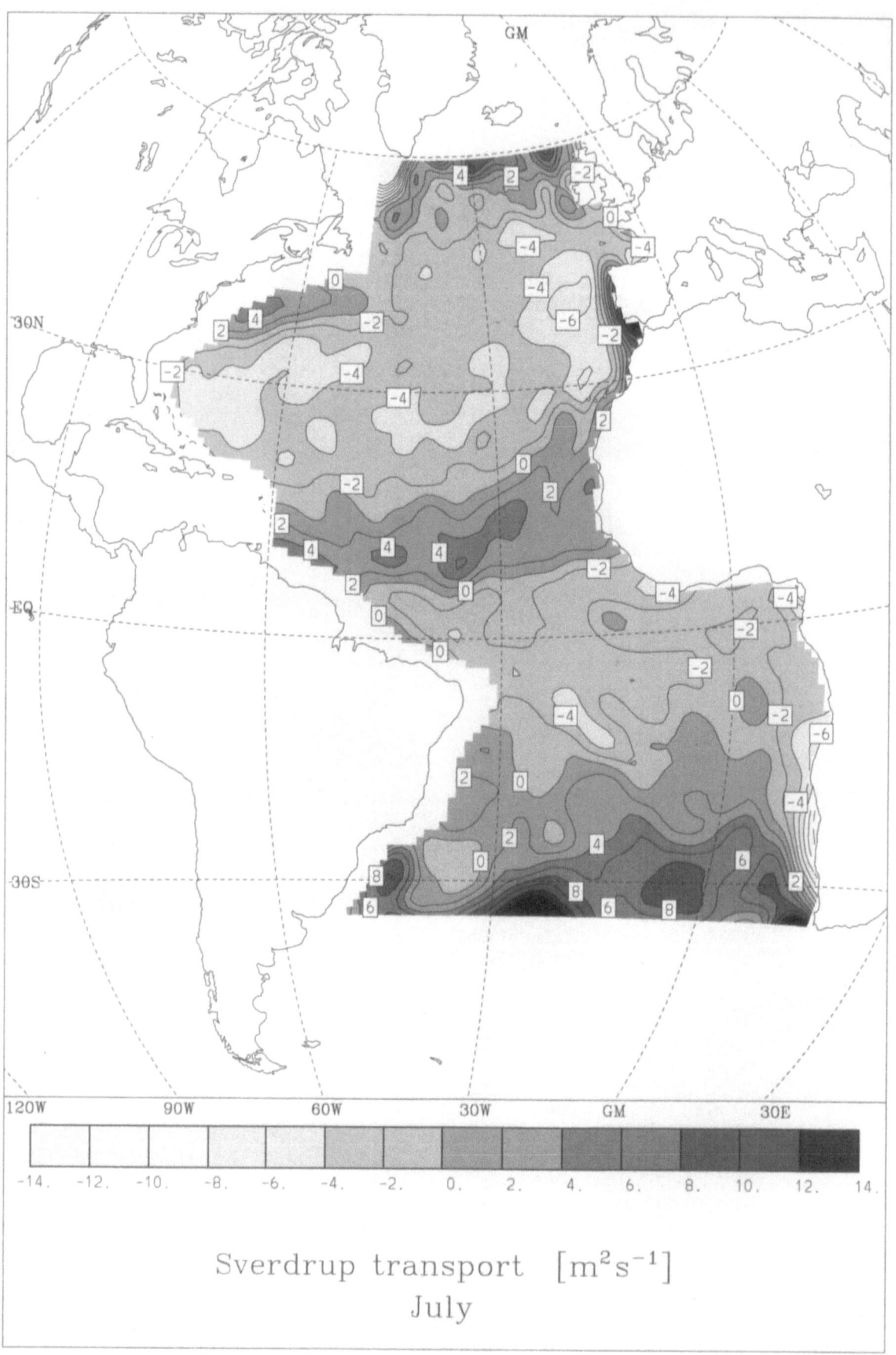

Sverdrup transport $[m^2 s^{-1}]$
July

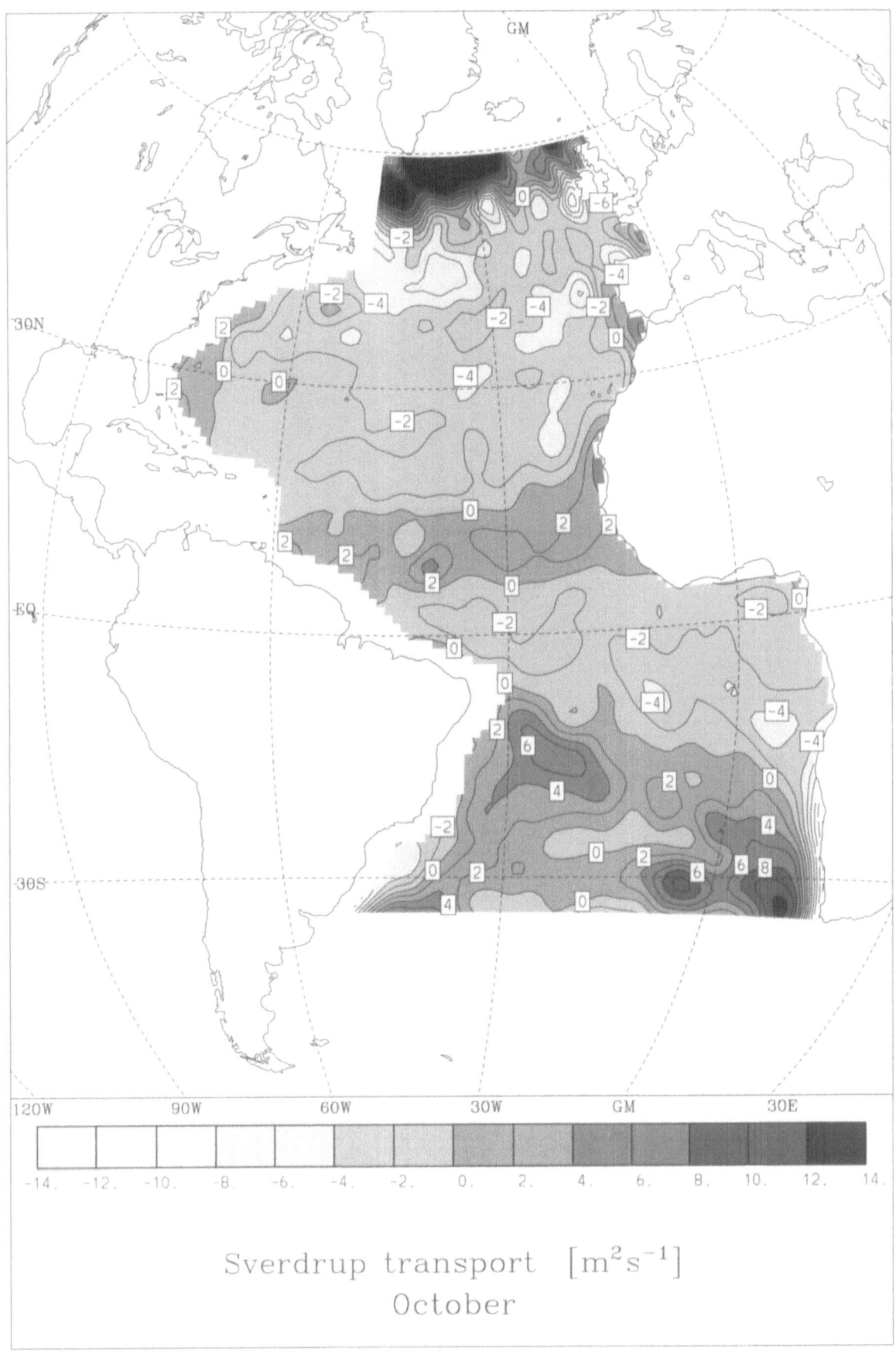

Sverdrup transport $[m^2 s^{-1}]$
October

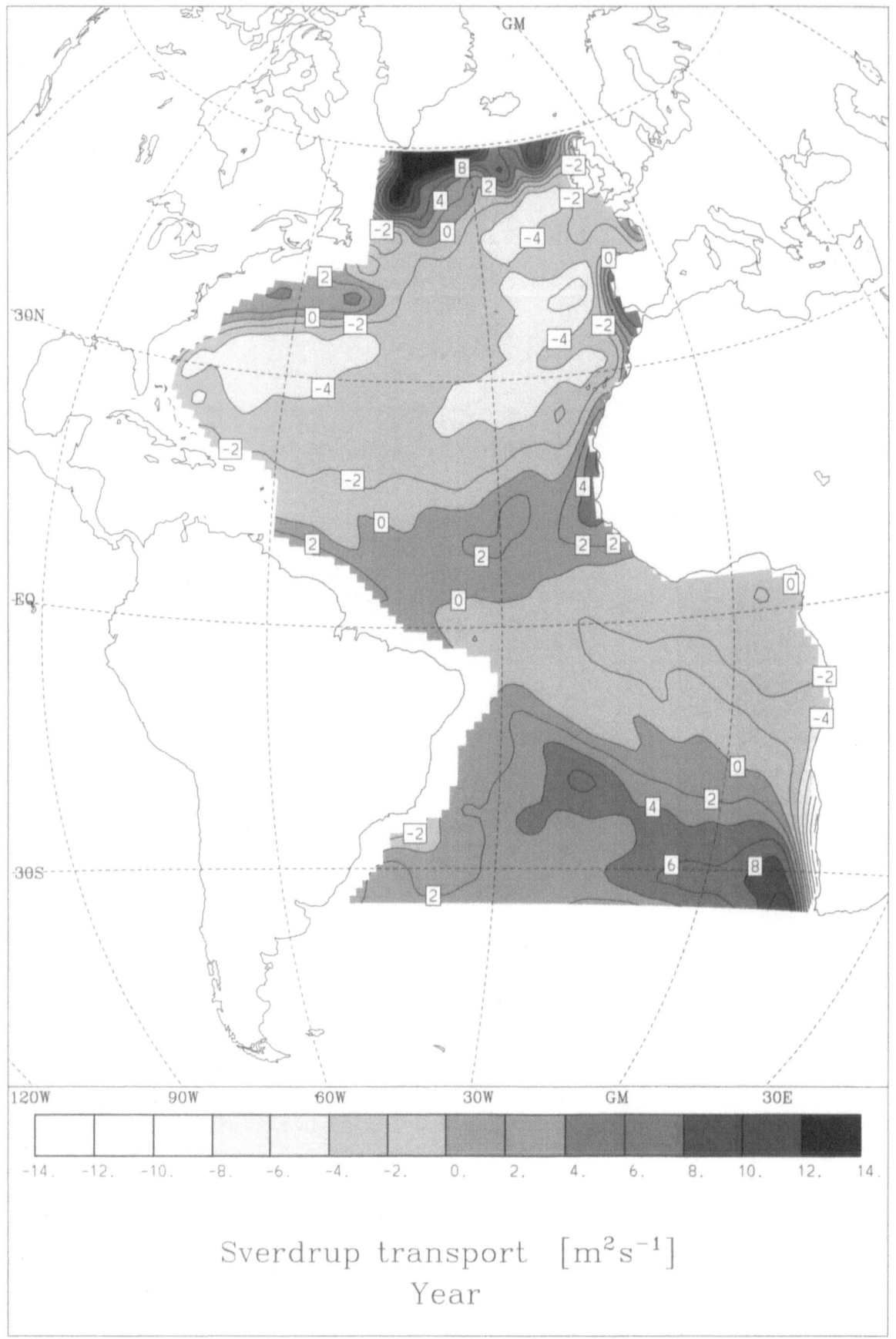

Sverdrup transport $[m^2s^{-1}]$
Year

Streamfunction of Sverdrup Transport

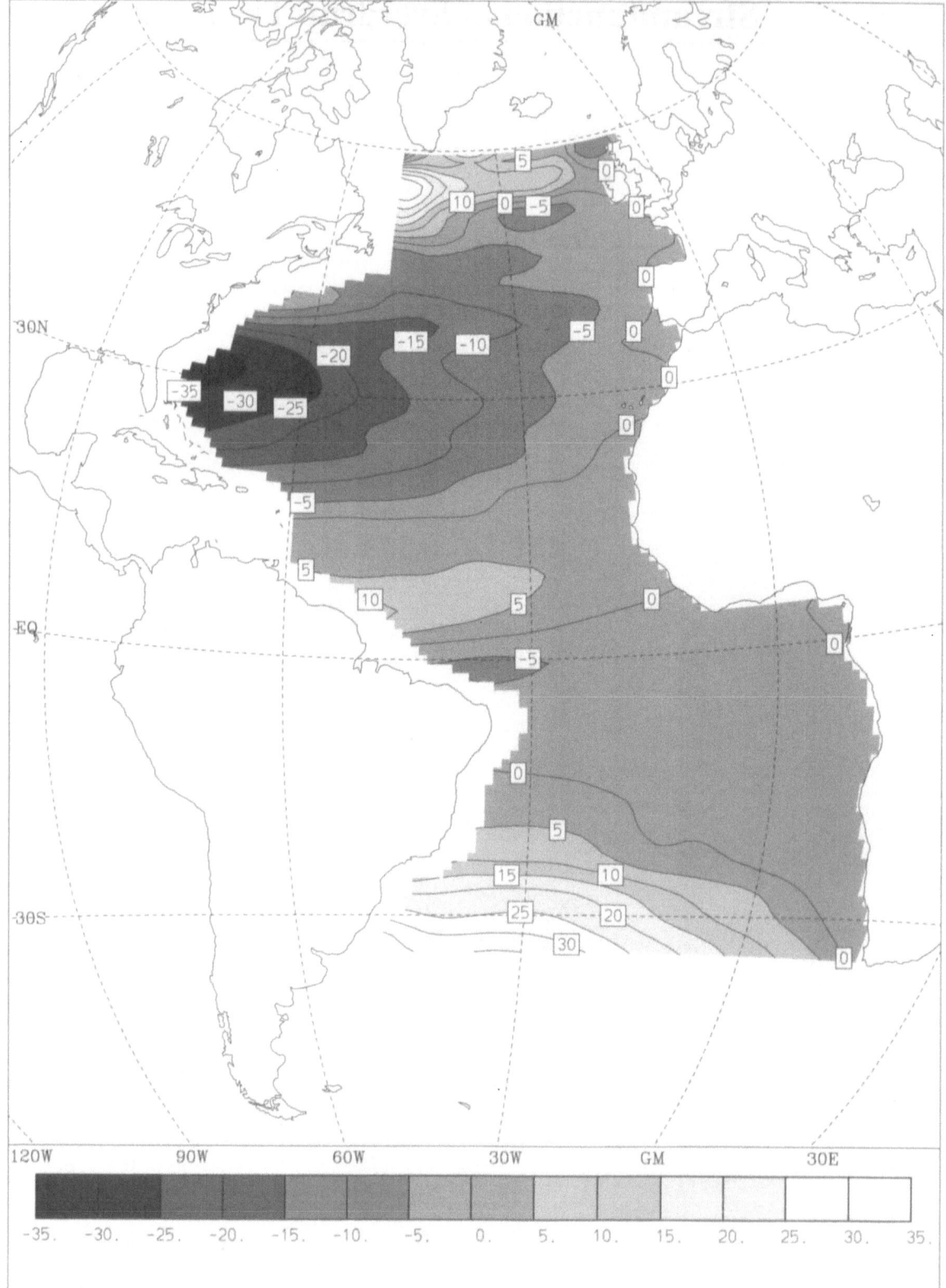

Streamfunction of Sverdrup transport $[10^6 m^3 s^{-1}]$
January

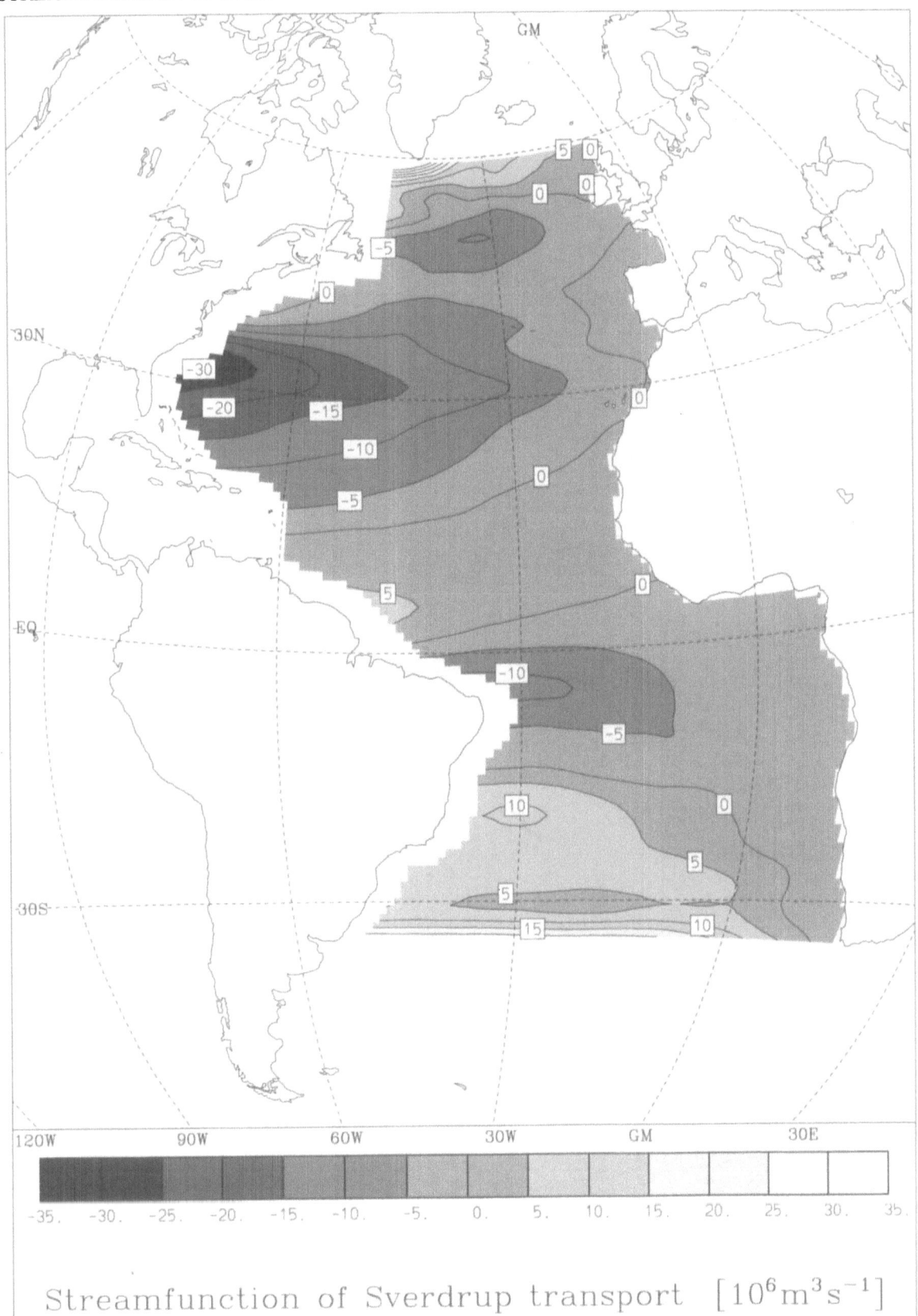

Streamfunction of Sverdrup transport $[10^6 m^3 s^{-1}]$

April

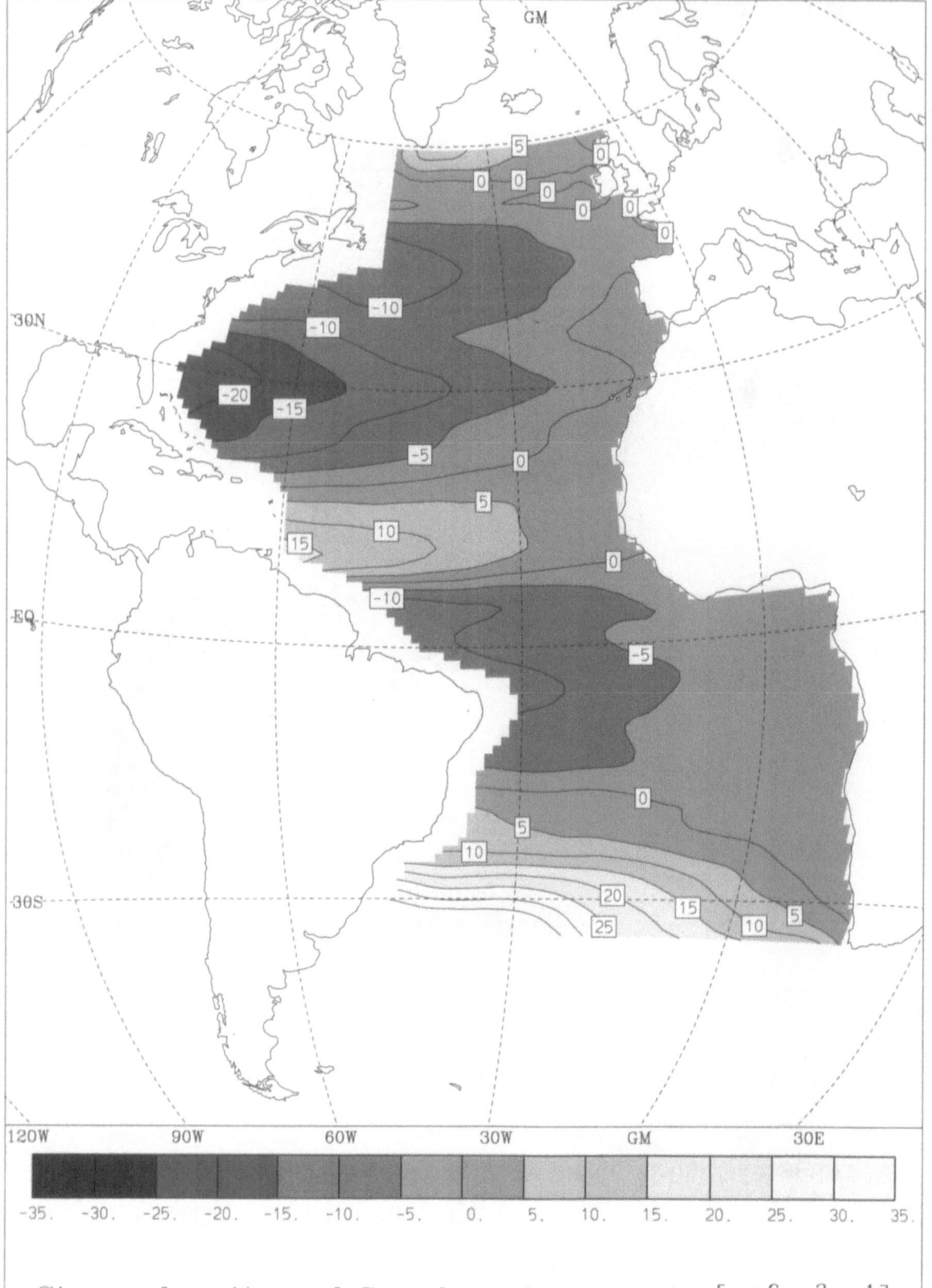

Streamfunction of Sverdrup transport $[10^6 \mathrm{m}^3\mathrm{s}^{-1}]$
July

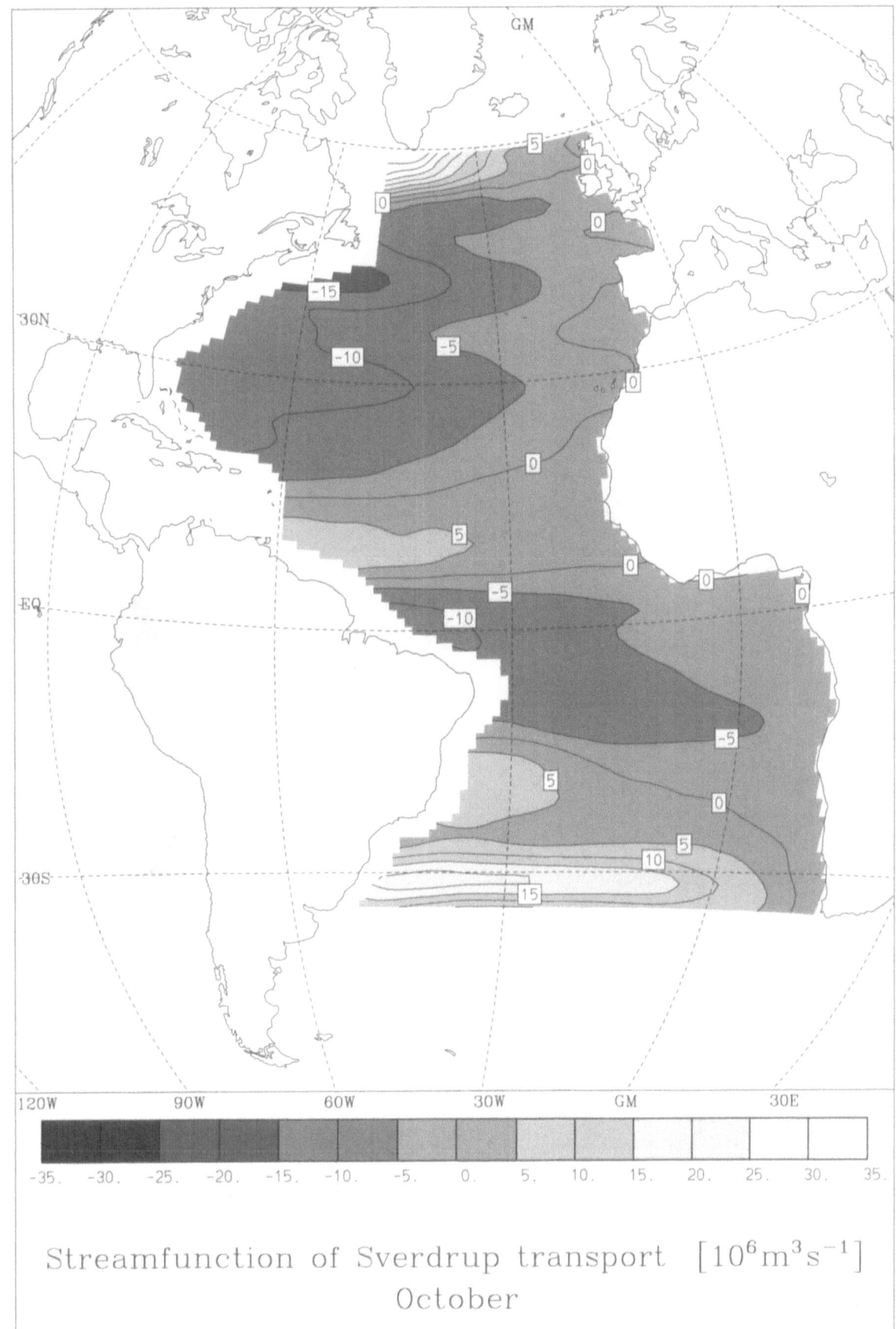

Streamfunction of Sverdrup transport $[10^6 \mathrm{m}^3\mathrm{s}^{-1}]$
October

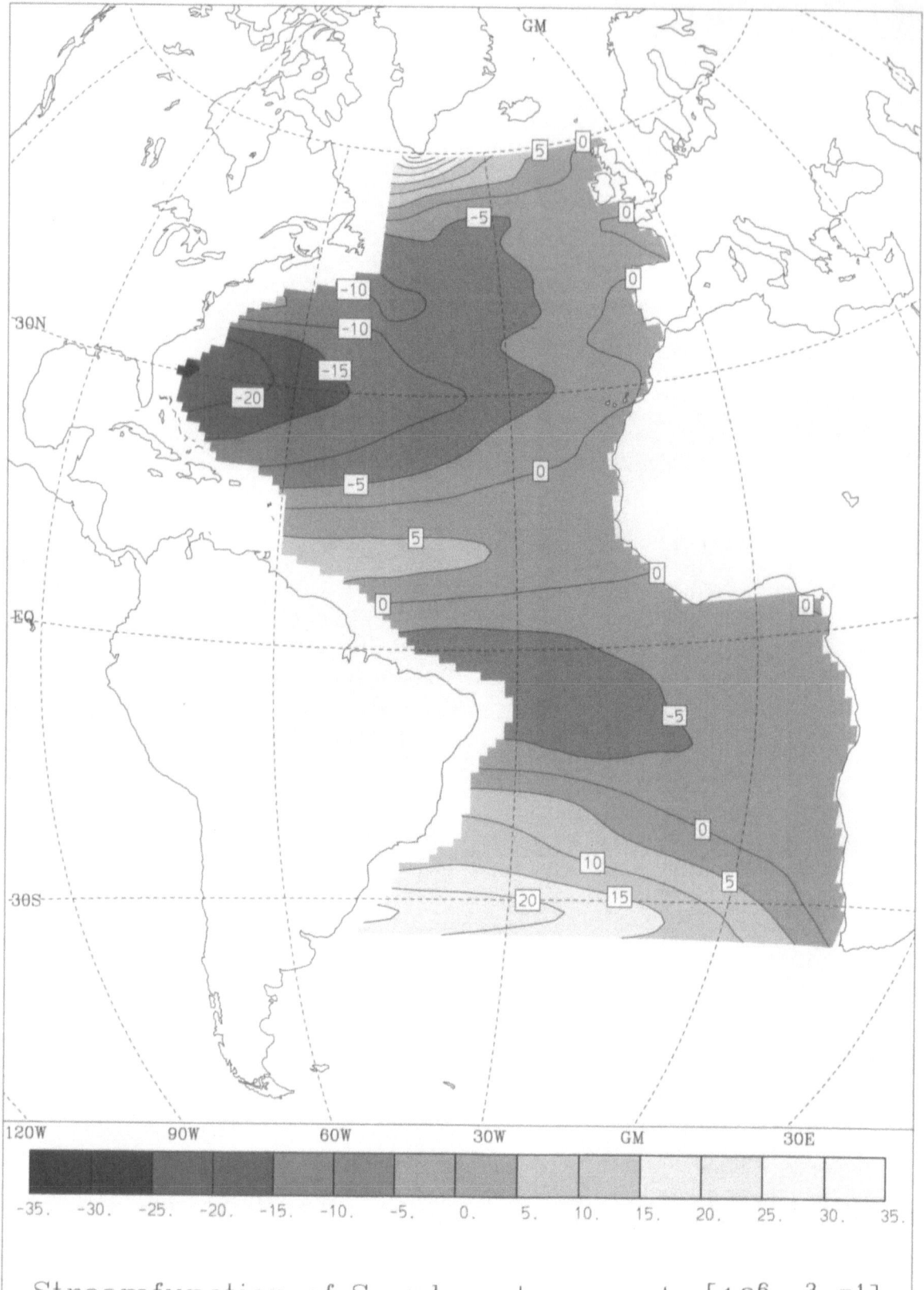

Streamfunction of Sverdrup transport $[10^6\,\mathrm{m}^3\mathrm{s}^{-1}]$
Year